THE GENETICS
of Cattle

THE GENETICS
of Cattle

Edited by

R. FRIES

Lehrstuhl für Tierzucht
Technische Universität München
Freising-Weihenstephan
Germany

and

A. RUVINSKY

Animal Sciences, SRSNR
University of New England
Armidale
Australia

CABI *Publishing*

CABI *Publishing* is a division of CAB *International*

CABI Publishing	CABI Publishing
CAB International	10 E 40th Street
Wallingford	Suite 3203
Oxon OX10 8DE	New York, NY 10016
UK	USA
Tel: +44 (0)1491 832111	Tel: +1 212 481 7018
Fax: +44 (0)1491 833508	Fax: +1 212 686 7993
Email: cabi@cabi.org	Email: cabi-nao@cabi.org

© CAB *International* 1999. All rights reserved. No part of this publication may be reproduced in any form or by any means, electronically, mechanically, by photocopying, recording or otherwise, without the prior permission of the copyright owners.

A catalogue record for this book is available from the British Library, London, UK

Library of Congress Cataloging-in-Publication Data
The genetics of cattle/edited by R. Fries and A. Ruvinsky.
    p.   cm.
   Includes bibliographical references and index.
   ISBN 0-85199-258-7 (alk. paper)
   1. Cattle--Genetics.   2. Cattle--Breeding.   I. Fries, R. (Ruedi).
II. Ruvinsky, Anatoly.   III. C.A.B. International.
SF201.G45   1999
636.2′08′21--dc21   98-45203
                                                                                                                                             CIP

ISBN 0 85199 258 7

Typeset in Garamond by AMA DataSet Ltd
Printed and bound in the UK by Biddles Ltd, Guildford and King's Lynn

Contents

Preface · vii
R. Fries and A. Ruvinsky

1. Systematics and Phylogeny of Cattle · 1
   J.A. Lenstra and D.G. Bradley
2. Genetic Aspects of Domestication · 15
   D.G. Bradley and E.P. Cunningham
3. Genetics of Colour Variation · 33
   T.A. Olson
4. Genetics of Morphological Traits and Inherited Disorders · 55
   F.W. Nicholas
5. Blood Groups and Biochemical Polymorphisms · 77
   H.C. Hines
6. The Molecular Genetics of Cattle · 123
   D. Vaiman
7. Molecular Genetics of Molecules with Immunological Functions: Major Histocompatibility Complex, Immunoglobulins, T-cell Receptors, Cytokines and their Receptors · 163
   H.A. Lewin, M. Amills and V.K. Ramiya
8. Genetics of Disease Resistance · 199
   A.J. Teale
9. Molecular Biology and Genetics of Bovine Spongiform Encephalopathy · 229
   N. Hunter
10. Cytogenetics and Physical Chromosome Maps · 247
    R. Fries and P. Popescu
11. Genetic Linkage Mapping, the Gene Maps of Cattle and the Lists of Loci · 329
    W. Barendse and R. Fries
12. Genetics of Behaviour in Cattle · 365
    D. Buchenauer

13.	Genetics and Biology of Reproduction in Cattle *B.W. Kirkpatrick*	391
14.	Reproductive Technologies and Transgenics *N.L. First, M. Mitalipova and M. Kent First*	411
15.	Developmental Genetics *A. Ruvinsky and L.J. Spicer*	437
16.	Genetic Resources and Conservation *D.L. Simon*	475
17.	Marker-assisted Selection *M.R. Dentine*	497
18.	Genetic Improvement of Dairy Cattle *M.E. Goddard and G.R. Wiggans*	511
19.	Molecular Genetics of Milk Production *W.S. Bawden and K.R. Nicholas*	539
20.	Genetic Improvement of Beef Cattle *B.P. Kinghorn and G. Simm*	577
21.	Genetics of Meat Quality *D.M. Marshall*	605
22.	Genetic Aspects of Cattle Adaptation in the Tropics *S. Newman and S.G. Coffey*	637
23.	Standardized Genetic Nomenclature for Cattle *C.H.S. Dolling*	657
24.	Breeds of Cattle *D.S. Buchanan and S.L. Dolezal*	667
Index		697

Preface

Since the time of domestication about 10,000 years ago, cattle have played an increasingly important role in the development of human civilizations around the world. It is not easy to find a country that does not have a more or less significant population of cattle. Certainly, traditions and climatic and economic conditions modify the way cattle are used, bred and kept. However, cattle have always provided essential human requirements, such as food, clothing, draught power, soil improvement and many others, including cultural and religious necessities. The unique features of cattle as ruminants, with their ability to effectively digest rough plant mass, allow cattle to occupy a special ecological position in the human environment. The current number of cattle worldwide exceeds 1200 million and continues to grow.

Cattle are among the largest domesticated species and certainly the process of domestication was a great challenge for neolithic communities. It seems likely that the traction power of the ox in the earliest stages of its domestication marked a turning-point in the development of agriculture. Over time, cow's milk steadily became a staple source of food in many geographical areas. This process is continuing and milk and numerous milk products are spreading into countries where they were not traditional. The total world production of cow's milk was 467 million t in 1997. Another very important product is beef. Worldwide production of beef and veal is approaching 50 million t year^{-1}.

Progress in cattle breeding and selection during the last century has been impressive. Breeding programmes based on principles of quantitative genetics, artificial insemination and embryo transfer, as well as the computerization of the industry, are the main reasons for the tremendous intensification in milk production. However, a genomic revolution and biotechnology promise new developments. Just a decade ago, the location of only a few genes on the cattle chromosomes were known. At the time of publication of this book, this number has reached 2200 loci. Previously separated, quantitative and molecular genetics are now taking a united approach to the identification of loci underlying important cattle traits, so called quantitative trait loci (QTL), and are expected to provide new tools for cattle breeding in the near future. Cloning and other new reproduction technologies will also benefit the cattle industry enormously.

The main purpose of this book is to collect essential data concerning cattle genetics and bring together previously non-united areas of research. The 24 chapters of this book can be partitioned into four sections. The first four chapters cover the systematics and phylogeny of cattle, domestication and factorial genetics. Chapters 5–11 present information about biochemical polymorphism, immunogenetics and disease resistance, genome structure and gene mapping. The third section discusses genetics of reproduction, development and behaviour. Finally, Chapters 16–22 are devoted to genetics applied to cattle improvement. Standard genetic nomenclature for cattle is presented in Chapter 23 and cattle breeds in Chapter 24.

The authors of this book have made every attempt to highlight the most important publications in the area of cattle genetics for the last several decades, with emphasis on the most recent papers, reviews and books. However, we realize that omissions and errors are very difficult to avoid and apologize for possible mistakes. This book is addressed to a broad audience, which includes researchers, lecturers, students, farmers and specialists working in the industry. *The Genetics of Cattle* is the third book in a series of monographs on mammalian genetics published by CAB *International*. Two other books, *The Genetics of Sheep* (1997) and *The Genetics of the Pig* (1998), are based on similar ideas and structure. *The Genetics of the Horse* will continue this series.

This book is a result of truly international cooperation. Scientists from several European countries, USA and Australia put a lot of effort into this book. The editors are very grateful to all of them. It is our pleasure and debt to thank many people who helped tremendously in reviewing the book: A.J. Ball, J.S.F. Barker, T.K. Bell, A.M. Crawford, M. Enns, I.R. Franklin, D. Gallagher, H.-U. Graser, G.N. Hinch, F. Hughes, J.K. Lunney, C. Moran, L. Piper, B.C. Powell, J.E.O. Rege, H.W. Radsma, G. Rogers, D.P. Sponenberg and P. Wynn.

It is our hope that this book will be useful for many people throughout the world interested in cattle genetics and perhaps will support consolidation and further progress of this field of science.

Ruedi Fries
Anatoly Ruvinsky

Systematics and Phylogeny of Cattle

J.A. Lenstra[1] and D.G. Bradley[2]
[1]*Institute of Infectious Diseases and Immunology, Faculty of Veterinary Medicine, Utrecht University, Yalelaan 1, 3584 CL Utrecht, The Netherlands;* [2]*Department of Genetics, Trinity College, Dublin 2, Eire*

Introduction	1
The Phylogenetic Position of the Tribe of the Bovini	2
The order Artiodactyla and the suborder of the ruminants	2
The family Bovidae	3
The subfamily Bovinae and the tribe Bovini	5
The Bovini	5
Buffalo species	5
Bison species	6
Yak	7
South-East Asian cattle	8
Zebu and taurine cattle	8
Phylogeny of Bovini	9
References	11

Introduction

The mammalian tribe Bovini (subfamily Bovinae, family Bovidae) contains all the most important of the world's larger domestic species. The domestications of these cattle and cattle-like taxa were among the most significant advances of the neolithic transition. With their specialized digestive system, these species can utilize cellulose as an energy source, generating the benefits of dairy, meat and hide production. In addition, cattle provided, for several millennia, the animal draught power that underpinned agriculture in the Old World and have assumed a cultural and even religious importance, which is preserved to the present day in some societies.

In this chapter we describe the wild and domestic cattle species and their systematics, and we review the phylogenetic analyses that have been performed on this closely related and poorly resolved group.

The Phylogenetic Position of the Tribe of the Bovini

The order Artiodactyla and the suborder of the ruminants

Cattle belong to the zoological order of the artiodactyls, or even-toed ungulates, the oldest fossils of which are 50–60 million years old. Artiodactyls were commonly divided into Suiformes (pigs, hippopotamuses and peccaries), Tylopoda (camels and llamas) and Ruminantia (see, for example, Franklin, 1997; Ruvinsky and Rothschild, 1998; http://www.pathcom.com/~dhuffman/artiodactyla.html). One of the most striking aspects of recent higher-mammalian phylogeny has emerged from the synthesis of genetics and morphological analyses and indicated that cetaceans (whales, porpoises and dolphins) and artiodactyls possessed a common ancestor not shared by any other group. This close relationship may seem counter-intuitive but has been deduced from very different suites of data: morphological analysis, sequence comparisons and interspersed repeat element (SINE) distributions (Thewissen and Hussain, 1993; Graur and Higgins, 1994; Buntjer *et al.*, 1997).

A recent and more controversial assertion, also based on protein and deoxyribonucleic acid (DNA) sequences (Graur and Higgins, 1994; Gatesy, 1997; Montgelard *et al.*, 1997), has been that cetaceans are deeply nested within the artiodactyl cluster. Surprisingly, not only may cattle be more closely related to whales than to horses, but cattle and whales may even form a clade to the exclusion of other artiodactyls, such as pigs and camels. This provocative finding has been strengthened by an elegant analysis of the species distributions of SINEs (Shimamura *et al.*, 1997). An insertion of one of these elements into a new location in an ancestral genome may be considered to be a unique event, with a negligible probability of independent recurrence. Thus a shared SINE insertion may be a powerful phylogenetic marker for the grouping of taxa (Fig. 1.1). In particular, the finding of a SINE uniquely shared by ruminants, cetaceans and the hippopotamus has suggested a clustering of these animals to the exclusion of pigs and camels (Shimamura *et al.*, 1997) and challenges the existence of a Suiformes clade (see also Montgelard *et al.*, 1997).

The distinguishing feature of the ruminants is their special digestive system, with rumination and consecutive digestions in three or four subdivisions of the stomach, enabling an efficient utilization of grass and leaves as food. The chevrotains of the Tragulidae family, with a three-chambered stomach, are the most primitive ruminants. The other ruminants form the separated infraorder of the Pecora, which comprises the Antilocapridae, Cervidae, Moschidae (with *Moschus* as the only extant species), Giraffidae and Bovidae (Janis and Scott, 1987). The relatively small interfamilial divergence indicates that these families were formed about 25 million year ago within a period of 5 million years (Kraus and Miyamoto, 1991). Phylogenetic studies (Kraus and Miyamoto, 1991; Cronin *et al.*, 1996; Montgelard *et al.*, 1997; see also Chikuni *et al.*, 1995) have suggested an early split-off of the

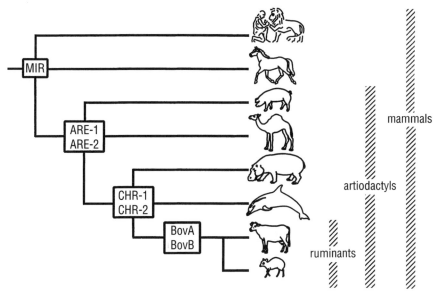

Fig. 1.1. Phylogeny of the order of the artiodactyls, as based on the sharing of interspersed repeat elements: the mammalian-wide interspersed repeat (MIR; Jurka et al., 1995); the artiodactyl ARE-1B and ARE-2B repeats (Alexander et al., 1995; Buntjer et al., 1997); the cetacean/hippopotamus/ruminant CHR-1 and CHR-2 repeats (Shimamura et al., 1997); and the ruminant SINE repeat elements Bov-A and Bov-B (Jobse et al., 1995).

Antilocapridae but do not resolve the branching order of the Cervidae, Giraffidae and Bovidae (Fig. 1.2).

The family Bovidae

Bovidae are distinguished by the presence of permanent hollow horns. After their emergence about 20 million years ago, their fast radiation was probably inititated by the emergence of the African savannah 18 to 23 million years ago (Allard et al., 1992) and has created a great variety of morphologies. Most of the more than 100 species, several of which are represented by large populations, originate from Africa, the rest from Eurasia or North America. Many bovids live in herds, which in open areas reduces their vulnerability to carnivore predators. Usually, the dominance of one or a few sires imposes a strong selection of the male lineage, a phenomenon that has been reinforced by selective breeding practices in domestic species.

Extant bovids are divided into between five and nine subfamilies (Morris, 1965; Gentry, 1978; Franklin, 1997; see Fig. 1.2), with the higher number being supported by molecular studies. Undisputed divisions include the Bovinae (such as cattle, nilgai and eland), Cephalophinae (duikers), Caprinae (sheep, goats and related animals), Hippotraginae (e.g. roan antelope) and Antilopinae (gazelles, dik-diks). Reduncinae (reedbucks) and Ancelaphinae (gnu, hartebeest, etc.) are often classified as tribes within the Hippotraginae (Reduncini

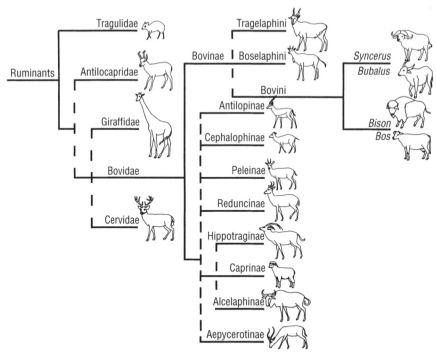

Fig. 1.2. Phylogeny of the order of ruminants, the family of Bovidae and the subfamily of the Bovinae and the main division within the tribe of the Bovini (Buntjer et al., 1997; Gatesy et al., 1997; Montgelard et al., 1997).

and Ancelaphini, respectively), but comparative sequencing of mitochondrial DNA (mtDNA) (Gatesy et al., 1997) supports a subfamily status for both taxa. The mitochondrial phylogeny further supports an assignment of impala to a separate Aepycerotinae subfamily rather than a classification within the Ancelaphinae or Antelopini. Likewise, reebok does not fit readily in one of the other subfamilies and is the sole member of the Peleinae (Gatesy et al., 1997).

The radiation of the Bovidae seems to have been faster than, for example, in the Cervidae (Buntjer, 1997) and has defied attempts to arrive at an unambiguous phylogeny. However, mtDNA sequence analyses (Allard et al., 1992; Gatesy et al., 1997), as well as a study of the genes for nuclear ribonuclease (Beintema et al., 1986), indicate that the first split was of the Bovinae subfamily, comprising the tribes Tragelaphini, Boselaphini and Bovini (Fig. 1.2). Within the other subfamilies, mitochondrial sequences indicate a clustering of the Caprinae, Hippotraginae and Alcelaphinae. The reebok may be considered as sister group of the Reduncinae, but the position of the impala remains unclear (Gatesy et al., 1997). The classification of the Neotragini and Antilopini tribes within the Antilopinae subfamily is confirmed by the similarity of their mtDNA (Allard et al., 1992; Gatesy et al., 1997). However, molecular data disbanded two tribes within the Caprinae tribes, the Saigini (saiga and Tibetan antelopes; Gatesy et al., 1997) and the Ovibovini (musk ox and takin; Groves and Shields, 1996).

The subfamily Bovinae and the tribe Bovini

The subfamily of the Bovinae is divided into the tribes of Tragelaphini or Strepsocerotini (spiral-horned antelopes), Boselaphini (nilgai, four-horned antelope) and large-sized Bovini. This tribe comprises the wild and domesticated buffalo and cattle species and started to diverge more than 4 million years ago. Bovini are sometimes divided in the genera Bubalina (*Bubalus* species, water-buffalo, arni, tamarao and anoa), Syncerina (African buffalo) and Bovina (*Bos* and *Bison* species). However, molecular data (see Phylogeny of Bovini) suggest a dichotomy of buffalo (*Bubalus, Syncerus*) vs. non-buffalo (*Bos, Bison*, Fig. 1.2). Within the Bovinae, *Bison* (bison, wisent) and *Bos* are considered as subgenera. In other classifications the *Bos* species are split into the subgenera *Poephagus* (yak), *Bos* (taurine and indicine cattle) and *Bibos* (gaur, banteng, kouprey), but this is not consistent with a suggested clustering of yak with the bison species (Geraads, 1992; Buntjer, 1997).

The *Bos* and *Bison* bovines have karyotypes of 2n = 60 with acrocentric or telocentric autosomes, except gaur with 2n = 58 (Gallagher and Womack, 1992). In contrast, African buffalo (2n = 52) and water-buffalo (2n = 50 for the river type, 48 for the swamp type) have several metacentric autosomes. These differences in chromosome number probably explain why interspecies hybridizations are restricted to the non-buffalo bovines. Within this group, interspecies hybridization occurs spontaneously or as a result of domestic activity. The crosses of *Bos taurus* (common cattle) × *Bos indicus* (zebu) yield completely fertile offspring. This indicates that the divergence of these species has not yet led to a complete speciation. Other crosses usually yield sterile bulls and fertile cows. The offspring of gayal–zebu crosses and of yak–taurine crosses (see below) may be considered as bovine equivalents of the mule and the hinny, being better suited for specific tasks than their parents.

The Bovini

Buffalo species

Although in America often synonymous with bison, the zoological denotation of buffalo refers to the African buffalo (*Syncerus caffer*), the water-buffalo (*Bubalus bubalis*), the Indian wild buffalo or arni (*Bubalus arnee*), the anoa (*Bubalus* or *Anoa depressicornis*) or the tamarao (*Bubalus* or *Anoa mindorensis*).

The African buffalo lives on the African savannah south of the Sahara and is one of the most dangerous game animals. Although it can be tamed, African buffalo has never been domesticated. The anoa is the smallest bovine species and lives in the forests of Sulawesi, Indonesia. The two types of lowland and mountain anoa (the latter sometimes denoted *Bubalus quarlesi*) appear to have quite divergent mtDNA sequences (Tanaka *et al.*, 1996; Kikkawa *et al.*,

1997). The tamarao is also smaller than the water-buffalo and lives on the southern Philippine island, Mindoro.

About 2000 Indian wild buffaloes live in parks in Assam, Nepal and Burma. This Indian wild buffalo is the progenitor of the domestic water-buffalo. Typically, both species seek protection against heat and insects by wallowing in water. Domestication took place 5000 years BC in China (Chen and Li, 1989) and/or 2500 BC in Mesopotamia (Cockrill, 1984; see also Lau *et al.*, 1998) and resulted in tranquil and docile animals. Two types of water-buffalo have been discerned: the swamp type in South-East Asia (sometimes denoted *Bubalus carabanesis*) and the river type on the Indian subcontinent and farther west. The swamp type resembles the arni progenitor most closely. The two types differ in karyotype (see previous section) and at the level of protein and DNA sequences (Beintema *et al.*, 1986; Amano *et al.*, 1994; Barker *et al.*, 1997; Kikkawa *et al.*, 1997; Lau *et al.*, 1998). Estimates on the divergence time of the river and swamp types entirely depend on the calibration of the molecular clock. Extrapolation of recent rates of microsatellite variation and of mitochondrial sequence divergence yielded estimates of 10,000–15,000 years (Barker *et al.*, 1997) and 28,000–87,000 years (Lau *et al.*, 1998), respectively. However, interpolation of mitochondrial sequence variation on the basis of the buffalo–cattle divergence indicated a divergence time of 1.7 million years (Tanaka *et al.*, 1996). The latter used only one sequence per taxon, which may introduce wide errors into the estimates (see also Loewe and Scherer, 1997).

About 148 million water-buffalo (an estimate from 1994; see http://ww2.netnitco.net/users/djligda/wbfacts/html) are kept around the world. Most of these (95%) live in Asia, but substantial populations have been established in Egypt, Italy, the Balkans and Brazil. In India and Pakistan, well-defined breeds of the river type are kept (Cockrill, 1984). Swamp buffalo are mainly used as draught animals, while river buffalo are dairy animals. In Italy, buffalo milk, rich in butterfat, is used for the preparation of mozzarella cheese, an essential component of the traditional Neapolitan pizza. Water-buffalo is also an excellent beef animal, yielding tender meat with a low fat content.

Mating with normal cattle occurs, but there are no reliable reports of the birth of hybrids.

Bison species

The American bison (*Bison bison*) are the wild cattle of the North American prairie. A separate subspecies (*Bison bison athabasca*) lives in woodland. The prairie bison underwent a narrow population bottleneck as a result of excessive hunting. In 1894 only 800 animals remained, most of which carried bullets in their bodies. Conservation efforts have resulted in a modern population of about 25,000 animals in national parks. The detection of taurine mtDNA sequences in one modern population (Polziehn *et al.*, 1995) indicates that, in at least in one case, *B. taurus* cows have been used to restore the bison population.

In addition to these animals, 15,000 bison are reared as a source of meat, which has reputedly a low content of fat and cholesterol (see http://bigweb.com/mall/buffalo/index.html). Hybridization with normal cattle has also resulted in the registered beefalo breed, which comprises about 8000 animals (Felius, 1995), although not much is known about its genetic composition. Beefalo meat also has the reputation of being low in fat and cholesterol.

The wisent (*Bison bonasus*) is the European relative of the bison and suffered an even more severe population bottleneck. All 3332 animals (1990 count), which are now kept in zoos and parks, descend from only 23 bulls and 23 cows, which, in 1923, were available for breeding. Interestingly, in a phylogenetic analysis of mitochondrial COII gene sequences from banteng, gaur, cattle, yak, bison and wisent, the latter two did not cluster. Rather, the bison sample was substantially divergent from all other non-buffalo sequences, including the sequence of American bison (Janecek *et al.*, 1996). This anomalous phylogenetic situation may be the result of lineage sorting in an ancestral species and is a caution against overinterpreting data from a single locus in comparisons of closely related species. American bison mtDNA variants have been detected in two wisents and are possibly a consequence of crossbreeding in the species' recent regeneration (R.T. Loftus, personal communication).

Yak

The most typical characteristic of the yak is their long and woolly hair, necessary for their natural habitat at altitudes of 3000–5000 m. Their distinctive grunt gave the animal the systematic name of *Bos grunniens*. Since wild yaks only grunt during the mating season, these animals were baptized *Bos mutus*. In 1983 only a few hundred wild animals still lived in Tibet. Like most wild bovine species (including the extinct aurochs), cows have only one-third of the weight of the bulls, a sexual dimorphism that is always reduced by domestication (Bonnemaire, 1984; Felius, 1995).

Domestication was early and perhaps took place when agriculture first appeared (Zeuner, 1963). At present, about 12 million animals are kept at regions above 2000 m in China (mainly in Tibet), Pamir, Tien Shan, Mongolia and neighbouring regions in Siberia. Domestic animals are smaller and have a greater variety of coat colour than in the wild. The yak is a sturdy animal and may negotiate and graze in terrain inaccessible to normal cattle. They provide many products, such as unusually rich milk (up to 7% fat) for the production of butter, meat, dung as fuel, leather and hair for ropes, tents and felt (Bonnemaire, 1984; Lensch, 1996).

Yak–taurine hybrids, or yakows, are mostly derived from taurine bulls and yak cows at altitudes of 1500–2000 m (Bonnemaire, 1984; Felius, 1995). Male hybrids are invariable sterile. By hybrid vigour, the F_1 hybrid offspring is larger than the parents and the cows yield more milk. In Mongolia, the F_1 hybrid, or khainag, is traditionally used as a pack and draught animal. Female khainags are used to produce B_1 and B_2 backcrosses with both species (Tumennasan *et al.*, 1997).

South-East Asian cattle

Small populations of wild banteng (*Bos* or *Bibos javanicus*) number together a few thousand animals in hilly woodlands below 2000 m in South-East Asia, Java and Kalimantan. Bantengs are shy animals, with, again, a large size difference between bulls and cows.

Banteng is the wild progenitor of Bali cattle, and was domesticated before 3500 BC in Indonesia or Indochina. About 1.5–2 million animals are kept on the island of Bali and other islands in Indonesia (Rollinson, 1984) and are used for tillage and slaughter. Bali cattle tolerate heat, but tend to be nervous and shy. Their meat is reputed to be lean and exceptionally tender.

The Madura zebu breed on the island of Madura ('the racing bull') is supposed to be a hybrid of zebu and banteng, and there are some molecular data that suggest an admixture in cattle of that region (Namikawa, 1981; Kikkawa *et al.*, 1995). However, experimental crossing of taurine or indicine breeds with banteng has met with little success.

From the gaur, or Indian bison (*Bos* or *Bibos gaurus*), only a few thousand animals still exist in tropical forests and woodlands in India, Indochina and the Malay peninsula. The gaur has a stubborn disposition, with a typical sexual size dimorphism, which is reduced in their domestic relatives, the gayal or mithan (*Bos* or *Bibos frontalis*). Domestication probably dates back to 500 BC (Felius, 1995) and may have exploited the animal's fondness for salt (Simoons, 1984). About 150,000 gayals are kept near the eastern border of India. Gayals are friendly animals, used primarily for traditional sacrifice and increasing the social status of their owners. Animals are often permitted to range around the village, and matings with gaur bulls probably take place frequently.

The selembu is a gayal–zebu hybrid in Bhutan and India, which is valued as a dairy animal, while the sterile males are used for draught. Accidental hybrids of a gaur bull with taurine or indicine cattle display considerable heterosis and may also have potential for meat production (Bongso *et al.*, 1988).

The kouprey or Cambodian wild ox (*Bos* or *Bibos sauveli*) is a severely threatened species, of which only a few hundred animals may persist in Cambodia and Vietnam (see http://www.wcmc.org.uk/infoserv/countryp/vietnam/app4.html). It lives in open parts of the jungle, occasionally in mixed herds with banteng (Felius, 1995).

Zebu and taurine cattle

Common cattle (*B. taurus*) and zebu (*B. indicus*) both descend from the aurochs (*Bos primigenius*). This was a large animal that once lived in North Africa and Eurasia, ranging from the Atlantic to the Pacific coast. However, it has been extinct in Europe since 1627 and from earlier times in other continents. Domestication took place as early as 7000 BC and more than 1200 million cattle are kept today, chiefly as sources of milk, meat and hides.

Analyses of microsatellite variation and mtDNA sequences (Loftus *et al.*, 1994; MacHugh *et al.*, 1997) have revealed that the ancestors of zebu and taurine cattle diverged some hundreds of thousands of years ago and must therefore be the result of at least two biologically independent domestication events. Evidence that separate strains of aurochs have been incorporated into African and European *B. taurus* has emerged from a more extensive mtDNA analysis (Bradley *et al.*, 1996). Successful extraction and polymerase chain reaction (PCR) amplification of mitochondrial DNA from 12,000-year-old British auroch bones have been achieved and have revealed sequences of divergent, but unmistakably *B. taurus* type (Bailey *et al.*, 1996).

About 800 different cattle breeds are recognized (Felius, 1995; http://dad.fao.org; http://www.ansi.okstate.edu/breeds/), many of which are genetic isolates (see Chapter 24). Selection has been directed primarily towards production of milk and/or meat, but often also towards their appearance (coat colour, horns). There has been inherent selection for local disease resistance and docility, except in breeds used for bullfighting. Most developed taurine breeds have superior production, but zebu breeds are better able to survive in dry and warm climates. Dairy taurine breeds have been commonly crossed to local zebu breeds in the tropics, but the favourable effects of heterosis are mainly in the first generation (McDowell *et al.*, 1996).

The classification of breeds may be based on geography, history and morphology (Epstein and Mason, 1984; Felius, 1995). The primary distinction is between humped (*B. indicus*) and humpless (*B. taurus*) cattle, although many breeds are of intermediate morphology and genetically admixed origin, particularly those in the greater part of Africa (Felius, 1995; Frisch *et al.*, 1997; MacHugh *et al.*, 1997). Further taxonomic division into long-horned and short-horned types is sometimes employed, but it is perhaps unlikely that this subdivision is clearly reflected at the genomic level.

Phylogeny of Bovini

The phylogeny of the Bovini has been studied by morphological analysis (Groves, 1981; Geraads, 1992), nuclear ribosomal DNA (rDNA) restriction fragment length polymorphism (RFLP) analysis (Wall *et al.*, 1992), sequencing of the mitochondrial cytochrome oxidase gene (Janecek *et al.*, 1996) and PCR-generated fingerprinting (Buntjer, 1997). This latter study is based on the variation of many nuclear loci distributed over the genome; 69 markers

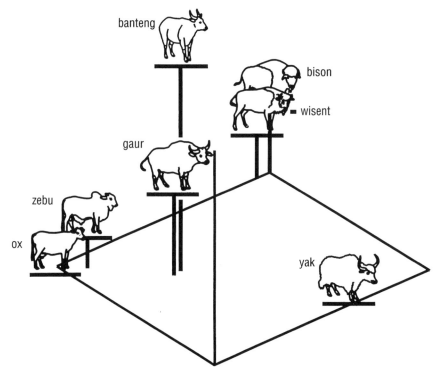

Fig. 1.3. Principal-coordinate analysis representing 72% of the interspecies variation of 85 AFLP markers informative for the non-buffalo Bovini species (Buntjer, 1997).

generated by a PCR specific for the mammalian-wide interspersed repeat (MIR) and 361 amplified fragment length polymorphisms (AFLP).

With the exception of Geraads (1992), all studies consistently cluster together the buffalo taxa (*Bubalus*, *Anoa* and *Syncerus*) relative to the *Bos* and *Bison* species. On a lower taxonomic level, the clustering of *Bubalus* and/or *Anoa* species (arni, water-buffalo, tamarao, anoa) is indicated by morphology (Groves, 1981), rDNA analysis (Wall *et al.*, 1992) and MIR-specific fingerprinting (Buntjer *et al.*, 1997). Tanaka *et al.* (1996) found that the tamarao is more closely related to the water-buffalo and the arni than to the anoa.

Data on the phylogeny of the other Bovini are largely conflicting. Morphological studies (Groves, 1981; Geraads, 1992) do not agree on the separate position of the *Bison* species. Ribosomal DNA restriction sites (Wall *et al.*, 1992) are hardly informative at this level, while the large divergence of the mitochondrial sequences of bison and wisent, together with a miniscule difference between yak and cattle (Janecek *et al.*, 1996), may indicate anomalies in maternal lineages. Also, mtDNA chromosome data, which are based on a single segregating unit, may not be representative of whole genomes, due to stochastic lineage survival and loss in ancestral species. In the study of phylogeny in closely related species, surveys of variation at a range of separately segregating sites are likely to yield more accurate reconstructions. In comparisons of nuclear fingerprints generated by PCR, bison and wisent are clearly similar and different from the *Bos* species (Buntjer, 1997). MIR-specific PCR, and AFLP analysis (Buntjer, 1997) also indicate a clustering of yak with both *Bison* species, in agreement with Geraads (1992).

The gaur and the banteng, often designated as *Bibos* species, are clustered by the rDNA data as well as the mitochondrial sequences. These species have no stable positions in the trees constructed on the basis of the nuclear fingerprints. However, a principal-coordinate plot of the AFLP data (Fig. 1.3) does indicate the phylogenetic proximity of gaur and banteng (Buntjer, 1997).

Buntjer (1997) proposed that the retention of ancestral polymorphisms and recombination events interfere with the reconstruction of phylogenetic trees. This may explain the lack of agreement between the different phylogenetic reconstructions. Additionally, the principal-coordinate plot, which represents 66–72% of the total variation in the nuclear fingerprints, suggests an interesting correspondence between genetic distance and geographical origin of the species. The positions of the taxa stretch from *B. taurus* (Middle East), through *B. indicus* (India) to the South-East Asian species *B. javanicus* and *B. gaurus*. The *Bison* species and the yak, both adapted to colder climates, are clearly different from the other non-buffalo Bovini. It is a compelling hypothesis that, at this taxonomic level, geographical factors may have determined the degree of exchange of genetic material between the ancestral populations and hence the genetic similarity of extant species.

References

Alexander, L.J., Rohrer, G.A., Stone, R.T. and Beattle, C.W. (1995) Porcine SINE-associated microsatellite markers: evidence for new artiodactyl SINEs. *Mammalian Genome* 6, 464–468.

Allard, M.W., Miyamoto, M.M., Jarecki, L., Kraus, F. and Tennant, M.R. (1992) DNA systematics and evolution of the artiodactyl family Bovidae. *Proceedings of the National Academy of Sciences of the USA* 89, 3972–3976.

Amano, T., Miyakoshi, Y., Takada, T., Kikkawa, Y. and Suzuki, H. (1994) Genetic variants of ribisomal DNA and mitochondrial DNA between swamp and river buffaloes. *Animal Genetics* 25(Suppl. 1), 29–36.

Bailey, J.F., Richards, M.B., Macaulay, V.A., Colson, I.B., James, I.T., Bradley, D.G., Hedges, R.E.M. and Sykes, B.C. (1996) Ancient DNA suggests a recent expansion of European cattle from a diverse wild progenitor species. *Proceedings of the Royal Society of London B* 23, 1467–1473.

Barker, J.S.F., Moore, S.S., Hetzel, D.J.S., Evans, D., Tan, S.G. and Byrne, K. (1997) Genetic diversity of Asian water buffalo *(Bubalus bubalis)*: microsatellite variation and a comparison with protein-coding loci. *Animal Genetics* 28, 103–115.

Beintema, J.J., Fitch, W.M. and Carsana, A. (1986) Molecular evolution of pancreatic-type ribonucleases. *Molecular Biology and Evolution* 3, 262–275.

Bongso, T.A., Hilmi, M., Sopian, M. and Zulkilfi, S. (1988) Chromosomes of gaur cross domestic cattle hybrids. *Research in Veterinary Science* 44, 251–254.

Bonnemaire, J. (1984) Yak. In: Mason, I.L. (ed.) *Evolution of Domesticated Animals.* Longman, New York, pp. 39–45.

Bradley, D.G., MacHugh, D.E., Cunningham, P. and Loftus, R.T. (1996) Mitochondrial diversity and the origins of African and European cattle. *Proceedings of the National Academy of Sciences of the USA* 93, 5131–5135.

Buntjer, J.B. (1997) DNA repeats in the vertebrate genome as probes in phylogeny and species identification. Academic thesis, Utrecht University.

Buntjer, J.B., Hoff, I.A. and Lenstra, J.A. (1997) Artiodactyl interspersed DNA repeats in cetacean genomes. *Journal of Molecular Evolution* 45, 66–69.

Chen, Y.C. and Li, X.H. (1989) New evidence of the origin and domestication of the Chinese swamp buffalo *(Bubalus bubalis)*. *Buffalo Journal* 1, 51–55.

Chikuni, K., Mori, Y., Tabata, M., Monma, M. and Kosugiyama, M. (1995) Molecular phylogeny based on the κ-casein and cytochrome b sequences in the mammalian suborder Ruminantia. *Journal of Molecular Evolution* 41, 859–866.

Cockrill, W.R. (1984) Water buffalo. In: Mason, I.L. (ed.) *Evolution of Domesticated Animals.* Longman, New York, pp. 52–63.

Cronin, M.A., Stuart, R., Pieson, B.J. and Patton, J.C. (1996) κ-casein gene phylogeny of higher ruminants (pecora, artiodactyla). *Molecular Phylogenetics and Evolution* 6, 295–311.

Epstein, H. and Mason, I.L. (1984) Cattle. In: Mason, I.L. (ed.) *Evolution of Domesticated Animals.* Longman, New York, pp. 6–34.

Felius, M. (1995) *Encyclopedia of Cattle Breeds.* Misset, Doetinchem, the Netherlands.

Franklin, I.R. (1997) Systematics and phylogeny of sheep. In: Pier, L. and Ruvinsky, A. (eds) *The Genetics of Sheep.* CAB International, Wallingford, UK.

Frisch, J.E., Drinkwater, R., Harrison, B. and Johnson, S. (1997) Classification of the southern African sanga an East African shorthorned zebu. *Animal Genetics* 28, 77–83.

Gallagher, D.S., Jr and Womack, J.E. (1992) Chromosome conservation in the Bovidae. *Journal of Heredity* 83, 287–298.

Gatesy, J. (1997) More DNA support for a Cetacea/Hippopotamidae clade: the blood-clotting protein gene γ-fibrinogen. *Molecular Biology and Evolution* 14, 537–543.

Gatesy, J., Amato, G., Vrba, E., Schaller, G. and DeSalle, R. (1997) A cladistic analysis of mitochondrial ribosomal DNA from the Bovidae. *Molecular Phylogenetics and Evolution* 7, 303–319.

Gentry, A.W. (1978) Bovidae. In: Maglio, V.J. and Cooke, H.B.S. (eds) *Evolution of African Mammals.* Harvard University Press, Cambridge, Massachusetts, pp. 540–572.

Geraads, D. (1992) Phylogenetic analysis of the tribe Bovini (Mammalia: Artiodactyla). *Zoological Journal of the Linnean Society* 104, 193–207.

Graur, D. and Higgins, D.G. (1994) Molecular evidence for the inclusion of the Cetaceans within the order Artiodactyla. *Molecular Biology and Evolution* 11, 357–364.

Groves, C.P. (1981) Systematic relationships in the Bovini (Artiodactyla, Bovidae). *Zeitschrift für Zoologie und Systematische Evolution* 19, 264–278.

Groves, P. and Shields, G.F. (1996) Phylogenetics of the Caprinae based on cytochrome b sequence. *Molecular and Phylogenetic Evolution* 5, 467–476.

Janecek, L.L., Honeycutt, R.L., Adkins, R.M. and Davis, S.K. (1996) Mitochondrial gene sequences and the molecular systematics of the artiodactyl subfamily Bovinae. *Molecular Phylogenetics and Evolution* 6, 107–119.

Janis, C.M. and Scott, K.M. (1987) The inter-relationships of higher ruminant families with special emphasis of the members of the Cervoidea. *American Musea Novitates* 2893, 1–85.

Jobse, C., Buntjer, J.B., Haagsma, N., Breukelman, H.J., Beintema, J.J. and Lenstra, J.A. (1995) Evolution and recombination of bovine DNA repeats. *Journal of Molecular Evolution* 41, 277–283.

Jurka, J., Zietkiewicz, E. and Labuda, D. (1995) Ubiquitous mammalian-wide interspersed repeats (MIRs) are molecular fossils from the mesozoic era. *Nucleic Acids Research* 23, 170–175.

Kikkawa, Y., Amano, T. and Suzuki, H. (1995) Analysis of genetic diversity of domestic cattle in east and Southeast Asia in terms of variations in restriction sites and sequences of mitochondrial DNA. *Biochemical Genetics* 33, 51-60.

Kikkawa, Y., Yonekawa, H., Suzuki, H. and Amano, T. (1997) Analysis of genetic diversity of domestic water buffaloes and anoas based on variations in the mitochondrial gene for cytochrome b. *Animal Genetics* 28, 195–201.

Kraus, F. and Miyamoto, M.M. (1991) Rapid cladogenesis among the pecoran ruminants: evidence from mitochondrial DNA sequences. *Systematic Zoology* 40, 117–130.

Lau, C.H., Drinkwater, R.D., Yusoff, K., Tan, S.G., Hetzel, D.J.S. and Barker, J.S.F. (1998) Genetic diversity of Asian water buffalo *(Bubalus bubalis)*: mitochondrial DNA D-loop and cytochrome b sequence variation. *Animal Genetics*, 29, 253–264.

Lensch, J. (1996) Yaks – Asian mountain cattle – in science and practice. *International Yak Newsletter* 2, 1–11.

Loewe, L. and Scherer, S. (1997) Mitochondrial Eve: the plot thickens. *Trends in Ecology and Evolution* 12, 422–423.

Loftus, R.T., MacHugh, D.E., Bradley, D.G., Sharp, P.M. and Cunningham, E.P. (1994) Evidence for two independent domestications of cattle. *Proceedings of the National Academy of Sciences of the USA* 91, 2757–2761.

McDowell, R.E., Wilk, J.C. and Talbott, C.W. (1996) Economic viability of crosses of *Bos taurus* and *Bos indicus* for dairying in warm climates. *Journal of Dairy Sciences* 79, 1292–1303.

MacHugh, D.E., Shriver, M.D., Loftus, R.T., Cunningham, P. and Bradley, D.G. (1997) Microsatellite DNA variation and the evolution, domestication and phylogeography of taurine and zebu cattle (*Bos taurus* and *Bos indicus*). *Genetics* 146, 1071–1086.

Montgelard, C., Catzeflis, F.M. and Douzery, E. (1997) Phylogenetic relationships of artiodactyles and cetaceans as deduced from the comparison of cytochrome *b* and 12S rRNA mitochondrial sequences. *Molecular Biology and Evolution* 14, 550–559.

Morris, D. (1965) *The Mammals. A Guide to the Living Species*. Hodder & Stoughton, London.

Namikawa, T. (1981) Geographical distribution of bovine hemoglobin-beta (Hbb) alleles and the phylogenetic analysis of the cattle in Eastern Asia. *Zeitschrift für Tierzuchtung und Zuchtungsbiologie* 98, 17.

Polziehn, R.O., Strobeck, C., Sheraton, J. and Beech, R. (1995) Bovine mtDNA discovered in North American bison populations. *Conservation Biology* 9, 1638–1643.

Rollinson, D.H.L. (1984) Bali cattle. In: Mason, I.L. (ed.) *Evolution of Domesticated Animals*. Longman, New York, pp. 28–34.

Ruvinsky, A. and Rothschild, M.F. (1998) Systematics and evolution of the pig. In: Rothschild, M.F. and Ruvinsky, A. (eds) *The Genetics of the Pig*, CAB International, Wallingford, UK, pp. 1–16.

Shimamura, M., Yasue, U., Ohshima, K., Abe, H., Kato, H., Kishiro, T., Goto, M., Munechka, I. and Okada, N. (1997) Molecular evidence from retroposons that whales form a clade within the even-toed ungulates. *Nature* 388, 666–669.

Simoons, F.J. (1984) Gayal or mithan. In: Mason, I.L. (ed.) *Evolution of Domesticated Animals*. Longman, New York, pp. 34–39.

Tanaka, K., Solis, C.D., Masangkay, J.S., Maeda, K., Kawamoto, Y. and Namikawa, T. (1996) Phylogenetic relationship among all living species of the genus *Bubalus* based on DNA sequences of the cytochrome *b* gene. *Biochemical Genetics* 34, 443–452.

Thewissen, J.G.M. and Hussain, S.T. (1993) Origin of underwater hearing in whales. *Nature* 361, 444–445.

Tumennasan, K., Tuya, T., Hotta, Y.M., Takase, H., Speed, R.M. and Chandley, A.C. (1997) Fertility investigations of the F1 hybrid and backcross progeny of cattle (*Bos taurus*) and yak (*B. grunniens*) in Mongolia. *Cytogenetics and Cell Genetics* 78, 69–73.

Wall, D.A., Davis, S.K. and Read, B.M. (1992) Phylogenetic relationships in the subfamily Bovinae (Mammalia: Artiodactyla) based on ribosomal DNA. *Journal of Mammalogy* 73, 262–275.

Zeuner, F.E. (1963) *A History of Domesticated Animals*. Hutchinson, London.

Genetic Aspects of Domestication

D.G. Bradley and E.P. Cunningham
Department of Genetics, Trinity College, Dublin 2, Eire

Introduction	15
Present Variations in Cattle Populations	16
Europe	17
Asia	19
Africa	20
Australasia and the Americas	21
The Genetic Effects of the Domestication Process	22
Phenotype	22
Genotype	24
Archaeological Inferences from Patterns of Genetic Variation	26
DNA variation	26
Microsatellites and African cattle origins	28
References	29

Introduction

The cultural, economic and sometimes religious importance that cattle assume in modern pastoral societies is undoubtedly reflective of that in the earliest herding communities. Although predated by dogs, sheep, pigs and goats, cattle were the largest and subsequently most important of the domesticated animal species. The domestication of an animal of the dimensions of the now-extinct aurochs, or wild ox *(Bos primigenius)*, must have been a formidable task and was one of the definitive points of human development. In addition to products such as milk, meat, hides and dung, it is likely that the traction power represented by the domestic ox would have been instrumental in opening up the heavy soils of Europe for the expansion of agriculture from the Near East.

Cattle species today include: humpless cattle *(Bos taurus)*, which predominate in Europe, northern Asia, West Africa and the Americas; humped cattle *(Bos indicus)*, which thrive in more arid climes; Bali cattle of South-East Asia *(Bos javanicus)*; gayal or mithan *(Bos gaurus)* of north-west India; and the domestic yak *(Bos grunniens)*. From morphology and limits to interfertility, it is clear that the wild progenitors of the latter three were distinct species and, as such, must have involved separate domestications in Asia. However, a distinction between humped (zebu) and humpless (taurine) cattle has been less obvious, and several authors (Payne, 1970; Epstein and Mason, 1984) have favoured a single domestic origin for *B. indicus* and *B. taurus*, centred on the taurine cattle of the early neolithic centres of the Near East. This view holds that zebu morphology developed as a derivative type on the eastern margin of the region. However, the aurochs had a wide range, stretching over much of Eurasia and parts of North Africa, and some geographical variation is detectable in osteological remains. Some archaeologists (Zeuner, 1965; Grigson 1980; Meadow, 1993) have held that the domestication of two or more geographically distinct strains of wild oxen is a more likely explanation of modern and ancient patterns of variation.

In this review, we shall assess the effects of the domestication process on the genetics of the aurochs. Additionally, we shall describe what the study of modern genetic variation implies about the nature of the different *B. primigenius* strains involved.

Present Variations in Cattle Populations

With a global total of almost 1300 million head, cattle are the most numerous of domesticated mammals. This dominant position can be attributed to two factors. As efficient and relatively unselective herbivores, they occupy a complementary position to humans in exploiting plant foods, subsisting largely on inedible forages. Secondly, they have multiple utility in evolving farming systems, providing dung for fuel, hides, work, currency and saving and security functions, as well as milk and meat. Their role is particularly important in pastoral areas, which comprise two-thirds of all agriculturally managed land and which are not suitable, for climatic or other reasons, for crop production.

As agriculture has evolved from subsistence through to commercial systems, the role of cattle has changed considerably. In developed farming systems, cattle now have clearly defined and specialized roles as either milk or meat producers. With the growing demand for livestock products in the diet, as disposable incomes have increased, and the steadily lower cost of grains, as a result of technical advances, the dietary base for cattle production has shifted to a considerable degree from forage to grains, particularly in milk production systems.

These developments have led to sharpening differences between the developed and developing world in the numbers, nature and function of cattle. In the developed world, with low population growth, high consumption levels of cattle products and steady increases in productivity per animal, total animal numbers are decreasing. This is particularly the case in dairy cattle. In developing countries, on the other hand, demand, driven by rising population numbers and income levels, is growing. While productivity per animal is increasing in many countries, cattle numbers are also tending to increase. At present, over 70% of the world's cattle are in developing countries, with a ratio of one bovine to every 4.7 people, while in the developed world, with under 30% of the global population of cattle, the ratio is one bovine to 3.3 people.

The broad variation in today's global cattle populations is most conveniently classified on a geographical and breed basis. To a large degree, these criteria run parallel. The breeds prevalent today in Europe, Asia and Africa are mainly ones native to those continents, while cattle in Australasia and the Americas are derivatives from Old World populations.

Europe

The most comprehensive survey of cattle breeds in Europe is given by Simon and Buchenauer (1993). They list 277 separate breeds within national boundaries, although in many cases these represent national populations of a widely used breed, such as the Holstein Friesian. Using estimates of effective population size and generation interval, they classified breeds according to their degree of endangerment. A breed was regarded as normal if the expected increase in inbreeding over a 50-year period was under 5%. With this criterion, 128 breeds were classified as normal. The remainder, representing more than half of the total, were in some degree endangered, with 57 being rated as critical.

Formal breed definition, with herd-books and rules of registration, began about 200 years ago with Coates' Shorthorn herd-book in Britain. For most breeds, the herd-book is of much more recent origin, with many established in the late 19th century. While herd-books included only a minority of animals, most bulls in use would have been registered, and therefore, over a number of generations, genes from the herd-book section would predominate in the wider population. This process has tended to sharpen the differences between breed types in Europe over the past two centuries. As a result, almost all cattle in Europe can be assigned to clearly defined breed groups.

A contrary trend has been systematic migration of genes from one population to another. With the advent of artificial insemination in the 1950s, this process has become easier, cheaper and more rapid. Much of this gene flow has been between breeds of similar colour. Thus, various Scandinavian Red populations have been systematically exchanging genetic material for some decades, as have the different groups of brown dairy cattle. All populations of black and white cattle now make extensive use of an internationally available pool of bulls.

A third factor affecting breed variability is the rapid process of breed replacement. A century ago, Holstein Friesian-type cattle were essentially limited to their home area of the Netherlands and northern Germany. Today, they constitute almost two-thirds of the dairy cow population of Europe. In the process, many local breeds have disappeared. On a smaller scale, the NRF breed in Norway has replaced seven local populations. This process is driven by ever more effective selection both between and within breeds for economic performance within a specialized production system.

To offset this genetic erosion at breed level, most countries in Europe now have put in place specific programmes to conserve endangered breeds. This is supported by actions at the level of the European Union.

The principal European breed groups are as follows.

Holstein Friesian

Holstein Friesian originated in the Netherlands and northern Germany, where, by the middle of the 19th century, the characteristic black and white colour pattern was predominant. Substantial exports from 1860 onwards led to the establishment of black and white cattle in many European countries and in the USA, where it became the principal dairy breed. With an estimated 70 million head worldwide, Holstein Friesian represents more than one-third of the world's dairy cattle (Cunningham and Syrstad, 1987).

Following trial importations and a major experiment in Poland (Jasiorowski *et al.*, 1988), which demonstrated genetic superiority for milk production in American Holsteins, widespread importations have now largely converted all European black and white populations to American genotypes. There is also considerable gene flow between European populations. European, North American and other Holstein Friesian groups can therefore now be regarded as effectively a single global breed.

Red or red and white

In many European countries, there are significant populations of red or red and white dairy cattle (Montbeliard and Normande in France, MRY in the Netherlands, Red Dane, Finnish Ayrshire, Rotbunte and Angler in Germany). There has been less gene flow between these populations than in the case of Holstein Friesian. However, in most cases, significant genetic change has begun. The Red Dane has largely incorporated American Brown Swiss genes, while many of the other populations have made significant use of American Red Holstein sources.

Brown breeds

These have their origin in native Alpine breeds. They are significant in Germany, Switzerland, Austria, Italy and France. In all cases, the pressures for high-output dairy performance have led to extensive use of American Brown Swiss genes.

Simmental or Fleckvieh

This group of breeds, with origins in southern Germany and Switzerland, has also been extensively used in Hungary, Ukraine and Russia. While used for milk production, it is more heavily muscled than the other dairy breeds. This dual-purpose character has been maintained, and the breed has not been subject to much inward gene flow.

Beef breeds

Specialized beef breeds represent less than 20% of Europe's cattle. Significant purebred groups are important in some countries, notably in France, while crossbred animals are the norm elsewhere. The principal French breeds, Charolais, Limousin and Blonde d'Aquitaine, are all heavily muscled, but differ in some other characteristics. Charolais are white/cream in colour and have a docile temperament and a large adult body size. Limousin are somewhat smaller, brown, less docile and of finer bone and slightly higher meat content. Blonde d'Aquataine are white in colour and large-framed animals that have evolved from a base of several breeds in south-west France.

The principal British beef breeds Angus and Hereford have a worldwide distribution, although their relative position in their home region of Britain and Ireland has declined. They mature at lower adult weights than the French beef breeds and with a higher fat level in the carcass, and are therefore somewhat less suitable for intensive production systems.

Many other local breeds (e.g. Alentejana in Portugal, Retinto in Spain, Maremmana and Marchegianna in Italy, Salers in France) subsist as beef producing breeds in their home regions. Generally they are declining in numbers and involved in conservation programmes.

Asia

Of the approximately 800 identified cattle breeds in the world, one-quarter are found in Asia. However, a large proportion of Asian cattle can be assigned to these breed groups only in the most general sense. Bhat and Taneja (1987) concluded that only 18% of the cattle population of the Indian subcontinent belonged to a well-defined breed.

Most Asian cattle breeds are *B. indicus* types. These are characterized by a pronounced thoracic or cervicothoracic hump and are often collectively called zebu cattle. The most recent and thorough description of Asian breeds is given by Payne and Hodges (1997). Their coverage extends to some 80 of the approximately 200 Asian breeds. Cattle are not numerous in the drier west Asian countries. They are generally varied in colour, shorthorn and often known by the Arabic term *Baladi*. Most of these breeds are declining in number or involved in crossbreeding.

From Iran eastward, humped cattle predominate. The hump consists of muscle, connective tissue and variable amounts of fat. The size and shape may vary by breed according to the sex and age of the animal. The function of the hump is not clearly understood, and earlier theories that it was a fat storage depot are not now generally accepted. Zebu cattle tend to have a narrow body, sloping rump and long legs. The hide is loosely attached and the brisket and dewlap are usually well developed, particularly in males. Size varies over a wide range, with breed averages ranging from below 200 to above 400 kg for mature cows.

The body characteristics can be seen as adapted to heat dissipation. In addition to heat tolerance, *B. indicus* cattle have better resistance to ticks than *B. taurus* types, and are better able to withstand periods of low feed intake. Reflecting their different selection history since domestication, particularly in recent centuries, they have a lower milk potential than *B. taurus,* and in particular require the presence of the calf for milk let-down.

The most widespread zebu breed type is the Hariana, predominant in northern India. The bullocks are widely used for draught purposes and cows are milked, giving up to 1000 kg per lactation.

Kankrej (Gujrati) and Gir, both from the western regions of the country, combine good draught ability with acceptable milk yields, and also have a high potential for meat production. The same can be said for Ongole (Nellore), a breed originating from southern India. All three breeds have been exported to Central and South America, are the basis for the zebu breeds that predominate in beef production in Brazil and have also been important in milk production.

Sahiwal and Red Sindhi are zebu breeds originating in Pakistan. Both are reddish in colour, have short horns and have an established reputation as milk-producing cattle. Tharparkar is a third breed from the same region, also with dairy potential; it is usually somewhat larger than Sahiwal or Red Sindhi, and white or light grey in colour.

Zebu cattle are also found in south China, Taiwan and Indonesia, although in all cases some input from Indian breeds may be involved.

Bali cattle, from the island of Bali and neighbouring islands of Indonesia, differ from both *B. taurus* and *B. indicus* types, and are said to originate from a separate domestication. They are classified as *B. javanicus*, and have 2n = 60 chromosomes, similar to *B. indicus* and *B. taurus*. Their colour is reddish brown, with characteristic reddish-white markings on the rear and underparts. Bali cattle are used primarily for work, but are good meat producers.

In the sub-Himalayan areas of India, Bhutan and China, related taxa are also found in both wild and domesticated circumstances: Gaur (*B. gaurus*), Mithun and Dulong. All three have 2n = 58 chromosomes and are apparently cross-fertile with cattle, although the offspring, particularly males, may not be fertile.

Africa

Some 120 cattle breeds have been classified for Africa. As in Asia, most cattle can be assigned to breed type only in a general way, and formal herd-books and registration structures are rare. In their discussion of African breeds, Payne and Hodges (1997) offer brief descriptions of 72 breeds.

African cattle fall into three broad groups, though with considerable gene flow between them. In North and much of West Africa, *B. taurus* types predominate. In Egypt, the local *Baladi* cattle can be compared morphologically to similar types in west Asia. A more distinctive group is the Atlas Brown, found in Tunisia, Algeria and Morocco. Their colour is brown to dark, and the animals are described as sturdy. They are used for work, as well as milk and meat production, although productivity is generally low.

B. taurus breeds of West Africa are found primarily in the more humid areas of the countries ranging from Senegal to Nigeria. The best-known breed types are the N'Dama, centred on Guinea but with populations in several other countries, and West African Shorthorn, including Namchi and Kapsiki in Cameroon and Baoulé in the Ivory Coast and Burkina Faso. All are used for meat and milk production, and to some extent for draught purposes. They are of considerable interest because of their demonstrated tolerance of trypanosomiasis, a blood-parasitic disease transmitted by tsetse flies and generally fatal to introduced breeds.

In the drier regions, north of the trypanosomiasis belt, most cattle breeds are of *B. indicus* type. Ranging from west to east, the group includes Maure in Mauritania, Mali, Burkina Faso; Fulani ranging throughout the western Sahel; Baggara in western Sudan; Kenana and Butana in Central Sudan; and Boran in Ethiopia, Somalia and Kenya.

Many of these breeds are kept in nomadic herding systems, and vary considerably in size, colour and other characteristics. Some, including Kenana, Butana and Boran, have a considerable reputation as beef cattle.

The native breeds of Central and Southern Africa generally display a phenotype intermediate between *B. taurus* and *B. indicus*, with, in particular, reduced hump and dewlap. They are collectively called Sanga breeds, and are believed to be the result of crossing between *B. taurus* and *B. indicus* types. Most are involved in pastoral systems with domestic milk production and limited use for draught purposes. Some of the better-known breed types within this group are Ankole in Tanzania, Uganda and neighbouring countries; Barotse in Zambia and Zimbabwe; Tuli and Mashona in Zimbabwe; Nguni and Tswana in the Republic of South Africa, Botswana and Mozambique. Each breed has local subtypes.

Australasia and the Americas

These continents have very large cattle populations, all of postcolonial origin. In Australasia, the initial importations were exclusively of British breeds. In New Zealand, this has largely remained the case, with Angus predominant in beef-producing areas, and Jersey and Holstein Friesian in dairy-farming areas.

In Australia, dairy farming is almost exclusively with Holstein Friesian, but the great bulk of the cattle population is involved in beef production in pastoral areas. In temperate regions, Hereford and Angus types were once dominant. Today, these areas also include numbers of Simmental, Charolais and other types, and crosses between all of these beef breeds. In more tropical parts of the country, *B. indicus* types and crosses have largely replaced European breeds. The *B. indicus* strains involved are largely of North American origin, which in turn derive from Brazilian stock, in their turn based on importations from India in the last century.

The cattle of Latin America were derived from Spanish and Portugese breeds, possibly with some West African influence. Locally adapted descendants of these importations are generally called Criollo. In Brazil, very large numbers of descendants of Indian cattle (Nellore, Guzera, Gir) are used in beef production. In Argentina, the very large beef herd is mainly derived from Hereford and Angus.

North American dairy cattle are today almost exclusively Holstein Friesian. Beef cattle are four times more numerous and include derivatives of all European beef breeds, though with Angus and Hereford still very important. In addition, in southern parts of the country, *B. indicus* types derived from Brazilian imports and their crosses with European breeds, predominate.

The Genetic Effects of the Domestication Process

Phenotype

The main criteria used to decide whether animal remains at an archaeological site are wild or domestic are: the presence of a species outside its normal geographical range; an age and sex structure in the remains which would be atypical in the wild; a sudden increase in the frequency of a species between different time horizons; and evidence for reduction in animal size (Grigson, 1989; Meadow, 1989). The latter is likely to be a consequence, not only of the changed environment in which the animals exist, but of genetic change in the managed herds.

A convincing archaeological record of changing cattle size over the time in which domestication took place is available from two regions: the Near East (Grigson, 1989) and Baluchistan (Meadow, 1993). In each case, when bone dimensions are normalized by comparisons with those of a reference animal, definite and consistent decrease is evident in collections from later levels. For example, the box plot in Fig. 2.1 summarizes the distributions of bone measurements from the western part of the Near East in collections spanning a period of five millennia (based on Grigson, 1989). This type of data from a defined region, coupled with other changes in the patterns of faunal remains, provides strong support for local domestication of the aurochs.

The restriction of cattle foraging and the diseases, which are a consequence of confinement, could have been environmental factors leading to the reduction of animal size that occurs after domestication. However, it is likely that the major source of change would have been genetic. Selection for docility and manageability of animals would have been a strong motivation for early herders. Additionally, the time differences between archaeological levels may be several hundreds of years, representing many animal generations with ample opportunity for selective effects. Size decrease is most pronounced in

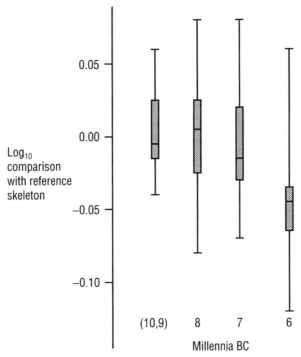

Fig. 2.1. The size of *Bos* remains in the western part of the Near East from the tenth to the sixth millennium BC. Adapted from Near Eastern bovine data presented and analysed by Grigson (1989). The data shown consist of logarithmic comparisons of bone measurements with those taken from the elements of a standard skeleton, in this case that of a female auroch found in Denmark. A clear diminution in size in sixth-millennium cattle is apparent and is taken to be a result of, and also evidence for, the domestication process.

male cattle and the marked sexual dimorphism of the aurochs, typical of *Bovini*, is almost absent in domestic cattle.

Genetic change seems to have been a continuing process in cattle domestic history. The earliest cattle of Europe were relatively large long-horned beasts and have mostly been replaced by the smaller short-horned varieties, which predominate today. Overall, we can see today an enormous range of local breeds, strains or types. In the course of human history, these have been moved with the migration of peoples, crossed with other local strains, and selected continuously both by human intervention and by the forces of natural selection operating in the particular farming system of which they are a part.

Change in the farming system involves changed breeding objectives for livestock. In traditional systems, number of livestock rather than output per head is often the main consideration. In these circumstances, traits related to survival in the face of nutritional, health and climatic stress predominate. As farming systems become more market-orientated, volume and value of saleable product take over. In these circumstances, selection goals often change from multipurpose use to much narrower targets. They shift to such traits as prolificacy, early maturity, individual growth rate and milk yield, and aspects of milk composition and carcass quality.

Selection for such traits has transformed the genetic constitution of many developed breeds. For example, in many non-market traditional cattle systems, cows first calve at 4–5 years and thereafter at 2-year intervals, producing perhaps 500 l of milk year^{-1}. In modern, highly developed, dairy production systems, cows first calve at 2 years, and thereafter at yearly intervals, and produce up to 10,000 l of milk year^{-1}. The genetic change in the animal has made possible the intensification of the system, while the evolution of the system of feed supply, management and health care has made possible the support for high-producing genotypes. The net result is a dramatic improvement in efficiency of resource use.

In addition to the direct effect of selection on the genetic constitution of livestock populations, changes in breeding objectives can also promote inter-population gene flow. Thus, in West Africa, demand for beef animals of higher body weight has led to substantial crossbreeding of larger *B. indicus* on the smaller *B. taurus* populations. Such patterns of crossbreeding can, over a few generations, effectively lead to total replacement of the domestic pool of genetic diversity. In some cases – where artificial insemination is widely used or where domestic strains are simply bypassed by the importation of totally new genotypes – this can happen very rapidly.

Genotype

The principal product of a domestic animal is its progeny. Domestication of cattle may have involved the capture of a relatively small number of aurochs, from which the 1300 million head alive today are descended. One question is whether such an initial restricted sampling of the genetic variation from the wild progenitor, together with subsequent population bottlenecks experienced in early herds, has reduced genetic variability in the species.

One of the most valuable indicators of genetic variation in a species is the examination of protein polymorphisms through electrophoretic assay. A similar array of enzymatic and non-enzymatic proteins is accessible for study in a range of mammalian species, and the amount of heterozygosity observed at these loci is a useful comparative indicator of total variability. Estimates for average heterozygosity in *B. taurus* populations have varied between 4 and 8% (Baccus *et al.*, 1983), and compare well with the mammalian mean of $4.14 \pm 0.25\%$. Figure 2.2 illustrates the distribution of average heterozygosity estimates obtained from studies of 183 mammalian taxa (after Bancroft *et al.*, 1995; see also Nevo *et al.*, 1984). The relatively high level of estimates from cattle are indicated. Figures from *B. taurus* are comparable to, or greater than, those from other ungulates. Notably, those species that are known to have undergone historical population bottlenecks, such as *Bison* sp., have yielded lower heterozygosity values (Hartl and Pucek, 1994).

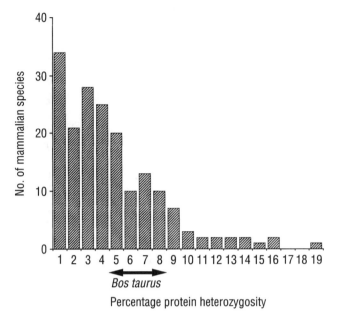

Fig. 2.2. Distribution of average protein heterozygosity for 183 mammalian species or subspecies (adapted from Bancroft et al., 1995). The range of a series of estimates calculated for domestic cattle populations is denoted by an arrow on the horizontal axis.

The persistence of significant genotypic variation in modern cattle is also evidenced by deoxyribonucleic acid (DNA) studies. Extensive typing of microsatellite markers has revealed substantial variability both within and among populations (MacHugh et al., 1997), although such data are difficult to compare across species because of a bias due to the initial selection of microsatellites for high polymorphism in the species of origin.

Many DNA studies of intraspecific variation have centred on highly polymorphic sequences of the major histocompatibility complex. This gene-rich genomic segment contains several genes important in the immune response, and it is well established that deeply divergent alleles or haplotypes are maintained by balancing selective forces, sometimes even across species boundaries (Takahata, 1990). Similarly to sheep and goat, cattle exhibit many alleles, high heterozygosities and substantial divergence between alleles in sequence studies of class II *DRB* loci (Mikko, 1997). This is in contrast to the situation in American bison where the number of alleles seems to have been noticeably reduced by the severe hunting-induced bottleneck of the last century, although genetic divergences between alleles themselves are similar to those within cattle (Mikko et al., 1997).

Thus it seems there is little evidence from autosomal genetic systems for any significant reduction in bovine variability as a result of the domestication process. This is perhaps explainable in several ways. Firstly, the undoubted population bottleneck which would have resulted from the initial capture of a limited number of wild aurochs may have been short-lived. The genetic effects of a bottleneck are dependent on a number of factors, including the number of generations for which it is endured. It may be that wild oxen were initially adopted into systems in which cattle numbers were expanding, and thus little loss of genetic variation was effected. Secondly, the process of incorporating aurochs into the ancestral domestic pool may have been an often-repeated one, resulting in a high proportion of the genetic variation in the wild species being harnessed. That variation may have been considerable, the aurochs itself being a successful, wide ranging species. Lastly, more sensitive molecular tools may be required to detect any diversity loss induced by domestication.

The effects of a population bottleneck, while not detectable using autosomal markers, may in fact be discerned through examination of mitochondrial DNA (mtDNA) haplotypes. These possess an effective population size, which is one-quarter that of nuclear DNA, and are more susceptible to lineage loss in reduced populations. Indeed, extensive study of mtDNA variation in widely sampled populations has revealed the imprint of the domestication process and allowed novel inference about the nature of events some 10,000 years ago (Bradley *et al.*, 1998).

Archaeological Inferences from Patterns of Genetic Variation

DNA variation

An initial survey of mtDNA control region sequence variation, using 26 samples from Western Europe, Africa and India (origins shown in Fig. 2.3), revealed the existence of two substantially divergent clades (Loftus *et al.*, 1994). The sequences of European, *B. taurus* origin showed approximately 5% difference from those of Indian *B. indicus* provenance. Sequences from Africa, sampled from both taurine and zebu breeds, were much more similar to those from Europe but showed some consistent difference (averaging 0.76%). The right-hand side of Fig. 2.4 includes a simplified phylogenetic tree, which summarizes these data.

Further analysis of a shorter region of the mtDNA chromosome has brought the number of sequences analysed to over 100 and this basic pattern of variation has been confirmed (Bradley *et al.*, 1996). The deep phylogenetic division between Indian mtDNA haplotypes and the rest has been interpreted as reflecting that between the progenitor of *B. indicus* and the ancestor of *B. taurus* cattle. A number of analyses attest that the grouping of these sequences into continental groups is justifiable (Bradley *et al.*, 1996) and this allows an inference to be drawn from comparisons based on quantitative divergence. The time depth of the divergence between the two major clades may be estimated by either applying the mtDNA control-region molecular clock, which has been calibrated in other species, or by using comparison with the equivalent *Bison* sequence and a corresponding palaeontological estimate of at least 1 million years for that interspecific separation. Several calculations

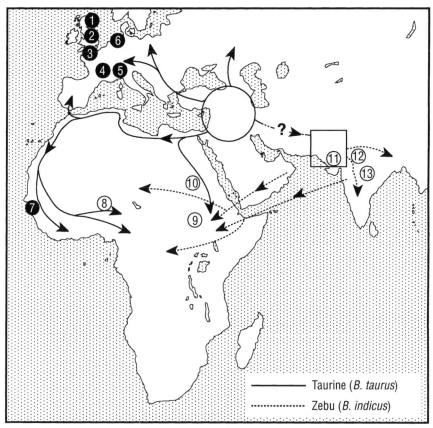

Fig. 2.3. Postulated migratory routes of cattle across western Asia, Africa and Europe (reprinted from Loftus et al., 1994). Geographical origins of breeds sampled for mtDNA analysis are represented by numbered circles (taurine, black; zebu, white).1, Aberdeen Angus; 2, Hereford; 3, Jersey; 4, Charolais; 5, Simmental; 6, Friesian; 7, N'Dama; 8, White Fulani; 9, Kenana; 10, Butana; 11, Tharparkar; 12, Sahiwal; 13, Hariana. The large circle represents the original domestication event and the square represents the formation of Asian zebu. The data discussed here are not consistent with zebu having developed as post-domestic derivatives of Middle Eastern taurines – indicated with a dotted line broken by a question mark.

have yielded figures for the Indian vs. Europe/Africa bifurcation as occurring in the order of hundreds of thousands of years BP (Bradley *et al.*, 1998). This is consistent with other data, such as that from microsatellites and biochemical polymorphisms, which emphasize zebu/taurine divergence (Manwell and Baker, 1980; MacHugh *et al.*, 1997) and leave it difficult to imagine that all modern cattle might have come from a single strain of wild ox domesticated only approximately 10,000 BP (see Fig. 2.3). There is strong support from these data for assertions that *B. indicus* cattle are descended from eastern *B. primigenius* variants, which may have been domesticated in centres such as Mehrgarh, Baluchistan (Meadow, 1993).

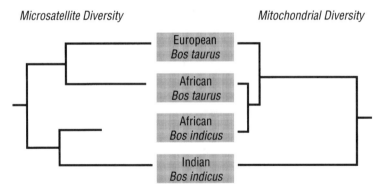

Fig. 2.4. Summary of phylogenetic tree topologies derived from microsatellite diversity (left-hand side) and mitochondrial sequence diversity (right-hand side) (reprinted from Bradley et al., 1998). Both types of data reveal the deepest phylogenetic division within cattle, that between *Bos indicus* and *Bos taurus*. However, the positioning of the African zebu is ambiguous. These cluster with Indian zebu (with truncated branch lengths) in microsatellite analyses and with the African taurines in mtDNA-derived phylogenies. Thus, African zebu seem to be hybrids, with the majority of their genome derived from introgressing *Bos indicus* but with maternally inherited mtDNA variation that is representative of the original *Bos taurus* domesticates of that continent. Both phylogenetic reconstructions illustrate some divergence between African and European taurines.

Microsatellites and African cattle origins

One surprising feature of mtDNA variation is that the zebu cattle of Africa, which are unmistakably *B. indicus* in morphology and physiology, exhibit sequence types which show no distinction from those of neighbouring *B. taurus* breeds. This is likely to reflect a hybrid origin for African zebu. An investigation of these and some additional cattle populations, using 20 microsatellite markers, has given a phylogenetic reconstruction that ties more closely with breed morphology (MacHugh *et al.*, 1997). The simplified left-hand phylogeny in Fig. 2.4 shows African *B. indicus* clustering with Indian cattle and a grouping of the *B. taurus* populations from Europe and Africa. Branch lengths of African zebu in neighbour-joining trees, which are constructed from allele frequency-derived genetic distances, are truncated and this is a further suggestion that these populations may be hybrid in origin.

Particularly strong inferences concerning hybridization may be drawn from frequencies of microsatellite alleles that are zebu-specific. Certain loci show allele length distributions in pure taurine (such as European and certain West African populations) and pure zebu (such as those of southern Indian provenance) cattle which are separate and distinctive. Consequently, it is possible to identify private alleles in approximately half of the markers tested which are specific to *B. indicus* and their hybrids, and which allow the estimation of the extent of admixture in breed ancestry. The descendants of primeval African cattle persist in parts of humid West Africa and are pure *B. taurus* in nature. Due to their long association with the continent, they display a suite of adaptations to the prevalent disease challenges, including, notably, a tolerance for trypanosomiasis, which is arguably the most important African livestock disease (Murray et al., 1982). Microsatellite-based estimation of zebu genetic influence in a range of breeds has illustrated a cline of admixture which stretches from the exclusively taurine N'Dama cattle of Guinea both northward to Sahelian zebu cattle, such as the Mauritanian Maure breed, and eastward, to the Sudanese Butana and Kenana breeds, which have displayed the highest *B. indicus* influence (MacHugh et al., 1997).

A Y-chromosome polymorphism that distinguishes between the two taxa has also been described (Bradley et al., 1994). In an examination of the frequency of the zebu form in the same series of breeds, stretching north–south from Mauritania to Guinea, it was striking that the paternally inherited Y variant seemed to be the most aggressively introgressing genetic element. These data, together with the apparently total absence of *B. indicus* mtDNA chromosomes on the continent, suggest that most African cattle are the products of a progressive, male-driven, admixture between the indigenous taurine breeds and immigrating zebu cattle, which have made major incursions within the last millennium and a half (Bradley et al., 1998).

The time depth of the ancestral divergence between African and European *B. taurus* cattle has been estimated, using both microsatellite and mtDNA variation (Bradley et al., 1996; MacHugh et al., 1997). These estimates (the lowest of which is 22,000 BP), are suggestive of the input of separate aurochs strains into these two continental gene pools. However, it must be noted that calibrations of the genetic divergence in both these genetic systems are difficult and potentially subject to wide errors. Further evidence from cattle breeds of the Near East and Egypt should do much to clarify the nature of the early domestication process in the Levant and, potentially, the Nile valley.

References

Baccus, R., Ryman, N., Smith, M.H., Reurterwall, C. and Cameron, D. (1983) Genetic variability and differentiation of large mammals. *Journal of Mammalogy* 64, 109-120.

Bancroft, D.R., Pemberton, J.M. and King, P. (1995) Extensive protein and microsatellite variability in an isolated, cyclic ungulate population. *Heredity* 74, 326–336.

Bhat, P.N. and Taneja, V.K. (1987). Principles of indigenous animal improvement in the tropics – the programme for India. In: Hodges, J. (ed.) *Animal Genetic Resources. Strategies for Improved Use and Conservation.* FAO Animal Production and Health Paper No. 44/1, FAO, Rome, pp. 44–52.

Bradley, D.G., MacHugh, D.E., Loftus, R.T., Sow, R.S., Hoste, C.H. and Cunningham, E.P. (1994) Zebu–taurine variation in Y chromosomal DNA: a sensitive assay for genetic introgression in West African trypanotolerant cattle populations. *Animal Genetics* 25, 7–12.

Bradley, D.G., MacHugh, D.E., Cunningham, P. and Loftus, R.T. (1996) Mitochondrial diversity and the origins of African and European cattle. *Proceedings of the National Academy of Sciences, USA* 93, 5131–5135.

Bradley, D.G., Loftus, R.T., Cunningham, P. and MacHugh, D.E. (1998) Genetics and domestic cattle origins. *Evolutionary Anthropology* 6, 79–86.

Cunningham, E.P. and Syrstad, O. (1987) *Cross-breeding* Bos indicus *and* Bos taurus *for Milk Production in the Tropics.* FAO Animal Production and Health Paper No. 68, FAO, Rome.

Epstein, H. and Mason, I.L. (1984) Cattle. In: Mason, I.L. (ed.) *Evolution of Domesticated Animals.* Longman, London, pp. 6–27.

Grigson, C. (1980) The craniology and relationships of four species of *Bos.* 5. *Bos indicus* L. *Journal of Archaeological Science* 7, 3–32.

Grigson, C. (1989) Size and sex: the evidence for the domestication of cattle in the Near East. In: Milles, A., Williams, D. and Gardner, G. (eds) *The Beginnings of Agriculture.* BAR International Series 496. British Archaeological Reports, Oxford, pp. 77–109.

Hartl, G.B. and Pucek, Z. (1994) Genetic depletion in the European Bison (*Bison bonasus*) and the significance of electrophoretic heterozygosity for conservation. *Conservation Biology* 8, 167–174.

Jasiorowski, H.A., Stoltzman, M. and Reklewski, Z. (1988) *The International Friesian Comparison Trial. A World Perspective.* FAO, Rome.

Loftus, R.T., MacHugh, D.E., Bradley, D.G., Sharp, P.M. and Cunningham, E.P. (1994) Evidence for two independent domestications of cattle. *Proceedings of the National Academy of Sciences, USA* 91, 2757–2761.

MacHugh, D.E., Shriver, M.D., Loftus, R.T., Cunningham, P. and Bradley, D.G. (1997) Microsatellite DNA variation and the evolution, domestication and phylogeography of taurine and zebu cattle (*Bos taurus* and *Bos indicus*). *Genetics* 146, 1071–1086.

Manwell, C. and Baker, C.M.A. (1980) Chemical classification of cattle. 2. Phylogenetic tree and specific status of the zebu. *Animal Blood Groups and Biochemical Genetics* 11, 151–162.

Meadow, R.H. (1989) Osteological evidence for the process of animal domestication. In: Clutton-Brock, J. (ed.) *The Walking Larder: Patterns of Domestication, Pastoralism and Predation.* Unwin Hyman, London, pp. 80–90.

Meadow, R.H. (1993) Animal domestication in the Middle East: a revised view from the Eastern Margin. In: Possehl, G. (ed.) *Harappan Civilisation.* Oxford and IBH, New Delhi, pp. 295–320.

Mikko, S. (1997) *A Comparative Analysis of Genetic Diversity at Mhc DRB Loci in Some Ruminant Species.* Swedish University of Agricultural Sciences, Uppsala.

Mikko, S., Spencer, M., Morris, B., Stabile, S., Basu, T., Stormont, C. and Andersson, L. (1997) A comparative analysis of Mhc *DRB3* polymorphism in the American Bison (*Bison bison*). *Journal of Heredity* 88, 499–503.

Murray, M., Morrison, W.I. and Whitelaw, D.D. (1982) Host susceptibility to African trypanosomiasis: trypanotolerance. *Advances in Parasitology* 21, 1–68.

Nevo, E., Beiles, A. and Ben-Shlomo, R. (1984) The evolutionary significance of genetic diversity: ecological, demographic and life history correlates. In: Mani, G.S. (ed.) *Evolutionary Dynamics of Genetic Diversity.* Springer-Verlag, Berlin, pp. 13–213.

Payne, W.J.A. (1991) Domestication: a forward step in civilisation. In: Hickman, C.G. (ed.) *Cattle Genetic Resources.* World Animal Science Vol. B7, Elsevier, Amsterdam, pp. 51–72.

Payne, W.J.A. and Hodges, J. (1997) *Tropical Cattle: Origins, Breeds and Breeding Policies.* Blackwell Science, Oxford.

Scherf, B. (ed.) (1995) *World Watch List for Domestic Animal Diversity*, 2nd edn. FAO/UNEP, Rome.

Simon, D.L. and Buchenauer, D. (1993) *Genetic Diversity of European Livestock Breeds*. EAAP Publication 66, Wageningen Press, Wageningen, the Netherlands.

Takahata, N. (1990) A simple genealogical structure of strongly balanced allelic lines and trans-species evolution of polymorphism. *Proceedings of the National Academy of Sciences, USA* 87, 2419–2423.

Zeuner, F.E. (1963) *A History of Domesticated Animals*. Hutchinson, London.

Genetics of Colour Variation

3

T.A. Olson

Animal Science Department, University of Florida, 202-B Animal Science Building, Gainesville, FL 32611, USA

Introduction	33
The Basis of Pigmentation in Cattle	35
The Wild Type	35
Variations from the Wild-type Coat Colour	36
The *Extension* locus (black/red)	36
Influence of *Agouti* locus mutants	40
The *dun* locus	41
The *Albino* locus	42
Dilution mutants of the Charolais, Simmental, Gelbvieh and other breeds	42
White-spotting Mutants	44
The *S* locus mutants	46
Blaze mutants	47
Roan	48
Colour-sided	48
Belted	50
Brockling/pigmented legs	50
Minimal white spotting	51
White-spotting patterns in African and zebu breeds	51
References	51

Introduction

Variation in coat colour and spotting patterns of cattle have been of interest for many centuries, as indicated by the Lascaux cave drawings of cattle, which depict cattle with white-spotting patterns. In the 1800s, when some of the English breeds were being developed, breeders attempted to produce a reasonably uniform coat colour and spotting pattern within most breeds, as an aid to breed identity. The red colour and white spotting pattern of the Hereford breed is an example. In addition to the primarily aesthetic aspects of colour,

©CAB International 1999. *The Genetics of Cattle* (eds R. Fries and A. Ruvinsky)

there is evidence that, under tropical conditions with high levels of solar radiation, animals with a light-coloured hair coat and darkly pigmented skin are better adapted (Finch and Western, 1977; Finch *et al.*, 1984). Most zebu breeds, which are well adapted to tropical conditions, have such a colour. The Italian breeds that have been imported into the USA, the Chianina, Marchigiana, Romagnola and Piemontese, also show this colour.

Several studies have indicated that the percentage of white on Holstein cows can have an impact on milk yield and reproductive traits in regions of high solar radiation. King *et al.* (1988) reported that cows with over 60% white that had calved in February and March required fewer services per conception and had fewer days open than cows with a higher percentage of black in their coats. Cows with a higher percentage of white showed a lowered depression in milk yield following exposure to intense solar radiation without shade (Hansen, 1990). The white cows also showed smaller alterations in physiological variables than did cows with less white. Becerril *et al.* (1994) reported a modest advantage in productivity of Holstein cows with greater quantities of white during the summer months in Florida. They also observed an advantage (not significant) for reproductive traits in these 'whiter' cows. Any advantage of the white hair coat is due to its lowered absorption of incident solar radiation (Stewart, 1953), which results in reduced heat stress.

Another association of a colour-related trait and an economically important trait is the relationship between eyelid pigmentation and the susceptibility to eye lesions, leading to 'cancer eye' in Hereford and other cattle breeds (Anderson, 1991). It was found that increasing eyelid pigmentation in Hereford cattle resulted in a decreased incidence of lesion development. In addition, several studies were cited by Brown *et al.* (1994) that reported an association between colour and horn fly counts on cattle, with greater numbers of flies being observed on darker than lighter cattle.

Changes in the breeding programmes of the US cattle industry in recent years have led to additional interest in the genetic control of coat colour in cattle. For example, the development of a composite breed of cattle in which it may be desired to fix a certain coat colour, such as solid red, from a foundation population that includes both black and spotted animals will require a knowledge of the inheritance patterns involved if it is to be easily achieved. Such information is also useful in the establishment of new colour and spotting patterns within established breeds, which allow upgrading (e.g. black, non-spotted Simmental). The existence of price discounts or premiums for feeder calves of various colours also leads to greater interest in colour genetics. For example, black or black, white-faced calves may receive a premium, regardless of actual breed composition; on the other hand, calves with zebu breeding that express grey or dark brown colours may receive a price dock, whereas black or red calves with the same zebu breeding may not suffer one.

The study of the genetic control of coat colour in cattle dates back over 90 years to the study of the inheritance of coat colour in Shorthorns and their crosses by Barrington and Pearson (1906). Reviews of the subject have been published periodically since then, including those of Ibsen (1933), Lauvergne

(1966), Searle (1968) and Olson and Willham (1982). This chapter will incorporate the results of recent observations and research to update the conclusions of Olson and Willham.

The Basis of Pigmentation in Cattle

The basis of coat colour in cattle and all mammals is the presence or absence of melanins in the hair (Searle, 1968). The melanin is found in the melanosomes of the cytoplasm of the melanocytes. These melanosomes are transferred to the hair as it grows through a process of exocytosis. The melanocytes migrate from the neural crest during embryonic development and only areas of the body in which they are found are pigmented. White spotting occurs in areas where the skin or hair lacks melanocytes. Pigmentation in all or part of the body may also be diminished by reduced activity of the melanocytes. Melanins of two types, eumelanin and phaeomelanin, are found in mammals. Eumelanin is responsible for black and brown colours and phaeomelanin for reddish brown, reds, tans and yellows. A more detailed discussion of the production of coat pigmentation can by found in Searle (1968) and the review of sheep colour genetics by Sponenberg (1997).

The Wild Type

To discuss most effectively the variation in coat colour and spotting patterns in any species, it is useful to explain the effects of mutants relative to the wild type. For spotting patterns in cattle, the wild type is simply a solidly pigmented animal or simply the lack of any spotting. Choice of a wild type for pigmentation is somewhat more difficult, but the colour of the aurochs of Europe, the wild ancestor of most (or all) *Bos taurus* breeds, seems appropriate (Olson, 1980). Aurochs were essentially reddish brown to brownish black with a tan muzzle ring. There was apparently some variation in the proportion of the body that was black and bulls were darker than cows. This coat colour or a similar one is occasionally observed in certain breeds today. Some Jersey, Brown Swiss and Longhorn purebreds, as well as crosses of these breeds with red breeds and some crosses of the Brahman with red breeds, produce the wild-type pigmentation pattern. Animals with wild-type coat colour tend to be darker at their extremities (head and neck, feet and hindquarters), similar to bay colour in the horse. Cattle with this type of brownish-black colour at maturity are born a reddish brown and darken when the calves shed out for the first time.

Adult bulls of several wild relatives of cattle, namely bison and banteng, have a similar dark brown coat colour. In banteng, adult cows are much lighter than bulls, having more of a tan colour, whereas, in the bison, cows are coloured like bulls.

Variations from the Wild-type Coat Colour

The most commonly observed variants from wild-type coat colour in cattle are red and solid black. Other coat colours of cattle are simply modifications of three basic colours: black, wild type (brown-black) and red. Most variations from these basic colours involve lightening or removal of pigmentation. Good examples are the light red colour of Limousin, the tan colour of many Jerseys and the almost complete removal of pigmentation of Chianina, some zebu and some Brown Swiss. Other mutant genes are responsible for the dilute colours of the Charolais and Simmental dilute pigmentation uniformly over the entire body, and for those found in Limousin, Jersey, Brown Swiss, Chianina (and the other white Italian breeds) and Brahman (and other zebu breeds), which tend to have differential effects on different parts of the body, especially the underline, the poll and along the back. Mutants thought to influence coat colour of cattle are shown in Table 3.1.

The Extension *locus (black/red)*

The *Extension* (*E*) locus is responsible for most of the variation in cattle coat colour. Three alleles present at this locus include: E^D, dominant black, E^+, the wild-type allele responsible for most combinations of red or reddish brown and black; and *e*, recessive red. The order of dominance of these alleles is $E^D > E^+ > e$ and is complete. This locus has been identified as the *melanocyte-stimulating hormone (MSH) receptor* locus (Robbins *et al.*, 1993) and commercial companies have deoxyribonucleic acid (DNA) probes to identify whether phenotypically black animals carry E^+ or *e*. A point mutation of a single base substitution (thymine (T) for cytosine (C)) was found in the E^D allele, which resulted in an amino acid change from leucine to proline (Klungland *et al.*, 1995). A base deletion resulting in a frame shift has been found in the *e* allele (Klungland *et al.*, 1995; Joerg *et al.*, 1996).

Microsatellite analysis by Klungland *et al.* (1995) indicated that the *Extension* locus is located on chromosome 18. The *E* locus alleles regulate the level of tyrosinase production, which, in turn, determines whether eumelanin (black) or phaeomelanin (red) pigment is produced. Low levels of tyrosinase lead to the production of phaeomelanin, while high levels lead to eumelanin production. Tyrosinase levels are determined by the activity of the MSH receptor. If the E^D allele is present, an MSH receptor is produced that is active regardless of the activity of any alleles of the *Agouti* (*A*) locus, resulting in increased tyrosinase within the melanocytes and eumelanin production (Robbins *et al.*, 1993). The colour of animals with E^+ has been shown to be influenced by genes at the *Agouti* locus in other species. When the *wild-type* allele, A^+, is present at the *Agouti* locus together with E^+ at the *Extension* locus, the combination of areas of black along with red pigmentation of the wild-type colour is produced. A low level of tyrosinase (resulting in red pigment

Table 3.1. Mutants influencing the colour of cattle.

Locus symbol	Locus name	Allele symbol	Allele name/description	Mode of inheritance relative to wild type	Breed distribution
E	Extension	E^D	Dominant black/uniformly black at birth	Dominant	Holstein, Angus, etc.
E	Extension	E^+	Brown-black with darker extremities, bulls are darker than cows and calves are born a reddish brown (wild type)	—	Jersey, Brown Swiss, Brahman
E	Extension	e	Red/red without any dark pigmentation	Recessive to E^D and E^+	Hereford, Red Angus, Guernsey, Simmental and other red breeds
Br	Brindle	Br	Brindle/alternating stripes black and red pigmentation	Dominant to lack of brindling	Most solid red and black breeds
A	Agouti	A^{bp}	Patterned blackish/a modifier of wild type that is similar to the wild-type pattern, but nearly entirely black and not influenced by sex	Dominant in the presence of E^+, hypostatic to E^D	Holstein, Jersey, Brown Swiss, Brahman
A	Agouti	a^w	White-bellied agouti/removal of red pigment and a part of the black pigment while causing more uniform distribution of black pigmentation, especially across the sides of the animal	Recessive	Brown Swiss, Grey Steppe
A	Agouti	a^i	Fawn/removal of red and black pigmentation, particularly red along the underline, along the back (dorsal strip) resulting in tan to fawn colour	Recessive	Limousin, Jersey, Brahman, Chianina, Grey Steppe
D_c	Charolais dilution	Dc	Charolais dilution/heterozygotes: strong dilution of black to light grey, red to light cream; homozygotes are white or nearly white	Nearly completely dominant	Charolais
D_s	Simmental dilution	Ds	Simmental dilution/heterozygotes: moderate dilution of black to light grey, red to light red; homozygotes are lighter	Incompletely dominant	Simmental, Scottish Highland, Murrey Grey, some Gelbvieh
D_n	Dun	Dn	Dun/removal of red pigmentation with a reduced effect upon black pigment	Incompletely dominant	Brown Swiss, Brahman, Chianina

production only) is present in the melanocytes of individuals homozygous for *e*, probably as a result of non-functional MSH receptors in such animals (Klungland *et al.*, 1995).

The E^D allele is found in animals born solid black (with or without white spotting) and is responsible for the black colour of Angus and Holstein cattle. Some Texas Longhorns and a number of other breeds worldwide carry E^D. Animals with E^D do not change colour with age (at least not until advanced age, when greying about the face occurs in a small proportion of animals).

The E^+, or *wild-type* allele, at the *Extension* locus produces a reddish brown with varying amounts of black. The black pigmentation may be restricted to the head and neck, feet, hindquarters and tail or may cover nearly the entire body, with only an area of reddish brown over the ribs, a tan dorsal stripe and a tan muzzle ring and poll. Bulls with wild-type colour are generally darker than cows. Breeds that possess E^+ are Jersey, Brown Swiss, Texas Longhorn and Brahman and other zebu breeds. It is also found in Holstein cattle, where the colour may be referred to as red-black to differentiate it from 'true' red, that produced by *e/e*. The E^+ gene was increased in frequency within the Holstein breed through use of a popular sire, Hanover-Hill Triple Threat-Red, which was E^+/e in genotype. Calves that are E^+– (see below) in genotype are born red or reddish brown, perhaps with minimal black pigmentation on the feet and switch. There is complete dominance of E^D over both E^+ and *e*. Thus, it may be necessary to use DNA probes to differentiate if phenotypically black animals are heterozygous for either E^+ or *e*.

The red colour of Hereford, Simmental, Red Angus and other red breeds is due to homozygosity for the recessive gene, *e*. There is considerable variation in the intensity of red in cattle, from the dark red of Red Danish, Shorthorn and Maine-Anjou cattle to the lighter shades of some Herefords and Guernseys. While there may be a major (single) gene influencing the darker red colour and certainly there are single genes that lighten red colour, intensity of red is, in general, quantitative (Koger and Mankin, 1952). The desirable aspect of the red coat colour, from a genetic standpoint relative to composite breed development, is that, as a recessive, it will always 'breed true'. Exceptions would be the segregation of very light red or cream-coloured animals from some light red parents or wild-type animals from cattle that may appear red but on closer examination are brindles with minimal expression of black pigmentation.

The brindle coat colour is characterized by narrow alternating stripes of black and red arranged vertically on the entire body or confined to the head, neck and rear quarters. As mentioned above, some cattle may express the brindle colour only minimally, for example, only on the face. The base colour (the colour of the areas between the black stripes) may range from light red or fawn to dark brown or even nearly white, depending on the effects of alleles at other loci. Brindle coat colour is observed in the Texas Longhorn and Normande breeds, and is often produced in crossbreeding programmes, especially those including zebu breeds. The symbol *Br* will be used here to identify the gene responsible for producing the brindle coat colour. The gene responsible for brindling, however, requires the wild-type coat colour (as produced by animals that are E^+/E^+ or E^+/e in genotype) to be expressed (Bjarnadottir, 1993, as cited by Adalsteinsson *et al.*, 1995). Therefore, brindle animals are E^+– Br– in genotype, where the – following the E^+ and Br indicates that the other allele at each locus may be either another copy of the dominant gene or its recessive counterpart.

The Br gene is apparently hypostatic to both the E^D and the e alleles. Since e/e animals have no black pigment, they cannot express the black stripes of the brindle pattern. Also, it appears that the brindle gene is unable to modify the expression of the black coat colour produced by E^D. As a result, E^D– Br– animals are black and e/e Br– animals, red. The interaction of the *wild-type* coloration and Br is seen even in interspecific crosses, as bison × Hereford crosses are often brindle in colour. An explanation for the variation in the quantity of brindling expressed on animals is that brindled areas are apparently restricted to areas that would have been dark brown to black had the animal not possessed Br. Certainly it is true that E^+– br^+/br^+ (wild-type or brown) animals show great variation in the amount of black pigmentation they express, from limited amounts on the head or head and neck to animals that may appear to possess an E^D allele as they are essentially solid black. In some adult animals, the only indication that wild-type animals do not possess E^D, that is that they are not true black, is that they may have a tan muzzle ring. Thus, it appears likely that brindling is only expressed in areas that would have been black in E^+– br^+/br^+ animals.

The brindle coat colour is produced in a high percentage of the offspring of the cross of Hereford or Red Angus with Brahman (or other white zebu breeds). The Jersey × Hereford cross also nearly always produces brindle progeny. An explanation for these results appears to be that both the Hereford and Angus breeds carry the Br gene in quite high frequency, but do not express it as they are E^D– e/e in genotype. The reason that e/e animals do not express brindling is that they lack the black pigment, which is, in part, removed to produce the striped pattern of brindling. The other piece of this puzzle is that most grey/white zebu breeds, as well as the Jersey and Brown Swiss, carry the E^+ allele at a high frequency. The reason that the zebu breeds do not appear to express the wild-type (brown/black) pattern is that alleles at other loci (see discussions of the *Agouti* and *dun* loci) generally remove all of the red pigment present and most of the black pigment. The alleles that remove pigment in zebu breeds tend to be recessive and therefore are usually not expressed in the zebu × Hereford or zebu × Angus crosses. Thus, the F_1 animals from this cross (E^+/e Br/br^+ in genotype) express the wild-type coat colour and it is modified by Br to produce the brindle coat colour. Because of this epistatic relationship between the E and Br loci, the F_2 generation of a zebu × Red Angus cross would be expected to produce a 9 : 3 : 4 ratio of brindle, brown (wild type) and red.

If it were desired to produce a composite breed of cattle based on a zebu × Hereford/Angus base that was solid red (non-brindle), the E^+ gene would need to be eliminated from the population. This could be done in one generation of selection if only red F_2 animals were used as sires and dams of the F_3 generation. Care would need to be taken in the evaluation of these cattle, as some brindle animals show only a very limited expression of the pattern, being essentially red, and yet these animals would carry E^+. The heads, necks and rear quarters of all apparently red animals would need to be carefully examined for brindle striping to ensure that all animals carrying E^+ were eliminated.

Influence of Agouti *locus mutants*

While various authors have discussed possible mutants at the *A* locus in cattle that are homologous to those found at the *A* locus of other species (Lauvergne, 1966; Searle, 1968; Olson and Willham, 1982; Adalsteinsson et al., 1995), the *A* locus mutants in cattle remain incompletely understood. Mutants at the *Agouti* locus in other species have an impact on the expression of the wild-type pattern. It has been speculated that the lighter belly of the Limousin and Jersey cattle is due to a mutant at this locus (a^i) and that the modifications of the wild-type pattern observed in Brown Swiss, Brahman and Chianina breeds may be due to another allele (a^w). Adalsteinsson et al. (1995) have presented data that support the existence of a *recessive black* allele at the *A* locus, symbolized by *a*, which, when homozygous, modifies the wild-type coat colour to solid black. The wild-type allele of the *A* locus, A^+, allows the expression of the combination of black and red pigmentation of the wild-type coat colour. Homozygosity for *a* does not have an impact on animals that are e/e or E^D– in genotype. Alleles at the *Agouti* locus with the same type of effect are found in mice, dogs and horses. The cattle that Adalsteinsson et al. studied were Icelandic cattle and there is little evidence to date that the a allele exists in other, more populous breeds. The *a* gene seems to be found only at low frequency even in the Icelandic cattle.

A modification of the wild-type pattern, extending black pigment and resulting in a nearly black animal, has been described in the literature and named 'patterned blackish' (Majeskie, 1970). This pattern of black coat colour does not seem to be affected by the sex of the animal and is sometimes seen in progeny of crosses of Jersey, Brown Swiss and Brahman with red breeds. The *patterned blackish* gene is probably responsible for the occasional nearly black Jersey cows and Brown Swiss cows with greater proportions of black pigmentation. Perhaps the proper symbol to use to describe the mutant responsible for patterned blackish is A^{pb}, given its apparent modifying of the wild-type pattern.

Red cattle with black ear tips have been observed in zebu cattle in the USA (Olson, 1975) and southern China (Chen et al., 1994). It would seem plausible that such animals are produced by modifying the distribution of black pigment in E^+ animals, but no confirmation of this is possible at this time.

The dun *locus*

The genes that are responsible for the coat colours of Jersey, Brown Swiss, Brahman (and other grey or white zebu breeds), Grey Steppe and Chianina (and similar-coloured Italian breeds) are not well understood, due in part to the great variation in coat colour within these breeds. The coat colours of these breeds are related, as all involve a lightening of the basic coat pigmentation to a greater or lesser degree. In general, the pigmentation tends to be removed to a greater degree on the underline and red pigmentation is removed to a greater degree than black. The removal of red pigmentation is essentially complete in many Brown Swiss and most Brahman and Chianina cattle.

It is clear that these 'lightened' breeds carry similar colour mutants, based on the coat colours of crosses between them. Crosses between Jersey and Brahman resemble Jerseys and crosses between Brown Swiss and both Brahman and Chianina resemble Brown Swiss. Also, in Brown Swiss, grey Brahman, Chianina and crosses among these breeds, there is little to no red pigmentation expressed, indicating the presence of a gene that acts similarly to the '*Chinchilla*' mutant, c^{ch}, in other mammalian species. The *dun* gene, which has been studied by S. Adalsteinsson (personal communication) and Berge (1949, 1961) seems to produce the effects produced by this proposed '*chinchilla-like*' gene. Perhaps the symbol *dn*, for *dun*, is appropriate to identify this mutant. It seems to be incompletely recessive to its normal allele, Dn^+. When heterozygous, there is some effect of *dn* on red pigment, but its effect on black pigmentation is minimal. The *A* locus (a^i and a^w) and *dun* mutants, at least when heterozygous, have little or no effect on animals carrying E^D, except for a slight lightening to brown on the poll and along the back. Homozygosity for both a^w and *dn* acting upon black produced by E^D may produce a dilute coat colour, similar to the dilute-coloured animals from crosses of Simmental with Angus or Holstein. The *dn* mutant is probably present in the Brown Swiss, Chianina, Brahman and similarly coloured breeds, in addition to the Scandinavian cattle discussed by Berge.

A hypothesis regarding these colours is that a recessive allele, which could be called *fawn* and be symbolized as a^i, is responsible for the lightened underline and overall lightening resulting in tan to light red coat colours in Limousin, Guernsey and Jersey cattle. The darker extremities of Jersey are due to E^+, whereas Limousin is *e/e*, as are most Guernseys. Grey Brahman and Chianina also carry a^i in its homozygous state but, in addition, are homozygous for *dn*, which removes the rest of the red pigment. This results in silver-grey in the case of Brahman, which carries E^+, and white in Chianina, which is *e/e*. Since the Chianina and some zebu breeds may not carry E^+ and as a result produce only red pigment, they may also often be nearly white. Jersey and Limousin, which retain some red pigmentation, are probably Dn^+/Dn^+ at the *dun* locus.

The usual coat colour of Brown Swiss differs from that of Brahman (grey) in that its grey pigmentation is more uniformly distributed across the body, except for the underline, and is not usually confined to the extremities. Whether this difference is caused by a different allele at the *A* locus (a^w) or an independent dominant mutant is not known. For the purposes of this chapter, the coat colour of the Brown Swiss breed will be assumed to be due to homozygosity for a^w. Most US Brown Swiss also seem to be homozygous for *dn* acting together with a^w on a wild-type background coat colour. Cream-coloured animals, occasionally seen in crossbreeding programmes involving Brahman and Brown Swiss with Red Angus, may be the result of heterozygosity for both a^i or a^w and *dn*. Nearly white animals are sometimes produced by backcrosses of Brown Swiss or Brahman with their F_1 crosses with Angus. Such 3/4 animals are 'whiter' than their Brown Swiss or Brahman parents. They resemble Chianina cattle in coat colour. This is somewhat confusing, unless one considers the effect of *Br* on E^+. The effect of the brindle gene is to remove black pigment and sometimes it removes nearly all the black pigment on an animal, with the brindle colour being shown only on the face. With the eumelanin restricted by *Br* and the red pigmentation removed by *dn* acting together with a^i or a^w, perhaps a nearly white animal can be produced.

The Albino *locus*

Recessive albinism has occurred periodically in cattle of a number of breeds, including the Brown Swiss (Petersen *et al.*, 1944). In some cases, the animals reported had no pigment in either the eyes or skin and so could be considered as being 'true albinos' and would appear to be *c/c* in genotype, the allelic symbolism for true albinos in rodents and other species. In other cases, such as the recent case of a Florida Cracker calf born of a dam–son mating, the calf involved seemed to have some residual pigmentation. This may have been the same type of albinism that was discussed by Cole *et al.* (1934) and referred to as a 'ghost pattern' in white cattle of Holstein origin. Lauvergne (1966) also described a similar coat colour in a black and white spotted German breed.

Partial albinism, which, in some cases, was associated with dwarfism, was reported in Hereford cattle by Hafez *et al.* (1958). Since dwarfism has not been associated with *C* locus mutants in other species, another locus may have been involved in the dwarf individuals. It appears likely that none of the *C* locus mutants involved in these cases have been maintained and certainly none are established in any current breeds. Thus, the *C* locus mutants are not indicated in Table 3.1.

Dilution mutants of the Charolais, Simmental, Gelbvieh and other breeds

There are at least two additional loci in cattle that result in a diluted coat colour. Nearly all Charolais cattle are homozygous for a dilution mutation that is nearly completely dominant in mode of inheritance. The Simmental,

Gelbvieh and some other breeds carry a second, less dominant dilution mutation at varying frequencies. Both of these mutations uniformly dilute pigment over the entire body. These dilutions affect black at least as much as they do red pigmentation.

The dilution mutation carried by the Charolais breed that is responsible for the white coat colour of Charolais is symbolized by Dc, for *Charolais dilution*. The mutant approaches complete dominance over its *wild-type* (non-diluting) allele, dc^+. Most Charolais are homozygous for both e and Dc, with the red coat that would have been produced by the e/e genotype being modified to an essentially white one by Dc/Dc. Some US Charolais that have been upgraded from Angus may, however, carry E^D. Crosses (F_1) of Angus and Charolais are generally light grey, due to the effect of heterozygosity for Dc (Dc/dc^+) acting upon black (E^D/e). Similarly, crosses of red cattle with Charolais are cream to light yellow. In recent years, there has been an interest in producing red Charolais which would not carry Dc and thus be dc^+/dc^+ in genotype. It is not surprising that such animals should be produced occasionally, as all animals are not homozygous for Dc. It is not clear what other breeds of cattle possess Dc; the Cacereño breed of Spain, however, does appear to possess an identical coat colour to that of the Charolais.

A number of other breeds (Simmental, Gelbvieh, Scottish Highland, Texas Longhorn and others) carry a different dilution mutation, Ds, for *Simmental dilution*. This mutation is incompletely dominant to its wild-type allele, ds^+. Thus, red animals heterozygous for Ds (i.e. $e/e\ Ds/ds^+$) are light red and black animals (i.e. $E^D-\ Ds/ds^+$) exhibit varying intensities of grey colour. The Australian breed, the Murrey Grey, appears to be homozygous for Ds acting upon a black background. Red (e/e) animals homozygous for Ds are light yellow and black (E^D-) animals homozygous for Ds are light grey, similar to black cattle heterozygous for Dc. None of the breeds listed above carry Ds at a high frequency like that of Dc in Charolais. It seems likely that its frequency has decreased in the Simmental breed in the USA, as it is now desirable to have Simmental sires that will not produce dilute calves in crosses with black cows. On the other hand, the frequency of Ds in Gelbvieh cattle seems to have increased in the USA, due to the influence of certain popular sires.

The presence of Ds in the heterozygous state in animals with black hair coats is always obvious. This is not the case with red animals and this fact may have increased the popularity of the black animals now found in Simmental, Gelbvieh and other European breeds that utilized upgrading programmes in their development in the USA. Through the use of a black Simmental bull on black cows, for example, a commercial cattle producer can be assured that no dilute progeny will be born.

The desire to avoid the production of dilute black cattle can be traced, at least in part, to the phenomenon of the so-called 'rat-tailed' calves. Occasionally, the cross of a Simmental (and some other breeds) with a black mate (often Angus, or Angus crossbred) will result in a very dark, charcoal-coloured calf. Such calves are generally darker than the dilute colour produced by Ds, and apparently the rat-tailed condition is not related to the *Ds* dilution gene. Rat-tailed animals frequently show hair coats that are very dark, but still slightly diluted, and are also abnormal (short and curly) with the hair on the switch being reduced or completely absent. Because of the lack of switch, such animals are referred to as rat-tailed and are often severely discriminated against when such calves are sold as feeder calves. The price discounts are due to the perception that such animals have reduced cold tolerance, and there are reports of death losses in rat-tailed calves in Nebraska, when severe early winter results in extremely rapid temperature drops. Also, rat-tailed calves gain weight more slowly during the winter than calves of comparable genetic make-up with normal hair coats (R.R. Schalles and L.V. Cundiff, personal communication). The bizarre situation with regard to rat-tailed animals is that, when such animals show white spotting, the hair in the white areas is completely normal. It should also be emphasized that the abnormal hair coat is never observed in red animals.

While the mechanism of inheritance of the rat-tailed condition is not completely clear, a procedure to reduce its incidence is available. Since the condition is apparently always expressed in animals with black hair coats, black Simmental do not carry the gene/genes responsible for the condition. Therefore, the use of only black Simmental, or red Simmental that are the progeny of black parents, should eliminate the possibility of producing rat-tailed (or normal-haired dilute) progeny when subsequent crosses are made with black mates. Another approach would be the use of test matings of red sires to homozygous black cows. Bulls that produce seven or more black calves (no rat-tails or dilutes) have less than a 1% chance of carrying the gene/genes responsible for the rat-tailed condition or *Ds*. After several generations of use of bulls tested in this fashion, the gene frequencies of *Ds* and of whatever genes are responsible for rat-tailed calves would be reduced substantially.

In addition to the colours that have been mentioned, other colour patterns caused by additional mutants surely exist in cattle but have not been well documented to date. A partial list of such colours includes black ear rims and black on the feet of basically red zebu and zebu crossbred cattle, irregular spots of light red pigment on a basically dark red or brown zebu (generally found in animals with Gyr breeding), and a type of red in Holsteins that appears not to be recessive in mode of inheritance.

White-spotting Mutants

Since the wild type for white spotting is a lack of spotting, any white spotting on cattle is due to a mutant or combination of several mutants (Olson, 1980). In general, the understanding of the genetic control of white spotting is complete, except for a few, infrequently observed patterns to be discussed later. Major mutant genes affecting spotting patterns in cattle are listed in Table 3.2.

Table 3.2. Common white-spotting mutants in cattle.

Locus symbol	Locus name	Allele symbol	Mutant name/description	Inheritance relative to wild type	Breed distribution
S	Spotting	S^H	*Hereford pattern*/white face, belly, feet and tail, often with white stripe over shoulders when homozygous. Only white face is present in S^H/S^+	Incompletely dominant	Hereford, Braford, Beefmaster
S	Spotting	S^P	*Pinzgauer pattern*/sides of body pigmented; variable amounts of white appear along dorsal and ventral areas extending forward from tail and rump	Incompletely dominant	Pinzgauer, Charolais, Longhorn, Florida Cracker
S	Spotting	s	*Recessive spotting*/piebald: irregular areas of pigmented and white; feet, belly and tail usually white	Recessive	Holstein, Guernsey, Jersey, Simmental, Ayrshire, Maine-Anjou, Belgian Blue and others
R	Roan	R	*Roan*/homozygote: nearly white except for small amounts of pigmentation on the edges of the ears; heterozygote: a combination of pigmented and white hairs	Incompletely dominant	Shorthorn, Belgian Blue
Bt	Belting	Bt	*Belting*/belt of white of various widths around paunch	Dominant	Dutch Belted, Galloway
Bl	Blaze	Bl	*Blaze*/white head, often a blaze when heterozygous, without associated white areas on other parts of body produced by Hereford pattern	Incompletely dominant	Simmental, Holstein(?), Gronigen
Bc	Brockling	Bc	*Brockling*/areas of pigmentation within areas of white spotting produced by other mutants	Dominant	Nearly all solid-coloured breeds plus the Shorthorn, Ayrshire and Normande
Cs	Colour-sided	Cs	*Colour-sided*/ homozygote: white body with pigmented ears, muzzle and feet (white park pattern); heterozygote: colour-sided pattern, white dorsal stripe with irregular edges (roaned) and white roaning on head; roaning may be confined to head, rump and tail	Incompletely dominant	Texas Longhorn, White Park, British White, Florida Cracker, English Longhorn and Belgian Blue

The S locus mutants

Much of the variation in white spotting among US breeds of cattle is due to a multiple allelic series at the *S* locus (Olson, 1981). In mice, this locus is referred to as *piebald* and has been documented to affect the differentiation of melanocytes at the neural crest, as well as their migration from the neural crest to the rest of the body (Jackson, 1994). The action of the *S* locus mutants is to produce mice with various types of white spotting patterns in combination with any pigmentation. The *S* locus in cattle contains at least three mutants, in addition to the wild-type, non-spotting allele, S^+. These mutants are S^H, which is responsible for the Hereford pattern when homozygous, S^P, which is responsible for the Pinzgauer-type line-backed pattern (sometimes referred to as the Gloucester pattern, after the rare English breed), and *s*, *recessive spotting*, responsible for the irregular white spotting of the Holstein, Guernsey, Ayrshire, Jersey and Simmental breeds.

The order of dominance at the *S* locus is $S^H = S^P > S^+ > s$. The mutants, S^H and S^P, could be considered as codominant, as animals of genotype S^H/S^P, such as Pinzgauer × Hereford crossbreds, express both white face due to S^H and a white dorsal stripe and white across the underline due to S^P. Both S^H and S^P are incompletely dominant to S^+. Animals that are S^H/S^+, such as Angus × Hereford crossbreds, express a restricted Hereford pattern, in that they have less white on the head than S^H/S^H and generally have little or no white on other parts of the body, except on the switch and underline. Likewise, animals of genotype S^P/S^+, such as Pinzgauer × Angus crossbreds, generally have much less white than animals with genotype S^P/S^P. The white on S^P/S^+ animals resulting from Angus crosses can be restricted to a small amount on the tail or on the tail head extending along the spine across the rump. In other cattle, heterozygous animals may possess the full Pinzgauer pattern. The Charolais breed also possesses S^P in low frequency. Spotting patterns produced by S^P can be seen in animals with Charolais breeding but lacking the *Dc* (dilution) gene. Texas Longhorns and the related Florida Cracker cattle also possess S^P (Fig. 3.1).

Recessive spotting, *s*, is completely recessive to S^H, S^P and S^+. Matings between animals with perfect Hereford markings have produced spotted (*s/s*) progeny (Franke *et al.*, 1975). Similarly, Angus (non-spotted) bred to Holstein have produced spotted calves, indicating that some Angus were S^+/s but did not show excessive white, due to dominance of S^+ over *s*. The amount of white on animals that are *s/s* varies considerably. Some Holstein cattle are 90–95% white, whereas others are 90–95% black. Such differences are due to highly heritable (~0.9), quantitative, modifying factors (Briquet and Lush, 1947; Becerril *et al.*, 1993). Modifying genes that are quantitative also influence the degree of expression of all other white-spotting patterns. For example, the amount of white on S^P/S^P or S^P/S^+ animals may be increased from that usually observed in the Pinzgauer to cover nearly the entire posterior and part of the anterior half of the body, resulting in pigmentation only on the feet, head, sides of the neck and shoulders. Some Texas Longhorn and Florida Cracker

Fig. 3.1. A Florida Cracker cow with the white-spotting pattern produced by the $S^P/-$ genotype and her nearly white calf, which demonstrates the pattern often produced by the genotype $S^P/-$ Cs/cs^+.

cattle display such a spotting pattern. Other Longhorn and Cracker cattle that are S^P/S^+ in genotype may only express a limited amount of white speckling on the rear quarters. Apparently, there has been selection within the Pinzgauer breed for the degree of expression of S^P seen in most Pinzgauer cattle. Similarly, the amount of white could be increased or decreased on Herefords if breed standards of acceptable amounts and locations of white would allow them. The s gene may also exist in yaks, as some yak × cattle crosses show a spotting pattern similar to that produced by s/s animals.

Blaze *mutants*

The Simmental breed and perhaps a few Holsteins carry a gene that produces a white spotting pattern on the face that is distinct from S^H. The symbol used for this gene is Bl, for the blaze pattern it usually produces when heterozygous and in combination with S^+. Since full-blood (of 100% European origin) Simmental cattle are all spotted and must be s/s in genotype, the white facial spotting of the breed must be due to a gene at a locus independent of S. The genotype (for white spotting) of many Simmental × Angus crosses is Bl/bl^+ S^+/s. Such animals are solid-coloured with a white blaze on their face that usually does not include the eyes. In combination with s/s, both Bl/Bl and Bl/bl^+ will usually have a solid white face and head. Photographs of the Gronigen breed of Holland indicate that the white facial spotting of this breed may also be due to Bl.

Roan

Shorthorns, Belgian Blues, Texas Longhorns and Florida Crackers carry a gene, R, responsible, when heterozygous, for roan colour. The allele to R is r^+, a normal or wild-type gene that does not restrict pigmentation. Roan colour is a mixture of pigmented and white hairs and is produced in animals that are heterozygous, R/r^+. When homozygous for R, an almost white animal is produced, with the exception of pigmentation of the hair within the ears. While the most often observed roan is a red roan, as found in the Shorthorn breed, the *Roan* gene acts equally effectively in the removal of any pigment. The roan colour is frequently used as an example of incomplete dominance in genetic textbooks, with the three phenotypes involved being identified as white, roan and red. This leads to the misconception that there are red and 'white' genes that are allelic. Obviously, what is ignored in these texts is that all Shorthorn cattle are e/e in genotype and, in this genetic background, R/R, R/r^+ and r^+/r^+ animals are white, roan and red, respectively. When an animal is heterozygous for both the E and R loci, as in the cross of a white Shorthorn with an Angus, a 'blue roan', $E^D/e\ R/r^+$ is produced. Ibsen (1933) cited the results of Evvard *et al.* (1930), who reported the segregation of colours and roaning patterns from F_2 progeny of blue roan F_1 parents that produced 11 blue roans, one black, three whites (with black points, i.e. black hairs in the ears), one red roan and five reds. These results demonstrate the independence of the *roan* and E loci. Just as roaning can occur in both black and red animals, brindle roans, dilute black roans and many other colours in combination with roan could be produced. The expression of the *roan* gene when heterozygous is highly variable, with some animals being roan over the entire body, while, in others, roaning may be restricted to just the centre of the forehead. Recent studies have confirmed the independence of the *roan* and *Extension* (located on chromosome 18 (Klungland *et al.*, 1995)) loci, with the *roan* locus having been determined to be located on chromosome 5 (Charlier *et al.*, 1996).

The white colour caused by homozygosity for R has been associated with 'White heifer disease' in the Shorthorn and Belgian Blue breeds. Hanset (1969) has determined that the reproductive problems associated with the white coat colour have both multifactorial genetic and environmental causes. The R allele appears to act in conjunction with auxiliary genes to produce the reproductive problems associated with white heifer disease. Systematic selection against the auxiliary genes in the Belgian Blue breed seems to have been effective in lowering the incidence of reproductive problems in these cattle (Hanset, 1969).

Colour-sided

Texas Longhorn, Florida Cracker, English Longhorn, some Scandinavian cattle and apparently some African breeds possess what has been called the colour-sided pattern, the gene for which is symbolized as Cs (Wriedt, 1925). What appears to be the colour-sided pattern is also found in yaks. The Cs gene is partially dominant and continues to be expressed in Florida commercial cattle after many generations of crossing with non-spotted breeds. Animals carrying *Cs* in the heterozygous state show extreme variation in its expression. A pattern commonly seen in animals heterozygous for *Cs* includes a very irregular white strip along the dorsal and ventral parts of the animal, with roan areas along the edges and a roan or 'dappled' pattern of white on the head (Fig. 3.2). In other heterozygotes, the white stripe may be restricted to the rump and tail along with a little roaning on the head. The spotting patterns produced by animals heterozygous for *Cs* may be differentiated from those produced by S^P in that the spotting produced by *Cs* generally has a ragged or roan-like edge, whereas the edges of spots produced by S^P are clearly defined. The nearly white colour of other breeds, such as the British White and American White Park, is also apparently due to the action of *Cs*. Homozygotes for *Cs* generally exhibit the 'white park' pattern, that is, a nearly solid white animal with pigmented ears and a pigmented muzzle and often with some pigmentation just above the feet. It appears that the Blanco Orejinegro (translates as white with black ears) breed of Colombia owes its pattern to homozygosity for *Cs*. It also appears, based on limited data, that a 'white park' pattern may be able to be produced in cattle that are heterozygous for *Cs* in some breeds, such as the White Galloway and American White Park cattle. Other heterozygotes for *Cs* may appear similar to 'blue roans' in colour if they also are E^D–. Charlier *et al.* (1996) report that the *colour-sided* allele also appears to be present in low frequency in the Belgian Blue breed.

Fig. 3.2. A Florida Cracker cow with the colour-sided pattern produced in animals with the Cs/cs^+ genotype.

It has been observed in Florida Cracker cattle that animals which carry both R and Cs but cannot be homozygous for either are white park in phenotype. Allelism between R and Cs has been suggested; however, Florida Cracker cows that are white park in phenotype (due to carrying both R and Cs) have produced calves without any white spotting. This would indicate a lack of allelism of these two genes. Animals with both Cs and S^P may also show a predominantly white coat. The Florida Cracker calf show in Fig. 3.1 is of the type frequently observed in animals that possess both Cs and S^P. Similarly, Shorthorn × Hereford crosses that possess both S^H and R may be predominantly white. Thus, it appears that there is an additive effect involved in the reduction of pigmentation in animals heterozygous for several spotting loci.

Belted

One of the most striking white-spotting mutants, Bt, produces the belted pattern of the Dutch Belted and Belted Galloway breeds. Belting is dominant and expresses itself with a white belt of varying widths around the midsection. Crosses of belted cattle with Holsteins (s/s) produce belted animals resembling the belted parent. As is the case in most spotting patterns, modifying genes are responsible for influencing the width and uniformity of this belt. Allelism with other spotting mutants has not been documented.

Brockling/pigmented legs

A major gene, referred to by previous authors as the *brockling* gene (Olson, 1975) or *'pigmented legs'* gene (Ibsen, 1933), Bc, interacts with apparently any white-spotting mutant, producing areas of pigmentation within areas that would be white had the Bc gene not been present. The most commonly observed expression of the *brockling* gene is in Hereford × Angus crossbreds, where Bc from the Angus produces pigmented spots on the face (the brockled-face pattern), which otherwise would be white, due to S^H. In s/s animals, legs are usually white, but when an s/s animal carries Bc as well, legs are pigmented to varying degrees. Ayrshire cattle with white-spotted sides and legs which are pigmented are s/s $Bc-$ in genotype. Only a few spotted breeds appear to carry Bc. These include the Ayrshire, Jersey, Norwegian Red and Normande. In addition to the Angus, most other non-spotted breeds also carry Bc at a reasonably high frequency. The exception to this rule is the Brown Swiss (Olson, 1975). A desirable function of Bc in Hereford crossbreds carrying S^H is that it usually results in pigmented areas surrounding the eyes, which has been shown to reduce the likelihood of cancer eye (Anderson, 1991). The so-called red-eyed condition in Hereford and Simmental cattle is very probably due to a different gene(s), which may be dominant, but this has not been well documented.

Minimal white spotting

In some solid-coloured breeds, white spotting along the underline, especially in front of the navel, can disqualify an animal from registration. Such spotting may be due to the presence of s. In many cases, however, such spotting is not caused by s and it is unclear as to the genetic mechanism involved. Selection against animals with white areas on the underline should reduce the incidence of such spotting, but this reduces the selection intensity possible for traits related to productivity.

White-spotting patterns in African and zebu breeds

A number of striking spotting patterns can be observed in several of the African Sanga and zebu breeds. These include the Pakistani zebu breed, the Dhanni, which is predominantly white but with small black body spotting and with black ears and a black muzzle. The Landim cattle of Mozambique possess a similar pattern. It is possible that this pattern is produced by modification of the line-backed pattern produced by S^p. Many Nguni cattle of South Africa show similar spotting. The Bapedi cattle, which were developed from Nguni cattle, often show a type of mottled grey-black, which is not frequently observed in other breeds. The Lohani cattle of Pakistan have pronounced speckling throughout a black body. This spotting pattern is also observed in some American Red Brahman cattle and has been observed in Texas Longhorn cattle. The mode of inheritance of this pattern is somewhat unclear although it does occur in Red Brahman × Angus crosses. Another Pakistani breed, the Rojhan, appears to have the recessive spotting pattern (s/s). The American Brahman does carry s at a modest frequency. The cattle referred to as West African Shorthorns and some other African breeds also appear to possess recessive spotting but modified by Bc, resulting in pigmentation on the legs and other modifications to recessive spotting.

References

Adalsteinsson, S., Bjarnadottir, S., Vage, D.I. and Jonmundsson, J.V. (1995) Brown coat colour produced by the loci Extension and Agouti. *Journal of Heredity* 86, 395–398.
Anderson, D.E. (1991) Genetic study of eye cancer in cattle. *Journal of Heredity* 82, 21–26.
Barrington, A. and Pearson, K. (1906) On the inheritance of coat color in cattle. I. Shorthorn crosses and pure Shorthorn. *Biometrica* 4, 427–437.
Becerril, C.M., Wilcox, C.J., Wiggins, G.R. and Sigmon, K.N. (1994) Transformation of measurements of percentage of white coat colour for Holsteins and estimation of heritability. *Journal of Dairy Science* 77, 2651–2657.

Becerril, C.M., Wilcox, C.J., Lawlor, T.J., Wiggins, G.R. and Webb, D.W. (1993) Effects of percentage of white coat colour on Holstein production and reproduction in a subtropical environment. *Journal of Dairy Science* 76, 2286–2291.

Berge, S. (1949) Inheritance of dun, brown and brindle colour in cattle. *Heredity* 3, 195.

Berge, S. (1961) Influence of dun on brown and brindle cattle. *Zeitschrift für Tierzüchtung und Züchtungsbiologie* 75, 298.

Briquet, R., Jr and Lush, J.L. (1947) Heritability of amount of white spotting in Holstein-Friesian cattle. *Journal of Heredity* 38, 98.

Brown, A.H., Jr, Johnson, Z.B., Simpson, R.B., Brown, M.A., Steelman, C.D. and Rosenkrans, C.F., Jr (1994) Relationship of horn fly to face fly infestation in beef cattle. *Journal of Animal Science* 72, 2264–2269.

Charlier, C., Denys, B., Belanche, J.I., Coppieters, W., Grobet, L., Mni, M., Womack, J., Hanset, R. and Georges, M. (1996) Microsatellite mapping of the bovine roan locus: a major determinant of White Heifer Disease. *Mammalian Genome* 7, 138–142.

Chen, Y., Wang, Y., Cao, H., Pang, Z. and Yang, G. (1994) Black-ear gene and blood polymorphism in four southern Chinese cattle groups. *Animal Genetics* 25 (Suppl. 1), 89–90.

Cole, L.J., Van Lone, E.E. and Johannsson, I. (1934) Albinotic dilution of colour in cattle. *Journal of Heredity* 25, 145–156.

Evvard, J.M., Shearer, P.S., Lindstrom, E.W. and Smith, A.D.B. (1930) The inheritance of colour and horns in blue gray cattle. II. *Iowa Agricultural Experiment Station Research Bulletin* 133, 1–16.

Finch, V.A. and Western, D. (1977) Cattle colours in pastoral herds: natural selection or social preference. *Ecology* 58, 1384.

Finch, V.A., Bennett, I.L. and Holmes, C.R. (1984) Coat colour in cattle: effect on thermal balance, behaviour and growth, and relationship with coat type. *Journal of Agricultural Science (Cambridge)* 102, 141.

Franke, D.E., Burns, W.C. and Koger, M. (1975) Variation in the coat colour pattern of Hereford cattle. *Journal of Heredity* 66, 147–150.

Hafez, E.S.E., O'Mary, C.C. and Ensminger, M.E. (1958) Albino-dwarfism in Hereford cattle. *Journal of Heredity* 49, 111–115.

Hansen, P.J. (1990) Effects of coat colour on physiological responses to solar radiation in Holsteins. *Veterinary Record* 127, 333.

Hanset, R. (1969) La White Heifer Disease dans la race bovine de Moyenne et Haute Belgique: un bilan de dix années. *Annales de Medécine Vétérinaire* 113, 12–21.

Ibsen, H.L. (1933) Cattle inheritance. I. Colour. *Genetics* 18, 441–480.

Jackson, I.J. (1994) Molecular and developmental genetics of mouse coat colour. *Annual Review of Genetics* 28, 189–217.

Joerg, H., Fries, H.R., Meijerink, E. and Stranzinger, G.F. (1996) Red coat colour in Holstein cattle is associated with a deletion in the MSHR gene. *Mammalian Genome* 7, 317–318.

King, V.L., Denise, S.K., Armstrong, D.V., Torabi, M. and Weirsma, F. (1988) Effects of a hot climate on the performance of first lactation cows grouped by coat colour. *Journal of Dairy Science* 71, 1093–1096.

Koger, M. and Mankin, J.D. (1952) Heritability of intensity of red colour in Hereford cattle. *Journal of Heredity* 43, 15.

Klungland, H., Våge, D.I., Gomez-Raya, L., Adalsteinsson, S. and Lien, S. (1995) The role of melanocyte-stimulating hormone (MSH) receptor in bovine coat colour determination. *Mammalian Genome* 6, 636–639.

Lauvergne, J.J. (1966) Génétique de la couleur de pelage dex boivins domestiques. *Bibliographia Genetica* 20, 1–168.
Majeskie, J.L. (1970) Characteristics and inheritance of certain coat colours and patterns in cattle. PhD thesis, Kansas State University, Manhattan, Kansas.
Olson, T.A. (1975) An analysis of the inheritance of coat colour in cattle. MS thesis, Iowa State University, Ames.
Olson, T.A. (1980) Choice of a wild-type standard in colour genetics of domestic cattle. *Journal of Heredity* 71, 442–444.
Olson, T.A. (1981) The genetic basis for piebald patterns in cattle. *Journal of Heredity* 72, 113–116.
Olson, T.A. and Willham, R.L. (1982) *Inheritance of Coat Colour in Cattle: a Review*. Iowa State University and Agricultural Experiment Station Bulletin 595, Ames, Iowa.
Petersen, W.E., Gilmore, L.O., Fitch, J.B. and Winters, L.M. (1944) Albinism in cattle. *Journal of Heredity* 35, 135–144.
Robbins, L.S., Nadeau, J.H., Johnson, K.R., Kelly, M.A., Roselli-Rehfuss, L., Baack, E., Mountjoy, K.G. and Cone, R.D. (1993) Pigmentation phenotypes of variant extension locus alleles result from point mutations that alter MSH receptor function. *Cell* 72, 827–834.
Searle, A.G. (1968) *Comparative Genetics of Coat Colour in Mammals*. Logos Press, London.
Sponenberg, D.P. (1997) The genetics of colour and hair texture. In: Piper, L. and Ruvinsky, A. (eds) *The Genetics of Sheep*. CAB International, Wallingford, pp. 51–86.
Stewart, R.E. (1953) Absorption of solar radiation by the hair of cattle. *Agricultural Engineering* 34, 235.
Wriedt, C. (1925) Colour-sided cattle. *Journal of Heredity* 16, 51–56.

Genetics of Morphological Traits and Inherited Disorders

4

F.W. Nicholas
Department of Animal Science, University of Sydney, NSW 2006, Australia

Introduction	55
The Range of Possibilities	56
Previous Reviews	56
Current Sources of Information	57
An Overview	57
Conclusion	68
Acknowledgement	68
References and List of Reviews	68

Introduction

Morphological traits and inherited disorders are of considerable importance in cattle, from an agricultural point of view and in terms of animal models for inherited human diseases. The main aim of this chapter is to present a summary of bovine morphological traits and inherited disorders for which there is substantial evidence of single-gene inheritance. A secondary aim is to provide a list of other traits and disorders for which the genetic basis has not yet been determined.

With the molecular revolution now in full swing, and, in particular, with the development of gene markers covering all regions of all bovine chromosomes (see Chapter 11), knowledge of morphological traits and inherited disorders in cattle will increase rapidly in the decades ahead. As described below, regularly updated information on the subject-matter of this chapter is available on the Internet. By this means, it is possible for readers throughout the world to obtain the latest information on any single-locus trait or inherited disorder in cattle.

The Range of Possibilities

The spectrum of morphological traits and inherited disorders ranges from those that are definitely due to the action of just one gene, to those that are due to the combined action of many genes and many non-genetic (environmental) factors. In between these two extremes are many traits and disorders which appear to run in families, but for which there is insufficient information to enable a conclusion to be drawn about whether one or more genes are involved. Unfortunately, the literature abounds with examples of traits and disorders that have been claimed to be due to just one gene, despite the data being so sparse that such a claim cannot be justified. Similar problems exist with claims of inheritance being recessive or dominant; in most cases, there is insufficient information to justify the claims that have been made. In the fullness of time, of course, additional data might support the initial claims. But we must be careful not to jump the gun.

This scarcity of reliable data on the inheritance of traits and disorders poses a challenge to those who are asked to compile lists of such traits – as required for this chapter. No two reviewers will interpret the evidence in exactly the same way, so we must expect that lists of single-locus traits and disorders compiled by different authors will differ at the margins. In the fullness of time, as more data become available, these differences will be resolved.

Previous Reviews

Several comprehensive reviews of inherited traits and disorders in cattle have been published over the years. The first major summary was by Shrode and Lush (1947). Since then, there have been surveys of inherited disorders by Gilmore (1957), Lauvergne (1968), Leipold *et al.* (1972), Jolly and Leipold (1973), Leipold and Schalles (1977) and Herzog (1992). A full list of reviews is presented in the References. It should be noted that some of these reviews are concerned with congenital traits and disorders, i.e. traits and disorders that are present at birth. Not all such traits and disorders are inherited.

No discussion of inherited disorders in cattle can be complete without a special mention of the pioneering work of Dr Horst Leipold, whose name appears regularly in the list of reviews. His pioneering research into the inheritance of disorders and his encylopaedic knowledge of inherited disorders will continue to earn him the gratitude of those who follow in his footsteps. Mention must also be made of the mammoth 'retirement' project of Dr Keith Huston, a former colleague of Dr Leipold. In reviewing all the published information on inherited disorders, Dr Huston has compiled an annotated list of inherited disorders in cattle (category 2 in COGNOSAG, 1998) that will be a very important source of information for many years to come.

Current Sources of Information

While a list of reviews is useful, it is even more useful to have a single catalogue of morphological traits and inherited disorders that is regularly updated, and which is made available both electronically (on the Internet) and in hardcover (book) form on a regular basis. Human geneticists have long had access to such a resource – McKusick's *Mendelian Inheritance in Man* (MIM). Now in its 12th edition as a book (McKusick, 1998) and accessible on the Internet as Online Mendelian Inheritance in Man (OMIM) at http://www3.ncbi.nlm.nih.gov:80/omim/, this catalogue contains a wealth of information on thousands of morphological traits and inherited disorders in humans. It also contains a surprising quantity of information on cattle, because McKusick has always been interested in potential animal models of human disorders.

As described in Chapter 23 of the present book, the Committee on Genetic Nomenclature of Sheep and Goats (COGNOSAG) has been assembling a catalogue of genes and inherited traits in cattle. As mentioned above, category 2 of this catalogue (describing loci for visible traits other than coat colour) has been compiled by Dr Keith Huston. An electronic version of this catalogue is accessible on the Internet at http://www.angis.su.oz.au/Databases.

In 1978, the present author commenced compiling a catalogue of inherited traits and disorders in a wide range of animal species. Being modelled on, and complementary to, McKusick's catalogue, this catalogue is called Mendelian Inheritance in Animals (MIA). It is accessible on the Internet as Online Mendelian Inheritance in Animals (OMIA) at http://www.angis.su.oz.au/Databases/BIRX/omia/ in the same format as OMIM, and at http://tetra.gig.usda.gov in a different format.

Entries for all inherited disorders in cattle are included in OMIA, together with other traits in cattle for which single-locus inheritance has been claimed, however dubiously. Each entry consists of a list of references arranged chronologically, so as to present a convenient history of knowledge about each disorder or trait. For some entries, there is additional information on inheritance or molecular genetics. If the disorder or trait has a human homologue, the relevant MIM numbers are included, providing a direct hyperlink to the relevant entry in McKusick's online catalogue OMIM.

Online MIA is updated regularly.

An Overview

Table 4.1 provides a list of single-locus morphological traits and inherited disorders that have been reported in cattle – a total of 39 – together with the earliest report plus the most recent reference, extracted from MIA. The entries have been divided into two categories: those that have been characterized at the

Continued on p. 68

Table 4.1. A list of single-locus morphological traits and inherited disorders in cattle (from Online Mendelian Inheritance in Animals).

Molecular defect known	Summary	Earliest reference; latest reference (if different from earliest reference)
Citrullinaemia	The clinical signs of this disorder result from ammonia poisoning, due to a fault in the urea cycle. This cycle is the biochemical process by which potentially toxic ammonia (a by-product of catabolism of proteins) is converted to urea, which is excreted in urine. The fault in the cycle arises from a deficiency of one of the enzymes involved in the cycle, namely argininosuccinate synthetase (ASS). The absence of this enzyme leads to a build-up of citrulline and, more seriously, of ammonia. To date, all cases of this lethal disorder in cattle appear to be due to the same mutation, namely a nonsense mutation in the fifth of nine exons of the ASS gene. Normal bovine ASS is a peptide containing 412 amino acids; the mutation occurs in the 86th codon	Harper et al. (1986b); Grupe et al. (1996)
Coat colour, extension	The extension locus encodes the melanocyte-stimulating hormone receptor (MSHR; also known as MC1R). This receptor controls the level of tyrosinase within melanocytes. Tyrosinase is the limiting enzyme involved in synthesis of melanins: high levels of tyrosinase result in the production of eumelanin (dark colour, e.g. brown or black), while low levels result in the production of phaeomelanin (light colour, e.g. red or yellow). When melanocyte-stimulating hormone (MSH) binds to its receptor, the level of tyrosinase is increased, leading to production of eumelanin. The wild-type allele at the extension locus corresponds to a functional MSHR, and hence to dark pigmentation in the presence of MSH. Klungland et al. (1995) presented molecular evidence for three alleles at the MSHR locus in Norwegian and Icelandic cattle. The wild-type allele (E^+) encodes the normal functional receptor for MSH. The E^D allele contains a mis-sense mutation, changing the 99th amino acid from leucine to proline. The resultant MSHR molecule is constitutively expressed, i.e. it is expressed without the need for MSH binding. This results in continuously high levels of tyrosinase, and hence production of eumelanin (black coat colour). The e allele contains a single base deletion (a frame-shift mutation), which gives rise to a non-functional receptor, and hence to low levels of tyrosinase, resulting in production of phaeomelanin (red coat colour). Because one copy of the E^D allele is sufficient to produce functional MSHR molecules and hence to produce black coat colour, this allele is dominant, while the e allele is recessive. Thus black is dominant to red. As expected, the coat colour associated with wild-type (E^+) allele varies, depending, among other things, on the level of MSH produced. Joerg et al. (1996) extended these results by showing that red coat colour in Friesians is also due to a deletion in the MSHR gene (see Chapter 3 for more details on coat colour)	Wright (1917); Joerg et al. (1996)

Morphological Traits and Inherited Disorders 59

Deficiency of uridine monophosphate synthase	Uridine monophosphate synthase (UMPS) is the enzyme responsible for converting orotic acid to uridine monophosphate (UMP), which is an essential component of pyrimidine nucleotides. This enzyme actually has two enzymatic functions: orotic phosphoribosyltransferase (OPRTase) and orotidine monophosphate decarboxylase (OMPDCase), corresponding to the last two steps in pyrimidine synthesis. Since pyrimidines are vital ingredients of nucleic acids, it follows that a deficiency of UMPS can have quite severe consequences. This disorder is of particular interest in cattle, because it is one of the few cases in which an embryonic lethal has been identified. (By their very nature, embryonic lethals are difficult to identify; their only manifestation is a return to service.) To date, the only known cause of this disorder in cattle is a nonsense mutation in codon 405 of the UMPS gene, resulting in a complete deficiency of functional UMPS (Schwenger *et al*., 1993). Since nucleotides are required in such vast quantities during embryonic development, it is not surprising that homozygosity for the nonsense mutation results in embryonic death around 40 days *in utero*. The practical effect of this disorder is that carrier cows show a higher rate of return to service, because some of their pregnancies end in early natural abortion. Given that returns to service can have so many different causes, it would not have been possible to identify the mutation from reproductive records. How, then, was this embryonic lethal mutation detected initially? The answer is: by chance! As part of a nutrition study at the University of Illinois, the level of orotic acid was determined in the milk of cows in the university herd. Some cows had exceptionally high levels of this acid, and one possible explanation was that they were deficient in UMPS. Subsequent biochemical tests showed that these cows had only 50% of the normal activity of this enzyme (reviewed by Shanks and Robinson, 1990). Inheritance and molecular studies finally brought the whole story to light	Robinson *et al*. (1983); Harlizius *et al*. (1996)
Glycogen storage disease V	The first occurrence of this disorder in any domestic species was reported by Angelos *et al*. (1995), in Charolais cattle. Tsujino *et al*. (1996) showed that the disorder in these cattle is due to a mis-sense mutation in codon 489 of the gene for myophosphorylase	Angelos *et al*. (1995); Tsujino *et al*. (1996)
Goitre, familial	Enlargement of the thyroid gland, causing a swelling in the front of the neck. Familial forms of this disorder have been identified in several species. But only in cattle and goats has the molecular basis been determined. In Afrikander cattle, inherited goitre is due to a nonsense mutation in the thyroglobulin gene. The nonsense mutation occurs in exon 9. Interestingly, both normal and affected cattle produce a second, shortened thyroglobulin (TG) transcript that lacks exon 9. The sequences obtained from normal (European) and mutant (Afrikander) genes also differ by a mis-sense mutation at codon 699 (A → G; Ala → Gly; a conservative amino acid change) and an A → G transition 25 bases into intron 9. It is not known whether these polymorphisms exist within the Afrikander breed. The intronic transition generates a Pst recognition site	Jamieson *et al*. (1945); Wither (1997)

continued overleaf

Table 4.1. *Continued.*

Molecular defect known	Summary	Earliest reference; latest reference (if different from earliest reference)
Leucocyte adhesion deficiency	Affected animals die because of extreme susceptibility to infections, caused by an inability of white blood cells (leucocytes) to pass from the bloodstream into infected tissue. This inability is due to the lack of a membrane glycoprotein called CD18, resulting from a mis-sense mutation in the CD18 gene	Kehrli *et al.* (1990); Jorgensen and Madsen (1997)
Mannosidosis, alpha	A lysosomal storage disease in which there is a build-up (storage) of mannose-rich compounds, due to the lack of the enzyme α-mannosidase, whose task is to cleave mannose from such compounds. Clinical signs include ataxia, head tremor, aggression and finally paralysis and death. This was one of the first inherited lysosomal storage disorders to be studied extensively in animals. The pioneering work was done by Jolly and colleagues at Massey University in New Zealand, who developed an enzyme assay that enabled carriers to be distinguished from homozygote normals with a high degree of accuracy (Jolly *et al.*, 1973). More than 20 years later, the bovine gene was cloned and characterized by Tollersrud *et al.* (1997), who identified a different causative base substitution (giving rise to an amino acid substitution) in each of the Angus and Galloway breeds. Berg *et al.* (1997) extended these results to several other derivative breeds. In summary, a G662A transition is the mutation responsible in Galloways; and a T961C transition is the causative mutation in Angus, Murray Grey and Brangus from Australia, and in Red Angus cattle imported into Australia from Canada as embryos. Given that both Galloways and Angus originated in Scotland, and that the other breeds studied are Angus derivatives, it seems that the two mutations arose in Scotland, and have then been disseminated to other parts of the world. Prior to the introduction of the control programme in 1974 in New Zealand, approximately 3000 Angus calves were affected each year in that country. As reported by Tollersrud *et al.* (1997), the bovine gene spans 16 kb and comprises 24 exons. The open reading frame comprises 2997 bases encoding a signal peptide of 50 amino acids plus a further 949 amino acids, which are cleaved into five peptides in the mature enzyme. In Angus cattle, a T961 → C transition results in a Phe321 → Leu substitution. In Galloway cattle, a G662 → A transition results in a Arg221 → His substitution	Whittem and Walker (1957); Berg *et al.* (1997)

Morphological Traits and Inherited Disorders

Maple syrup urine disease	A progressive neurological dysfunction that occurs in the first few hours of life, accompanied by an odour of maple syrup in the urine. Within 12–48 h, affected calves develop central nervous system (CNS) depression, lethargy and a 'scruffy' unclean coat, progressing to coma and death within 48–72 h. This disorder is due to a build-up of branched-chain keto acids. In normal animals, the catabolism (breakdown) of the three branched-chain amino acids (leucine, isoleucine and valine) involves transamination to branched-chain α-keto acids, which are then oxidatively decarboxylated by the enzyme branched-chain α-keto acid dehydrogenase. Maple syrup urine disease results from the deficiency of this enzyme. In Poll Herefords, the deficiency is due to a nonsense mutation in codon 6 of the gene for the E1-α subunit of branched-chain α-keto acid dehydrogenase.	Harper et al. (1986a); Healy and Dennis (1995)
Muscular hypertrophy	The double-muscle trait in cattle is characterized by an increase in muscle mass of approx. 20%, resulting in substantially higher meat yield, a higher proportion of expensive cuts of meat, and lean and very tender meat, for which a substantial premium is paid. The trait is autosomal recessive, and the locus has been given the symbol mh. It occurs at such a high frequency in Piedmontese and Belgian Blue cattle that it is characteristic of these breeds. However, it also occurs in other breeds. Along with its obvious advantages, double-muscling also has one major drawback – a greatly increased incidence of calving difficulties, to the extent that Caesarean sections are the rule for deliveries within these breeds. However, its advantages are sufficient for double-muscled cattle to play a major role in animal agriculture in several countries, and they are found in many countries. Double-muscling in Belgian Blues is due to an 11-bp deletion (of nucleotides 821 to 831), resulting in a frame shift and subsequent premature termination in the bioactive carboxy-terminal domain of the gene (Grobet et al., 1997; Kambadur et al., 1997; McPherron and Lee, 1997). This is a region that is very highly conserved in the transforming growth factor (TGF) family of peptides. The same mutation is responsible for double-muscling in the Asturiana breed (Georges et al., 1998; Grobet et al., 1998). In contrast, double-muscled Piedmontese cattle have a G–A transition that changes a cysteine residue to a tyrosine in the same highly conserved region of the gene (Kambadur et al., 1997; McPherron and Lee, 1997). In a screen of 35 double-muscled cattle from ten European breeds, seven different sequence variants (alleles) were discovered in the coding region of the myostatin gene (Georges et al., 1998; Grobet et al., 1998). Five of these could cause a deficiency of myostatin: the 11-bp deletion already described; an insertion/deletion in which ten unrelated bases are inserted in the place of seven bases that have been deleted at nucleotide 418; a C–T transition at nucleotide 610; a G–T transversion at nucleotide 676; and a G–A transition at nucleotide 938. Two other mutants were unlikely to cause a deficiency: a C–A transversion at nucleotide 282 (resulting in a conservative Phe–Leu amino acid substitution);	Fischer (1953); Grobet et al. (1998)

continued overleaf

Table 4.1. *Continued.*

Molecular defect known	Summary	Earliest reference; latest reference (if different from earliest reference)
	and a silent C–T transition at nucleotide 414. For most of the breeds, double-muscled animals were homozygous for one of the five harmful mutations, or were compound heterozygotes for two mutants. Obviously, there is considerable genetic heterogeneity in the cause of double-muscling. Furthermore, the mutations are not all unique to one breed: two are shared by more than one breed. In addition, two breeds (Limousin and Blonde d'Aquitaine) have double-muscling but do not have any of the five harmful mutations. Clearly, there are more harmful mutations to be discovered. More importantly, the discovery that mutations in the myostatin gene have a profound effect on meat yield and quality opens the way for elucidating the role of myostatin in meat production, which in turn will suggest novel (possibly non-genetic) ways of enhancing meat yield and quality	
Protoporphyria	Protoporphyrin is the last intermediate in the seven-step biosynthesis of haem from aminolaevulinic acid (ALA). The conversion of protoporphyrin to haem is catalysed by the enzyme ferrochelatase. Protoporphyria is a disorder resulting from the build-up of protoporphyrin (and the other five, earlier intermediates (porphyrins) in this pathway due to the lack of this enzyme. Protoporphyrin is extremely photoreactive. Because of this, photosensitivity is the main clinical sign of this disorder. No details have yet been published. However, the report of the use of a deoxyribonucleic acid (DNA) genotyping test by Healy *et al.* (1995a), citing a personal communication from G.S. Johnson at the University of Missouri, implies that the molecular basis of this disorder within the gene for ferrochelatase has been determined by Dr Johnson	Ruth *et al.* (1977); Shibuya *et al.* (1995)
Sex reversal: XY female	A polymerase chain reaction (PCR) test failed to detect the SRY gene in three XY female Japanese cattle (Kawakura *et al.*, 1996). The authors concluded that these cases of sex reversal were due to a deletion of the SRY gene	Matejka *et al.* (1989); Kawakura *et al.* (1997)
Spherocytosis	In a population of Japanese cattle, this disorder is due to a nonsense mutation (CGA → TGA; Arg → Stop) in the gene for band 3 of the red-cell membrane, at the position corresponding to codon 646 of the human gene (Inaba *et al.*, 1996). The lack of this protein produced very unstable red-cell membranes, resulting in anaemia and retarded growth	Inaba *et al.* (1996)

Other single-locus traits and disorders

Ceroid lipofuscinosis	A disorder characterized by blindness, mental dullness and abnormal behaviour, caused by abnormal storage (build-up) of lipofuscins, which are fatty pigments (lipopigments or ceroid bodies) formed by the solution of a pigment in fat. The brain atrophies, showing signs of extensive accumulation of ceroid bodies in nerve tissue and later in viscera	Martinus *et al.* (1991); Jolly *et al.* (1992)
Chediak–Higashi syndrome	The mouse homologue of this disorder is the beige coat-colour mutation, which is a mutation in the gene for the lysosomal trafficking regulator, *Lyst*. As its name suggests, this gene is involved in lysosomal functioning – lysosomes being the digestive system of the cell. Thus Chediak–Higashi syndrome is a lysosomal disorder	Ament and O'Mary (1963); Ogawa *et al.* (1997)
Coat colour, agouti	This locus encodes a peptide antagonist of the melanocyte-stimulating hormone receptor (MC1R), which is the product of the extension locus (see Coat colour, extension). In cattle, there are two alleles at this locus: the dominant A^+ allele, conferring brown coat colour, and the recessive *a* allele, resulting in black (i.e. non-agouti). However, the effect of these alleles is seen only if the E^+ allele is present at the *E* (*extension*) locus (see Chapter 3 for more details on coat colour)	Adalsteinsson *et al.* (1995)
Coat colour, albinism	Congenital lack of pigment in all parts of the body. Due to a non-functional form of the enzyme tyrosinase (see Chapter 3 for more details on coat colour)	Detlefsen (1920); Foreman *et al.* (1994)
Coat colour, roan	Roan coat colour in cattle is due to heterozygosity at a locus for which the two homozygous phenotypes are red and white (in breeds like the Shorthorn) or black and white (in breeds like the Belgian Blue). Unfortunately, the white coat colour appears to be associated with a reproductive disorder (imperforate hymen) in females, known as white heifer disease. Charlier *et al.* (1996a) identified a microsatellite marker linked to this locus (see Chapter 3 for more details on coat colour)	Charlier *et al.* (1996a)
Dwarfism, Dexter	The Dexter is an ancient breed of cattle characterized by its small size. Part of this small size appears to be due to heterozygosity for a dwarfism gene. When homozygous, this gene results in the so-called Dexter bulldog dwarf, which is characterized by disproportionate dwarfism, a short vertebral column, marked micromelia, a relatively large head and retruded muzzle, cleft palate and protruding tongue, and a large abdominal hernia	Wilson (1909); Harper *et al.* (1998)

continued overleaf

Table 4.1. *Continued.*

Molecular defect known	Summary	Earliest reference; latest reference (if different from earliest reference)
Dwarfism, growth-hormone-receptor deficiency	Homozygotes for this mutation have a mature weight and size approximately 70% that of normal Brahman cattle. In a comparison of ovarian growth and development between dwarfs and normals, Chase et al. (1998) showed that the dwarfs have abnormally high levels of circulating growth hormone, and abnormally low levels of insulin-like growth factor I, resulting in failure to maintain follicular growth during the midluteal phase of the oestrous cycle. However, follicular growth resumes following luteolysis when secretion of luteinizing hormone increases. In being due to a deficiency of growth hormone receptor, GHRD in Brahman cattle is homologous with sex-linked dwarfism in chickens. The molecular basis of this disorder has still to be determined. This mutation occurred in the Brahman herd at the US Department of Agriculture (USDA) Subtropical Agricultural Research Station (STARS), near Beltsville, Florida (Elsasser et al., 1990)	Elsasser et al. (1990); Chase et al. (1998)
Dwarfism, snorter	An autosomal recessive form of dwarfism that reached an alarming frequency in Herefords in the USA in the 1960s. At one stage, it was thought that the dwarfism was associated with a lysosomal storage disease, and that snorter dwarves were a model for Hurler's syndrome (mucopolysaccharidosis I). But this is now known not to be the case. Various techniques of carrier detection, including the use of X-rays and a device for measuring the extent of concavity in the dorsal line of the head (a profilometer), proved to be less effective than initially hoped. The disorder gradually decreased in frequency as breeders avoided blood lines known to be carrying the gene	Johnson et al. (1950); Hurst et al. (1975)
Dyserythro-poiesis	An inability to produce the normal quantity of red blood cells; a slow but progressive disease seen in Polled Hereford calves, usually resulting in death	Steffen et al. (1992)
Ehlers–Danlos syndrome, type VII	Although evidence is incomplete, it appears that this disorder in cattle is due to a mutation in the gene for the enzyme procollagen-I amino proteinase, which is the enzyme responsible for removing surplus amino acids from the N-terminal end of procollagen-I molecules	Hanset and Ansay (1967); Lapiere and Nusgens (1993)
Epitheliogenesis imperfecta	Congenital absence of skin. Also known as aplasia cutis	Hadley (1927); Agerholm et al. (1993)

Morphological Traits and Inherited Disorders 65

Factor XI deficiency	A heritable bleeding disorder, due to the deficiency of blood clotting factor XI. Also known as plasma thromboplastin antecedent PTA deficiency and Rosenthal syndrome	Kociba et al. (1969); Coomber et al. (1997)
Gangliosidosis, GM1	A lysosomal storage disease in which there is a build-up (storage) of GM1 gangliosides (a type of glycolipid) in various tissues, due to the lack of the enzyme β-galactosidase, whose task is to break down the GM1 ganglioside into its constituents. Characterized by progressive neuromuscular dysfunction and impaired growth from an early age	Donnelly et al. (1972); Figueras et al. (1992)
Glycogen storage disease II	A lysosomal storage disease in which there is a build-up (storage) of glycogen, due to the lack of the enzyme α-glucosidase, whose task is to break down glycogen into its constituent glucose molecules. Characterized by poor growth, incoordination, muscle weakness and eventual recumbency. Also known as Pompe's disease, cardiac form of generalized glycogenosis, cardiomegalia glycogenica diffusa, acid maltase deficiency and glycogenosis type II	Richards et al. (1977); Healy et al. (1995b)
Horns	There is substantial variation in the extent of horn growth, making classification difficult. However, in general, the presence or absence of horns can be attributed to the action of two alleles at an autosomal locus, with the polled condition being dominant to horned. The absence of horns (polledness) is of substantial benefit in cattle, from an economic and welfare point of view: bruising due to horns is eliminated and the stress associated with dehorning is avoided. Compared with horned cattle, polled cattle have a narrower skull and the frontal eminence is more pronounced (Brenneman et al., 1996). Microsatellites TGLA49 and BM6438 show complete linkage to the horns locus (Schmutz et al., 1995)	Spillman (1906); Harlizius et al. (1997)
Hypotrichosis	Also known as hairlessness	Mohr and Wriedt (1928); Stober et al. (1995)
Lethal trait A46	An autosomal recessive disorder in which there is reduced ability to absorb zinc from the intestine, resulting in loss of hair, skin lesions and parakeratosis, atrophy of the thymus and decreased resistance to infections. Affected calves soon die, but treatment with regular oral doses of zinc solution completely eliminate the clinical signs, enabling homozygotes to be kept alive and well for years. Also known as acrodermatitis enteropathica, zinc-deficiency type; hereditary parakeratosis; hereditary thymus hypoplasia; and hereditary zinc deficiency	McPherson et al. (1964); Machen et al. (1996)
Mannosidosis, β	An autosomal recessive lysosomal storage disease due a deficiency of the enzyme β-mannosidase. This deficiency results in a build-up of oligosaccharides, especially in the nervous system, causing a gradual and fatal neural deterioration	Jolly et al. (1990); Schmutz et al. (1996)

continued overleaf

Table 4.1. *Continued.*

Molecular defect known	Summary	Earliest reference; latest reference (if different from earliest reference)
Myoclonus	An autosomal recessive disorder in which affected calves experience uncontrolled muscle tetany even before birth, sometimes resulting in the embryonic fracture of bones. Due to a deficiency of inhibitory glycine receptors in the spinal cord. Since these are the same receptors to which strychnine attaches, the tetany is similar to that seen in cases of strychnine poisoning. Also known as myoclonus epilepsy of Unverricht and Lundborg, Baltic myoclonus epilepsy and progressive myoclonus epilepsy	Blood and Gay (1971); Lummis et al. (1990)
Porphyria, congenital erythropoietic	Porphyrins are a class of organic compounds characterized by four pyrrole nuclei connected in a ring structure. When combined with iron, porphyrins form haem, which is a component of haemoglobin, cytochromes, catalases and peroxidases. Thus, porphyrins are constituents of many compounds that play a vital role in biological systems. The biosynthesis of porphyrins involves a six-step process, starting with aminolaevulinic acid (ALA) and ending with protoporphyrin. Each step is catalysed by an enzyme. A deficiency of any one of these enzymes results in a build-up of intermediates prior to the step for which the enzyme is lacking, and a deficiency of intermediates after that step. In much of the literature, these intermediates are loosely called porphyrins. Congenital erythropoietic porphyria results from a deficiency of the third enzyme, uroporphyrinogen III cosynthetase, and a consequent build-up of intermediates (i.e. a build-up of porphyrins) that have been synthesized in the three previous steps. Also known as pink tooth	Fourie (1936); Franco et al. (1992)
Progressive degenerative myeloencephalopathy	A wasting or 'dying back' of the peripheral nerves, known as weaver disease. There is some evidence that heterozygotes may have higher milk production, which could result in selection favouring heterozygotes. The identification of a linked microsatellite marker by Georges et al. (1993) has been very useful for carrier detection	Leipold (1988); Millonig et al. (1996)
Protamine-2 deficiency	In most mammals, protamine-2 constitutes the major component of basic protein in sperm nuclei. It binds to DNA during the elongation of spermatids. Pigs and cattle have the protamine-2 gene, but produce very little protamine-2. It appears that all cattle are homozygous for a mutation in the protamine-2 gene. Presumably, the mutation occurred before the splitting of the bovine evolutionary lineage	Maier et al. (1990)

Morphological Traits and Inherited Disorders

Spinal muscular atrophy	An autosomal recessive neuronal degeneration resulting in weakness, trembling and inability to stand. The clinical signs appear within the first 6 weeks of life, and death results within a few weeks	Elhamidi et al. (1989); Lidauer and Essl (1994)
Syndactyly	Fusion of the claws (cleats or digits). Occurs in one, two, three or (rarely) four legs. Syndactyly has been reported in many breeds of cattle in many countries. Most of the documentation, however, concerns its occurrence in US Holsteins, where, as a result of the siring of more than 60,000 calves by a bull who was subsequently shown to be a carrier, the disorder attracted considerable attention (Anon., 1967). The possibility that artificial selection favouring heterozygotes may have contributed to the unacceptably high frequency of the disorder was suggested by data showing that carrier females produced on average 0.95 kg of butterfat and 135.29 kg of milk more than homozygotes (Leipold et al., 1973). Progeny testing of males (Johnson et al., 1980) and females (Leipold and Peeples, 1981) was the main means of identifying carriers until Charlier et al. (1996b) used a genome-wide set of DNA markers in an identity-by-descent linkage study to show that the syndactyly gene is located on chromosome 15. Also known as mule foot	Sing and Tandon (1949); Fazili (1997)
Testicular feminization	This is an abnormality of sexual development in which affected individuals have an XY chromosomal constitution, undescended testes and female secondary sexual characteristics (including female external genitalia). Also, instead of normally developed Mullerian-duct derivatives (Fallopian tubes, uterus, cervix and upper portion of the vagina), they have underdeveloped Wolffian-duct derivatives (epididymis, vas deferens and seminal vesicle). In all species so far investigated, the inheritance is X-linked recessive. In humans and mice, this disorder is known to be due to a deficiency of an androgen receptor encoded by a gene on the X chromosome. The presence of a Y chromosome induces the undifferentiated embryonic gonads to develop as testes, but, in the absence of androgen receptor, the androgens produced by the testes cannot exert any effect. The result is that the embryo follows the default path of development, which is female. Also known as androgen insensitivity syndrome, androgen receptor deficiency and dihydrotestosterone receptor (DHTR) deficiency	Nes (1966); Peter et al. (1993)
Tibial hemimelia	An autosomal recessive disorder reported in Galloways, characterized by a gross distortion of the hind legs, a resultant inability to stand and subsequent death. The breed society's laudable attempts to control this disorder consisted of the creation of a herd of carrier cows for the progeny-testing of young bulls, an insurance scheme for all bulls sold at the annual yearling bull sale and the publication of a list of known carriers	Young (1951); Salako and Abdullahi (1982)
Vertical-fibre hide defect	An autosomal recessive, non-lethal defect of skin, resulting in downgrading during the tanning process	Peters and Duffy (1985); Kronick and Sacks (1991)

molecular level, and the rest. Obviously, each of these lists will expand as new discoveries are made. An up-to-date list of single-locus traits and disorders (including references) can be obtained by accessing OMIA on the Internet. At the time of writing, OMIA also contains information on a further 280 traits and inherited disorders for which the evidence of single-locus inheritance is insufficient to justify inclusion in Table 4.1. Information about these traits and disorders, including a full list of references, can be found by searching for 'cattle' in OMIA.

It is readily acknowledged that the information in OMIA is incomplete, and that it includes errors of omission and commission. However, one of the advantages of having this type of information stored in a database is that errors can be rectified easily as soon as they are spotted. The author would therefore be very grateful to any readers who identify errors in the information supplied in this chapter.

Conclusion

The list of inherited morphological traits and disorders presented in this chapter provides an indication of the range of such traits and disorders that have been observed and studied in cattle. The molecular and gene-mapping revolutions now under way will lead to an explosion of knowledge in this area in the years ahead. To exploit fully the genetic variation that does occur, breeders and researchers need to be continually on the lookout for unusual animals, saving them where possible. If DNA can be sampled from several generations of a family in which a particular morphological trait or disorder occurs and if careful records on the occurrence of the trait or disorder in that family have been kept, it will be an increasingly straightforward matter to identify the gene responsible.

Acknowledgement

Some of the information on horns was supplied by Ulrika Tjäldén and Vanja Kinch, Swedish University of Agricultural Sciences, Uppsala.

References and List of Reviews (marked with an asterisk)

Adalsteinsson, S., Bjarnadottir, S., Vage, D.I. and Jonmundsson, J.V. (1995) Brown coat color in Icelandic cattle produced by the loci extension and agouti. *Journal of Heredity* 86, 395–398.

*Agerholm, J.S., Basse, A. and Christensen, K. (1993) Investigations on the occurrence of hereditary diseases in the Danish cattle population 1989–1991. *Acta Veterinaria Scandinavica* 34, 245–253.

Ament, D.L. and O'Mary, C.C. (1963) Albinism and related blood disorders in Herefords. *Journal of Animal Science* 22, 815.

*Andresen, E. and Christensen, K. (1977) Genetics of disease in relation to preventive veterinary medicine. In: *Husdyrsykdommer i Relasjon Tilary. NK Vet-Symposium, Voksenasen, Oslo*. Nordisk Komite for Veterinaervitenskapelig Samarbeid, Oslo, pp. 15–53.

Angelos, S., Valberg, S.J., Smith, B.P., Mcquarrie, P.S., Shanske, S., Tsujino, S., Dimauro, S. and Cardinet, G.H. (1995) Myophosphorylase deficiency associated with rhabdomyolysis and exercise intolerance in 6 related Charolais cattle. *Muscle and Nerve* 18, 736–740.

*Anon. (1958) Incidence of deformities among A.I. calves. In: *Thirty Fourth Annual Report*, New Zealand Dairy Board, Wellington, p. 83.

Anon. (1967) The A.I. dilemma. *Holstein-Friesian World* 64, 1394–1395.

*Ballarini, G. (1977) Hereditary diseases in veterinary practice. *Rivista di Zootecnia e Veterinaria* 2, 177–186.

*Bauer, H. (1955) Hereditary defects of the udder. *Zuchtungskunde* 27, 256–266.

Berg, T., Healy, P.J., Tollersrud, O.K. and Nilssen, O. (1997) Molecular heterogeneity for bovine alpha-mannosidosis – PCR based assays for detection of breed-specific mutations. *Research in Veterinary Science* 63, 279–282.

Blood, D.C. and Gay, C.C. (1971) Hereditary neuraxial oedema of calves. *Australian Veterinary Journal* 47, 520.

Brenneman, R.A., Davis, S.K., Sanders, J.O., Burns, B.M., Wheeler, T.C., Turner, J.W. and Taylor, J.F. (1996) The polled locus maps to BTA1 in a *Bos indicus* × *Bos taurus* cross. *Journal of Heredity* 87, 156–161.

Charlier, C., Denys, B., Belanche, J.I., Coppieters, W., Grobet, L., Mni, M., Womack, J., Hanset, R. and Georges, M. (1996a) Microsatellite mapping of the bovine roan locus – a major determinant of white heifer disease. *Mammalian Genome* 7, 138–142.

Charlier, C., Farnir, F., Berzi, P., Vanmanshoven, P., Brouwers, B., Vromans, H. and Georges, M. (1996b) Identity-by-descent mapping of recessive traits in livestock – application to map the bovine syndactyly locus to chromosome 15. *Genome Research* 6, 580–589.

Chase, C.C., Kirby, C.J., Hammond, A.C., Olson, T.A. and Lucy, M.C. (1998) Patterns of ovarian growth and development in cattle with a growth hormone receptor deficiency. *Journal of Animal Science* 76, 212–219.

*Cho, D.Y. and Leipold, H.W. (1977) Congenital defects of the bovine central nervous system. *Veterinary Bulletin* 47, 489–504.

*COGNOSAG (1998). *Mendelian Inheritance in Cattle*. University of Camerino, Matalica, Italy.

Coomber, B.L., Galligan, C.L. and Gentry, P.A. (1997) Comparison of *in vitro* function of neutrophils from cattle deficient in plasma factor XI activity and from normal animals. *Veterinary Immunology and Immunopathology* 58, 121–131.

*Desnick, R.J., Patterson, D.F. and Scarpelli, D.G. (1982) *Animal Models of Inherited Metabolic Diseases*, 1st edn. Alan R. Liss, New York.

Detlefsen, J.A. (1920) A herd of albino cattle. *Journal of Heredity* 11, 378–379.

Donnelly, W.J.C., Hannan, J., Sheahan, B.J. and O'Connor, P.J. (1972) Cerebrospinal lipidosis in Friesian calves. *Veterinary Record* 91, 225–226.

Elhamidi, M., Leipold, H.W., Vestweber, J.G.E. and Saperstein, G. (1989) Spinal muscular atrophy in Brown Swiss calves. *Journal of Veterinary Medicine Series A – Animal Physiology Pathology and Clinical Veterinary Medicine – Zentralblatt für Veterinarmedizin Reihe a* 36, 731–738.

Elsasser, T.H., Hammond, A.C., Kozak, A.S. and Olson, T.A. (1990) Genetic and physiological consequences of a miniature condition in Brahman cattle [abstract]. *Journal of Animal Science* 68, 304.

Fazili, M.U.R. (1997) Congenital unidactylism along with contracture of flexor tendon in a calf. *Indian Veterinary Journal* 74, 445.

Figueras, A., Moralesolivas, F.J., Capella, D., Palop, V. and Laporte, J.R. (1992) Bovine gangliosides and acute motor polyneuropathy. *British Medical Journal* 305, 1330–1331.

Fischer, H. (1953) The genetics of the 'doppellender' condition in cattle. *Deutsche Tierarztliche Wochenschrift* 60, 25–27.

*Foote, R.H., Henderson, C.R. and Bratton, R.W. (1956) Testing bulls in artificial insemination centres for lethals, type and production. In *Proceedings of the Third International Congress of Animal Reproduction, Cambridge*, International Congress of Animal Reproduction, Cambridge, pp. 49–53.

Foreman, M.E., Lamoreux, M.L., Kwon, B. and Womack, J.E. (1994) Mapping the bovine albino locus. *Journal of Heredity* 85, 318–320.

Fourie, P.J.J. (1936) The occurrence of congenital porphyrinuria (pink tooth) in cattle in South Africa (Zwaziland). *Onderstepoort Journal of Veterinary Science and Animal Industry* 7, 535–566.

Franco, D.A., Lin, T.L. and Leder, J.A. (1992) Bovine congenital erythropoietic porphyria – bovine review article. *Compendium on Continuing Education for the Practicing Veterinarian* 14, 822–826.

Georges, M. (1997) A deletion in the bovine myostatin gene causes the double-muscled phenotype in cattle. *Nature Genetics* 17, 71–74.

Georges, M., Dietz, A.B., Mishra, A., Nielsen, D., Sargeant, L.S., Sorensen, A., Steele, M.R., Zhao, X.Y., Leipold, H.W., Womack, J.E. and Lathrop, M. (1993) Microsatellite mapping of the gene causing weaver disease in cattle will allow the study of an associated quantitative trait locus. *Proceedings of the National Academy of Sciences of the USA* 90, 1058–1062.

Georges, M., Grobet, L., Poncelet, D., Royo, L.J., Pirottin, D. and Brouwers, B. (1998) Positional candidate cloning of the bovine *mh* locus identifies an allelic series of mutations disrupting the myostatin function and causing double-muscling in cattle. *Proceedings of the 6th World Congress on Genetics Applied to Livestock Production* 26, 195–204.

*Gilman, J.P.W. (1956) Inherited developmental defects in cattle. In *Proceedings of the Ninety Second Annual Meeting of the American Veterinary Medical Association*, American Veterinary Medical Association, Ithaca, New York, pp. 49–53.

*Gilmore, L.O. (1957) Inherited defects in cattle. *Journal of Dairy Science* 40, 593–595.

*Gotink, W.M., de Groot, T. and Stegenga, T. (1955) Hereditary defects of cattle in the Netherlands. *Tijdschrift voor Diergeneeskunde* 80 (Suppl.), 1–45.

*Greene, H.J., Leipold, H.W., Huston, K., Noordsdy, J.L. and Dennis, S.M. (1973) Congenital defects in cattle. *Irish Veterinary Journal* 27, 37–45.

Grobet, L., Martin, L.J.R., Poncelet, D., Pirottin, D., Brouwers, B., Riquet, J., Schoeberlein, A., Dunner, S., Menissier, F., Massabanda, J., Fries, R., Hanset, R., Grobet, L., Poncelet, D., Royo, L.J., Brouwers, B., Pirottin, D., Michaux, C., Menissier, F., Zanotti, M., Dunner, S. and Georges, M. (1997) Molecular definition of an allelic series of mutations disrupting the myostatin function and causing double-muscling in cattle. *Mammalian Genome* 9, 210–213.

*Grubbs, S.T. and Olchowy, T.W.J. (1997) Bleeding disorders in cattle – a review and diagnostic approach. *Veterinary Medicine* 92, 737–743.

Grupe, S., Dietl, G. and Schwerin, M. (1996) Population survey of citrullinemia on German Holsteins. *Livestock Production Science* 45, 35–38.

Hadley, F.B. (1927) Congenital epithelial defects of calves. Epitheliogenesis imperfecta neonatorum bovis – a recessive brought to light by inbreeding. *Journal of Heredity* 18, 487–495.

*Hadley, F.B. and Warwick, B.L. (1927) Inherited defects of livestock. *Journal of the American Veterinary Medical Association* 70, 492–504.

*Hamori, D. (1959) Genetic pathological investigations. I. Hereditary deformities of the limbs in calves involving the tendons. *Magyar Allatorvosok Lapja* 14, 53–56.

*Hamori, D. (1966) Studies on genetical defects in herds of cattle, pigs and sheep. *Hungarian Agriculture Review* 15, 29–30.

*Hamori, D. (1973) Breeding hygiene examination of cattle twins. *Monatshefte für Veterinarmedizin* 28, 857–861.

Hanset, R. and Ansay, M. (1967) [Dermatosparaxia (fragility of the skin) – a new hereditary defect of connective tissues in cattle]. *Annales de Médecine Vétérinaire* 111, 451–470.

Harlizius, B., Schober, S., Tammen, I. and Simon, D. (1996) Isolation of the bovine uridine monophosphate synthase gene to identify the molecular basis of DUMPS in cattle. *Journal of Animal Breeding and Genetics – Zeitschrift für Tierzuchtung und Zuchtungsbiologie* 113, 303–309.

Harlizius, B., Tammen, I., Eichler, K., Eggen, A. and Hetzel, D.J.S. (1997) New markers on bovine chromosome 1 are closely linked to the polled gene in Simmental and Pinzgauer cattle. *Mammalian Genome* 8, 255–257.

Harper, P.A.W., Healy, P.J. and Dennis, J.A. (1986a) Maple syrup urine disease as a cause of spongiform encephalopathy in calves. *Veterinary Record* 119, 62–65.

Harper, P.A.W., Healy, P.J., Dennis, J.A., O'Brien, J.J. and Rayward, D.H. (1986b) Citrullinaemia as a cause of neurological disease in neonatal Friesian calves. *Australian Veterinary Journal* 63, 373–379.

Harper, P.A.W., Latter, M.R., Nicholas, F.W., Cook, R.W. and Gill, P.A. (1998) Chondrodysplasia in Australian Dexter cattle. *Australian Veterinary Journal* 76, 199–202.

*Healy, P.J. (1996) Testing for undesirable traits in cattle – an Australian perspective. *Journal of Animal Science* 74, 917–922.

*Healy, P.J. and Dennis, J.A. (1993) Inherited enzyme deficiencies in livestock. *Veterinary Clinics of North America – Food Animal Practice* 9, 55–63.

Healy, P.J. and Dennis, J.A. (1995) Heterozygote detection for maple syrup urine disease in cattle. *Australian Veterinary Journal* 72, 346–348.

Healy, P.J., Nicholls, P.J., Martiniuk, F., Tzall, S., Hirschhorn, R. and Howell, J.M. (1995a) Evidence of molecular heterogeneity for generalised glycogenosis between and within breeds of cattle. *Australian Veterinary Journal* 72, 309–311.

Healy, P.J., Dennis, J.A. and Moule, J.F. (1995b) Use of hair root as a source of DNA for the detection of heterozygotes for recessive defects in cattle. *Australian Veterinary Journal* 72, 392.

*Heizer, E.E. (1958) Genetic abnormalities in cattle. *Veterinary Science News* 12, 6–9.

*Herzog, A. (1992) Genetic defects in cattle and the possibilities of control. *Wiener Tierarztliche Monatsschrift* 79, 142–148.

*Hiraga, T. and Dennis, S.M. (1993) Congenital duplication. *Veterinary Clinics of North America – Food Animal Practice* 9, 145–161.

Hurst, R.E., Cezayirli, R.C. and Lorincz, A.E. (1975) Nature of the glycosaminoglycanuria (mucopolysacchariduria) in brachycephalic 'Snorter' dwarf cattle. *Journal of Comparative Pathology* 85, 481–486.

*Huston, K. (1993) Heritability and diagnosis of congenital abnormalities in food animals. *Veterinary Clinics of North America – Food Animal Practice* 9, 1–9.

*Huston, K., Chase, R. and Waller, R. (1974) Identifying genes with visible effects in cattle, full, three quarter, and half sib, other methods. In: *Proceedings of the First World Congress on Genetics Applied to Livestock Production, Madrid*, pp. 39–46.

*Hutt, F.B. (1934) Inherited lethal characters in domestic animals. *Cornell Veterinarian* 24, 1–25.

*Hutt, F.B. (1968) Genetic defects of bones and joints in domestic animals. *Cornell Veterinarian* 58 (Suppl.), 104–113.

Inaba, M., Yawata, A., Koshino, I., Sato, K., Takeuchi, M., Takakuwa, Y., Manno, S., Yawata, Y., Kanzaki, A., Sakai, J., Ban, A., Ono, K. and Maede, Y. (1996) Defective anion transport and marked spherocytosis with membrane instability caused by hereditary total deficiency of red cell band 3 in cattle due to a nonsense mutation. *Journal of Clinical Investigation* 97, 1804–1817.

Jamieson, S., Simpson, B.W. and Russell, J.B. (1945) Bovine congenital goitre. *Veterinary Record* 57, 429–431.

Joerg, H., Fries, H.R., Meijerink, E. and Stranzinger, G.F. (1996) Red coat color in Holstein cattle is associated with a deletion in the MSHR gene. *Mammalian Genome* 7, 317–318.

Johnson, J.L., Leipold, H.W., Snider, G.W. and Baker, R.D. (1980) Progeny testing for bovine syndactyly. *Journal of the American Veterinary Medical Association* 176, 549–550.

Johnson, L.E., Harshfield, G.S. and McCone, W. (1950) Dwarfism, a hereditary defect in beef cattle. *Journal of Heredity* 41, 177–181.

*Jolly, R.D. (1993) Lysosomal storage diseases in livestock. *Veterinary Clinics of North America – Food Animal Practice* 9, 41–53.

*Jolly, R.D. and Hartley, W.J. (1977) Storage diseases of domestic animals. *Australian Veterinary Journal* 53, 1–8.

*Jolly, R.D. and Healy, P.J. (1986) Screening for carriers of genetic diseases by biochemical means. *Veterinary Record* 119, 264–267.

*Jolly, R.D. and Leipold, H.W. (1973) Inherited diseases of cattle – a perspective. *New Zealand Veterinary Journal* 21, 147–155.

Jolly, R.D., Tse, C.A. and Greenway, R.M. (1973) Plasma alpha mannosidase activity as a means of detecting mannosidosis heterozygotes. *New Zealand Veterinary Journal* 21, 64–69.

*Jolly, R.D., Dodds, W.J., Ruth, G.R. and Trauner, D.B. (1981) Screening for genetic diseases, principles and practice. *Advances in Veterinary Science and Comparative Medicine* 25, 245–276.

Jolly, R.D., Thompson, K.G., Bayliss, S.L., Vidler, B.M., Orr, M.B. and Healy, P.J. (1990) Beta-mannosidosis in a Salers calf – a new storage disease of cattle. *New Zealand Veterinary Journal* 38, 102–105.

Jolly, R.D., Gibson, A.J., Healy, P.J., Slack, P.M. and Birtles, M.J. (1992) Bovine ceroid-lipofuscinosis – pathology of blindness. *New Zealand Veterinary Journal* 40, 107–111.

Jorgensen, J.N. and Madsen, P. (1997) Genetic parameters for and BLAD effects on beef production traits and disease frequency. *Acta Agriculturae Scandinavica Section A – Animal Science* 47, 1–8.

Kambadur, R., Sharma, M., Smith, T.P.L. and Bass, J.J. (1997) Mutations in myostatin (GDF8) in double-muscled Belgian Blue and Piedmontese cattle. *Genome Research* 7, 910–916.

Kawakura, K., Miyake, Y.I., Murakami, R.K., Kondoh, S., Hirata, T.I. and Kaneda, Y. (1996) Deletion of the SRY region on the Y chromosome detected in bovine gonadal hypoplasia (XY female) by PCR. *Cytogenetics and Cell Genetics* 72, 183–184.

Kawakura, K., Miyake, Y.I., Murakami, R.K., Kondoh, S., Hirata, T.I. and Kaneda, Y. (1997) Abnormal structure of the Y chromosome detected in bovine gonadal hypoplasia (XY female) by FISH. *Cytogenetics and Cell Genetics* 76, 36–38.

Kehrli, M.E., Schmalstieg, F.C., Anderson, D.C., Vandermaaten, M.J., Hughes, B.J., Ackermann, M.R., Wilhelmsen, C.L., Brown, G.B., Stevens, M.G. and Whetstone, C.A. (1990) Molecular definition of the bovine granulocytopathy syndrome – identification of deficiency of the Mac-1 (CD11b/CD18) glycoprotein. *American Journal of Veterinary Research* 51, 1826–1836.

Klungland, H., Vage, D.I., Gomezraya, L., Adalsteinsson, S. and Lien, S. (1995) The role of melanocyte-stimulating hormone (msh) receptor in bovine coat color determination. *Mammalian Genome* 6, 636–639.

Kociba, G.J., Ratnoff, O.D., Loeb, W.F., Wall, R.L. and Heider, L.E. (1969) Bovine plasma thromboplastin antecedent (factor XI) deficiency. *Journal of Laboratory and Clinical Medicine* 74, 37–41.

Kronick, P.L. and Sacks, M.S. (1991) Quantification of vertical-fiber defect in cattle hide by small-angle light scattering. *Connective Tissue Research* 27, 1–13.

*Kuhn, C. (1997) Molecular genetic background of inherited defects in cattle [German]. *Archiv für Tierzucht – Archives of Animal Breeding* 40, 121–127.

*Ladds, P.W. (1993) Congenital abnormalities of the genitalia of cattle, sheep, goats, and pigs. *Veterinary Clinics of North America – Food Animal Practice* 9, 127–144.

Lapiere, C.M. and Nusgens, B.V. (1993) Ehlers–Danlos syndrome type VII-C, or human dermatosparaxis, the offspring of a union between basic and clinical research. *Archives of Dermatology* 129, 1316–1319.

*Larsson, E.L. (1952) *Inherited Defects in Swedish Cattle: a Brief Survey*. Lts Forlag 88, Stockholm.

*Lauvergne, J.J. (1968) *Catalogue des Anomalies Héréditaires des Bovines*. Bulletin Technique du Département de Génétique Animale. Jouy-en-Josas, France.

*Lauvergne, J.J. (1978) Hereditary and non-hereditary congenital abnormalities in Normandy cattle. *Annales de Génétique et de Sélection Animale* 10, 131–134.

*Lauvergne, J.J. and Winzenreid, H.U. (1972) Genetic mutations with visible effects in Swiss cattle breeds. *Annales de Génétique et de Sélection Animale* 4, 523–535.

*Leipold, H.W. (1974a) Hereditary diseases of the bovine neurologic system. *Journal of the American Veterinary Medical Association* 165, 726.

*Leipold, H.W. (1974b) Congenital defects in cattle. *Giessener Beitrage zur Erbpathologie und Zuchthygiene* 6, 8–23.

*Leipold, H.W. (1980a) Diagnosis and control of undesirable genetic diseases and lethal factors in cattle. In: *Reports and Summaries. Eleventh International Congress on Diseases of Cattle, Tel-Aviv*, Vol. 1, pp. 543–557.

*Leipold, H.W. (1980b) Diagnosis of congenital defects in cattle in the A.I. industry. *Advanced Animal Breeder* 28, 5–11.

Leipold, H.W. (1988) Straight talk on weavers. *Brown Swiss Bulletin* January 30–32.

*Leipold, H.W. and Dennis, S.M. (1987) Cause, nature, effect and diagnosis of bovine congenital defects. *Irish Veterinary News* 9, 11–19.

Leipold, H.W. and Peeples, J.G. (1981) Progeny testing for bovine syndactyly. *Journal of the American Veterinary Medical Association* 179, 69–70.

*Leipold, H.W. and Schalles, R. (1977) Genetic defects in cattle, transmission and control. *Veterinary Medicine and Small Animal Clinician* 45, 80–85.

*Leipold, H.W., Dennis, S.M. and Huston, K. (1972) Congenital defects of cattle: nature, cause and effect. *Advances in Veterinary Science and Comparative Medicine* 16, 103–150.

Leipold, H.W., Dennis, S.M. and Huston, K. (1973) Syndactyly in cattle. *Veterinary Bulletin* 43, 399–403.

*Leipold, H.W., Hiraga, T. and Dennis, S.M. (1993a) Congenital defects of the bovine central nervous system. *Veterinary Clinics of North America – Food Animal Practice* 9, 77–91.

*Leipold, H.W., Hiraga, T. and Dennis, S.M. (1993b) Congenital defects of the bovine musculoskeletal system and joints. *Veterinary Clinics of North America – Food Animal Practice* 9, 93–104.

Lidauer, M. and Essl, A. (1994) Estimation of the frequencies for recessive lethal genes for spinal muscular atrophy, arachnomelia and weaver in the Austrian Braunvieh population. *Zuchtungskunde* 66, 54–65.

*Lojda, L. (1986) [Control of inherited disorders in domestic livestock in Czechoslovakia]. *Tierhygiene – Information* 18, 55–64.

Lummis, S.C.R., Gundlach, A.L., Johnston, G.A.R., Harper, P.A.W. and Dodd, P.R. (1990) Increased gamma-aminobutyric acid receptor function in the cerebral cortex of myoclonic calves with an hereditary deficit in glycine strychnine receptors. *Journal of Neurochemistry* 55, 421–426.

Machen, M., Montgomery, T., Holland, R., Braselton, E., Dunstan, R., Brewer, G. and Yuzbasiyangurkan, V. (1996) Bovine hereditary zinc deficiency – lethal trait A46. *Journal of Veterinary Diagnostic Investigation* 8, 219–227.

McKusick, V.A. (1996) *Mendelian Inheritance in Man,* 12th edn. Johns Hopkins University Press, Baltimore.

McPherron, A.C. and Lee, S.J. (1997) Double muscling in cattle due to mutations in the myostatin gene. *Proceedings of the National Academy of Sciences of the USA* 94, 12457–12461.

McPherson, E.A., Beattie, I.S. and Young, G.B. (1964) An inherited defect in Friesian calves. *Nordisk Veterinaer-medicin* 16 (Suppl.1) 533–540.

Maier, W.M., Nussbaum, G., Domenjoud, L., Klemm, U. and Engel, W. (1990) The lack of protamine-2 (p2) in boar and bull spermatozoa is due to mutations within the P2 gene. *Nucleic Acids Research* 18, 1249–1254.

Martinus, R.D., Harper, P.A.W., Jolly, R.D., Bayliss, S.L., Midwinter, G.G., Shaw, G.J. and Palmer, D.N. (1991) Bovine ceroid-lipofuscinosis (Batten disease) – the major component stored is the DCCD-reactive proteolipid, subunit-C, of mitochondrial ATP synthase. *Veterinary Research Communications* 15, 85–94.

Matejka, M., Berland, H.M., Darre, R., Delverdier, M. and Valognes, J. (1989) 1 case of XY gonadic dysgenesia in a heifer. *Revue de Médécine Vétérinaire* 140, 1011–1014.

Millonig, J.H., Millen, K.J. and Hatten, M.E. (1996) A high-density molecular genetic map around the *Weaver* locus. *Mammalian Genome* 7, 616–618.

Mohr, O.L. and Wriedt, C. (1928) *Hairless*, a new recessive lethal in cattle. *Journal of Genetics* 19, 315–336.

Nes, N. (1966) [Testicular feminisation in cattle]. *Nordisk Veterinaermedicin* 18, 19–29.

*Nordlund, S. (1962) Lethal defects and their combating in Swedish Friesian cattle. *Proceedings of the Fourth International Congress on Animal Reproduction, The Hague* 4, 765–767.
Ogawa, H., Tu, C.H., Kagamizono, H., Soki, K., Inoue, Y., Akatsuka, H., Nagata, S., Wada, T., Ikeya, M., Makimura, S., Uchida, K., Yamaguchi, R. and Otsuka, H. (1997) Clinical, morphologic, and biochemical characteristics of chediak-higashi-syndrome in fifty-six Japanese black cattle. *American Journal of Veterinary Research* 58, 1221–1226.
Peter, A.T., Scheidt, A.B., Campbell, J.W. and Hahn, K.A. (1993) Male pseudohermaphroditism of the testicular feminization type in a heifer. *Canadian Veterinary Journal – Revue Vétérinaire Canadienne* 34, 304–305.
Peters, D.E. and Dufty, J.H. (1985) The effectiveness of selective breeding for reducing the incidence of the vertical fibre hide condition in Hereford cattle. *Journal of the American Leather Chemists Association* 80, 42–46.
*Priester, W.A. (1972) Congenital ocular defects in cattle, horses, cats and dogs. *Journal of the American Veterinary Medical Association* 160, 1504–1511.
*Queinneg, G. (1977) Diseases of genetic origin in calves. In: Mornet, P. and Espinasse, J. (eds) *Le Veau*. Maloine, Paris, pp. 419–465.
*Rasbech, N.O. (1964) Fertility and reproductive disorders of various species of farm livestock in Denmark. *British Veterinary Journal* 120, 415–430.
Richards, R.B., Edwards, J.R., Cook, R.D. and White, R.R. (1977) Bovine generalized glycogenosis. *Neuropathology and Applied Neurobiology* 3, 45–56.
Robinson, J.L., Drabik, M.R., Dombrowski, D.B. and Clark, J.H. (1983) Consequences of UMP synthase deficiency in cattle. *Proceedings of the National Academy of Sciences of the USA* 80, 321–323.
*Rousseaux, C.G. (1994) Congenital defects as a cause of perinatal mortality of beef calves. *Veterinary Clinics of North America – Food Animal Practice* 10, 35–51.
Ruth, G.R., Schwartz, S. and Stephenson, B. (1977) Bovine protoporphyria, the first nonhuman model of this hereditary photosensitizing disease. *Science* 198, 199–201.
Salako, M.A. and Abdullahi, U.S. (1982) Tibial hemimelia in a Bunaji calf. *Veterinary Record* 110, 430.
*Saperstein, G. (1993) Congenital abnormalities of internal organs and body cavities. *Veterinary Clinics of North America – Food Animal Practice* 9, 115–125.
*Scherenberg, G. (1958) Inherited defects in domestic animals: their removal by operative treatment. *Tierzüchter* 10, 457.
Schmutz, S.M., Marquess, F.L.S., Berryere, T.G. and Moker, J.S. (1995) DNA marker-assisted selection of the polled condition in Charolais cattle. *Mammalian Genome* 6, 710–713.
Schmutz, S.M., Moker, J.S., Leipprandt, J.R. and Friderici, K.H. (1996) Beta-mannosidase maps to cattle chromosome 6. *Mammalian Genome* 7, 474.
Schwenger, B., Schober, S. and Simon, D. (1993) DUMPS cattle carry a point mutation in the uridine monophosphate synthase gene. *Genomics* 16, 241–244.
*Scott, D. and Tizard, I. (1993) Primary immunodeficiencies of food animals. *Veterinary Clinics of North America – Food Animal Practice* 9, 65–75.
Shanks, R.D. and Robinson, J.L. (1990) Deficiency of uridine monophosphate synthase among Holstein cattle. *Cornell Veterinarian* 80, 119–122.
Shibuya, H., Nonneman, D., Tamassia, M., Allphin, O.L. and Johnson, G.S. (1995) The coding sequence of the bovine ferrochelatase gene. *Biochimica et Biophysica Acta – Bioenergetics* 1231, 117–120.

*Shrode, R.R. and Lush, J.L. (1947) The genetics of cattle. *Advances in Genetics* 1, 209–261.

Sing, S. and Tandon, R.K. (1949) A short note on calves with unbifurcated hooves. *Indian Journal of Veterinary Sciences* 12, 61.

Spillman, W.J. (1906) Mendel's law in relation to animal breeding. *Proceedings of the American Breeders Association* 1, 171–177.

*Steffen, D.J. (1993) Congenital skin abnormalities. *Veterinary Clinics of North America – Food Animal Practice* 9, 105–114.

Steffen, D.J., Elliott, G.S., Leipold, H.W. and Smith, J.E. (1992) Congenital dyserythropoiesis and progressive alopecia in polled Hereford calves – hematologic, biochemical, bone marrow cytologic, electrophoretic, and flow cytometric findings. *Journal of Veterinary Diagnostic Investigation* 4, 31–37.

*Stegenga, T. (1964) Congenital anomalies in cattle. *Tijdschrift voor Diergeneeskunde* 89, 286–293.

Stober, M., Weitze, K.F., Hoedemaker, M., Pohlenz, J., Liebler, E., Wurm, S., Harlizius, B., Treviranus, A. and Sissoko, J. (1995) Congenital hypotrichosis in Holstein–Friesian calves. *Tierarztliche Umschau* 50, 224.

*Strobl, M. (1956) Anomalies in Spotted Mountain cattle. *Tierzuchter* 8, 31–32.

Tollersrud, O.K., Berg, T., Healy, P., Evjen, G., Ramachandran, U. and Nilssen, O. (1997) Purification of bovine lysosomal alpha-mannosidase, characterization of its gene and determination of two mutations that cause alpha-mannosidosis. *European Journal of Biochemistry* 246, 410–419.

*Toombs, R.E., Wikse, S.E. and Kasari, T.R. (1994) The incidence, causes, and financial impact of perinatal mortality in North American beef herds. *Veterinary Clinics of North America – Food Animal Practice* 10, 137–146.

*Trotter, D.M., Huston, K., Leipold, H.W., Adrian, R.A. and Merriam, J.G. (1971) Anatomic variations of normal calves. *Giessener Beitrage zur Erbpathologie und Zuchthygiene* 1(2), 31–54.

Tsujino, S., Shanske, S., Valberg, S.J., Cardinet, G.H., Smith, B.P. and Dimauro, S. (1996) Cloning of bovine muscle glycogen phosphorylase cDNA and identification of a mutation in cattle with myophosphorylase deficiency, an animal model for McArdle's disease. *Neuromuscular Disorders* 6, 19–26.

*van der Plank, G.M. and Hoiting, H. (1954) Inherited diseases in calves of the Meuse-Rhine-Yassel breed. *Tijdschrift voor Diergeneeskunde* 79, 149–150.

Whittem, J.H. and Walker (1957) Neuronopathy and pseudolipidosis in Angus calves. *Journal of Pathology and Bacteriology* 74, 281–288.

*Wikse, S.E., Kinsel, M.L., Field, R.W. and Holland, P.S. (1994) Investigating perinatal calf mortality in beef herds. *Veterinary Clinics of North America – Food Animal Practice* 10, 147–166.

Wilson, J. (1909) The origin of the Dexter-Kerry breed of cattle. *Proceedings of the Royal Dublin Society* 12, 1–17.

Wither, S.E. (1997) Congenital goiter in cattle. *Canadian Veterinary Journal – Revue Vétérinaire Canadienne* 38, 178.

*Wriedt, C. (1930) *Heredity in Livestock*. Macmillan, London.

Wright, S. (1917) Colour inheritance in mammals. VI. cattle. *Journal of Heredity* 8, 521–527.

Young, G.B. (1951) A case of tibial hemimelia in cattle. *British Veterinary Journal* 107, 23–28.

Blood Groups and Biochemical Polymorphisms

H.C. Hines

Animal Genetics Laboratory, Department of Animal Sciences, Ohio State University, 2027 Coffey Road, Columbus, OH 43210, USA

Introduction	77
Early blood-group studies	77
The Wisconsin contribution	78
Nomenclature and definitions	79
Blood Groups	81
Blood-group systems	81
Standardization of typing results	87
Monoclonal antibodies: utility and instructiveness	87
Nature and function of blood-group antigens	88
Biochemical Polymorphisms	90
Plasma proteins	90
Red-cell proteins	95
White-cell proteins	98
Milk proteins	99
Utilization of Blood Groups and Biochemical Polymorphisms	100
Parentage verification	101
Twin determinations	101
Breed structure and phylogenetic studies	102
Association of blood group and biochemical polymorphism markers with traits of economic importance	103
Neonatal Isoerythrolysis	105
Summary	106
References	106

Introduction

Early blood-group studies

As the 20th century burst upon the world, it was greeted by two momentous discoveries, which ushered in the era of immunogenetics. One was the rediscovery in 1900 of Mendel's principles of genetics and the other was the discovery by Landsteiner (1900, 1901) of human blood-group variability. Now, as we stand on the threshold of the 21st century, we can pause to consider where we have been with cattle immunogenetics and where we are heading.

Simultaneous with Landsteiner's human ABO discoveries, though not as famous, was the work of Ehrlich and Morgenroth (1900) with farm animals. They demonstrated individual differences in the blood of goats and introduced the technique of isoimmunization, injecting blood from donors into recipients to produce isoantisera for use as typing reagents. The first cattle blood-group work was reported by Todd and White (1910). They not only reported some of the same results in cattle as Ehrlich and Morgenroth had described in goats, but they also contributed the important insight that blood-group variation might be vast and complicated. In contrast to the then generally held view that blood groups were under the control of only a few loci and alleles, the immunization and absorption experiments of Todd and White suggested that cattle blood-group variation was complex and extensive.

Despite such promising results, systematic animal blood-group research languished for the next 30 years before surfacing in the early 1940s in the immunogenetics laboratory of M.R. Irwin at the University of Wisconsin. The immunogenetic investigations undertaken there culminated in a profusion of seminal papers on cattle and chicken blood groups published throughout the 1940s and early 1950s.

The Wisconsin contribution

In the 1930s M.R. Irwin began immunogenetic studies at the University of Wisconsin, working initially with pigeon–dove hybrids. In a 1936 paper describing this work (Irwin and Cole, 1936), the authors first used the term immunogenetics. In about 1940, L.C. Ferguson, working in Irwin's laboratory, began immunizations of cattle by transfusing blood between closely related cattle (e.g. daughter into dam). Two factors contributed to the success of this venture for Ferguson and Irwin. First, they utilized immunizations to raise the desired antibodies, instead of relying upon naturally occurring antibodies, as often existed in humans. Secondly, they found early on that cattle red cells were not very agglutinable, and they quickly adopted the haemolytic test as their means of determining reactivity.

In order for the reaction between the antibody and the red-cell antigen to culminate in cell lysis, a complex of serum enzymes collectively termed complement must be present. The workers in Irwin's laboratory also determined that, although guinea-pig serum was a good complement source for haemolytic tests in some other species, it was only suitable in cattle following absorption to remove cattle haemolytic antibodies (Ferguson, 1941; Irwin, 1956). Rabbit serum proved to be the complement source of choice.

In the initial cattle blood-group report from the Wisconsin immunogenetics laboratory, Ferguson (1941) presented evidence for the existence of seven red-cell antigens. Two of them appeared to be related, while the others seemed to be independent. In the second paper, Ferguson et al. (1942) reported on 23 more. Again, one pair of the cellular antigens appeared to be related to each other, while the rest behaved independently. The seemingly independent antigenic factors, as they came to be called, conformed to the expectations of products of dominant genes, while their individual absence was controlled by recessive alternatives. In each of the related factor pairs, the factors seemed to represent a subtype series. For example, for the pair C and E, C could appear alone but E was observed only when C was also present (Ferguson, 1941). The observed phenotypes were C, CE and '–' (absence of both C and E). Factor C was later termed C_2 and E (CE) was renamed C_1 (Stormont, 1950), in keeping with subtype nomenclature practices.

Other relationships among detected factors were soon apparent. In 1945 Stormont et al. reported that factors B, G and K were not products of independent genes. Only five of the eight possible factor combinations were observed: B, G, BG, BGK and –. Factor K only occurred in the presence of both B and G. By the late 1940s the ever-growing list of factors was beginning to be sorted into genetic systems (Stormont, 1950), and in 1951 evidence of the B and C systems was published (Stormont et al., 1951). Information on the F–V and Z systems followed quickly (Stormont, 1952).

Two other significant discoveries originating in the Wisconsin immunogenetics laboratory should be noted. The first was the recognition that the J antigenic factor was fundamentally different from all of the other detected factors (Stormont, 1949). Stormont showed that the J factor was not inherently a red cell membrane constituent, but instead a serum component, which is adsorbed on to erythrocytes. Also notable was the discovery by Owen (1945) that fraternal twins commonly exchange blood-forming components *in utero* and that, at this stage, they readily accept and maintain these contributions from their co-twins. His findings markedly contributed to the development of Burnet and Fenner's (1949) hypothesis of self-recognition.

Several reviews of the early days of cattle immunogenetics are available (Irwin, 1956, 1974, 1976; Stormont, 1978). They make fascinating reading for those interested in learning more about the beginnings of this field.

Nomenclature and definitions

Some of the terms and designations routinely used in blood-typing work merit description and definition to make them more meaningful to others. The fundamental component of blood-group determination has been variously called a cellular antigen, antigenic determinant, antigenic factor, blood-group factor, blood factor, or simply factor. Sometimes the term specificity is used. In any of these aliases, the subject is the entity identified by a functionally monospecific antibody source (a blood-typing reagent), which is usually an absorption-purified isoimmune serum. Some would argue with characterizing factors as unit antigens or epitopes and prefer to define them simply as the targets (perhaps a collection) recognized by a specific reagent. However, although all of the structures identified by a particular factor designation may not be absolutely identical, the fact that monoclonal antibodies satisfactorily distinguish many of these factors (Tucker *et al.*, 1986, 1987; Metenier *et al.*, 1991; Honberg and Larsen, 1992) seems to argue strongly for regarding them as epitopes.

As factors were discovered, they were assigned letter designations, beginning with A and extending through the alphabet. Following the assignment of Z, the alphabetic series was begun again, with a prime (′) being appended to each letter (i.e. A′, B′, etc.). At the next turn of the alphabetic cycle, a double prime (″) was appended (i.e. A″, B″, etc.). No relationship was normally implied among factors possessing the same letter but different prime designations (e.g. there is no relationship between A, A′ and A″).

Groups of factors for which the genetic control is almost invariably transmitted as a unit are termed phenogroups. The term was coined with the intention of conveying the concept of unit genetic transmission, while remaining silent about the nature of the control or about the characteristics of the product(s) (Stormont, 1972a). In light of the considerable current evidence of tightly linked genes as the basis of most B and C blood-group variation (Bouw and Fiorentini, 1970; Oosterlee and Bouw, 1974; Ruiterkamp *et al.*, 1977; Grosclaude *et al.*, 1979, 1983), it is appropriate to think of phenogroups as the products of haplotypes. In fact, phenogroup is also commonly used instead of haplotype to represent the genetic unit that codes for the phenogroup product. Within phenogroups, the usual order of factors is alphabetical and implies nothing about spatial relationships of factors or products or about the chromosomal order of the controlling genes.

Blood-group system, or simply system, refers to the entire array of products coded for by the alleles at one locus. System or complex is also used to refer to the group of linked loci controlling complex blood groups, such as the B or C, just as phenogroup or haplotype seems more appropriate than allele as a designation for a linked allelic entity, so system or complex seems more appropriate than locus as a designator of a linked locus region.

Blood-group loci are identified by the designation of the alphabetically first factor of that system. When used in other than an exclusively blood-group context, the blood-group locus designators are prefaced by 'EA', signifying erythrocyte antigen. The 11 cattle blood-group loci/systems are thus: *EAA*, *EAB*, *EAC*, *EAF*, *EAJ*, *EAL*, *EAM*, *EAS*, *EAZ*, *EAR′*, and *EAT′*. When the blood-group context is clear the EA prefix is usually omitted. Current blood-group locus and allele designations may be found in Anon. (1985). Current nomenclature for biochemical polymorphisms may be found in Larsen *et al.* (1992).

Blood Groups

Blood-group systems

The 11 currently recognized cattle blood-group systems are given in Table 5.1, and the locations of the controlling loci are shown in Table 5.2. The systems range in complexity from one-factor, two-allele systems, such as L and T′, to the highly polymorphic B system, with approximately 50 factors and more than 1000 phenogroups. Stormont (1962) and more recently Bell (1983) have reviewed the area. Some details about each of the systems will be considered in turn.

The A system

The A system was recognized by Stormont *et al.* (1960) when they realized that the factors formerly assigned to the separate systems A–H and D (Stormont and Suzuki, 1956) were under single locus control. Larsen (1961) confirmed that factor D belonged to the A system, and Stormont (1961) promptly added Z′ to form the current list.

The six A-system factors occur in at least 11 phenogroups. The Z′ factor, occurring in phenogroups A_1D_2Z' and Z′, is virtually absent in northern European breeds. The rare situations in which it has been detected in those breeds are probably evidence of crossbreeding. It is found in low frequency in Channel Island breeds and in somewhat higher frequencies in *Bos indicus* cattle. The null phenogroup is not encountered in cattle, but is seen in bison (Stormont *et al.*, 1960).

Table 5.1. Cattle blood-group systems and factors.

System	Factors	No. of alleles/phenogroups
A	$A_1, A_2, D_1, D_2, H, Z'$	≥ 11
B	$B_1, B_2, G_1, G_2, G_3, I_1, I_2, K, O_1, O_2, O_3, O_x, P_1, P_2, Q, T_1, T_2, Y_1, Y_2, A', B', D', E'_1, E'_2, E'_3, E'_4, F'_1, F'_2, G', I'_1, I'_2, J'_1, J'_2, K', O'_1, O'_2, P'_1, P'_2, Q', Y', A'', B'', D'', F'' G'', I'', J'', K'', O''$	≥ 1000
C	$C_1, C_2, E, R_1, R_2, W, X_0, X_1, X_2, C', L', X', C''_1, C''_2$	≥ 100
F	$F_1, F_2, V_1, V_2, N', V'$	≥ 8
J	J, Oc	≥ 4
L	L	2
M	M_1, M_2, M'	3
S	$S, U, U'_1, U'_2, H', H'', S'', U''$	≥ 15
Z	Z_1, Z_2	3
R′	R′, S′	3
T′	T′	2

The A locus, *EAA*, has been mapped to chromosome 15 (Bishop *et al.*, 1994), where it is located about 4 centimorgans (cM) towards the telomere from the locus for haemoglobin beta, *HBB* (US Department of Agriculture (USDA)/Agriculture Research Service (ARS)/Meat Animal Research Center (MARC) Cattle Genome Maps web site*).

The B system

The B system (Stormont *et al.*, 1951) is probably the most complex blood group system so far reported for any species. It has been compared to the human rhesus (Rh) system, and even scrutinized as a possible Rh orthologue (Lewin *et al.*, 1994), but in fact far exceeds it in complexity. Its 50 or so factors, occurring in more than 1000 phenogroups, are powerful discriminators of breeds and agents for the resolution of parentage questions.

The plethora of B-system diversity has raised questions about the nature of the genetic control of this system. In fact, a small percentage but significant number of cases of irregular inheritance of phenogroups has accumulated over the last half-century of blood typing (Stormont, 1954; Datta, 1959; Bouw, 1962; Moustgaard and Neimann-Sorensen, 1962; Stormont and Suzuki, 1962; Lie and Braend, 1963; Bouw *et al.*, 1964; Larsen, 1964; Stormont *et al.*, 1964; Bouw and Fiorentini, 1970; Sellei and Rendel, 1970; Glasnak and Sulc, 1971; Wegrzyn *et al.*, 1971; Oosterlee and Bouw, 1974; Dorynek and Kaczmarek, 1975; Ruiterkamp *et al.*, 1977; Grosclaude *et al.*, 1979, 1983). These have enabled tentative B-system maps to be created, showing relative positions of the genetic elements controlling the different factors (Green, 1966; Bouw and Fiorentini, 1970; Oosterlee and Bouw, 1974; Ruiterkamp *et al.*, 1977; Grosclaude *et al.*, 1979, 1983). A generalized, composite map based upon the common features of the published maps is presented in Fig. 5.1.

Ruiterkamp *et al.* (1977) presented evidence supporting the interesting notion that the controlling genetic elements are not evenly spaced, but occur in clusters. Two gene clusters reported were one centred about E'_4, comprising factors P, E'_1, E'_2, F' and G'', and a second centred about O_x, comprising factors O_1, O_2, O_3, A', J'_1, K', O', H_{10} and NF_7.

```
                <---E'₁---> <-BGK->    O₁-Oₓ
  Q-YY'-D'--G₁----G" G' P'-B--Q'---G₃ E'₃--A' J' O'-----I₁--B'---I"--------I'
           G₂     E'₃ F' P'            T  I₂ K'
```

Fig. 5.1. Composite B-system map (based upon results of Bouw and Fiorentini, 1970; Ruiterkamp *et al.*, 1977; Grosclaude *et al.*, 1979). *Doubtful locations: (i) P and P' may be to the right of B. Placed to the left on the basis of reported inclusion in E' cluster group; (ii) A' may be to the right of I_1. Placed to the left on the basis of reported inclusion in O cluster group.

---

*The USDA/ARS/MARC Cattle Genome Maps web site is:
http://sol.marc.usda.gov/genome/cattle/cattle.html

From their data Grosclaude *et al.* (1979) estimated the genetic distance between the Q and I′ factors on opposite ends of the map to be 1.34 cM, but felt that this was an overestimation. Based upon other data, they suggested that a more conservative estimate of about one-half that amount (0.7 cM) might be more correct. Even with the more conservative estimate, assuming that conventional genetic conditions exist, the distance circumscribes a length of deoxyribonucleic acid (DNA) sequence that could contain several tens of loci. The recombination rate is high enough for most workers to believe that a series of linked genes must be involved in the control of B-system factors. Conversely, Stormont (1981) argues for intragenic crossing-over. Since almost nothing is known about either the nature of the genetic control region or the B-system gene product(s), it is conceivable that aspects of both viewpoints may be correct. Multiple genes may be involved in B-system control and simultaneously some genes may produce products carrying multiple epitopes. Additionally, similar or even identical epitopes may exist on the products of different genes. Such situations would help to explain suspicions of differing rates of crossing-over in different breeds (Stormont, 1972a).

In an attempt to clarify the picture, two blocking-antibody studies were conducted. Unfortunately, they produced contradictory results. Ostrand-Rosenberg (1976) found no evidence that antibodies against one factor blocked those against another, regardless of whether the two factors were in the same or different phenogroups. Conversely, Auditore and Stormont (1978) found that antibodies against one factor blocked those against another when the two factors were in the same phenogroup, but not when they were in a different one. While we shall not know the true story at either the gene or product levels until the sequence of the relevant DNA segment and the physicochemical nature of the product(s) have been determined, it is hard to conceive that multiple genes are not a major part of the explanation.

It may be instructive to note that, as the molecular nature of the *Rh* genes and the biochemical composition of their products were elucidated, it became evident that neither the one-locus theory of Wiener (1944) nor the three-locus theory of Fisher (cited by Race, 1944) was strictly correct. Instead, there are two genes and three products (Mouro *et al.*, 1993; Colin *et al.*, 1994; Cartron and Agre, 1995). When present, the *D* gene encodes the D protein; the antigenically D-negative state ensues when the *D* gene is deleted from the genome. The *CcEe* gene encodes two products, probably through alternative splicing events. The *C* and *c* alleles differ by six nucleotides, *E* and *e* by only one. On chromosomes containing the *D* gene, the *CcEe* gene is closely linked to it. Furthermore, the Rh antigenic complex in the red-cell membrane includes a mix of *Rh* gene products, Rh-related glycoproteins and several apparently unrelated products (Cartron and Agre, 1995).

The B 'locus', *EAB*, has been mapped to chromosome 12 (Kappes *et al.*, 1994). It is in the vicinity of 12q24 (USDA/ARS/MARC Cattle Genome Maps web site).

The C system

The C system (Stormont *et al.*, 1951) is the second most complex cattle blood group system. It was reviewed by Stormont in 1962, and since that time several additional factors have been identified (Nasrat, 1965; Duniec *et al.*, 1973; Grosclaude *et al.*, 1981; Larsen, 1981). The C system now consists of 14 factors and more than 100 phenogroups. In northern European breeds, C'' (C''_2) seems to form a closed system with C_1 and C_2 (Duniec *et al.*, 1973), but in French breeds this does not appear to be the case (Grosclaude *et al.*, 1981).

The C system differs subtly from the B system in that a much higher proportion of the possible factor combinations are actually observed in phenogroups. The practical result of this difference is that it is much harder in the C system to deduce probable phenogroups from phenotypes.

As in the B system, a small but significant number of cases of irregular phenogroup inheritance has accumulated (Bouw, 1962; Nasrat, 1965; Bouw *et al.*, 1974; Guérin *et al.*, 1981; Buys, 1990). Based upon the observed apparent crossovers, three C-system maps have been created (Bouw *et al.*, 1974; Guérin *et al.*, 1981; Buys, 1990). They are in generally good agreement. At two points in the map of Guérin *et al.* (1981) there are gene clusters. One cluster consists of factors C_1, C_2, C''_1, and C''_2; the other of factors X_1, X_2, C' and F10. A composite C-system map is presented in Fig. 5.2.

The C 'locus', *EAC*, has been mapped to chromosome 18, where it lies near the telomeric end (Kappes *et al.*, 1994; USDA/ARS/MARC Cattle Genome Maps web site).

The F system

Constituent F-system factors F (Stormont, 1950) and V (Ferguson *et al.*, 1942), initially considered to be under independent control, were reported in 1952 to comprise the F–V system (Stormont, 1952). Both F and V factors were later found to have linear subtypes (Rendel, 1958a; Stormont, 1962), and the system name was shortened to F (Anon., 1985). Factors V' (Hall and Ross, 1981) and N' (Grosclaude, 1966; Miller, 1966) were added, and a null allele was reported (Dabczewski, 1969; Osterhoff and Politzer, 1970; Carr *et al.*, 1974), although not all available reagents were used in those studies (Bell, 1983). In European breeds, the null allele is at best extremely rare and the system is effectively closed (Bell, 1983). Some of the complexities of the system have been described by Larsen (1982). The F locus, *EAF*, has been mapped near

```
                    C₁              F10
       L'---W-R₁---C₂---E-----X₁-C'
              R₂   C"₁          X₂
                   C"₂
```

Fig. 5.2. Composite C-system map (based upon the results of Guérin *et al.*, 1971; Bouw *et al.*, 1974, Buys, 1990).

The J system

As mentioned earlier, the J antigenic factor is fundamentally different from all of the other detected factors (Stormont, 1949). Stormont showed that the J factor is not inherently a red-cell membrane constituent, but instead a serum component which is adsorbed – or, perhaps more accurately, transferred – on to erythrocytes (Thiele *et al.*, 1975). The red-cell membrane substance has been shown to be a glycolipid, while the serum J substance consists of both glycolipid and glycoprotein components. Surprisingly, however, efficient *in vitro* coating of red cells is only accomplished by using the protein-containing, lipid-free residue obtained by lipid extraction of J-positive serum (Thiele *et al.*, 1975). These workers have hypothesized that the carbohydrate component of serum J glycoprotein may become detached and transferred to red-cell membrane glycosphingolipid.

Cattle may be divided into three classes on the basis of their red-cell and serum J status: J^{CS} individuals have J substance on their cells and in their sera, J^S cattle have J substance in the sera but not on the cells, and J^a cattle have no J substance in either place (Stone and Irwin, 1954). It appears that whether or not cattle are classified as J^{CS} is largely determined by the amount of J substance in the serum; J^{CS} cattle generally have a much higher serum concentration of J substance than do J^S cattle (Stone, 1962). Even within J^{CS} and J^S classes there are variations in J substance strength; these variations are under the control of multiple alleles, although the number of alleles is not certain (Stone, 1962).

The J factor is the only cattle blood-group antigen for which antibodies are regularly found in the serum of antigenically negative animals. Isoimmunization of cattle has generally failed to induce anti-J production (Stormont, 1949; Stone, 1956) and, while immunization of rabbits has been successful (Stormont and Suzuki, 1960; Bednekoff *et al.*, 1962, 1963), natural anti-J from J^a cattle is the usual source of typing reagent. Utilizing the system for parentage verification is not entirely satisfactory because of the J quantitative variation and the inability to find an anti-J source which will lyse all J^{CS} intergrades.

Depending upon the season and other factors, J^a cattle may or may not have anti-J in their serum (Stone, 1956). It has been conclusively demonstrated that season strongly affects anti-J titre, with highest titres occurring in summer and lowest titres in winter in both northern and southern hemispheres (Stone, 1956; Osterhoff, 1962; Rendel and Sellei, 1967; Mzee and Braend, 1979). Mzee and Braend (1979) also demonstrated that the amount of seasonal change in anti-J titres is directly related to the magnitude of seasonal change in the animal's environment.

A further interesting property of the J system is that it contains a soluble substance termed Oc (Sprague, 1958). The Oc (O of cattle) substance is identified by its ability to inhibit sheep anti-O reactivity. The data of Sprague indicate that it is the product of a recessive allele, J^{Oc}, in the J^{CS}, J^S, J^a series. The J factor also exhibits wide-ranging species cross-reactivity. It is serologically related to the human A factor (Stormont, 1949), the sheep R factor (Stormont, 1951) and the pig A factor (Sprague, 1958).

The J locus, *EAJ*, is linked to that for lactoglobulin beta, *LGB*, at a distance of from 4 cM (Larsen, 1970) to 18 cM (Hines *et al.*, 1981). The locus *EAJ* has been mapped to the telomeric end of BTA 11 (Ma *et al.*, 1996).

The L system
The L system is one of only two systems which remain one-factor, two-allele systems (Bell, 1983). The system was described by Stormont (1959). The L locus, *EAL*, has been mapped to BTA 3 (Kappes *et al.*, 1994), where it is located near the centromeric end of the chromosome (USDA/ARS/MARC Cattle Genome Maps web site).

The M system
The M system is a three-factor system with the three alleles M^{M1}, $M^{M2M'}$ and M^m. The M system was initially described by Stormont (1959). Non-linear subtype M′ was recognized by Miller *et al.* in 1962. Since M is a relatively simple system with weak haemolytic reactions, it was a surprise to most workers to find that it was associated with the bovine major histocompatibility complex (MHC), bovine leucocyte antigens (BoLA) (Leveziel and Hines, 1984). First considered to be simply genetic linkage, the relationship has now been shown to be much more fundamental (Hines and Ross, 1987; Honberg *et al.*, 1995). Immunization, immunoprecipitation and absorption experiments indicate that the M1 factor and the BoLA A24 antigen share the same structure; M′ and BoLA A16 are similarly related to each other. None of the many other BoLA class I antigens have been found to have red-cell counterparts. The situation may be similar to the Bennett–Goodspeed (Bg) phenomenon in humans (Morton *et al.*, 1969), where some but not all of the human leucocyte antigens (HLA) have red-cell counterparts. The M locus, *EAM*, has been mapped to BTA 23 (Bishop *et al.*, 1994), where it is part of the bovine MHC, BoLA.

The S system
The S system was described by Stormont *et al.* (1961). It is a nine-factor system with more than 15 phenogroups known. It is the third most complex bovine blood-group system. The S locus, *EAS*, is linked to that for protease inhibitor 2, *PI2*, at a distance of about 20 cM (Georges *et al.*, 1987). The locus *EAS* has been mapped to BTA 21 (Kappes *et al.*, 1994), where it is found at or near q15 (USDA/ARS/MARC Cattle Genome Maps web site).

The Z system
The existence of the Z system was reported by Stormont (1952). Although a Z_2 subtype is recognized, reagents that differentiate Z_1 from Z_2 cells are not widely available. With the inclusion of Z_2, Z is a two-factor, three-allele system. Monospecific typing sera, termed dosage reagents, have also been developed which discriminate between animals homozygous and heterozygous for Z

alleles (Stormont, 1952). Such reagents are used in some blood-typing laboratories, but they require careful management, as the discriminating reactions are quantitative rather than qualitative. Using electron microscopy, Ostrand-Rosenberg (1975) has demonstrated that EAZ^Z/EAZ^Z cells have twice as many Z factors on their membranes as EAZ^Z/EAZ^z cells. The Z locus, *EAZ*, has been mapped to BTA 10 (Kappes *et al.*, 1994), where it lies toward the telomeric end (USDA/ARS/MARC Cattle Genome Maps web site).

The R' system
The R' system was recognized by Miller (1966). Originally termed the R'–S' system, it is a two-factor system, which in most breeds is closed. A null allele has been encountered rarely (H.C. Hines, unpublished observations). The R' locus, *EAR'*, has been mapped to BTA 16 (Kappes *et al.*, 1994). It is found at the telomeric end of the chromosome (USDA/ARS/MARC Cattle Genome Maps web site).

The T' system
The T' system remains a one-factor, two-allele system (Bell, 1983). It was described by Grosclaude (1965). The T' locus, *EAT'*, has been mapped to BTA 19, where it is linked to that for post-transferrin 2 (*PTF2*) (Kappes *et al.*, 1994) at a distance of approximately 9 cM. The growth hormone locus, *GH*, which lies between *PTF2* and *EAT'*, is mapped to the q26–qter region (USDA/ARS/MARC Cattle Genome Maps web site).

Standardization of typing results

As more laboratories became engaged in cattle blood typing and as there was more international shipment of animals, semen and eventually embryos, it became increasingly important to ensure that the typing results were consistent across laboratories. To compare and standardize results, the Dairy Cattle Research Branch of the USDA, ARS, instituted a cooperative programme of comparative testing in 1959 (Kiddy and Hooven, 1961). The percentage of laboratory agreement in the biennial tests quickly rose to nearly 100%. As biochemical polymorphisms became part of blood typing, they were added to the test. Eventually the responsibility for conducting the tests was assumed by the International Society for Animal Blood Group Research and its successor, the International Society for Animal Genetics.

Monoclonal antibodies: utility and instructiveness

Many monoclonal antibodies with specificity for cattle blood-group factors have been produced and several are in routine use in typing laboratories (Tucker *et al.*, 1981, 1986, 1987; Méténier *et al.*, 1991; Hønberg and Larsen, 1992). More usable specificites have been produced from murine/bovine hybridomas, but murine/murine hybridomas are more stable (Tucker *et al.*, 1987; Hønberg and Larsen, 1992). Many of the specificities produced by the murine/murine hybridomas do not parallel any recognized specificities of monospecific, polyclonal typing sera, while most of the murine/bovine hybridomas do (Hønberg and Larsen, 1992). In behaving exactly like polyclonal typing sera, monoclonal products of the latter group of murine/bovine hybridomas confirm the unit-entity view of blood-group factors.

At the same time, some of the reactions of the monoclonals have been quantitatively variable in reactions with the same factor in different phenogroups, suggesting factor cross-reactivity rather than identity in these situations (Hønberg and Larsen, 1992). A still unanswered question is whether it is impossible – or simply difficult – to produce monoclonal antibodies against some recalcitrant factors.

Nature and function of blood-group antigens

Not surprisingly, in comparison to human blood-group antigens, very little is known about the bovine counterparts. In the late 1960s, Spooner and Maddy began investigations of cattle red-cell membranes, reporting that butanol extraction yielded a protein fraction in which F, V, J, L, R′, S′ and T′ antigens were retained, while all others were lost (Spooner and Maddy, 1970, 1971). All of these antigens were found in the sialic-acid-rich protein fraction. Maddy and Spooner (1970) also found that cattle erythrocytes fell into two different categories of membrane protein, which differed in their content of sialic acid. These two types of cells also differed in their ability to be agglutinated in an antiglobulin test (Spooner *et al.*, 1970). Cowpertwait and Spooner (1971) demonstrated that the agglutination classes were under the genetic control of a locus they designated *Agg*. The allele for high agglutinability, Agg^l, is recessive to that for low agglutinability, Agg^L.

Meanwhile, Hatheway *et al.* (1969) found factors from the A, B, C, F, L, S and Z systems in an ether–alkali extract of red-cell ghosts. Treatment of this extract with papain eliminated reactivity for all factors except F, V and C. The C reactivity was weak and was not studied further. The V factor was released into the soluble fraction, while F specificity remained with the residue. Neuraminidase treatment of ghosts removed F reactivity, but did not weaken V reactivity.

Hines (1971) examined the agglutination of cattle erythrocytes with several plant extracts (lectins). Although there were some individual lectin differences, the striking observation was that there was a general pattern of agglutination for all lectins. Some major membrane characteristic seemed to override individual lectin specificities and determine whether the animal would fall into the agglutinator or non-agglutinator class. The agglutinators were not invariably EAF V animals, but there was a strong association with presence of the V factor.

Hines *et al.* (1972a) showed that EAF F, FV and V phenotypic classes differed greatly in amounts of sialic acid, ghost material per ml of packed red cells, phenol-extracted mucoid per ml of red cells, and pronase-released mucoprotein per ml of red cells. In all of these characteristics, F values were the highest, FV intermediate and V lowest. Complicating the picture, *Phaseolus vulgaris* lectin non-agglutinators also had markedly higher values for red-cell sialic acid, ghost material per ml of packed red cells, phenol-extracted mucoid per ml of red cells, and pronase-released mucoprotein per ml of red cells. It was impossible to disentangle the *EAF* and *Agg* effects. Paradoxically, although pronase removed more sialic acid from F than from intact V red cells, it destroyed the V factor but did not affect F reactivity. It appears that EAF F, Agg L cells have more glycoprotein on their membranes, but the specific part that the F factor plays in the picture is not known. Slight differences in red-cell size parallel the membrane glycoprotein differences (Stämpfli and Ittig, 1983).

Other insights into the nature of blood-group antigens have been afforded by proteolytic enzyme and lectin studies. Hines *et al.* (1972b) found that the concanavalin A agglutinability of trypsin-treated bovine erythrocytes closely paralleled blood-group A reactivity. Although there is a definite connection between EAA A and concanavalin A agglutinability of the trypsin-treated red cells, the blood-group A factor and the concanavalin receptor do not share identity (Ostrand-Rosenberg, 1976).

Multiple lines of evidence suggest that the complex blood groups are aptly described. The structures do not seem to be simple. Shaw and Stone (1962) investigated the time of appearance of antigenic factors on bovine fetal erythrocytes. They found not only different times of appearance among the factors of the same phenogroup, but also different times of appearance of the same factor occurring in different phenogroups. These results were supported by the results of proteolytic enzyme treatments reported by Hines *et al.* (1976). They found differential effects upon factors of the same phenogroup and also differential effects upon the same factor occurring in different phenogroups. Auditore *et al.* (1979) also obtained mixed results when they used proteolytic enzymes to study subtypic linearity of bovine B blood-group factors. Additionally the observation that factors of the E' series uniquely routinely elicit the production of immunoglobulin M (IgM) antibodies in immunizations (Hines, 1967) suggests that they are in some respect different from the rest.

Increasingly, the picture of the complex bovine blood groups looks similar to that of rhesus (Rh), where the membrane complex is now known to be composed of Rh gene products, Rh-related glycoproteins and several unrelated products (Cartron and Agre, 1995). As with the Rh situation, the real picture cannot be elucidated until pertinent basic molecular and biochemical studies are conducted.

Almost nothing is known about the functions of bovine blood-group antigens. The EAM M antigen appears to be a residual component of the MHC structure (Hines and Ross, 1987; Hønberg *et al.*, 1995). The *EAS* antigenic products may be involved in some way with the sodium–potassium pump, although apparently not as directly as the related *EAM* locus is in sheep (Ellory and Tucker, 1970; Ellory *et al.*, 1974; Rasmusen *et al.*, 1974).

Biochemical Polymorphisms

The biochemical polymorphisms of cattle consist of variations in the proteins of blood plasma/serum, red cells, white cells and milk. Most of these variations are detected electrophoretically, but some are serologically determined. One of the most important milestones along the road of biochemical polymorphism detection was the introduction by Smithies (1955) of the technique of starch-gel electrophoresis. This truly revolutionized detection methods, providing a very sophisticated tool for detecting differences in charge and/or size among macromolecules. When used with supporting media, such as starch, agarose or polyacrylamide, electrophoresis is capable of differentiating minute, genetically controlled variations in proteins of plasma, milk or intracellular components. When combined with specific enzyme-staining techniques, it permits visualization of differences in enzymes present in very small quantities.

The known genomic locations of the loci coding for cattle biochemical polymorphisms are listed in Table 5.2.

Plasma proteins

Albumin

Albumin (ALB) polymorphism was first reported by Ashton (1964) and by Braend and Efremov (1965), identifying alleles ALB^A, ALB^B and ALB^C. In African cattle, Carr (1966) observed three additional variants, and he assigned their controlling alleles the designations ALB^D, ALB^E and ALB^F. The last of the currently recognized alleles, ALB^G, was assigned by Spooner and Oliver (1969). In almost all breeds, ALB^A greatly predominates, and in northern European breeds it is virtually fixed.

The *ALB* and vitamin D-binding protein (*GC*) loci are extremely closely linked (Bouquet *et al.*, 1986). They have been mapped to chromosome 6, where they are separated by less than 1 cM (Kappes *et al.*, 1994). The *ALB/GC* linkage group maps very close to the casein cluster (*CSN1S1–CSN2–CSN1S2–CSN3*) (USDA/ARS/MARC Cattle Genome Maps web site). The casein cluster is found at BTA 6q31 (BovMap Genome Mapping Site web site*).

Alkaline phosphatase

Alkaline phosphatase (ALP) polymorphism was first described by Gahne (1963a). Detection following electrophoresis is by a specific enzyme-staining technique. There are two alleles, ALP^A and ALP^O. In all of the breeds reported, ALP^O predominates (Gahne, 1963a; Haenlein *et al.*, 1980). In contrast to the

---

*The BovMap Genome Mapping web site is http://locus.jouy.inra.fr/ cgi-bin/bovmap/Bovmap/main.pl

Table 5.2. Genomic locations* of loci of blood groups and biochemical polymorphisms

Locus	Locus symbol	Chromosomal location (BTA no.)	Comments
Blood group A	EAA	15	~4 cM telomeric from HBB
Blood group B	EAB	12	In vicinity of 12q24
Blood group C	EAC	18	Near the telomeric end of BTA 18
Blood group F	EAF	17	Near centromeric end of BTA 17
Blood group J	EAJ	11	At telomeric end of BTA 11
Blood group L	EAL	3	Near centromeric end of chromosome
Blood group M	EAM	23	Part of BoLA-A region
Blood group S	EAS	21	In vicinity of 21q15
Blood group Z	EAZ	10	Toward telomeric end of BTA 10
Blood group R'	EAR'	16	At telomeric end of BTA 16
Blood group T'	EAT'	19	9 cM from PTF2, near GH
Albumin	ALB	6	Closely linked to GC and casein genes
Ceruloplasmin	CP	1	20 cM from TF
GC protein	GC	6	Closely linked to ALB and casein genes
Protease inhibitor alpha	PI2	21	About 20 cM from EAS
Transferrin	TF	1	Near BTA 1 telomere; 20 cM from CP
Carbonic anhydrase	CA†	14	Near the telomeric end of BTA 14
Haemoglobin beta	HBB	15	Linked to EAA, maps to BTA q13-q23
Casein gene cluster	–	6	CSN1S1–CSN2–CSN1S2–CSN3 form a cluster which is closely linked to ALB and GC
Lactalbumin alpha	LAA	5	Maps to BTA 5q21
Lactoglobulin beta	LGB	11	Linked to EAJ; maps to BTA 11q28

*In addition to original publication information, supplemental current data on the chromosomal locations of these markers were obtained from two web sites:
http://sol.marc.usda.gov/genome/cattle/cattle.html
http://locus.jouy.inra.fr/cgi-bin/bovmap/Bovmap/main.pl
†The erythrocyte carbonic anhydrase locus is sometimes referred to as CA2, the notation of the corresponding human locus.
BTA, *Bos taurus*.

situation in almost all other electrophoretically determined biochemical polymorphism systems, the alleles are not codominant. Genotypes ALP^A/ALP^A and ALP^A/ALP^O produce electrophoretic phenotypes that are qualitatively identical. Whether or not they are quantitatively the same is uncertain (Gahne, 1963a).

Several factors in addition to the *ALP* status influence the ALP phenotype. The ALP A band is not seen until animals are about 10 months old (Gahne, 1967b). Furthermore, the level of serum alkaline phosphatase activity exhibits a gradual, steady decline from soon after birth to approximately 3 years, where it plateaus (Agergaard and Larsen, 1974). Once it reaches this age-independent stage, serum alkaline phosphatase level is most strongly influenced by the *ALP* genotype, with animals possessing the ALP^A allele having significantly higher levels than those lacking it (Gahne, 1967a).

There is also a relationship between EAJ and ALP phenotypes. There are significantly higher than expected numbers of animals in EAJ J-positive (EAJ J^{CS} and EAJ J^S), ALP O and EAJ J-negative (EAJ J^a), ALP A classes. Concomitantly, there are deficits in the EAJ J-positive, ALP A and EAJ-negative, ALP O classes (Rendel and Gahne, 1963; Haenlein *et al.*, 1980). The relationship is also manifest quantitatively. Animals that are EAJ J-positive, ALP A have lower levels of serum J substance and higher levels of alkaline phosphatase activity than J-positive, ALP O animals (Rendel and Gahne, 1963). Similar relationships exist between alkaline phosphatase activity and cattle J-related blood-group substances in humans (Arfors *et al.*, 1963), sheep (Rendel and Stormont, 1964) and dogs (Symons and Bell, 1992). An association between the ALP phenotype and the lactoglobulin beta (*LGB*) genotype has also been observed (Agergaard and Larsen, 1974; Haenlein *et al.*, 1980), but this is probably attributable to *EAJ–LGB* disequilibrium (see The J system, Lactoglobulin beta and Table 5.2).

Amylase 1
Amylase 1 (AMY1) was simultaneously described by Ashton (1965a) and by Hesselholt and Moustgaard (1965). Alleles *AMY1^A*, *AMY1^B* and *AMY1^C* are recognized. The *AMY1^A* allele has low frequency in almost all breeds. In southern European breeds, *AMY1^B* predominates slightly, while, in northern European and British breeds, *AMY1^C* is in a slight majority. In early publications amylase 1 was referred to as thread protein. For both *AMY1* and *AMY2*, the stain *p*-phenylenediamine dihydrochloride is commonly employed to aid in visualizing the translucent amylase bands on starch gels.

Amylase 2
Amylase 2 (AMY2) was recognized by Ebertus (1968) and by Mazumder and Spooner (1970). Two alleles, *AMY2^A* and *AMY2^B*, have been identified. The *AMY2^A* allele occurs in low frequency in all of the breeds that have been examined.

Ceruloplasmin
Ceruloplasmin (CP) polymorphism was recognized by Schröffel *et al.* (1970). Three alleles, *CP^A*, *CP^B* and *CP^C*, have been identified. In the breeds reported, *CP^B* occurs in low frequency. Instead of general protein dye, *p*-phenylenediamine is commonly employed to stain the variant zones on electrophoretic gels. Compared with CP A, variants CP B and CP C stain less intensely and their mobilities are more variable. The locus *CP* has been shown to be linked to the transferrin locus (*TF*) at a distance of about 20 cM (Larsen, 1977), and it is therefore mapped to chromosome 1.

Vitamin D-binding protein
Polymorphism of the group-specific component (GC) protein was first reported by Ashton (1963) and Gahne (1963b). Both referred to it as post-albumin. Discovery of the relationship to the human counterpart GC gave it the locus designation *GC*, which was retained even when it was recognized as

vitamin D-binding protein. Three alleles, GC^A, GC^B and GC^C, are known. The GC^C allele was reported by Masina *et al.* (1980) in very low frequency in the Apulian breed of Italy and by Van de Weghe *et al.* (1982) in relatively high frequency in White and Red East Flemish cattle. The locus is closely linked to that for albumin (Bouquet *et al.*, 1986; Kappes *et al.*, 1994), and both are now mapped to chromosome 6 in close proximity to the casein cluster (*CSN1S1–CSN2–CSN1S2–CSN3*) (USDA/ARS/MARC Cattle Genome Maps web site).

Immunoglobulin light-chain antigen

Polymorphic variation of some soluble macromolecules is detected by immunological rather than by electrophoretic means. Such soluble substances are termed allotypes. The term was defined by Faber and Stone (1976) as follows:

> An allotype is an inherited variant of molecules common to the normal individuals of a species; it is detected by immunological methods, and is not ordinarily distinguished by the usual physical and chemical methods; particulate antigens such as blood groups are not considered allotypes.

Allotypes are usually detected by alloimmune sera, although heteroimmune sera and normal cattle sera containing naturally occurring antibodies have been used.

Antigenic differences in immunoglobulin (Ig) light chains were detected by Blakeslee *et al.* (1971a). Immunoglobulin light-chain antigen (IGL) has two detectable antigenic specificities, IGL B1 and IGL B2, and a null phenotype, IGL Bo, which is unreactive with either anti-B1 or anti-B2. The system is under the control of three alleles, IGL^{B1}, IGL^{B2} and IGL^{Bo}. The IGL^{B1} and IGL^{B2} alleles are codominant with respect to each other, and both are dominant to IGL^{Bo} (Faber and Stone, 1975).

Immunoglobulin G2 heavy-chain antigen

Blakeslee *et al.* (1971a) also reported IgG2 heavy-chain antigen (*IGHG2*). They initially found only one specificity, IgG2 BVA1, which was controlled by a dominant gene, $IgG2^{BVA1}$. Later, Blakeslee *et al.* (1971b) found a sister specificity, IgG2 BVA2, controlled by a second allele, $IgG2^{BVA2}$. This appears to be a closed, codominant system.

Macroglobulin antigen

An allotypic macroglobulin polymorphism was discovered by Rapacz *et al.* (1968). They detected one specificity, encoded by MC^1. They found that the alternative null allele produced no detectable product. Later, Iannelli and Masina (1978) produced an alloantiserum that detected a second specificity, MC 2. It is encoded by allele MC^2, which is codominant and closes the system. The nature of the macromolecule carrying the MC specificities is not known.

Post-transferrin 1

The function of the protein termed post-transferrin 1 (PTF1) is not known. It was named for the fact that on alkaline electrophoretic gels it migrated immediately behind transferrin. It stains much more faintly than transferrin, presumably because it is present in the serum in low concentration. Detection of PTF1 protein was reported by Gahne *et al.* (1977), although they were not confident that they had observed polymorphism. They stated, 'The post-transferrin Ptf1 comprised of two or three weakly stained bands and no clear variation was observed between the samples.' Thinnes *et al.* (1976) detected variation in post-transferrins (three phenotypes, two alleles), but they did not indicate whether this was PTF1 or PTF2. Juneja and Gahne (1980) clearly identified PTF1 variation, and stated that it corresponded to the protein identified by Thinnes *et al.* (1976) as post-transferrin. The two alleles, originally termed *PTF1^A* and *PTF1^B*, are now designated *PTF1^F* and *PTF1^S* (Larsen *et al.*, 1992).

Post-transferrin 2
Post-transferrin 2 was so named because on alkaline electrophoretic gels it migrated more cathodally to transferrin than did post-transferrin 1. Like PTF1, its function is also not known. There are two alleles, *PTF2^F* and *PTF2^S*. A major hybrid product exists for the heterozygous genotype. This polymorphism was detected by Gahne *et al.* (1977) and Juneja and Gahne (1980). The *PTF2^F* allele predominates in the Swedish Red and White, Swedish Friesian, Charolais, Hereford and Simmental breeds studied by Gahne *et al.* (1977). Allelic frequencies have also been reported for East Asian and additional European breeds (Komatsu *et al.*, 1979).

The PTF2 locus, *PTF2*, has been mapped to chromosome 19, where it is linked to the T' locus, *EAT'* (Kappes *et al.*, 1994). The two loci are approximately 9 cM apart. The growth hormone locus, *GH*, which maps at q26–qter, lies between *PTF2* and *EAT'* (USDA/ARS/MARC Cattle Genome Maps web site).

Protease inhibitor alpha
Juneja and Gahne (1980) reported *PI2* polymorphism, using two-dimensional gel electrophoresis. Of the three alleles studied in the Swedish Red and White, Swedish Friesian, Charolais, Swedish Jersey and Simmental breeds, *PI2^S* always occurred with a frequency of at least 0.5, *PI2^F* was generally of intermediate frequency and *PI2^I* was of low frequency, being entirely absent from the samples examined from the Swedish Red and White and Swedish Jersey breeds. The PI2 locus, *PI2*, is linked to that for S, *EAS*, at a distance of about 20 cM (Georges *et al.*, 1987). By virtue of this linkage, *PI2* is thus assigned to BTA 21.

Transferrin
Transferrin was one of the first cattle polymorphisms recognized, following Smithies' introduction of the starch-gel technique (Smithies, 1955), undoubtedly because it formed such a prominent, clearly visualized zone of variation on the gels. Initially, alleles *TF^A*, *TF^D* and *TF^E* were detected (Ashton, 1958; Smithies and Hickman, 1958). Modifications of the technique permitted recognition of subdivisions of the *TF^D* allele into *TF^{D1}* and *TF^{D2}* (Ashton, 1965b; Jamieson, 1965; Kristjansson and Hickman, 1965). Other identified alleles

include TF^B and TF^F, found in zebu cattle (Ashton, 1959), TF^N, found as a rare variant in Swedish cattle (Gahne, 1961), TF^{GSA}, detected as a rare variant in South Africa (Osterhoff and van Heerden, 1965), TF^{GKE}, found in East African cattle (Ashton and Lampkin, 1965), TF^H, found in the Piedmont breed (Sartore and Bernoco, 1966), TF^J, found in East Asian and Japanese Black cattle (Abe *et al.*, 1968; Tsuji *et al.*, 1981), and TF^I (Soos *et al.*, 1973).

The *TF* locus has been shown to be linked to *CP* at a distance of about 20 cM (Larsen, 1977). The linkage group has been mapped to chromosome 1 (Bishop *et al.*, 1994), where it lies near the telomere (USDA/ARS/MARC Cattle Genome Maps web site). Transferrin has been heavily utilized for parentage verification and extensively investigated for associations with economically important traits.

Red-cell proteins

Acid phosphatase
Variation in acid phosphatase (ACP) was reported by Dogrul (1969). Although he observed two different ACP phenotypes, he was not confident that the variation had a genetic basis. Indeed, the fact that he found no differences between breeds in the frequency of the two phenotypes raised doubts that the detected differences had a genetic origin.

Carbonic anhydrase
Carbonic anhydrase (CA) polymorphism was first reported by Sartore *et al.* (1969). They detected two alleles, CA^F and CA^S, that gave rise to three phenotypes, F, FS and S. In Piedmont cattle, Sartore (1970) found a third allele, CA^{SPI}. Stormont *et al.* (1972) found a new zone of staining in three Angus samples, which they postulated to correspond to the product of an allele termed CA^C. Penedo *et al.* (1982) detected allele CA^Z at moderately low but not marginal frequencies in three breeds of zebu cattle.

The human isozyme to which cattle erythrocyte CA appears to correspond most closely is carbonic anhydrase II (*CA2*), and in some publications bovine *CA* is referred to as *CA2*. The CA locus, *CA* (*CA2*), has been mapped to chromosome 14 (Bishop *et al.*, 1994). It is found near the telomeric end of the chromosome (USDA/ARS/MARC Cattle Genome Maps web site).

Catalase
Catalase is reported to exhibit genetic variation under the control of two alleles, CAT^F and CAT^S. More information is awaited on this system.

Erythrocytic protein antigen
Using alloimmune serum and an immunodiffusion test, Rapacz *et al.* (1975) detected an erythrocyte protein allotypic factor, termed EC 1, which is encoded by the EC^1 allele. The alternative null allele, EC^0, produces no

detectable product. However, in the immunodiffusion test there is a dosage effect, which permits the EC^1/EC^1 homozygous genotype to be differentiated from the EC^1/EC^o heterozygote. The allele EC^1 occurs with widely varying frequencies in different breeds.

Erythrocytic protein 1

Thinnes *et al.* (1976) detected both erythrocyte protein 1 and 2 (*EP1* and *EP2*) variation in the intracellular contents of erythrocytes after precipitating the predominant haemoglobin. Nothing is known about the biochemical nature or function of either protein. Using general protein staining, the investigators found two codominant alleles, $EP1^A$ and $EP1^B$, for EP1. In their sample of the Deutsche Schwarzbunte breed, $EP1^A$ had a frequency of about 0.7.

Erythrocytic protein 2

As stated above, this protein polymorphism was also discovered by Thinnes *et al.* (1976). The two alleles, $EP2^A$ and $EP2^B$, showed codominant inheritance and were found in frequencies of 0.64 and 0.36 in their sample of the Deutsche Schwarzbunte breed.

Esterase D

Del Lama *et al.* (1989) first detected genetic variation in esterase D (ESD). There are two alleles, ESD^{D1} and ESD^{D2}. The predominant allele is ESD^{D2} in all breeds, and it appears to be fixed in *B. taurus* breeds.

Haemoglobin beta

Owing largely to its predominance in the erythrocyte and its ease of detection (no staining necessary), HBB was the first cattle protein polymorphism to be recognized (Cabannes and Serain, 1955; Bangham, 1957). In both of these reports, alleles HBB^A and HBB^B were detected. In almost all breeds, HBB^A is the predominant allele and is virtually fixed in northern European and British cattle. In southern European, Channel Island, African and *B. indicus* cattle, HBB^B is found in moderate frequencies, approaching equality in the Jersey breed. Eight other *HBB* alleles have been reported; almost all of them are rare. These alleles are: HBB^C (Crockett *et al.*, 1963; Carr, 1964), HBB^{DZAM} (Carr, 1965), HBB^D (Efremov and Braend, 1965), HBB^G (Braend, 1971), HBB^E (Khanna *et al.*, 1972), HBB^I (Osterhoff, 1975; Schwellnuss and Guérin, 1977), HBB^H (Han and Suzuki, 1976) and HBB^{XBAL} (Namikawa *et al.*, 1983). An allele, $HBB^{Khillary}$, reported by Naik *et al.* (1965), may be the same as HBB^{DZAM}. In a 1988 paper, Braend reported that by using isoelectric focusing he was able in Norwegian Red cattle to separate the HBB^A allele into HBB^{A4} and HBB^{A6} subtypes. The frequency of HBB^{A4} was 0.94. Whether or not there is a similar division in other breeds is not known. The locus *HBB* has been found to be linked to *EAA* (Larsen, 1966) and to the locus encoding parathyroid hormone,

PTH (Fries *et al.*, 1988). The loci *HBB* and *PTH* have been mapped to BTA 15q13–q23 (Fries *et al.*, 1988).

Peptidase B
Genetic variation in peptidase B (PEPB) enzyme was reported by Del Lama *et al.* (1992). There are two alleles, *PEPB¹* and *PEPB²*, giving rise to three phenotypes, PEPB 1, PEPB 2 and PEPB 12. The allele *PEPB¹* was found only in zebu cattle and in the Chianina and Marchigiana breeds. There was a marked difference in peptidase allele frequencies between the *B. taurus* and *B. indicus* breeds examined. In the five zebu breeds studied, *PEPB¹* was present in majority frequencies, while, in the European breeds, *PEPB²* was at least present in high frequency (Chianina and Marchigiana) or was fixed.

Peptidase D
Polymorphism of peptidase D (PEPD) was reported by Saison (1973). On electrophoresis, she observed three phenotypes: a fast band only (PEPD FO), a fast and a slower one (PEPD FS-1), and a fast together with the slowest one (PEPD FS-2). The fast band was always present; the variation was in the slower bands, which were either present or absent. Phenotype PEPD FS-2 was very rare, being seen in only three animals in the Simmental breed. No segregation studies were presented, so the type of control of these phenotypes is in doubt. Saison theorized control by three alleles, *PEPD^O*, *PEPD^{S-1}* and *PEPD^{S-2}*.

Phosphoglucomutase
Phosphoglucomutase (PGM1) polymorphism has been reported by Ansay *et al.* (1971b) and by Probeck and Geldermann (1976). Similar results are obtained with red cells and with leucocytes. Two alleles, *PGM1^A* and *PGM1^B*, have been detected. The B allele predominates in Belgian and German breeds that have been examined (Probeck and Geldermann, 1976).

Purine nucleoside phosphorylase
Ansay and Hanset (1972c) described variation in purine nucleoside phosphorylase (NP). They observed an age-dependent decline in enzymatic activity, which was particularly pronounced during the first year of life. Two electrophoretic phenotypes, NP H and NP l, were observed and determined to be controlled by two alleles, *NP^H* and *NP^l*. The *NP^H* allele was associated with higher enzymatic activity and was dominant to *NP^l*. The heterozygotes, *NP^H/NP^l*, have only half the enzymatic activity of *NP^H/NP^H* homozygotes, but this determination is complicated by the age dependency. Later, Ansay (1975) observed that the *NP^H* allele could be subdivided into a faster migrating variant, *NP^{HF}*, and a more slowly migrating one, *NP^{HS}*. These both had high enzymatic activity, were codominant with respect to each other and were dominant with respect to *NP^l*.

White-cell proteins

Adenosine deaminase
Ansay and Hanset (1972a) reported polymorphism in adenosine deaminase (ADA). They found four alleles, ADA^A, ADA^B, ADA^C and ADA^D.

Alkaline ribonuclease
The polymorphism at the alkaline ribonuclease locus (*ALR*) was first reported by Thinnes *et al.* (1976), who referred to the protein as leucocyte protein 1. There are two alleles, ALR^A and ALR^B. In Deutsche Schwarzbunte (Thinnes *et al.*, 1976) and Polish black and white cattle (Walawski and Prusinowska, 1981), the A allele predominates.

Glutamic oxalacetic transaminase
Ansay and Hanset (1972b) found limited polymorphism in glutamic oxalacetic transaminase (GOT1). The $GOT1^A$ allele overwhelmingly predominates, $GOT1^B$ being found only in the Central and Upper Belgium breed, at a frequency of about 1%.

Leucocytic protein 2
Polymorphism at the leucocytic protein 2 locus, *LEP2*, was reported by Thinnes *et al.* (1976). They found alleles $LEP2^A$ and $LEP2^B$, with the A allele occurring at a frequency of about 70% in the Deutsche Schwarzbunte breed.

Malate dehydrogenase
Malate dehydrogenase (MDH1) polymorphism was detected by Ansay *et al.* (1971a). They found evidence of two alleles, $MDH1^A$ and $MDH1^B$.

Mannose phosphate isomerase
Genetically controlled variation in mannose phosphate isomerase (MPI) was discovered by Ansay and Hanset (1973). Segregation data verify that the three phenotypes, MPI B, MPI C and MPI BC, are under the control of two codominant alleles, MPI^B and MPI^C. A fourth phenotype, MPI AB, was observed in Charolais cattle, but evidence for a postulated third allele, MPI^A, has not been confirmed. The MPI^C allele has a frequency of around 0.10 in most breeds studied.

Phosphoglucomutase
Phosphoglucomutase (PGM) polymorphism was first reported by Ansay *et al.* (1971b) and expanded upon by Probeck and Geldermann (1976). There are two codominant alleles, PGM^A and PGM^B. Similar phenotypes are also observed in erythrocytes. Probeck and Geldermann (1976) observed an excess of heterozygotes from segregating matings.

Phosphogluconate dehydrogenase

Probeck and Geldermann (1977) reported polymorphism of phosphogluconate dehydrogenase (PGD). Control at the *PGD* locus was vested in two alleles, PGD^A and PGD^B, with approximately equal frequencies in Deutsche Schwarzbunte (German Friesian) cattle. As with PGM, a marked excess of heterozygotes was observed from segregating matings.

Milk proteins

The four casein genes, *CSN1S1*, *CSN1S2*, *CSN2* and *CSN3*, are tightly linked on less than 200 kb of chromosome 6 in the apparent order *CSN1SN–CSN2–CSN1S2–CSN3* (Mercier and Vilotte, 1993). They have been assigned to BTA 6q31 (BovMap Bovine Genome Mapping web site). They are located very close to the *ALB* and *GC* loci. As expected, the casein loci are almost invariably transmitted as haplotypes; within breeds definite haplotypic combinations exist. Some haplotype frequencies have been reported (Hines *et al.*, 1977).

Casein alpha S1

Variation in α_{s1}-casein (now designated casein alpha S1 (*CSN1S1*)) was discovered by Thompson *et al.* (1962), who detected alleles $CSN1S1^A$, $CSN1S1^B$ and $CSN1S1^C$. Grosclaude *et al.* (1966) observed the additional allele $CSN1S1^D$, and Grosclaude *et al.* (1976a) recognized allele $CSN1S1^E$. In virtually all European breeds, $CSN1S1^B$ is the major allele and $CSN1S1^C$ is the minority variant. The allele $CSN1S1^A$ has been found only in US Holsteins and in Red Danish (RDM) cattle in Denmark (Aschaffenburg, 1968), and $CSN1S1^D$ has been detected only in the Flamande breed (Grosclaude *et al.*, 1966). In zebu breeds, $CSN1S1^C$ predominates (Aschaffenburg, 1968).

Casein alpha S2

Casein alpha S2 (*CSN1S2*) polymorphism was first reported by Grosclaude *et al.* (1976b) and elaborated upon by Grosclaude *et al.* (1978). Four codominant alleles exist, $CSN1S2^A$, $CSN1S2^B$, $CSN1S2^C$ and $CSN1S2^D$.

Casein beta

Of all the casein genes, *CSN2* exhibits the greatest polymorphism. Ten alleles are known. The first to be reported were $CSN2^A$, $CSN2^B$ and $CSN2^C$ (Aschaffenburg, 1961). Kiddy *et al.* (1966) and Peterson and Kopfler (1966), using new electrophoretic techniques, showed how subdivisions of the $CSN2^A$ allele could be detected. They reported existence of $CSN2^{A1}$, $CSN2^{A2}$ and $CSN2^{A3}$. Aschaffenburg *et al.* (1968) described the rare variant $CSN2^D$ in Indian and African zebu cattle. Voglino (1972) found $CSN2^E$ in Piedmont cattle. Creamer and Richardson (1975) reported the existence of $CSN2^{B2}$. Abe *et al.* (1975) described $CSN2^{A4}$ in Japanese cattle, and Grosclaude (1975) reported $CSN2^{A3M}$ in Mongolian cattle.

Casein kappa

The most common casein kappa alleles, $CSN3K^A$ and $CSN3^B$ were described in 1964 (Neelin, 1964; Woychick, 1964). Di Stasio and Merlin (1979) discovered $CSN3^C$ and Seibert *et al.* (1987) confirmed it. Erhardt (1989) detected $CSN3^E$.

Lactalbumin alpha

Lactalbumin alpha (LAA), formerly called α-lactalbumin, scarcely varies, except in zebu cattle (Aschaffenburg, 1968). Except for the report by Osterhoff and Pretorius (1966) of its existence in Brown Swiss, Holsteins and Jerseys in South Africa, the A variant has not been found in European breeds. In all breeds, LAA^B predominates. The codominant alternative in zebu breeds is LAA^A. The initial report is credited to Blumberg and Tombs (1958). Bell *et al.* (1981b) detected LAA^C in Bali cattle. The locus LAA has been mapped to BTA 5q21 (BovMap Bovine Genome Mapping web site).

Lactoglobulin beta

Lactoglobulin beta (LGB), previously termed β-lactoglobulin, has two main alleles, LGB^A and LGB^B (Aschaffenburg and Drewry, 1955), which are responsible for the bulk of the variation at this locus. Although in most breeds the A allele predominates slightly, there are some breeds in which it is in the minority (Aschaffenburg, 1968). All other alleles are rare. Bell (1962) described LGB^C in Australian Jerseys. Thymann and Larsen (1965) and Grosclaude *et al.* (1966) found LGB^D in Danish and French cattle. Bell *et al.* (1970) described LGB^{Dr} in the Australian Droughtmaster breed. Later, Grosclaude *et al.* (1976a) found LGB^{DYAK} in yaks. Most recently, Bell *et al.* (1981a) reported on the existence of three new alleles, LGB^E, LGB^F and LGB^G, in Bali cattle.

The LGB locus is linked to that for blood-group J, EAJ, at a distance of from 4 cM (Larsen, 1970) to 18 cM (Hines *et al.*, 1981). This linkage group has been mapped to chromosome 11 (Bishop *et al.*, 1994). The MARC web site map shows it at BTA 11q28, close to the telomeric end of this chromosome (USDA/ARS/MARC Cattle Genome Mapping web site).

Utilization of Blood Groups and Biochemical Polymorphisms

When the first reports on cattle blood groups emanated from Irwin's immunogenetics laboratory at the University of Wisconsin in the early 1940s, they were greeted with eager anticipation of their possible usefulness in verifying parentage and assisting in the selection of animals with superior production. The former potential was promptly realized, but the latter has proved much more elusive, as we shall see.

Parentage verification

Irwin and his students and co-workers had barely produced a collection of antisera and established the basic rules of inheritance of bovine blood-group antigens when the first client was knocking on their door. In 1940 the laboratory conducted the first tests for the Holstein Friesian Association, and in 1941 the American Guernsey Cattle Club also requested their services (Stormont, 1967). By 1946 the five breeds represented by the Purebred Dairy Cattle Association had contracted with them. It was soon established that blood typing was a powerful tool for excluding incorrect parentage and for corroborating (though never proving) the accuracy of recorded pedigree information (Braend, 1956; Irwin, 1956; Rasmusen, 1959; Stormont, 1959, 1967; Spooner, 1967; Rendel, 1968; Bell and Francis, 1970). Using blood groups alone, in most breeds the well-stocked laboratories could detect 95% of cases of incorrect parentage (Rasmusen, 1959; Bell and Francis, 1970). The addition of electrophoretic systems to the typing battery significantly raised the detection rate.

Stormont has pointed out the contribution of the blood-typing tests to the development of the artificial insemination (AI) industry (Stormont, 1967). Blood typing provided assurances that the intended sires were being used and it provided a means of correcting mistakes when such were made. Spot-checking programmes by breed associations have also contributed to reduced errors – partly through direct error detection, but more generally through the promotion of more accurate record-keeping by virtue of knowledge that such error-detection mechanisms exist.

A great advantage to those working with cattle blood groups accrues from the existence in the species of complex phenogroups such as are found in the B and C systems. Knowledge of these helps to guard against spurious laboratory reactions – or non-reactions – where there is either an extra reaction or a missed reaction in the apparent phenogroup reactivity pattern. Utilization of phenogroups also markedly magnifies the power of the test for excluding erroneous parentage (Stormont, 1959). This power needs to be wielded with some caution, however; it is a two-edged sword. Because factors in phenogroups can recombine, the blood-group analyst must constantly use his/her knowledge of system maps to determine whether a single crossover event could account for the observed irregularity. As a cautionary measure, many parentage-verification labs require exclusions in at least two systems before they issue a parentage-exclusion verdict. The availability of additional marker systems of biochemical polymorphisms and DNA markers not offered in the routine parentage test provides the opportunity for reaching a conclusive verdict in such situations.

Twin determinations

Blood groups and biochemical polymorphisms can also be useful for discriminating among different kinds of twins (Irwin, 1956; Rasmusen, 1959; Stormont,

1959, 1967; Rendel, 1963, 1968; Spooner, 1967; Bell and Francis, 1970). In research studies, it is often important to know with certainty whether an animal is a monozygotic or dizygotic twin. In the case of heifers born twins to bulls, it is important to ascertain whether exchange of blood and tissue-forming elements has occurred. In both of these situations, blood typing is useful. Identical twins will, of course, have identical blood types. Fraternal twins may have identical genotypes, but because of the extensive polymorphism that exists they usually will not.

Fraternal twins will, however, often have identical blood-group phenotypes by virtue of blood admixture resulting from exchange of blood-forming tissues. Such chimeras can usually be differentiated from identical twins because the chimeras will exhibit partial haemolytic reactions arising from the existence of two populations of cells. They will often also display more than two phenogroups in the complex systems (Stormont, 1959, 1967). In all kinds of twin determinations, biochemical polymorphisms, as well as the J system, have their own particular usefulness, because the products of most of these loci do not arise from haemopoietic tissues and are not exchanged (Stormont, 1967). Rendel (1963) has estimated that, by using all systems of blood groups and biochemical polymorphisms, dizygosity can correctly be discriminated from monozygosity about 98% of the time, and Bell and Francis (1970) estimate that by additionally utilizing morphological criteria the resolution approaches 100%.

Breed structure and phylogenetic studies

Fortunately, the very considerable genetic variation found in the blood groups and biochemical polymorphisms of cattle is divided rather well into variation among breeds, as well as into variation among animals within breeds. It is this division which not only permits the use of genetic-marker data for parentage verification within breeds, but also allows for interpretations regarding breed structure, breed relationships and the breed composition of individual animals. These breed characteristics have all been examined with genetic markers (Neimann-Sorensen, 1956; Rendel, 1958b, 1967; Maijala and Lindström, 1965).

In some cases, genetic markers are breed-specific, but more often, although there are breed overlaps in genetic-marker existence, there are marked breed differences in the frequencies of the controlling genes. For each breed, a gene-frequency profile and a genetic-marker phenotypic frequency profile may be established. These profiles may show some degree of similarity among related breeds, but they are also distinctive for each breed. Such profiles can be used to help flag reportedly purebred animals for further checking of crossbreeding (Hines, 1977).

It is also possible to utilize gene-frequency information to obtain quantitative estimates of breed relationships. Assuming all of the breed groups being evaluated have evolved from a common source and diverged in gene-frequency because of mutation and genetic drift, one would expect that breeds with more similar gene-frequency profiles should be more closely related than those with greater differences. Although it is also likely that selection pressures play some role in shaping frequency profiles, for phylogenetic analysis purposes the markers being studied are usually assumed to be selectively neutral. A second assumption is that, once the breed groups have diverged, there is no further interaction between them. To the extent that crossbreeding between the two groups occurs, the conclusions are biased. From such analyses one may construct phylogenetic trees, which represent the degree of relationship among breeds.

Phylogenetic analyses have sometimes involved complex mathematical algorithms to evaluate relationships and formulate phylogenetic trees (Kidd, 1971; Kidd and Sgaramella-Zonta, 1971). Such investigations have been undertaken for a variety of purposes, including helping to guide breeding-programme decisions (Kidd and Pirchner, 1971), helping to guide decisions about merging breeds (Kidd *et al.*, 1974), helping to set priorities regarding breed preservation (Kidd, 1974; Moazami-Goudarzi *et al.*, 1997) and helping to understand human and livestock history and migration patterns (Kidd and Pirchner, 1971; Kidd, 1971; Kidd and Cavalli-Sforza, 1974; Baker and Manwell, 1980).

Association of blood group and biochemical polymorphism markers with traits of economic importance

In the 1940s and 1950s, when cattle blood typing was a rapidly expanding field, hopes were high that some of the blood-group alleles would be found to be markers for superior production. Some studies seemed to confirm that hope, but too often they were contradicted by the following report. Even now, more than 50 years and many studies later, the relationships of most blood-group genotypes to quantitative trait differences are far from clear.

The earliest attempts to find associations between blood groups and milk-production traits yielded mixed results (McClure, 1952; Nair, 1957), but as additional blood-group systems and factors continued to be discovered throughout the 1950s and early 1960s (Stormont, 1962), euphoric expectations persisted. The development of gel electrophoretic methods fuelled these hopes by opening the door to the discovery of many additional polymorphic systems – primarily proteins of red cells, serum and milk. Many studies, involving many marker loci and several lactation traits, have resulted (see Rocha (1994) for a summary).

The early studies were almost entirely for 'direct' marker effects, i.e. pleiotropic or very closely linked gene effects. Some have questioned the rationale for this type of study, but Soller (1991) has argued that, because effective population size in many livestock populations is rather small, population-wide marker–quantitative trait locus (QTL) disequilibrium may exist for markers and QTL up to 5 cM apart. Several of the later studies have included within-family analyses of the type outlined by Geldermann (1975, 1976). Such analyses evaluate marker–QTL linkage at greater distances.

Reviewing the results from the association studies, one may initially be overwhelmed by all of the inconsistencies and ask in frustration whether there is any consistency at all hidden beneath all of the 'noise'. Indeed, for almost every marker locus investigated, there have been one or more studies in which it has been found to be associated with lactation trait differences and a number in which it has not!

There are several explanations for the chaos. First, with so many markers and traits typically being examined, the contribution of chance to variable results must be recognized. Next, the contributions of different environments and different genetic complements must be recognized. In different investigations, marker-linked QTL may differ, as may other QTL which interact epistatically with them. And then there is 'noise', which should be but is not completely eliminated by the analytical methodology. Kennedy *et al.* (1992) have pointed out the analytical shortcomings of many of the earlier studies, emphasizing the importance of utilizing analytical methods that adequately account for the residual correlations among the phenotypes of related animals.

Finally, one should keep in mind that most of the examined markers were not chosen because of any known connection with the target trait; they were chosen simply because they comprised the known repertoire of readily detectable polymorphic variants in the species. Therefore, it should come as no surprise that the consistently strongest results have involved those loci for which the lactation connection is obvious, namely the milk-protein loci.

Within the last decade, several large investigations have examined associations between blood groups and/or biochemical polymorphisms (Gonyon *et al.*, 1987; Haenlein *et al.*, 1987; Bovenhuis *et al.*, 1992; Andersson-Eklund and Rendel, 1993; Rocha, 1994). The results of these, supplemented by judicious evaluation of earlier studies, reveal some consistent, informative patterns.

A general observation is that marker relationships are more commonly found with milk-composition traits, particularly percentage traits, than with milk yield itself. Generalities aside, the single most striking association is that of the milk-protein loci with milk-component differences (Marziali and Ng-Kwai-Hang, 1986; Gonyon *et al.*, 1987; Haenlein *et al.*, 1987; Bovenhuis *et al.*, 1992). This is currently the premier example of genetic-marker–lactation-trait relationship in dairy cattle. The amount of the effect that is due to pleiotropy of the milk-protein genes themselves and the amount attributable to other linked genes remains to be elucidated. Bovenhuis *et al.* (1992) indicate that *CSN3* and *LGB* probably have direct effects upon protein and fat percentage, respectively, while there is more probability that the effects of *CSN2* upon fat percentage, and the effects of *CSN2* and *LGB* upon milk production and protein yield may involve linked genes.

A very generally found relationship is that of *EAM* upon milk yield and composition (Mitscherlich *et al.*, 1959; Rendel, 1959; Tolle, 1959; Hogreve, 1965; Bernikova, 1974; Gonyon *et al.*, 1987; Haenlein *et al.*, 1987; Andersson-Eklund *et al.*, 1990; Rocha, 1994). Almost invariably, the null allele, EAM^m, is associated with greater milk yield than that of the antigen-encoding allele (EAM^{M1} or $EAM^{M2M'}$) in the particular breed being studied. The depressive effect may be associated with the reported *EAM/BoLA* relationship to mastitis (Solbu *et al.*, 1982; Larsen *et al.*, 1985; Kaartinen *et al.*, 1988a, b).

A third generally consistent finding has been of EAB system effects upon fat percentage (Nair, 1957; Andresen *et al.*, 1959; Mitscherlich *et al.*, 1959; Rendel, 1959, 1960, 1961; Tolle, 1959; Neimann-Sorensen and Robertson, 1961; Conneally and Stone, 1965; Maijala, 1966; Brum *et al.*, 1968; Rausch *et al.*, 1968; Gonyon *et al.*, 1987; Andersson-Eklund *et al.*, 1990). Most commonly, these effects have been manifest as an association of $EAB^{BO1Y2D'}$ with increased fat percentage.

The nature of additional reported relationships is less certain. The most marked finding, although examined in only one study, is the indication of a QTL with a large effect upon fat percentage near *AMY1* (Andersson-Eklund and Rendel, 1993). Haenlein *et al.* (1987) found evidence of direct effects and Gonyon *et al.* (1987) and Andersson-Eklund *et al.* (1990) of linkage effects of *EAJ* upon lactation traits. Gonyon *et al.* (1987) also reported $EAJ \times EAL$ interaction effects upon milk yield and composition traits. While not confirming the interaction, Rocha (1994) found large linkage effects of *EAL* upon all lactation traits studied. Many earlier investigations reported *TF* effects (Ashton, 1960; Ashton *et al.*, 1964; Jamieson and Robertson, 1967; Ashton and Hewetson, 1969), but more recent studies have yielded mixed results (Geldermann *et al.*, 1985; Gonyon *et al.*, 1987; Haenlein *et al.*, 1987; Andersson-Eklund *et al.*, 1990).

A few non-lactation marker associations have also been noted. Three reports have cited significant deviations from expectation in segregation of *EAF* alleles from particular matings (Fowler *et al.*, 1963; Kraay, 1970; Rocha, 1994). The cause(s) of the deviations could not be determined. Brum *et al.* (1970) found little evidence of direct effects of blood and milk polymorphisms upon size measures in dairy heifers, but Beever *et al.* (1990) noted significant *EAB*-linked QTL effects upon preweaning growth and lean-muscle content in beef cattle.

In addition to the examination of specific effects of genetic marker loci, some have evaluated the degree of marker heterozygosity and used this measure as a relative estimate of the amount of overall genotypic heterozygosity. It is assumed that greater heterozygosity will contribute to improved fitness, which will in turn be manifest in superior survival, production and reproduction. Some such effects have been found for lactation traits (Rocha, 1994) and for measures of reproductive performance (Pirchner *et al.*, 1971; Schleger *et al.*, 1974, 1977; Hierl, 1976).

Neonatal Isoerythrolysis

Naturally occurring haemolytic disease of the newborn seems almost never to occur in cattle (Stormont, 1972b, 1975; Hines, 1973). Cows routinely develop anti-MHC antibodies against paternally inherited MHC class I antigens of their fetuses (Newman and Hines, 1980; Hines and Newman, 1981), but anti-erythrocyte antibodies are not found. A problem may occur, however, when blood-origin vaccines are used to vaccinate the pregnant cow (Dimmock and Bell, 1970; Hines *et al.*, 1973). Cows may develop antibodies against blood-group antigens in the products, and the antibodies may be further concentrated in the colostrum. Obviously, those producing blood-origin vaccines need to take care that blood-group antigens are not present in their preparations.

Summary

Polymorphic variation has been detected for 11 cattle blood-group systems, 13 plasma-protein systems, 12 red-cell-protein systems, eight white-cell-protein systems and six milk-protein systems. All of the blood-group and milk-protein loci, together with several of the plasma-protein and red-cell-protein loci, have been mapped to chromosomes. Some information has been learned about the structural nature of the blood-group antigens, but little about their function. Neonatal isoerythrolysis is not a naturally occurring phenomenon, but may arise in calves of dams receiving vaccines containing blood-group substances.

The blood group and blood biochemical polymorphism markers have had their greatest utility as vehicles for accomplishing parentage verification programmes. The milk-protein polymorphisms are associated with lactation trait differences. Other markers probably have effects upon economic traits also, but the effects appear to be small and in most cases not very clear. A deleterious association of M-system antigens with lactation traits may be effected through an increase in mastitis susceptibility.

References

Abe, T., Oishi, T., Amano, T., Kondoh, K., Nozama, K., Namikawa, T., Kumazaki, K., Koga, O., Hayashida, S. and Otsuka, J. (1968) Studies on native farm animals in Asia. I. On blood groups and serum protein polymorphism of East Asian cattle. *Japanese Journal of Zootechnical Science* 39, 523–535.

Abe, T., Komatsu, M., Oishi, T. and Kageyma, A. (1975) Genetic polymorphism in milk proteins in three Japanese cattle and four European breeds in Japan. *Japanese Journal of Zootechnical Science* 46, 591–599.

Agergaard, N. and Larsen, B. (1974) Bovine plasma alkaline phosphatase activity in relation to age, J substance and beta-lactoglobulin phenotypes. *Animal Blood Groups and Biochemical Genetics* 5, 11–19.

Andersson-Eklund, L. and Rendel J. (1993) Linkage between amylase-1 locus and a major gene for milk fat content in cattle. *Animal Genetics* 21, 101–103.

Andersson-Eklund, L., Danell, B. and Rendel, J. (1990) Associations between blood groups, blood protein polymorphisms and breeding values for production traits in Swedish Red and White Dairy bulls. *Animal Genetics* 21, 361–376.

Andresen, E., Hojgaard, N., Julling, B., Larsen, B., Moller, P., Moustgaard, J. and Neimann-Sorensen, A. (1959) Blood and serum group investigations on cattle, pig, and dog in Denmark. In: *Report of the VIth International Blood Group Congress (Munich)*, Munich, West Germany, p. 24.

Anon. (1985) Notice from the standing committee on cattle blood groups and biochemical polymorphisms. *Animal Genetics* 16, 249–252.

Ansay, M. (1975) Note on a third allele in the erythrocytic NP system of cattle. *Animal Blood Groups and Biochemical Genetics* 6, 121–124.

Ansay, M. and Hanset, R. (1972a) Polymorphisme de l'adénosine déaminase (ADA) dans l'éspèce bovine. *Annales de Génétique et de Sélection Animale* 4, 505–514.

Ansay, M. and Hanset, R. (1972b) Soluble glutamic oxalacetic transaminase (GOT) polymorphism in cattle. *Animal Blood Groups and Biochemical Genetics* 3, 163–168.

Ansay, M. and Hanset, R. (1972c) Purine nucleoside phosphorylase (NP) of bovine erythrocytes: genetic control of electrophoretic variants. *Animal Blood Groups and Biochemical Genetics* 3, 219–227.

Ansay, M. and Hanset, R. (1973) Polymorphism of mannose-6-phosphate isomerase in cattle. *Animal Blood Groups and Biochemical Genetics* 4, 169–173.

Ansay, M., Hanset, R. and Esser-Coulon, J. (1971a) La malate déshydrogénase mitochondriale: variants électrophorétiques de nature héréditaire dans l'espèce bovine. *Annales de Génétique et de Sélection Animale* 3, 235–243.

Ansay, M., Hanset, R. and Esser-Coulon, J. (1971b) Polymorphisme de la phosphoglucomutase (PGM) dans l'espèce bovine. *Annales de Génétique et de Sélection Animale* 3, 413–418.

Arfors, K.E., Beckman, L. and Lundin, L.G. (1963) Genetic variations of human serum phosphatases. *Acta Genetica Statistica Medica* 13, 89–94.

Aschaffenburg, R. (1961) Inherited casein variants in cow's milk. *Nature (London)* 192, 431–432.

Aschaffenburg, R. (1968). Reviews of the progress of dairy science. Section G. Genetics. Genetic variants of milk proteins: their breed distribution. *Journal of Dairy Research* 35, 447–460.

Aschaffenburg, R. and Drewry, J. (1955) Occurrences of different lactoglobulins in cow's milk. *Nature (London)* 176, 218–219.

Aschaffenburg, R., Sen, A. and Thompson, M.P. (1968) Genetic variants of casein in Indian and African Zebu cattle. *Comparative Biochemistry and Physiology* 25, 177–184.

Ashton, G.C. (1958) Genetics of beta-globulin polymorphism in British cattle. *Nature (London)* 182, 370–372.

Ashton, G.C. (1959) Beta-globulin alleles in some Zebu cattle. *Nature (London)* 184, 1135–1136.

Ashton, G.C. (1960) Beta-globulin polymorphism and economic factors in dairy cattle. *Journal of Agricultural Science* 54, 321.

Ashton, G.C. (1963) Polymorphism in the serum post-albumins of cattle. *Nature (London)* 198, 1117–1118.

Ashton, G.C. (1964) Serum albumin polymorphism in cattle. *Genetics* 50, 1421–1426.

Ashton, G.C. (1965a) Serum amylase (thread protein) polymorphism in cattle. *Genetics* 51, 431–437.

Ashton, G.C. (1965b) Serum transferrin D alleles in Australian cattle. *Australian Journal of Biological Science* 18, 665–670.

Ashton, G.C. and Hewetson, R.W. (1969) Transferrins and milk production in dairy cattle. *Animal Production* 11, 533.

Ashton, G.C. and Lampkin, G.H. (1965) Serum albumin and transferrin polymorphism in East African cattle. *Nature (London)* 205, 209–210.

Ashton, G.C., Fallon, G.R. and Sutherland, D.N. (1964) Transferrin (β-globulin) type and milk and butterfat production in dairy cows. *Journal of Agricultural Science* 62, 27.

Auditore, K.J. and Stormont, C. (1978) Cell surface distribution of alloantigens on bovine erythrocytes. *Immunogenetics* 6, 547.

Auditore, K.J., Morris, B.G., Suzuki, Y. and Stormont, C. (1979) Subtypic linearity in the bovine B blood group system. *Vox Sanguinis* 36, 236–239.

Baker, C.A., and Manwell, C. (1980) Chemical classification of cattle. 1. Breed groups. *Animal Blood Groups and Biochemical Genetics* 11, 127.

Bangham, A.D. (1957) Distribution of electrophoretically different haemoglobins among cattle breeds of Great Britain. *Nature (London)* 179, 467–468.

Bednekoff, A.G., Datta, S.P. and Stone, W.H. (1962) The J substance of cattle. VII. Production of immune anti-J in rabbits. *Journal of Immunology* 89, 408–413.

Bednekoff, A.G., Tolle, A., Datta, S.P., Friedman, J. and Stone, W.H. (1963) The J substance of cattle. VIII. The J-like substance of rabbits and the production of the immune anti-J. *Journal of Immunology* 91, 369–373.

Beever, J.E., George, P.D., Fernando, R.L., Stormont, C.J. and Lewin, H.A. (1990) Associations between genetic markers and growth and carcass traits in a paternal half-sib family of Angus cattle. *Journal of Animal Science* 68, 337–344.

Bell, K. (1962) One-dimensional starch-gel electrophoresis of bovine skim milk. *Nature (London)* 195, 705–706.

Bell, K. (1983) The blood groups of domestic mammals. In: Agar, N.S. and Board, P.G. (eds) *Red Blood Cells of Domestic Mammals*. Elsevier Science Publishers B.V., Amsterdam, pp. 133–164.

Bell, K. and Francis, J. (1970) Practical applications of cattle blood groupings. *Australian Veterinary Journal* 46, 119–120.

Bell, K., McKenzie, H.A., Murphy, W.H. and Shaw, D.C. (1970) β-lactoglobulin$_{\text{Droughtmaster}}$: a unique protein variant. *Biochimica et Biophysica Acta* 214, 427–436.

Bell, K., McKenzie, H.A. and Shaw, D.C. (1981a) Bovine beta-lactoglobulin E, F, and G of Bali (Banteng) cattle *Bos (Bibos) javanicus*. *Australian Journal of Biological Science* 34, 133–147.

Bell, K., Hoper, K.E. and McKenzie, H.A. (1981b) Bovine alpha-lactalbumin C and alpha$_{s1}$-, beta-, and kappa-caseins of Bali (Banteng) cattle, *Bos (Bibos) javanicus*. *Australian Journal of Biological Science* 34, 149–159.

Bernikova, N.N. (1974) Productivity of Black Pied cows at the Lesnoe Breeding Farm in relation to differences in their blood groups. *Animal Breeding Abstracts* 42, 4224 (abstract).

Bishop, M.D., Kappes, S.M., Keele, J.W., Stone, R.T., Sunden, S.L.F., Hawkins, G.A., Toldo, S.S., Fries, R., Grosz, M.D., Yoo, J. and Beattie, C.W. (1994) A genetic linkage map for cattle. *Genetics* 136, 619–639.

Blakeslee, D., Butler, J.E. and Stone, W.H. (1971a) Serum antigens of cattle. II. Immunogenetics of two immunoglobulin allotypes. *Journal of Immunology* 107, 227–235.

Blakeslee, D., Rapacz, J. and Butler, J.E. (1971b) Bovine immunoglobulin allotypes. *Journal of Dairy Science* 54, 1319–1320.

Blumberg, B.S. and Tombs, M.P. (1958) Possible polymorphism of bovine alpha-lactalbumin. *Nature (London)* 181, 683–684.

Bouquet, Y., Van de Weghe, A., VanZeveren, A. and Varewyck, H. (1986) Evolutionary conservation of the linkage between the structural loci for serum albumin and vitamin D binding protein (Gc) in cattle. *Animal Genetics* 17, 175.

Bouw, J. (1962) Some irregularities in the inheritance of B and C groups in Dutch cattle. In: *Proceedings of the 8th European Conference on Animal Blood Groups, Ljubljana, Yugoslavia*.

Bouw, J. and Fiorentini, A. (1970) Structure of loci controlling complex blood group systems in cattle. *European Congress Animal Blood Groups* 11, 109.

Bouw, J, Nasrat, G.E. and Buys, C. (1964) The inheritance of blood groups in the blood group system B in cattle. *Genetica* 35, 47–58.

Bouw J., Buys, C. and Schreuder, I. (1974) Further studies on the genetic control of the blood group system C of cattle. *Animal Blood Groups and Biochemical Genetics* 5, 105–114.

Bovenhuis, H., van Arendonk, J.A. and Korver, S. (1992) Associations between milk protein polymorphisms and milk production traits. *Journal of Dairy Science* 75, 2549.

Braend, M. (1956) The use of blood groups in bovine disputed parentage cases. *Cornell Veterinarian* 46, 83–87.

Braend, M. (1971) Haemoglobin variants of cattle. *Animal Blood Groups and Biochemical Genetics* 2, 15–21.

Braend, M. (1988) Haemoglobin polymorphism in Norwegian Red Cattle. *Animal Genetics* 19, 59–62.

Braend, M. and Efremov, G. (1965) Polymorphism of cattle serum albumin. *Nordisk Veterinaermedicin (Copenhagen)* 11, 585–588.

Brum, E.W., Rausch, W.H., Hines, H.C. and Ludwick, T. M. (1968) Association between milk and blood polymorphism types and lactation traits of Holstein cattle. *Journal of Dairy Science* 51, 1031–1038.

Brum, E.W., Hines, H.C., Ludwick, T.M. and Rader, E.R. (1970) Relationship between blood or milk polymorphisms and size measures of Holstein heifers. *Animal Blood Groups and Biochemical Genetics* 1, 247–252.

Burnet, F.M. and Fenner, F. (1949) *The Production of Antibodies*. Macmillan, Melbourne.

Buys, C. (1990) Additional informational on the linear order of the genes encoding the red blood group C-system antigens in cattle. *Animal Genetics* 21, 333.

Cabannes, R. and Serain, C. (1955) Hétérogénéité de l'hémoglobine des bovidés. Identification électrophorétique de deux hémoglobines bovines. *Comptes rendus Séances Société Belge de Biologie* 149, 7–10.

Carr, W.C. (1966) Serum albumin polymorphism of some breeds of cattle in Zambia. In: *Proceedings of the 10th European Conference on Animal Blood Groups and Biochemical Polymorphism, Paris, France*, pp. 293–297.

Carr, W.C., Macleod, J., Woolf, B. and Spooner, R.L. (1974) A survey of the relationship of genetic markers, tick-infestation level and parasitic diseases in Zebu cattle in Zambia. *Tropical Animal Health Production* 6, 203–214.

Carr, W.R. (1964) The haemoglobins of indigenous breeds of cattle in central Africa. *Rhodesian Journal of Agricultural Research* 2, 93–94.

Carr, W.R. (1965) A new haemoglobin variant. *Rhodesian Journal of Agricultural Research* 3, 62.

Cartron, J.-P. and Agre, P. (1995) RH blood groups and Rh-deficiency syndrome. In: Cartron, J.-P. and Rouger, P. (eds) *Blood Cell Chemistry. 6. Molecular Basis* of *Human Blood Group Antigens*. Plenum Press, New York and London, pp. 189–225.

Colin, Y., Bailly, P. and Cartron, J.-P. (1994) Molecular basis of RH and LW blood groups. *Vox Sanguinis* 67 (Suppl. 3), 67–72.

Conneally, P.M. and Stone, W.H.. (1965) Association between a blood group and butterfat production in dairy cattle. *Nature (London)* 206, 115.

Cowpertwait, J. and Spooner, R.L. (1971) Ox erythrocyte agglutinability. III. A comparison of agglutinability in various anti-globulin systems and its mode of inheritance. *Vox Sanguinis* 18, 251–256.

Creamer, L.K. and Richardson, B.C. (1975) A new variant of β-casein. *New Zealand Journal of Dairy Science Technology* 10, 170–171.

Crockett, J.R., Koger, M. and Chapman, H.L., Jr (1963) Genetic variations in hemoglobins of beef cattle. *Journal of Animal Science* 22, 173–176.

Dabczewski, Z. (1969) Studies on hereditary blood properties in Kerry cattle. *Irish Journal of Agricultural Research* 8, 279–284.

Datta, S.P. (1959) A possible heritable exception in cattle blood groups. *Genetics* 44, 504.

Del Lama, M.A., Del Lama, S.N., Mestriner, M.A. and Mortari, N. (1989) Esterase D: a new polymorphism in Zebu cattle. *Animal Genetics* 20 (Suppl. 1), 57–58.

Del Lama, S.N., Del Lama, M.A., Mestriner, M.A. and Mortari, N. (1992) Peptidase B polymorphism in cattle erythrocytes. *Biochemical Genetics* 30, 247–255.

Dimmock C.K., and Bell, K. (1970) Haemolytic disease of the newborn in calves. *Australian Veterinary Journal* 46, 44–47.

Di Stasio, L. and Merlin, P. (1979) Polimorfismi biochimici del latte nella razza bovina Grigio Alpina. *Rivista di Zootecnia e Veterinaria* 2, 64–67.

Dogrul, F. (1969) Investigations on erythrocyte acid phosphatase in cattle. In: *Proceedings of the 11th European Conference on Animal Blood Groups and Biochemical Polymorphisms, Warsaw, 1968*, pp. 223–226.

Dorynek, Z. and Kaczmarek, A. (1975) Recombination of blood group factors in the B system in cattle. In: *Proceedings of the 26th Meeting of the European Association for Animal Production, Warsaw, 1968*, pp. 117–122.

Duniec, M., Stawarz, K., Buys, C. and Bouw, J. (1973) A closed system within blood group locus C of cattle. *Animal Blood Groups and Biochemical Genetics* 4, 185–186.

Ebertus, R. (1968) Ein bisher nicht beschreibner Polymorphismos einer zweiten Amylase im Blutserum des Rindes. *Fortpflanzung, Besamung und Aufzucht der Haustiere* 4, 283–288.

Efremov, G. and Braend, M. (1965) A new hemoglobin in cattle. *Acta Veterinaria Scandinavica* 6, 109–111.

Ehrlich, P. and Morgenroth, J. (1900) Ueber Haemolysine. *Berliner Klinische Wochenschrift* 37, 453–458.

Ellory J.C., and Tucker, E.M. (1970) High potassium type red cells in cattle. *Journal of Agricultural Science* 74, 595–596.

Ellory, J.C., Tucker, E.M. and Rasmusen, B.A. (1974) The effect of sheep anti-L and certain cattle S-system reagents on active potassium transport in sheep and cattle red blood cells. *Animal Blood Groups and Biochemical Genetics* 5, 159–165.

Erhardt, G. (1989) K-Kaseine in Rindermilch – Nachweis eines weiteren Allels (K-Cn^E) in verschiedenen Rassen. *Journal of Animal Breeding and Genetics* 106, 225–231.

Faber, H.E. and Stone W.H. (1975) Serum antigens of cattle. V. Genetics of immunoglobulin light chain allotypes. *Immunogenetics* 3, 167–176.

Faber, H.E. and Stone, W.H. (1976) Cattle allotypes: a review and suggested nomenclature. *Animal Blood Groups and Biochemical Genetics* 7, 39–50.

Ferguson, L.C. (1941) Heritable antigens in the erythrocytes of cattle. *Journal of Immunology* 40, 213–242.

Ferguson, L.C., Stormont, C. and Irwin, M.R. (1942) On additional antigens in the erythrocytes of cattle. *Journal of Immunology* 44, 147–164.

Fowler, A.K., Ludwick, T.M., Weseli, D.F., Brum, E.W., Rader, E.R., Hines, H.C. and Plowman, D. (1963) Blood group segregation and possible effects on fertility in cattle. *Journal of Dairy Science* 46, 629 (abstract).

Fries, R., Hediger, R. and Stranzinger, G. (1988) The loci for parathyroid hormone and beta-globin are closely linked and map to chromosome 15 in cattle. *Genomics* 3, 302–307.

Gahne, B. (1961) Studies of transferrins in serum and milk of Swedish cattle. *Animal Production* 3, 135–145.

Gahne, B. (1963a) Genetic variation of phosphatase in cattle serum. *Nature (London)* 199, 305–306.

Gahne, B. (1963b) Inherited variations in the post-albumins of cattle serum. *Hereditas* 50, 126–135.

Gahne, B. (1967a) Inherited high alkaline phosphatase activity in cattle serum. *Hereditas* 57, 83–99.

Gahne, B. (1967b) Alkaline phosphatase isoenzymes in serum, seminal plasma, and tissues of cattle. *Hereditas* 57, 100–114.

Gahne, B., Juneja, R.R. and Grolmus, J. (1977) Horizontal polyacrylamide gradient gel electrophoresis for the simultaneous phenotyping of transferrin, post-transferrin, albumin and post-albumin in the blood plasma of cattle. *Animal Blood Groups and Biochemical Genetics* 8, 127–137.

Geldermann, H. (1975) Investigations on inheritance of quantitative characters in animals by gene markers. I. Methods. *Theoretical and Applied Genetics* 46, 319.

Geldermann, H. (1976). Investigation on inheritance of quantitative characters in animals by gene markers. II. Expected effects. *Theoretical and Applied Genetics* 46, 1.

Geldermann, H., Pieper, U. and Roth, B. (1985) Effects of marked chromosome sections on milk performance in cattle. *Theoretical and Applied Genetics* 70, 138–146.

Georges, M., Swillens, S., Bouquet, Y., Lequarre, A.S. and Hanset, R. (1987) Genetic linkage between the bovine plasma protease inhibitor 2 (Pi-2) and S blood group loci. *Animal Genetics* 18, 311–316.

Glasnak, V. and Sulc, J. (1971) A case of irregular inheritance in the B system of cattle. *Animal Blood Groups and Biochemical Genetics* 2, 185–187.

Gonyon, D.S., Mather, R.E., Hines, H.C., Haenlein, G.F., Arave, C.W. and Gaunt, S.N. (1987) Associations of bovine blood and milk polymorphisms with lactation traits: Holsteins. *Journal of Dairy Science* 70, 2585–2598.

Green, P. (1966) Towards a tentative map of the genetic determinants of the B blood group system of cattle. *Immunogenetics Letter* 4, 188–191.

Grosclaude, F. (1965) Note préliminaire sur un locus supplémentaire de groupes sanguins des bovins, le locus T'. *Annales de Biologie Animale, Biochimie, Biophysique (Paris)* 5, 403–406.

Grosclaude, F. (1966) Note sur le mode d'hérédite d'un 'nouveau' facteur de groupes sanguins des bovins. *Immunogenetics Letter* 4, 171–172.

Grosclaude, F. (1975) Variants génétiques des protéines du lait de vaches mongoles. *Etudes mongoles, Cahier du Centre d'Etudes Mongoles, Laboratoire d'Ethnologie, Université de Paris X, Nanterre* 6, 81–83.

Grosclaude, F., Pujolle, J., Garnier, J. and Ribadeau-Dumas. (1966) Mise en évidence de deux variants supplémentaires des protéines du lait de vache: Alpha$_{s1}$-Cn^D et Lg^D. *Annales de Biologie Animale, Biochimie, Biophysique (Paris)* 6, 215–222.

Grosclaude, F., Mahé, M.-F., Mercier, J.C., Bonnemarie, J. and Teissier, J.H. (1976a) Polymorphisme des lactoprotéins de bovinés Népalais. I. Mise en evidence, chez le Yak, et caractérisation biochimique de deux nouveaux variants: beta-lactoglobuline D$_{Yak}$ et caséine Alpha$_{s1}$. *Annales de Génétique et de Sélection Animale* 8, 461–479.

Grosclaude, F., Mahé, M.-F., Mercier, J.C., Bonnemarie, J. and Teissier, J.H. (1976b) Polymorphisme des lactoprotéins de bovinés Népalais. II. Polymorphisme des caséines 'Alpha$_s$ mineures'; le locus Alpha$_{s2}$-Cn est-il lié aux loci Alphs$_{s1}$-Cn, Beta-Cn et Kappa-Cn. *Annales de Génétique et de Sélection Animale* 8, 481–491.

Grosclaude, F., Joudrier, P. and Mahé, M.-F. (1978) Polymorphisme de la caséine alpha$_{s2}$ bovine: étroite liason du locus alpha$_{s2}$-Cn avec les loci alphs$_{s1}$-Cn, beta-Cn et kappa-Cn; mise en évidence d'une délétion dans le variant alpha$_{s2}$-Cn D. *Annales de Génétique et de Sélection Animale* 10, 313–327.

Grosclaude, F., Guerin, G. and Houlier, G. (1979) The genetic map of the B system of cattle blood groups as observed in French breeds. *Animal Blood Groups and Biochemical Genetics* 10, 199–218.

Grosclaude, F., Alaux, M.T., Houlier, G. and Guerin, G. (1981) The C system of cattle blood groups. 1. Additional factors in the system. *Animal Blood Groups and Biochemical Genetics* 12, 7–14.

Grosclaude, F., Lefebvre, J. and Noe, G. (1983) Nouvelles précisions sur la carte génétique du système de groupes sanguins B des bovins. *Génétique, Sélection, Evolution* 15, 45–54.

Guérin, G., Grosclaude, F. and Houlier, G. (1981) The C system of cattle blood groups. 2. Partial genetic map of the system. *Animal Blood Groups and Biochemical Genetics* 12, 15–21.

Haenlein, G.F., Hines, H.C. and Zikakis J.P. (1980) Frequency distribution of genetic markers in Guernsey cattle. *Journal of Dairy Science* 63, 1145–1153.

Haenlein G.F., Gonyon, D.S., Mather, R.E. and Hines H.C. (1987) Associations of bovine blood and milk polymorphisms with lactation traits: Guernseys. *Journal of Dairy Science* 70, 2599–2609.

Hall, J.G. and Ross, D.S. (1981) Evidence for the presence of an additional allele in the F system of British Friesian cattle blood. *Animal Blood Groups and Biochemical Genetics* 12, 229–240.

Han, S.K. and Suzuki, S. (1976) Studies on hemoglobin variants in Korean cattle. *Animal Blood Groups and Biochemical Genetics* 7, 21–25.

Hatheway, C.L., Weseli, D.F., Ludwick, T.M. and Hines, H.C. (1969) Separation and differential enzymic inactivation of the F and V isoantigens of the bovine erythrocyte. *Vox Sanguinis* 17, 204–216.

Hesselholt, M. and Moustgaard, J. (1965) Serum amylase polymorphism in cattle and its application in parentage control. (In Danish, with English summary.) In: *Årsberetning*. Institut for Sterilitetsforskning, Copenhagen, pp. 175–182.

Hierl, H.F. (1976) Beziehungen zwishen dem Heterozygotiegrad, geschazt aus Markengenen, und der Fruchtbarkeit beim Rind. II. Heterozygotie und Fruchtbarkeit. *Theoretical and Applied Genetics* 47, 77.

Hines, H.C. (1967) An examination of bovine blood typing antibodies by gel-filtration analysis. *Vox Sanguinis* 13, 263–269.

Hines, H.C. (1971) The agglutination of cattle erythrocytes by plant extracts. *Animal Blood Groups and Biochemical Genetics* 2, 221–228.

Hines, H.C. (1973) Isohemolytic disease – a review. *Proceedings of the National Anaplasmosis Conference* 6, 77–80.

Hines, H.C. (1977) Blood group examination of cattle breed purity. *Animal Blood Groups and Biochemical Genetics* 8 (Suppl. 1), 31–32.

Hines, H.C. and Newman, M.J. (1981) Production of foetally stimulated lymphocytotoxic antibodies by multiparous cows. *Animal Blood Groups and Biochemical Genetics* 12, 201–206.

Hines, H.C. and Ross, M.J. (1987) Serological relationships among antigens of the BoLA and the bovine M blood group systems. *Animal Genetics* 18, 361–369.

Hines, H.C., Salfner, B., Uhlenbruck, G. and Schmid, D.O. (1972a) Bovine erythrocyte membrane composition: association with the FV and other glycoprotein systems. *Vox Sanguinis* 22, 488–500.

Hines, H.C., Uhlenbruck, G. and Schmid, D.O. (1972b) Possible relationship between Concanavalin A reactivity and the bovine A blood group system. *Vox Sanguinis* 22, 529–531.

Hines, H.C., Bedell, D.M., Kliewer, I.O. and Hayat, C.S. (1973) Some effects of blood antigens in A. marginale and other vaccines. *Proceedings of the National Anaplasmosis Conference* 6, 82–85.

Hines, H.C., Trowbridge, C.L. and Zink, G.L. (1976) Removal of blood group determinants from bovine erythrocyte membranes. 3. Action of proteolytic enzymes on intact cells. *Animal Blood Groups and Biochemical Genetics* 7, 91–99.

Hines, H.C., Haenlein, G.F., Zikakis, J.P. and Dickey H.C. (1977) Blood antigen, serum protein, and milk protein gene frequencies and genetic interrelationships in Holstein cattle. *Journal of Dairy Science* 60, 1143–1151.

Hines, H.C., Zikakis, J.P., Haenlein, G.F.W., Kiddy, C.A. and Trowbridge, C.L. (1981) Linkage relationships among loci of polymorphisms in blood and milk of cattle. *Journal of Dairy Science* 64, 71–76.

Hogreve, F. (1965) Lifetime performance and blood group factors in cattle. investigations on Black Pied Lowland cows. *Tierrztliche Umschau (Constance)* 20, 17.

Hønberg, L.S. and Larsen, B. (1992) Bovine monoclonal alloantibodies to blood group antigens prepared by murine × bovine or (murine × bovine) × bovine interspecies fusions. *Animal Genetics* 23, 497–508.

Hønberg, L.S., Larsen, B., Koch, C., Ostergard, H. and Skjodt, K. (1995) Biochemical identification of the bovine blood group M' antigen as a major histocompatibility complex class I-like molecule. *Animal Genetics* 26, 307–313.

Iannelli, D. and Masina, P. (1978) Immunogenetics of a macroglobulin allotype in cattle. *Genetics Research, Cambridge* 31, 265–271.

Irwin, M.R. (1956) Blood grouping and its utilization in animal breeding. In: *7th International Congress of Animal Husbandry, Madrid*, Vol. 2, pp. 7–41.

Irwin, M.R. (1974) Comments on the early history of immunogenetics. [Review, 89 references.] *Animal Blood Groups and Biochemical Genetics* 5, 65–84.

Irwin, M.R. (1976) The beginnings of immunogenetics. *Immunogenetics* 3, 1–13.

Irwin, M.R. and Cole, L.J. (1936) Immunogenetic studies of species and species hybrids in doves and the separation of species-specific substances in the backcross. *Journal of Experimental Zoology* 73, 85–108.

Jamieson, A. (1965) The genetics of transferrins in cattle. *Heredity* 20, 419–441.

Jamieson, A. and Robertson, A. (1967) Cattle transferrins and milk production. *Animal Production* 9, 491.

Juneja, R.K. and Gahne, B. (1980) Two-dimensional gel electrophoresis of cattle plasma proteins: genetic polymorphisms of an α1-protease inhibitor. *Animal Blood Groups and Biochemical Genetics* 11, 215–228.

Kaartinen, L., Ali-Vehmas, T., Larsen, B., Jensen, N.E. and Sandholm, M. (1988a) Relation of the bovine M blood group system with growth of *Streptococcus agalactiae* and *Staphylococcus aureus* in whey. *Zentralblatt für Veterinarmedizin (B) (Berlin)* 35, 77–83.

Kaartinen, L., Ali-Vehmas, T., Larsen, B., Jensen, N.E. and Sandholm, M. (1988b) *In vitro* bacterial growth in whey from M' blood group positive and negative cows – growth of *Streptococcus agalactiae* is dependent on protease inhibitor level. *Zentralblatt für Veterinarmedzin (B)* 35, 681–687.

Kappes, S.M., Bishop, M.D., Keele, J.W., Penedo, M.C.T., Hines, H.C., Grosz, M.D., Hawkins, G.A., Stone, R.T., Sunden, S.L.F. and Beattie, C.W. (1994) Linkage of bovine erythrocyte antigen loci B, C, L, S, Z, R', and T' and the serum protein loci post-transferrin 2 (PTF2), vitamin D binding protein (GC), and albumin to DNA microsatellite markers. *Animal Genetics* 25, 133–140.

Kennedy, B.W., Quinton, M. and van Arendonk, J.A. (1992) Estimation of effects of single genes on quantitative traits. *Journal of Animal Science* 70, 2000.

Khanna, N.D., Singh, H, Bhatia, S.S. and Bhat, P.N. (1972) A rare hemoglobin variant in Afghan cattle and crosses. *Animal Blood Groups and Biochemical Genetics* 3, 59–60.

Kidd, K.K. (1971) Applications to man and cattle of methods of reconstructing evolutionary histories. In: Hodson, F.R., Kendell, D.G. and Tautu, P. (eds) *Mathematics in the Archaeological and Historical Sciences*. Edinburgh University Press, Edinburgh, pp. 356–360.

Kidd, K.K. (1974) Biochemical polymorphisms, breed relationships, and germ plasm resources in domestic cattle. In: *Proceedings of the First World Congress of Genetics Applied to Livestock Production, Madrid, Spain*, Vol. I, pp. 321–328.

Kidd, K.K. and Cavalli-Sforza, L.L. (1974) The role of genetic drift in the differentiation of Icelandic and Norwegian cattle. *Evolution* 28, 381.

Kidd, K.K. and Pirchner, F. (1971) Genetic relationships of Austrian cattle breeds. *Animal Blood Groups and Biochemical Genetics* 2, 145–158.

Kidd, K.K. and Sgaramella-Zonta, L.A. (1971) Phylogenetic analysis: concepts and methods. *American Journal of Human Genetics* 23, 235–252.

Kidd, K.K., Osterhoff, D., Erhard, L. and Stone, W.H. (1974) The use of genetic relationships among cattle breeds in the formulation of rational breeding policies: an example with South Devon (South Africa) and Gelbvieh (Germany). *Animal Blood Groups and Biochemical Genetics* 5, 21–28.

Kiddy, C.A. and Hooven, N.W. (1961) *Results of Repeatability and Standardization Tests in Cattle Blood Typing*. USDA-ARS Publication. 44–94, 11 pp.

Kiddy, C.A., Peterson, R.F. and Kopfler, F.C. (1966) Genetic control of the variants of beta-casein A. *Journal of Dairy Science* 49, 742.

Komatsu, M., Abe, T. and Oishi, T. (1979) Genetic variation of serum post-albumin and post-transferrin in nine East Asian and European cattle breeds. *Animal Blood Groups and Biochemical Genetics* 10, 185–188.

Kraay, G.J. (1970) Gene segregation at the F–V and Tf loci in some cattle breeds. *Canadian Journal of Animal Science* 50, 371–376.

Kristjansson, F.K. and Hickman, C.G. (1965) Subdivision of the TF^D for transferrins in Holstein and Ayrshire cattle. *Genetics* 52, 627–630.

Landsteiner, K. (1900) Zur Kenntnis der antifermentativen, lytischen und agglutinierenden Wirkungen des Blutserums und der Lymphe. *Zentralblatt für Bakteriologie* 27, 357–362.

Landsteiner, K. (1901) Über Agglutinations-erscheinungen normalen menschlichen Blutes. *Wiener Klinische Wochenschrift* 14, 1132–1134.

Larsen, B. (1961) Additional blood group factors of the A and B systems of cattle. *Acta Veterinaria Scandinavica* 11, 242–256.

Larsen, B. (1964) Irregular inheritance of blood groups in the B system of cattle. In: *Aarsberetning*. Institut for Sterilitetsforskning, Copenhagen, pp. 125–130.

Larsen, B. (1966) Test for linkage of the genes controlling hemoglobin, transferrin and blood types in cattle. *Royal Veterinary and Agricultural University Yearbook, Copenhagen* 41, 8.

Larsen, B. (1970) [Linkage relations of blood groups and polymorphic protein systems in cattle. Koblingsrelationer mellem blod-og polymorfe proteintypesystemer hos kvæg. *Aarsberetning, Institut for Sterilitetsforskning, Copenhagen* 13, 165–194.

Larsen, B. (1977) On linkage relations of ceruloplasmin polymorphism (Cp) in cattle. *Animal Blood Groups and Biochemical Genetics* 8, 111–113.

Larsen, B. (1981) Studies on the C blood group system in three Danish cattle breeds. *Animal Blood Groups and Biochemical Genetics* 12, 133–138.

Larsen, B. (1982) On the bovine F blood group system. *Animal Blood Groups and Biochemical Genetics* 13, 115–121.

Larsen B., Jensen, N.E., Madsen, P., Nielsen, S.M., Klastrup, O. and Madsen, P.S. (1985) Association of the M blood group system with bovine mastitis. *Animal Blood Groups and Biochemical Genetics* 16, 165–173.

Larsen, B., di Stasio, L., and Tucker, E.M. (1992) List of alleles for blood and milk polymorphisms in cattle, sheep, and goats. *Animal Genetics* 23, 188–192.

Leveziel, H. and Hines, H.C. (1984) Linkage in cattle between the major histocompatibility complex (BoLA) and the M blood group system. *Génétique, Sélection, Evolution* 16, 405–416.

Lewin, H.A., Beever, J.E., Da, Y., Hines, H.C. and Faulkner, D.B. (1994) The bovine B and C blood group systems are not likely to be the orthologues of human RH: an interesting twist in the comparative map. *Animal Genetics* 25 (Suppl. 1), 13–18.

Lie, H. and Braend, M. (1963) Irregularity in transmission of B alleles in cattle. *Immunogenetics Letter* 3, 23–26.

Ma, R.Z., Beever, J.E., Da, Y., Green, C.A., Russ, I., Park, C., Heyden, D.W., Everts, R.E., Fisher, S.R., Overton, K.M., Teale, A.J., Kemp, S.J., Hines, H.C., Guérin, G. and Lewin, H.A. (1996) A male linkage map of the cattle (*Bos taurus*) genome. *Journal of Heredity* 87, 261–271.

McClure, T.J. (1952) Correlation study of bovine erythrocyte antigen A and butterfat test. *Nature (London)* 170, 327.

Maddy, A.H. and Spooner, R.L. (1970) Ox erythrocyte agglutinability. 1. Variation in membrane protein. *Vox Sanguinis* 18, 34–41.

Maijala, K. (1966) On the possibility of predicting the success of a bull's daughters from his blood type. *Annales Agriculturae Fenniae* 5, 65.

Maijala, K. and Lindström, G. (1965) The inheritance of the new blood group factor SF3 in cattle. *Annales Agriculturae Fenniae* 4, 207–214.

Marziali, A.S. and Ng-Kwai-Hang, N.K. (1986) Relationships between milk protein polymorphisms and cheese yielding capacity. *Journal of Dairy Science* 69, 1193–1201.

Masina, P., Ramunno, L. and Iannelli, D. (1980) A new electrophoretic variant of vitamin D binding protein (post-albumin) in cattle serum. *Animal Blood Groups and Biochemical Genetics* 11, 271–273.

Mazumder, N.K. and Spooner, R.L. (1970) Studies on bovine serum amylase; evidence for two loci. *Animal Blood Groups and Biochemical Genetics* 1, 145–156.

Mercier, J.-C. and Vilotte, J.-L. (1993) Structure and function of milk protein genes. *Journal of Dairy Science* 76, 3079–3098.

Méténier, L., Nocart, M. and Alaux, M.T. (1991) Mouse monoclonal antibodies to bovine blood group antigens: additional results. *Animal Genetics* 22, 155–163.

Miller, W.J. (1966) Evidence for two new systems of blood groups in cattle. *Genetics* 54, 151–158.

Miller, W.J., Braend, M. and Stormont, C. (1962) A new reagent for the M blood group system of cattle. *Immunogenetics Letter* 2, 78–79.

Mitscherlich, E., Tolle, J. and Walter, E.. (1959) Untersuchungen ueber das Bestehen von Beziehungen zwischen Blutgruppenfaktoren und der Milchleistung des Rindes. *Zeitschrift für Tierzüchtung und Züchtungsbiologie* 72, 289–301.

Moazami-Goudarzi, K., Laloë, D., Furet, J.P. and Grosclaude, F. (1997) Analysis of genetic relationships between 10 cattle breeds with 17 microsatellites. *Animal Genetics* 28, 338–345.

Morton, J.A., Pickles, M.M. and Sutton, L. (1969) The correlation of the Bg blood group with the HL-A7 leucocyte group: demonstration of antigenic sites on red cells and leucocytes. *Vox Sanguinis* 17, 536–547.

Mouro, I., Colin, Y., Cherif-Zahar, B., Cartron, J.-P. and Van Kim, C.L. (1993) Molecular genetic basis of the human Rhesus blood group system. *Nature Genetics* 5, 62–65.

Moustgaard, J. and Neimann-Sorensen, A. (1962) Possible inter-allelic crossing over in the bovine B blood group system. *Immunogenetics Letter* 2, 62–64.

Mzee, R.M. and Braend, M. (1979) Seasonal variation of bovine anti-J in the tropics. *Animal Blood Groups and Biochemical Genetics* 10, 137–140.

Naik, S.N., Sukumaran, P.K. and Sanghvi, L.D. (1965) A note on blood groups and haemoglobin variants in Zebu cattle. *Animal Production* 7, 275–277.

Nair, P.G. (1957) Studies on associations between cellular antigens and butterfat percentage in dairy cattle. *Dairy Science Abstracts* 19, 369.

Namikawa, T., Takanaka, O. and Takhashi, K. (1983) Haemoglobin Bali (Bovine): Beta18(Bl)Lys-His: one of the 'missing links' between beta and beta^B of domestic cattle exists in the Bali cattle (Bovinae, *Bos banteng*). *Biochemical Genetics* 21, 787–796.

Nasrat, G.E. (1965) The inheritance of blood groups in the blood group system C in cattle. In: *Proceedings of the 9th European Conference on Animal Blood Groups, Prague, 1964*, pp. 69–74.

Neelin, J.M. (1964) Variants of kappa-casein revealed by improved starch gel electrophoresis. *Journal of Dairy Science* 47, 506–509.

Neimann-Sorensen, A. (1956) Blood groups and breed structure as exemplified by three Danish breeds. *Acta Agriculturae Scandinavica* 6, 115–137.

Neimann-Sorensen, A. and Robertson, A. (1961) The association between blood groups and several production characteristics in three Danish cattle breeds. *Acta Agriculturae Scandinavica* 11, 163–196.

Newman, M.J. and Hines, H.C. (1980) Stimulation of maternal anti-lymphocyte antibodies by first gestation bovine fetuses. *Journal of Reproduction and Fertility* 60, 237–241.

Oosterlee, C.C. and Bouw, J. (1974) Structure of loci in animals. In: *Proceedings of the 1st World Congress of Genetics Applied to Livestock Production, Madrid*, Vol. 1, pp. 243–252.

Osterhoff, D.R. (1962) Blood groups in bovines. II. Normally occurring isoantibodies in cattle blood. *Onderstepoort Journal of Veterinary Research* 29, 89–106.

Osterhoff, D.R. (1975) Hemoglobin types in African cattle. *Journal of the South African Veterinary Medical Association* 46, 186–189.

Osterhoff, D.R. and Politzer, N. (1970) F^f – a new allele in the FV blood group system. In: *Proceedings of the 11th Conference on Animal Blood Groups and Biochemical Polymorphisms, Warsaw, 1968*, pp. 135–142.

Osterhoff, D.R. and Pretorius, A.M. (1966) *Proceedings of the South African Society of Animal Production* 5, 166.

Osterhoff, D.R. and van Heerden, J.R.H. (1965) Tf^G – a new transferrin allele in cattle. In: *Proceedings of the 9th European Conference on Animal Blood Groups and Biochemical Polymorphisms, Prague, 1964*, 311–312.

Ostrand-Rosenberg, S. (1975) Gene dosage and antigenic expression on the surface of bovine erythrocytes. *Animal Blood Groups and Biochemical Genetics* 6, 81–99.

Ostrand-Rosenberg, S. (1976) Topology of bovine red cell antigens as a function of the organization of their genes. *Immunogenetics* 3, 53–64.

Owen, R.D. (1945) Immunogenetic consequences of vascular anastomoses between bovine twins. *Science* 102, 400–401.

Penedo, M.C.T., Mortari, N. and Magalhaes, L.E. (1982) Carbonic anhydrase polymorphism in Indian Zebu cattle. *Animal Blood Groups and Biochemical Genetics* 13, 141–143.

Peterson, R.F. and Kopfler, F.C. (1966) Detection of new types of β-casein by polyacrylamide gel electrophoresis at acid pH: a proposed nomenclature. *Biochemical and Biophysical Research Communications* 22, 388–392.

Pirchner, F., Chakrabarti, S., Erlacher, J., Schleger, W. and Rohrbacher, H. (1971) Heterozygotie und Fruchtbarkeit bei Rindern. [Heterozygosis and fertility in cattle.] *Deutsche Tierärztliche Wochenschrift* 78, 111–114.

Probeck, H.D. and Geldermann, H. (1976) Polymorphism of 6-phosphoglucmutase in cattle leucocytes. *Animal Blood Groups and Biochemical Genetics* 7, 33–37.

Probeck, H.D. and Geldermann, H. (1977) Polymorphism of 6-phosphogluconate dehogenase in the leucocytes of cattle. *Animal Blood Groups and Biochemical Genetics* 8, 157–160.

Race, R.R. (1944) An 'incomplete' antibody in human serum. *Nature (London)* 153, 771–772.

Rapacz, J., Korda, N. and Stone, W.H. (1968) Serum antigens of cattle. I. Immunogenetics of a macroglobulin allotype. *Genetics* 58, 387–398.

Rapacz, J., Korda, N. and Stone, W.H. (1975) Serum antigens of cattle. IV. Immunogenetics of a soluble antigenic determinant derived from lysed erythrocytes. *Genetics* 80, 323–329.

Rasmusen, B.A. (1959). Blood groups, disputed parentage, twins, and freemartins in cattle. *Illinois Veterinarian* 2, 31–34.

Rasmusen, B.A., Tucker, E.M., Ellory, J.C. and Spooner R.L. (1974) The relationship between S system of blood groups and potassium levels in red blood cells of cattle. *Animal Blood Groups and Biochemical Genetics* 5, 95–104.

Rausch, W.H., Brum, E.W. and Ludwick, T.M. (1968) Relationship between blood type and predicted differences in production of Holstein sires in artificial insemination. *Journal of Dairy Science* 51, 445–451.

Rendel, J. (1958a) Studies of cattle blood groups. I. Production of cattle iso-immune sera and the inheritance of 4 antigenic factors. *Acta Agriculturae Scandinavica* 8, 40–61.

Rendel, J. (1958b) Studies of cattle blood groups. IV. The frequency of blood group genes in Swedish cattle breeds, with special reference to breed structure. *Acta Agriculturae Scandinavica* 8, 191–215.

Rendel, J. (1959) A study on relationships between blood group and production characters in cattle. In: *Report of the VIth International Bloodgroup Congress (Munich)*, pp. 8–23.

Rendel, J. (1960) Untersuchungen uber Zusammenhange zwischen Blutgruppen und Leistungscharakteren des Rindes. *Journal of Zuchtungskunde* 32 (Suppl.), 291.

Rendel, J. (1961) Relationships between blood groups and the fat percentage of the milk in cattle. *Nature (London)* 189, 408–409.

Rendel, J. (1963) A study of the variation in cattle twins and pairs of single-born animals. I. Description of the material and the diagnosis of zygosity. *Zeitschrift für Tierzüchtung und Züchtungsbiologie* 79, 75–85.

Rendel, J. (1967) Studies of blood groups and protein variants as a means of revealing similarities and differences between animal populations. *Animal Breeding Abstracts* 35, 371–383.

Rendel, J. (1968) Inheritance of blood characteristics: basic results and practical applications. In: Johansson, I. and Rendel, J. (eds) *Genetics and Animal Breeding*. Oliver & Boyd, Edinburgh, pp. 185–213.

Rendel, J. and Gahne, B. (1963) Interaction between phosphatases and the J blood groups in cattle. *Immunogenetics Letter* 3, 38–43.

Rendel, J. and Sellei, J. (1967) The genetics of antibody production. II. A study of the anti-J variation in monozygous cattle twins. *European Conference on Animal Blood Groups* 10, 85–90.

Rendel, J. and Stormont, C. (1964) Variants of ovine alkaline serum phosphatases and their association with the R-O blood groups. *Proceedings of the Society for Experimental Biology and Medicine* 115, 853–856.

Rocha, J.L. (1994) Blood group polymorphisms and production and type traits in dairy cattle: after forty years of research. PhD dissertation, Texas A&M University, College Station, Texas, 315 pp.

Ruiterkamp, W.A., Spek, C.M. and Bouw, J. (1977) Gene clusters in the blood group system B of cattle. *Animal Blood Groups and Biochemical Genetics* 8, 231–240.

Saison, R. (1973) Red cell peptidase polymorphism in pigs, cattle, dogs, and mink. *Vox Sanguinis* 25, 173–181.

Sartore, G. (1970) Carbonic anhydrase types of cattle red cells. In: *Proceedings of the 11th Conference on Animal Blood Groups and Biochemical Polymorphisms, Warsaw, 1968*, pp. 211–216.

Sartore, G. and Bernoco, D. (1966) Research on biochemical polymorphisms in the indigenous cattle of Piedmont. In: *Proceedings of the 10th Conference on Animal Blood Groups and Biochemical Polymorphisms, Paris*, pp. 283–287.

Sartore, G., Stormont, C., Morris, B.G. and Grunder, A.A. (1969) Multiple electrophoretic forms of carbonic anhydrase in red cells of domestic cattle (*Bos taurus*) and American buffalo (*Bison bison*). *Genetics* 61, 823–831.

Schleger, W., Mayerhofer, G. and Pirchner, F. (1974) Relationship between heterozygosity as estimated from genetic markers and performance and average effects of marker genes. *Animal Blood Groups and Biochemical Genetics* 5 (Suppl.), 37 (abstract).

Schleger, W., Mayrhofer, G. and Stur, I. (1977) Relationships between marker gene heterozygosity and fitness in dairy cattle. *Animal Blood Groups and Biochemical Genetics* 8 (Suppl.), 42 (abstract).

Schröffel, J.A., Kubek, A. and Glasnak, V. (1970) Serum ceruloplasmin polymorphism in cattle. In: *Proceedings of the 11th European Conference on Animal Blood Groups and Biochemical Polymorphisms, Warsaw*, pp. 207–210.

Schwellnus, M. and Guérin, G. (1977) Difference between the HB C variants in Brahman and in indigenous Southern African cattle breeds. *Animal Blood Groups and Biochemical Genetics* 8, 161–169.

Seibert, B., Erhardt, G. and Senft, B. (1987) Detection of a new K-casein variant in cow's milk. *Animal Genetics* 18, 269–272.

Sellei J. and Rendel, J. (1970) A probable crossing-over between two B-alleles of cattle blood groups. *European Conference on Animal Blood Groups* 11, 115–116.

Shaw, D.H. and Stone, W.H. (1962) Time of appearance of antigenic factors on cattle erythrocytes. *Proceedings of the Society for Experimental Biology and Medicine* 111, 104–111.

Smithies, O. (1955) Zone electrophoresis in starch gels: group variations in the serum proteins of normal human adults. *Biochemistry Journal* 61, 629.

Smithies, O. and Hickman, C.G. (1958) Inherited variations in the serum proteins of cattle. *Genetics* 43, 374–385.

Solbu, H., Spooner, R.L. and Lie, O. (1982) A possible influence of the major histocompatibility complex (BoLA) on mastitis. *Proceedings of the Second World Congress of Genetics Applied to Livestock Production* 7, 368–371.

Soller, M. (1991) Mapping quantitative trait loci affecting traits of economic importance in animal populations using molecular markers. In: Schook, L.B., Lewin, H.A. and McLaren, D.G. (eds) *Gene-Mapping Techniques and Applications*. Marcel Dekker, New York, p. 21.

Soos, P., Stukovszky, J., Csontos, G. and Gippert, E. (1973) A further type of cattle serum transferrin. *Acta Veterinaria Academiae Scientiarium Hungaricae (Budapest)* 23, 303–305.

Spooner, R.L. (1967) Blood groups in animals and their practical application. with special reference to cattle. *Veterinary Record* 81, 699–708.

Spooner, R.L. and Maddy, A.H. (1970) Studies on ox red cell membranes: detection of red cell antigens in ox red cell protein. In: *Proceedings of the 11th European Conference on Animal Blood Groups and Biochemical Polymorphisms, Warsaw, 1968*, pp. 115–116.

Spooner, R.L. and Maddy, A.H. (1971) The isolation of ox red cell membrane antigens: antigens associated with sialoprotein. *Immunology* 21, 809–816.

Spooner, R.L. and Oliver, R.A. (1969) Albumin polymorphism in British cattle. *Animal Production* 11, 59–63.

Spooner, R.L. Cowpertwait, J. and Maddy, A.H. (1970) Ox erythrocyte agglutinability. 2. Differential agglutinability. *Vox Sanguinis* 18, 251–256.

Sprague, L.M. (1958) On the recognition and inheritance of the soluble blood group property 'Oc' of cattle. *Genetics*. 43, 906–912.

Stämpfli, G. and Ittig, H.P. (1983) FV blood group and haemoglobin type versus haematological and blood chemical parameters in young Swiss bulls. *Animal Blood Groups and Biochemical Genetics* 14, 181–189.

Stone, W.H. (1956) The J substance of cattle. III. Seasonal variation of the naturally occurring isoantibodies for the J substance. *Journal of Immunology* 77, 369.

Stone, W.H. (1962) The J substance of cattle. *Annals of the New York Academy of Science* 98, 269–280.

Stone, W.H. and Irwin, M.R. (1954) The J substance of cattle. 1. Developmental and immunogenetic studies. *Journal of Immunology* 73, 397–406.

Stormont, C. (1949) Acquisition of the J substance by the bovine erythrocyte. *Proceedings of the National Academy of Science* 35, 232–237 (abstract).

Stormont, C. (1950) Additional gene-controlled antigenic factors in the bovine erythrocyte. *Genetics* 35, 76–94.

Stormont, C. (1951) An example of a recessive blood group in sheep. *Genetics* 36, 577–578.

Stormont, C. (1952) The F–V and Z systems of bovine blood groups. *Genetics* 37, 39–48.

Stormont, C. (1954) On the genetics and serology of the B system of bovine blood groups. In: *Proceedings of the 9th International Congress of Genetics, Bellagio, 1953*, Vol. 2, pp. 1205–1206.

Stormont, C. (1959) On the applications of blood groups in animal breeding. *Proceedings of the 10th International Congress of Genetics, The Hague, 1959*, Vol. 1, pp. 206–224.

Stormont, C. (1961) Further expansion of the A system of bovine blood groups. *Federation Proceedings* 20, 66 (abstract).

Stormont, C. (1962) Current status of blood groups in cattle. *Annals of the New York Academy of Science* 98, 251–268.

Stormont, C. (1967) Contributions of blood typing to dairy science progress. *Journal of Dairy Science* 50, 253–259.

Stormont, C. (1972a) The language of phenogroups. *Haematologia (Budapest)* 6, 73–79.

Stormont, C. (1972b) The role of maternal effects in animal breeding. I. Passive immunity in newborn animals. [Review.] *Journal of Animal Science* 35, 1275–1279.

Stormont C. (1975) Neonatal isoerythrolysis in domestic animals: a comparative review. *Advances in Veterinary Science and Comparative Medicine* 19, 23–45.

Stormont, C. (1978) The early history of cattle blood groups. *Immunogenetics* 6, 1–15.

Stormont, C. (1981) The B and C systems of cattle revisited. In: Hildemann, W.H. (ed.) *Frontiers in Immunogenetics*. Elsevier–North Holland, New York, pp. 31–43.

Stormont, C. and Suzuki, Y. (1956) The 'D' system of bovine blood groups. *Journal of Animal Science* 15, 1208–1209 (abstract).

Stormont, C. and Suzuki, Y. (1960) On the 'J' classification of rabbits and production of anti-J in 'J-negative' rabbits. *Proceedings of the Society for Experimental Biology and Medicine* 105, 123–126.

Stormont, C. and Suzuki, Y. (1962) A possible duplication of the B locus in cattle. *Immunogenetics Letter* 2, 80–81.

Stormont, C., Irwin, M.R. and Owen, R.D. (1945). A probable series of alleles affecting cellular antigens in cattle. *Genetics* 30, 25–26 (abstract).

Stormont, C., Owen, R.D. and Irwin, M.R. (1951) The B and C systems of bovine blood groups. *Genetics* 36, 134–161.

Stormont, C., Miller, W.J. and Suzuki, Y. (1960) The convergence of the A–H , and D systems of bovine blood groups. *Genetics* 45, 1013 (abstract).

Stormont, C., Miller, W.J. and Suzuki, Y. (1961) The S system of bovine blood groups. *Genetics* 46, 541–551.

Stormont, C., Morris, B. and Gregory, P.W. (1964) On the origin of a new B allele in cattle. *Immunogenetics Letter* 3, 130–131.

Stormont, C., Morris, B.G. and Suzuki, Y. (1972) A new phenotype in the carbonic anhydrase system of cattle red cells. In: *Proceedings of the 12th Conference on Animal Blood Groups and Biochemical Polymorphisms, Budapest, 1970*, pp. 187–189.

Symons, M. and Bell, K. (1992) Canine plasma alkaline phosphatase polymorphism and its relationship with the canine Tr blood group system. *Animal Genetics* 23, 315–324.

Thiele, O.W., Krotlinger, F. and Ohl, C. (1975) Transfer of blood-group determinants from plasma onto erythrocytes. *Naturwissenschaften* 62, 586.

Thinnes, F., Geldermann, H. and Wens, U. (1976) New protein polymorphisms in cattle. *Animal Blood Groups and Biochemical Genetics* 7, 73–89.

Thompson, M.P., Kiddy, C.A., Pepper, L. and Zittle, C.A. (1962) Variations in the $alpha_s$-casein fraction of individual cow's milk. *Nature (London)* 195, 1001–1002.

Thymann, M. and Larsen, B. (1965) Milk protein polymorphism in Danish cattle. (In Danish, with English summary.) In: *Årsberetning*. Institut for Sterilitetsforskning, Copenhagen, pp. 225–250.

Todd, C. and White, R.G. (1910) On the haemolytic immune isolysins of the ox and their relation to the question of individuality and blood relationship. *Journal of Hygiene* 10, 185–195.

Tolle, A. (1959) [Principles and experimental results pertaining to relationships between blood group factors and heifer lactation.] Grundlagen und Untersuchungsergebnisse von Beziehungen zwischen Blutgruppenfaktoren und Faersenlaktation. In: *Report of the VIth International Blood Group Congress (Munich)*, p. 40.

Tsuji, S., Fukushima, T., Shiomi, M. and Abe, T. (1981) A new serum transferrin phenotype observed in Japanese Black Cattle. *Animal Blood Groups and Biochemical Genetics* 12, 299–305.

Tucker, E.M., Wright, L.J. and Varden, S.E. (1981) Monoclonal antibodies to blood group antigens on ruminant red cells. *Immunology Letters*. 3, 121–123.

Tucker, E.M., Metenier, L., Grosclaude, J., Clarke, S.W. and Kilgour, L. (1986) Monoclonal antibodies to bovine blood group antigens. *Animal Genetics* 17, 3–13.

Tucker, E.M., Clarke S.W. and Metenier, L. (1987) Murine/bovine hybridomas producing monoclonal alloantibodies to bovine red cell antigens. *Animal Genetics* 18, 29–39.

Van de Weghe, A., VanZeveren, A. and Bouquet, Y. (1982) The vitamin D-binding protein in domestic animals. *Comparative Biochemistry and Physiology* 73B, 977–982.

Voglino, G.F. (1972) A new beta-casein variant in Piedmont cattle. *Animal Blood Groups and Biochemical Genetics* 3, 61–62.

Walawski, K. and Prusinowska, I. (1981) Polymorphism of alkaline ribonuclease in the leucocytes of Black and White cattle. *Animal Blood Groups and Biochemical Genetics* 12, 167–169.

Wegrzyn J., Duniec, M.J., Duniec, M. and Trela, J. (1971a) Irregularities in the inheritance of alleles of B group system in cattle. *Genetica Polonica* 12, 471–475.

Wiener, A.S. (1944) The Rh series of allelic genes. *Science* 100, 595–597.

Woychick, J.H. (1964) Polymorphism in kappa-casein of cow's milk. *Biochemical and Biophysical Research Communications* 16, 267–271.

The Molecular Genetics of Cattle

D. Vaiman
Laboratoire de Génétique Biochimique et de Cytogénétique, Département de Génétique Animale, INRA, 78352 Jouy-en-Josas, France

Introduction	123
Molecular Anatomy of the Cattle Genome	124
General considerations	124
Basic constituents of cattle repetitive sequences	125
The New Molecular Tools Available for Studying the Cattle Genome	130
Large-insert DNA libraries	131
Production of genetic markers from definite chromosome fractions	133
Examples of Applications of Molecular Genetic Techniques to Cattle	135
Molecular markers and positional cloning	136
An example of gene structure and function: the casein locus	138
Molecular diagnoses in cattle embryos	144
Conclusions	147
Acknowledgement	147
References	148

Introduction

Genetics has been revolutionized in recent years by molecular biology techniques, making it possible to elucidate complex physiological functions at the gene level. Molecular genetics has enabled a wealth of information to be collected on the structure and function of complex eukaryotic genomes, mostly obtained from mice and humans. Comparisons between these two species indicate that, while mammalian genomes differ in their chromosome numbers and genomic organization, a high degree of similarity is often observed at the gene-sequence level. In contrast, global data about the specific features of cattle molecular genetics remain to be obtained. Obviously, while gene analyses in humans or mice are a prominent issue for public health and understanding of diseases, in domestic animals, genes are interesting when they are relevant for economically important functions. Although the tracking of genetic resistance to diseases is of major interest in animal farming, economically relevant loci will most of the time be connected to normal physiological functions of the healthy animal, such as reproduction, lactation and growth. This has led to the definition of economic trait loci (ETLs), which most probably will be searched for only in domestic animals. Thus, specific molecular tools have to be built to analyse the genome of these species.

To address the different aspects of cattle molecular genetics, this chapter will be divided into three sections.

- The specific features of cattle genomic organization will be presented, focusing mainly on bovine repetitive deoxyribonucleic acid (DNA).
- The molecular tools now available for molecular genetic studies in cattle will be described, with special emphasis given to the latest facilities available for positional cloning projects.
- Finally, applications of molecular genetics to the cattle genome will be outlined, and three applications will be further developed, i.e. the use of molecular genetics to map and identify genes and ETLs, the structural analysis of one of the most thoroughly studied loci in ruminants, the casein complex, and the possibilities raised by molecular genetic techniques for performing antenatal diagnoses of sex and transgene integration in embryos.

Molecular Anatomy of the Cattle Genome

General considerations

The cattle genome is composed of 29 pairs of autosomes and two sex chromosomes. All chromosomes except for the X chromosome are acrocentric, which makes their identification difficult and even impossible without banding. Numerous abnormalities, especially Robertsonian translocations, involving nearly all the chromosomes, have been reported in the literature, as for instance in the 14;20, 1;29, 21;27, 4;8 (for a review see Logue and Greig, 1985; Popescu, 1990), and are very helpful for identifying the chromosomes or for automatic flow sorting. In cattle, the total amount of DNA per diploid cell is evaluated at 6×10^{-12} g. This is similar to what is generally found in mammalian cells.

Like the genome of almost all vertebrates (with the notable exception of tetrodontoid fish (Brenner *et al.*, 1993)), the genome of ruminants, and among them cattle, is replete with various repetitive sequences, displaying different characteristics and origins. In this respect, it was one of the first genomes studied by analysing its renaturation in liquid phase (Britten and Kohne, 1968). These pioneer studies revealed that at least 50% of the cattle genome is composed of repetitive sequences, about half of which are highly repeated. The biological role of these non-coding elements is no longer downplayed, as repetitive sequences may play an essential part in the exon shuffling processes, which leads to the evolution of new functions by randomly assembling exons corresponding to protein domains (Dorit *et al.*, 1990). In vertebrates, non-coding repetitive sequences, which are far more numerous than coding repetitive sequences, have globally been categorized into two broad classes: (i) tandem repeats, mainly satellites, minisatellites or microsatellites, according to the size of the repeated unit, but also telomeric sequences; and (ii) dispersed elements. Apparently, no specific characteristics have been observed in cattle telomeric repeats. As for all the vertebrates studied up to now, they consist of arrays of the TTAGGG hexanucleotide, as shown by *in situ* hybridization (Meyne *et al.*, 1989). This has also been confirmed by the primed *in situ* synthesis (PRINS) technique (Gu and Hindkjær, 1996).

Eukaryotic genomes also contain coding repetitive sequences, such as ribosomal ribonucleic acid (RNA) genes, transfer RNA (tRNA) genes or multigene families. In the latter, up to dozens of sequences can be derived from an ancestral gene by gene duplications. Numerous examples of gene families are found in the mammalian genomes, such as the globin, lysozyme or immunoglobulin family, generated by tandem gene duplication followed by independent mutation events, leading to new functions. In gene families, some derived functions happen to be only distantly related to the original function of the ancestral polypeptide. Tandem gene duplications can also lead to the formation of pseudogenes, degenerated gene copies which accumulate non-sense mutations, frame shifts or stop codons, and to a rapid function loss. A typical bovine chromosome is composed of all these diverse elements and can be schematically represented as in Fig. 6.1.

Basic constituents of cattle repetitive sequences

Short interspersed elements, long interspersed elements, microsatellites and short interspersed element-associated microsatellites

In eukaryotes, RNA is occasionally reverse-transcribed in DNA, which is subsequently integrated into the chromosomes. Such integrated transposed elements are called retroposons (Rogers, 1985; Weiner *et al.*, 1986). Different highly repetitive eukaryotic DNA elements – short interspersed elements (SINEs), long interspersed elements (LINEs) and processed retropseudogenes – belong to this broad category. In most cases, but with the noticeable exception of *Alu* repeats in primates and the B1 family in rodents, SINEs from different species, including plants, are derived from tRNAs (Okada and Ohshima, 1995). While SINEs are short (less than 400 bp), LINEs, in contrast, are several kilobases long and contain two open reading frames (Scott *et al.*, 1987), one of which encodes a reverse transcriptase (Mathias *et al.*, 1991). Mammalian genomes are pervaded by LINEs, with about 100,000 elements in humans (Hwu *et al.*, 1986). Most of them, however, are truncated and are

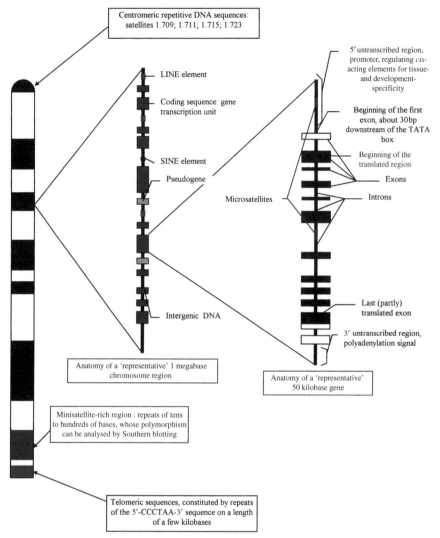

Fig. 6.1. Schematic representation of a 'typical' acrocentric mammalian R-banded chromosome. SINEs and LINEs can be present in the intergenic regions as well as inside the genes.

thus functionally inactive. Long terminal repeat (LTR)-type retrotransposons (derived from LINEs by addition of LTR elements) and retroviruses (derived from LINEs by gaining *env* genes) (Eickbush, 1992; Eickbush *et al.*, 1995) are also included in reverse transcriptase-containing retroposons. One example of such a retrovirus is the jaagsiekte sheep retrovirus (JSRV), which is also found in the genome of various ungulates and mammals from different orders (Hecht *et al.*, 1996). Retroviruses are integrated in the host genome, but structural defects render them unable to replicate; they can, however, be induced to replicate by viral infections.

At the sequence level, the distinction between satellite and dispersed DNA elements in the bovine genome is not clear-cut, because the same common set of elementary construction units comprises either dispersed or satellite bovine repetitive DNA (Zelnick et al., 1987; Lenstra et al., 1993). Among these basic blocks, there is one key 115 bp DNA element called Bov-A in the nomenclature of Lenstra et al. (1993), but previously designated BCS (Spence et al., 1985) and BMF (Zelnick et al., 1987). This DNA element is never found alone, but is either duplicated (and called in this case Bov-A2) or associated with a 73 bp pseudogene of tRNA^{Gly} (called in this case Bov-tA). Bov-A2 and Bov-tA SINEs account for 1.8 and 1.6% of the bovine genome, respectively (Lenstra et al., 1993). Another frequent (0.5% of the genome) dispersed element (Bov-B) corresponds to a 560 bp *PstI* repeated band visible after digestion of total bovine genomic DNA. This element was first called art-2 (Zelnick et al., 1987) and classified as a bovine SINE (Lenstra et al., 1993). Szemraj and co-workers demonstrated that the complete element, now called Bov-B (3.1 kb), encompasses an open reading frame that encodes a reverse transcriptase (Groenen et al., 1993; Szemraj et al., 1995). Thus, Bov-B falls into the LINE category. Lenstra and co-workers showed that Bov-A and Bov-B share a homologous region corresponding to about one-half of Bov-A. Furthermore, the other half of Bov-A is present at the other extremity of the complete 3.1 kb Bov-B element (Szemraj et al., 1995). Recently, a model has been proposed to account for these sequence similarities (Okada and Hamada, 1997). In this model, each half of the Bov-A element would contain a specific activity: a promoter-like activity for the 5' half, while the 3' half would contain a target for the reverse transcriptase. The Bov-A element in this case would result from the complete deletion of the central part of Bov-B. The very frequent duplication of Bov-A to form Bov-A2 could have been selected because of a more efficient promoter activity. As well as Bov-B, other LINEs have been found in the bovine genome, such as BATPS, located in the second intron of the adenosine triphosphate (ATP) synthase gene (Dyer et al., 1989). Part of this LINE has also been recognized in an α-lactalbumin pseudogene (Vilotte et al., 1993).

These different repeated elements are terminated by particular tandem repeats, called microsatellites (Hamada and Kakunaga, 1982; Tautz and Renz, 1984). Microsatellites or simple sequence repeated polymorphisms (SSRPs) are iterations of a simple sequence motif composed of 1–5 bp. The most frequent in mammalian genomes are poly-A tracts, probably originating from retrotransposition events (Aitman et al., 1991). However, poly-(TG/CA), the second most numerous, are the best-studied microsatellites, because the variability of their number of tandem repeats renders them incredibly powerful genetic markers. Microsatellites of the TG/CA type have considerably boosted the construction of genetic maps in mammals. Together with their very high level of polymorphism (ten alleles or more often being observed in a given population), prominent characteristics of TG/CA microsatellites are their number (several tens of thousands, the actual number varying according to the species under study), their relatively even distribution along the chromosomes and their easy polymerase chain reaction (PCR) analysis.

As detailed previously, in the cattle genome, the 3' half of the Bov-A element is always terminated by a microsatellite sequence, meaning that many constituents of bovine dispersed elements (Bov-A2, Bov-tA and Bov-B) are closely associated with a powerful genetic marker. Reciprocally, it has been estimated that 45% of the bovine microsatellites are associated with SINEs (Kaukinen and Varvio, 1992; Vaiman et al., 1994). This high level of association between microsatellites and SINEs seems to be a characteristic of Bovidae (in contrast, only 12–24% of the pig microsatellites are associated with previously described SINEs (Miller and Archibald, 1993; Alexander et al., 1995)). In the study of Alexander et al. (1995), one new microsatellite-associated porcine SINE (antiodactyl repetitive element (ARE)) has been identified and shown to share homology with bovine and ovine counterparts. The retroposition of these SINEs therefore predates the Suidae/Bovidae radiation (over 40 million years).

This SINE/microsatellite association has been used to design SINE-PCR primers (similar to *Alu*-PCR primers in humans), which can then be used in PCR reactions to amplify ruminant genomic DNA in a non-specific fashion. This has meant that new microsatellites could be isolated from sorted chromosomes (Band et al., 1997) or somatic cell hybrids (Kostia et al., 1997). One of the particularities of the ruminant genetic maps (Crawford et al., 1995; Vaiman et al., 1996b; Barendse et al., 1997; Kappes et al., 1997), when compared with the mouse, pig or human map, is the apparent non-clustering of microsatellites. In the non-ruminant species, the distribution of markers is not as even as could be expected for homogeneously covering the genome, and, often, clumps of microsatellites are found that are apparently inseparable by recombination. Microsatellites could be derived from degenerated CpG islands, which are genomic regions that are also called *Hpa*II tiny fragment (HTF) islands and are particularly rich in C and G nucleotides. These islands, often located in the vicinity of housekeeping gene promoters, are classically clustered in mammalian genomes (Ohno and Yomo, 1991). In contrast, the relatively even distribution of microsatellites observed in cattle, as well as in other ruminants, could be explained by the high proportion of SINE-associated microsatellites in these species. A given SINE-associated microsatellite could be subjected to the same genome dynamic as the SINE and to a rapid shuffling by retrotransposition over the whole genome.

Centromeric satellite deoxyribonucleic acid
Description   The other elementary units of bovine repetitive DNA are three short DNA elements of 23 bp, 23 bp and 31 bp (Pech et al., 1979). While no obvious sequence similarity can be observed in these elements, a common origin by duplication or triplication of an original 12 bp DNA stretch has been hypothesized (Jobse et al., 1995).

It has been shown in many organisms that the constitution of active kinetochores is made possible by the presence of heterochromatic repetitive DNA sequences (in large amounts in ruminants, as revealed by C-banding techniques, which specifically label centromeres). Satellite DNAs are essential elements of the centromere, the chromosome region enabling the correct attachment of achromatic spindle fibres during cellular division (Pluta et al., 1995). In primates, the main constituent of centromeric DNA belongs to the alphoid family, repetitions of a more or less altered basic unit of about 170 bp, displaying a species-specific variation. In ruminants, sequence conservation of satellite DNA also seems to be restricted to one or a few closely related species, as revealed by the relatively inefficient interspecific hybridization of bovine satellite sequences to sheep or goat metaphases. The very weak interspecific conservation indicates that the precise sequence of centromeric satellites does not seem to be very relevant for the correct biological function. However, a minimum number of repeats is almost certainly indispensable in the structure of this chromosome region.

Bovine satellites represent more than a quarter of the bovine cell DNA content. The analysis of bovine satellites has evolved with the available techniques. They were first separated by density gradient centrifugation as early as 1978 (Macaya et al., 1978). This made it possible to distinguish eight major different satellites, according to their position in the gradient (1.706, 1.709, 1.711a, 1.711b, 1.715, 1.720a, 1.720b and 1.723 g cm^{-3}) and 11 minor components, accounting for 23 and 4% of the genome, respectively. The first to be completely sequenced was the 1.706 component, representing 4.6% of the genome (Pech et al., 1979), and the study showed that the long unit of 2350 bp is actually composed of complex and altered repetitions of a very simple 23 bp motif, consisting of two basic related oligonucleotides of 12 and 11 bp.

Jobse et al. (1995) have clearly classified the different satellite sequences, and have demonstrated that some of them share similar sequences. The most simple component, i.e. the 1.715 (a basic unit of 1402 bp (Pluciennicak et al., 1982), 5.1% of the bovine genome, corresponding to about 110,000 copies), is only composed of repeats of the basic 31 bp element (also shared with other artiodactyls, such as cervids (Lee and Lin, 1996)). Two different satellites have been identified at 1.711 g cm^{-3} and for this reason are known as 1.711a and 1.711b (Macaya et al., 1978). A part of the 1.711b satellite is identical to the 1.715 satellite but it also contains a 1198 bp insertion, called INS-1.711b, closely related to retroviral LTR elements. The 1.711a satellite (a basic unit of 1360 bp, comprising 1.7 % of the genome (Kopecka et al., 1978)) contains a sequence called INS-1.711a (602 bp long), which shares some parts with INS-1.711b (bases 116 to 358 and 1041 to 1198). The rest of 1.711a is composed of 23 bp repeats.

The 1.723 component (0.5% of the genome, basic unit of 645 bp) is not related to the other satellites and does not contain simple short repeated units. The very frequent 1.709 component (4.3% of the genome) consists of a basic unit of 3200 bp (Skowronski et al., 1984). This last satellite displays a very peculiar structure, containing the two different dispersed elements Bov-A2 and Bov-B. The presence of a $(CA)_n$ microsatellite between these two elements resulted in the frequent unwanted isolation of the microsatellite contained in the 1.709 satellite sequence while screening for anonymous microsatellites (Vaiman et al., 1992).

Chromosomal localization   The chromosomal localization of several bovine satellites has been studied by fluorescence *in situ* hybridization (FISH) (Modi *et al.*, 1993). Strong hybridization signals have been observed on all the autosomal centromeres for 1.715. In contrast, 1.711a and 1.709 hybridize strongly to the majority, but not all, of the chromosome. In the same study, however, 1.711b-INS was shown to hybridize to interstitial chromosome regions. As could be foreseen, hybridization of the BovB LINE only revealed an interstitial signal.

Evolutionary perspective   The comprehensive work of Jobse *et al.* (1995) elucidated the relationship between different repetitive elements in even-toed ruminants. These authors searched for these repeats by Southern blot hybridization in Bovidae (cattle, buffaloes and goats), Cervidae (deer), Giraffidae (giraffe), Tragulidae (chevrotain), as well as outside the Ruminantia in Tylopoda (camel) and Suidae (pig). The most ancestral motifs – Bov-A, Bov-B, the 31-mer motif, as well as the 1.723 ancestor – predate the divergence of ruminants (30 to 50 million years ago). The 1.715 satellite is restricted to the members of the Bovidae family, while the other satellites result in clear hybridization patterns only in Bovinae. The well-known sequence similarities between bovine satellites illustrate the dynamic and complex evolution of eukaryotes, with some elementary repeated motifs early duplicated and shared by many species and some lately duplicated, constituting the species-specific centromeric satellites.

The New Molecular Tools Available for Studying the Cattle Genome

The 1990s have seen the development of many new tools for the molecular analysis of complex eukaryotic genomes. It is now evident that the aphorism implying that what is true for bacteria is also true for elephants is widely inaccurate. Indeed, the degree of complexity of mammalian genomes is three orders of magnitude greater than that of bacteria. The intricacy of introns and exons, the multitude of repetitive sequences, the centromeric and telomeric structures have all proved to need precisely adapted tools in order to be studied. One of the major challenges of eukaryotic genetics is how to identify all the genes needed to make a multicellular organism work. This work is now in progress, with the projects of massive systematic sequencing of complementary DNA (cDNA) in humans (Adams *et al.*, 1992; 1993; Weinstock *et al.*, 1994; Korenberg *et al.*, 1995). In databases, lists of human and mouse coding sequences are now available, making it possible to identify many anonymous exonic sequences (Rawlings and Searls, 1997). These lists undoubtedly benefit the whole community of mammalian geneticists, as they make it possible to practically exploit the interspecific similarity of many coding sequences (Nadeau *et al.*, 1995; Lyons *et al.*, 1997). However, a function has been assigned to only about 6000 genes among the estimated 75,000 needed for the functioning of eukaryotic organism (8%). Therefore, the knowledge of a small fragment of the sequence of every gene, generated by the expressed sequence tags (ESTs) projects, is only a first step towards an overall understanding of transcription in mammalian genomes.

The identification of new genes is a major endeavour for animal geneticists, and, in most cases, the function studied in domestic mammals is only attainable by positional cloning techniques. Schematically, these approaches are divided into three successive steps.

- Primarily, the meiotic segregation of highly informative, evenly distributed genetic markers has to be studied by genotyping informative families, i.e. families originating from a heterozygous animal for the character under scrutiny. This first step should be pursued until a small recombination fraction (less than 2 centimorgans (cM)) is observed between the character and the marker.
- This makes it possible to reach the second step, involving the constitution of a series of overlapping DNA clones (contig), in which the gene will be contained.
- The third and last step necessitates the identification of coding sequences in this contig. At this step, comparative database information from humans or mice can be a spectacular short cut towards the identification of the implicated gene. To reach this step and be able to test candidate genes, specific molecular tools have to be used. All these tools now exist in cattle.

Large-insert DNA libraries

An essential phase of positional cloning programmes is the possibility of enumerating all the coding sequences present in a DNA region previously delineated by genetic mapping. Genetic localizations have an accuracy of around 1 cM, although this precision level can be improved if the informative families used for discovering the linkage are very large. A 1 cM genetic map distance corresponds approximately to an average of 10^6 bp of DNA. This relationship can vary according to the sex, the chromosomal region studied, and the species. In cattle, the autosomal genetic size has recently been estimated at about 3500 cM (Barendse *et al.*, 1997), for a total physical size of 3.3×10^9 bp, showing that 1 cM satisfactorily approximates to 10^6 bp in this species.

Until 1989, the largest DNA insert that could be cloned was limited to about 45 kb, inserted in cosmid vectors (Meyerowitz *et al.*, 1980). Consequently, to cover a million base-pair region with overlapping DNA fragments, a very large number of cosmid clones had to be aligned (70 on average to obtain a 95% probability of the presence of every sequence of the contig). The development of yeast artificial chromosome (YAC) vectors at the end of the 1980s was therefore a considerable boon in analysing pertinent chromosome regions. The YAC vectors contain all the DNA elements necessary for propagation in the yeast cell: a centromere, telomeres and eukaryote origin of replication. Furthermore, they contain three selectable markers, one on each vector arm, and one which is interrupted by the cloning site. The YAC vectors have been shown to accommodate up to 2000 kb of foreign DNA. They are largely used to perform long walks along mammalian chromosomes (Silverman et al., 1989), to construct physical contigs (see, for example, Antoch et al., 1997; Carstea et al., 1997; Chandrasekharappa et al., 1997) and to analyse the fine structure of large genes. Libraries of YACs have been constructed in several species, including humans (Brownstein et al., 1989; Traver et al., 1989; Anand et al., 1990; Larin et al., 1991), mice (Larin et al., 1991; Chartier et al., 1992), invertebrates such as nematodes (Coulson et al., 1988) and Drosophila (Garza et al., 1989). In cattle, at least two YAC libraries have been constructed (Libert et al., 1993; Smith et al., 1996), corresponding to six and five genomic equivalents, respectively. In both cases, the average insert size comprises between 700 and 750 kb. A complementary resource is constituted by a sheep YAC library of 29,000 clones (Broom and Hill, 1994), which could be used to find genes absent from the bovine YAC libraries.

One major drawback of YAC cloning is the relatively frequent occurrence of chimeric genomic inserts (Banfi and Zoghbi, 1996). Chimerism most probably occurs by recombination rather than co-ligation (Wada et al., 1994). The level of chimerism of the bovine YAC libraries was estimated at 18–21% for one of the libraries (Smith et al., 1996) and at 33% for the other library (Libert et al., 1993). The high frequency of non-contiguous DNA clones, together with the fact that yeast manipulations are time-consuming, prompted some researchers to develop large-DNA insert cloning systems in Escherischia coli, based on the F factor. This resulted in the development of bacterial artificial chromosomes (BACs) (Shizuya et al., 1992), or P1-derived artificial chromosomes (PACs) (Ioannou et al., 1994). They accept much shorter inserts than YACs, with an average size limited to about 150 kb, although some recent results suggest that a larger insert size could be achieved (Zimmer and Verrinder Gibbins, 1997). On the other hand, the libraries constructed in BACs or PACs display a very low level of chimerism (less than 5%), and the DNA preparation is extremely simplified, consisting of a slightly modified plasmid purification. These advantages have made BACs the most appropriate candidates for the large-scale sequencing projects under way in several chromosome regions. In cattle, a BAC library has been constructed (Cai et al., 1995) and is now starting to be exploited (Ponce de Leon et al., 1996; Sun et al., 1997). This library is composed of 23,000 clones. A complementary resource for studying the ruminant genomes is constituted by a 64,000-clone goat BAC library (Schibler et al., 1998a). For both these libraries, the level of chimerism was proved to be very low. They thus constitute adequate molecular tools for positional cloning projects.

Unlike cosmid libraries, a reasonably complete BAC or YAC DNA library can consist of a rather limited number of clones (some tens of thousands). This makes it relatively easy to handle and individually grow all the clones from the library, and to array them in a superpool and pool format. Thus, the screening of the library can be performed by PCR on the pool DNA, and with a three-dimensional arrangement of the library. The address of a specific clone in the library can be obtained in less than 100 PCR reactions (Barillot *et al.*, 1991; Strauss *et al.*, 1992).

Large-insert DNA libraries for different mammalian species such as humans, mice, cattle, pig, goat and sheep, constitute enormously promising resources for the comparative and evolutionary analysis of their genomes. In mammals, while the total amount of DNA is fairly constant from one species to another, differences in karyotypes can be explained by translocations of relatively long chromosomal segments. Therefore, such libraries will initially make it possible to identify the break points differentiating the chromosomes of groups separated 70 million years ago, at the fountain-head of the mammalian evolutionary radiation, to evaluate the dynamics of chromosomal segments and their preferential sites of fusion and fission, and to precisely compare centromeric and telomeric structures.

Production of genetic markers from definite chromosome fractions

With the development of fairly complete second-generation genetic maps for the bovine genome (Barendse *et al.*, 1997; Kappes *et al.*, 1997), it clearly appears that any gene or even quantitative trait locus (QTL) is now attainable by genetic analysis of segregation. The next challenge will be to get closer to the gene in question, permitting its final identification and cloning. The chromosomal genetic coverage attained in humans for the whole genome is justified by the fact that a mutation in any human gene is the putative cause of a human genetic disease. Such global map development would be very expensive in domestic animals, where only some definite regions will probably be thoroughly studied. Therefore, the possibility of enriching some precise chromosomal regions in genetic markers is particularly tantalizing. For this purpose, genetic markers have been successfully isolated from three distinct DNA sources: flow-sorted chromosomes, microdissected chromosomes and somatic cell hybrids.

Production from sorted chromosomes
It has long been presumed that sorting bovine chromosomes would constitute an impossible challenge (Dixon *et al.*, 1991). This was due to the similar DNA content of successive bovine chromosome pairs, the difference often being less than 5% (Cribiu and Popescu, 1974). However, progress in flow sorting technology has finally made it possible to sort most bovine chromosome pairs. The specificity of the sorting could be controlled by the chromosome painting technique. This technique can be used for homologous painting (Yerle *et al.*, 1993; Schmitz *et al.*, 1995), as well as for heterologous painting, using sorted chromosome probes from one species (mainly from humans) to 'paint' bovine, pig, cat, horse or seal chromosomes (Scherthan *et al.*, 1994; Hayes, 1995; Rettenberger *et al.*, 1995a, b; Solinas-Toldo *et al.*, 1995; Fronicke *et al.*, 1996; Goureau *et al.*, 1996; Raudsepp *et al.*, 1996; Richard *et al.*, 1996; Hameister *et al.*, 1997). This consists in isolating a peak of sorted chromosomes on the

flow karyotype, amplifying it in a non-specific manner by PCR (using either degenerate primers or non-specific PCR conditions), labelling it by the incorporation of a fluorescent dye and hybridizing it on to metaphases. With this technique, 22 peaks could be analysed from the cattle flow karyotype. In some cases, it was even possible to separate the two chromosomes from a single pair (Schmitz et al., 1995). Recently, this sorted material was used to isolate specific microsatellite markers from a peak of bovine chromosomes 1 + X (Vaiman et al., 1997a, b) and from bovine chromosome 25 (Band et al., 1997).

Production from microdissected chromosomes
Recently, the density of some genetic chromosome maps has been considerably increased by isolating microsatellites from microdissected bovine chromosomes. The technique has been very successful for chromosome 1 (Sonstegard et al., 1997b) and X (Ponce de Leon et al., 1996; Sonstegard et al., 1997a). It is based upon the specific PCR amplification of the microdissected chromosomes, previously digested to completion and to which oligonucleotide linkers have been ligated. The main drawback with the technique is the identification of the chromosome of interest, because most banding techniques imply the use of acetic acid fixation, inducing a depurination of the chromosomes and probably lessening the maximum insert size. It is still possible to easily scrape recognizable chromosomes, such as the X or one circumscribed arm of a translocated chromosome. For instance, the enrichment of the chromosome 1 map has been performed after microdissection of the long arm of a 1;29 chromosome from a cell-cultured line harbouring this translocation (Sonstegard et al., 1997b). In this case, the translocated chromosome is a large submetacentric chromosome, easily recognizable from the other autosomes, which are acrocentric, and from the submetacentric X chromosome. With the localization of new genes, it seems obvious that other translocations will be of great utility for the future specific enrichment of chromosome maps.

An alternative to the direct isolation of microsatellite markers from microdissected chromosomes is the scraping of a very limited number of chromosomes or chromosome bands followed by a random PCR amplification, using degenerate oligonucleotides (Viersbach et al., 1994; Goldammer et al., 1996). While most of the DNA fragments obtained by this method are probably too small to enable recovery of a large number of microsatellite markers, they could be used to screen genomic cosmid or large-insert DNA libraries. This strategy of hybridization has already been used to constitute chromosome-specific BAC DNA libraries for human chromosome 22 (Kim et al., 1994, 1995), and could be applied to the cattle genome.

Production from somatic-cell hybrids
There are at least three different rodent/cattle somatic-cell hybrid panels available (Heuertz and Hors-Cayla, 1981; Womack and Moll, 1986; Heriz et al., 1994). These tools have been widely used to help build the nascent genetic and physical bovine maps in the last few years (Georges et al., 1991; Guérin et al., 1994; Vaiman et al., 1994). Recently, it has been shown that PCR

amplification with different bovine specific primers corresponding to short dispersed elements (SINEs) from somatic cell hybrid DNA resulted in different clones, which could be screened for microsatellite sequences. This strategy was recently employed to improve the genetic map of cattle chromosome 11, from two somatic cell lines, 8.1C and 8.1R, containing either bovine chromosomes 9, 11 and 18 or bovine chromosomes 9 and 11 in the hamster background (Kostia et al., 1997). Such a technique would make it possible to increase the genetic density of markers using hybrids containing only one or a few bovine chromosomes. As the SINE sequence is species-specific, no rodent DNA is amplified. Moreover, as 45% of the bovine microsatellites have been shown to be associated with SINE sequences, this technique could theoretically yield many different markers.

In conclusion, at least three different technical approaches are now available to specifically produce bovine genetic markers from specific chromosomal regions. This constitutes a major advantage for getting close to interesting genes. With the large-insert DNA genomic libraries and the development of databases of coding sequences, all the elements seem to be in place for discovering and further analysing genes relevant to economic agricultural traits in cattle.

Examples of Applications of Molecular Genetic Techniques to Cattle

The power of molecular genetic tools is such that no field of agricultural research can escape from their potential applications. In cattle, many new uses of molecular genetics are dependent on molecular genetic markers. Gene- or chromosome-specific molecular markers make possible early embryonic sex identification and disease detection, such as the bovine leucocyte adhesion deficiency (BLAD) (Mirck et al., 1995). On the other hand, anonymous DNA markers, mainly microsatellites, have been shown to be abundant in the bovine genome (Fries et al., 1990), similarly to all other eukaryotes studied today (Hamada and Kakunaga, 1982; Tautz and Renz, 1984). They are efficiently used for individual identification and parentage analysis, population genetics, construction of linkage maps and reverse genetics analysis. Molecular genetics is also involved in the setting up of many genetic improvement programmes for farm animals, such as marker-assisted selection and transgenesis. It is not feasible to comprehensively develop all the applications of molecular genetics to cattle in this chapter. Only a few specific points will thus be further developed.

- The progress towards gene identification and cloning by reverse genetics in cattle will be described.
- The recent inputs of molecular genetics to the structural analysis of the very important casein gene family, will be discussed.

- Finally, the use of molecular genetic methods for identifying transgene-carrying embryos and embryo sexing will be presented.

Molecular markers and positional cloning

The reverse genetics approach, inconceivable not long ago, has been exceptionally fecund in recent years, with more than ten gene localizations in ruminants (Fig. 6.2). In cattle, the first genes to be mapped were the 'polled' gene, localized in the peri-centromeric region of bovine chromosome 1 (Georges *et al.*, 1993a; Schmutz *et al.*, 1995; Harlizius *et al.*, 1997), and the gene responsible for the 'weaver disease' (progressive degenerative myeloencephalopathy), mapped to chromosome 4 (Georges *et al.*, 1993b), which has been shown to be associated with a gene controlling milk production (Hoeschele and Meinert, 1990). After these first successes, protocols were designed for identifying milk ETLs in cattle. The structure of the pre-existing selection schemes in dairy cattle was particularly adapted to the so-called 'granddaughter design' (Weller *et al.*, 1990). It is based upon the fact that, in dairy farming, a few males produce many male offspring, the genetic value of which can be very precisely calculated from the genetic value of their numerous daughters. It is thus possible to study the allelic segregation of genetic markers in the two-generation families constituted by grandfathers and fathers of cows. This approach has made it possible to identify a locus responsible for the quantity of milk and protein on chromosome 21, linked to the ETH131 marker (Ron *et al.*, 1994). Using the same design, the cattle genome was subsequently systematically screened by Georges and co-workers for ETLs involved in milk production. This pioneer work has provided strong statistical evidence (Logarithm of the odds (LOD) scores > 3.0) for the presence of five regions influencing different features of milk production on cattle chromosomes 1 (milk and protein yield), 6 (fat percentage, protein percentage, milk yield), 9 (fat yield and protein yield), 10 (fat yield) and finally 20 (protein percentage) (Georges *et al.*, 1995). Three other genes were mapped in cattle: the White Heifer disease locus (Charlier *et al.*, 1996a), the muscular hypertrophy gene found in the Blanc-Bleu Belge breed (Charlier *et al.*, 1995; Dunner *et al.*, 1997) and the syndactyly locus (Charlier *et al.,* 1996b).

These impressive results should not mask the achievements obtained in other ruminant species thanks to the overall genomic conservation of the three major ruminant species of economic interest (Hediger *et al.*, 1991), and to the possibility of amplifying by PCR a large proportion of genetic markers developed for cattle in sheep and goats (Moore *et al.*, 1991; Pépin *et al.*, 1995). The cross-species conservation of microsatellites, although imperfect, made it possible to map various genes in sheep and goats (Montgomery *et al.*, 1993, 1994, 1996; Cockett *et al.*, 1994, 1996; Vaiman *et al.*, 1996a).

The tremendous speed of ruminant gene localizations indicates that a new epoch has begun in research in animal genetics. Numerous programmes are under way to achieve the next steps, i.e. the cloning of these new genes. As

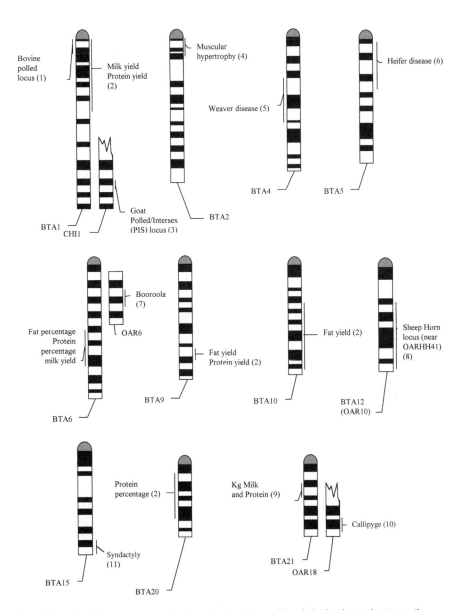

Fig. 6.2. The different gene localizations obtained by positional cloning in ruminants on the G-banded karyotype. BTA refers to bovine chromosomes, OAR to ovine chromosomes and CHI to goat chromosomes. The number in parentheses refers to the first reported localization in the literature: (1) Georges *et al.*, 1993a; (2) Georges *et al.*, 1995; (3) Vaiman *et al.*, 1996a; (4) Charlier *et al.*, 1995; (5) Georges *et al.*, 1993b; (6) Charlier *et al.*, 1996a; (7) Montgomery *et al.*, 1993; (8) Montgomery *et al.*, 1996; (9) Ron *et al.*, 1994; (10) Cockett *et al.*, 1994.

we have seen before, all the tools necessary to attain this goal already exist, and, in the next few years, we shall certainly see the first genes molecularly characterized in cattle by positional cloning.

Once a locus is identified as being of particular importance for agriculture, molecular genetics makes it possible to carry out fine-scale analysis. This has been the case in the molecular analysis of the casein family, a gene complex encoding the four major ruminant milk proteins.

An example of gene structure and function: the casein locus

This chapter will focus on the structural aspects of the casein locus as a specific example of chromosome structure in a well-studied gene-family region. The more physiological aspects will be treated further in the relevant chapter.

In mammals, milk is the only source of nutrients for the newborn. In a young mammalian organism, a highly energetic protein and mineral supply is necessary to make the rapid growth of the young warm-blooded animal possible. A large amino acid supply is required for constructing the body proteins, while minerals, such as calcium and phosphorus, are essential for building and consolidating the growing skeleton, as well as for the synthesis of cellular nucleic acids. Therefore, milk composition is of crucial importance for the survival of the young mammal, even though huge variations in milk compositions are frequently encountered in different mammalian orders. For instance, some mammals living under extreme conditions, such as whales or seals, (Nowak, 1991), need a milk which is very rich in fatty components (30% for seals compared with 3–4% in cow's milk).

Economic reasons led to the thorough study of the artiodactyl milk long before the development of molecular biology techniques. Since 1990, the development of molecular probes, PCR analysis of molecular polymorphisms, large-insert DNA libraries in BACs or YACs, and long-range analysis of DNA fragments has made it possible to study the fine molecular structure of the major milk-protein family: the casein family. This study has been performed recently (Rijnkels *et al.*, 1997a, b, c) in different species (human, mouse, cattle), and a comparative analysis gives interesting evolutionary insights into the evolution of the gene family. The structure and function of the four genes composing the ruminant casein locus provide a good example of concerted physiology. Many aspects have been studied, such as ontogenetic and tissue-specific regulation, evolution of new genes and gene function by gene duplication, exon skipping, alternative splicing and distant regulatory control elements.

Characteristics of cattle caseins
First evidence of genetic linkage: existence of a casein gene cluster   Four different casein genes have been found in cattle (and also in goats and sheep), and the primary structure of these four proteins was elucidated as early as the end of the 1970s (Mercier *et al.*, 1971, 1973; Ribadeau-Dumas *et al.*, 1972; Brignon *et al.*,

1977). The high level of polymorphism of these four genes made it possible to identify multiple variants, arising from genetic polymorphism; some differ in their phosphorylation patterns, resulting from amino acid substitutions. Different post-translational modifications not due to genetic polymorphisms are also observed, such as partial O-glycosylation in the κ-casein and partial phosphorylation in αs1- and αs2-caseins. The protein variants could be used for analysing the Mendelian segregation of the different genes. The first very tight linkage between αs1- and β-casein was demonstrated in 1964 (Grosclaude *et al.*, 1964). Later, it was shown that the four caseins were clustered in a single chromosome locus (Grosclaude *et al.*, 1978). This locus was mapped on cattle chromosome 6q31-33 (Threadgill and Womack, 1990). There has been a long-standing polemic about which chromosome was the casein chromosome in domestic ruminants because of the resemblance of cattle chromosomes 4 and 6. But recent cattle chromosome nomenclature standardization eventually brought this controversy to an end (Popescu *et al.*, 1996). Similarly, the casein locus has been mapped on chromosome 6 in goats and sheep, homologous to bovine chromosome 6 (Hayes *et al.*, 1992; Broad *et al.*, 1994). Recent progress in mapping techniques could specify the chromosome band 6q32 (Hayes *et al.*, 1993). Similarly, isolation and hybridization of a large-insert clone from a goat BAC library could demonstrate the same localization in goats (Schibler *et al.* 1998b).

Protein and DNA sequences of caseins: casein origin The primary amino acid sequence of the four caseins made it possible to screen cDNA libraries and obtain the complete sequences of the bovine transcripts corresponding to αs1- and κ-casein (Stewart *et al.*, 1984), and then to αs2- and β-casein (Stewart *et al.*, 1987). Sequence comparisons with homologous caseins from other species could then be performed, even though not all four caseins are present in all mammalian species. Sequence comparisons revealed, for instance, that, despite many base substitutions observed between rat and bovine β-casein, it is essentially the same polypeptide (Stewart *et al.*, 1987), while similar observations have been made for κ-casein. In contrast, for αs1- and αs2-caseins, many points of divergence are found between rat and bovine sequences, with a high rate of point mutation on the one hand, and several insertion/deletion events on the other. Basically, these two contrasting observations can probably be related to the fact that β- and κ-caseins have been found in all eutherian milks studied (Jenness, 1979), presumably demonstrating a very essential role of these two proteins in milk assimilation by the newborn mammal. The three calcium-sensitive caseins (αs1, αs2 and β, which precipitate at low calcium concentrations) were very early supposed to be derived from the duplication of a unique ancestral gene. However, this could not be deduced from most of their coding sequences, which are in fact highly divergent. The more revealing fact was that they share similar multiple phosphorylation sites, together with the high sequence similarity observed in their signal peptides (Gaye *et al.*, 1977; Mercier, 1981). On the other hand, κ-casein is largely different from the other three caseins, suggesting a different evolutionary origin. When searches for similarities are performed, it is found to be related to the γ chain of fibrinogen (FGG), (Jollès *et al.*, 1974).

The three calcium-sensitive caseins are maintained in micelles by interactions with the κ-casein, which seems to stabilize the micellar structure (Waddy and Mackinlay, 1971). The formation of curd (due to micelle aggregation) is governed mainly by β- and κ-casein, while the others two caseins probably determine the ability of the casein micelles to transport colloidal $CaPO_4$ (Stewart *et al.*, 1984, 1987).

After identification and sequencing of the cDNAs encoding the different caseins, studies were pursued at the gene level. Bovine casein genes were successfully cloned and sequenced for β-casein (Bonsing *et al.*, 1988), αs1-casein (Koczan *et al.*, 1991), αs2-casein (Groenen *et al.*, 1993) and κ-casein (Alexander *et al.*, 1988).

Common regulatory sequences important in casein expression have been found in the 5′ region of the calcium-sensitive caseins, making it possible to define a 'milk-box' encompassing a 18 bp motif (YNCCYYAGAATTTYTNRR) around −140 bp, of which 12 bp (AGAA...ATTTYCTA) are repeated at −110 bp before the starting-point of the transcription (Schmitt-Ney *et al.*, 1991). Between −90 and −50 bp, a 14 bp sequence (GANTTCTTRGAATT) is the putative binding site for a nuclear factor specific for the mammary gland (Watson *et al.*, 1991). This last sequence is common to other milk proteins (α-lactalbumin, β-lactoglobulin and whey acidic protein). At −30 bp, less specific *cis*-binding sites are found, with a Simian virus 40 (SV40)-type enhancer and a nuclear factor octamer-1 (NF Oct.1) site. Most of the data were obtained on β-casein, because the size of this gene facilitates the handling of the complete transcription unit and its microinjection, completely or in part as a transgene. However, the sharing of common motifs between αs1-, αs2- and β-casein genes, as well as their physical closeness in the same chromosome region, is strongly in favour of a concerted regulation of these three genes. Knowledge about the casein promoters is of utmost importance for producing transgenic animals in which expression of the synthesized peptide is targeted to the mammary tissue.

Polymorphisms of the ruminant caseins
An example of polymorphism at the gene level in a ruminant casein Like many other eukaryotic genes, casein genes, in particular αs1 and αs2, present a complex exon/intron structure, with mainly short exons sharing at least partial sequence similarity. The average exon size for these two caseins is below 60 bp, with 19 and 16 exons, respectively, in the transcription unit (Mercier and Vilotte, 1993), ten of which are less than 30 bp long. This illustrates the fact that casein genes are extremely fragmented genes. Moreover, the sequence relatedness of some exons suggests the possibility that unequal crossing- over can lead to the appearance of new variants, which occurs frequently (Ng-Kwai-Hang and Grosclaude, 1992). This high level of allelic variation has been particularly studied in αs1-casein in the goat species (Grosclaude *et al.*, 1987; Brignon *et al.*, 1989). The structure of the bovine polypeptide is similar to that in goats, with

199 amino acid residues for the most common allele. In contrast, the ovine polypeptide is eight residues shorter, and only one type of polypeptide is found (Mercier et al., 1985). In the goat, seven allelic forms are found, and the analysis of transcripts from casein allele F revealed the occurrence of at least ten different RNA types arising from exon skipping (Leroux et al., 1992). Single nucleotide insertions, insertions of nucleotide stretches, or deletions have been shown to be responsible for these exon skipping events. Comparison of differentially spliced variants of goat αs1-casein allele F revealed that exons 13 and 16 (with 24 nucleotides each) are skipped in this form. Interestingly, these two exons are constitutively skipped in sheep (Passey et al., 1996). As these two exons display over 58% sequence similarity at the DNA level, and taking into account the weak conservation of the casein sequences, it can be relatively confidently assumed that they are derived from a single ancestral exon. These two exons are also very similar to the fourth exon of the bovine or sheep β-casein gene, although the latter is one codon larger (27 nucleotides instead of 24). Globally, these results substantiate the evolutionary relationships between the different calcium-sensitive caseins, and support the hypothesis of a common evolutionary rooting from a unique ancestral gene (Mercier et al., 1971; Gaye et al., 1977; Jones et al., 1985). The results regarding αs1-casein were obtained in goats, and the same association between protein variants and protein quantity in the milk was not observed in cows. However, the variant A of bovine αs1-casein could be recovered from a mammary gland cDNA library constructed from a cow homozygous for αs1-casein variant B. The difference between the two variants was shown to be due to an internal 13 amino acid deletion (McKnight et al., 1989). This result clearly illustrates that unusual splicing events sometimes occur during the maturation of casein messenger RNA (mRNA), and have the potential to generate new variants.

Polymorphism of non-coding DNA inside the casein locus The existence of specific probes for the four caseins made it possible to study their polymorphism by restriction fragment length polymorphism (RFLP) (Seibert et al., 1985; Levéziel et al., 1991; Schlieben et al., 1991; Miranda et al., 1993) and to reveal the existence of numerous allelic variants. Later, two different microsatellite sequences were described in the ruminant caseins: one in cattle in the third intron of κ-casein (Alexander et al., 1988; Moore et al., 1992), and another one in the 3′ untranslated region of the goat β-casein gene (Pépin et al., 1994). This microsatellite is conserved in sheep with at least five alleles, while more than 14 alleles are found in goats. In contrast, it could not be amplified from bovine DNA. Similarly, the bovine κ-casein microsatellite is not present or is monomorphic in goats or sheep. These genetic markers were used to genetically map the casein locus on the bovine (Barendse et al., 1994, 1997) and goat chromosomes (Vaiman et al., 1996b). These localizations are perfectly consistent with the previously reported physical localizations (Hayes et al., 1992; Schibler, 1998b, for the goats). One interesting feature of these two polymorphic microsatellites is their physical associations with coding sequences for which the polymorphism has been extensively studied at the protein and DNA level. This situation provides an unprecedented opportunity to analyse linkage disequilibrium between a highly polymorphic genetic marker and a highly variable genetic system (Levéziel *et al.*, 1994). Similarly, linkage disequilibrium has been observed in goats between the β-casein microsatellite and the different αs1-casein alleles (Pépin *et al.*, 1994).

The study concerning the κ-casein microsatellite is more complete, because its polymorphism could be studied relatively to protein or 'classical' DNA polymorphisms, such as RFLPs, which were thoroughly studied in the past (Seibert *et al.*, 1985; Levéziel *et al.*, 1991; Schlieben *et al.*, 1991; Miranda *et al.*, 1993). In the κ-casein microsatellite, six alleles were identified (Levéziel *et al.*, 1994); some were sequenced, and 330 animals belonging to nine different purebred *Bos taurus* French breeds and a *B. taurus* × *Bos indicus* crossbreed were genotyped, showing a very strong linkage disequilibrium between the two types of polymorphism. The disequilibrium was also studied at the level of the complete locus, including variants of αs1- and β-casein; this clearly indicated the presence of preferential non-random allelic associations, and showed that polymorphic microsatellites associated with coding variants could be potent markers for marker-assisted selection schemes. An analogous linkage disequilibrium has been described in cattle between an intronic bovine leucocyte antigen (BoLA) DRB3 microsatellite and multiple variants observed at this locus (Ellegren *et al.*, 1993).

Evolutionary comparisons of the casein family in mammals
In all the mammal species studied so far, the whole casein family is contained in a fragment of less than 300 kb (Fig. 6.3), and the overall organization seems to be conserved (Mercier and Vilotte, 1993). The casein family has been analysed in bovines by two independent pulse-field analysis experiments (Ferretti *et al.*, 1990; Threadgill and Womack, 1990). Both have shown that the four casein genes (αs1, β, αs2 and κ) are present in this order, separated by relatively short distances, 20 kb, 105 kb and 55 kb according to the study of Threadgill and Womack (1990). According to the study of Ferretti *et al.* (1990), these genes are located in a larger region (250 kb instead of 200 kb), and κ-casein is located further away than the others. A more recent study, combining pulse-field analysis and the isolation of phage and cosmid clones corresponding to 60% of the locus (140 kb out of 250 kb) was performed by Rijnkels *et al.* (1997b). The intergenic distances are largely consistent with those described by Ferretti *et al.* (1990). In this study transcriptional orientation could be defined, showing that αs1, αs2 and κ-casein display the same orientation, while the transcriptional unit of β-casein is in reverse orientation. Interestingly, the same authors presented similar studies of the casein locus in humans (Rijnkels *et al.*, 1997c) and mice (Rijnkels *et al.*, 1997a), making an evolutionary comparison possible among three different mammalian orders. Independently, the organization of the human casein locus as well as the murine casein locus has recently been studied by other authors (Fujiwara *et al.*, 1997; George *et al.*, 1997). Recently, the goat BAC library described previously was used to build a contig of five BACs encompassing the complete locus. The goat locus

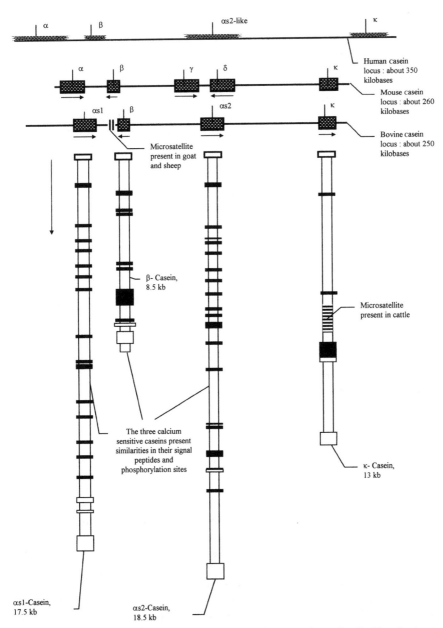

Fig. 6.3. Comparative structure of the casein locus in humans, mice and cattle. The structure of the bovine casein genes is detailed below the bovine locus, with the complete size of the transcription units. The untranslated exons are represented by white boxes, whereas the translated ones are represented by solid black boxes. The position of ruminant polymorphic microsatellites is indicated. The vertical arrow represents the orientation of the transcription units.

organization is similar to the cattle one (L. Schibler *et al.*, unpublished observations). It has been assumed until recently that human milk was devoid of α-like casein. However, the presence of an αs1-like protein has been reported (Cavaletto *et al.*, 1994; Rasmussen *et al.*, 1995). Hybridization with an αs2-casein bovine probe also revealed the existence of an αs2-like casein in the human gene family (Rijnkels *et al.*, 1997c). This αs2-like human casein gene is located between β- and κ-casein, similarly to where it is found in ruminants. In mice, an additional gene has been described, and the position of αs2-casein is occupied by γ- and δ- (also called ε-) casein genes. These two murine loci are strongly related to αs2-casein at the sequence level, although their transcriptional orientation is inverted. The presence of two αs2-like casein genes is consistent with the supposed situation in rabbits, where two different αs2-like casein cDNAs have also been discovered (Dawson *et al.*, 1993), although the structure of the gene family is largely unknown in this species. The inverted orientation of αs1, β, and αs2 caseins (which display sequence similarity) seems to be a consistent feature of the gene family. One explanation of this organization can be the subsequent stabilization provided by such a structure, which prevents the instability provoked by unequal crossing-over, but also prevents homogenization of the implied genes by gene conversion mechanisms (Rijnkels *et al.*, 1997a). It can be hypothesized that the impediment to gene conversion mechanisms by the inverted orientation of casein genes enables them to evolve relatively freely.

Molecular diagnoses in cattle embryos

Introduction

In cattle, the identification of genetic traits in preimplantation embryos is of high economic interest. Indeed, due to the fact that cattle are costly animals and have a long pregnancy, it is of particular interest to be able to determine the genotype status of precious embryos before implanting them in foster mothers. Two especially important traits exemplify the specific problems linked to bovine embryo transfer; one is the determination of the embryonic sex, the other is the implantation of embryos demonstrated to be transgenics. While the production of transgenic cows is a tedious, inefficient and very expensive process (the cost of one transgenic cow has been estimated to be at least $300,000 (Wall *et al.*, 1992; Wall, 1997)), success has been obtained – for instance, cattle producing human lactoferrin under the control of an αs1-casein promoter (Krimpenfort *et al.*, 1991). The two problems of transgenesis verification and sex determination at the embryo level are intimately linked: the implantation of multiple modified embryos is an efficient way of overcoming the very low transgenesis rate in cattle – 0.02–0.2%, compared with 3.5% in mice (Krimpenfort *et al.*, 1991; Wall *et al.*, 1992). However, in this case, the embryos must have the same genetic sex in order to prevent freemartinism (Wilmut *et al.*, 1991; see Chapter 15). Furthermore, the determination of the sex of the embryo is in itself of great practical interest, since dairy production relies mainly on females, while the production of males is the major objective for meat breeders and insemination centres.

Embryo sexing

In most mammals, sex is determined at fertilization by the presence or absence of a Y chromosome in spermatozoa. The obvious cytogenetic difference between the X and Y chromosomes has motivated a long quest for a primary male sex-determining gene carried by the Y chromosome. The first candidates for the role of testis determining factor was the H-Y antigen (Wachtel *et al.*, 1975), a male-specific minor histocompatibility antigen described primarily by graft rejection experiments in mice (Eichwald and Silmser, 1955). Later on, the determination of a critical 140 kb region, resulting from the analysis of human patients with chromosome anomalies, and the further identification of a zinc finger protein in this region (ZFY) made this latter protein a serious candidate for the function of testis determining factor (Page *et al.*, 1987; Schneider-Gädicke *et al.*, 1989). Further delimitation of the sex-determining region to a 35 kb region finally showed that the testis determining factor was not ZFY, but a different, very close gene, called the sex region of the Y chromosome SRY (Gubbay *et al.*, 1990; Sinclair *et al.*, 1990). Later on, it was shown that a 14 kb DNA fragment containing SRY was sufficient to reverse XX mice to the male sex (Koopman *et al.*, 1991). The SRY function is conserved across mammals but its sequence conservation varies according to the part of the gene under study. The highly conserved high mobility group (HMG) box, corresponding to the DNA-binding domain, is the most preserved, while the rest of the sequence has no function assigned and is greatly variable (Tucker and Ludigran, 1993). Cattle, sheep and goat SRY has been cloned and sequenced (Payen and Cotinot, 1993, 1994). This last study has demonstrated that, even within the Bovidae, the sequence of SRY is highly variable outside the HMG box.

Most of the length of the Y chromosome is devoid of coding sequences, but it is particularly rich in repetitive sequences. Such sequences are very easy to amplify by PCR and have been used to sex mice and humans (Bradbury *et al.*, 1990; Handyside *et al.*, 1990). Various repetitive DNA fragments of this type have been identified in cattle, such as BRY.4a (Matthews and Reed, 1991), BRY.1 (Matthews and Reed, 1991, 1992), BOV97M (Miller and Koopman, 1990), BC1.2 and BC1.34 (Cotinot *et al.*, 1991). The copy number of the BC1.2 probe was estimated at 2000 to 2500 copies on the Yq arm of chromosome Y (Popescu *et al.*, 1988; Cotinot *et al.*, 1991). Recently, the BRY.4a probe was used in FISH experiments and shown to label three-quarters of the chromosome (Thomsen and Jorgensen, 1994). The specific advantages of the PCR technique, compared with classical Southern blot analyses, have made this the most extensively used tool in the identification of embryonic sex (Aasen and Medrano, 1990; Schroder *et al.*, 1990; Peura *et al.*, 1991; Horvat *et al.*, 1993). The sensitivity of the PCR technique, however, is in itself a drawback, the slightest contamination being able to lead to a PCR artefact, when working from a very limited number of cells. Reciprocally, the absence of a specific

Y-chromosome PCR amplification in an experiment can be due to either a female embryo or an undetermined impairment of the normal polymerase function. To make the technique fully workable, various technical solutions were proposed: co-amplification of an autosomal fragment in the reaction (Rao et al., 1993); confirmation using several different Y-chromosome sequences, or nested PCR, combined with RFLP, making it possible to distinguish between ZFX (the X homologue of ZFY) and ZFY (Pollevick et al., 1992). These different techniques are now operational and several commercial detection kits enable embryonic sex to be diagnosed before implantation. These probes also function in some cases in other species, such as the water buffalo (*Bubalus bubalis*) (Rao et al., 1993), the bison (Cotinot et al., 1991), sheep and goats (Matthews and Reed, 1991; Rao and Totey, 1992).

Molecular diagnoses of transgenesis
Transgenic animals are obtained by the development of embryos into which foreign DNA has been introduced. More precisely, the foreign DNA is microinjected into the male pronucleus before the fusion with the female pronucleus. Transgenic mice are currently one of the major models for studying human diseases (Palmiter and Brinster, 1986; Gordon, 1989). In large domestic animals, the expression of transgenes governed by specific milk-gene promoters is an attractive alternative to the production of pharmaceutical proteins by microorganisms. Post-translational modifications rendering the peptide active can be obtained in eukaryotic cells only. This motivated the vision of large domestic transgenic animals, and among them cattle, as 'transgenic bioreactors' (Jänne et al., 1994), while transgenic cattle could also be obtained for modifying or improving some of their genetic characteristics.

Transgenesis, although widely used, at least in rodents, largely exploits the egg as a 'black box', and little is actually known about the fate of the microinjected DNA (Bishop and Smith, 1989). Some copies are integrated, sometimes in long head-to-tail arrays, at different places in the egg DNA, while other unintegrated copies remain present even at the morula and blastula stage of the developing embryo (Bowen et al., 1994; Jänne et al., 1994). This unintegrated DNA is a permanent problem for screening embryos by PCR before implantation. One recent idea for overcoming this problem consists in using the difference between methylation patterns between bacterial and chromosomal eukaryotic DNA. Eukaryotic cells do not methylate their adenine residues, and demethylation is thought to occur while the cell replicates. The fate of a methylated transgene after microinjection differs according to whether it is integrated or not. When it remains extrachromosomal, it seems to escape demethylation and can thus be degraded by *Dpn*I, a restriction enzyme that cleaves the DNA at the GATC site, but only when the adenine residue is methylated. In contrast, integrated copies of the transgene behave like other genomic DNA, lose their methylation after a few cellular cycles and cannot be degraded by the restriction enzyme. This ingenious technique was successfully used to discriminate transgenic embryos from non-transgenic ones in cattle (Hyttinen *et al.*, 1994, 1996). The problem with all these applications is that a biopsy or a bisection of the putative transgenic embryos has to be carried out and many embryos can be lost. Jänne *et al.* (1994) reported a result of 0.6–3 pregnancies for 600 microinjected oocytes, but the PCR test of integration made it possible to transfer only six embryos checked previously for sex and transgene integration. Recently, transgenic mice have been obtained using a construct merging the green fluorescent protein (GFP) with the elongation factor 1α promoter (Takada *et al.*, 1997). The authors injected bovine embryos with the same construct and could detect GFP in the bovine blastocysts, suggesting that this method could constitute a promising alternative to disturbing biopsies.

Transgenesis could potentially accelerate the improvement of cattle breeds, once the uncertainties of the choice of the embryos to be implanted are efficiently controlled. The application of molecular biology to verify transgene integration will probably be of extreme importance in the future, in order to shorten the time necessary for the genetic improvement of the bovine species.

Conclusions

During the writing of this chapter, the first successful positional cloning experiment was achieved in cattle, with the identification of the gene responsible for the double-muscling phenotype (Grobet *et al.*, 1997). This gene was first localized in 1995 (Charlier *et al.*, 1995). The discovery of a mutation causing an analogous phenotype in the mouse (McPherron *et al.*, 1997) made it possible to identify the homologous human gene. Comparative mapping data supported this candidate, as the double-muscling phenotype mapped to the appropriate position. Mutation analysis of the normal and mutant animals revealed a deletion of 11 bases systematically found in carrier animals (Grobet *et al.*, 1997). Clearly, this success was made possible by the synchronous development of the complete tool-box of molecular genetics in cattle, i.e. dense genetic maps, YAC libraries and comparative genomic analysis of humans, mice and cattle. The discovery of other genes and QTL can now be foreseen. Together with the possibility of altering the germplasm of bovine embryos, new vistas are now opened for animal geneticists.

Acknowledgement

The critical reading of the manuscript by Edmond Paul Cribiu and François Grosclaude is greatly appreciated.

References

Aasen, E. and Medrano, J.F. (1990) Amplification of the ZFY and ZFX genes for sex identification in humans, cattle, sheep and goats. *Biotechnology* 8, 1279–1281.

Adams, M.D., Dubnick, M., Kerlavage, A.R., Moreno, R., Kelley, J.M., Utterback, T.R., Nagle, J.W., Fields, C. and Venter, J.C. (1992) Sequence identification of 2,375 human brain genes. *Nature* 355, 632–634.

Adams, M.D., Kerlavage, A.R., Fields, C. and Venter, J.C. (1993) 3,400 new expressed sequence tags identify diversity of transcripts in human brain. *Nature Genetics* 4, 256–267.

Aitman, T.J., Hearne, C.M., McAleer, M.A. and Todd, J.A. (1991) Mononucleotide repeats are an abundant source of length variants in mouse genomic DNA. *Mammalian Genome* 1, 206–210.

Alexander, L.J., Rohrer, G.A., Stone, R.T. and Beattie, C.W. (1995) Porcine SINE-associated microsatellite markers: evidence for new artiodactyl SINEs. *Mammalian Genome* 6, 464–468.

Alexander, L.J., Stewart, F., Mackinlay, A.G., Kapelinskaya, T.V., Tkach, T.M. and Gorodetski, S.I. (1988) Isolation and characterization of the bovine k-casein gene. *European Journal of Biochemistry* 178, 395–401.

Anand, R., Riley, J.H., Butler, R., Smith, J.C. and Markham, A.F. (1990) A 3.5 genome equivalent multi access YAC library: construction, characterisation, screening and storage. *Nucleic Acids Research* 18, 1951–1956.

Antoch, M.P., Song, E.J., Chang, A.M., Vitaterna, M.H., Zhao, Y., Wilsbacher, L.D., Sangoram, A.M., King, D.P., Pinto, L.H. and Takahashi, J.S. (1997) Functional identification of the mouse circadian Clock gene by transgenic BAC rescue. *Cell* 16, 655–667.

Band, M., Vaiman, D., Cribiu, E.P. and Ron, M. (1997) Four bovine microsatellites derived from a sorted chromosome 25 library. *Animal Genetics* 28, 239–240.

Banfi, S. and Zoghbi, H.Y. (1996) Detection of chimerism in YAC clones. *Methods in Molecular Biology* 54, 115–121.

Barendse, W., Armitage, S.M., Kossarek, L.M., Shalom, A., Kirkpatrick, B.W., Ryan, A.M., Clayton, D., Li, L., Neibergs, H.L., Zhang, N., Grosse, W.M., Weiss, J., Creighton, P., McCarthy, F., Ron, M., Teale, A.J., Fries, R., McGraw, R.A., Moore, S.S., Georges, M., Soller, M., Womack, J.E. and Hetzel, D.J.S. (1994) A genetic linkage map of the bovine genome. *Nature Genetics* 6, 227–234.

Barendse, W., Vaiman, D., Kemp, S., Sugimoto, Y., Armitage, S.M., Williams, J., Sun, S., Eggen, A., Agaba, M., Aleyasin, A., Band, M., Bishop, M., Buitkamp, J., Byrne, K., Collins, F., Cooper, L., Coupettiers, W., Denis, B., Drinkwater, R., Easterday, K., Ennis, S., Erhardt, G., Ferretti, L., Gao, Q., Georges, M., Gurung, R., Harlizius, B., Hawkins, G., Hetzel, D.J.S., Hirano, T., Joergensen, C., Kessler, M., Kirkpatrick, B., Konfortov, B., Kuhn, C., Lenstra, H., Leveziel, H., Lewin, H., Leyhe, B., Li, L., Martin Burriel, I., McGraw, R.A., Miller, J.R., Moody, D., Moore, S., Nakane, S., Nijman, I., Olsaker, I., Pomp, D., Rando, A., Ron, M., Shalom, A., Soller, M., Teale, A., Thieven, U., Vage, D., Varvio, S., Velmala, R.V.J., Weikard, R., Woodside, C., Womack, J., Zanotti, M. and Zaragoza, P. (1997) A medium-density genetic linkage map of the bovine genome. *Mammalian Genome* 8, 21–28.

Barillot, E., Lacroix, B. and Cohen, D. (1991) Theoretical analysis of library screening using a N-dimensional pooling strategy. *Nucleic Acids Research* 19, 6241–6247.

Bishop, J.O. and Smith, P. (1989) Mechanism of chromosomal integration of microinjected DNA. *Molecular Biology and Medicine* 6, 283–298.

Bonsing, J., Ring, J.M., Stewart, A.F. and Mackinlay, A.G. (1988) Complete nucleotide sequence of the bovine beta-casein gene. *Australian Journal of Biological Sciences* 41, 527–537.

Bowen, R.A., Reed, M.L., Schnieke, A., Seidel, G.E.J., Stacey, A., Thomas, W.K. and Kajikawa, O. (1994) Transgenic cattle resulting from biopsied embryos: expression of c-ski in a transgenic calf. *Biology of Reproduction* 50, 664–668.

Bradbury, M.W., Isola, L.M. and Gordon, J.W. (1990) Enzymatic amplification of a Y chromosome repeat in a single blastomere allows identification of the sex of preimplantation mouse embryos. *Proceedings of the National Academy of Sciences of the USA* 87, 4053–4057.

Brenner, S., Elgar, G., Sandford, R., Macrae, A., Venkatesh, B. and Aparicio, S. (1993) Characterization of pufferfish (Fugu) genome as a compact model vertebrate genome. *Nature* 366, 265–268.

Brignon, G., Mahé, M.-F., Grosclaude, F. and Ribadeau-Dumas, B. (1989) Sequence of caprine alpha s1-casein and characterization of those of its genetic variants which are synthesized at a high level, alpha s1-CnA, B and C. *Protein Sequence and Data Analysis* 2, 181–188.

Brignon, G., Ribadeau-Dumas, B., Mercier, J.-C., Pélissier, J.-P. and Das, B.-C. (1977) Complete amino acid sequence of bovine alphaS2-casein. *FEBS Letters* 76, 274–279.

Britten, R.J. and Kohne, D.E. (1968) Repeated sequences in DNA: hundreds of thousands of copies of DNA sequences have been incorporated into the genomes of higher organisms. *Science* 161, 529–540.

Broad, T.E., Burkin, D.J., Cambridge, L.M., Maher, D.W., Lewis, P.E., Ansari, H.A., Pearce, P.D. and Jones, C. (1994) Seven loci on human chromosome 4 map onto sheep chromosome 6: a proposal to restore the original nomenclature of this sheep chromosome. *Mammalian Genome* 5, 429–433.

Broom, M.F. and Hill, D.F. (1994) Construction of a large-insert yeast-artificial chromosome library from sheep DNA. *Mammalian Genome* 5, 817–819.

Brownstein, A.H., Silverman, G.A., Little, R.D., Burke, D.T., Korsmeyer, S.J., Schlessinger, D. and Olson, M.V. (1989) Isolation of single-copy human genes from a library of yeast artificial chromosome clones. *Science* 244, 1348–1351.

Cai, L., Taylor, J.F., Wing, R.A., Gallagher, D.S., Woo, S.-S. and Davis, S.K. (1995) Construction and characterization of a bovine bacterial artificial chromosome library. *Genomics* 29, 413–425.

Carstea, E.D., Morris, J.A., Coleman, K.G., Loftus, S.K., Zhang, D., Cummings, C., Gu, J., Rosenfeld, M.A., Pavan, W.J., Krizman, D.B., Nagle, J., Polymeropoulos, M.H., Sturley, S.L., Ioannou, Y.A., Higgins, M.E., Comly, M., Cooney, A., Brown, A., Kaneski, C.R., Blanchette-Mackie, E.J., Dwyer, N.K., Neufeld, E.B., Chang, T.Y., Liscum, L. and Tagle, D. (1997) Niemann–Pick C1 disease gene: homology to mediators of cholesterol homeostasis. *Science* 277, 228–231.

Cavaletto, M., Cantisani, A., Giuffrida, G., Napolitano, L. and Conti, A. (1994) Human alpha S1-casein like protein: purification and N-terminal sequence determination. *Biological Chemistry Hoppe Seyler* 375, 149–151.

Chandrasekharappa, S.C., Guru, S.C., Manickam, P., Olufemi, S.E., Collins, F.S., Emmert-Buck, M.R., Debelenko, L.V., Zhuang, Z., Lubensky, I.A., Liotta, L.A., Crabtree, J.S., Wang, Y., Roe, B.A., Weisemann, J., Boguski, M.S., Agarwal, S.K., Kester, M.B., Kim, Y.S., Heppner, C., Dong, Q., Spiegel, A.M., Burns, A.L. and Marx, S.J. (1997) Positional cloning of the gene for multiple endocrine neoplasia-type 1. *Science* 276, 404–407.

Charlier, C., Coppieters, W., Farnir, F., Grobet, L., Leroy, P.L., Michaux, C., Mni, M., Schwers, A., Vanmanshoven, P. and Hanset, R. (1995) The mh gene causing double-muscling in cattle maps to bovine chromosome 2. *Mammalian Genome* 6, 788–792.

Charlier, C., Denys, B., Belanche, J.I., Coppieters, W., Grobet, L., Mni, M., Womack, J., Hanset, R. and Georges, M. (1996a) Microsatellite mapping of the bovine roan locus: a major determinant of White Heifer disease. *Mammalian Genome* 7, 138–142.

Charlier, C., Farnir, F., Berzi, P., Vanmanshoven, P., Brouwers, B., Vromans, H. and Georges, M. (1996b) Identity-by-descent mapping of recessive traits in livestock: application to map the bovine *syndactyly* locus to chromosome 15. *Genome Research* 6, 580–589.

Chartier, F.L., Keer, J.T., Sutcliffe, M.J., Henriques, D.A., Mileham, P. and Brown, S.D. (1992) Construction of a mouse yeast artificial chromosome library in a recombination-deficient strain of yeast. *Nature Genetics* 1, 132–136.

Cockett, N.E., Jackson, S.P., Shay, T.L., Nielsen, D., Moore, S.S., Steele, M.R., Barendse, W., Green, R.D. and Georges, M. (1994) Chromosomal localization of the callipyge gene in sheep (*Ovis aries*) using bovine DNA markers. *Proceedings of the National Academy of Sciences of the USA* 91, 3019–3023.

Cockett, N.E., Jackson, S.P., Shay, T.L., Farnir, F., Berghmans, S., Snowder, G.D., Nielsen, D.M. and Georges, M. (1996) Polar overdominance at the ovine callipyge locus. *Science* 273, 236–238.

Cotinot, C., Kirszenbaum, M., Leonard, M., Gianquinto, L. and Vaiman, M. (1991) Isolation of bovine Y-derived sequence: potential use in embryo sexing. *Genomics* 10, 646–653.

Coulson, A., Waterston, R., Kiff, J., Sulston, J. and Kohara, Y. (1988) Genome linking with yeast artificial chromosomes. *Nature* 335, 184–186.

Crawford, A.M., Dodds, K.G., Pierson, C.A., Ede, A.J., Montgomery, G.W., Garmonsway, H.G., Beattie, A.E., Davies, K., Maddox, J.F., Kappes, S.W., Stone, R.T., Nguyen, T.C., Penty, J.M., Lord, E.A., Broom, J.E., Buitkamp, J., Schwenger, W., Epplen, J.T., Matthew, P., Matthews, M.E., Hulme, D.J., Beh, K.J., McGraw, R.A. and Beattie, C.W. (1995) An autosomal genetic linkage map of the sheep genome. *Genetics* 140, 703–724.

Cribiu, E.P. and Popescu, P. (1974) L'idiogramme de *Bos taurus* L. *Annales de Génétique et de Sélection animale* 6, 291–296.

Dawson, S.P., Wilde, C.J., Tighe, P.J. and Mayer, R.J. (1993) Characterization of two novel casein transcripts in rabbit mammary gland. *Biochemical Journal* 15, 777–784.

Dixon, S.C., Miller, N.G.A., Tucker, E.M. and Carter, N.P. (1991) Flow sorting of farm animal chromosomes. *Animal Genetics* 22 (Suppl.), 87.

Dorit, R.L., Schoenbach, L. and Gilbert, W. (1990) How big is the universe of exons? *Science* 250, 1377–1382.

Dunner, S., Charlier, C., Farnir, F., Brouwers, B., Canon, J. and Georges, M. (1997) Towards interbreed IBD fine mapping of the mh locus: double-muscling in the Asturiana de los Valles breed involves the same locus as in the Belgian Blue cattle breed. *Mammalian Genome* 8, 430–435.

Dyer, M.R., Gay, N.J. and Walker, J.E. (1989) DNA sequences of a bovine gene and of two related pseudogenes for the proteolipid subunit of mitochondrial ATP synthase. *Biochemical Journal* 260, 249–258.

Eichwald, E.J. and Silmser, C.R. (1955) Untitled communication. *Transplantation Bulletin* 2, 148–149.

Eickbush, D.G., Lathe, W.C., Francino, M.P. and Eickbush, T.H. (1995) R1 and R2 retrotransposable elements of *Drosophila* evolve at rates similar to those of nuclear genes. *Genetics* 139, 685–695.

Eickbush, T.H. (1992) Transposing without ends: the non-LTR retrotransposable elements. *New Biology* 4, 430–440.

Ellegren, H., Davies, C.J. and Andersson, L. (1993) Strong association between polymorphisms in an intronic microsatellite and in the coding sequence of the *BoLA-DRB3* gene: implications for microsatellite stability and PCR-based *DRB3* typing. *Animal Genetics* 24, 269–275.

Ferretti, L., Leone, P. and Sgamarella, V. (1990) Long range restriction analysis of the bovine casein genes. *Nucleic Acids Research* 18, 6829–6833.

Fries, R., Eggen, A. and Stranzinger, G. (1990) The bovine genome contains polymorphic microsatellites. *Genomics* 8, 403–406.

Fronicke, L., Chowdhary, B.P., Scherthan, H. and Gustavsson, I. (1996) A comparative map of the porcine and human genomes demonstrates ZOO-FISH and gene mapping-based chromosomal homologies. *Mammalian Genome* 7, 285–290.

Fujiwara, Y., Miwa, M., Nogami, M., Okumura, K., Nobori, T., Suzuki, T. and Ueda, M. (1997) Genomic organization and chromosomal localization of the human casein gene family. *Human Genetics* 99, 368–373.

Garza, D., Ajioka, J.W., Burke, D.T. and Hartl, D.L. (1989) Mapping the *Drosophila* genome with yeast artificial chromosomes. *Science* 246, 641–646.

Gaye, P., Gautron, J.-P., Mercier, J.-C. and Haze, G. (1977) Amino terminal sequences of the precursors of ovine caseins. *Biochemical and Biophysical Research Communications* 79, 903–911.

George, S., Clark, A.J. and Archibald, A.L. (1997) Physical mapping of the murine casein locus reveals the gene order as alpha-beta-gamma-epsilon-kappa. *DNA and Cell Biology* 16, 477–484.

Georges, M., Gunawardana, A., Threadgill, D., Lathrop, M., Olsaker, I., Mishra, A., Sargeant, L.L., Schoerbelein, A., Steele, M.R., Terrie, C., Threadgill, D.S., Zhao, X., Holm, T., Fries, R. and Womack, J.E. (1991) Characterization of a set of variable number of tandem repeat markers conserved in Bovidae. *Genomics* 11, 24–32.

Georges, M., Drinkwater, R., King, T., Mishra, A., Moore, S.S., Nielsen, D., Sargeant, L.S., Sorensen, A., Steele, M.R., Zhao, X., Womack, J.E. and Hetzel, J. (1993a) Microsatellite mapping of a gene affecting horn development in *Bos taurus*. *Nature Genetics* 4, 206–210.

Georges, M., Dietz, A.B., Mishra, A., Nielsen, D., Sargeant, L.S., Sorensen, A., Steele, M.R., Zhao, X.Y., Leipold, H., Womack, J.E. and Lathrop, M. (1993b) Microsatellite mapping of the gene causing weaver disease in cattle will allow the study of an associated quantitative trait locus. *Proceedings of the National Academy of Sciences of the USA* 90, 1058–1062.

Georges, M., Nielsen, D., Mackinnon, M., Mishra, A., Okimoto, R., Pasquino, A.T., Sargeant, L.S., Sorensen, A., Steele, M.R., Zhao, X., Womack, J.E. and Hoeschele, I. (1995) Mapping quantitative trait loci controlling milk production in dairy cattle by exploiting progeny testing. *Genetics* 139, 907–920.

Goldammer, T., Weikard, R., Brunner, R.M. and Schwerin, M. (1996) Generation of chromosome fragment specific bovine DNA sequences by microdissection and DOP-PCR. *Mammalian Genome* 7, 291–296.

Gordon, J.W. (1989) Transgenic animals. *International Review of Cytology* 115, 171–229.

Goureau, A., Yerle, M., Schmitz, A., Riquet, J., Milan, D., Pinton, P., Frelat, G. and Gellin, J. (1996) Human and porcine correspondence of chromosome segments using bidirectional chromosome painting. *Genomics* 36, 252–262.

Grobet, L., Royo Martin, L.J., Poncelet, D., Pirottin, D., Brouwers, B., Riquet, J., Schoeberlein, A., Dunner, S., Ménissier, F., Massabanda, J., Fries, R., Hanset, R. and Georges, M. (1997) A deletion in the bovine *myostatin* gene causes the double-muscling phenotype in cattle. *Nature Genetics* 17, 71–74.

Groenen, M.A., Dijkhof, R.J., Verstege, A.J. and van der Poel, J.J. (1993) The complete sequence of the gene encoding bovine alpha s2-casein. *Gene* 123, 187–193.

Grosclaude, F., Garnier, J., Ribadeau-Dumas, B. and Jeunet, R. (1964) Etroite dépendance des loci contrôlant le polymorphisme des caséines αs et β. *Comptes Rendus de l'Académie des Sciences* 259, 1569.

Grosclaude, F., Joudrier, P. and Mahe, M.-F. (1978) Polymorphisme de la caséine alpha-S2 bovine: étroite liaison du locus alpha-S2 avec les loci alpha S1, beta, et kappa; mise en évidene d'une délétion dans le variant alpha-S2 D. *Annales de Génétique et de Sélection Animale* 10, 313–327.

Grosclaude, F., Mahé, M.F., Brignon, G., Di Stasio, L. and Jeunet, R. (1987) A mendelian polymorphism underlying quantitative variations of goat alpha S1 casein. *Genetics Selection Evolution* 19, 399–412.

Gu, F. and Hindkjær, J. (1996) Primed in situ labeling (PRINS) detection of the telomeric (CCCTAA)n sequences in chromosomes of domestic animals. *Mammalian Genome* 7, 231–232.

Gubbay, J., Collignon, J., Koopman, P., Capel, B., Economou, A., Mounsterberg, A., Vivian, N., Goodfellow, P. and Lovell-Badge, R. (1990) A gene mapping to the sex-determining region of the mouse Y chromosome is a member of a novel family of embryonically expressed genes. *Nature* 346, 245–250.

Guérin, G., Eggen, A., Vaiman, D., Nocart, M., Laurent, P., Béchet, D. and Ferrara, M. (1994) Further characterization of a somatic cell hybrid panel: ten new assignments to the bovine genome. *Animal Genetics* 25, 31–35.

Hamada, H. and Kakunaga, T. (1982) Potential Z-DNA forming sequences are highly dispersed in the human genome. *Nature* 302, 396–398.

Hameister, H., Klett, C., Bruch, J., Dixkens, C., Vogel, W. and Christensen, K. (1997) Zoo-FISH analysis: the American mink (*Mustela vison*) closely resembles the cat karyotype. *Chromosome Research* 5, 5–11.

Handyside, A.H., Kontogianni, E.H., Hardy, K. and Winston, R.M. (1990) Pregnancies from biopsied human preimplantation embryos sexed by Y-specific DNA amplification. *Nature* 344, 768–770.

Harlizius, B., Tammen, I., Eichler, K., Eggen, A. and Hetzel, D.J. (1997) New markers on bovine chromosome 1 are closely linked to the polled gene in Simmental and Pinzgauer cattle. *Mammalian Genome* 8, 255–257.

Hayes, H. (1995) Chromosome painting with human chromosome specific DNA libraries reveals the extent and the distribution of conserved segments in bovine chromosomes. *Cytogenetics and Cell Genetics* 71, 168–174.

Hayes, H., Petit, E., Bouniol, C. and Popescu, P. (1993) Localization of the alpha-S2-casein gene (CASAS2) to the homoeologous cattle, sheep, and goat chromosomes 4 by *in situ* hybridization. *Cytogenetics and Cell Genetics* 64, 281–285.

Hayes, H., Petit, E., Lemieux, N. and Dutrillaux, B. (1992) Chromosomal localization of the ovine beta-casein gene by non-isotopic *in situ* hybridization and R-banding. *Cytogenetics and Cell Genetics* 61, 286–288.

Hecht, S.J., Stedman, K.E., Carlson, J.O. and DeMartini, J.C. (1996) Distribution of endogenous type B and type D sheep retrovirus sequences in ungulates and other mammals. *Proceedings of the National Academy of Sciences of the USA* 93, 3297–3302.

Hediger, R., Ansari, H.A. and Stranzinger, G.F. (1991) Chromosome banding and gene localizations support extensive conservation of chromosome structure between cattle and sheep. *Cytogenetics and Cell Genetics* 57, 127–134.

Heriz, A., Arruga, M.V., Monteagudo, L.V., Tejedor, M.T., Pitel, F. and Echard, G. (1994) Assignment of the transition protein 1 (TNP1) gene to U17 bovine synteny group by PCR. *Mammalian Genome* 5, 742.

Heuertz, S. and Hors-Cayla, M.C. (1981) Cattle gene mapping by somatic cell hybridization study of 17 enzyme markers. *Cytogenetics and Cell Genetics* 30, 137–145.

Hoeschele, I. and Meinert, T.R. (1990) Association of genetic defects with yield and type traits: the weaver locus effect on yield. *Journal of Dairy Science* 73, 2503–2515.

Horvat, S., Medrano, J.F., Behboodi, E., Anderson, G.B. and Murray, J.D. (1993) Sexing and detection of gene construct in microinjected bovine blastocysts using the polymerase chain reaction. *Transgenic Research* 2, 134–140.

Hwu, H.R., Roberts, J.W., Davidson, E.H. and Britten, R.J. (1986) Insertion and/or deletion of many repeated DNA sequences in human and higher ape evolution. *Proceedings of the National Academy of Sciences of the USA* 83, 3875–3879.

Hyttinen, J.M., Peura, T., Tolvanen, M., Aalto, J., Alhonen, L., Sinervirta, R., Halmekyto, M., Myohanen, S. and Jänne, J. (1994) Generation of transgenic dairy cattle from transgene-analyzed and sexed embryos produced *in vitro*. *Biotechnology* 12, 606–608.

Hyttinen, J.M., Peura, T., Tolvanen, M., Aalto, J. and Jänne, J. (1996) Detection of microinjected genes in bovine preimplantation embryos with combined DNA digestion and polymerase chain reaction. *Molecular Reproduction and Development* 43, 150–157.

Ioannou, P.A., Amemiya, C.T., Garnes, J., Kroisel, P.M., Shizuya, H., Chen, C., Batzer, M.A. and de Jong, P.J. (1994) A new bacteriophage P1-derived vector for the propagation of large human DNA fragments. *Nature Genetics* 6, 84–89.

Jänne, J., Hyttinen, J.M., Peura, T., Tolvanen, M., Alhonen, L., Sinervirta, R. and Halmekyto, M. (1994) Transgenic bioreactors. *International Journal of Biochemistry* 26, 859–870.

Jenness, R. (1979) Comparative aspects of milk proteins. *Journal of Dairy Research* 46, 197–210.

Jobse, C., Buntjer, J.B., Haagsma, N., Breukelman, H.J., Beintema, J.J. and Lenstra, J.A. (1995) Evolution and recombination of bovine DNA repeats. *Journal of Molecular Evolution* 41, 277–283.

Jollès, J., Schoentgen, F., Hermann, J., Alais, C. and Jolles, P. (1974) The sequence of sheep kappa-casein: primary structure of para-kappa A-casein. *European Journal of Biochemistry* 46, 127–132.

Jones, W.K., Yu-Lee, L.Y., Clift, S.M., Brown, T.L. and Rosen, J.M. (1985) The rat casein multigene family: fine structure and evolution of the beta-casein gene. *Journal of Biological Chemistry* 260, 7042–7050.

Kappes, S.M., Keele, J.W., Stone, R.T., McGraw, R.A., Sonstegard, T.S., Smith, T.P., Lopez-Corrales, N.L. and Beattie, C.W. (1997) A second-generation linkage map of the bovine genome. *Genome Research* 7, 235–249.

Kaukinen, J. and Varvio, S.-L. (1992) Artiodactyl retroposons: association with microsatellites and use in SINEmorph detection by PCR. *Nucleic Acids Research* 20, 2955–2958.

Kim, U.-J., Shizuya, H., Birren, B., Slepak, T., De Jong, P. and Simon, M.I. (1994) Selection of chromosome 22-specific clones from human genomic BAC library using a chromosome-specific cosmid library pool. *Genomics* 22, 336–339.

Kim, U.J., Shizuya, H., Deaven, L., Chen, X.N., Korenberg, J.R. and Simon, M.I. (1995) Selection of a sublibrary enriched for a chromosome from total human bacterial artificial chromosome library using DNA from flow sorted chromosomes as hybridization probes. *Nucleic Acids Research* 23, 1838–1839.

Koczan, D., Hobom, G. and Seyfert, H.M. (1991) Genomic organization of the bovine alpha-S1 casein gene. *Nucleic Acids Research* 19, 5591–5596.

Koopman, P., Gubbay, J., Vivian, N., Goodfellow, P. and Lovell-Badge, R. (1991) Male development of chromosomally female mice transgenic for Sry. *Nature* 351, 117–121.

Kopecka, H., Macaya, G., Cortadas, J., Thiery, J.P. and Bernardi, G. (1978) Restriction enzyme analysis of satellite DNA components from the bovine genome. *European Journal of Biochemistry* 84, 189–195.

Korenberg, J.R., Chen, X.N., Adams, M.D. and Venter, J.C. (1995) Toward a cDNA map of the human genome. *Genomics* 29, 364–370.

Kostia, S., Vilkki, J., Pirinen, M., Womack, J.E., Barendse, W. and Varvio, S.L. (1997) SINE targeting of bovine microsatellites from bovine/rodent hybrid cell lines. *Mammalian Genome* 8, 365–367.

Krimpenfort, P., Rademakers, A., Eyestone, W., Schans, A. van der, Broek, S. van der, Kooiman, P., Kootwijk, E., Platenburg, G., Pieper, F., Strijker, R. and Boer, H. de (1991) Generation of transgenic dairy cattle using *in vitro* embryo production. *Biotechnology* 9, 844–847.

Larin, Z., Monaco, A.P. and Lehrach, H. (1991) Yeast artificial chromosome libraries containing large inserts from mouse and human DNA. *Proceedings of the National Academy of Sciences of the USA* 88, 4123–4127.

Lee, C. and Lin, C.C. (1996) Conservation of a 31-bp bovine subrepeat in centromeric satellite DNA monomers of *Cervus elaphus* and other cervid species. *Chromosome Research* 4, 427–435.

Lenstra, J.A., Van Boxtel, J.A.F., Zwaagstra, K.A. and Schwerin, M. (1993) Short interspersed nuclear element (SINE) sequences of the Bovidae. *Animal Genetics* 24, 33–39.

Leroux, C., Mazure, N. and Martin, P. (1992) Mutations away from splice site recognition sequences might cis-modulate alternative splicing of goat alpha s1-casein transcripts: structural organization of the relevant gene. *Journal of Biological Chemistry* 267, 6147–6157.

Levéziel, H., Méténier, L., Guérin, G., Cullen, P., Provot, C., Bertaud, M. and Mercier, J.-C. (1991) Restriction fragment length polymorphism of ovine casein genes: close linkage between the alpha s1-, alpha s2-, beta- and kappa-casein loci. *Animal Genetics* 22, 1–10.

Levéziel, H., Rodellar, C., Leroux, C., Pepin, L., Grohs, C., Vaiman, D., Mahé, M.-F., Martin, P. and Grosclaude, F. (1994) A microsatellite within the bovine kappa-casein gene reveals a polymorphism correlating strongly with polymorphisms at protein and DNA levels *Animal Genetics* 25, 223–228.

Libert, F., Lefort, A., Okimoto, R., Womack, J. and Georges, M. (1993) Construction of a bovine genomic library of large yeast artificial chromosome clones. *Genomics* 18, 270–276.

Logue, D. and Greig, A. (1985) Infertility in the bull, ram and boar: 1. Failure to mate. *In Practice* 7, 185–191.

Lyons, L.A., Laughlin, T.F., Copeland, N.G., Jenkins, N.A., Womack, J.E. and O'Brien, S.J. (1997) Comparative anchor tagged sequences (CATS) for integrative mapping of mammalian genomes. *Nature Genetics* 15, 47–56.

Macaya, G., Cortadas, J. and Bernardi, G. (1978) An analysis of the bovine genome by density-gradient centrifugation. Preparation of the dG+dC-rich DNA components. *European Journal of Biochemistry* 84, 179–188.

McKnight, R.A., Jimenez-Flores, R., Kang, Y., Creamer, L.K. and Richardson, T. (1989) Cloning and sequencing of a complementary deoxyribonucleic acid coding for a bovine alpha s1-casein A from mammary tissue of a homozygous B variant cow. *Journal of Dairy Science* 72, 2464–2473.

McPherron, A.C., Lawler, A.M. and Lee, S.J. (1997) Regulation of skeletal muscle mass in mice by a new TGF-beta superfamily member. *Nature* 387, 83–90.

Mathias, S.L., Scott, A.F., Kazazian, H.H., Boeke, J.D. and Gabriel, A. (1991) Reverse transcriptase encoded by a human transposable element. *Science* 254, 1808–1810.

Matthews, M.E. and Reed, K.C. (1991) A DNA sequence that is present in both sexes of *Artiodactyla* is repeated on the Y chromosome of cattle, sheep, and goats. *Cytogenetics and Cell Genetics* 56, 40–44.

Matthews, M.E. and Reed, K.C. (1992) Sequences from a family of bovine Y-chromosomal repeats. *Genomics* 13, 1267–1273.

Mercier, J.-C. (1981) Phosphorylation of caseins, present evidence for an amino acid triplet code posttranslationally recognized by specific kinases. *Biochimie* 63, 1–17.

Mercier, J.-C. and Vilotte, J.-L. (1993) Structure and function of milk protein genes. *Journal of Dairy Science* 76, 3079–3098.

Mercier, J.-C., Grosclaude, F. and Ribadeau-Dumas, B. (1971) Primary structure of bovine s1 casein: complete sequence. *European Journal of Biochemistry* 23, 41–51.

Mercier, J.-C., Brignon, G. and Ribadeau-Dumas, B. (1973) Primary structure of bovine kappa B casein: complete sequence. *European Journal of Biochemistry* 35, 222–235.

Mercier, J.-C., Gaye, P., Soulier, S., Hue-Delahaie, D. and Vilotte, J.-L. (1985) Construction and identification of recombinant plasmids carrying cDNAs coding for ovine alpha S1-, alpha S2-, beta-, kappa-casein and beta-lactoglobulin: nucleotide sequence of alpha S1-casein cDNA. *Biochimie* 67, 959–971.

Meyerowitz, E.M., Guild, G.M., Prestidge, L.S. and Hogness, D.S. (1980) A new high-capacity cosmid vector and its use. *Gene* 11, 271–282.

Meyne, J., Ratliff, R.L. and Moyzis, R.K. (1989) Conservation of the human telomere sequence (TTAGGG)n among vertebrates. *Proceedings of the National Academy of Sciences of the USA* 86, 7049–7053.

Miller, J.R. and Archibald, A.L. (1993) 5' and 3' SINE-PCR allows genotyping of pig families without cloning and sequencing steps. *Mammalian Genome* 4, 243–246.

Miller, J.R. and Koopman, M. (1990) Isolation and characterization of two male-specific DNA fragments from the bovine gene. *Animal Genetics* 21, 77–82.

Miranda, G., Anglade, P., Mahe, M.F. and Erhardt, G. (1993) Biochemical characterization of the bovine genetic kappa-casein-C and E-variants. *Animal Genetics* 24, 27–31.

Mirck, M.H., Von Bannisseht-Wijsmuller, T., Timmermans-Besselink, W.J., Van Luijk, J.H., Buntjer, J.B. and Lenstra, J.A. (1995) Optimization of the PCR test for the mutation causing bovine leukocyte adhesion deficiency. *Cellular and Molecular Biology* 41, 695–698.

Modi, W.S., Gallagher, D.S. and Womack, J.E. (1993) Molecular organization and chromosomal localization of six highly repeated DNA families in the bovine genome. *Animal Biotechnology* 4, 143–161.

Montgomery, G.W., Crawford, A.M., Penty, J.M., Dodds, K.G., Ede, A.J., Henry, H.M., Pierson, C.A., Lord, E.A., Galloway, S.M., Schmack, A.E., Sise, J.A., Swarbrick, P.A., Hanrahan, V., Buchanan, F.C. and Hill, D.F. (1993) The ovine Booroola fecundity gene (FecB) is linked to markers from a region of human chromosome 4q. *Nature Genetics* 4, 410–414.

Montgomery, G.W., Lord, E.A., Penty, J.M., Dodds, K.G., Broad, T.E., Cambridge, L., Sunden, S.L.F., Stone, R.T. and Crawford, A.M. (1994) The booroola fecundity (*FecB*) gene maps to sheep chromosome 6. *Genomics* 22, 148–153.

Montgomery, G.W., Henry, H.M., Dodds, K.G., Beattie, A.E., Wuliji, T. and Crawford, A.M. (1996) Mapping the Horns (Ho) locus in sheep: a further locus controlling horn development in domestic animals. *Journal of Heredity* 87, 358–363.

Moore, S.S., Sargeant, L.L., King, T.J., Mattick, J.S., Georges, M. and Hetzel, D.J.S. (1991) The conservation of dinucleotide microsatellites among mammalian genomes allows the use of heterologous PCR primer pairs in closely related species. *Genomics* 10, 654–660.

Moore, S.S., Barendse, W., Berger, K.T., Armitage, S.M. and Hetzel, D.J.S. (1992) Bovine and ovine DNA microsatellites from the EMBL and GENBANK databases. *Animal Genetics* 23, 463–467.

Nadeau, J.H., Grant, P.L., Mankala, S., Reiner, A.H., Richardson, J.E. and Eppig, J.T. (1995) A Rosetta stone of mammalian genetics. *Nature* 373, 363–365.

Ng-Kwai-Hang, K.F. and Grosclaude, F. (1992) Genetic polymorphism of milk proteins. In: Fox, P.F. (ed.) *Advanced Dairy Chemistry.* Vol. 1, *Proteins.* Elsevier Applied Science, London and New York, pp. 405–455.

Nowak, R.M. (1991) *Walker's Mammals of the World.* Johns Hopkins University Press, London.

Ohno, S. and Yomo, T. (1991) The grammatical rule for all DNA: junk and coding sequences. *Electrophoresis* 12, 103–108.

Okada, N. and Hamada, M. (1997) The 3′ ends of tRNA-derived SINEs originated from the 3′ ends of LINEs: a new example from the bovine genome. *Journal of Molecular Evolution* 44, S52–S56.

Okada, N. and Ohshima, K. (1995) Evolution of tRNA-derived SINEs. In: Maraia, R. J. (ed.) *The Impact of Short Interspersed Elements (SINEs) on the Host Genome.* R.G. Landes, Austin, Texas, pp. 61–79.

Page, D.C., Mosher, R., Simpson, E.M., Fisher, E.M., Mardon, G., Pollack, J., McGillivray, B., de la Chapelle, A. and Brown, L.G. (1987) The sex-determining region of the human Y chromosome encodes a finger protein. *Cell* 51, 1091–1104.

Palmiter, R.D. and Brinster, R.L. (1986) Germ-line transformation of mice. *Annual Review of Genetics* 20, 465–499.

Passey, R., Glenn, W. and Mackinlay, A. (1996) Exon skipping in the ovine alpha s1-casein gene. *Comparative Biochemical Physiology B Biochemistry and Molecular Biology* 114, 389–394.

Payen, E.J. and Cotinot, C.Y. (1993) Comparative HMG-box sequences of the SRY gene between sheep, cattle and goats. *Nucleic Acids Research* 21, 2772.

Payen, E.J. and Cotinot, C.Y. (1994) Sequence evolution of SRY gene within Bovidae family. *Mammalian Genome* 5, 723–725.

Pech, M., Streeck, R.E. and Zachau, H.G. (1979) Patchwork structure of a bovine satellite DNA. *Cell* 18, 883–893.

Pépin, L., Amigues, Y., Mahé, M.-F., Persuy, M.-A. and Leroux, C. (1994) Linkage disequilibrium between two multi-allelic markers at the cluster of caseins in goat. In: *International Conference on Animal Genetics*. Prague, Abstract G-27.

Pépin, L., Amigues, Y., Lépingle, A., Berthier, J.-L., Bensaïd, A. and Vaiman, D. (1995) Sequence conservation of microsatellites between cattle (*Bos taurus*) and goat (*Capra hircus*), and related species: examples of use in parentage testing and phylogeny analysis. *Heredity* 74, 53–61.

Peura, T., Hyttinen, J.M., Turunen, M., Aalto, J., Rainio, V. and Janne, J. (1991) Birth of calves developed from embryos of predetermined sex. *Acta Veterinaria Scandinavia* 32, 283–286.

Plucienniczak, A., Shownovki, J. and Jaworski, B. (1982) Nucleotide sequence of bovine 1. 715 satellite DNA and its relation to other bovine satellite sequences. *Journal of Molecular Biology* 158, 293–304.

Pluta, A.F., Mackay, A.M., Ainsztein, A.M., Goldberg, I.G. and Arnshaw, W.C. (1995) The centromere: hub of chromosomal activities. *Science* 270, 1591–1594.

Pollevick, G.D., Giambiagi, S., Mancardi, S., de Luca, L., Burrone, O., Frasch, A.C. and Ugalde, R.A. (1992) Sex determination of bovine embryos by restriction fragment polymorphisms of PCR amplified ZFX/ZFY loci. *Biotechnology* 10, 805–807.

Ponce de Leon, F.A., Ambady, S., Hawkins, G.A., Kappes, S.M., Bishop, M.D., Robl, J.M. and Beattie, C.W. (1996) Development of bovine X chromosome linkage group and painting probes to assess cattle, sheep, and goat X chromosome segment homologies. *Proceedings of the National Academy of Sciences of the USA* 93, 3450–3454.

Popescu, C.P. (1990) Consequences of abnormalities of chromosome structure in domestic animals. *Reproduction Nutrition Development* Suppl. 1, 105S–116S.

Popescu, C.P., Cotinot, C., Boscher, J. and Kirszenbaum, M. (1988) Chromosomal localization of a bovine male specific probe. *Annals of Genetics* 31, 39–42.

Popescu, C.P., Long, S., Riggs, P., Womack, J., Schmutz, S., Fries, R. and Gallagher, D.S. (1996) Standardization of cattle karyotype nomenclature: report of the committee for the standardization of the cattle karyotype. *Cytogenetics and Cell Genetics* 74, 259–261.

Rao, K.B. and Totey, S.M. (1992) Sex determination in sheep and goats using bovine Y-chromosome specific primers via polymerase chain reaction: potential for embryo sexing. *Indian Journal of Experimental Biology* 30, 775–777.

Rao, K.B., Pawshe, C.H. and Totey, S.M. (1993) Sex determination of *in vitro* developed buffalo (*Bubalus bubalis*) embryos by DNA amplification. *Molecular Reproduction and Development* 36, 291–296.

Rasmussen, L.K., Due, H.A. and Petersen, T.E. (1995) Human alpha s1-casein: purification and characterization. *Comparative Biochemistry and Physiology B, Biochemistry and Molecular Biology* 111, 75–81.

Raudsepp, T., Fronicke, L., Scherthan, H., Gustavsson, I. and Chowdhary, B.P. (1996) Zoo-FISH delineates conserved chromosomal segments in horse and man. *Chromosome Research* 4, 218–225.

Rawlings, C.J. and Searls, D.B. (1997) Computational gene discovery and human disease. *Current Opinion in Genetics and Development* 7, 416–423.

Rettenberger, G., Klett, C., Zechner, U., Bruch, J., Just, W., Vogel, W. and Hameister, H. (1995a) ZOO-FISH analysis: cat and human karyotypes closely resemble the putative ancestral mammalian karyotype. *Chromosome Research* 3, 479–486.

Rettenberger, G., Klett, C., Zechner, U., Kunz, J., Vogel, W. and Hameister, H. (1995b) Visualization of the conservation of synteny between humans and pigs by heterologous chromosomal painting. *Genomics* 26, 372–378.

Ribadeau-Dumas, B., Brignon, G., Grosclaude, F. and Mercier, J.-C. (1972) Primary structure of bovine beta casein: complete sequence. *European Journal of Biochemistry* 25, 505–514.

Richard, F., Lombard, M. and Dutrillaux, B. (1996) ZOO-FISH suggests a complete homology between human and capuchin monkey (Platyrrhini) euchromatin. *Genomics* 36, 417–423.

Rijnkels, M., Wheeler, D.A., de Boer, H.A. and Pieper, F.R. (1997a) Structure and expression of the mouse casein gene locus. *Mammalian Genome* 8, 9–15.

Rijnkels, M., Kooiman, P.M., de Boer, H.A. and Pieper, F.R. (1997b) Organization of the bovine casein gene locus. *Mammalian Genome* 8, 148–152.

Rijnkels, M., Meershoek, E., de Boer, H.A. and Pieper, F.R. (1997c) Physical map and localization of the human casein gene locus. *Mammalian Genome* 8, 285–286.

Rogers, J.H. (1985) The origin and evolution of retroposons. *International Review of Cytology* 93, 187–279.

Ron, M., Band, M., Yanai, A. and Weller, J.I. (1994) Mapping quantitative trait loci with DNA microsatellites in a commercial dairy cattle population. *Animal Genetics* 25, 259–264.

Scherthan, H., Cremer, T., Arnason, U., Weier, H.U., Lima-de-Faria, A. and Fronicke, L. (1994) Comparative chromosome painting discloses homologous segments in distantly related mammals. *Nature Genetics* 6, 342–347.

Schibler, L., Vaiman, D., Oustry, A., Guinec, N., Dangy-Caye, A.L., Billault, A. and Cribiu, E.P. (1998a) Construction and extensive characterization of a goat bacterial artificial chromosome library with threefold genome coverage. *Mammalian Genome* 9, 119–124.

Schibler, L., Vaiman D., Oustry, A., Giraud-Delville, C. and Cribiu, E.P. (1998b) Comparative gene mapping: a fine-scale survey of chromosome rearrangements between ruminants and humans. *Genome Research* 8, 901–915.

Schlieben, S., Erhardt, G. and Senft, B. (1991) Genotyping of bovine kappa-casein (kappa-CNA, kappa-CNB, kappa-CNC, kappa-CNE) following DNA sequence amplification and direct sequencing of kappa-CNE PCR product. *Animal Genetics* 22, 333–342.

Schmitt-Ney, M., Doppler, W., Ball, R.K. and Groner, B. (1991) Beta-casein gene promoter activity is regulated by the hormone-mediated relief of transcriptional repression and a mammary-gland-specific nuclear factor. *Molecular and Cellular Biology* 11, 3745–3755.

Schmitz, A., Oustry, A., Chaput, B., Yerle, M., Milan, D., Frelat, G. and Cribiu, E.P. (1995) The bovine bivariate flow karyotype and peak identification by chromosome painting with PCR-generated probes. *Mammalian Genome* 6, 415–420.

Schmutz, S.M., Marquess, F.L., Berryere, T.G. and Moker, J.S. (1995) DNA marker-assisted selection of the polled condition in Charolais cattle. *Mammalian Genome* 6, 710–713.

Schneider-Gädicke, A., Beer-Romero, P., Brown, L.G., Nussbaum, R. and Page, D.C. (1989) ZFX has a gene structure similar to ZFY, the putative human sex determinant, and escapes X inactivation. *Cell* 57, 1247–1258.

Schroder, A., Miller, J.R., Thomsen, P.D., Roschlau, K., Avery, B., Poulsen, P.H., Schmidt, M. and Schwerin, M. (1990) Sex determination of bovine embryos using the polymerase chain reaction. *Animal Biotechnology* 1, 121–133.

Scott, A.F., Schmeckpeper, B.J., Abdelrazik, M., Comey, C.T., O'Hara, B., Rossiter, J.P., Cooley, T., Heath, P., Smith, K.D. and Margolet, L. (1987) Origin of the human L1 elements: proposed progenitor genes deduced from a consensus DNA sequence. *Genomics* 1, 113–125.

Seibert, B., Erhardt, G. and Senft, B. (1985) Procedure for simultaneous phenotyping of genetic variants in cow's milk by isoelectric focusing. *Animal Blood Groups and Biochemical Genetics* 16, 183–191.

Shizuya, H., Birren, B., Kim, U.-J., Mancino, V., Slepak, T., Tachiri, Y. and Simon, M. (1992) Cloning and stable expression of 300-kilobase-pair fragments of human DNA in *Escherischia coli* using an F-factor-based vector. *Proceedings of the National Academy of Sciences of the USA* 89, 8794–8797.

Silverman, G.A., Ye, R.D., Pollock, K.M., Sadler, J.E. and Korsmeyer, S.J. (1989) Use of yeast artificial chromosome clones for mapping and walking within human chromosome segment 18q21.3. *Proceedings of the National Academy of Science of the USA* 86, 7485–7489.

Sinclair, A.H., Berta, P., Palmer, M.S., Hawkins, J.R., Griffiths, B.L., Smith, M.J., Foster, W., Frischauf, A.-M., Lovell-Badge, R. and Goodfellow, P.N. (1990) A gene from the human sex-determining region encodes a protein with homology to a conserved DNA-binding motif. *Nature* 346, 240–244.

Skowronski, J., Plucienniczak, A., Bednarek, A. and Jaworski, J. (1984) Bovine 1.709 satellite: recombination hotspots and dispersed repeated sequence *Journal of Molecular Biology* 177, 399–416.

Smith, T.P.L., Alexander, L.J., Sonstegard, T.S., Yoo, J., Beattie, C.W. and Broom, M.F. (1996) Construction and characterization of a large insert bovine YAC library with fivefold genomic coverage. *Mammalian Genome* 7, 155–156.

Solinas-Toldo, S., Lengauer, C. and Fries, R. (1995) Comparative genome map of human and cattle. *Genomics* 27, 489–496.

Sonstegard, T.S., Lopez-Corrales, N.L., Kappes, S.M., Stone, R.T., Ambady, S., Ponce de Leon, F.A. and Beattie, C.W. (1997a) An integrated genetic and physical map of the bovine X chromosome. *Mammalian Genome* 8, 16–20.

Sonstegard, T.S., Ponce de Leon, F.A., Beattie, C.W. and Kappes, S.M. (1997b) A chromosome-specific microdissected library increases marker density on bovine chromosome 1. *Genome Research* 7, 76–80.

Spence, S.E., Young, R.M., Garner, K.J. and Lingrel, J.B. (1985) Localization and characterization of members of a family of repetitive sequences in the goat beta globin locus. *Nucleic Acids Research* 13, 2171–2186.

Stewart, A.F., Willis, I.M. and Mackinlay, A.G. (1984) Nucleotide sequences of bovine alpha S1- and kappa-casein cDNAs. *Nucleic Acids Research* 12, 3895–3907.

Stewart, A.F., Bonsing, J., Beattie, C.W., Shah, F., Willis, I.M. and Mackinlay, A.G. (1987) Complete nucleotide sequences of bovine alpha S2- and beta-casein cDNAs: comparisons with related sequences in other species. *Molecular Biology and Evolution* 4, 231–241.

Strauss, W.M., Jaenisch, E. and Jaenisch, R. (1992) A strategy for rapid production and screening of yeast artificial chromosome libraries. *Mammalian Genome* 2, 150–157.

Sun, H.S., Cai, L., Davis, S.K., Taylor, J.F., Doud, L.K., Bishop, M.D., Hayes, H., Barendse, W., Vaiman, D., McGraw, R.A., Hirano, T., Sugimoto, Y. and Kirkpatrick,

B.W. (1997) Comparative linkage mapping of human chromosome 13 and bovine chromosome 12. *Genomics* 39, 47–54.

Szemraj, J., Plucienniczak, G., Jaworski, J. and Plucienniczak, A. (1995) Bovine Alu-like sequences mediate transposition of a new site-specific retroelement. *Gene* 152, 261–264.

Takada, T., Lida, K., Awaji,T., Itoh, K., Takahashi, R., Shibui, A., Yoshida, K., Sugano, S., and Tsujimoto, G. (1997) Selective production of transgenic mice using green fluorescent protein as a marker. *Nature Biotechnology* 15, 458–461.

Tautz, D. and Renz, M. (1984) Simple sequences are ubiquitous repetitive components of eukaryotic genomes. *Nucleic Acids Research* 12, 4127–4138.

Thomsen, P.D. and Jorgensen, C.B. (1994) Distribution of two conserved, male-enriched repeat families on the *Bos taurus* Y chromosome. *Mammalian Genome* 5, 171–173.

Threadgill, D.W. and Womack, J.E. (1990) Genomic analysis of the major bovine milk protein genes. *Nucleic Acids Research* 18, 6935–6942.

Traver, C.N., Klapholz, S., Hyman, R.W. and Davis, R.W. (1989) Rapid screening of a human genomic library in yeast artificial chromosomes for single-copy sequences. *Proceedings of the National Academy of Sciences of the USA* 86, 5898–5902.

Tucker, P.K. and Ludigran, B.L. (1993) Rapid evolution of the sex determining locus in old world mice and rats. *Nature* 364, 715–717.

Vaiman, D., Osta, R., Mercier, D., Grohs, C. and Leveziel, H. (1992) Characterization of five new bovine dinucleotide repeats. *Animal Genetics* 23, 537–541.

Vaiman, D., Mercier, D., Moazami-Goudarzi, K., Eggen, A., Ciampolini, R., Lépingle, A., Velmala, R., Kaukinen, J., Varvio, S.-L., Martin, P., Levéziel, H. and Guérin, G. (1994) A set of 99 cattle microsatellites: characterization, synteny mapping and polymorphism. *Mammalian Genome* 5, 288–297.

Vaiman, D., Koutita, O., Oustry, A., Elsen, J.-M., Manfredi, E., Fellous, M. and Cribiu, E.P. (1996a) Genetic mapping of the autosomal region involved in XX sex-reversal and horn development in goats. *Mammalian Genome* 7, 133–137.

Vaiman, D., Schibler, L., Bourgeois, F., Oustry, A., Amigues, Y. and Cribiu, E.P. (1996b) A genetic linkage map of the male goat genome. *Genetics* 144, 279–305.

Vaiman, D., Pailhoux, E., Schmitz, A., Giraud-Delville, C., Cotinot, C. and Cribiu, E.P. (1997a) Mass production of genetic markers from a limited number of sorted chromosomes. *Mammalian Genome* 8, 153–156.

Vaiman, D., Schibler, L., Oustry, A., Furet, J.-P., Barendse, W. and Cribiu, E.P. (1997b) A cytogenetically anchored high resolution linkage map of bovine chromosome 1. *Cytogenetics and Cell Genetics* 79, 204–207.

Viersbach, R., Schwanitz, G. and Nothen, M.M. (1994) Delineation of marker chromosomes by reverse chromosome painting using only a small number of DOP-PCR amplified microdissected chromosomes. *Human Genetics* 93, 663–667.

Vilotte, J.-L., Soulier, S. and Mercier, J.-C. (1993) Complete sequence of a bovine alpha-lactalbumin pseudogene: the region homologous to the gene is flanked by two directly repeated LINE sequences. *Genomics* 16, 529–532.

Wachtel, S.S., Koo, G.C. and Boyse, E.A. (1975) Evolutionary conservation of H-Y ('male') antigen. *Nature* 254, 270–272.

Wada, M., Abe, K., Okumura, K., Taguchi, H., Kohno, K., Imamoto, F., Schlessinger, D. and Kuwano, M. (1994) Chimeric YACs were generated at unreduced rates in conditions that suppress coligation. *Nucleic Acids Research* 22, 1651–1654.

Waddy, C.T. and Mackinlay, A.G. (1971) Protein kinase activity from lactating bovine mammary gland. *Biochimica Biophysica Acta* 250, 491–500.

Wall, R.J. (1997) A new lease on life for transgenic livestock. *Nature Biotechnology* 15, 416–417.

Wall, R.J., Hawk, H.W. and Nel, N. (1992) Making transgenic livestock: genetic engineering on a large scale. *Journal of Cellular Biochemistry* 49, 113–120.

Watson, C.J., Gordon, K.E., Robertson, M. and Clark, A.J. (1991) Interaction of DNA-binding proteins with a milk protein gene promoter *in vitro*: identification of a mammary gland-specific factor. *Nucleic Acids Research* 19, 6603–6610.

Weiner, A.M., Deininger, P.L. and Efstratiadis, A. (1986) Nonviral retroposons: genes, pseudogenes, and transposable elements generated by the reverse flow of genetic information. *Annual Review of Biochemistry* 55, 631–661.

Weinstock, K.G., Kirkness, E.F., Lee, N.H., Earle-Hughes, J.A. and Venter, J.C. (1994) cDNA sequencing: a means of understanding cellular physiology. *Current Opinion in Biotechnology* 5, 599–603.

Weller, J.I., Kashi, Y. and Soller, M. (1990) Power of daughter and granddaughter designs for determining linkage between marker loci and quantitative trait loci in dairy cattle. *Journal of Dairy Science* 73, 2525–2537.

Wilmut, I., Hooper, M.L. and Simons, J.P. (1991) Genetic manipulation of mammals and its application in reproductive biology. *Journal of Reproduction and Fertility* 92, 245–279.

Womack, J.E. and Moll, Y.D. (1986) Gene map of the cow: conservation of linkage with mouse and man. *Journal of Heredity* 77, 2–7.

Yerle, M., Schmitz, A., Milan, D., Chaput, B., Monteagudo, L., Vaiman, M., Frelat, G. and Gellin, J. (1993) Accurate characterization of porcine bivariate flow karyotype by PCR and fluorescence *in situ* hybridization. *Genomics* 16, 97–103.

Zelnick, C.R., Burks, D.J. and Duncan, C.H. (1987) A composite transposon 3' to the cow fetal globin gene binds a sequence specific factor. *Nucleic Acids Research* 15, 10437–10453.

Zimmer, R. and Verrinder Gibbins, A.M. (1997) Construction and characterization of a large-fragment chicken bacterial artificial chromosome library. *Genomics* 42, 217–226.

Molecular Genetics of Molecules with Immunological Functions: Major Histocompatibility Complex, Immunoglobulins, T-cell Receptors, Cytokines and their Receptors

H.A. Lewin*, M. Amills and V.K. Ramiya

Laboratory of Immunogenetics, Department of Animal Sciences, 1201 W. Gregory Drive, University of Illinois at Urbana-Champaign, IL 61801, USA

Introduction	164
Major Histocompatibility Complex	165
Structure–function of major histocompatibility complex molecules	166
Genomic organization of the cattle major histocompatibility complex	167
Class I genes and proteins	168
Class IIa genes and proteins	170
Class IIb genes and proteins	171
Class I peptide-binding motifs	171
The Immunoglobulins	172
Immunoglobulin genetics	173
Cattle C_H and C_L genes	175
Cattle V_λ genes	176
Cattle V_H genes	176
T-cell Receptors	177
T-cell subtypes and their receptors	177

*Corresponding author: tel. (217) 333-5998; fax: (217) 244-5617; email: h-lewin@ux1.cso.uiuc.edu

©CAB International 1999. *The Genetics of Cattle* (eds R. Fries and A. Ruvinsky)

T-cell receptor structure, gene organization and
diversification 179
Cattle T-cell receptor gene segments 179
Cytokines and Cytokine Receptors 180
Cytokine genes of cattle 181
Interferons 184
Tumour necrosis factors 185
Transforming growth factors 185
Colony-stimulating factors 186
Cytokine receptors 186
Summary 186
Acknowledgments 187
References 188

Introduction

The immune system of vertebrates consists of diverse organs, tissues and cells. Despite the system's complexity, its components are interrelated and act in a highly coordinated and specific manner to recognize, eliminate and recall foreign macromolecules without destroying the host. In doing so, the immune system must distinguish between self and non-self molecules. Acquired immunity, whether cellular or humoral, permits the elimination of all substances (living and non-living) within the body that are not recognized as self. The major characteristics of acquired immunity are: (i) specificity; (ii) diversity; (iii) memory; (iv) distinguishing self from non-self; and (v) autoregulation of responses. These characteristics are rendered by the system's major components; lymphocytes (primarily T and B cells) and phagocytic cells (macrophages and dendritic cells) which work in concert in mounting effective immune responses.

Recognition of non-self antigens is a primary event in the induction of immune responses. Unlike B lymphocytes, T cells do not recognize antigens from invading pathogens in native form. T cells depend on specialized cells (dendritic cells, macrophages and B cells) for processing and display of peptides bound to major histocompatibility complex (MHC) antigens on their surface. Additionally, a set of costimulatory molecules must interact with their ligands simultaneously with MHC–peptide (antigen) stimulus to achieve a successful T-cell activation signal (Jenkins et al., 1987; Mondino et al., 1996). The B cells depend on helper T (T_H) cells to expand clonally and differentiate into antibody-producing plasma cells (humoral response), while macrophages acquire their enhanced killing capability through activated T cells. In general, functional communication among cells of the immune system is carried out through their specific cell-surface receptors and soluble regulatory polypeptides called cytokines. Upon activation, T_H cells may differentiate into T-cell subsets that secrete cytokines, thus promoting cellular immune responses (T_H1 responses) or humoral responses (T_H2 responses), depending on host genetics and various factors in peripheral lymphoid microenvironments. The T-cell subsets that produce cytokine patterns associated with these responses are termed T_H1 and T_H2 cells, but clearly distinct patterns of cytokine production by individual cells *in vivo* do not exist for most species. However, *in vitro,* cloned T cells can be distinguished as T_H1 or T_H2 subtypes by their cytokine expression pattern (reviewed by Kelso, 1995). In this review, we limit our presentation to the 'classical' molecules that function in the immune system: MHC, immunoglobulins (Igs), T-cell receptors (TCRs) for antigen, cytokines and their receptors, as they are known for cattle, with presentation of comparative structure–function relationships. As research on the immune systems of cattle and other vertebrates progresses, new molecules will be identified that will be relevant to animal health and disease. The greatest challenge in immunogenetics during the next decade will be to understand how this complex system of hundreds of genes is regulated and ultimately might be manipulated to improve the yield and efficiency of milk and meat production.

Major Histocompatibility Complex

The MHC was originally named for its role in genetically determining the fate of allografts. The true function of MHC molecules eluded immunologists for nearly 40 years until the Nobel Prize-winning discovery by Zinkernagel and Doherty (1974), who showed that T cells require antigens be presented to cytotoxic T lymphocytes (CTLs) in the context of self MHC antigens. The advent of molecular biology led to an explosion of knowledge on the fine structure of genes encoding MHC molecules and the chromosomal organization of literally hundreds of other genes contained within the neighbourhood of the complex, some associated with immunological function and others not (for a general review see Janeway and Travers, 1997). Moreover, the high degree of conservation of MHC genes permitted the use of cloned probes for studying the evolutionary history of these genes across phylogenetically distant species. For example, MHC genes have been identified in all vertebrates except jawless fish (agnatha) using mammalian gene probes. More recently, comprehensive sequencing of alleles and the structural determination of MHC molecules have provided a detailed understanding of the role of MHC-encoded polymorphic molecules in antigen presentation. The elucidation of the three-dimensional structure of MHC molecules (Bjorkman *et al.*, 1987; Brown *et al.*, 1993) revealed precisely how the enormous polymorphism of MHC molecules relates to genetic control of immune responses and disease susceptibility.

The promise of manipulating immune responses for the purpose of improving resistance to infectious diseases attracted immunologists and geneticists to study the MHC of cattle. Accounts of the discovery and characterization of the cattle MHC have been reviewed elsewhere and will not be recounted here, nor will the voluminous literature dealing with disease association studies (for a review see Lewin, 1996). Below, the essential features of the classical MHC molecules (class I and class II) are reviewed, and a brief overview of the bovine lymphocyte antigen (BoLA) system is presented, with attention given to recent information. Discussion of the organization and function of other genes located within the MHC, including many with immunological function, will be limited, necessarily. For a listing of the class III genes (a catch-all terminology for genes located between the class I and class II genes in mice and humans) and other genes found within the MHC, refer to Campbell and Trowsdale (1993).

Structure–function of major histocompatibility complex molecules

Histocompatibility antigens are grouped into two types, class I and class II, on the basis of sequence similarity, structure and function. Class I molecules are heterodimers, consisting of an α (heavy) chain with molecular mass of 45 kDa, associated non-covalently with β2-microglobulin (12 kDa). The class I heavy chains are assembled in the lumen of the endoplasmic reticulum (ER), where they are loaded with peptides produced in the cytosol by the 20S multicatalytic proteasome. Two of the proteasome subunits, LMP-2 and LMP-7, are encoded within the MHC, proximal to the class II genes. These peptides are actively transported from the cytosol to the ER by the MHC-encoded transporters associated with antigen processing (TAP-1 and TAP-2 heterodimers) before they are directed to the cell surface. The class I molecules are expressed on all nucleated cells, and their main function is to present peptides to $CD8^+$ T lymphocytes, which kill virus-infected and neoplastic cells.

In contrast to class I molecules, both chains of class II molecules (mol. mass 33 + 28 kDa) are encoded by genes within the MHC. The α chains form non-covalently-associated heterodimers with β chains within the lumen of the ER. An invariant chain (Ii), also known as class II-associated invariant chain peptide (CLIP), binds non-covalently to nascent class II molecules to prevent premature loading with peptides. The class II–CLIP complexes are exported from the ER, where they come into contact with peptides produced in low-pH endosomes. These peptides are derived from intracellular parasites, such as viruses and bacteria, which replicate in the cytosol. Although the precise intracellular trafficking of class II molecules is still being defined, it appears that CLIP is removed in the endosomes by proteolytic degradation, and peptide is then loaded into class II molecules with the assistance of another class II-like molecule, DM, also encoded within the MHC. Class II molecules are expressed on 'professional' antigen-presenting cells (APCs), such as dendritic cells and macrophages. Class II molecules on APCs present peptides derived from extracellular pathogens to $CD4^+$ T cells, which, once stimulated, activate macrophages and B cells to generate inflammatory and antibody responses, respectively.

The three-dimensional structure of class I molecules and class II molecules determined by X-ray crystallography revealed sites for peptide binding and interaction with the TCR for antigen (Bjorkman *et al.*, 1987; Garrett *et al.*, 1989;

Madden et al., 1991; Brown et al., 1993). Primary sequence data from all vertebrate species suggest that the structure of MHC molecules is highly conserved. The membrane distal portion of MHC class I molecules consists of an eight-stranded β-pleated sheet, topped by two α helices. The peptide-binding region is formed from two subdomains (α1 and α2, class I; α1 and β1, class II), each composed of a four-stranded β sheet and a single helix. The two helices are separated by an extended groove approximately 30 Å long and 12 Å at its widest point. The peptide-binding groove of class I molecules tapers at both ends to a width of approximately 5 Å and is 'closed' by an amino acid residue with a bulky side-chain, usually Tyr 84 and Trp 167, conserved in all class I molecules (Saper et al., 1991). In the class II molecules, the blocking side-chains at the ends of the cleft are replaced by smaller ones and/or repositioned by changes in secondary structure (Madden, 1995). The class I antigen-binding groove contains six pockets, into which amino acid residues of processed peptides are non-covalently embedded. In general, class I molecules bind peptides of 8–11 amino acid residues, one or two of which are allele-specific (anchor residues). Naturally processed peptides bound to class II molecules are 13–17 amino acids long and also share similar structural features among peptides that bind to the same allele. Peptides bind in an elongated fashion along the length of the groove of both class I and class II molecules, with the exposed region(s) formed by a conformational kink(s) in the peptide. These exposed regions, intimately associated with MHC molecules, are what is recognized by specific receptors on T lymphocytes.

Genomic organization of the cattle major histocompatibility complex

The BoLA system is located on two distinct segments of cattle chromosome 23 ((*Bos taurus*) (BTA) 23; Table 7.1). This organization distinguishes BoLA from the MHC of rodents and primates (see Lewin, 1996; Fig. 7.1). The BoLA class IIa, class III and class I regions map to BTA 23q22 approximately 35 centimorgans (cM) from the centromere, whereas the class IIb region is located 15–30 cM proximal to the centromere (Lewin, 1996; Band et al., 1998). The class I region spans about 770–1650 kb and contains 10–20 different genes (Bensaid et al., 1991; reviewed in Lewin, 1996). The class IIa cluster contains the *DQ* and *DR* regions tightly linked to the class III and the class I genes (Gwakisa et al., 1994; van Eijk et al., 1995), similar in organization to the human orthologues. Physical mapping of the class II region (Groenen et al., 1990; reviewed in Lewin, 1996) has allowed the estimation of the distance between *DQ* and *DR* regions (400 kb), *DRA* and *DRB* (340 kb) and *DQA* and *DQB* (170 kb). The class IIb gene cluster contains *LMP2, LMP7, DNA, DOB, DYA, DYB* and *DIB* (reviewed in Lewin, 1996; Band et al., 1998). Recent data suggest that the organization of the class II genes in cattle arose by a chromosomal inversion in an ancestral mammal (Band et al., 1998).

Significant variations in the recombination rate between different bulls in the interval between *DRB3* and *DYA* have been observed (Park et al., 1995;

Table 7.1. Chromosomal locations of some molecules important in immune function.

Locus name	Chromosomal location		
	Mouse	Human	Cattle
IGH@	12F1–2	14q32.33	21q23–q24
IGK@	6C	2p12	nd
IGL@	16A–B1	22q11.2	17
IGJ	5	4q21	6
CD3E	9B	11q23.3	15
CD3D	9B	11q23.3	15
CD3G	9B	11q23.3	nd
CD3Z	1G–H	1q22–q23	3q11–q14
CD4	6F	12pter–p12	nd
CD8A	6C	2p12	11
CD8B	6C	2p12	nd
CD28	1C1–C3	2q33	nd
CD80 (B7.1)	16B5	3q13.3–q21	nd
TCRA	14D1–D2	14q11.2	10
TCRB	6A–C	7q35	4
TCRG	13A2–A3	7p15–p14	4
TCRD	14D1–D2	14q11.2	10
MHC	17B–D	6p21.3	23q22

nd, not determined; @ indicates a cluster of loci.

Park and Lewin, 1999), suggesting the existence of a polymorphic recombination hot spot or suppressor in the region containing the inversion break point. Further evidence for unusual recombination activity in this interval was obtained from the genetic analysis of secondary oocytes and first polar bodies (Jarrell *et al.*, 1995). The coming onslaught of deoxyribonucleic acid (DNA) sequence data will reveal interesting information on the comparative organization and evolution of the cattle MHC.

Class I genes and proteins

Southern blot and restriction fragment length polymorphism (RFLP) analysis of the cattle class I region provided evidence for the existence of about 10–20 class I genes (Lindberg and Andersson, 1988). Molecular cloning and sequencing of bovine class I complementary DNAs (cDNAs) from a single bull heterozygous for BoLA class I serological antigens yielded six different sequences, demonstrating the presence of at least three different transcribed loci (Garber *et al.*, 1994). Interestingly, one of the cDNA clones (BS1b) did not contain the typical features of a classical class I gene (Garber *et al.*, 1994). The predicted amino acid of BS1b sequence lacks a phosphorylation site in the cytoplasmic (CY) domain and the aliphatic amino acid Leu 294 was replaced by the basic

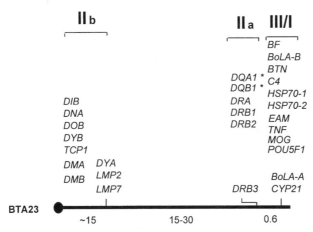

Fig. 7.1. A map of structural genes located within the cattle MHC. Approximate map distances are compiled from multiple sources and expressed in centiMorgans. *DQA* and *DQB* genes are duplicated in a subset of haplotypes. Several loci have been more precisely mapped using radiation hybrid analysis (Band *et al.*, 1998). Genes mapping distal to the class I region on BTA23 are not shown.

amino acid His, a feature that perturbs the strongly hydrophobic 287–308 region. This sequence may correspond to a non-classical class I (class Ib) gene, as such unusual amino acid replacements have also been observed in the mouse *Qa* and *TL* genes.

The International Society of Animal Genetics (ISAG) BoLA Nomenclature Committee (Davies *et al.*, 1997) summarized 21 different class I sequences, grouping them into two main clusters according to the length of the transmembrane (TM) domain (35 or 37 amino acids). However, the exact number of transcribed classical class I genes in each haplotype has not been clearly established (Garber *et al.*, 1994; Ellis *et al.*, 1996).

The predicted amino acid sequences of cattle class I genes have the typical features of the class I molecules: Cys at positions 101 and 164 of the $\alpha 2$ domain and at positions 203 and 259 of the $\alpha 3$ domain for the formation of disulphide bridges, a single conserved site for N-linked glycosylation at Asn 86 of the $\alpha 1$ domain, and a conserved site for phosphorylation at Ser 333 of the CY domain (Garber *et al.*, 1994).

Current evidence suggests that there are at least three class I molecules expressed on the cell surface (al-Murrani *et al.*, 1994). The multi-band patterns observed using one-dimensional isoelectric focusing cannot be explained by the occurrence of alternative messenger ribonucleic acid (mRNA) splicing or by post-translational modifications. The serological analysis of class I polymorphism has led to the identification of 27 internationally accepted BoLA-A specificities and 23 others that have not been given a locus designation (Davies *et al.*, 1994; reviewed in Lewin, 1996).

Class IIa genes and proteins

BoLA-DRA and -DRB

The bovine *DRA* gene encodes a mature protein of 230 amino acids and is monomorphic (Davies *et al.*, 1997). The *BoLA-DRA* gene is the orthologue of human leucocyte antigen (*HLA*)- *DRA*. There are three *DRB* genes but only *DRB3* appears to be functional. The *DRB3* gene is transcribed at high levels in the periphery, and two different RNA transcripts (1970 and 1460 bp) have been described (Muggli-Cockett and Stone, 1989). The *BoLA-DRB1* gene is a pseudogene; it has stop codons in exons 2 and 4 and a 2 bp deletion at positions 141–142 resulting in a frame-shift mutation (Muggli-Cockett and Stone, 1988; Groenen *et al.*, 1990). Moreover, a single base deletion at the 3′ end of the TM-encoding exon negates the consensus sequence for intron splicing. The *DRB1* pseudogene shows a biallelic polymorphism (Muggli-Cockett and Stone, 1988). The *DRB2* gene has the typical features of a functional *DRB* gene: Cys residues at positions 15 and 79 of β1 and 117 and 173 of β2 domains. However, it also has some unusual characteristics: the splice junction of the 3′ end of exon 3 of β2 is GA instead of GT, and there is no glycosylation site at position 19 of the β1-coding exon (Muggli-Cockett and Stone, 1989). The *DRB2* gene appears to be transcribed at a low level, if at all, in the periphery. The *DRB3* gene encodes a functional DRβ chain and has an overall identity of 83.1% and 86.3% at the nucleotide level with *DRB1* and *DRB2*, respectively. The *DRB3* gene displays a high degree of polymorphism, with at least 66 different alleles. Most variation is concentrated in the second exon, which encodes peptide-binding residues. Comparison of *DRB3* polymorphism in African and European cattle has revealed only a few novel sequence motifs. This patchwork polymorphism may be generated by the occurrence of parallel point mutations due to selective pressure, or to the exchange of short sequences by intragenic recombination (Mikko and Andersson, 1995).

BoLA-DQA and -DQB

Expression of BoLA-DQ antigens has been demonstrated using monoclonal antibodies and isoelectric focusing (reviewed in Lewin, 1996). They are termed 'DQ' because the genes that encode *BoLA-DQ* have strong sequence similarity with *HLA-DQ* genes. The cattle genome contains two distinct *DQA* genes (*DQA1* and *DQA2*), which differ by about 25% in their second exon sequences (van der Poel *et al.*, 1990; Sigurdardóttir *et al.*, 1991). Haplotypes carrying one or both *DQA* genes have been observed (Sigurdardóttir *et al.*, 1991). To date, about 47 *DQA* sequences have been reported. There may be a third *DQA* gene in cattle of African origin (Ballingall *et al.*, 1997).

The *DQB* genes are duplicated in some haplotypes in *Bos taurus* (Sigurdardóttir *et al.*, 1992) and *B. indicus* (Marello *et al.*, 1995). In duplicated haplotypes, both *DQB* genes are transcriptionally active (Xu *et al.*, 1994; Marello *et al.*, 1995). The *DQB1* gene is present in all haplotypes, whereas *DQB2, DQB3* and *DQB4* are present only in duplicated haplotypes

(Sigurdardóttir *et al.*, 1992). The *DQB3* and *DQB4* genes are very divergent compared with *DQB1* and *DQB2*. The *DQB3* gene has multiple substitutions at positions that are conserved in β-chain genes and thus may be a pseudogene. The *DQB* genes are highly polymorphic, with 40 different sequences identified to date (Davies *et al.*, 1997). The variability indices calculated for each polymorphic site are, on average, lower for *DQB* compared with *DRB*.

Class IIb genes and proteins

As discussed above, the class IIb region is physically distant from the rest of the cattle MHC. Genes within the cluster have human and mouse orthologues, but several genes (*DYA*, *DIB*) do not, probably arising by gene duplication after the divergence of ruminants.

The class II-like gene *BoLA-DMA*, a chaperon for loading class II molecules with peptide, displays a high degree of overall similarity (> 70%) to its human and mouse orthologues (Niimi *et al.*, 1995). The *DYA* gene was first detected by Southern blot analysis as a cross-hybridizing fragment with a human *DQA* cDNA probe (Andersson *et al.*, 1988) and later confirmed as a novel class II gene by DNA sequencing (van der Poel *et al.*, 1990). It is possible *DYA* diverged from an ancestral *DQA*-like gene, or it may constitute a distinct group of class II genes (van der Poel *et al.*, 1990). The *DYA* gene does not appear to be expressed; thus its function is unknown. Another 'novel' BoLA class IIb gene is *DIB*, which has about 75% sequence similarity to *BoLA-DQB* (Stone and Muggli-Cockett, 1990). The presence of the *DIB* gene has been demonstrated in cattle, gaur, bison, sheep, gazelle, wapiti, serow, muntjak and giraffe. Northern analysis suggested that neither *DIB* nor *DOB* genes were expressed in the periphery (Stone and Muggli-Cockett, 1993).

Class I peptide-binding motifs

A ripe area for investigation is the identification of peptide-binding motifs of BoLA class I and class II alleles. This information will ultimately be important for the production of broadly effective peptide subunit vaccines. The sequencing of peptides bound to BoLA-A11 revealed that the majority of eluted peptides are nonamers (Hegde *et al.*, 1995). Position 2 (P2) was predominantly occupied by Pro and P9 by Ile or Val, while Ala and Pro were the preferred residues for P3 and P5, respectively (Table 7.2). These two positions may constitute auxiliary anchor residues. Peptides bound to BoLA-A31 and -A18 have also been characterized (Gaddum *et al.*, 1996). The majority of peptides were nonamers, but evidence was obtained for the presence of dodecamers bound to BoLA-A31. No specific anchor motif was detected for an A31-associated class I molecule at P2, but a hydrophobic Ile anchor at P3 was found (Table 7.2). For BoLA-A18, Gln was a strong anchor at P2. Surprisingly, there were three strong anchors at P2 (Val), P7 (Ile) and P10 (Tyr) for another class I

Table 7.2. Motifs of MHC class I-bound peptides in humans and cattle.*

Allele	Peptide sequence									References
	1	2	3	4	5	6	7	8	9	
HLA-A2.1		<u>L</u>							<u>V</u>	Rammensee et al., 1993
HLA-A2.5		M	E/K	V	K				<u>L</u>	
	V	Y	G	V	I			K		
		L	P	E	Y					
		I	F	D	L					
		Q	I		I					
HLA-B27		<u>R</u>								
BoLA-A11		<u>P</u>	A	I	P	V	L	Q	<u>I</u>	Hegde et al., 1995
		F	P		G	I	I	V	<u>V</u>	
BoLA-A18		<u>Q</u>		P						Gaddum et al., 1996
		I								
		L								
BoLA-A31		V	P	P	F			I		Bamford et al., 1995
		I	I							
		A								
		Q								

*Amino acids are in standard single-letter code. All known anchor residues are underlined. Other residues are either auxiliary anchors or not an anchor site.

protein associated with BoLA-A31-bearing haplotype. No anchor motifs at P9 were detected for any of these three class I molecules. Peptides eluted from BoLA-A20 MHC class I molecules isolated from bovine muscle-derived fibroblastoid cells persistently infected with the parainfluenza type-3 virus showed that P2 and P9 were occupied mostly by Lys and Arg, respectively (Bamford et al., 1995).

The Immunoglobulins

Humoral immunity is mediated by the Ig class of proteins. The Igs, commonly referred to as antibodies, are produced by B lymphocytes. Activated B cells differentiate to antibody-secreting plasma cells, which localize in the lamina propria of the gut and other mucosal tissues, as well as in specialized compartments of lymph nodes, spleen and bone marrow. Each B cell expresses surface Ig of unique specificity (the B-cell receptor (BCR)), allowing for 'clonal selection' by antigen. The BCR molecules recognize specific antigenic determinants (epitopes) on proteins, polysaccharides or nucleic acids in their native state. In humans and rodents, B cells are generated in bone marrow, whereas in ruminants, the predominant site of B-cell maturation is ileal Peyer's patch (IPP) (Reynaud et al., 1991). Mammals typically express some or all of the five known Ig heavy chain isotypes in serum and other body fluids: IgM, IgD, IgG, IgA and IgE. The different isotypes are associated with different Ig functions, e.g. Fc (fragment crystallizable) receptor binding and complement activation.

Ruminants have several features that distinguish their humoral immune system from that of other mammals. For example, ruminant IgG is passively acquired from colostrum via the gut after birth, whereas, in primates, IgGs are acquired by the fetus via transport across the placenta. Another apparent difference from primates and rodents is that ruminant B cells do not express IgD on their surface (Naessens, 1997). In this section, a general review of comparative Ig genetics is provided, followed by a summary of what is known for cattle. A detailed review of Ig structure and function in vertebrates can be found elsewhere (for example, see Janeway and Travers, 1997).

Immunoglobulin genetics

Immunoglobulins are large globular proteins shaped to a form a Y-like structure. Each Ig monomer molecule has four polypeptides, two identical light chains and two identical heavy chains. Heavy and light chains have N-terminal variable (V) and C-terminal constant (C) regions, as defined by their primary amino acid sequences. While the V regions determine antibody specificity the C regions carry out effector functions, such as binding to macrophages and activation of the classical pathway of complement activation. Amino acid variability in V regions is concentrated in areas called hypervariable or complementarity-determining regions (CDRs). There are three CDRs: CDR1, CDR2 and CDR3. The less variable framework residues comprise the classical Ig folds that expose the CDRs on the outer surface of the molecule. The CDRs, formed by association of heavy and light chains, make up a functional antigen-binding site. On the B-cell surface, Igs are non-covalently associated with additional molecules, such as Ig-α, and Ig-β heterodimers, which play an essential role in intracellular Ig transport and signal transduction. The Ig-α and Ig-β are encoded by *mb-1* and *B29* genes, respectively, and are structurally very similar to CD3γ, δ, and ϵ chains of T cells.

Mice and humans have three multigene families encoding Ig chains (κ, δ and H). Each family contains multiple V gene segments (~ 300 bp long), joining (J) gene segments (~ 50 bp long) and C gene segments (~ 300 bp long), all separated by introns of variable length. In addition to V and J segment genes, the H chain cluster contains short diversity (D) gene segments (~ 5–50 bp long). Among species, there are differences in the number and arrangement of genes in each cluster, which are reflected in the genes used for Ig production.

In differentiating B cells, genetic rearrangements of Ig gene segments are required to form a functional BCR. For light-chain genes, a rearrangement of germline DNA leads to V gene segments joining with J gene segments. Within the heavy-chain cluster, D gene segments first join with J gene segments, followed by D–J joining to V gene segments. The DNA sequences that mediate rearrangements lie at the immediate 3' ends of variable light-chain (V_L) and variable heavy-chain (V_H) gene segments, 5′ and 3′ to each D_H gene segment, and at the immediate 5′ ends of each J_L or J_H segment. The recombination signals are heptamer and nonamer oligonucleotides, which are identical in all V, D and J gene segments. Heptamers and nonamers are separated by 12 or 23 bp of random sequence (spacers) in the germline, with recombination occurring only between an element with a 12 bp spacer and an element with a 23 bp spacer. For joining to occur, the heptamer–nonamer must be orientated in opposite directions. This interaction brings light-chain V and J gene segments together, forming loops of non-coding intervening DNA, which are excised. In heavy-chain germline DNA, the V and J gene segments are flanked by 23 bp spacer elements, and the D gene segments are flanked on both sides by a 12 bp spacer, thus permitting recombination with upstream V and downstream J elements.

The RAG-1 and RAG-2 proteins mediate Ig gene rearrangement by cleaving double-stranded DNA adjacent to a joining signal sequence (Gellert, 1996). The termini of rearranging elements may undergo additional processing by two mechanisms. First, nucleotides can be randomly removed by an exonuclease. Second, terminal deoxynucleotidyl transferase can add DNA sequences at D_H and J_H junctions, creating 'N regions' at critical functional sites of an Ig molecule. These N regions may vary considerably in size and nucleotide sequence. The D segment encodes the CDR3 region of BCR, and preferential reading-frame usage is considered to stabilize the structure, thus avoiding rigid hydrophobic residues and an abrupt stop codon.

The genetic recombination of Ig gene segments occurs sequentially during B-cell differentiation. The heavy-chain V, D and J segments rearrange first. In general, a productive rearrangement of heavy-chain gene segments on one chromosome prevents rearrangement on the homologous chromosome (allelic exclusion). Abortive rearrangements at both heavy-chain loci will result in cell death. In humans and mice, the presence of μ heavy-chain stimulates the rearrangement at the κ light-chain locus by an unknown mechanism. Abortive rearrangement of both κ-chain alleles leads to rearrangement at the λ locus. Failure to produce a productive rearrangement at one of the κ or λ alleles will also result in cell death.

Mature B cells alter their Ig isotype expression in response to antigens with a highly regulated, deletional process, called isotype switch recombination. Switch (S) regions are located in introns directly upstream of C_H genes (except $C_δ$). The S regions generally contain a number of G-rich pentamer motifs, which are conserved among species. Switch recombination differs from V(D)J recombination, in that different signals and enzymes are used in the process; all isotype switches are productive, and switching is a highly regulated process, which occurs after antigen stimulation.

Six mechanisms may contribute to the enormous potential diversity of antibodies. They are: (i) multiplicity of V, D and J germline gene segments; (ii) the combinatorial freedom of these gene segments to associate; (iii) junctional diversity arising from both imprecise DNA rearrangement and insertion of N regions; (iv) somatic hypermutation in V_H and V_L regions; (v) gene conversion;

and (vi) combinatorial association of light-chain and heavy-chain polypeptides. Theoretically, 1.3×10^8–1.6×10^9 BCR specificities are possible in the mouse (Pascual and Capra, 1991).

Different species may employ different strategies for generation of the Ig repertoire. For example, the chicken heavy-chain gene repertoire is generated from a single set of functional V_H, D_H and J_H segments, which is then followed by a gene conversion-like process. Gene conversion uses nearby pseudogenes as a source of donor nucleotides for non-reciprocal exchange to generate diversity (Reynaud et al., 1987, 1989). In ruminants, relatively few heavy-chain VDJ rearrangements are possible, because of the limited number of V_H segments, but diversity in the repertoire is most probably generated by somatic hypermutation in V regions rather than by gene conversion (Reynaud et al., 1991).

Cattle C_H and C_L genes

The cattle immunoglobulin heavy-chain gene cluster is located on chromosome 21q23–q24 (Tobin-Janzen and Womack, 1992), spanning about 400 kb (Rabbani et al., 1997). Genes identified in the cluster include *IGHM, IGHG1, IGHG2, IGHG3, IGHG4, IGHA* and *IGHE* (Table 7.1). Cattle IgM is the predominant isotype found in serum. The *IGHM* gene is separated from the most proximal *IGHG* gene by 6 kb (Knight and Becker, 1987). The gene encoding the Ig J chain (*IGJ*) has also been cloned, sequenced and mapped to cattle chromosome 6 (Zhang et al., 1992; Kulseth and Rogne, 1994). The J chain plays a role in molecular assembly and selective transport of IgM and IgA across epithelial cell layers.

There are three expressed cattle Cγ genes, each differing in structure and implied function (Symons et al., 1987, 1989). The major isotype found in colostrum is IgG1, whereas IgG2 constitutes approximately half of the serum IgG. As in rabbit, an extra intrachain disulphide bond occurs in cattle IgG molecules at C_H1 (Symons et al., 1989). Both *IGHG1* and *IGHG2* have four exons corresponding to three C_H domains and the hinge region. The hinge region and C_H2 domain of *IGHG2* are shorter than the other heavy-chain genes, which might affect the Fc-receptor function of the IgG2 molecule (Symons et al., 1989). The IgG2 (A1) and IgG2 (A2) allotypes differ at the site of the light- and heavy-chain bond in the C_H1 domain and vary in amino acids in the middle of the hinge region and in the intradomain loop of C_H3 (Kacskovics and Butler, 1996). The *IGHG3* gene differs from *IGHG1* and *IGHG2* at the hinge region, in that the *IGHG3* hinge is encoded by two exons (Rabbani et al., 1997). Analysis of cattle populations using pulsed-field gel electrophoresis indicated the existence of two putative *IGHG3* alleles (Rabbani et al., 1997).

The IgA isotype is the major Ig in exocrine secretions of cattle, with the exception of lacteal secretions (for a review of cattle IgA genes and gene products, see Butler, 1998). Like mice and pigs, cattle have one gene encoding

C_α (*IGHA*), whereas humans have two and rabbits have 13 genes encoding C_α. There are at least two *IGHA* alleles in cattle; IgGA appears to have an additional N-linked glycosylation site at position 282 in the C_H2 domain. The hinge region in C_α is only five amino acids long.

Immunoglobulin E is commonly associated with allergies and parasitic infections. Cattle have one gene encoding C_ε (*IGHE*), separated by approximately 15 kb from *IGHA*. The *IGHE* gene contains four C_H domains; the C_H4 domain is associated with the membrane-bound form of IgE (Mousavi *et al.*, 1997).

In general, much less is known about the organization of light-chain C-region genes in cattle. The λ chain is encoded on cattle chromosome 17 (Tobin-Janzen and Womack, 1996). Lambda chains represent the predominant light-chain isotype found on Igs (Butler, 1983). A recent study of the cattle λ light-chain repertoire revealed evidence for at least four C_λ and two J_λ genes (Parng *et al.*, 1996). Cattle C_λ cDNA encodes a polypeptide of 107 amino acids, differing by 24 amino acids from that of human C_λ chain (Ivanov *et al.*, 1988).

Cattle V_λ genes

On the basis of Southern hybridization and nucleotide sequence analysis, there are two families of V_λ genes (Parng *et al.*, 1995; Sinclair *et al.*, 1995). Analysis of cDNA sequences and Southern blotting experiments indicate that one V_λ family is preferentially utilized (> 90%) and that only one J gene and one C gene is selected in the germline. There are at least 14 genes in the $V_\lambda 1$ group, which can be further subdivided into three subgroups on the basis of the length of CDR1. The CDR3 regions also show considerable variation in length among the subgroups (10–15 amino acids). Comparison of germline V_λ sequences with cDNA sequences indicates that a large number of the germline sequences are pseudogenes. On the basis of the relatively small number of germline V_λ sequences and the location of V_λ pseudogenes, Parng *et al.* (1996) suggested that diversification of V_λ occurs by gene conversion followed by somatic mutation.

Cattle V_H genes

Analysis of V_H encoded sequences from fetal and adult splenocytes and heterohybridomas led to the assignment of V_H sequences to one family, *Bov VH1*, corresponding to the murine *Q52* and human V_HII families (Berens *et al.*, 1997; Saini *et al.*, 1997; Sinclair *et al.*, 1997). Southern blotting and sequencing suggest 13–15 members of the *Bov VH1* family exist (Saini *et al.*, 1997; Sinclair *et al.*, 1997). Although cattle orthologues of other murine V_H families are present in the genome these genes do not appear to be expressed. The limited usage of V_H segments in cDNAs suggests that heavy-chain diversity in cattle is not generated through DNA rearrangements.

How then is diversity generated in cattle Igs? Analysis of CDRs encoded by V_H segments revealed extreme variability in the length of CDR3s (four to 26 amino acids). It is unclear whether this observation represents differences in the length of D segments or possibly D–D fusions (Sinclair *et al.*, 1997). In addition, extensive variation in CDR1 and CDR2 in cDNA compared with germline sequences was observed, suggesting a high substitution rate (transitions rather than transversions) in the CDRs compared with framework regions. These results suggest that somatic hypermutation is the primary mechanism for generating heavy-chain diversity in cattle.

T-cell Receptors

Thymus-derived (T) lymphocytes regulate antibody production by B cells, cellular immune responses, and the killing of virus-infected cells and tumour cells. Unlike B cells, T cells are incapable of recognizing antigen in native conformation (reviewed in Janeway and Travers, 1997). As discussed in the previous section on MHC, the cognate receptor for antigen on T cells recognizes antigen-derived peptides that are complexed with self MHC molecules. When TCRs engage MHC + peptide in the presence of additional costimulatory molecules, T cells are activated to carry out their effector functions (Fig. 7.2).

T-cell subtypes and their receptors

There are three major types of functional T cells in the vertebrate immune system: (i) T_H; (ii) CTL; and (iii) γδ T cells (Janeway and Travers, 1997). The TCR of CD4$^+$ (MHC class II restricted) and CD8$^+$ (MHC class I restricted) T cells consists of an αβ heterodimer encoded by two distinct genes (*TCRA* and *TCRB*). T cells bearing αβ TCR constitute > 90% of all T cells in human and mouse peripheral blood. Unlike mice and humans, cattle and other ruminants have a high frequency of T cells with γδ receptors in peripheral blood (discussed below).

The TCR γ and δ chains are encoded by distinct genes (*TCRG* and *TCRD*). The physiological function of γδ T cells is unclear, as their mode of antigen recognition is different from conventional αβ T cells. Their TCR seem to accommodate a larger variety of ligands and may bind antigen independently from MHC (Rock *et al.*, 1994; Schild *et al.*, 1994). The γδ T cells may also be stimulated by peptides in the context of class I, class Ib and class II antigens (Holoshitz *et al.*, 1989; Kozbor *et al.*, 1989; Vidović *et al.*, 1989; Guo *et al.*, 1995), and have been implicated in the regulation of αβ T cells (Kaufmann *et al.*, 1993), macrophage activation (Nishimura *et al.*, 1995) and tissue-repair mechanisms (Boismenu and Havran, 1994; Komano *et al.*, 1995). The ligands of γδ T cells include heat-shock protein-60 peptides (Imani and Soloski, 1991),

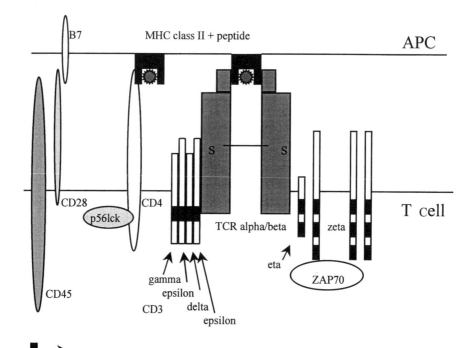

■ → ITAM

Fig. 7.2. A simplified view of APC–T-cell interaction. Various components involved in APC–T-cell interaction are shown. The intracellular src-related tyrosine kinase (p56lck), and syk-related kinase (ZAP70) bind to co-receptor CD4 and signal-transducing ζ chain. Also shown is the distribution of the immunoglobulin receptor-related tyrosine-associated activation motif (ITAM), which is essential for signalling in lymphocytes (Cambier, 1995). A simultaneous signalling via CD28-B7 (costimulatory molecules) is crucial for successful induction of immune response.

isopentenylpyrophosphate, monomethylphosphate and pyrophosphate (collectively known as phospholigands; De Libero, 1997).

In contrast to mice and humans, ruminant γδ T cells constitute 15–50% of ruminant peripheral-blood mononuclear cells, and are also present at high frequency in lymphoid and non-lymphoid tissues, such as skin, intestinal epithelium and lamina propria (Hein and Dudler, 1997). A 215 kDa transmembrane protein termed WC1 in cattle and T19 in sheep is encoded by a novel multigene family, which has not been found in other mammals (except pigs) or rodents (Hein and Mackey, 1991; Carr *et al.*, 1994). Interestingly, ovine γδ T cells are faster in their response to mitogens *in vitro* (Evans *et al.*, 1994), strengthening the idea that γδ T cells could override αβ T-cell response *in vivo* by their early, active response. Given the chronic stimulation of the mucosal immune system by rumen microbial antigens and the relatively low incidence of autoimmune diseases, ruminants appear to have a powerful mechanism for immune suppression. It would seem that the relatively high frequency of γδ T cells in ruminant lymphoid tissues and peripheral blood is intimately related to this powerful system of suppression. Thus, ruminants will be useful models in understanding the role of γδ T cells in maintaining immune system homoeostasis during adaptive responses to various pathogens.

T-cell receptor structure, gene organization and diversification

The αβ TCR is a heterodimer, comprised of 49 kDa and 44 kDa polypeptides, respectively (for a review of TCR structure and genetic organization, see Janeway and Travers, 1997). The γδ TCR consists of a γ chain (55 kDa) and a δ chain (42 kDa). Each chain has loops of about 70 amino acids, and, like Ig molecules, the α and δ chains of the TCR have V, J and C regions. The β and δ chains also have a D segment (Davis *et al.*, 1984). As in Ig molecules, there are three CDR regions that determine TCR specificity. Recent crystallographic analysis of TCR αβ heterodimer and of TCRs binding to their MHC–peptide ligands (Garboczi *et al.*, 1996; Garcia *et al.*, 1996) suggests that CDR1 and CDR2 of the TCR α chain interact with the N-terminal region of the MHC-bound peptide, whereas CDR3 of the α chain interacts with the central portion of the peptide, and CDR3 of the β chain lies over the C-terminal region of the peptide. These results suggest that the TCR is aligned diagonally across the MHC–peptide surface during T-cell interaction with a cell. For activation, T cells require expression of coreceptors, such as CD4 (binds monomorphic region of MHC class II molecules) and CD8 (binds monomorphic region of MHC class I molecules), and the signal-transducing CD3 complex, consisting of CD3 γ, δ, ε, ξ, existing as either a homodimer or heterodimer with η chain (Cantrell, 1996; Fig. 7.2).

Mice and humans have four multigene families encoding α, β, γ and δ chains (*TCRA, TCRB, TCRG* and *TCRD*). Each gene family is found on a different chromosome (Table 7.1). The number and organization of genes in each family will not be reviewed here (for a review see Janeway and Travers, 1997). The TCR germline DNA undergoes rearrangement by mechanisms similar to Ig gene rearrangement (Toyonaga and Mak, 1987; reviewed in Hay, 1993).

The TCR multigene families use four main strategies for diversification: (i) multiple germline segments; (ii) combinatorial joining of these segments; (iii) junctional diversity (including junctional flexibility, random N-region addition and D-segment joining); and (iv) combinatorial association of the α and β or γ and δ polypeptide subunits. The potential diversity of human TCRs was calculated to be 10^{15-18} (Davis and Bjorkman, 1988).

Cattle T-cell receptor gene segments

Both Southern blot and cDNA sequencing established that cattle have a single C_α gene segment (Ishiguro *et al.*, 1990). There are at least five bovine V_α families which share 16–47% identity at the amino acid level. Several different J_α regions, ranging in length between 20 and 24 codons, were identified in cDNA clones (Ishiguro and Hein, 1994).

Cattle express at least two C_β gene segments (*CB1* and *CB2*; Tanaka *et al.*, 1990) but the exact number of *TCRB* gene segments is unknown. The junctional sequences of cattle *TCRB* chains are highly diverse. However, a clear-cut distinction has not been made between D_β segments and N additions at junctional regions of transcripts.

Ruminant genes encoding C_γ genes have distinct features compared with their human and mouse counterparts. Four functional C_γ genes in cattle have been identified (Takeuchi *et al.*, 1992; Hein and Dudler, 1997). Some ruminant C_γ chains have a variable number of repeats of a five amino acid motif, TTEPP, near the 5′ end of the hinge. These motifs, which are absent from human and mouse sequences, indicate the presence of additional C_γ exons (Ishiguro and Hein, 1994). Ten bovine V_γ sequences have been found by cDNA sequencing (Takeuchi *et al.*, 1992; Hein and Dudler, 1997). Thus, cattle *TCRG* gene segments may resemble the genomic organization in humans, with a few J_γ segments (< 5) localized at the 5′ end of the C region (Ishiguro *et al.*, 1993).

Cattle have a single C_δ gene (Hein *et al.*, 1990; Takeuchi *et al.*, 1992). Twenty-one distinct V_δ sequences have been identified by cDNA cloning, all falling into one major family ($V_\delta 1$; Hein and Dudler, 1997). Three distinct J_δ sequences have been described (Ishiguro *et al.*, 1993; Hein and Dudler, 1997). Thus, unlike humans and mice, cattle and other ruminants can express an unusually large number V_δ gene segments.

Invariant genes of the TCR complex, *CD3G*, *CD3D* and *CD3E*, are conserved and closely related. Sheep and cattle sequences are very similar (Clevers *et al.*, 1990), sharing 70–80% overall identity with their human and mouse counterparts. However, the extracellular protein sequence of sheep CD3 shares only 46–59% identity with cattle. The CD3 molecule is required for signal transduction and cell surface expression of the TCR (Fig. 7.2). The *CD3E* and *CD3D* genes map to cattle chromosome 15 (Table 7.1).

Cytokines and Cytokine Receptors

Cytokines are a family of low-molecular-weight regulatory molecules involved in cell growth, inflammation, immunity, differentiation and tissue repair. They are characterized by pleiotropic activity, acting on a wide range of cell types. Unlike hormones, cytokines generally act in a local manner through autocrine and paracrine mechanisms. Cytokines are produced constitutively by many cell types and, once activated, they are tightly regulated. Furthermore, cytokines function singly or, more often in networks, positively and negatively influencing cell responses, and in some instances acting synergistically with other cytokines to amplify a particular cellular response. In humans and mice, several cytokine genes map in a conserved cluster (e.g. interleukin *(IL)-3, IL-4, IL-13,* granulocyte–macrophage colony-stimulating factor *(GM-CSF)*). On the basis of structural features, cytokines may be classified into six families (Fig.

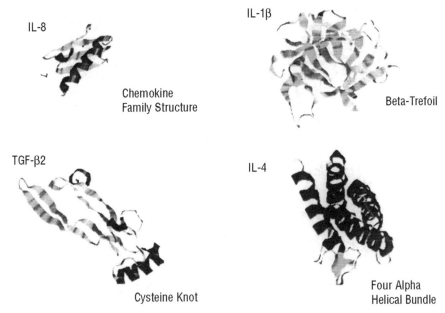

Fig. 7.3. Structural families of cytokines. Structural data obtained using X-ray crystallography and nuclear magnetic resonance (NMR) have resulted in a new scheme of cytokine classification. Cytokines IL-2, -3, -4, -5, -6, -7, -9, -13, G-CSF, GM-CSF belong to a four α-helical bundle family (haematopoietins). The β-trefoil family members, IL-1α, β and IL-1Rα, form a bowl-like structure. Chemokine family members, including IL-8, consist of a triple-stranded antiparallel pairs of β-sheet, with a long C-terminal helix. The TGF-β2 monomers consist of two antiparallel pair of β-strands, which form a flat curved surface and a separate long α-helix, stabilized by disulphide bonds (cysteine knot). Representative structures of the EGF and TNF families are not shown.

7.3). For the following presentation, only bovine cytokine genes or cDNAs that have been cloned will be discussed.

Cytokine genes of cattle

Interleukin-1 (IL-1) is primarily an inflammatory cytokine, with a wide range of biological activities on many different cell types, including B cells, T cells and monocytes. *In vivo*, IL-1 induces hypotension, fever, weight loss, neutrophilia and inflammatory responses. In humans and mice, there are two distinct molecular forms of IL-1 (IL-1α and IL-1β), derived from two different genes (Dinarello, 1988). Despite the low sequence similarity between IL-1α and IL-1β, these molecules bind to the same receptor and have very similar, if not identical, biological properties. In general, IL-1α is cell-associated and IL-1β is secreted. Cattle *IL1A* and *IL1B* cDNAs have been cloned (Leong *et al.*, 1988; Maliszewski *et al.*, 1988a), and the proteins share 23% amino acid similarity. In cattle alveolar macrophages, there is tenfold more IL-1β mRNA expression than IL-1α mRNA. Cattle IL-1α shares 73% and 62% identity with the human and mouse proteins, respectively, while cattle IL-1β shares 61% and 58% with human and mouse protein sequences, respectively. A conserved tetrapeptide (Ile–Thr–Asp–Phe) near the carboxy terminus is known be responsible for cross-species biological activity (Mosley et al., 1987). Both forms of IL-1 are stable tetrahedral globular proteins, formed by an antiparallel six-stranded barrel closed at one end by a six-stranded β-sheet to form a bowl-like structure (Graves et al., 1990). Cattle *IL1A* maps to BTA 11 (ArkDB, 1998).

Interleukin-2 (IL-2) is produced mainly by activated T cells and is both an autocrine and a paracrine growth factor for $CD4^+$ and $CD8^+$ T cells (reviewed by Theze et al., 1996). Interleukin-2 can activate natural killer (NK) cells and is a proliferation signal for anti-IgM costimulated B cells. Cattle IL-2 cDNAs have been cloned (Cerretti et al., 1986a; Reeves et al., 1986), and the mature protein shares approximately 65% and 50% identity with human and mouse IL-2 molecules, respectively. Structurally, IL-2 is a globular protein composed of six α-helical regions (A–F), with a disulphide bridge between Cys 58 and Cys 105 near a bent loop between α-helices A and B. The α-helices C, D, E and F form an antiparallel α-helical bundle (Brandhuber et al., 1987). A site on the B α-helix is thought to bind the p55 chain of the IL-2 receptor (Bazan, 1992). The IL-2 gene is located on cattle chromosome 17q22–q23 (Chowdhary et al., 1994).

Interleukin-3 (IL-3) is a heavily glycosylated haemopoietic growth factor that stimulates colony formation of human erythroid, megakaryocyte, neutrophil, eosinophil, basophil, mast-cell and monocytic lineages (Ihle, 1992). Interleukin-3 is considered to be more important in committing progenitor cells to a differentiation pathway rather than self-renewal of primitive progenitor cells. The predominant sources of IL-3 are activated T cells, mast cells and eosinophils. Many of its activities are influenced by other cytokines. Cattle IL-3 shares 85%, 35% and 28% similarity at protein level with sheep, human and mouse IL-3, respectively (Mwangi et al., 1995). On the basis of comparative mapping information, *IL3* is predicted to map to BTA 7.

Interleukin-4 (IL-4) is produced by activated T_H2 cells. Interleukin-4 induces secretion of IgG1, IgG4 and IgE by mouse, human and bovine B cells (Callard et al., 1991; Estes, 1996). It also affects macrophage function and is therefore involved in inflammatory responses (Crawford et al., 1988). The IL-4 gene of cattle is 92% similar to the human gene, but cattle *IL4* is shorter than mouse *IL4*, due to a deletion of 51 nucleotides (Heussler et al., 1992). The three-dimensional structure of IL-4 shows a compact globular organization, with a predominantly hydrophobic core. A four α-helix bundle, with the helices arranged in a left-handed antiparallel bundle and with two overhand connections containing a two-stranded antiparallel β-sheet, make up most of the molecule (Walter et al., 1992). The gene *IL4* is located on BTA 7q15–q21 (Buitkamp et al., 1995).

Interleukin-5 (IL-5) is a T-cell-derived glycoprotein that stimulates eosinophil colony formation and is an eosinophil differentiation factor in humans and mice. It is also a growth and differentiation factor for mouse but not human B cells (Takatsu et al., 1988). Bovine IL-5 has been cloned and shown to have

97%, 66%, 59% and 58% identity at amino acid level with the sheep, human, mouse and rat IL-5 sequences, respectively (Mertens *et al.*, 1996). Interleukin-5 is an antiparallel disulphide-linked homodimer, which has a novel two-domain structure, with each domain showing significant homology to the cytokine fold in GM-CSF, IL-2, IL-4 and growth hormone. The IL-5 structure consists of two left-handed bundles of four α-helices laid end to end and two short β-sheets on opposite sides of the molecule (Milburn *et al.*, 1993). The monomer is biologically inactive. The gene encoding bovine IL-5 has not yet been mapped, but is predicted to be located on BTA 7, on the basis of comparative mapping information.

Interleukin 6 (IL-6) is a glycoprotein produced by both lymphoid and non-lymphoid cells. Interleukin-6 regulates immune responses, acute phase reactions and haemopoiesis (Sehgal, 1990). Bovine IL-6 has 65%, 53%, 42% and 42% sequence similarity to porcine, human, mouse and rat proteins, respectively. The sequence immediately upstream to the RNA CAP-site is conserved, and this region contains consensus sequences for NF-kB, CREB, NF-IL6 and AP-1 factors (Droogmans *et al.*, 1992). Structural analysis suggests that IL-6 contains five α-helices interspersed with variable-length loops; four of these helices constitute a classical four-helix bundle, with a fifth helix located in the C–D loop (Xu *et al.*, 1997). Cattle *IL6* has been mapped to BTA 4 (Barendse *et al.*, 1997).

Interleukin 7 (IL-7) is a stromal cell-derived growth factor for progenitor B cells and T cells. The main population in the thymus that is responsive to IL-7 is CD4⁻CD8⁻ thymocytes. It also stimulates proliferation and differentiation of mature T cells (Henney, 1989; Welch *et al.*, 1989). Cattle IL-7 has 75% and 65% sequence similarity to human and mouse IL-7 proteins, respectively (Cludts *et al.*, 1992). The cattle *IL7* gene has been mapped to BTA 14 (Barendse *et al.*, 1997).

Interleukin 8 (IL-8) is a neutrophil and T-lymphocyte chemotactic and activating inflammatory cytokine produced by monocytes, lymphocytes, fibroblasts, endothelial cells, epithelial cells and granulocytes. Interleukin-8 is also a potent angiogenic factor. Many cell types, including macrophages, can produce IL-8 in response to microbial and non-microbial agents in cattle. Bovine IL-8 protein shares 76% and 87% similarity with the human and porcine IL-8 proteins, respectively (Morsey *et al.*, 1996). The three-dimensional structure of an IL-8 dimer revealed a general architecture that is similar to that of the α1 and α2 domains of class I histocompatibility molecules. The two α-helices were predicted from the binding site for the cellular receptor for IL-8 (Clore *et al.*, 1990). The gene *IL8* has been mapped to chromosome 4q12–q21 in humans (Modi *et al.*, 1990) and, on the basis of comparative mapping information, should map to BTA 6.

Interleukin-10 (IL-10) is an acid-labile non-disulphide-linked homodimer expressed by many cell types, including activated T cells, B cells, monocytes, macrophages and keratinocytes. Interleukin-10 generally inhibits cytokines (specifically, gamma interferon (IFN-γ)) produced by T_H1 cells, NK cells and macrophages (Moore *et al.*, 1993). It can also stimulate or enhance proliferation of B cells, thymocytes and mast cells (Howard and O'Garra, 1992). While considered as a T_H2 cytokine in the mouse, its expression is not restricted to T_H2 cells in humans or cattle. Cattle IL-10 has 77% and 71% similarity with human and mouse proteins, respectively (Hash et al., 1994). The molecule is a tight dimer made of two interpenetrating subunits, forming a V-shaped structure. Each IL-10 molecule consists of six α-helices, four originating from one subunit and two from the other. Four of the α-helices form a classical 'up–up–down–down' bundle observed in all other helical cytokines (Zdanov et al., 1995). The *IL10* gene has been assigned to cattle chromosome 16 (Beever et al., 1997).

Interleukin-12 (IL-12) induces IFN-γ production by T_H1 cells and NK cells. The bioactivity of IL-12 is mediated by heterodimers of disulphide-linked p35 and p40 protein subunits. Bovine p40 has 84% and 68% sequence similarity, whereas p35 has 82% and 57% similarity to human and mouse proteins, respectively (Zarlenga et al., 1995). On the basis of comparative mapping data, the genes encoding cattle p35 (*IL12A*) and p40 (*IL12B*) will be located on BTA 1 and BTA 7, respectively.

Interleukin-15 (IL-15) shares biological activities and receptor components with IL-2 (β and γ chains). Like IL-2, it is also a member of the four-helix bundle cytokine family (Giri et al., 1995). Interleukin-15 mRNA is expressed by a variety of cell types, including mononuclear cells, epithelial cells and skeletal muscle. Cattle IL-15 cDNA has been cloned and predicted to have 78% and 74% sequence similarity to human and mouse proteins, respectively (Canals et al., 1997). Purified IL-15 stimulated proliferation of bovine lymphoblasts and its expression was upregulated by endotoxin in macrophages (Canals et al., 1997). The *IL15* gene has not yet been mapped in cattle but is predicted to map to BTA 17 on the basis of comparative mapping data.

Interferons

The alpha interferons (IFN-α) are a family of inducible secreted proteins, which confer resistance to viruses in target cells, inhibit cell proliferation and regulate MHC class I expression. The IFN-α family is produced predominantly by leucocytes and fibroblasts (type I IFNs). In cattle, IFN-α is encoded by an 'intronless' multigene family. The similarity in amino acid sequence of cattle IFN-α to human IFN-α is 60% (Velan et al., 1985, 1986), whereas mouse and human IFN-α subtypes share only 40% similarity. Omega interferon (IFN-ω) is a trophoblast-specific IFN. At least four IFN-ω genes have been identified in cattle (with only one known to be functional), having 63% similarity to human IFN-ω (Hansen et al., 1991). In cattle, IFN gene numbers have been estimated from the number of polymorphic loci (ten *IFNA*, six *IFNB*, 20 *IFNW* and six *IFNT*; Ryan and Womack, 1993). The Cys residues are conserved in cattle *IFNB* genes and are essential for antiviral activity (Shepard et al., 1981). Type I IFNs possess common structural features with respect to the chain- folding topology and the hydrogen-bond networks between various polypeptide segments. The *IFNA1*, *IFNB1* and *IFNW1* genes map to BTA 8q15 (Ryan *et al.*, 1992).

Ruminants also express another type I IFN, called IFN-τ, which is structurally related to IFN-α. Tau interferon is considered as the major antiluteolytic factor, and is secreted by the conceptus of ruminant species before trophoblast attachment and implantation (Martal *et al.*, 1997). The gene encoding IFN-τ (*IFNT*) is also located on BTA 8.

Gamma interferon (IFN-γ; a type II IFN) is produced by activated T_H1 cells. Bovine IFN-γ has 63% similarity to the human protein (Cerretti *et al.*, 1986b). Human IFN-γ associates to form a dimer, with each subunit consisting of six α-helices (A–F), which make up 62% of the structure, and it is stabilized by multiple intersubunit interactions (Ealick *et al.*, 1991). The gene for *IFNG* is located on cattle chromosome 5q22–q24 (Threadgill *et al.*, 1990).

Tumour necrosis factors

Tumour necrosis factors (TNFs) are also known as cachectin and necrosin. Alpha tumour necrosis factor (TNF-α) is a potent paracrine and endocrine mediator of inflammatory immune functions and is selectively cytotoxic for many tumour cells. Many of its actions occur in combination with other cytokines. The bovine TNF α protein sequence (Cludts *et al.*, 1993) has 79% identity with the human protein. It exists as a homotrimer characterized by an edge-to-face association of antiparallel sandwich structure of wedge-shaped monomers (Jones *et al.*, 1989). Also known as lymphotoxin, TNF-β has 35% similarity to TNF-α and binds to the same receptor. Lymphotoxin forms a heteromeric complex with a dimer of p33 proteins on the cell surface (Androlewicz *et al.*, 1992). While TNF-α is produced predominantly by accessory cells, TNF-β is produced by activated lymphocytes. In cattle, *TNFA* has been mapped to BTA 23 (Barendse *et al.*, 1997).

Transforming growth factors

The transforming growth factor (TGF) family of cytokines regulate cell growth, differentiation, matrix production and apoptosis. While TGFs have important functions during embryonal development in pattern formation and tissue specification, in adults they are involved in tissue repair and modulation of the immune system (Letterio and Roberts, 1997). Beta TGF (TGF-β) is comprised of three related dimeric proteins, TGF-β1–3, all which are members of a superfamily. The expressed proteins are biologically inactive disulphide-linked dimers, which are cleaved to active forms. Most nucleated cell types express all TGFs, or a combination of the three forms. Bovine TGF-β1 has 92% similarity to human TGF-β at the protein level (Van Obberghen-Schilling *et al.*, 1987). Transforming growth factor is one of the predominant growth factors present in cow's milk (Rogers *et al.*, 1996).

The monomer TGF-β2 assumes an extended conformation that lacks a well-defined hydrophobic core. However, in dimer form, 56% of the non-polar surface is buried, with four disulphides and the extended β-strand region constituting the conserved structural motif of the TGF-β family (Daopin et al., 1993). The gene for cattle TGF (*TGF1*) was mapped to BTA 5 (Kappes et al., 1997).

Colony-stimulating factors

Colony-stimulating factors (CSFs) induce growth and differentiation of haemopoietic cells. The GM-CSF, granulocyte CSF (G-CSF) and monocyte CSF (M-CSF) are well-known members of this family. A beneficial effect of G-CSF in reducing the incidence and severity of bovine mastitis has been reported (Kehrli et al., 1991). Cattle GM-CSF shares 80% and 57% similarity with the human and mouse proteins, respectively (Maliszewski et al., 1988b; Leong et al., 1989). The gene encoding GM-CSF (*CSF2*) has been assigned to chromosome 7 in cattle (Zhang and Womack, 1992). On the basis of comparative mapping information, it is likely that *CSF1* and *CSF3* map to BTA 3 and BTA 19, respectively.

Cytokine receptors

On the basis of structural similarities, cytokine receptors have been classified into superfamilies (Bazan, 1990). Examples include: cytokine receptor superfamily (CKR-SF), immunoglobulin superfamily (IG-SF), tyrosine kinase receptor superfamily (PTKR-SF), complement control protein superfamily (CCP-SF), rhodopsin superfamily (Rho-SF) and interferon receptor superfamily (IFNR-SF). The CKR-SF is also known as the haematopoietic superfamily. Cytokine receptors fold into distinct structural domains in their tertiary structures. Examples include immunoglobulin domains (IG), cytokine receptor domain (CK), fibronectin type III (F3) and nerve growth factor receptor type (N). Binding of cytokine to its receptor results in the transduction of intracellular signals, which leads to bioactivity of that cytokine. The cytokine receptors identified in cattle and their properties are summarized in Table 7.3.

Summary

The past 10 years have witnessed impressive progress in characterizing genes affecting immune function in cattle. Conservation in DNA sequence of genes among vertebrate species has resulted in rapid elucidation of orthologues and aided in studies of comparative immunobiology. However, orthology at the gene level is not always accompanied by identity in chromosomal organization

Table 7.3. Available data for cattle cytokine receptors

Receptor (R)	Mol. mass (kDa)	Human chromosome location	Types of domains/ superfamily	Cattle chromosome location	References
IL-1R (Type I)	80	2q12	IG	11	Yoo et al., 1994
(Type II)	60	2q12–q22	IG	11	Yoo et al., 1994
IL-2R α	55	10p15–p14	IG	13q13–q14	Yoo et al., 1995b
γ	64	Xq13	CK, F3	Xq23	Yoo et al., 1996
IL-6R β pseudogene	130	17p11	IG, CK, F3	19	Barendse et al., 1997
IL-8R (low-affinity)	32	2q34–q35	Rho	cDNA sequenced	GenBank (U19937)
IFN-α and -β	102	21q22.1	F3	1	Langer et al., 1992
G-CSFR	150	1p35–p34.3	IG, F3	3	Yoo et al., 1995a
M-CSFR	150	5q33–q35	IG	7	Zhang and Womack, 1992

IG, immunoglobulin superfamily; CK, cytokine receptor or haematopoietic superfamily; F3, fibronectin type II domain; Rho, rhodopsin superfamily.

or gene function. This is most clearly evident in the case of the unique architecture of the cattle MHC. Another prominent example is the TCR genes and T-cell function. Clearly, the γδ T cells of cattle must serve a more front-line function than in primates or rodents. There are also gross differences in lymphoid architecture and sites of lymphoid development between primates, rodents and ruminants. Immunologists studying cattle must not be coaxed into blindly accepting any paradigm that exists for primates or rodents without extensive experimental analysis. This is particularly true for cytokines and their networks. For example, major differences in expression pattern and function have been identified for cattle IL-10. Unfortunately, molecular genetic analysis is now far easier to perform than detailed functional studies of immune-system molecules. As a consequence, the complex interacting networks that regulate innate and adaptive immunity in cattle remain to be elucidated. Although gaining a complete understanding of the immunobiology of cattle is going to be a long and difficult task, insights into the adaptation and evolution of the vertebrate immune system gained from such studies will be well worth the effort. Moreover, new vaccines, treatments for common and costly diseases, such as mastitis, and selection for genetic resistance to parasites will probably result from continued research in cattle genomics and immunology.

Acknowledgements

This work was supported in part by grants to H.A.L. from the US Department of Agriculture (USDA) (96-35204-3314 and 97-35205-4738). M.A. is a Fellow of the of the Spanish Ministry of Education and Culture. We thank Colleen

Olmstead for technical assistance with preparation of the manuscript and Roger Sayle for RasMol software.

References

al-Murrani, S.W.K., Glass, E.J. and Hopkins, J. (1994) BoLA class I charge heterogeneity reflects the expression of more than two loci. *Animal Genetics* 25, 165–172.

Andersson, L., Lundèn, A., Sigurdardóttir, S., Davies, C.J. and Rask, L. (1988) Linkage relationships in the bovine MHC region: high recombination frequency between class II subregions. *Immunogenetics* 27, 273–280.

Androlewicz, M.J., Browning, J.L. and Ware, C.F. (1992) Lymphotoxin is expressed as a heteromeric complex with a distinct 33-kDa glycoprotein on the surface of an activated human T cell hybridoma. *Journal of Biological Chemistry* 267, 2542–2547.

ArkDB (1998) The cattle genome database. Roslin Institute Bioinformatics Group. http://bos.cvm.tamu.edu/bovgbase.html

Ballingall, K.T., Luyai, A. and McKeever, D.J. (1997) Analysis of genetic diversity at the *DQA* loci in African cattle: evidence for a *BoLA-DQA3* locus. *Immunogenetics* 46, 237–244.

Bamford, A.I., Douglas, A., Friede, T., Stevanovic, S., Rammensee, H.G. and Adair, B.M. (1995) Peptide motif of a cattle MHC class I molecule. *Immunology Letters* 45, 129–136.

Band, M., Larson, J.H., Womack, J.E. and Lewin, H.A. (1998) A radiation hybrid of map BTA23: identification of a chromosomal rearrangement leading to separation of the cattle MHC class II subregions. *Genomics* 53, 269–275.

Barendse, W., Vaiman, D., Kemp, S.J., Sugimoto, Y., Armitage, S.M., Williams, J.L., Sun, H.S., Eggen, A., Agaba, M., Aleyasin, S.A., Band, M., Bishop, M.D., Buitkamp, J., Byrne, K., Collins, F., Cooper, L., Coppettiers, W., Denys, B., Drinkwater, R.D., Easterday, K., Elduque, C., Ennis, S., Erhardt, G., Ferretti, L., Flavin, N., Gao, Q., Georges, M., Gurung, R., Harlizius, B., Hawkins, G., Hetzel, J., Hirano, T., Hulme, D., Jorgensen, C., Kessler, M., Kirkpatrick, B.W., Konfortov, B., Kostia, S., Kuhn, C., Lenstra, J.A., Leveziel, H., Lewin, H.A., Leyhe, B., Lil, L., Martin Buriel, I., McGraw, R.A., Miller, R.J., Moody, D.E., Moore, S.S., Nakane, S., Nijman, I.J., Olsaker, I., Pomp, D., Rando, A., Ron, M., Shalom, A., Teale, A.J., Thieven, U., Urquhart, B.G.D.,Vage, D.-I., Van de Weghe, A., Varvio, S., Velmala, R., Vilkki, J., Weikard, R., Woodside, C., Womack, J.E., Zanotti, M. and Zaragoza , P. (1997) A medium density genetic linkage map of the bovine genome. *Mammalian Genome* 8, 21–28.

Bazan, J.F. (1990) Haemopoietic receptors and helical cytokines. *Immunology Today* 11, 350–354.

Bazan, J.F. (1992) Unraveling the structure of IL-2. *Science* 257, 410–413.

Beever, J.E., Fisher, S.R., Guérin, G. and Lewin, H.A. (1997) Mapping of eight human chromosome 1 orthologs to cattle chromosomes 3 and 16. *Mammalian Genome* 8, 533–536.

Bensaid, A., Kaushal, A., Baldwin, C.L., Clevers, H., Young, J.R., Kemp, S.J., MacHugh, N.D., Toye, P.G. and Teale, A.J. (1991) Identification of expressed bovine class I MHC genes at two loci and demonstration of physical linkage. *Immunogenetics* 33, 247–254.

Berens, S.J., Wylie, D.E. and Lopez, O.J. (1997) Use of a single V_H family and long CDR3s in the variable region of cattle Ig heavy chains. *International Immunology* 9, 189–199.

Bjorkman, P.J., Saper, M.A., Samraoui, B., Bennett, W.S., Strominger, J.L. and Wiley, D.C. (1987) Structure of the human class I histocompatibility antigen, HLA-A2. *Nature* 329, 506–512.

Boismenu, R. and Havran, W.L. (1994) Modulation of epithelial cell growth by intra-epithelial γδ T cells. *Science* 266, 1253–1255.

Brandhuber, B.J., Boone, T., Kenney, W.C. and McKay, D.B. (1987) Crystals and a low resolution structure of interleukin-2. *Journal of Biological Chemistry* 262, 12306–12308.

Brown, J.H., Jardetzky, T.S., Gorga, J.C., Stern, L.J., Urban, R.G., Strominger, J.L. and Wiley, D.C. (1993) Three dimensional structure of the human class II histo-compatibility antigen HLA-DR1. *Nature* 364, 33–39.

Buitkamp, J., Schwaiger, F.-W., Solinas-Toldo, S., Fries, R. and Epplen, J.T. (1995) The bovine interleukin-4 gene: genomic organization, localization, and evolution. *Mammalian Genome* 6, 350–356.

Butler, J.E. (1983) Bovine immunoglobulins: an augmented review. *Veterinary Immunology and Immunopathology* 4, 43–152.

Butler, J.E. (1998) Immunoglobulin diversity, B-cell and antibody repertoire development in large farm animals. *Revue Scientifique et Technique, Office International des Épizooties* 17, 42–70.

Callard, R.E., Smith, S.H. and Scott, K.E. (1991) The role of interleukin 4 in specific antibody responses by human B cells. *International Immunology* 3, 157–163.

Cambier, J.C. (1995) Antigen and Fc receptor signaling: the awesome power of the immunoreceptor tyrosine-based activation motif (ITAM). *Journal of Immunology* 155, 3281–3285.

Campbell, R.D. and Trowsdale, J. (1993) Map of the human MHC. *Immunology Today* 14, 349–352.

Canals, A., Gasbarre, L.C., Boyd, P.C., Almeria, S. and Zarlenga, D.S. (1997) Cloning and expression of bovine interleukin-15: analysis and modulation of transcription by exogenous stimulation. *Journal of Interferon and Cytokine Research* 17, 473–480.

Cantrell, D. (1996) T cell antigen receptor signal transduction pathways. *Annual Review of Immunology* 14, 259–274.

Carr, M.M., Howard, C.J., Sopp, P., Manser, J.M. and Parsons, K.R. (1994) Expression on porcine γδ lymphocytes of a phylogenetically conserved surface antigen previously restricted in expression to ruminant γδ T lymphocytes. *Immunology* 81, 36–40.

Cerretti, D.P., McKereghan, K., Larsen, A., Cantrell, M.A., Anderson, D., Gillis, S., Cosman, D. and Baker, P.E. (1986a) Cloning, sequence, and expression of bovine interleukin 2. *Proceedings of the National Academy of Sciences of the USA* 83, 3223–3227.

Cerretti, D.P., McKereghan, K., Larsen, A., Cosman, D., Gillis, S. and Baker, P.E. (1986b) Cloning, sequence, and expression of bovine interferon-γ. *Journal of Immunology* 136, 4561–4564.

Chowdhary, B.P., Hassanane, M.S. and Gustavsson, I. (1994) Regional localization of the bovine interleukin-2 (IL2) gene to chromosome 17q22→q23 by *in situ* hybridization. *Cytogenetics and Cell Genetics* 65, 166–168.

Clevers, H., MacHugh, N.D., Bensaid, A., Dunlap, S., Baldwin, C.L., Kaushal, A., Iams, K., Howard, C.J. and Morrison, W.I. (1990) Identification of a bovine surface antigen uniquely expressed on CD4⁻CD8⁻T cell receptor γ/δ⁺T lymphocytes. *European Journal of Immunology* 20, 809–817.

Clore, G.M., Appella, E., Yamada, M., Matsushima, K. and Gronenborn, A.M. (1990) Three dimensional structure of interleukin 8 in solution. *Biochemistry* 29, 1689–1696.

Cludts, I., Droogmans, L., Cleuter, Y., Kettmann, R. and Burny, A. (1992) Sequence of bovine interleukin 7. *DNA Sequence* 3, 55–59.

Cludts, I., Cleuter, Y., Kettmann, R., Burny, A. and Droogmans, L. (1993) Cloning and characterization of the tandemly arranged bovine lymphotoxin and tumour necrosis factor-α genes. *Cytokine* 5, 336–341.

Crawford, R.M., Finbloom, D.S., Ohara, J., Paul, W.E. and Meltzer, M.S. (1988) Regulation of macrophage effector function by B cell stimulatory factor-1 (BSF-1). *Advances in Experimental Medicine and Biology* 239, 223–229.

Daopin, S., Li, M. and Davies, D.R. (1993) Crystal structure of TGF-β_2 refined at 1.8 Å resolution. *Proteins* 17, 176–192.

Davies, C.J., Joosten, I., Bernoco, D., Arriens, M.A., Bester, J., Ceriotti, G., Ellis, S., Hensen, E.J., Hines, H.C., Horin, P., Kristensen, B., Lewin, H.A., Meggiolaro, D., Morgan, A.L.G., Morita, M., Nilsson, Ph.R., Oliver, R.A., Orlova, A., Østergård, H., Park, C.A., Schuberth, H.-J., Simon, M., Spooner, R.L. and Stewart, J.A. (1994) Polymorphism of bovine MHC class I genes. Joint report of the fifth international bovine lymphocyte antigen (BoLA) workshop, Interlaken, Switzerland, 1 August 1992. *European Journal of Immunology* 21, 239–258.

Davies, C.J., Andersson, L., Ellis, S.A., Hensen, E.J., Lewin, H.A., Mikko, S., Muggli-Cockett, N.E., van der Poel, J.J. and Russell, G.C. (1997) Nomenclature for factors of the BoLA system, 1996: report of the ISAG BoLA nomenclature committee. *Animal Genetics* 28, 159–168.

Davis, M.M. and Bjorkman, P.J. (1988) T-cell antigen receptor genes and T-cell recognition. *Nature* 334, 395–402.

Davis, M.M., Chien, Y.H., Gascoigne, N.R.J. and Hedrick, S.M. (1984) A murine T cell receptor gene complex: isolation, structure and rearrangement. *Immunological Reviews* 81, 235–258.

De Libero, G. (1997) Sentinel function of broadly reactive human γδ T cells. *Immunology Today* 18, 22–26.

Dinarello, C.A. (1988) Biology of interleukin 1. *FASEB Journal* 2, 108–115.

Droogmans, L., Cludts, I., Cleuter, Y., Kettmann, R. and Burny, A. (1992) Nucleotide sequence of the bovine *interleukin*-6 gene promotor. *DNA Sequence* 3, 115–117.

Ealick, S.E., Cook, W.J., Vijay-Kumar, S., Carson, M., Nagabhushan, T.L., Trotta, P.P. and Bugg, C.E. (1991) Three-dimensional structure of recombinant human interferon-γ. *Science* 252, 698–702.

Ellis, S.A., Staines, K.A. and Morrison, W.I. (1996) cDNA sequence of cattle MHC class I genes transcribed in serologically defined haplotypes A18 and A31. *Immunogenetics* 43, 156–159.

Estes, D.M. (1996) Differentiation of B cells in the bovine: role of cytokines in immunoglobulin isotype expression. *Veterinary Immunology and Immunopathology* 54, 61–67.

Evans, C.W., Lund, B.T., McConnell, I. and Bujdoso, R. (1994) Antigen recognition and activation of ovine γδ T cells. *Immunology* 82, 229–237.

Gaddum, R.M., Willis, A.C. and Ellis, S.A. (1996) Peptide motifs from three cattle MHC (BoLA) class I antigens. *Immunogenetics* 43, 238–239.

Garber, T.L., Hughes, A.L., Watkins, D.I. and Templeton, J.W. (1994) Evidence for at least three transcribed *BoLA* class I loci. *Immunogenetics* 39, 257–265.

Garboczi, D.N., Ghosh, P., Utz, U., Fan, Q.R., Biddison, W.E. and Wiley, D.C. (1996) Structure of the complex between human T-cell receptor, viral peptide and HLA-A2. *Nature* 384, 134–141.

Garcia, K.C., Degano, M., Stanfield, R.L., Brunmark, A., Jackson, M.R., Peterson, P.A., Teyton, L. and Wilson, I.A. (1996) An αβ T cell receptor structure at 2.5 Å and its orientation in the TCR-MHC complex. *Science* 274, 209–219.

Garrett, T.P.J., Saper, M.A., Bjorkman, P.J., Strominger, J.L. and Wiley, D.C. (1989) Specificity pockets for the side chains of peptide antigens in HLA-Aw68. *Nature* 342, 692–696.

Gellert, M. (1996) A new view of V(D)J recombination. *Genes to Cells* 1, 269–275.

Giri, J.G., Anderson, D.M., Kumaki, S., Park, L.S., Grabstein, K.H. and Cosman, D. (1995) IL-15, a novel T cell growth factor that shares activities and receptor components with IL-2. *Journal of Leukocyte Biology* 57, 763–766.

Graves, B.J., Hatada, M.H., Hendrickson, W.A., Miller, J.K., Madison, V.S. and Satow, Y. (1990) Structure of interleukin 1 α at 2.7-Å resolution. *Biochemistry* 29, 2679–2684.

Groenen, M.A.M., van der Poel, J.J., Dijkhof, R.J.M. and Giphart, M.J. (1990) The nucleotide sequence of bovine MHC class II *DQB* and *DRB* genes. *Immunogenetics* 31, 37–44.

Guo, Y., Ziegler, H.K., Safley, S.A., Niesel, D.W., Vaidya, S. and Klimpel, G.R. (1995) Human T-cell recognition of *Listeria monocytogenes*: recognition of listeriolysin O by TcR αβ$^+$ and TcR γδ$^+$ T cells. *Infection and Immunity* 63, 2288–2294.

Gwakisa, P., Mikko, S. and Andersson, L. (1994) Close genetic linkage between *DRBP1* and *CYP21* in the MHC of cattle. *Mammalian Genome* 5, 731–734.

Hansen, T.R., Leaman, D.W., Cross, J.C., Mathialagan, N., Bixby, J.A. and Roberts, R.M. (1991) The genes for the trophoblast interferons and the related interferon-$α_{II}$ possess distinct 5′-promoter and 3′-flanking sequences. *Journal of Biological Chemistry* 266, 3060–3067.

Hash, S.M., Brown, W.C. and Rice-Ficht, A.C. (1994) Characterization of a cDNA encoding bovine interleukin 10: kinetics of expression in bovine lymphocytes. *Gene* 139, 257–261.

Hay, F. (1993) The generation of diversity. In: Roitt, I., Brostoff, J. and Male, D. (eds) *Immunology*. Mosby, Chicago, Illinois, pp. 5.1–5.14.

Hegde, N.R., Ellis, S.A., Gaddum, R.M., Tregaskes, C.A., Sarath, G. and Srikumaran, S. (1995) Peptide motif of the cattle MHC class I antigen BoLA-A11. *Immunogenetics* 42, 302–303.

Hein, W.R. and Dudler, L. (1997) TCR γδ$^+$ cells are prominent in normal bovine skin and express a diverse repertoire of antigen receptors. *Immunology* 91, 58–64.

Hein, W.R. and MacKay, C.R. (1991) Prominence of γδ T cells in the ruminant immune system. *Immunology Today* 12, 30–34.

Hein, W.R., Dudler, L., Marcuz, A. and Grossberger, D. (1990) Molecular cloning of sheep T cell receptor γ and δ chain constant regions: unusual primary structure of γ chain hinge segments. *European Journal of Immunology* 20, 1795–1804.

Henney, C.S. (1989) Early events in lymphopoiesis: the role of interleukins 1 and 7. *Journal of Autoimmunity* 2 (Suppl.), 155–161.

Heussler, V.T., Eichhorn, M. and Dobbelaere, D.A.E. (1992) Cloning of a full-length cDNA encoding bovine interleukin 4 by the polymerase chain reaction. *Gene* 114, 273–278.

Holoshitz, J., Koning, F., Coligan, J.E., De Bruyn, J. and Strober, S. (1989) Isolation of CD4⁻ CD8⁻ mycobacteria-reactive T lymphocyte clones from rhematoid arthritis synovial fluid. *Nature* 339, 226–229.

Howard, M. and O'Garra, A. (1992) Biological properties of interleukin 10. *Immunology Today* 13, 198–200.

Ihle, J.N. (1992) Interleukin-3 and hematopoiesis. *Chemical Immunology* 51, 65–106.

Imani, F. and Soloski, M.J. (1991) Heat shock proteins can regulate expression of the Tla region-encoded class Ib molecule Qa-1. *Proceedings of the National Academy of Sciences of the USA* 88, 10475–10479.

Ishiguro, N. and Hein, W.R. (1994) The T cell receptor. In: Goddeeris, B.M.L. and Morrison, I.W. (eds) *Cell-mediated Immunity in Ruminants*. CRC Press, Boca Raton, Florida, pp. 59–73.

Ishiguro, N., Tanaka, A. and Shinagawa, M. (1990) Sequence analysis of bovine T-cell receptor α chain. *Immunogenetics* 31, 57–60.

Ishiguro, N., Aida, Y., Shinagawa, T. and Shinagawa, M. (1993) Molecular structures of cattle T-cell receptor gamma and delta chains predominantly expressed on peripheral blood lymphocytes. *Immunogenetics* 38, 437–443.

Ivanov, V.N., Karginov, V.A., Morozov, I.V. and Gorodetsky, S.I. (1988) Molecular cloning of a bovine immunoglobulin lambda chain cDNA. *Gene* 67, 41–48.

Janeway, C.A. and Travers, P. (1997) *Immunobiology: the Immune System in Health and Disease*, 3rd edn. Current Biology, New York, and Garland Publishing, New York.

Jarrell, V.L., Lewin, H.A., Da, Y. and Wheeler, M.B. (1995) Gene-centromere mapping of bovine *DYA, DRB3*, and *PRL* using secondary oocytes and first polar bodies: evidence for four-strand double crossovers between *DYA* and *DRB3*. *Genomics* 27, 33–39.

Jenkins, M.K., Pardoll, D.M., Mizuguchi, J., Chused, T.M. and Schwartz, R.H. (1987) Molecular events in the induction of a nonresponsive state in interleukin 2-producing helper T- lymphocyte clones. *Proceedings of the National Academy of Sciences of the USA* 84, 5409–5413.

Jones, E.Y., Stuart, D.I. and Walker, N.P. (1989) Structure of tumour necrosis factor. *Nature* 338, 225–228.

Kacskovics, I. and Butler, J.E. (1996) The heterogeneity of bovine IgG2–VIII. The complete cDNA sequence of bovine IgG2a (A2) and IgG1. *Molecular Immunology* 33, 189–195.

Kappes, S.M., Keele, J.W., Stone, R.T., McGraw, R.A., Sonstegard, T.S., Smith, T.P., Lopez-Corrales, N.L. and Beattie, C.W. (1997) A second-generation linkage map of the bovine genome. *Genome Research* 7, 235–249.

Kaufmann, S.H., Blum, C. and Yamamoto, S. (1993) Crosstalk between α/β T cells and γ/δ T cells *in vivo*: activation of α/β T-cell responses after γ/δ T-cell modulation with the monoclonal antibody GL3. *Proceedings of the National Academy of Sciences of the USA* 90, 9620–9624.

Kehrli, M.E., Jr, Goff, J.P., Stevens, M.G. and Boone, T.C. (1991) Effects of granulocyte colony-stimulating factor administration to periparturient cows on neutrophils and bacterial shedding. *Journal of Dairy Science* 74, 2448–2458.

Kelso, A. (1995) Th1 and Th2 subsets: paradigms lost? *Immunology Today* 16, 374–379.

Knight, K.L. and Becker, R.S. (1987) Isolation of genes encoding bovine IgM, IgG, IgA and IgE chains. *Veterinary Immunology and Immunopathology* 17, 17–24.

Komano, H., Fujiura, Y., Kawaguchi, M., Matsumoto, S., Hashimoto, Y., Obana, S., Mombaerts, P., Tonegawa, S., Yamamoto, H. and Itohara, S. (1995) Homeostatic regulation of intestinal epithelia by intraepithelial γδ T cells. *Proceedings of the National Academy of Sciences of the USA* 92, 6147–6151.

Kozbor, D., Trinchieri, G., Monos, D.S., Isobe, M., Russo, G., Haney, J.A., Zmijewski, C. and Croce, C.M. (1989) Human TCR-γ⁺/δ⁺, CD8⁺ T lymphocytes recognize tetanus toxoid in an MHC-restricted fashion. *Journal of Experimental Medicine* 169, 1847–1851.

Kulseth, M.A. and Rogne, S. (1994) Cloning and characterization of the bovine immunoglobulin J chain cDNA and its promoter region. *DNA and Cell Biology* 13, 37–42.

Langer, J.A., Puvanakrishnan, R. and Womack, J.E. (1992) Somatic cell mapping of the bovine interferon-alpha receptor. *Mammalian Genome* 3, 237–240.

Leong, S.R., Flaggs, G.M., Lawman, M. and Gray, P.W. (1988) The nucleotide sequence for the cDNA of bovine interleukin-1 beta. *Nucleic Acids Research* 16, 9054

Leong, S.R., Flaggs, G.M., Lawman, M.J.P. and Gray, P.W. (1989) Cloning and expression of the cDNA for bovine granulocyte-macrophage colony-stimulating factor. *Veterinary Immunology and Immunopathology* 21, 261–278.

Letterio, J.J. and Roberts, A.B. (1997) TGF-β: a critical modulator of immune cell function. *Clinical Immunology and Immunopathology* 84, 244–250.

Lewin, H.A. (1996) Genetic organization, polymorphism, and function of the bovine major histocompatibility complex. In: Schook, L.B. and Lamont, S.J. (eds) *The Major Histocompatibility Complex Region of Domestic Animal Species*. CRC Press, Boca Raton, Florida, pp. 65–98.

Lindberg, P.-G. and Andersson, L. (1988) Close association between DNA polymorphism of bovine major histocompatibility complex class I genes and serological BoLA-A specificities. *Animal Genetics* 19, 245–255.

Madden, D.R. (1995) The three-dimensional structure of peptide-MHC complexes. *Annual Review of Immunology* 13, 587–622.

Madden, D.R., Gorga, J.C., Strominger, J.L. and Wiley, D.C. (1991) The structure of HLA-B27 reveals nonamer self-peptides bound in an extended conformation. *Nature* 353, 321–325.

Maliszewski, C.R., Baker, P.E., Schoenborn, M.A., Davis, B.S., Cosman, D., Gillis, S. and Cerretti, D.P. (1988a) Cloning, sequence and expression of bovine interleukin 1α and interleukin 1β complementary DNAs. *Molecular Immunology* 25, 429–437.

Maliszewski, C.R., Schoenborn, M.A., Cerretti, D.P., Wignall, J.M., Picha, K.S., Cosman, D., Tushinski, R.J., Gillis, S. and Baker, P.E. (1988b) Bovine GM-CSF: molecular cloning and biological activity of the recombinant protein. *Molecular Immunology* 25, 843–850.

Marello, K.L., Gallagher, A., McKeever, D.J., Spooner, R.L. and Russell, G.C. (1995) Expression of multiple *DQB* genes in *Bos indicus* cattle. *Animal Genetics* 26, 345–349.

Martal, J., Chene, N., Camous, S., Huynh, L., Lantier, F., Hermier, P., L' Haridon, R., Charpigny, G., Charlier, M. and Chaouat, G. (1997) Recent developments and potentialities for reducing embryo mortality in ruminants: the role of IFN-τ and other cytokines in early pregnancy. *Reproduction, Fertility, and Development* 9, 355–380.

Mertens, B., Gobright, E. and Seow, H.F. (1996) The nucleotide sequence of the bovine interleukin-5-encoding cDNA. *Gene* 176, 273–274.

Mikko, S. and Andersson, L. (1995) Extensive MHC class II *DRB3* diversity in African and European cattle. *Immunogenetics* 42, 408–413.

Milburn, M.V., Hassell, A.M., Lambert, M.H., Jordan, S.R., Proudfoot, A.E.I., Graber, P. and Wells, T.N.C. (1993) A novel dimer configuration revealed by the crystal structure at 2.4 Å resolution of human interleukin-5. *Nature* 363, 172–176.

Modi, W.S., Dean, M., Seuanez, H.N., Mukaida, N., Matsushima, K. and O'Brien, S.J. (1990) Monocyte-derived neutrophil chemotactic factor (MDNCF/IL-8) resides in a gene cluster along with several other members of the platelet factor 4 gene superfamily. *Human Genetics* 84, 185–187.

Mondino, A., Khoruts, A. and Jenkins, M.K. (1996) The anatomy of T-cell activation and tolerance. *Proceedings of the National Academy of Sciences of the USA* 93, 2245–2252.

Moore, K.W., O' Garra, A., de Waal Malefyt, R., Vieira, P. and Mosmann, T.R. (1993) Interleukin-10. *Annual Review of Immunology* 11, 165–190.

Morsey, M.A., Popowych, Y., Kowalski, J., Gerlach, G., Godson, D., Campos, M. and Babiuk, L.A. (1996) Molecular cloning and expression of bovine interleukin-8. *Microbial Pathogenesis* 20, 203–212.

Mosley, B., Dower, S.K., Gillis, S. and Cosman, D. (1987) Determination of the minimum polypeptide lengths of the functionally active sites of human interleukins 1α and 1β. *Proceedings of the National Academy of Sciences of the USA* 84, 4572–4576.

Mousavi, M., Rabbani, H. and Hammarstrom, L. (1997) Characterization of the bovine ε gene. *Immunology* 92, 369–373.

Muggli-Cockett, N.E. and Stone, R.T. (1988) Identification of genetic variation in the bovine major histocompatibility complex DRβ-like genes using sequenced bovine genomic probes. *Animal Genetics* 19, 213–225.

Muggli-Cockett, N.E. and Stone, R.T. (1989) Partial nucleotide sequence of a bovine major histocompatibility class II DRβ-like gene. *Animal Genetics* 20, 361–370.

Mwangi, S.M., Logan-Henfrey, L., McInnes, C. and Mertens, B. (1995) Cloning of the bovine interleukin-3-encoding cDNA. *Gene* 162, 309–312.

Naessens, J. (1997) Surface Ig on B lymphocytes from cattle and sheep. *International Immunology* 9, 349–354.

Niimi, M., Nakai, Y. and Aida, Y. (1995) Nucleotide sequences and the molecular evolution of the *DMA* and *DMB* genes of the bovine major histocompatibility complex. *Biochemical and Biophysical Research Communications* 217, 522–528.

Nishimura, H., Emoto, M., Hiromatsu, K., Yamamoto, S., Matsuura, K., Gomi, H., Ikeda, T., Itohara, S. and Yoshikai, Y. (1995) The role of $\gamma\delta$ T cells in priming macrophages to produce tumor necrosis factor-α. *European Journal of Immunology* 25, 1465–1468.

Park, C. and Lewin, H.A. (1999) Fine mapping of a region of variation in recombination rate on BTA23 to the *D23S22-D23S23* interval using sperm typing and meiotic breakpoint analysis *Genomics* (in press).

Park, C., Russ, I., Da, Y. and Lewin, H.A. (1995) Genetic mapping of *F13A* to BTA23 by sperm typing: difference in recombination rate between bulls in the *DYA-PRL* interval. *Genomics* 27, 113–118.

Parng, C.L., Hansal, S., Goldsby, R.A. and Osborne, B.A. (1995) Diversification of bovine λ-light chain genes. *Annals of the New York Academy of Sciences* 764, 155–157.

Parng, C.L., Hansal, S., Goldsby, R.A. and Osborne, B.A. (1996) Gene conversion contributes to Ig light chain diversity in cattle. *Journal of Immunology* 157, 5478–5486.

Pascual, V. and Capra, J.D. (1991) Human immunoglobulin heavy-chain variable region genes: organization, polymorphism, and expression. *Advances in Immunology* 49, 1–74.

Rabbani, H., Brown, W.R., Butler, J.E. and Hammarstrom, L. (1997) Polymorphism of the *IGHG3* gene in cattle. *Immunogenetics* 46, 326–331.

Rammensee, H.G., Falk, K. and Rötzchke, O. (1993) Peptides naturally presented by MHC class I molecules. *Annual Review of Immunology* 11, 213–244.

Reeves, R., Spies, A.G., Nissen, M.S., Buck, C.D., Weinberg, A.D., Barr, P.J., Magnuson, N.S. and Magnuson, J.A. (1986) Molecular cloning of functional bovine interleukin 2 cDNA. *Proceedings of the National Academy of Sciences of the USA* 83, 3228–3232.

Reynaud, C.A., Anquez, V., Grimal, H. and Weill, J.C. (1987) A hyperconversion mechanism generates the chicken light chain preimmune repertoire. *Cell* 48, 379–388.

Reynaud, C.A., Dahan, A., Anquez, V. and Weill, J.C. (1989) Somatic hyperconversion diversifies the single V_H gene of the chicken with a high incidence in the D region. *Cell* 59, 171–183.

Reynaud, C.A., Mackay, C.R., Müller, R.G. and Weill, J.C. (1991) Somatic generation of diversity in a mammalian primary lymphoid organ: the sheep ileal Peyer's patches. *Cell* 64, 995–1005.

Rock, E.P., Sibbald, P.R., Davis, M.M. and Chien, Y.-H. (1994) CDR3 length in antigen-specific immune receptors. *Journal of Experimental Medicine* 179, 323–328.

Rogers, M.-L., Goddard, C., Regester, G.O., Ballard, F.J. and Belford, D.A. (1996) Transforming growth factor β in bovine milk: concentration, stability and molecular mass forms. *Journal of Endocrinology* 151, 77–86.

Ryan, A.M. and Womack, J.E. (1993) Type I interferon genes in cattle: restriction fragment length polymorphisms, gene numbers and physical organization on bovine chromosome 8. *Animal Genetics* 24, 9–16.

Ryan, A.M., Gallagher, D.S. and Womack, J.E. (1992) Syntenic mapping and chromosomal localization of bovine alpha and beta interferon genes. *Mammalian Genome* 3, 575–578.

Saini, S.S., Hein, W.R. and Kaushik, A. (1997) A single predominantly expressed polymorphic immunoglobulin V_H gene family, related to mammalian group, I, Clan, II, is identified in cattle. *Molecular Immunology* 34, 641–651.

Saper, M.A., Bjorkman, P.J. and Wiley, D.C. (1991) Refined structure of the human histocompatibility antigen HLA-A2 at 2.6Å resolution. *Journal of Molecular Biology* 219, 277–319.

Schild, H., Mavaddat, N., Litzenberger, C., Ehrich, E.W., Davis, M.M., Bluestone, J.A., Matis, L., Draper, R.K. and Chien, Y.H. (1994) The nature of major histocompatibility complex recognition by γδ T cells. *Cell* 76, 29–37.

Sehgal, P.B. (1990) Interleukin-6: molecular pathophysiology. *Journal of Investigative Dermatology* 94 (Suppl.), 2s–6s.

Shepard, H.M., Leung, D., Stebbing, N. and Goeddel, D.V. (1981) A single amino acid change in IFN-$β_1$ abolishes its antiviral activity. *Nature* 294, 563–565.

Sigurdardóttir, S., Mariani, P., Groenen, M.A.M., van der Poel, J. and Andersson, L. (1991) Organization and polymorphism of bovine major histocompatibility complex class II genes as revealed by genomic hybridizations with bovine probes. *Animal Genetics* 22, 465–475.

Sigurdardóttir, S., Borsch, C., Gustafson, K. and Andersson, L. (1992) Gene duplications and sequence polymorphism of bovine class II *DQB* genes. *Immunogenetics* 35, 205–213.

Sinclair, M.C., Gilchrist, J. and Aitken, R. (1995) Molecular characterization of bovine V λ regions. *Journal of Immunology* 155, 3068.

Sinclair, M.C., Gilchrist, J. and Aitken, R. (1997) Bovine IgG Repertoire is dominated by a single diversified V_H gene family. *Journal of Immunology* 159, 3883–3889.

Stone, R.T. and Muggli-Cockett, N.E. (1990) Partial nucleotide sequence of a novel bovine major histocompatibility complex class II β-chain gene, *BoLA-DIB*. *Animal Genetics* 21, 353–360.

Stone, R.T. and Muggli-Cockett, N.E. (1993) *BoLA-DIB*: species distribution, linkage with *DOB*, and Northern analysis. *Animal Genetics* 24, 41–45.

Symons, D.B., Clarkson, C.A., Milstein, C.P., Brown, N.R. and Beale, D. (1987) DNA sequence analysis of two bovine immunoglobulin CHγ pseudogenes. *Journal of Immunogenetics* 14, 273–283.

Symons, D.B.A., Clarkson, C.A. and Beale, D. (1989) Structure of bovine immunoglobulin constant region heavy chain gamma 1 and gamma 2 genes. *Molecular Immunology* 26, 841–850.

Takatsu, K., Tominaga, A., Harada, N., Mita, S., Takahashi, T., Kikuchi, Y. and Yamaguchi, N. (1988) T cell-replacing factor (TRF)/interleukin 5 (IL-5): molecular and functional properties. *Immunological Reviews* 102, 107–135.

Takeuchi, N., Ishiguro, N. and Shinagawa, M. (1992) Molecular cloning and sequence analysis of bovine T–cell receptor γ and δ chain genes. *Immunogenetics* 35, 89–96.

Tanaka, A., Ishiguro, N. and Shinagawa, M. (1990) Sequence and diversity of bovine T–cell receptor β-chain genes. *Immunogenetics* 32, 263–271.

Theze, J., Alzari, P.M. and Bertoglio, J. (1996) Interleukin 2 and its receptors: recent advances and new immunological functions. *Immunology Today* 17, 481–486.

Threadgill, D.W., Adkison, L.R. and Womack, J.E. (1990) Syntenic conservation between humans and cattle. II. Human chromosome 12. *Genomics* 8, 29–34.

Tobin-Janzen, T.C. and Womack, J.E. (1992) Comparative mapping of *IGHG1*, *IGHM*, *FES*, and *FOS* in domestic cattle. *Immunogenetics* 36, 157–165.

Tobin-Janzen, T.C. and Womack, J.E. (1996) The immunoglobulin lambda light chain constant region maps to *Bos taurus* chromosome 17. *Animal Biotechnology* 7, 163–172.

Toyonaga, B. and Mak, T.W. (1987) Genes of the T–cell antigen receptor in normal and malignant T cells. *Annual Review of Immunology* 5, 585–620.

van der Poel, J.J., Groenen, M.A.M., Dijkhof, R.J.M., Ruyter, D. and Giphart, M.J. (1990) The nucleotide sequence of the bovine MHC class II alpha genes: *DRA*, *DQA* and *DYA*. *Immunogenetics* 31, 29–36.

van Eijk, M.J.T., Beever, J.E., Stewart, J.A., Nicholaides, G.E., Green, C.A. and Lewin, H.A. (1995) Genetic mapping of *BoLA-A*, *CYP21*, *DRB3*, *DYA* and *PRL* on BTA 23. *Mammalian Genome* 6, 151–152.

Van Obberghen-Schilling, E., Kondaiah, P., Ludwig, R.L., Sporn, M.B. and Baker, C.C. (1987) Complementary deoxyribonucleic acid cloning of bovine transforming growth factor-β1. *Molecular Endocrinology* 1, 693–698.

Velan, B., Cohen, S., Grosfeld, H., Leitner, M. and Shafferman, A. (1985) Bovine interferon α genes. structure and expression. *Journal of Biological Chemistry* 260, 5498–5504.

Velan, B., Cohen, S., Grosfeld, H. and Shafferman, A. (1986) Isolation of bovine IFN-α genes and their expression in bacteria. *Methods in Enzymology* 119, 464–474.

Vidović, D., Roglić, M., McKune, K., Guerder, S., MacKay, C. and Dembic, Z. (1989) Qa-1 restricted recognition of foreign antigen by a γδ T-cell hybridoma. *Nature* 340, 646–650.

Walter, M.R., Cook, W.J., Zhao, B.G., Cameron, R.P., Jr, Ealick, S.E., Walter, R.L., Jr, Reichert, P., Nagabhushan, T.L., Trotta, P.P. and Bugg, C.E. (1992) Crystal structure of recombinant human interleukin-4. *Journal of Biological Chemistry* 267, 20371–20376.

Welch, P.A., Namen, A.E., Goodwin, R.G., Armitage, R. and Cooper, M.D. (1989) Human IL-7: a novel T cell growth factor. *Journal of Immunology* 143, 3562–3567.

Xu, A., Park, C. and Lewin, H.A. (1994) Both *DQB* genes are expressed in BoLA haplotypes carrying a duplicated *DQ* region. *Immunogenetics* 39, 316–321.

Xu, G.Y., Yu, H.A., Hong, J., Stahl, M., McDonagh, T., Kay, L.E. and Cumming, D.A. (1997) Solution structure of recombinant human interleukin-6. *Journal of Molecular Biology* 268, 468–481.

Yoo, J., Stone, R.T., Kappes, S.M. and Beattie, C.W. (1994) Linkage analysis of bovine interleukin 1 receptor types I and II (IL-1R I, II). *Mammalian Genome* 5, 820–821.

Yoo, J., Kappes, S.M., Stone, R.T. and Beattie, C.W. (1995a) Linkage analysis assignment of bovine granulocyte colony stimulating factor receptor (G-CSFR) to chromosome 3. *Mammalian Genome* 6, 686–687.

Yoo, J., Ponce de Leon, F.A., Stone, R.T. and Beattie, C.W. (1995b) Cloning and chromosomal assignment of the bovine interleukin-2 receptor alpha (IL-2R alpha) gene. *Mammalian Genome* 6, 751–753.

Yoo, J., Stone, R.T., Solinas-Toldo, S., Fries, R. and Beattie, C.W. (1996) Cloning and chromosomal mapping of bovine interleukin-2 receptor gamma gene. *DNA and Cell Biology* 15, 453–459.

Zarlenga, D.S., Canals, A., Aschenbrenner, R.A. and Gasbarre, L.C. (1995) Enzymatic amplification and molecular cloning of cDNA encoding the small and large subunits of bovine interleukin 12. *Biochimica et Biophysica Acta* 1270, 215–217.

Zdanov, A., Schalk-Hihi, C., Gustchina, A., Tsang, M., Weatherbee, J. and Wlodawer, A. (1995) Crystal structure of interleukin-10 reveals the functional dimer with an unexpected topological similarity to interferon gamma. *Structure* 3, 591–601.

Zhang, N. and Womack, J.E. (1992) Synteny mapping in the bovine: genes from human chromosome 5. *Genomics* 14, 126–130.

Zhang, N., Threadgill, D.W. and Womack, J.E. (1992) Synteny mapping in the bovine: genes from human chromosome 4. *Genomics* 14, 131–136.

Zinkernagel, R.M. and Doherty, P.C. (1974) Restriction of *in vitro* T cell-mediated cytotoxicity in lymphocytic choriomeningitis within a syngeneic or semallogeneic system. *Nature* 248, 701–702.

Genetics of Disease Resistance

8

A.J. Teale*
Institute of Aquaculture, University of Stirling, Stirling FK9 4LA, UK

Introduction	199
Genetics of Disease Resistance	200
The nature of genetic control of immunity to disease	203
The goals of research into the genetics of disease resistance in cattle	204
Research approaches	206
Disease-resistance Traits in Cattle	207
Bovine leucosis	207
Brucellosis	208
Dermatophilosis	209
Mastitis	211
Trypanosomiasis	212
Helminthiasis	214
Tick resistance	216
Concluding Remarks	218
References	219

Introduction

In this chapter, the genetic basis of immunity in domestic cattle, *Bos taurus* and *Bos indicus*, is reviewed, and research advances which have contributed to development of current understanding are cited. It is not the intention here to review the genetics of the fundamental cellular and molecular aspects of innate immunity of cattle, or of immune responses which follow from antigen recognition in the form of specific acquired immunity. For these, the reader is referred to Chapter 7 of this volume and to other relatively recent and comprehensive reviews (Teale *et al.*, 1991; Andersson and Davies, 1994; Lewin, 1996).

---

*Formerly International Livestock Research Institute, PO Box 30709, Nairobi, Kenya.

The focus here is on immunity in terms of being refractory to infection or resistant to the deleterious consequences of the infected state and, more particularly, on the genetic basis of diversity within domestic cattle with respect to such traits. Non-infectious diseases are not considered.

Genetics of Disease Resistance

Studies in this field have been driven by two principal objectives. The first of these, as with all scientific endeavour, is to increase knowledge and understanding. However, as experimental animals cattle are far from ideal, and certainly as far as fundamental genetic processes are concerned, they are generally not considered a model species. Nevertheless, as they are studied, apparently unusual and possibly unique features are revealed. A good example in the current context is the organization of the class II region of the major histocompatibility complex (MHC), which, in genetic (Andersson *et al.*, 1988) and also in physical terms (Skow *et al.*, 1996), is somewhat larger than in other mammalian species where this has been investigated. The high proportion of γ/δ T cells in cattle in comparison with non-ruminant species is another example (MacKay and Hein, 1989; Clevers *et al.*, 1990).

Cattle also provide some rather unique diseases for study, such as East Coast fever. This disease is due to infection with the intracellular protozoan parasite *Theileria parva*. The parasite is able to induce its host cells, which are for the most part lymphocytes, to proliferate. As a model system for the study of control of cell division, and the consequences of loss of such control, this disease of cattle is arguably without equal (ole-MoiYoi, 1989)

The second major stimulus to research into disease resistance in cattle is the prospect of useful applications in agriculture. It is clear that there is significant variation in cattle in terms of resistance to diseases, and that this variation is of economic importance. Nevertheless, the field has been slow to develop for several reasons. First, there are alternative options for disease control, such as management changes, treatment, vaccination, vector control, movement control, test and slaughter and isolation and quarantine. Second, the genetic route to improving the disease resistance of meaningful numbers of animals has appeared to be a slow and arduous one, due to the long generation times involved and, in many parts of the world, to an inadequate animal breeding and distribution infrastructure. Third, the spectre of negative impacts of improving disease resistance on valuable productivity traits, themselves often the result of centuries of selection, has been raised and then periodically highlighted. Fourth, the logistics of experimentation in disease resistance in cattle can pose a considerable challenge, mainly because of the nature of the animal, which is expensive to obtain and maintain and relatively slow to reproduce.

However, the situation is changing and the stimulus to undertake research that will provide a new option for disease control is increasing. This is so for several reasons. First, resistance among pathogens to chemotherapeutic and chemoprophylactic drugs is apparently increasing. Compelling examples are resistance to anthelminthics (Waller, 1997) and to trypanocidal compounds (Peregrine, 1994). Certainly where sheep are concerned, the situation in the case of helminthiasis control is such that entire industries are threatened in some countries. In the case of trypanocide resistance, it can be argued that development of the livestock sector in some of the poorest countries of the world is jeopardized. Second, safe, effective and inexpensive vaccines have not been developed for some very important diseases. Again, gastrointestinal helminthiasis and trypanosomiasis provide good examples. The costs of non-genetic disease-control options are also a consideration. The investment required to bring a vaccine or drug to the market must inevitably be passed on to the farmer, and may preclude the development of some potential new products. There are also environmental concerns associated with some conventional approaches to disease control, and these are increasing. There is mounting resistance to the use of chemicals in agriculture, which applies to agents administered directly to animals, both internally and externally, as well as agents deployed in the environment of livestock, such as insecticides and disinfectants.

The genetic option does not suffer these negative features, with one possible exception. It would be naïve to suppose that parasite and pathogen genomes will remain passive while those of their hosts change in an effort to make their parasites' and pathogens' lives more difficult. However, the fact that some livestock populations are undoubtedly relatively resistant to certain diseases, and have remained so for thousands of years in some cases, suggests that new 'agreements' between pathogens and hosts can be brokered at new levels. This view of the genetic option raises another important point. It implies that it will usually be a means of disease control rather than a means of parasite control *per se* (although, as in the case of some forms of colibacillosis of pigs (Sellwood *et al.*, 1975; Sellwood, 1979), certain host genetic changes can totally prevent infection at the level of the individual animal). A subjective comparison of different disease-control options, in terms of a variety of features, is given in Fig. 8.1.

Host resistance can operate at different levels (outlined in Fig. 8.2). The first is at the level of invasion of potential pathogens. Most microorganisms, and even quite complex organisms, proceed no further than the first physical barrier they encounter, be it an epidermis or a mucosal or serosal surface. At the next level, once past this host–environment interface, many potential pathogens fail to establish a foothold. This may be due to innate or acquired immune responses or, much more commonly, to a range of innate non-immunological factors, such as lack of essential nutrients/substrates, lack of receptors for potential intracellular parasites, incompatible intracellular processing mechanisms, etc. These factors, among others, simply reflect fundamental incompatibilities in what would otherwise be a host–parasite or host–pathogen relationship. At the third level, potential pathogens may establish, but not cause significant disease. A good example here is provided by *Trypanosoma congolense* infection in resistant cattle types (Murray and Trail, 1984; Trail *et al.*, 1989).

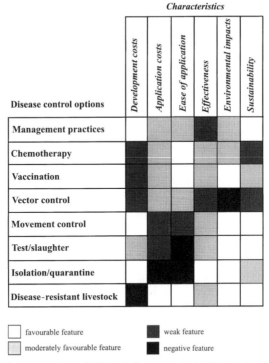

Fig. 8.1. A subjective comparison of different disease control options in terms of some of their more important characteristics.

There is yet a fourth level at which host resistance can operate, although in this case it is effectively more to the benefit of the population than to the benefit of the individual host. This occurs when resistance results in reduced environmental levels of the infective stages of parasites. The effect is most obvious in the case of the helminthiases and other parasite infections. In these examples, in a population, resistant individuals exert a controlling effect on levels of parasites and thus reduce challenge levels to the benefit of all members of the population, resistant or otherwise.

Of these different levels at which infectious disease resistance operates, the first (physical barrier) level is arguably the least tractable for the livestock geneticist and animal breeder. At the level of barriers to establishment of invading organisms, experience in various species suggests that aspects of innate and acquired immunity may offer opportunities for exploitation of genetically determined variability, and this may also be the case with non-immunological host–parasite incompatibility factors. There may also be opportunities at the level of resistance to the consequences of the infected or parasitized state, where again it is possible to envisage involvement of immunological and non-immunological mechanisms. Resistance of the kind that tolerates parasites and pathogens, but at levels or in ways which result in reduced environmental contamination and reduced challenge levels, is a

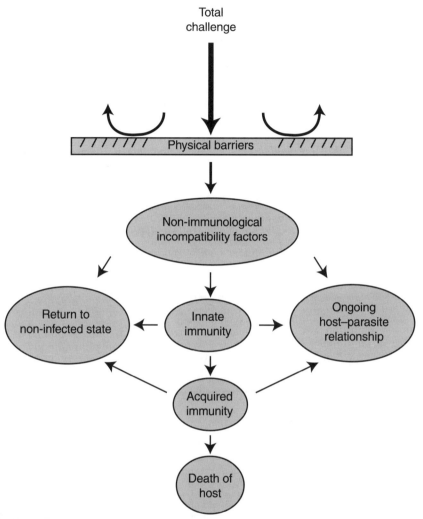

Fig. 8.2. The processes following from challenge with potentially pathogenic organisms, and their consequences.

particularly attractive target. This is because, as with resistance to effects of infection, it allows coexistence of parasites/pathogens and their hosts. It is reasonable to speculate that disease control achieved in these ways may be the most sustainable because it allows a continuing host–parasite relationship.

The nature of genetic control of immunity to disease

While it is not unlikely that, as in other species, major genes responsible for most of the genetic variation involved in some instances of disease resistance will be found in cattle, the likelihood is that most disease-resistance traits will be under multigene control. Moreover, the genetic background on which disease-resistance genes find themselves can be expected to exert effects. A practical implication of this is that those genes which bestow disease resistance on some animals or breeds may not do so on others. Conversely, it can be expected that apparently susceptible animals will carry genes capable of endowing other animals with disease resistance. By extension of these sorts of considerations, it is not difficult to foresee that some, perhaps rare, animals belonging to populations or breeds generally considered to be susceptible will be resistant.

A further level of complexity will be provided by significant environmental effects in most cases of disease resistance. Among these in cattle will be such factors as nutritional status, productivity level and pregnancy status. Among environmental factors, level of challenge will be critical. In some cases, the genotype of the pathogens effectively becomes an important 'environmental' effect. This is the case, for example, with bovine trypanosomiasis. The African trypanosomes have remarkably plastic genomes. One consequence of this is antigenic variation (Borst *et al.*, 1996; Cross, 1996). The process of gene switching, which is responsible for antigenic variation, is semi-random, and thus it follows that any two animals infected with cloned parasites expressing identical antigens at the point of infection will not necessarily experience the same series of parasite antigens subsequently. In the same way, there may also be changes in parasite populations in different host animals in terms of rate of multiplication, pathogenicity and various other properties that could have an impact on the host as infections proceed.

Consequently, it can be expected that many disease-resistance traits will be complex in nature. The complexity will result from multiple gene effects in both the host and the parasite, creating a dynamic interplay between the infected and infecting organisms. Moreover, this will be occurring against a background of changing environmental effects, some of which may be major, but most of which will be subtle and almost subliminal.

The goals of research into the genetics of disease resistance in cattle

As indicated at the outset, the major driver of research into the genetic basis of disease resistance in cattle is the prospect of increasing agricultural productivity. Before considering progress to date in this field it is perhaps helpful to consider what products are being targeted in order to achieve this goal.

The ultimate objective is animals that combine efficiency and productivity with disease resistance. Production traits have been intensively selected for by animal breeders, particularly in agriculturally advanced countries, for hundreds of years in some cases. Such selection has been spectacularly successful in terms of the quantities of end-products, such as meat and milk, which individual animals are able to produce. In cattle, however, improvements in the efficiency with which food is converted into these end-products may be less obvious.

Disease Resistance

Unlike productivity, disease resistance has largely been the product of natural selection over thousands of years. It is probable that to some extent, and especially where intensive human selection for productivity has operated, the gains made previously by natural selection for disease resistance have been lost. Intensive livestock systems have made considerable use of non-genetic means of disease control, and animal breeders have not often considered disease resistance as a factor in improvement programmes. It must be said, however, that this has not been the case universally, and a few enlightened national programmes have shown considerably more prescience than others. None the less, for the most part, the consequence is that high production potential and disease resistance are attributes of different individuals, and often of different populations and breeds. Given this situation, where there is a need for increased sustainability in disease control through an increase in the genetic capacity of populations to be resistant, the simplest and most obvious course is one of replacement of susceptible with resistant stock. That this can be done, and be successful, is illustrated by the beef industry of northern Australia, which switched from an emphasis on *B. taurus* cattle to an emphasis on *B. indicus* animals with their tick (Vercoe and Frisch, 1992) and haemoparasite resistance (Bock *et al.*, 1997). Such a strategy, however, can mean a step backwards in terms of production potential, but this could be avoided if disease resistance and production attributes could be combined. This requires new types of animals, and providing efficient means to produce them rapidly and in large numbers is one of the major challenges facing cattle geneticists at the present time.

The requirement for speed in the process at the selection and breeding stages is apparent if genetics is to have any impact in this area, and the development of genetic markers is currently seen as being pivotal to the entire process. As a first step, markers will be the tools for marker-assisted breeding. They are also the tools for genetic mapping of genes controlling target traits. If mapping results in the identification of such genes, the way is opened for transgenesis. Given that recent advances in cloning of ruminants (Wilmut *et al.*, 1997) continue into the future, even low-efficiency transgenesis will potentially have a very broad impact through the production of clonal transgenic populations comprising significant numbers of animals.

Whether or not this will transpire remains to be seen, but developments in transgenesis and cloning notwithstanding, identification of disease-resistance genes in cattle could have a potential impact in another way. If genes can be identified, underlying molecular mechanisms may become apparent. In the case of disease resistance, it is possible to foresee this leading to the development of new tools and strategies for disease control, such as new drugs, vaccines and immunodulators. The result will be non-genetic delivery of the results of genetic research into disease resistance.

The various products flowing from the development and application of genetic markers, with emphasis on disease resistance, are shown in Fig. 8.3.

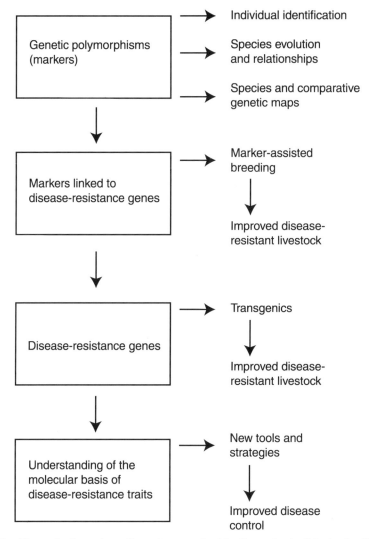

Fig. 8.3. The applications of genetic markers, emphasizing those of potential value for disease-control purposes.

Research approaches

By way of introduction to subsequent sections, a brief consideration of the general approaches which have been and are currently being taken in research on immune response and disease resistance in cattle helps to place the advances which have been made to date in context.

The most basic studies have involved breed or population comparisons. Many studies have been reported, and those which have been carefully conducted and properly controlled to minimize environmental noise have led to identification of potentially valuable genetic resources (populations of cattle which are valuable in their own right for the genetic-replacement approach to disease control (by breed substitution or crossbreeding), or which become the subject of further studies with a view to identification of markers or disease-resistance genes).

In turn, the search for disease-resistance markers has involved screening for associations between disease resistance and a variety of polymorphisms, often in candidate genes or genetic regions. The MHC has understandably received considerable attention in this regard. Many of these studies have involved polymorphism surveys in extant populations considered to constitute a disease-resistance resource, and comparison with polymorphisms in non-resistant populations. While this approach has met with some success, it suffers an important weakness in that the product is an association of unknown inheritance pattern.

Considerable power results from taking the extra step of searching for polymorphisms that are co-inherited with disease resistance. The step beyond this is genome-wide scanning for genetic markers in regions containing disease-resistance genes. For obvious reasons, this is the most demanding in terms of time and resources, but it is clearly an approach with great power and potential, and one that is likely to lead to considerable advances in the case of those disease-resistance traits which merit the commitment required.

Disease-resistance Traits in Cattle

There have been numerous attempts to identify, describe and analyse infectious disease-resistance traits in cattle. Some of the better-researched and understood examples will be reviewed here.

Bovine leucosis

Bovine leucosis (BL), due to infection with bovine leukaemia virus (BLV), is undoubtedly of considerable economic importance in dairy industries of North America and Europe. Cattle infected with BLV and suffering persistent lymphocytosis (PL) have reduced milk and fat yields in comparison with infected PL-negative peers (Da *et al.*, 1993). Da and colleagues (1993) estimated the loss to the American dairy industry due to the effects of BLV infection to be $42 million annually.

That susceptibility of cattle to BLV is influenced by genetic factors was suggested by Burridge and colleagues in 1979. It was then several years before evidence was reported that the source of a significant component of the genetic influence was associated with bovine lymphocyte antigen (BoLA) type (Lewin and Bernoco, 1986). The study reported was conducted in a single herd of Shorthorn cattle and revealed resistance and susceptibility (in terms of virus-dependent B-cell proliferation and lymphocytosis in seropositive animals) to be associated with alternative BoLA haplotypes defined by class I serology. Importantly, Lewin and Bernoco (1986) extended the association study at herd level to a study of the offspring of an individual bull carrying the alternative haplotypes associated with resistance and susceptiblity. They were able to show appropriate co-inheritance of the haplotypes and resistance/susceptibility in 33 offspring examined.

Lewin and colleagues (1988) and Stear and colleagues (1988c) extended studies to other breeds, as a result of which it became clear that several BoLA class I types could be associated with BL resistance and susceptibility, as defined by lymphocyte numbers in seropositive animals. As such, these studies provided an early indication that BL resistance/susceptibility genes might be linked to the genes encoding serologically defined class I polymorphisms, rather than being the class I genes *per se*.

Attention then shifted to the BoLA class II region, with the report in 1992 of an association between development of PL in cattle and DRB2 (van Eijk *et al.*, 1992). The attention to the class II region was subsequently focused on DRB3 (Xu *et al.*, 1993), and specifically on amino acid residues 70 and 71, with the report that Glu–Arg at 70 and 71 was a feature of haplotypes previously found to be associated with resistance to PL. This highly defined association was also supported in a case study.

Since the original report linking DRB3 with resistance/susceptibility to PL in American Holstein Friesians, several DRB3 types have been associated with the trait in Black Pied cattle (Sulimova *et al.*, 1995; Ernst *et al.*, 1997).

A recent study in Holstein Friesians in two herds in Italy suggested an association with BoLA haplotypes defined by A-locus serology and restriction fragment length polymorphisms (RFLPs) in the class II region, including the DRB3 gene (Zanotti *et al.*, 1996). It is notable that the resistant haplotype in the Italian study carries one of the DRB3 types associated with resistance in the studies of Black Pied cattle (Sulimova *et al.*, 1995; Ernst *et al.*, 1997). Likewise, the susceptible haplotype in the Italian Holstein Friesians carries a DRB3 allele associated with susceptibility in the Black Pied cattle.

As far as the mechanism(s) underlying resistance to the effects of BLV infection are concerned, little is known. However, a recent report (Mirsky *et al.*, 1996) suggests that factors affecting viral spread may be important, and that they may be under genetic control.

Brucellosis

Since Thimm reported evidence of natural resistance to brucellosis in East African shorthorn zebu cattle (Thimm, 1973), a considerable amount of research has been undertaken to confirm genetic control of variation of resistance in cattle to brucellosis, and to identify underlying cellular and molecular mechanisms. Breeding experiments confirming the heritability of variation in brucellosis resistance, and as a basis for numerous subsequent studies, were initiated by Templeton and Adams and colleagues in the 1970s (Templeton and Adams, 1995). In these studies, males and females received a standard challenge with *Brucella abortus* S2308 in mid-gestation. Resistant cows did not abort and *Brucella* organisms could not be recovered from cows or calves. Bulls were classified resistant if challenge organisms could not be recovered from semen at slaughter. Candidate immune responses were examined in these animals and, of these, two stand out. First, resistant and susceptible animals were found to differ with respect to anti-LPS immunoglobulin G2a (IgG2a) allotypes, with the A allotype over-represented in susceptible animals (Estes *et al.*, 1990). Second, significant differences emerged in macrophage function in terms of respiratory burst in response to *B. abortus* (Harmon *et al.*, 1989) and ability to control bacterial growth (Price *et al.*, 1990; Campbell and Adams, 1992; Qureshi *et al.*, 1996).

With the realization that the differences in antibacterial activity of macrophages derived from resistant and susceptible cattle extended to the other intracellular pathogens *Mycobacterium bovis* and *Salmonella dublin* (Qureshi *et al.*, 1996), close parallels with resistance to intracellular pathogens in mice under the control of the *Bcg/Ity/Lsh* gene were soon drawn. The murine gene was positionally cloned and designated *Nramp*I (Vidal *et al.*, 1993), and its product is postulated to function in the phagolysosomal membrane to concentrate oxidation products with antibacterial activity. Subsequently, Feng and colleagues (1996) reported cloning and analysis of the bovine homologue.

Interestingly, the polymorphism at position 169, which is associated with variation in resistance to intracellular pathogens in mice, has not so far been reported in other species. However, a single-strand conformational polymorphism (SSCP) has now been reported to be associated with resistance and susceptibility to brucellosis in cattle (Adams and Templeton, 1998).

Thus, there is some evidence that resistance to brucellosis is at least partly under the control of the bovine *Nramp*I gene, but a survey of the genes of a variety of bovid species and the mode of inheritance in cattle suggest that other genes are also involved (Adams and Templeton, 1998). Importantly, these studies have demonstrated the potential of selection for brucellosis resistance in cattle to achieve rapid gains, and thus have underscored the value of this important disease-resistance trait.

Dermatophilosis

The bacterial infection dermatophilosis (reviewed by Ambrose, 1996) is caused by *Dermatophilus congolensis*, and most commonly manifests as a skin infection which can be of varying severity in terms of pathology at a given infection site and in terms of total skin area affected. The disease occurs sporadically throughout the world and in a broad range of species, including humans. The principal economic importance of dermatophilosis arises from the disease in cattle in which production losses parallel severity of skin lesions and which, in many untreated cases, results in death of infected animals. This disease occurs in its severest form in cattle in the tropical humid regions, and especially in West Africa and the Caribbean. In some parts of West Africa, dermatophilosis precludes the keeping of exotic cattle types and is considered to be as serious as trypanosomiasis in terms of constraining livestock agriculture.

The aetiology of dermatophilosis is unclear. While controlled skin challenge can lead to development of lesions, concurrent tick infestation, particularly with the bont tick *Amblyomma variegatum*, is associated with dramatically increased severity of lesions and continuing disease progression in most cattle types (Ambrose, 1996). The tick effect is also a major factor in field-challenge situations (Hadrill and Walker, 1994; Koney *et al.*, 1996).

Laboratory-challenge experiments in sheep have shown that the effect of *A. variegatum* is systemic, in that ticks attached at sites remote from the site of bacterial challenge are able to exacerbate infection (Walker and Lloyd, 1993). It also appears that this effect is due to adult ticks, but not nymphal ticks (Lloyd and Walker, 1993).

Resistance to dermatophilosis has been described in West Africa, where it is most apparent in the N'Dama breed (Morrow *et al.*, 1996) and it is also a characteristic of the Creole cattle of Guadeloupe (Maillard *et al.*, 1993a). The N'Dama is a West African longhorn *B. taurus* breed descended from the earliest domestic cattle on the African continent. The Caribbean Creole cattle of Guadeloupe are an admixture of *B. taurus* and zebu types and, interestingly, there is some evidence that at least some of the *B. taurus* genetic component originated in West Africa (Maillard *et al.*, 1993a) and includes N'Dama genes.

As far as the aetiology of dermatophilosis is concerned, it seems clear that immunosuppression occurs in animals with severe and progressive skin lesions (Koney *et al.*, 1994a, 1996), and the extent of immunosuppression correlates with levels of susceptibility. Significantly, immunosuppression is a feature of tick infestation, especially infestation with *A. variegatum*. Moreover, disease severity correlates with levels of tick infestation (Koney *et al.*, 1994b; Morrow *et al.*, 1996). Thus, while *D. congolensis* can induce skin lesions in the absence of tick infestation, the disease only becomes a significant problem where ticks occur on animals. This in turn, where *A. variegatum* is involved, leads to generalized immunosuppression. The question therefore arises whether resistance is, in effect, resistance to ticks and/or to the effects of tick infestation. This is unclear at the present time.

Studies on the possible underlying genetic mechanism responsible for variation in dermatophilosis resistance have thus far taken a candidate-gene approach, which has resulted in identification of a strong association with polymorphisms in exon 2 of the *BoLA-DRB3* gene in Brahman cattle of Martinique (Maillard *et al.*, 1993b, 1996). Polymorphisms associated with both resistance and susceptibility have been found in *DRB3* exon 2. In the same study populations, a weaker, but significant, association with the BoLA-A8 class I type was found.

On the basis of these results, it is evident that a gene occurs in the MHC region of Brahman cattle in Martinique which contributes significantly to genetic control of dermatophilosis resistance. It is not yet clear whether this gene(s) is a class I and/or II gene, or another linked gene(s). Segregation studies should help clarify the situation.

Mastitis

Because of the economic importance of mastitis, the genetic basis of resistance and susceptibility has received considerable attention. The heritability of mastitis resistance and susceptibility has been established in several studies of various cattle breeds (Stanik and Vasil, 1986; Stavikova *et al.*, 1990; Philipsson *et al.*, 1995; Kelm *et al.*, 1997). The utility of somatic cell scores (SCS) in milk samples as a selection parameter has also been established, based on the high correlation between SCS and mastitis occurrence (Shook and Schutz, 1994; Philipsson *et al.*, 1995). While limited attention has been given to morphological indicator traits with useful heritabilities (Seykora and McDaniel, 1985), considerably more attention has been given to identification of potential immunological and molecular markers of mastitis resistance and susceptibility. Moreover, as far as molecular markers are concerned, most attention has been focused on the MHC.

The search for MHC markers of mastitis resistance has largely involved attempts to discern associations between MHC phenotype and/or genotype and mastitis incidence at the population and breed level. Various studies have been reported as definition of the bovine MHC has improved. Thus associations with MHC class I type (Oddgeirsson *et al.*, 1988; Lundén *et al.*, 1990; Simpson *et al.*, 1990; Weigel *et al.*, 1990; Aarestrup *et al.*, 1995; Simon *et al.*, 1995) and with class II type (Lundén *et al.*, 1990; Dietz *et al.*, 1997; Kelm *et al.*, 1997) have been reported. In some cases, associations have been supported by independent studies in different populations/breeds, but not in all cases (Mejdell *et al.*, 1994). Some studies have been reported in which no associations were found (Våge *et al.*, 1992). Few attempts have been made to identify associations in artificial challenge experiments, although Schukken and colleagues (1994) reported a serologically defined class I specificity associated with susceptibility to intramammary challenge with *Staphylococcus aureus*.

Few studies have gone beyond a simple search for associations to examine co-inheritance of potential markers with mastitis susceptibility, although in one such study, based on a progeny test of informative bulls, an MHC association emerged (Mejdell *et al.*, 1994).

The weakness of simple association studies in the form of marker comparisons between different disease classes is particularly evident where mastitis has been the focus. However, the evidence for a role for genes in the MHC region derives a degree of credibility from the number of independent studies that have revealed associations. The large progeny-test study reported by Mejdell and colleagues (1994) also indicated a role for genes in the MHC region. However, it is clear that genes other than MHC genes also contribute to mastitis resistance, among which, as indicated earlier, genes controlling morphological features may be important (Seykora and McDaniel, 1985). The precise nature of the important genes within or linked to the MHC region remains unclear, mainly because the types of experiments that go beyond association of resistance/susceptibility with entire haplotypes remain to be done.

Some attempts to link aspects of immune response with mastitis resistance have been reported (Mallard *et al.*, 1995; Kelm *et al.*, 1997) and, while weak associations have been found (Kelm *et al.*, 1997), there are no clear indications of the nature of controlling genes. The possible role for genes controlling IgG2 isotype and non-MHC leucocyte antigens, or genes linked to these (reported by Kelm and colleagues), emphasizes the polygenic nature of variation in mastitis resistance/susceptibility.

Trypanosomiasis

Resistance to African tsetse-transmitted trypanosomiasis in cattle, which is commonly referred to as trypanotolerance, is possibly the most researched example of genetic control of infectious-disease resistance in livestock species. The disease in cattle is caused by infection with *Trypanosoma brucei brucei*, *T. congolense* and *T. vivax*. These haemoprotozoa of the order Kinetoplastida cause extracellular infections in a range of wild and domestic species. However, the severest effects in economic terms occur in cattle, with losses to the livestock sector estimated in billions of US dollars annually (Murray and Gray, 1984; Hursey and Slingenbergh, 1995). The parasites are generally transmitted by the bites of infected tsetse flies (*Glossina* spp.), although mechanical transmission also occurs (particularly in the case of *T. vivax* infections). The disease is characterized by anaemia, lymphadenopathy, weight loss and abortion, and in susceptible animals advancing cachexia eventually leads to death after months or even years of infection. Several valuable reviews of trypanotolerance in cattle have been developed previously (Murray *et al.*, 1979, 1982, 1984).

Control of trypanosomiasis by conventional means is difficult. This, together with the very serious impacts that the disease has on human development through its effects on livestock agriculture, has focused attention on trypanotolerance as an option for disease control (Trail *et al.*, 1989).

It has been recognized for many decades that some *B. taurus* types (or breeds) of West African cattle appear to be resistant to trypanosomiasis in comparison with susceptible *B. indicus* breeds (Stewart, 1937). The trypanotolerance trait is particularly evident in the longhorn N'Dama and the small West African shorthorn breeds (Roberts and Gray, 1973; Roelants, 1986; Doko *et al.*, 1991). Notably, in the small West African shorthorn breeds, there is evidence of considerable variation in levels of resistance, especially at higher challenge levels (Roelants, 1986), when highly resistant and susceptible animals are differentiated.

It should be emphasized that, as with some other examples of resistance to infectious disease, trypanosomiasis resistance is not absolute. It is clearly influenced by level of challenge and other environmental factors, such as nutritional status and workload. There is also some evidence of variation in resistance within the generally susceptible Boran zebu breed of East Africa (Njogu et al., 1985). Moreover, N'Dama cattle appear to be as susceptible as Boran cattle to challenge with an unusual *T. vivax* which causes a particularly acute and highly haemorrhagic syndrome (Williams et al., 1992). However, in laboratory-challenge experiments, the majority of animals recover from a severe anaemic episode, irrespective of whether they are N'Damas or Borans.

Although considerable evidence had been accumulated indicating that trypanotolerance is a breed characteristic, and thus by inference under a degree of genetic control, unequivocal evidence for this being the case with the N'Dama breed had to await controlled challenge experiments in animals reared in a non-challenge environment (Logan et al., 1988; Paling et al., 1991a, b). These experiments confirmed the trypanotolerance of the N'Dama breed, and the detailed comparisons of responses to challenge with *T. congolense* established the key parameters which have been used subsequently to detect and measure trypanotolerance.

Studies of field challenges have also demonstrated genetic control of variation in trypanotolerance within the N'Dama breed (Trail et al., 1991) and, at the same time, emphasized the importance of resistance to anaemia development as a component of the trait (Trail et al., 1990a). Indeed, a field-challenge test, in which animals are ranked for selection purposes based on their ability to control anaemia, has been described (Trail et al., 1990b).

Other features of the response to trypanosome challenge have been extensively studied in controlled challenges of N'Dama and susceptible Boran cattle with the objective of understanding the mechanisms underlying the trait. Clearly, superior control of parasites in peripheral blood is a key feature of trypanotolerant animals (Logan et al., 1988; Paling et al., 1991a, b), and for this reason emphasis has been given to seeking differences in immune responses to challenge which differentiate trypanotolerant and trypanosusceptible cattle (Williams et al., 1991, 1996; Flynn et al., 1992; Authié et al., 1993a, b; Sileghem et al., 1993). Most notably, these studies have revealed breed differences in response to challenge with *T. congolense* in terms of antibody responses to cryptic and invariant trypanosome antigens, in early development of costimulatory cytokines, in T-cell responses and in circulating leucocyte populations. Despite this effort, which in terms of advancing knowledge of trypanosomiasis immunology in cattle has been very valuable, a causal relationship between immunological responses and trypanotolerance has not been clearly established (Taylor, 1998). However, at the present time, few would doubt that the immune response is an important component of trypanotolerance.

Interesting non-immunological factors which may be contributing to trypanotolerance are the level and type of sialic acid in erythrocyte membranes (Esievo et al., 1982, 1986, 1990). The importance of this system follows from the observation that *T. vivax* secretes a neuraminidase capable of hydrolysing sialic acid *in vitro* (Esievo, 1979). It seems that erythrocytes of N'Dama cattle not only have more sialic acid in the cell membrane than erythrocytes of zebus, but they may also have more sialic acid species.

A genetic approach to understanding the basis of trypanotolerance was proposed by Soller and Beckman (1988), requiring a genome scan for regions containing trypanotolerance genes. This approach, by making use of the developing genetic map of cattle, is seen as a way towards identification of genetic markers with potential value for marker-assisted breeding, and also as a step towards identification of trypanotolerance genes (Teale, 1993). Although a genome scan in a segregating population of cattle is a very demanding approach, it does not suffer the weaknesses of breed comparisons. A genome scan in a population of more than 200 F_2 N'Dama × Boran cattle is currently in progress (Teale, 1993).

While it is probable that genetic markers of trypanotolerance will be identified in scans of the bovine genome, it is unlikely that it will be possible to positionally clone in cattle. However, inbred mouse strains show considerable variation in resistance to infection with *T. congolense*, the major parasite of cattle (Jennings *et al.*, 1978; Morrison and Murray, 1979), and, given the opportunity to examine large numbers of meioses and utilize a very detailed genetic map in this species, prospects of identifying genes influencing response to *T. congolense* are greatly improved. Genome scans in independent murine F_2 populations have been reported (Kemp *et al.*, 1996, 1997; Kemp and Teale, 1998), which have revealed three regions of the C57BL/6 genome harbouring resistance genes. Further studies are ongoing in advanced intercross populations which are expected to refine genetic mapping preparatory to positional cloning. The utility of the murine model for this bovine disease is that, should trypanosomiasis resistance map to homologous genetic regions in the two species, it becomes likely that the bovine genes will be identified through their murine homologues (Teale, 1997).

Helminthiasis

Although there has been considerable research into helminthiasis resistance in small ruminant species, particularly sheep, rather less has been done in cattle. Considering the global importance of helminthiasis in cattle, this is perhaps surprising. The limited attention to the genetic aspects of the disease probably results from the fact that the effects of disease tend to be insidious rather than dramatic, and, in developed cattle industries, control by grazing management and anthelminthic use has been very effective. However, in economic terms, productivity losses due to helminthiasis can be significant. As far as anthelminthic use is concerned, the threat of resistance to the drugs developing in target parasites is very real, although the scale of the problem at the present time does not approach that facing many sheep industries.

The research that has been conducted in this area has largely focused on resistance to gastrointestinal nematodes, which are a worldwide problem. Trematode infections, which are a major problem in some areas of the world, have received less attention. In this section, only research on genetic aspects of resistance to gastrointestinal nematodes will be considered.

Differences in resistance to haemonchosis were reported in lines of zebu cattle in Nigeria more than 35 years ago (Ross *et al.*, 1959, 1960), and subsequently significant sire effects on aspects of resistance to other helminths were reported (Kloosterman *et al.*, 1978; Leighton *et al.*, 1989; Gasbarre *et al.*, 1990). Kloosterman and colleagues (1978) focused on Dutch Friesian cattle and infection with *Cooperia* spp., and their results concerning genetic influences were somewhat equivocal. When 14 sire groups were studied for responses to artificial challenges, only differences in antibody responses were detected. However, when two further sire groups were compared, significant differences in egg outputs were observed between groups. A subsequent pasture challenge of further calves derived from the two sires in the previous experiment did not reveal significant differences between sire groups. It is notable that, following these studies, Kloosterman and colleagues concluded that it was 'unlikely that genetic variability in resistance to nematodes can ever be used in cattle for breeding resistant stock'.

More encouraging results were obtained subsequently in studies in Angus cattle (Leighton *et al.*, 1989). In a large study involving natural challenge of calves in 26 sire groups (related sires), correcting for other effects Leighton and colleagues reported a heritability of 0.29 (SE 0.18) for faecal egg count. The complexity of this resistance began to emerge in a very comprehensive follow-up study (Gasbarre *et al.*, 1990). The genetic effect seen in sire groups was confirmed, but the effect was not observed until after 2–3 months of grazing, suggesting a genetic influence on acquired aspects of resistance. Moreover, with increasing challenge and changes in the range of challenge species towards the end of the season, the genetic effect was lost. Significantly, in this calf population, it was possible to show a correlation between earlier low faecal-egg output on pasture and worm burdens at post-mortem, although, in experimental challenges, this correlation did not apply to infections with *Ostertagia ostertagi*. Nevertheless, even in the case of the *O. ostertagi* infections, worm fecundity was correlated with overall resistance status expressed during pasture challenge.

A good example of a breed difference in helminthiasis resistance has been reported in cattle under natural challenge in village herds in the Gambia. N'Dama cattle appeared to shed lower numbers of eggs (during high challenge periods) and carry smaller worm burdens than zebu cattle (Claxton and Leperre, 1991). This breed difference applied to infections at all levels of the gut, including in the abomasum where *Haemonchus contortus* was the major worm species involved.

Little is known about the underlying mechanisms responsible for helminthiasis resistance in cattle. Serum antibody levels have received most attention (Gasbarre *et al.*, 1990), although their role in genetically determined resistance is unclear. Gasbarre and colleagues (1993) reported a high heritability for serum antibodies, but suggested this was under separate genetic control from control of faecal-egg output, although subsequently evidence has been reported to support a role for antibody response in terms of mediating genetically determined variation in egg output (Hammond *et al.*, 1997).

Similarly, little is known about the genes, or even genetic regions influencing this disease-resistance trait. In a preliminary study, Stear and colleagues found no association between BoLA class I type and faecal-egg output, although importantly they reported significant heritabilities in a large study population of crossbred cattle (Stear *et al.*, 1988a). In a follow-up study, a weak association between faecal-egg output and BoLA class I type was found (Stear *et al.*, 1988b).

Tick resistance

Tick resistance in cattle has been the subject of study over many decades in many breeds in a variety of locations. It has involved studies of infestations with tick species of all the major genera, e.g. *Boophilus microplus* (O'Kelly and Spiers, 1976; Stear *et al.*, 1984), *Boophilus decoloratus* (Rechav and Kostrzewski, 1991; Ali and de Castro, 1993), *Amblyomma americanum* (George *et al.*, 1985; Barnard, 1990a), *Amblyomma hebraeum* (Rechav *et al.*, 1991; Norval *et al.*, 1996), *A. variegatum* (Claxton and Leperre, 1991; Mattioli *et al.*, 1993; Morrow *et al.*, 1996), *Rhipicephalus appendiculatus* (Latif *et al.*, 1991), *Haemphysalis longicornis* (Dicker and Sutherst, 1981), *Ixodes rubicundus* (Fourie and Kok, 1995). These studies have involved both natural and artificial challenges, and resistance has generally been measured on the basis of an overall visual assessment of tick burdens (de Castro *et al.*, 1991), most commonly by tick counts, and also in terms of effects on ticks, such as survival (Utech and Wharton, 1982), weight and fecundity (Barnard, 1990b; Latif *et al.*, 1991). Effects on tick-population growth rates have also been determined (Barnard, 1990a) and have served to underscore the value of this particular disease-resistance trait.

In some cases, attempts have been made to correlate host morphological characteristics (de Castro *et al.*, 1991), and especially aspects of immune response, with tick resistance (George *et al.*, 1985; Rechav *et al.*, 1990; Claxton and Leperre, 1991), although causal relationships have not been revealed unequivocally. The MHC class I typing of cattle of defined resistance/susceptibility to *B. microplus* revealed only weak associations (Stear *et al.*, 1984).

However, some general conclusions can be drawn from the results of studies of tick resistance reported to date. First, there is evidence that where cattle are challenged by multiple tick species, resistance is generated to all of them (Rechav *et al.*, 1990; de Castro *et al.*, 1991; Ali and de Castro, 1993; Mattioli *et al.*, 1993; Solomon and Kaaya, 1996). Proportions of ticks of different species tend to be the same on susceptible and resistant animals.

Second, tick resistance tends to be acquired with exposure (O'Kelly and Spiers, 1976; George *et al.*, 1985; Spickett *et al.*, 1989; Morrow *et al.*, 1996). It appears that it is the extent to which this acquired resistance develops which determines the definitive resistance status. Perhaps not surprisingly, seasonal fluctuations in resistance apply to cattle across the resistance/susceptibility spectrum. In some cases, resistance differences are not apparent during periods of low challenge (Fourie and Kok, 1995), and a seasonal waning in resistance, which occurs across breeds and which appears to be independent of nutritional status, has been reported (Doube and Wharton, 1980).

The most commonly reported observation on tick resistance, which applies irrespective of tick species, is the relatively high resistance of *B. indicus* (zebu) cattle in comparison with *B. taurus* types. This has been reported to be the case with a variety of breeds (Dicker and Sutherst, 1981; Barnard, 1990a; Rechav *et al.*, 1990; Latif *et al.*, 1991). This difference between the major cattle subspecies is also reflected in their crosses (O'Kelly and Spiers, 1976; Dicker and Sutherst, 1981; Utech and Wharton, 1982; George *et al.*, 1985; Barnard, 1990a, b; Ali and de Castro, 1993; Fourie and Kok, 1995; Norval *et al.*, 1996). Thus crossbreds have generally been found to be intermediate in terms of resistance in comparison with pure *B. indicus* and *B. taurus* types.

However, there are exceptions to the general rule that *B. indicus* tend to be more tick-resistant than *B. taurus* breeds, most notably in the case of the N'Dama cattle (*B. taurus*) of West Africa, which in contemporary comparisons with *B. taurus* breeds of European origin and with *B. indicus* breeds, have shown the highest levels of resistance (Claxton and Leperre, 1991; Mattioli *et al.*, 1993; Morrow *et al.*, 1996). The predominant tick species in these comparisons was *A. variegatum*, which has implications for dermatophilosis resistance (Morrow *et al.*, 1996).

It is also notable that, where studied, so-called Sanga breeds, which are an established *B. taurus*–*B. indicus* admixture, have outperformed both *B. indicus* and *B. taurus* types. A clear example is provided by the Nguni breed of southern Africa (Spickett *et al.*, 1989; Rechav and Kostrzewski, 1991). The superiority of the Ethiopian Horro breed in comparison with Boran (*B. indicus*) and crosses with European breeds is a second example (Ali and de Castro, 1993).

Overall, however, it would seem that a history of exposure to ticks is the most important factor in determining apparent levels of resistance, irrespective of cattle type and breed (Latif *et al.*, 1991; Rechav and Kostrzewski, 1991; Ali and de Castro, 1993). Further, it seems that there is potential tick resistance in relatively tick-naïve populations of a wide variety of breeds of both *B. indicus* and *B. taurus* type (Utech and Wharton, 1982; Spickett *et al.*, 1989).

The precise nature of the genetic control of tick resistance in cattle is unknown, and, indeed, different mechanisms may operate in different circumstances. However, there are no convincing indications to date that the trait is heterogeneous between cattle populations and in the face of different tick challenges.

Given the report of Kerr and colleagues (1994) that tick resistance segregating in a *B. taurus* population in Australia may be under the control of a single major gene, it is quite possible that with further investigations a major tick-resistance gene will be identified. Its role in other populations can then be readily examined.

Concluding Remarks

The principal objectives of research into the genetic control of variation in disease resistance in cattle at the present time are, first, to assess the potential value of the traits for improvement of productivity, second, to obtain genetic markers for use in breeding programmes, and, third, if possible, to achieve an understanding of the molecular basis of the traits. Potentially useful disease resistance traits have been described where the diseases are due to infections with viruses, bacteria, protozoa and endo- and ectoparasites. Examples of all of these have been reviewed here.

Few would argue that many of the traits have potential value, but whether this is realized or not will depend on elucidation of the mode and extent of inheritance in different environments and on different genetic backgrounds. It will also depend in many cases on the identification of markers sufficiently closely linked to controlling genes to be of use in breeding programmes. Clearly, if specific controlling genes can be identified, prospects for making use of disease-resistance traits will be enhanced considerably.

Consideration of how research has developed and progressed over the decades since disease-resistance traits in cattle were first described in the scientific literature can lead to a conclusion that progress has been slow. But this has largely been due to the fact that research has had to await the development of tools and technologies. In the interim, effort has been focused on doing what was possible. The literature contains many anecdotal descriptions of possible disease-resistance traits and descriptions of breed and population comparisons. Over the past 20 years, as the means to define the bovine MHC have been developed, a major effort has been put into identifying genes in this region which influence these traits, based on the premise that, by virtue of function, the MHC is the probable location of such genes. This candidate-gene approach has been partially successful, but major genes in the MHC influencing economically important disease-resistance traits have not yet been forthcoming. Perhaps this is not surprising, and the situation could change in the relatively near future.

The technologies are now in place to expand the search for disease resistance genes beyond the MHC to encompass the entire genome. The major international effort made in the past few years to develop genetic maps for cattle is one of the great success stories in bovine genetics research. The challenge now is to make use of the maps to identify the markers required for animal-breeding purposes and for the purposes of identifying target disease-resistance genes. In some cases, where laboratory animal models exist, their use may facilitate the process considerably. Another important requirement for this is detailed comparative maps which allow identification of positional candidates across species. These are developing quite rapidly.

Society is increasingly demanding environmental friendliness, which in the livestock field means, among other things, reduced use of chemotherapeutic agents and of parasite and vector control chemicals. This strengthens the case for further research into the genetics of disease resistance in livestock species.

However, such research, especially when it is conducted in the target species, is very costly. It has also tended to be regarded as long-term, and has been viewed as having vaguely defined impacts. However, recent advances in animal-breeding technologies are bringing the day closer when new genotypes, produced by marker-assisted introgression, for example, and possibly by transgenesis, will be propagated and disseminated through the livestock industry in such a way that a real impact will be possible in the foreseeable future. But, if this is to happen, society will have to accept the technologies, their products and the costs involved. Whether this will transpire is not clear at present. What is clear is that the prospects for livestock geneticists achieving an understanding of the fundamental mechanisms responsible for genetic control of at least some disease-resistance traits will continue to improve if they are given the support that is needed.

References

Aarestrup, F.M., Jensen, N.E. and Ostergard, H. (1995) Analysis of associations between major histocompatibility complex (BoLA) class I haplotypes and subclinical mastitis of dairy cows. *Journal of Dairy Science* 78, 1684–1692.

Adams, L.G. and Templeton, J.W. (1998) Genetic resistance to bacterial diseases of animals. *Revue Scientifique et Technique de l'Office International des Epizooties* 17 (1), 200–219.

Ali, M. and de Castro, J.J. (1993) Host resistance to ticks (Acari:Ixodidae) in different breeds of cattle to Bako, Ethiopia. *Tropical Animal Health and Production* 25, 215–222.

Ambrose, N.C. (1996) The pathogenesis of dermatophilosis. *Tropical Animal Health and Production* 28, 29S–37S.

Andersson, L. and Davies, C.J. (1994) The major histocompatibility complex in cell mediated immunity in ruminants. In: Goddeeris, B.M. and Morrison, W.I. (eds) *Cell Mediated Immunity in Ruminants*. CRC Press, Boca Raton, Florida, pp. 37–46.

Andersson, L., Lundén, A., Sigurdardóttir, S., Davis, C.J. and Rask, L. (1988) Linkage relationships in the bovine MHC region: high recombination frequency between class II subregions. *Immunogenetics* 27, 273–280.

Authié, É., Muteti, D.K. and Williams, D.J. (1993a) Antibody responses to invariant antigens of *Trypanosoma congolense* in cattle of differing susceptibility to trypanosomiasis. *Parasite Immunology* 15, 101–111.

Authié, É., Duvallet, G., Robertson, C. and Williams, D.J.L. (1993b). Antibody responses to a 33 kDa cysteine protease of *Trypanosoma congolense*: relationship to 'trypanotolerance' in cattle. *Parasite Immunology* 15, 465–474.

Barnard, D.R. (1990a) Population growth rates for *Amblyomma americanum* (Acari: Ixodidae) on *Bos indicus*, *B. taurus* and *B. indicus* × *B. taurus* cattle. *Experimental and Applied Acarology* 9, 259–265.

Barnard, D.R. (1990b) Cattle breed alters reproduction in *Amblyomma americanum* (Acari:Ixodidae). *Experimental and Applied Acarology* 10, 105–109.

Bock, R.E., de Vas A.J., Kingston, T.G. and McClellan, D.J. (1997) Effect of breed of cattle on innate resistance to infection with *Babesia bovis*, *B. bigemina* and *Anaplsma marginale*. *Australian Veterinary Journal* 75, 337–340.

Borst, P., Rudenko, G., Taylor, M.C., Blundell, P.A., van Leeuwen, F., Bitter, W., Cross, M. and McCulloch, R. (1996) Antigenic variation in trypanosomes. *Archives of Medical Research* 27, 379–388.

Burridge, M.J., Wilcox, C.J. and Hennemann, J.M. (1979) Influence of genetic factors on the susceptibility of cattle to bovine leukaemia virus infection. *European Journal of Cancer* 15, 1395–1400.

Campbell, G.A. and Adams, L.G. (1992) The long-term culture of bovine monocyte-derived macrophages and their use in the study of intracellular proliferation of *Brucella abortus*. *Veterinary Immunology and Immunopathology* 34, 291–305.

Claxton, J. and Leperre, P. (1991) Parasite burdens and host susceptibility of Zebu and N'Dama cattle in village herds in Gambia. *Veterinary Parasitology* 40, 293–304.

Clevers, H., MacHugh, N., Bensaid, A., Dunlap, S., Baldwin, C.L., Kaushal, A., Iams, K., Howard, C.J. and Morrison, W.I. (1990) Identification of a bovine surface antigen uniquely expressed on CD4$^-$, CD8$^-$ T cell receptor gamma/delta$^+$ T lymphocytes. *European Journal of Immunology* 20, 809–817.

Cross, G.A. (1996) Antigenic variation in trypanosomes: secrets surface slowly. *Bioessays* 18, 283–291.

Da, Y., Shanks, P.D., Stewart, J.A. and Lewin, H.A. (1993) Milk and fat yields decline in leukaemia virus infected Holstein cattle with persistent lymphocytosis. *Proceedings of the National Academy of Sciences of the USA* 90, 6538–6541.

de Castro, J.J., Capstick, P.B., Nokoe, S., Kiara, H., Rinkanya, F., Slade, R., Okello, O. and Bennun, L. (1991) Towards the selection of cattle for tick resistance in Africa. *Experimental and Applied Acarology* 12, 219–227.

Dicker, R.W. and Sutherst, R.W. (1981) Control of the bush tick (*Haemaphysalis longicornis*) with Zebu × European cattle. *Australian Veterinary Journal* 57, 66–68.

Dietz, A.B., Cohen, N.D., Timms, L. and Kehrli, M.E., Jr (1997) Bovine lymphocyte antigen class II alleles as risk factors for high somatic cell counts in milk of lactating dairy cows. *Journal of Dairy Science* 80(2), 406–412.

Doko, A., Guedegbe, B., Baelmans, R., Demey, F., N'Diaye, A., Pandey, V.S. and Verhulst, A. (1991) Trypanosomiasis in different breeds of cattle from Benin. *Veterinary Parasitology* 40, 1–7.

Doube, B.M. and Wharton, R.H. (1980) The effect of locality, breed and previous tick experience on seasonal changes in the resistance of cattle to *Boophilus microplus* (Ixodidea:Ixodidae). *Experientia* 15, 1178–1179.

Ernst, L.K., Sulimova, G.E., Orlova, A.R., Udina, I.G. and Pavlenko, S.K. (1997) Features of the distribution of BoLA-A antigens and alleles of the BoLA-DRB3 gene in Black Pied cattle in relation to association with leukaemia. *Genetika* 33, 87–95.

Esievo, K.A.N. (1979) *In vitro* production of neuraminidase (sialidase) by *Trypanosoma vivax*. In: OAU/STRC (ed.) *Proceedings of the 16th Meeting of International Scientific Council for Trypanosomiasis Research and Control, Yaounde, Cameroun*. Eleza Service, Nairobi, pp. 205–210.

Esievo, K.A.N., Saror, D.I., Ilemobade, A.A. and Hallway, M.H. (1982) Variation in erythrocyte survase and free serum sialic acid concentrations during experimental *Trypanosoma vivax* infections in cattle. *Research in Veterinary Science* 32, 1–5.

Esievo, K.A.N., Saror, D.I., Kolo, M.N. and Eduvie, L.O. (1986) Erythrocytes surface sialic acid in N'Dama and Zebu cattle. *Journal of Comparative Pathology* 96, 95–99.

Esievo, K.A.N., Jaye, A., Andrews, J.N., Ukoha, A.I., Alafiatayo, R.A., Eduvie, L.O., Saror, D.I. and Njoku, C.O. (1990) Electrophoresis of bovine erythrocyte sialic acids: existence of additional band in trypanotolerant N'Dama cattle. *Journal of Comparative Pathology* 102, 357–361.

Estes, D.M., Templeton, J.W., Hunter, D.M. and Adams, L.G. (1990) Production and use of murine monoclonal antibodies reactive with bovine 1gM isotype and IgG subisotypes (IgGI, IgG2a and IgG2b) in assessing immunoglobulin levels in return of cattle. *Veterinary Immunology and Immunopathology* 25, 61–72.

Feng, J., Li, Y., Hashad, M., Schurr, E., Gros, P., Adams, L.G. and Templeton, J.W. (1996) Bovine natural resistance associated macrophage protein I (*Nramp* I) gene. *Genome Research* 6, 956–964.

Flynn, J.N., Sileghem, M. and Williams, D.J.L. (1992) Parasite-specific T-cell responses of trypanoloerant and trypanosusceptible cattle during infection with *Trypanosoma congolense*. *Immunology* 75, 639–645.

Fourie, L.J. and Kok, D.J. (1995) A comparison of *Ixodes rubicundus* (Acari:Ixodidae) infestations on Friesian and Bonsmara cattle in South Africa. *Experimental and Applied Acarology* 19, 529–531.

Gasbarre, L.C., Leighton, E.A. and Davies, C.J. (1990) Genetic control of immunity to gastrointestinal nematodes of cattle. *Veterinary Parasitology* 37, 257–272.

Gasbarre, L.C., Leighton, E.A. and Davies, C.J. (1993) Influence of host genetics upon antibody responses against gastrointestinal nematode infections in cattle. *Veterinary Parasitology* 46, 81–91.

George, J.E., Osburn, R.L. and Wikel, S.K. (1985) Acquisition and expression of resistance by *Bos indicus* and *Bos indicus* × *Bos taurus* calves to *Amblyomma americanum* infestation. *Journal of Parasitology* 71, 174–182.

Hadrill, D.J. and Walker, A.R. (1994) Effect of acaricide control of *Amblyomma variegatum* ticks on bovine dermatophilosis on nevis. *Tropical Animal Health and Production* 26, 28–34.

Hammond, A.C., Williams, M.J., Olson, T.A., Gasbarre, L.C., Leighton, E.A. and Menchaca, M.A. (1997) Effect of rotational vs continuous intensive stocking of bahiagrass on performance of Angus cows and calves and interaction with sire type on gastrointestinal nematode burden. *Journal of Animal Science* 75, 2291–2299.

Harmon, B.G., Adams, L.G., Templeton, J.W. and Smith, R., III (1989) Macrophage function in mammary glands of *Brucella abortus*-infected cows and cows that resisted infection after inoculation of *Brucella abortus*. *American Journal of Veterinary Research* 50, 459–465.

Hursey, B.S. and Slingenbergh, J. (1995) The tsetse fly and its effects on agriculture in sub-Saharan Africa. *World Animal Review* 3–4, 67–73.

Jennings, F.W., Whitelaw, D.D., Holmes, P.H. and Urquhart, G.M. (1978) The susceptibility of strains of mice to infection with *Trypanosoma congolense*. *Research in Veterinary Science* 25, 399–400.

Kelm, S.C., Detilleux, J.C., Freeman, A.E., Kehrli, M.E. Jr, Dietz, A.B., Fox, L.K., Butler, J.E., Kasckovics, I. and Kelly, D.H. (1997) Genetic association between parameters of inmate immunity and measures of mastitis in periparturient Holstein cattle. *Journal of Dairy Science* 80, 1767–1775.

Kemp, S.J. and Teale, A.J. (1998). The genetic basis of trypanotolerance in mice and cattle. *Parasitology Today* 14, 450–454.

Kemp, S.J., Iraqi, F., Darvasi, A., Soller, M. and Teale, A.J. (1996) Genetic control of resistance to trypanosomiasis. *Veterinary Immunology and Immunopathology* 54, 239–243.

Kemp, S.J., Iraqi, F., Darvasi, A., Soller, M. and Teale, A.J. (1997) Localisation of genes controlling resistance to trypanosomiasis in mice. *Nature Genetics* 16, 194–196.

Kerr, R.J., Frisch, J.E., Kinghorn, B.P., Smith, C., Gavora, J.S., Benkel, B., Chesnais, J., Fairfull, W., Gibson, J.P., Kennedy, B. and Burnside, E.B. (1994) Evidence for a major gene for tick resistance in cattle. *Proceedings of 5th World Congress, University of Guelph, Guelph, Ontario, Canada* 20, 7–12.

Kloosterman, A., Albers, G.A.A. and Van den Brink, R. (1978) Genetic variation among calves in resistance to nematode parasites. *Veterinary Parasitology* 4, 353–368.

Koney, E.B., Morrow, A.N., Heron, I., Ambrose, N.C. and Scott, G.R. (1994a) Lymphocyte proliferative responses and the occurrence of dermatophilosis in cattle naturally infested with *Amblyomma variegatum*. *Veterinary Parasitology* 55, 245–256.

Koney, E.B., Walker, A.R., Heron, I.D., Morrow, A.N. and Ambrose, N.C. (1994b) Seasonal prevalence of ticks and their association with dermatophilosis in cattle on the Accra plains in Ghana. *Revue d'Élevage et de Médecine Vétérinaire des Pays Tropicaux* 47, 163–167.

Koney, E.B., Morrow, A.N. and Heron, I.D. (1996) The association between *Amblyomma variegatum* and dermatophilosis: epidemiology and immunology. *Tropical Animal Health and Production* 28, 18S–25S.

Latif, A.A., Nokoe, S., Punyua, D.K. and Capstick, P.B. (1991) Tick infestations on Zebu cattle in Western Kenya: quantitative assessment of host resistance. *Journal of Medical Entomology* 28, 122–126.

Leighton, E.A., Murrell, K.D. and Gasbarre, L.C. (1989) Evidence for genetic control of nematode egg-shedding rates in calves. *Journal of Parasitology* 75, 498–504.

Lewin, H.A. (1996) Genetic organization, polymorphism, and function the bovine major histocompatibility complex. In: Schook, L.B. and Lamont, S.J. (eds) *The Major Histocompatibility Complex Region of Domestic Animal Species*. CRC Press, Boca Raton, Florida, pp. 65–98.

Lewin, H.A., Wu, M.C., Stewart, J.A. and Nolan, T.J. (1988) Association between *BoLA* and subclinical bovine leukaemia virus infection in a herd of Holstein-Friesian cows. *Immunogenetics* 27, 338–344.

Lewin, H.A. and Bernoco, D. (1986) Evidence for BoLA-linked resistance and susceptibility to subclinical progression of bovine leukaemia virus infection. *Animal Genetics* 17, 197–207.

Lloyd, C.M. and Walker, A.R. (1993) The systemic effect of adult and immature *Amblyomma variegatum* ticks on the pathogenesis of dermatophilosis. *Revue d'Élevage et de Médecine Vétérinaire des Pays Tropicaux* 46, 313–316.

Logan, L.L., Paling, R.W., Moloo, S.K. and Scott, J.R. (1988) Comparative studies on the responses of N'Dama and Boran cattle to experimental challenge with tsetse-transmitted *Trypanosoma congolense*. In: ILCA and ILRAD (eds) *Livestock Production in Tsetse Affected Areas of Africa*. ILCA/ILRAD, Nairobi, pp. 152–167.

Lundén, A., Sigurdardóttir, S., Edfors-Lilja, I., Danell, B., Rendel, J. and Andersson, L. (1990) The relationship between bovine major histocompatibility complex class II polymorphism and disease studied by use of bull breeding values. *Animal Genetics* 21, 221–232.

MacKay, C.R. and Hein, W.R. (1989). A large proportion of bovine T-cells express the gamma delta T-cell receptor and show a distinct tissue distribution and surface phenotype. *International Immunology* 1, 540.

Maillard, J.C., Kemp, S.J., Naves, M., Palin, C., Demangel, C., Accipe, A., Maillard, N. and Bensaid, A. (1993a) An attempt to correlate cattle breed origins and diseases associated with or transmitted by the tick *Amblyomma variegatum* in the French West Indies. *Revue d'Élevage et de Médecine Vétérinaire des Pays Tropicaux* 46, 283–290.

Maillard, J.C., Palin, C., Trap, I. and Bensaid, A. (1993b) An attempt to identify genetic markers of resistance or susceptibility to dermatophilosis in the zebu Brahman population of Martinique. *Revue d'Élevage et de Médecine Vétérinaire des Pays Tropicaux* 46, 291–295.

Maillard, J.C., Martinez, D. and Bensaid, A. (1996) An amino acid sequence coded by the exon 2 of the BoLA DRB3 gene associated with a BoLA class I specificity constitutes a likely genetic marker of resistance to dermatophilosis in Brahman zebu cattle of Martinique (FWI). *Annals of the New York Academy of Science* 791, 185–197.

Mallard, B.A., Leslie, K.E., Dekkers, J.C.M., Hedge, R., Bauman, M. and Stear, M.J. (1995) Differences in bovine lymphocyte antigen associations between immune responsiveness and risk of disease following intramammary infection with *Staphylococcus aureus*. *Journal of Dairy Science* 78, 1937–1944.

Mattioli, R.C., Bah, M., Faye, J., Kora, S. and Cassama, M. (1993) A comparison of field tick infestation on N'Dama, Zebu and N'Dama × Zebu crossbred cattle. *Veterinary Parasitology* 47, 139–148.

Mejdell, C.M., Lie, Ø., Solbu, H., Arnet, E.F. and Spooner, R.L. (1994) Association of major histocompatibility complex antigens (BoLA-A) with AI bull progeny test results for mastitis, ketosis and fertility in Norwegian cattle. *Animal Genetics* 25, 99–104.

Mirsky, M.L., Olmstead, C.A., Da, Y. and Lewin, H.A. (1996) The prevalence of proviral bovine leukaemia virus in peripheral blood mononuclear cells at two subclinical stages of infection. *Journal of Virology* 70, 2178–2183.

Morrison, W.I. and Murray, M. (1979) *Trypansoma congolense*: inheritance of susceptibility to infection in inbred strains of mice. *Experimental Parasitology* 48, 364–374.

Morrow, A.N., Koney, E.B. and Heron, I.D. (1996) Control of *Amblyomma variegatum* and dermatophilosis on local and exotic breeds of cattle in Ghana. *Tropical Animal Health and Production* 28, 44S–49S.

Murray, M. and Gray, A.R. (1984) The current situation on animal trypanosomiasis in Africa. *Preventive Veterinary Medicine* 2, 23–30.

Murray, M. and Trail, J.C.M. (1984) Genetic resistance to animal trypanosomiasis in Africa. *Preventive Veterinary Medicine* 2, 541–551.

Murray, M., Morrison, W.I., Murray, P.K., Clifford, D.J. and Trail, J.C.M. (1979) Trypanotolerance – a review. *World Animal Review* 31, 2–12.

Murray, M., Morrison, W.I. and Whitelaw, D.D. (1982) Host susceptibility to trypanosomiasis: trypanotolerance. *Advances in Parasitology* 21, 1–68.

Murray, M., Trail, J.C.M., Davies, C.E. and Black, S.J. (1984) Genetic resistance to African trypanosomiasis. *Journal of Infectious Diseases* 149, 311–319.

Njogu, A.R., Dolan, R.B., Wilson, A.J. and Sayer, P.D. (1985) Trypanotolerance in East African Orma Boran cattle. *Veterinary Record* 117, 632–636.

Norval, R.A.I., Sutherst, R.W. and Kerr, J.D. (1996) Infestations of the bont tick *Amblyomma hebraeum* (Acari:Ixodidae) on different breeds of cattle in Zimbabwe. *Experimental and Applied Acarology* 20, 599–605.

Oddgeirsson, O., Simpson, S.P., Morgan, A.L., Ross, D.S. and Spooner, R.L. (1988) Relationship between the bovine major histocompatibility complex (BoLA), erythrocyte markers and susceptibility to mastitis in Icelandic cattle. *Animal Genetics* 19, 11–16.

O'Kelly, J.C. and Spiers, W.G. (1976) Resistance to *Boophilus microplus* (Canestrini) in genetically different types of calves in early life. *Journal of Parasitology* 62, 312–317.

ole-MoiYoi, O.K. (1989) *Theileria parva*: an intracellular protozoan parasite that induces reversible lymphocyte transformation. *Experimental Parasitology* 69, 204–210.

Paling, R.W., Moloo, S.K., Scott, J.R., Gettinby, G., McOdimba, F.A. and Murray, M. (1991a). Susceptibility of N'Dama and Boran cattle to sequential challenges with tsetse-transmitted clones of *Trypanosoma congolense*. *Parasite Immunology* 13, 427–445.

Paling, R.W., Moloo, S.K., Scott, J.R., McOdimba, F.A., Logan-Henfrey, L.L., Murray, M. and Williams, D.J.L. (1991b). Susceptibility of N'Dama and Boran cattle to tsetse-transmitted primary and rechallenge infections with a homologous serodeme of *Trypanosoma congolense*. *Parasite Immunology* 13, 413–425.

Peregrine, A.S. (1994) Chemotherapy and delivery systems: haemoparasites. *Veterinary Parasitology* 54, 223–248.

Philipsson, J., Ral, G. and Berglund, B. (1995) Somatic cell count as a selection criterion for mastitis resistance in dairy cattle. *Livestock Production Science* 41, 195–200.

Price, R.E., Templeton, J.W. Smith, R., III and Adams, L.G. (1990) Ability of mononuclear phagocytes from cattle naturally resistant or susceptible to brucellosis to control *in vitro* intracellular survival of *Brucella abortus*. *Infection and Immunity* 58, 879–886.

Qureshi, T., Templeton, J.W. and Adams, L.G. (1996) Intracellular survival of *Brucella abortus*, *Mycobacterium bovis* (BCG), *Salmonella dublin* and *Salmonella typhimurium* in macrophages from cattle genetically resistant to *Brucella abortus*. *Veterinary Immunology and Immunopathology* 50, 55–66.

Rechav, Y. and Kostrzewski, M.W. (1991) Relative resistance of six cattle breeds to the tick *Boophilus decoloratus* in South Africa. *Onderstepoort Journal of Veterinary Research* 58, 181–186.

Rechav, Y., Dauth, J. and Els, D.A. (1990) Resistance of Brahman and Simmentaler cattle to southern African ticks. *Onderstepoort Journal of Veterinary Research* 57, 7–12.

Rechav, Y., Kostrzewski, M.W. and Els, D.A. (1991) Resistance of indigenous African cattle to the tick *Amblyomma hebraeum*. *Experimental and Applied Acarology* 12, 229–241.

Roberts, C.J. and Gray, A.R. (1973) Studies on trypanosome-resistant cattle II. The effects of trypanosomiasis on N'Dama, Muturu and Zebu cattle. *Tropical Animal Health and Production* 5, 220–233.

Roelants, G.E. (1986) Natural resistance to African trypanosomiasis. *Parasite Immunology* 8, 1–10.

Ross, J.G., Lee, R.P. and Armour, J. (1959) Haemonchosis in Nigeria zebu cattle: the influence of genetical factors in resistance. *Veterinary Record* 71, 27–31.

Ross, J.G., Armour, J. and Lee, R.P. (1960) Further observations on the influence of genetical factors in resistance to helminthiasis in Nigeria zebu cattle. *Veterinary Parasitology* 72, 119–122.

Schukken, Y.H., Mallard, B.A., Dekkers, J.C.M., Leslie, K.E. and Stear, M.J. (1994) Genetic impact on the risk of intramammary infection following *Staphylococcus aureus* challenge. *Journal of Dairy Science* 77, 639–647.

Sellwood, R. (1979) *Escherichia coli* diarrhoea in pigs with or without the K88 receptor. *Veterinary Record* 105, 228–230.

Sellwood, R., Gibbons, R.A., Jones, G.W. and Rutter, J.M. (1975) Adhesion of enteropathogenic *Escherichia coli* to pig intestinal brush borders: the existence of two pig phenotypes. *Journal of Medical Microbiology* 8, 405–411.

Seykora, A.J. and McDaniel, B.T. (1985) Udder and teat morphology related to mastitis resistance: a review. *Journal of Dairy Science* 68, 2087–2093.

Shook, G.E. and Schutz, M.M. (1994) Selection on somatic cell score to improve resistance to mastitis in the United States. *Journal of Dairy Science* 77, 648–658,

Sileghem, M.R., Flynn, J.N., Saya, R. and Williams, D.J. (1993) Secretion of co-stimulatory cytokines by monocytes and macrophages during infection with *Trypanosoma (Nannomonas) congolense* in susceptible and tolerant cattle. *Veterinary Immunology and Immunopathology* 37, 123–134.

Simon, M., Dusinsky, R. and Stavikova, M. (1995) Association between BoLA antigens and bovine mastitis. *Veterinary Medicine* 40, 7–10.

Simpson, S.P., Oddgeirsson, O., Jonmundsson, J.V. and Oliver, R.A. (1990) Associations between the bovine major histocompatibility complex (BoLA) and milk production in Icelandic dairy cattle. *Journal of Dairy Research* 57, 437–440.

Skow, L.C., Snaples, S.N., Davis, S.K., Taylor, J.F., Huang, B. and Gallagher, D.H. (1996) Localization of bovine lymphocyte antigen (BoLA) DYA and class I loci to different regions of chromosome 23. *Mammalian Genome* 7, 388–389.

Soller, M. and Beckman, J.S. (1988) *Mapping Trypanotolerance Loci of the N'Dama Cattle of West Africa*. Consultation Report, FAO, Rome.

Solomon, G. and Kaaya, G.P. (1996) Comparison of resistance in three breeds of cattle against African ixodid ticks. *Experimental and Applied Acarology* 20, 223–230.

Spickett, A.M., De Klerk, D., Enslin, C.B. and Scholtz, M.M. (1989) Resistance of Nguni, Bonsmara and Hereford cattle to ticks in a Bushveld region of South Africa. *Onderstepoort Journal of Veterinary Research* 56, 245–250.

Stanik, J. and Vasil, M. (1986) Mastitis in dairy cows in large-scale farming operating from the genetic aspect. *Veterinary Medicine* 31, 21–26.

Stavikova, M., Lojda, L. Zakova, M., Mach, P., Prikryl, S. and Pospisil, J. (1990) The genetic contribution of cows to the prevalence of mastitis in the following generation. *Veterinarni Medicina* 35, 257–265.

Stear, M.J., Newman, M.J., Nicholas, F.W., Brown, S.C. and Holroyd, R.G. (1984) Tick resistance and the major histocompatibility system. *Australian Journal of Experimental Biology and Medical Science* 62, 47–52.

Stear, M.J., Tierney, T.J., Baldock, F.C., Brown, S.C., Nicholas, F.W. and Rudder, T.H. (1988a) Failure to find an association between class I antigens of the bovine major histocompatibility system and faecal worm egg counts. *International Journal of Parasitology* 18, 859–861.

Stear, M.J., Tierney, T.J., Baldock, F.C., Brown, S.C., Nicholas, F.W. and Rudder, T.H. (1988b). Class I antigens of the bovine major histocompatibility system are weakly associated with variation in faecal worm egg counts in naturally infected cattle. *Animal Genetics* 19, 115–122.

Stear, M.J., Dimmock, C.K., Newman, M.J. and Nicholas, F.W. (1988c) BoLA antigens are associated with increased frequency of persistent lymphocytosis in bovine leukaemia virus infected cattle and with increased incidence of antibodies to bovine leukaemia virus. *Animal Genetics* 19, 151–158.

Stewart, J.L. (1937) The cattle of the Gold Coast. *Veterinary Record* 49, 1289–1297.

Sulimova, G.E., Udina, I.G., Shaikhaev, G.O. and Zakharov, I.A. (1995) DNA polymorphism of the BoLA-DRB3 gene in cattle in connection with resistance and susceptibility to leukaemia. *Genetika* 9, 1294–1299.

Taylor, K.A. (1998) Immune responses of cattle to African trypanosomes: protective or pathogenic? *International Journal for Parasitology* 28, 219–240.

Teale, A.J. (1993) Improving control of livestock disease: animal biotechnology in the Consultative Group on International Agricultural Research. *Biosciences* 43, 475–483.

Teale, A.J. (1997) Biotechnology: a key element in the CGIAR's livestock research programme. *Outlook on Agriculture* 26, 217–225.

Teale, A.J., Kemp, S.J. and Morrison, W.I. (1991) The major histocompatibility complex and disease resistance in cattle. In: Owen, J.B. and Axford,R.F.E. (eds) *Breeding for Disease Resistance in Farm Animals.* CAB International, Wallingford, UK, pp. 86–99.

Templeton, J.W. and Adams, L.G. (1995) Natural resistance to bovine brucellosis. In: Adams, L.G. (ed.) *Advances in Brucellosis Research.* Texas A & M University Press, College Station, Texas, pp. 144–150.

Thimm, B. (1973) Hypothesis of a higher natural resistance against brucellosis in East African shorthorn zebu. *Zentralblatt fur Veterinarmedizin [B]* 20, 490–494.

Trail, J.C.M., d'Ieteren, G. and Teale, A.J. (1989) Trypanotolerance and the value of conserving livestock genetic resources. *Genome* 31, 805–812.

Trail, J.C., d'Ieteren, G.D.M., Feron, A., Kakiese, O., Mulungo, M. and Pelo, M. (1990a) Effect of trypanosome infection, control of parasitaemia and control of anaemia development on productivity of N'Dama cattle. *Acta Tropica* 48, 37–45.

Trail, J.C.M., d'Ieteren, G.D.M., Colardelle, C., Maille, J.C., Ordner, G., Sauveroche, B. and Yangari, G. (1990b) Evaluation of a field test for trypanotolerance in young N'Dama cattle. *Acta Tropica* 48, 47–57.

Trail, J.C.M., d'Ieteren, G.D.M., Maille, J.C. and Yangari, G. (1991) Genetic aspects of control of anaemia development in trypanotolerant N'Dama cattle. *Acta Tropica* 48, 285–291.

Utech, K.B. and Wharton, R.H. (1982) Breeding for resistance to *Boophilus microplus* in Australian Illawarra Shorthorn and Brahman × Australian Illawarra Shorthorn cattle. *Australian Veterinary Journal* 58, 41–46.

Våge, D.I., Lingaas, F., Spooner, R.L., Arnet, E.F. and Lie, Ø. (1992) A study on association between mastitis and serologically defined class I bovine lymphocyte antigens (BoLA-A) in Norwegian cows. *Animal Genetics* 23, 533–536.

van Eijk, M.J., Stewart-Haynes, J.A., Beever, J.E., Fernando, R.L. and Lewin, M.A. (1992) Development of persistent lymphocytosis in cattle is closely associated with DRB2. *Immunogenetics* 37, 64–68.

Vercoe, J.E. and Frisch, J.E. (1992) Genotype (breed) and environment interaction with particular reference to cattle in the tropics. *Australasian Journal of Animal Science* 5, 401–409.

Vidal, S.M., Malo, D., Vogan, K., Skamene, E. and Gros, P. (1993) Natural resistance to infection with intracellular parasites: isolation of a candidate for *Bcg. Cell* 73, 469–485.

Walker, A.R. and Lloyd, C.M. (1993) Experiments on the relationship between feeding of the tick *Amblyomma variegatum* (Acari:Ixodidae) and dermatophilosis skin disease in sheep. *Journal of Medical Entomology* 30, 136–143.

Waller, P.J. (1997) Nematode parasite control of livestock in the tropics/subtropics: the need for novel approaches. *International Journal of Parasitology* 27, 1193–1201.

Weigel, K.A., Freeman, A.E., Kehrli, M.E., Jr, Stear, M.J. and Kelley, D.H. (1990) Association of class I bovine lymphocyte antigen complex alleles with health and production traits in dairy cattle. *Journal of Dairy Science* 73, 2538–2546.

Williams, D.J.L., Naessens, J., Scott, J.R. and McOdimba, F.A. (1991) Analysis of peripheral leucocyte populations in N'Dama and Boran cattle following a rechallenge infection with *Trypanosoma congolense*. *Parasite Immunology* 13, 171–185.

Williams, D.J.L., Logan-Henfrey, L.L., Authié, E., Seely, C. and McOdimba, F. (1992) Experimental infection with a haemorrhage-causing *Trypanosoma vivax* in N'Dama and Boran cattle. *Scandinavian Journal of Immunology* 36, 34–36.

Williams, D.J.L., Taylor, K., Newson, J., Gichuki, B. and Naessens, J. (1996) The role of anti-variable surface glycoprotein antibody responses in bovine trypanotolerance. *Parasite Immunology* 18, 209–218.

Wilmut, I., Schieke, A.E., McWhir, J., Kind, A.J. and Campbell, K.H. (1997) Viable offspring derived from foetal and adult mammalian cells. *Nature* 385(6619), 810–813.

Xu, A., van Eijk, M.J., Park, C. and Lewin, H.A. (1993) Polymorphism in *BoLA-DRB3* exon 2 correlates with resistance to persistent lymphocytosis caused by bovine leukaemia virus. *Journal of Immunology* 151, 6977–6985.

Zanotti, M., Poli, G., Ponti, W., Polli, M., Rocchi, M., Bolzani, E., Longeri, M., Russo, S., Lewin, H.A. and van Eijk, M.J. (1996) Association of BoLA class II haplotypes with subclinical progression of bovine leukaemia virus infection in Holstein-Friesian cattle. *Animal Genetics* 27, 337–341.

Molecular Biology and Genetics of Bovine Spongiform Encephalopathy

9

N. Hunter

Institute for Animal Health, BBSRC/MRC Neuropathogenesis Unit, West Mains Road, Edinburgh EH9 3JF, UK

Introduction	229
Clinical Signs and Pathology	230
The Importance of the Prion Protein	231
Pathogenesis of Bovine Spongiform Encephalopathy in Cattle	232
Natural bovine spongiform encephalopathy in cattle	232
Experimental bovine spongiform encephalopathy in cattle	233
Preclinical Diagnosis of Bovine Spongiform Encephalopathy in Cattle	233
Bovine Spongiform Encephalopathy Transmission Characteristics	233
Genetics of Transmissible Spongiform Encephalopathies in Sheep and Humans	234
The Bovine *PrP* Gene	235
PrP Gene Polymorphism Frequencies in Healthy and Bovine Spongiform Encephalopathy-affected Cattle	237
PrP coding region genotypes	237
Family studies	239
Is bovine spongiform encephalopathy not subject to host genetic control by the *PrP* gene?	240
Conclusions	241
References	242

Introduction

Bovine spongiform encephalopathy (BSE), although a disease currently in decline, is still a subject of much debate conerning its aetiology, epidemiology, mode of transmission and genetics. It was first recognized as a new neurological disease in cattle in the UK in 1986 and, since then, there have been approximately 170,000 UK cases. The economic impact of BSE and the associated control measures of culling of healthy but at-risk animals has been enormous.

BSE is one of a group of related diseases known as transmissible spongiform encephalopathies (TSEs), the oldest known of which is scrapie, which occurs in sheep and goats, but which also occur in humans. The TSEs are all slowly progressive, inevitably fatal, neurodegenerative disorders, characterized by vacuolated brain neurons and the deposition of an abnormal form of a host protein, prion protein (PrP). The TSEs are also experimentally transmissible, and are usually studied in laboratory rodents. The most likely source of infection for cattle was the use of a dietary protein supplement, meat and bone meal (MBM), which was regularly fed to cattle, particularly dairy cattle, and contained the rendered remains of animal offal and carcasses, principally from ruminants (Wilesmith *et al.*, 1991; Nathanson *et al.*, 1997). Such feeding practices were made illegal in July 1988 and BSE cases are now occurring less and less frequently. It is projected that the epidemic is likely to be coming to an end in the year 2001 (Anderson *et al.*, 1996).

There is a strong genetic component in the patterns of disease incidence of scrapie in sheep and of some forms of human TSE and there is overwhelming evidence that the genetic component is the *PrP* gene. However, although in mice, sheep, goats and humans there are polymorphisms and mutations of the *PrP* gene linked to disease incidence, such linkage has not so far been demonstrated for cattle.

This chapter will describe BSE and *PrP* genetics in cattle and will set this against the background of what is known about *PrP* genetics in sheep and humans.

Clinical Signs and Pathology

Cattle affected by BSE become very difficult to handle and show increasing signs of ataxia and altered behaviour, with fear and/or aggression and sensitivity to noises and to touch. Affected animals spend less time ruminating than healthy cattle (Austin and Pollin, 1993), although their physiological drive to eat appears to remain normal. Several studies have noted that BSE cattle have low heart rates (bradycardia), which may be related to the low food intake associated with reduced rumination or which may indicate that there is some damage to the vagus during disease development (Austin *et al.*, 1997). Cattle affected by BSE also show significant neuronal loss in the brain (Jeffrey and Halliday, 1994) and the appearance of vacuolar lesions in brain sections is very similar to that seen in sheep scrapie. Bovine spongiform encephalopathy was confirmed as a TSE by the demonstration of the diagnostic TSE-related PrP fibrils in brain extracts (Hope *et al.*, 1988) and by transmission of the disease to mice (Bruce *et al.*, 1994).

The Importance of the Prion Protein

The PrP is a normal host protein found in every mammal so far examined and consists of approximately 250 amino acids (the exact length depends on the species). It is glycosylated at either one, or both, of two possible glycosylation sites and is attached to the outside of the neuronal cell membrane by a glycophosphoinositol (GPI) anchor (Hope, 1993). The protein, in a conformationally altered form (PrP^{SC}), which is relatively resistant to protease digestion, is the major constituent of scrapie-associated fibrils (SAF), now known to be a hallmark of TSEs in general, and has a characteristic banded pattern when visualized on electrophoresis gels. The normal protein is designated PrP^C and is fully sensitive to protease digestion. It is thought that PrP^{SC} is formed directly from PrP^C by some means that induces a change to the three-dimensional structure of the molecule. The main physical differences between PrP^C and PrP^{SC} are shown in Table 9.1, but the actual difference in structure between the two isoforms is not fully understood. Theoretical analysis has suggested that PrP^C is made up mostly of α-helices and that PrP^{SC} is mostly β-sheets (Huang et al., 1994). However, nuclear magnetic resonance spectroscopy carried out on part of the PrP molecule (Riek et al., 1996) has provided direct evidence of a different structure for PrP^C, which would include three α-helices and an antiparallel β-sheet, which might act as the start site for the conformational change to PrP^{SC}, for which increased β-sheet content is expected. Molecules with high beta sheet content are more resistant to protease enzymatic digestion, probably because the structure gives regions of the protein protection from physical exposure to the enzyme.

The nature of TSE agents remains an unsettled question, despite the wide support for the prion hypothesis which proposes that the diseases are caused by an infectious protein. The PrP^{SC} isoform is so closely associated with TSE infectivity that it can be considered as a reliable marker for infection, and, indeed, the prion hypothesis proposes that it is PrP^{SC} which is itself the infectious agent, causing disease by acting as a seed for the conversion of the normal PrP^C into more of the abnormal isoform PrP^{SC} and thus appearing to multiply the infectivity (Prusiner et al., 1990). Variant forms (allotypes) of the PrP protein are associated with differences in incubation period of

Table 9.1. Differences between normal PrP (PrP^C) and its disease-associated isoform (PrP^{SC}).

	PrP^C	PrP^{SC}
Proteinase K (PK)	Sensitive	Partially resistant
Molecular mass (−PK)	33–35 kDa	33–35 kDa
Molecular mass (+PK)	Degraded	27–30 kDa
Detergent	Soluble	Insoluble
Location	Cell surface	Aggregates
Turnover	Rapid	Slow
Infectivity	Does not copurify	Copurifies

experimental scrapie both in laboratory mice (Carlson et al., 1986) and in sheep (Goldmann et al., 1991a). In addition, some of the human TSEs are familial and present excellent linkage between *PrP* gene mutations and the incidence of disease. The prion hypothesis therefore also accommodates the idea that TSEs can sometimes be simply genetic in origin, with the mutant protein being more likely spontaneously to adopt the disease-associated conformation, both causing disease and producing a seed for a new infection, should a transmission to another individual occur (Collinge and Palmer, 1994). An alternative view is that variant forms of PrP^C control susceptibility to an infecting agent that is made up of PrP^{SC} associated with another molecule specifying TSE strain characteristics, with the joint infectious structure being called a virino (Farquhar et al., 1998). Finally, there are also a few who believe that TSEs are viruses (Braig and Diringer, 1985; Manuelidis et al., 1995) with candidate viral nucleic acids being demonstrated in infectious preparations (Dron and Manuelidis, 1996).

Whatever the nature of the infectious agent, BSE, like many other TSEs, is very resistant to heat and to chemical methods of inactivation (Taylor et al., 1994; Taylor, 1995), making it difficult and expensive to decontaminate farms, abattoirs and laboratories.

Pathogenesis of Bovine Spongiform Encephalopathy in Cattle

Natural bovine spongiform encephalopathy in cattle

A single major strain of BSE has predominated throughout the epizootic and it is transmissible to mice at high efficiency and more rapidly than comparable transmissions of sheep scrapie. The mice that give the shortest incubation period with BSE are called RIII mice, but a characteristic pattern of incubation periods and pathology is produced by BSE when inoculated into a precise series of mouse lines (Fraser et al., 1992). The RIII mice have been used in a major investigation of tissues of cattle looking for evidence of infectivity and there has been no detectable infectivity found in any fluid (including milk; Taylor et al., 1995) or tissue, apart from brain and spinal cord (Fraser and Foster, 1993). The mouse bioassay may not be as sensitive as the homologous system of carrying out cattle-to-cattle transmissions; however, the mouse bioassay is sensitive enough to pick up BSE from the spleens of experimentally infected sheep (Foster et al., 1996). This strongly suggests that pathogenesis of BSE in sheep would be expected to be similar to scrapie in sheep and is not at all like pathogenesis of BSE in cattle. The very much lower levels of infectivity in peripheral tissues of cattle are confirmed by analysis of PrP^{SC} protein, which is detected in central nervous system (CNS) tissues in cattle but in both CNS and peripheral tissues in experimental BSE in sheep and scrapie in sheep (Somerville et al., 1997).

Experimental bovine spongiform encephalopathy in cattle

A large-scale experiment in which calves were dosed orally with BSE is giving interesting results, although as yet incomplete. Clinical signs appeared in the cattle from about 36 months postinoculation (mpi) but infectivity was found in the distal ileum of cattle killed at 6 and 10 mpi (Wells *et al.*, 1994, 1998). Infectivity was also demonstrated (by mouse bioassay) in the peripheral nervous system: in the cervical and dorsal-root ganglia 32–40 mpi and in the trigeminal ganglia 36 and 38 mpi. These tissues were negative in the naturally infected BSE cases, but this may be related to the initial dose of infection, which may very well have been less in the naturally infected cattle. The study is continuing.

Preclinical Diagnosis of Bovine Spongiform Encephalopathy in Cattle

Several research groups are looking for markers that would diagnose BSE by ante-mortem biopsy. The cerebrospinal fluid (CSF), for example, has been found to have elevated levels of apolipoprotein E (Hochstrasser *et al.*, 1997) and another protein called 14-3-3, which is of some use as a screening method for Creutzfeld–Jakob disease (CJD) in humans (Hsich *et al.*, 1996), may also be informative in cattle with BSE (Lee and Harrington, 1997a, b).

The identification of PrP^{SC} in affected individuals is also potentially of interest, although less of the disease-associated form of PrP is found in peripheral cattle tissues than in sheep with scrapie or experimental BSE (Somerville *et al.*, 1997). Humans with the new variant form of CJD have been found to have PrP^{SC} deposits in tonsil biopsies, something which is not the case with sporadic CJD (Arya, 1997; Collinge *et al.*, 1997) but which does occur in sheep scrapie (Schreuder *et al.*, 1996).

Bovine Spongiform Encephalopathy Transmission Characteristics

In mice, BSE produces a unique pattern of incubation periods and pathology, and by this means it has been possible to demonstrate that the new TSEs in zoo animals, domestic cats and humans were caused by the same BSE agent (Bruce *et al.*, 1994, 1997) which has affected cattle. The means by which these infections have occurred is uncertain, although contamination of food is implicated. Within cattle, there has been some argument about whether there is maternal transmission of disease from affected cow to calf. In a study on the offspring of BSE-affected pedigree beef suckler cows, much less likely than dairy cattle to be fed MBM-meal derived protein concentrates, none of 219 calves which had been suckled for at least a month went on to develop BSE themselves (Wilesmith and Ryan, 1997). As these animals would have consumed 111,500 l of milk, it suggests that milk is not a potential source of infection. A large-scale cohort study has also been carried out comparing animals born to BSE-affected cattle with animals whose mothers were healthy (Wilesmith et al., 1997). Of the offspring from BSE-affected mothers, 42 out of 301 (14%) developed BSE, whereas only 13 out of 301 (4.3%) offspring of BSE-unaffected mothers developed BSE. This places the calves from BSE-affected cows at greater risk of developing disease themselves ($P < 0.0001$) but does not distinguish between inheritance of genetic control of susceptibility and true maternal transmission of disease. Re-analysis of the data provided support for a genetic component (Ferguson et al., 1997), but the fact that a calf is even more likely to go on to develop BSE if it is born after the onset of symptoms in its mother argues for an element of direct maternal transmission of infection (Donnelly et al., 1997). However, such a low frequency of maternal transmission, if it occurs at all, is not thought to be able to sustain the epizootic beyond the year 2001(Anderson et al., 1996).

BSE has been experimentally transmitted to many species, including mice (Bruce et al., 1994), sheep and goats (Foster et al., 1994), marmosets (Baker et al., 1993), mink (Robinson et al., 1994) and macaques (Lasmezas et al., 1996). Because of the hypothesis that BSE may have originated from sheep scrapie, attempts have been made to transmit scrapie to cattle to make sure they were susceptible to scrapie and to observe any differences or similarities. A pool of nine scrapie brains from Suffolk and Hampshire sheep from five flocks in Iowa and Wisconsin was used to inoculate 18 calves intracerebrally. The challenged cattle did indeed become ill 14–18 months following challenge; however, the clinical signs and pathology were not the same as with BSE in cattle (Cutlip et al., 1994). A further transmission experiment, which used the scrapie-infected cattle brain tissue from the first-round study and transmitted that to yet more cattle, produced a disease that retained its characteristics and remained different from BSE (Cutlip et al., 1997). It is, however, likely that the US sheep scrapie brains were infected with only a limited range of scrapie strains compared with the number that would have entered into the rendering process in the UK. The strain which was the origin of BSE in the UK may have been very rare or simply able to produce different clinical symptoms from those produced by the US scrapie source. It is known in mice, for example, that different strains of scrapie target different brain regions and hence result in variations in clinical signs (Bruce et al., 1991). These small-scale cattle experiments would be useful if a positive BSE-like result had been generated; however, the distinct nature of the scrapie-affected cattle tells us very little and does not rule out the hypothesis that BSE came from sheep scrapie.

Genetics of Transmissible Spongiform Encephalopathies in Sheep and Humans

Studies of natural scrapie in sheep have confirmed the importance of three codons in the sheep *PrP* gene (136,154 and 171) (Belt et al., 1995; Clouscard *et al.*, 1995; Hunter *et al.*, 1996), originally shown to be associated with differing incubation periods following experimental challenge of sheep with different sources of scrapie and BSE (Goldmann *et al.*, 1991a, 1994), and, although there are breed differences in *PrP* allele frequencies and in disease-associated alleles, some clear rules have emerged from this work. The most resistant genotype in all sheep breeds is $AA_{136}RR_{154}RR_{171}$. This genotype is also resistant to experimental challenge with both scrapie and BSE (Goldmann *et al.*, 1994). Other genotypes encoding QQ_{171} are more susceptible to scrapie. For example, in Suffolk sheep the genotype $AA_{136}RR_{154}QQ_{171}$ is most susceptible, although not all animals of this genotype succumb to disease and it is quite a common genotype among healthy animals (Westaway *et al.*, 1994; Hunter *et al.*, 1997a). The *PrP* genetic variation in Suffolk sheep is much less than in some other breeds, the so-called 'valine breeds'. Breeds such as Cheviots, Swaledales and Shetlands encode *PrP* alleles with valine at codon 136, and the genotype $VV_{136}RR_{154}QQ_{171}$ is the most susceptible to scrapie (Hunter *et al.*, 1994a, 1996). The genotype $VV_{136}RR_{154}QQ_{171}$ is rare and, when it does occur, is almost always in scrapie-affected sheep, and so it has been suggested that scrapie may be simply a genetic disease (Ridley and Baker, 1995). However healthy animals of this genotype can live up to 8 years of age, well past the usual age at death from scrapie (2–4 years) (Hunter *et al.*, 1996, 1997b) and can be easily found in scrapie-free countries (Australia and New Zealand), and so the genetic-disease hypothesis seems less likely than an aetiology which involves host genetic control of susceptibility to an infecting agent.

In humans, sporadic forms of CJD are associated with a *PrP* gene codon 129 polymorphism (methionine/valine), in that homozygous individuals (either MM_{129} or VV_{129}) are over-represented in CJD cases and heterozygosity seems to confer some protection (Palmer *et al.*, 1991). New variant CJD, which, unlike sporadic CJD, is caused by an infectious agent indistinguishable from BSE, has also occurred so far only in MM_{129} genotypes (n > 30). Other forms of TSEs in humans appear to be genetic diseases – for example, Gerstmann–Straussler syndrome (GSS), which is linked to a codon 102 proline-to-leucine mutation (Hsiao *et al.*, 1989). There are many other human *PrP* gene mutations associated with disease; for example, one familial form of CJD is linked to an insert of 144 bp coding for six extra octapeptide repeats at codon 53 (Poulter *et al.*, 1992) and a codon 200 mutation (glutamic acid to lysine) is linked to CJD in Israeli Jews of Libyan origin, Slovaks in north central Slovakia, a family in Chile and a German family in the USA (Prusiner and Scott, 1997). However, the fact that sheep scrapie, which also demonstrates excellent linkage with PrP genotype, has been shown to be unlikely to be a genetic disease also has implications for the interpretation of the human data (Hunter *et al.*, 1997b).

The Bovine *PrP* Gene

When BSE was found in cattle, it was an obvious step to study the bovine *PrP* gene for markers of resistance or susceptibility to disease similar to those which had been found in sheep and humans. There is a great deal of allelic complexity in both the *PrP* coding region and in its flanking regions in the sheep (Hunter *et al.*, 1989, 1993; Goldmann *et al.*, 1990; Muramatsu *et al.*, 1992; Laplanche *et al.*, 1993; Bossers *et al.*, 1996) and also in the human *PrP* gene (Collinge and Palmer, 1994; Prusiner and Scott, 1997). In contrast, the bovine *PrP* gene is remarkably invariant, with very few polymorphisms described (Fig. 9.1).

The bovine *PrP* gene coding region, which has been mapped to bovine syntenic group U11 (Ryan and Womack, 1993), was first sequenced in 1991 (Goldmann *et al.*, 1991b) and this sequence was subsequently confirmed by two other groups (Yoshimoto *et al.*, 1992; Prusiner *et al.*, 1993). Allowing for various polymorphic forms of the gene in each species, there is very little difference (> 90% identity) between the cattle and sheep *PrP* gene. The bovine *PrP* gene has so far revealed only two polymorphisms of the coding region (Goldmann *et al.*, 1991b). The cattle tested have been of many different breeds but also include zebu and N'Dama cattle from the Gambia and feral scrub cattle from Florida (Ryan and Womack, 1993), so the lack of variation does not seem to result from studying related animals. The polymorphisms are firstly a silent *Hin*dII restriction fragment length polymorphism (RFLP) caused by a C to T transition at nucleotide (nt) 576 and secondly a difference in the number of octapeptide repeats found between nt 160 and nt 309 approximately. There are in this region a series of glycine-rich repeats, encoded by 24- or 27-nt G–C-rich elements. Two of the sequencing studies (Goldmann *et al.*, 1991b; Prusiner *et al.*, 1993) have designated these in the same manner, which gives

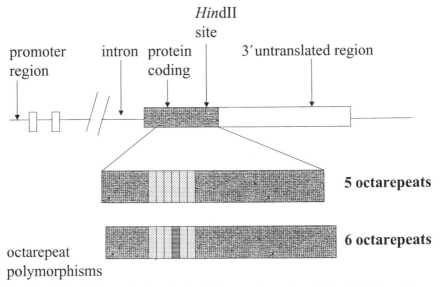

Fig. 9.1. Bovine *PrP* gene structure. Bovine *PrP* gene polymorphisms indicated in the protein-coding open reading frame (unbroken by introns) of the gene. Direction of transcription is from left to right. Each octapeptide repeat is distinguishable on the basis of DNA sequence, the extra repeat in the 6-repeat-encoding allele is therefore indicated in stripes rather than dots.

one allele with five copies of the repeat and another allele with six copies. This short region is so repetitive that other start points for counting the repeats are possible. This means that, whereas others (Yoshimoto et al., 1992) use a 4 : 5 system to describe the two bovine *PrP* alleles, the sequences are the same as in the 5 : 6 system. The 5 : 6 system has the advantage of allowing comparisons to be made easily with the human *PrP* gene octarepeat studies, which use the same repeat designation (Goldfarb et al., 1992). In humans many variations in the PrP octapeptide repeat number have been described, ranging from 4 to 14, some of which have clear linkage to the incidence of human TSE (Poulter et al., 1992).

PrP Gene Polymorphism Frequencies in Healthy and Bovine Spongiform Encephalopathy-affected Cattle

PrP *coding region genotypes*

There have been two main studies that have addressed the question of association of *PrP* genotype with incidence of BSE in cattle. One is discussed in the next section but in the other (Hunter et al., 1994b) *PrP* genotypes of BSE-affected cattle were compared with healthy animals and a case–control study of a single BSE-affected herd. The *Hin*dII genotype was not found to be informative and is not discussed further here. Genotype frequencies of the octapeptide repeat polymorphism are presented for 172 histopathologically confirmed BSE-affected cattle in Table 9.2. The majority of the cattle (91%) were of the 6 : 6 genotype, with 9% 6 : 5 and no 5 : 5 animals. For convenience, cattle were separated into five breed groups: Friesian (92), Friesian × Holstein (14), Friesian crosses (20), Ayrshire (16) and others (30). Friesian × Holsteins and Ayrshires had higher frequencies of the 6 : 5 genotype

Table 9.2. *PrP* gene octarepeat genotypes in a BSE case-study (data taken from Hunter et al., 1994b).

Breed group	Genotype frequency (%)			Number
	6 : 6*	6 : 5[†]	5 : 5[‡]	
All	91	9	0	172
Friesian	95	5	0	92
Fr × Holstein	79	21	0	14
Friesian crosses (others)	100	0	0	20
Ayrshire	69	31	0	16
Others	90	10	0	30

*6 : 6 – homozygous for the 6-octapeptide repeat-encoding allele.
[†]6 : 5 – heterozygous for the 6- and the 5-octapeptide repeat-encoding alleles.
[‡]5 : 5 – homozygous for the 5-octapeptide repeat-encoding allele.

(21 and 31%, respectively) than other breed groups (ranging from 0 to 10%), but these differences were not significant.

In the case–control single-herd study cattle (total 90 animals), shown in Table 9.3, the octapeptide repeat frequencies were 89% 6 : 6 and 11% 6 : 5 in the 85 healthy cattle. All five BSE cases were 6 : 6. (BSE case-study frequencies are also given in Table 9.3 for comparison). There were therefore no significant differences between BSE affected and healthy cattle in this herd in frequencies of the two known *PrP* polymorphisms.

The healthy (or unaffected) cattle group represented 108 animals from three herds with no history of BSE. Again, the majority of animals were of the genotype 6 : 6 (82%), with 17% being 6 : 5. A single animal of 5 : 5 genotype (representing 1% of this sample) was also found. From the Central Veterinary Laboratory (CVL) BSE database, information was available on the age of onset of disease in the BSE-affected UK animals. The youngest animal, an Ayrshire, was 29 months and the oldest, an Ayrshire cross, was 121 months. Table 9.4 gives the age/octarepeat genotype comparisons for all 172 cattle in the BSE case-study and for three breed groups large enough to analyse: Friesian, Ayrshire and Friesian × Holstein. There was no association between genotype and age. All the 6 : 5-genotype animals fell well within the range set by the greater numbers of 6 : 6 animals.

In the CVL BSE database, cattle are described as either home-bred (born on the same farm where they later became BSE-affected) or purchased (born elsewhere and transferred at some later date to the affected farm). This may or may not give an indication of where the animals contracted BSE. Of the BSE-affected cattle samples collected in 1991 (146 animals, all female), the majority (67%) were home-bred. There was no evidence that this frequency was related to genotype; for instance, the ten 6 : 5 animals in the 1991 group were 80% home-bred. The frequency was breed-dependent, however, in that, in this sample, more than 80% of Friesian, Ayrshire and Friesian × Holstein cattle were home bred but other Friesian crosses, Aberdeen Angus crosses,

Table 9.3. Cattle *PrP* octarepeat genotype frequencies (data taken from Hunter et al., 1994b).

Cattle group	No. of cattle	Frequency (%)		
		6 : 6*	6 : 5[†]	5 : 5[‡]
Case-study				
BSE	172	91	9	0
Herd study				
Healthy	85	89	11	0
BSE	5	100	0	0
Unaffected	108	82	17	1

*6 : 6 – homozygous for the 6-octapeptide repeat-encoding allele.
[†]6 : 5 – heterozygous for the 6- and the 5-octapeptide repeat-encoding alleles.
[‡]5 : 5 – homozygous for the 5-octapeptide repeat-encoding allele.

Table 9.4. BSE case-study: genotype comparison with age of onset of BSE (data taken from Hunter et al., 1994b).

Breed	Genotype*	No.	Mean age (months)	SD (months)	Range (months)
All	All	172	60	13	29–121
	6 : 6	156	60	13	29–121
	6 : 5	16	59	10	47–79
Friesian	All	92	59	12	38–110
	6 : 6	87	59	12	38–110
	6 : 5	5	60	11	49–72
Ayrshire	All	16	58	16	29–105
	6 : 6	11	61	19	29–105
	6 : 5	5	53	5	47–60
Friesian × Holstein	All	14	63	12	49–81
	6 : 6	11	62	12	49–81
	6 : 5	3	69	12	56–79

*Genotype designation as in Tables 9.2 and 9.3.

Simmental, Limousin, Herefords and other crosses were much more likely to have been purchased animals.

The genotype frequencies found in the above study (Hunter et al., 1994b) were similar to frequencies found in studies of healthy Belgian (Grobet et al., 1994) and US cattle (McKenzie et al., 1992) and in 210 Holstein and 46 Hereford bulls used actively in artificial insemination programmes in the USA, where the frequency of 6 : 6 was 97% and 99%, resepectively (Brown et al., 1993). The other bulls were 6 : 5, reducing the frequency of the 5 : 5 genotype to less than 0.5% in these US Holsteins.

Family studies

There is some evidence of heightened risk of BSE in offspring of BSE-affected cows (Wilesmith and Ryan, 1997), and several studies attempting to find evidence of inheritance of genetic control of susceptibility, rather than maternal transmission of disease itself, have not conclusively ruled out some element of genetic control of susceptibility in cattle (Curnow and Hau, 1996; Donnelly et al., 1997; Ferguson et al., 1997; Wilesmith et al., 1997).

It is normally the case that family studies are carried out in order to get an indication of whether there is genetic linkage with disease incidence. In the case of TSEs, however, the gene that controls disease incidence is known (the *PrP* gene) and it might be thought to be unnecessary to carry out family studies of cattle affected by BSE. However, in the light of the findings that the bovine *PrP* coding region does not show disease-linked variation, one such family study is of interest (Neibergs et al., 1994). This study used a technique known as single-strand conformational polymorphism (SSCP) analysis, which is designed to reveal the presence of a polymorphism within stretches of deoxyribonucleic acid (DNA) but does not provide details of exactly what that polymorphism is. Thus, the SSCP analysis revealed three possible alleles in the *PrP* gene region, generating (in the nomenclature used by Neibergs *et al.* (1994)) the genotypes AA, BB and AB. The source of the change in DNA which resulted in the A type as opposed to the B type is unknown. However, BSE-affected animals and their relatives were found to be more likely to have the AA genotype than the other animals analysed, with BSE-affected animals giving AA frequency of 48%, their relatives 58% and unrelated healthy animals 29%. Although the AA genotype cannot be regarded as a marker for BSE susceptibility in these cattle, it is suggestive that there may be some genetic linkage with disease incidence outside the *PrP* gene coding region itself. It is interesting that, in this study, non-UK cattle (Boran and N'Dama from Kenya, Friesian Sahiwal, Brahman/Brahman crosses from Australia and Brangus from the USA) had extremely low frequencies of AA (5%), suggesting that something is indeed genetically different about UK cattle; however, this has never been followed up.

Is bovine spongiform encephalopathy not subject to host genetic control by the PrP *gene?*

Despite the lack of evidence that the *PrP* gene coding region controls incidence of BSE in cattle, transmission studies of BSE to mice, to sheep and to goats strongly suggest that BSE infectivity does 'select' animals of certain *PrP* genotypes in these experimental models. In the mouse transmission studies of Fraser *et al.* (1992), BSE-affected cattle brain homogenate was injected into strains of mice which differed at amino acids 108 and 189 of the *PrP* gene (Hunter *et al.*, 1992), giving shorter incubation periods in the mice of $Prnp^{a/a}$ genotype (leucine and threonine at codons 108 and 189) than in mice of $Prn^{b/b}$ genotype (phenylalanine and valine). Transmission of BSE to Neuropathogenesis Unit (NPU) Cheviots (Goldmann *et al.*, 1994) also reveals an association of disease with the *PrP* genotype homozygous for glutamine at codon 171 and, when BSE is injected into goats, animals with isoleucine rather than methionine at codon 142 have longer incubation periods (Goldmann *et al.*, 1996). In mice, sheep and goats, BSE does associate with *PrP* variants, so why not in cattle?

It may be that there is a link between BSE incidence and an as yet unknown *PrP* polymorphism, as suggested indirectly by one study (Neibergs *et al.*, 1994). Sequencing of the *PrP* gene coding region from several individual cattle has revealed no new amino acid sequence variants (N. Hunter, W. Goldmann and G. Smith, unpublished); however, in other genes, both the 5' and 3' untranslated regions (UTRs) are attracting attention because of involvement in regulation of messenger ribonucleic acid (mRNA) function (Jackson, 1993). The *PrP* gene is present in the DNA of each cell in the body, but the amount of protein produced, or expressed, in each cell type is not the same.

The PrP expression is highest in brain, but there are variations in expression in peripheral tissues. Transgenic mouse studies have shown that PrP expression levels are vital to the development of disease following scrapie challenge, so there may also be disease-associated mutations in regions of the *PrP* gene which control expression. This has been shown already for other genes; for example, mRNA 3' UTR mutations are thought to be responsible for myotonic dystrophy (Jackson, 1993). The bovine *PrP* gene has, like the sheep, a 3' UTR of several kilobases and in the sheep 3' UTR there is a strong association of an *Eco*RI RFLP with incidence of both experimental and natural scrapie (Hunter *et al.*, 1989, 1991). This may be an indication that this region may have a role in control of gene expression and therefore, potentially, in disease control also.

It may be that, in cattle, the six-octarepeat *PrP* allele is dominant in conferring susceptibility, as all the BSE cases so far described have been 6 : 6 or 6 : 5. This could be tested by direct challenge of cattle and/or transgenic mice carrying different octarepeat alleles; however, if the six-octarepeat allele does confer susceptibility, why have there not been more BSE cases in a cattle population which apparently has very high frequencies of this allele?

Using BSE occurrence as sole measure of the frequency of susceptible cattle and 'absence of BSE' to estimate the numbers of resistant cattle could simply be wrong. John Wilesmith (1988), in suggesting that BSE resulted from feeding cattle infected ruminant material, also described the difficulties of carrying out case–control studies to confirm this (Wilesmith *et al.*, 1992). Most dairy-herd cattle would have been fed potentially contaminated concentrates before the order banning this use of MBM in 1988; however, most of these cattle did not develop BSE. There may therefore have been uneven distribution of infection in the food and only those cattle ingesting a large enough dose went on to develop BSE. This problem, which affects the epidemiology, may also apply to the *PrP* genetics. With essentially one form of the cattle *PrP* gene predominating and if this is the 'susceptible' allele, most cattle may have the potential to develop BSE if given a sufficiently high dose of infection.

Conclusions

In cattle, unlike sheep, the option to control TSE disease by breeding for resistance is not available – there are no genetic markers linked to BSE. In the UK, BSE is in decline, as a result of the physical measures taken to control cattle food, along with the slaughter of any animal considered at risk of disease. Because of this, it may be thought that there is no point trying to understand the genetics of BSE; however, BSE has apparently spread to other species, including humans, and it has the potential to spread into sheep, where its pathogenesis may be different from that in cattle. Our knowledge of BSE may therefore protect us from similar new diseases in the future.

References

Anderson, R.M., Donnelly, C.A., Ferguson, N.M., Woolhouse, M.E.J., Watt, C.J., Udy, H.J., Mawhinney, S., Dunstan, S.P., Southwood, T.R.E., Wilesmith, J.W., Ryan, J.B.M., Hoinville, L.J., Hillerton, J.E., Austin, A.R. and Wells, G.A.H. (1996) Transmission dynamics and epidemiology of BSE in British cattle. *Nature* 382, 779–788.

Arya, S. C. (1997) Diagnosis of new variant Creutzfeldt–Jakob disease by tonsil biopsy. *Lancet* 349, 1322–1323.

Austin, A. and Pollin, M. (1993) Reduced rumination in bovine spongiform encephalopathy and scrapie. *Veterinary Record* 132, 324–325.

Austin, A.R., Pawson, L., Meek, S. and Webster, S. (1997). Abnormalities of heart rate and rhythm in bovine spongiform encephalopathy. *Veterinary Record* 141, 352–357.

Baker, H.F., Ridley, R.M. and Wells, G.A. (1993) Experimental transmission of BSE and scrapie to the common marmoset. *Veterinary Record* 132, 403–406

Belt, P.B.G.M., Muileman, I.H., Schreuder, B.E.C., Bos-de Ruijter, J., Gielkens, A.L.J. and Smits, M.A. (1995) Identification of five allelic variants of the sheep *PrP* gene and their association with natural scrapie. *Journal of General Virology* 76, 509–517.

Bossers, A., Schreuder, B.E.C., Muileman, I.H., Belt, P.B.G.M. and Smits, M.A. (1996) *PrP* genotype contributes to determining survival times of sheep with natural scrapie. *Journal of General Virology* 77, 2669–2673.

Braig, H.R. and Diringer, H. (1985) Scrapie: concept of a virus-induced amyloidosis of the brain. *EMBO Journal* 4, 2309–2312.

Brown, D.R., Zhang, H.M., DeNise, S.K. and Ax, R.L. (1993) Bovine prion gene allele frequencies determined by AMFLP and RFLP analysis. *Animal Biotechnology* 4, 47–51.

Bruce, M.E., McConnell, I., Fraser, H. and Dickinson, A.G. (1991) The disease characteristics of different strains of scrapie in Sinc congenic mouse lines: implications for the nature of the agent and host control of pathogenesis. *Journal of General Virology* 72, 595–603.

Bruce, M., Chree, A., McConnell, I., Foster, J., Pearson, G. and Fraser, H. (1994) Transmission of bovine spongiform encephalopathy and scrapie to mice – strain variation and the species barrier. *Philosophical Transactions of the Royal Society of London Series B – Biological Sciences* 343, 405–411.

Bruce, M.E., Will, R.G., Ironside, J.W., McConnell, I., Drummond, D., Suttie, A., McCardle, L., Chree, A., Hope, J., Birkett, C., Cousens, S., Fraser, H. and Bostock, C.J. (1997) Transmissions to mice indicate that 'new variant' CJD is caused by the BSE agent. *Nature* 389, 488–501.

Carlson, G.A., Kingsbury, D.T., Goodman, P.A., Coleman, S., Marshall, S.T., DeArmond, S., Westaway, D. and Prusiner, S.B. (1986) Linkage of prion protein and scrapie incubation time genes. *Cell* 46, 503–511.

Clouscard, C., Beaudry, P., Elsen, J.M., Milan, D., Dussaucy, M., Bounneau, C., Schelcher, F., Chatelain, J., Launay, J.M. and Laplanche, J.L. (1995) Different allelic effects of the codons 136 and 171 of the prion protein gene in sheep with natural scrapie. *Journal of General Virology* 76, 2097–2101.

Collinge, J. and Palmer, M.S. (1994) Molecular-genetics of human prion diseases. *Philosophical Transactions of the Royal Society of London Series B – Biological Sciences* 343, 371–378.

Collinge, J., Hill, A., Ironside, J. and Zeidler, M. (1997) Diagnosis of new variant Creutzfeldt–Jakob disease by tonsil biopsy – Reply. *Lancet* 349, 1323.

Curnow, R.N. and Hau, C.M. (1996) The incidence of bovine spongiform encephalopathy in the progeny of affected sires and dams. *Veterinary Record* 138, 407–408.

Cutlip, R.C., Miller, J.M., Race, R.E., Jenny, A.L., Katz, J.B., Lehmkuhl, H.D., Debey, B.M. and Robinson, M.M. (1994) Intracerebral transmission of scrapie to cattle. *Journal of Infectious Diseases* 169, 814–820.

Cutlip, R.C., Miller, J.M. and Lehmkuhl, H.D. (1997) Second passage of a US scrapie agent in cattle. *Journal of Comparative Pathology* 117, 271–275.

Donnelly, C.A., Ferguson, N.M., Ghani, A.C., Wilesmith, J.W. and Anderson, R.M. (1997) Analysis of dam–calf pairs of BSE cases: confirmation of a maternal risk enhancement. *Proceedings Of the Royal Society Of London Series B – Biological Sciences* 264, 1647–1656.

Dron, M. and Manuelidis, L. (1996) Visualisation of viral candidate cDNAs in infectious brain fractions from Creutzfeldt–Jacob disease by representational difference analysis. *Journal of Neurovirology* 2, 240–248.

Farquhar, C., Somerville, R. and Bruce, M. (1998) Straining the prion hypothesis. *Nature* 391, 345–346.

Ferguson, N.M., Donnelly, C.A., Woolhouse, M.E.J. and Anderson, R.M. (1997) Genetic interpretation of heightened risk of BSE in offspring of affected dams. *Proceedings of the Royal Society of London Series B – Biological Sciences* 264, 1445–1455.

Foster, J.D., Hope, J., McConnell, I., Bruce, M. and Fraser, H. (1994).Transmission of bovine spongiform encephalopathy to sheep, goats, and mice. *Annals of the New York Academy of Sciences* 724, 300–303.

Foster, J.D., Bruce, M., McConnell, I., Chree, A. and Fraser, H. (1996) Detection of BSE infectivity in brain and spleen of experimentally infected sheep. *Veterinary Record* 138, 546–548.

Fraser, H. and Foster, J.D. (1993) Transmission of mice, sheep and goats and bioassay of bovine tissues. In: Bradley, R. and Marchant, B. (eds) *Transmissible Spongiform Encephalopathies: Proceedings of a Consulation on BSE with the Scientific Veterinary Committee of the Commission of the European Communities*. European Commission, Brussels, pp. 145–159.

Fraser, H., Bruce, M.E., Chree, A., McConnell, I. and Wells, G.A. (1992) Transmission of bovine spongiform encephalopathy and scrapie to mice. *Journal of General Virology* 73, 1891–1897.

Goldfarb, L.G., Brown, P.B. and Gajdusek, D.C. (1992) The molecular genetics of human transmissible spongiform encephalopathies. In: Prusiner, S.B., Powell, C.J.J. and Anderton, B. (eds) *Prion Diseases of Humans and Animals*. Ellis Horwood, Chichester, pp. 139–153.

Goldmann, W., Hunter, N., Foster, J.D., Salbaum, J.M., Beyreuther, K. and Hope, J. (1990) Two alleles of a neural protein gene linked to scrapie in sheep. *Proceedings of the National Academy of Sciences of the USA* 87, 2476–2480.

Goldmann, W., Hunter, N., Benson, G., Foster, J.D. and Hope, J. (1991a) Different scrapie-associated fibril proteins (PrP) are encoded by lines of sheep selected for different alleles of the Sip gene. *Journal of General Virology* 72, 2411–2417.

Goldmann, W., Hunter, N., Martin, T., Dawson, M. and Hope, J. (1991b) Different forms of the bovine *PrP* gene have five or six copies of a short, G-C-rich element within the protein coding exon. *Journal of General Virology* 72, 201–204.

Goldmann, W., Hunter, N., Smith, G., Foster, J. and Hope, J. (1994) PrP genotype and agent effects in scrapie – change in allelic interaction with different isolates of agent in sheep, a natural host of scrapie. *Journal of General Virology* 75, 989–995.

Goldmann, W., Martin, T., Foster, J., Hughes, S., Smith, G., Hughes, K., Dawson, M. and Hunter, N. (1996) Novel polymorphisms in the caprine *PrP* gene: a codon 142 mutation associated with scrapie incubation period. *Journal of General Virology* 77, 2885–2891.

Grobet, L., Vandevenne, S., Charlier, C., Pastoret, P.P. and Hanset, R. (1994) Polymorphism of the prion protein gene in Belgian cattle. *Annales de Médecine Vétérinaire* 138, 581–586.

Hochstrasser, D.F., Frutiger, S., Wilkins, M.R., Hughes, G. and Sanchez, J.C. (1997) Elevation of apolipoprotein E in the CSF of cattle affected by BSE. *FEBS Letters* 416, 161–163.

Hope, J. (1993) The biology and molecular biology of scrapie-like diseases. *Archives of Virology* 7, 201–214.

Hope, J., Reekie, L.J.D., Hunter, N., Multhaup, G., Beyreuther, K., White, H., Scott, A.C., Stack, M.J., Dawson, M. and Wells, G.A.H. (1988) Fibrils from brains of cows with new cattle disease contain scrapie-associated protein. *Nature* 336, 390–392.

Hsiao, K., Baker, H.F., Crow, T.J., Poulter, M., Owen, F., Terwilliger, J.D., Westaway, D., Ott, J. and Prusiner, S.B. (1989) Linkage of a prion protein missense variant to Gerstmann–Sraussler syndrome. *Nature* 338, 342–345.

Hsich, G., Kenney, K., Gibbs, C.J., Lee, K.H. and Harrington, M.G. (1996) The 14-3-3 brain protein in cerebrospinal fluid as a marker for transmissible spongiform encephalopathies. *New England Journal of Medicine* 335, 924–930.

Huang, Z.W., Gabriel, J.M., Baldwin, M.A., Fletterick, R.J., Prusiner, S.B. and Cohen, F.E. (1994) Proposed 3-dimensional structure for the cellular prior protein. *Proceedings of the National Academy of Sciences of the USA*, 91, 7139–7143.

Hunter, N., Foster, J.D., Dickinson, A.G. and Hope, J. (1989) Linkage of the gene for the scrapie-associated fibril protein (*PrP*) to the Sip gene in Cheviot sheep. *Veterinary Record* 124, 364–366.

Hunter, N., Foster, J.D., Benson, G. and Hope, J. (1991) Restriction fragment length polymorphisms of the scrapie-associated fibril protein (*PrP*) gene and their association with susceptibility to natural scrapie in British sheep. *Journal of General Virology* 72, 1287–1292.

Hunter, N., Dann, J.C., Bennett, A.D., Somerville, R.A., McConnell, I. and Hope, J. (1992) Are Sinc and the PrP gene congruent? Evidence from PrP gene analysis in *Sinc* congenic mice. *Journal of General Virology* 73, 2751–2755.

Hunter, N., Goldmann, W., Benson, G., Foster, J.D. and Hope, J. (1993) Swaledale sheep affected by natural scrapie differ significantly in *PrP* genotype frequencies from healthy sheep and those selected for reduced incidence of scrapie. *Journal of General Virology* 74, 1025–1031.

Hunter, N., Goldmann, W., Smith, G. and Hope, J. (1994a) The association of a codon 136 *PrP* gene variant with the occurrence of natural scrapie. *Archives of Virology* 137, 171–177.

Hunter, N., Goldmann, W., Smith, G. and Hope, J. (1994b) Frequencies of *PrP* gene variants in healthy cattle and cattle with BSE in Scotland. *Veterinary Record* 135, 400–403.

Hunter, N., Foster, J., Goldmann, W., Stear, M., Hope, J. and Bostock, C. (1996) Natural scrapie in a closed flock of Cheviot sheep occurs only in specific *PrP* genotypes. *Archives of Virology* 141, 809–824.

Hunter, N., Moore, L., Hosie, B., Dingwall, W. and Greig, A. (1997a) Natural scrapie in a flock of Suffolk sheep in Scotland is associated with *PrP* genotype. *Veterinary Record* 140, 59–63.

Hunter, N., Cairns, D., Foster, J., Smith, G., Goldmann, W. and Donnelly, K. (1997b). Is scrapie a genetic disease? Evidence from scrapie-free countries. *Nature* 386, 137.

Jackson, R.J. (1993) Cytoplasmic regulation of mRNA function: the importance of the 3′ untranslated region. *Cell* 74, 9–14.

Jeffrey, M. and Halliday, W.G. (1994) Numbers of neurons in vacuolated and non-vacuolated neuroanatomical nuclei in bovine spongiform encephalopathy-affected brains. *Journal of Comparative Pathology* 110, 287–293.

Laplanche, J.L., Chatelain, J., Westaway, D., Thomas, S., Dussaucy, M., Brugerepicoux, J. and Launay, J.M. (1993) PrP polymorphisms associated with natural scrapie discovered by denaturing gradient gel-electrophoresis. *Genomics* 15, 30–37.

Lasmezas, C., Deslys, J., Adjou, K., Lamoury, F. and Dormont, D. (1996) BSE transmission to macaques. *Nature* 381, 743–745.

Lee, K. and Harrington, M. (1997a) 14-3-3 and BSE. *Veterinary Record* 140, 206–207.

Lee, K.H. and Harrington, M.G. (1997b). The assay development of a molecular marker for transmissible spongiform encephalopathies. *Electrophoresis* 18, 502–506.

McKenzie, D.I., Cowan, C.M., Marsh, R.F. and Aiken, J.M. (1992) *PrP* gene variability in the US cattle population. *Animal Biotechnology* 3, 309–315.

Manuelidis, L., Sklaviadis, T., Akowitz, A. and Fritch, W. (1995) Viral particles are required for infection in neurodegenerative Creutzfeldt–Jakob-disease. *Proceedings of the National Academy of Sciences of the USA* 92, 5124–5128.

Muramatsu, Y., Tanaka, K., Horiuchi, M., Ishiguro, N., Shinagawa, M., Matsui, T. and Onodera, T. (1992) A specific RFLP type associated with the occurrence of sheep scrapie in Japan. *Archives in Virology* 127, 1–9.

Nathanson, N., Wilesmith, J. and Griot, C. (1997) Bovine spongiform encephalopathy (BSE): causes and consequences of a common source epidemic. *American Journal of Epidemiology* 145, 959–969.

Neibergs, H.L., Ryan, A.M., Womack, J.E., Spooner, R.L. and Williams, J.L. (1994) Polymorphism analysis of the prion gene in BSE-affected and unaffected cattle. *Animal Genetics* 25, 313–317.

Palmer, M.S., Dryden, A.J., Hughes, J.T. and Collinge, J. (1991) Homozygous prion protein genotype predisposes to sporadic Creutzfelt–Jakob disease. *Nature* 352, 340–342.

Poulter, M., Baker, H.F., Frith, C.D., Leach, M., Lofthouse, R., Ridley, R.M., Shah, T., Owen, F., Collinge, J. and Brown, J. (1992) Inherited prion disease with 144 base pair gene insertion. 1. Genealogical and molecular studies. *Brain* 115, 675–685.

Prusiner, S.B. and Scott, M.R. (1997) Genetics of prions. *Annual Review of Genetics* 31, 139–175.

Prusiner, S.B., Scott, M., Foster, D., Pan, K.-M., Groth, D., Mirenda, C., Torchia, M., Yang, S.-L., Serban, D., Carlson, G.A., Hoppe, P.C., Westaway, D. and DeArmond, S.J. (1990) Transgenetic studies implicate interactions between homologous PrP isoforms in scrapie prion replication. *Cell* 63, 673–686.

Prusiner, S.B., Miklos, F., Scott, M., Serban, H., Taraboulos, A., Gabriel, J.-M., Wells, G.A.H., Wilesmith, J.W., Bradley, R., DeArmond, S.J. and Kristensson, K. (1993) Immunologic and molecular biologic studies of prion proteins in bovine spongiform encephalopathy. *Journal of Infectious Diseases* 136, 602–613.

Ridley, R.M. and Baker, H.F. (1995) The myth of maternal transmission of spongiform encephalopathy. *British Medical Journal* 311, 1071–1075.

Riek, R., Hornemann, S., Wider, G., Billeter, M., Glockshuber, R. and Wuthrich, K. (1996) NMR structure of the mouse prion protein domain PrP (121–231). *Nature* 382, 180–182.

Robinson, M.M., Hadlow, W.J., Huff, T.P., Wells, G.A.H., Dawson, M., Marsh, R.F. and Gorham, J.R. (1994) Experimental-infection of mink with bovine spongiform encephalopathy. *Journal of General Virology* 75, 2151–2155.

Ryan, A.M. and Womack, J.E. (1993) Somatic cell mapping of the bovine prion protein gene and restriction fragment length polymorphism studies in cattle and sheep. *Animal Genetics* 24, 23–26.

Schreuder, B.E.C., van Keulen, L.J.M., Vromans, M.E.W., Langeveld, J.P.M. and Smits, M.A. (1996) Pre clinical test for prion diseases. *Nature* 381, 563.

Somerville, R.A., Birkett, C.R., Farquhar, C.F., Hunter, N., Goldmann, W., Dornan, J., Grover, D., Hennion, R., Percy, C., Foster, J. and Jeffrey, M. (1997) Immuno-detection of PrP^{SC} in spleens of some scrapie-infected sheep but not BSE-infected cows. *Journal of General Virology* 78, 2389–2396.

Taylor, D.M. (1995) Survival of mouse-passaged bovine spongiform encephalopathy agent after exposure to paraformaldehyde-lysine-periodate and formic-acid. *Veterinary Microbiology* 44, 111–112.

Taylor, D.M., Fraser, H., McConnell, I., Brown, D.A., Brown, K.L., Lamza, K.A. and Smith, G.R.A. (1994) Decontamination studies with the agents of bovine spongiform encephalopathy and scrapie. *Archives of Virology* 139, 313–326.

Taylor, D.M., Ferguson, C.E., Bostock, C.J. and Dawson, M. (1995) Absence of disease in mice receiving milk from cows with bovine spongiform encephalopathy. *Veterinary Record* 136, 592.

Wells, G.A.H., Dawson, M., Hawkins, S.A.C., Green, R.B., Dexter, I., Francis, M.E., Simmons, M.M., Austin, A.R. and Horigan, M.W. (1994) Infectivity in the ileum of cattle challenged orally with bovine spongiform encephalopathy. *Veterinary Record* 135, 40–41.

Wells, G.A.H., Hawkins, S.A.C., Green, R.B., Austin, A.R., Dexter, I., Spencer, Y.I., Chaplin, M.J., Stack, M.J. and Dawson, M. (1998) Preliminary observations on the pathogenesis of experimental bovine spongiform encephalopathy (BSE): an update. *Veterinary Record* 142, 103–106.

Westaway, D., Zuliani, V., Cooper, C.M., Dacosta, M., Neuman, S., Jenny, A.L., Detwiler, L. and Prusiner, S.B. (1994) Homozygosity for prion protein alleles encoding glutamine-171 renders sheep susceptible to natural scrapie. *Genes and Development* 8, 959–969.

Wilesmith, J. and Ryan, J. (1997) Absence of BSE in the offspring of pedigree suckler cows affected by BSE in Great Britain. *Veterinary Record* 141, 250–251.

Wilesmith, J.W., Wells, G.A.H., Cranwell, M.P. and Ryan, J.B.M. (1988). Bovine spongiform encephalopathy – epidemiological studies. *Veterinary Record* 123, 638–644.

Wilesmith, J.W., Ryan, J.B.M. and Atkinson, M.J. (1991) Bovine spongiform encephalopathy – epidemiologic studies on the origin. *Veterinary Record* 128, 199–203.

Wilesmith, J.W., Ryan, J.B. and Hueston, W.D. (1992) Bovine spongiform encephalopathy: case–control studies of calf feeding practices and meat and bonemeal inclusion in proprietary concentrates. *Research in Veterinary Science* 52, 325–331.

Wilesmith, J., Wells, G., Ryan, J., Gavier-Widen, D. and Simmons, M. (1997) A cohort study to examine maternally-associated risk factors for bovine spongiform encephalopathy. *Veterinary Record* 141, 239–243.

Yoshimoto, J., Iinumo, T., Ishiguro, N., Imamura, M. and Shinagawa, M. (1992) Comparative sequence analysis and expression of bovine PrP gene in mouse-L929 cells. *Virus Genes* 6, 343–356.

Cytogenetics and Physical Chromosome Maps

10

R. Fries[1] and P. Popescu[2]

[1]*Lehrstuhl für Tierzucht, Technische Universität München, 85350 Freising-Weihenstephan, Germany;* [2]*Laboratoire de Génétique Biochimique et de Cytogénétique, Départment de Génétique INRA, 78352 Jouy-en-Josas, France*

Introduction	247
The Normal Cattle Chromosomes	248
Standardization of the cattle karyotype	249
Location of nucleolar organizer regions	253
The meiotic chromosomes	254
Chromosome Abnormalities	256
Numerical chromosome abnormalities	256
Structural chromosome abnormalities	257
The 1/29 translocation	259
The impact of the 1/29 translocation and other chromosome abnormalities	260
Approaches to Chromosome Mapping in Cattle	260
Rationale of chromosome mapping	262
Somatic-cell hybrid mapping	262
Isotopic and fluorescence *in situ* hybridization	262
Radiation-hybrid panels	264
Comparative mapping	265
Chromosome sorting and microdissection	265
The Chromosome Map of Cattle	266
Acknowledgements	302
References	302

Introduction

Cattle cytogenetics had a comparatively strong start due to the early detection of the famous 1/29 translocation in 1964 in Sweden. The 1/29 translocation not only had a major impact on cattle cytogenetics but also fostered this discipline in other domestic species. The investigation of the 1/29 translocation was the driving force of cattle cytogenetics until the mid-1980s, when gene mapping and molecular genetics awoke a new interest in cytogenetics. Today, cytogenetics and gene mapping are largely interdependent. By considering the gene content of chromosomes, the cytogenetic anatomy of an organism adopts a functional dimension. The chromosomes are no longer merely the visual germplasm but represent strategically important units in genome analysis The functional anatomy is the basis for the systematic investigation of the substrate of animal genetic improvement, i.e. the genetic variation, at a molecular level. Consequently, cytogenetics and physical gene mapping are dealt with in one chapter in this book.

The Normal Cattle Chromosomes

Domestic cattle (*Bos taurus* L.), like many species belonging to the genus *Bos*, have 60 chromosomes: 58 autosomes and two sex chromosomes. All autosomes are acrocentric and the X and Y are submetacentric. Without bands, a cattle karyotype stained by Giemsa is presented as a decreasing series, arbitrarily divided into several groups. The only criterion for the karyotype arrangement is the relative length of each chromosome, expressed as a fraction of the total haploid autosome length.

The sex chromosomes are easily identifiable because they are biarmed. The X chromosome is similar in length to the largest autosome, being about 5% of the total haploid complement. The Y chromosome is one of the smallest chromosomes of the complement, with a relative length close to that of the smallest autosome pair.

Since the early 1970s, several banding techniques, such as C-bands by barium hydroxide using Giemsa (CBG), G-bands by trypsin using Giemsa (GTG), Q-bands by fluorescence using Hoechst 33258 (QFH), Q-bands by fluorescence using quinacrine (QFQ), R-bands by 5-bromo-2′-deoxyuridine (BudR) using acridine orange (RBA), R-bands by BudR using Giemsa (RBG) and staining of nucleolar organizer regions (NORs), have been applied to cattle chromosomes. These methods produced different systems of cytogenetic nomenclature of cattle chromosomes.

The first international conference for the standardization of the karyotypes of domestic animals, including cattle, is usually referred to as the Reading Conference, since it took place at Reading University, England, in 1976 (Ford *et al.*, 1980). The Reading Conference proposed a GTG-band standard karyotype for cattle chromosomes and created the basis for all subsequent nomenclature efforts. The Reading nomenclature system did not include a diagrammatic representation of the bovine karyotype.

Standardization of the cattle karyotype

In 1989, the second international conference on standardization of domestic animal karyotypes, establishing an International System for Cytogenetic Nomenclature of Domestic Animals (ISCNDA; Di Berardino et al., 1990), took place in Jouy-en-Josas, France. During this conference, the G-band karyotype of the Reading standard was improved by using prometaphase chromosomes, with a better resolution of bands. Ideograms with band numbering for each chromosome were established, based on sequential G- and R-banding of chromosomes of the same cell. The resulting schematic representations consisted of 410 GTG-bands and 404 RBA/RBG-bands.

Inconsistencies of the cattle chromosome nomenclature were clarified during the Ninth North American Colloquium on Domestic Animal Cytogenetics and Gene Mapping, held at Texas A&M University, College Station, in 1995 (Popescu et al., 1996). This new nomenclature ('Texas standard') correlates the

Table 10.1. The Texas standard chromosome nomenclature (Popescu et al., 1996) correlated with previous nomenclatures, synteny group designations and marker gene assignments. The homologies with human and sheep chromosomes and chromosome arms are also indicated. Marker gene symbols are explained in Table 10.4.

Texas standard	U group	Marker gene	Human chromosome	Length % *	Reading (GTG)	ISCNDA (GTG)	ISCNDA (QFQ)	ISCNDA (RBA/RBG)	Sheep chromosome
1	U 10	SOD1	3, 21	5.87	1	1	1	1	1q
2	U 17	VIL	1p, 2q	5.12	2	2	2	2	2q
3	U 6	HSD3B	1p	4.71	3	3	3	3	1p
4	U 13	INHBA	7p	4.67	4	4	4	4	4
5	U 3	IFNG	1q, 12, 22	4.48	5	5	5	5	3q
6	U 15	CSN@	4	4.33	6	6	6	6	6
7	U 22	RASA	5q, 19p	4.18	7	7	7	7	5
8	U 18	IFNA	8p, 9q	4.13	8	8	8	8	2p
9	U 2	IGF2R	6q	3.86	9	9	9	9	8
10	U 5	CYP19	5q, 14, 15	3.67	10	10	10	10	7
11	U 16	LGB	2, 9q	3.94	11	11	11	11	3p
12	U 27	RB1	13	3.29	12	12	12	12	10
13	U 11	IL2RA	10p, 20	3.09	13	13	13	13	13
14	U 24	TG	8q	3.15	14	14	14	14	9
15	U 19	FSHB	5, 11p	3.11	15	15	15	15	15
16	U 1	PIGR	1q	3.07	16	16	16	16	12
17	U 23	FGG	4q, 12q, 22	2.83	17	17	17	17	17
18	U 9	GPI	16q, 19q	2.60	18	18	18	18	14
19	U 21	GH	17	2.54	19	19	19	19	11
20	U 14	MAP1B	5	2.75	20	20	20	20	16
21	U 4	IGH@	14, 15	2.72	21	21	21	21	18
22	U 12	LTF	3	2.51	22	22	22	22	19
23	U 20	BOLA	6p	2.09	23	23	23	23	20
24	U 28	DSCI	18	2.37	24	24	24	24	23
25	U 8	ELN	7q, 16p	1.97					24
26	U 26	APT1	10q	1.96	26	26	26	26	22
27	U 25	DEFB@	8	1.83					26
28	U 29	CGN1	10q	1.73	28	27	27	28	25
29	U 7	LDHA	11	1.99					21
X		PGK1	X	5.45	X	X	X	X	X
Y		ZFY	Y	2.13	Y	Y	Y	Y	Y

*Chromosome measurement expressed as relative length of the haploid genome.
@ indicates a cluster of loci.

Reading G-band and the ISCNDA GTG-, QFQ- and RBA/RBG-band nomenclatures with marker genes mapped to each chromosome (Table 10.1). This table also presents the relative length of each chromosome and indicates the comparative relationships of the cattle chromosomes with the human and sheep

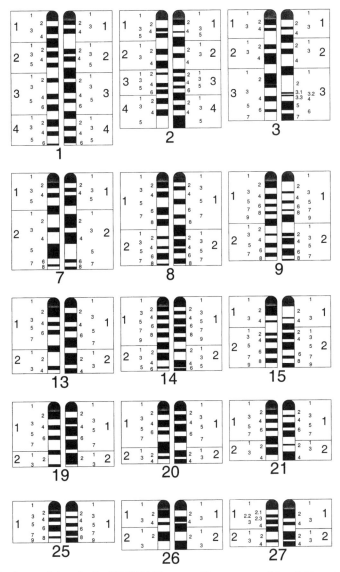

Fig. 10.1 (and opposite). Revised ISCNDA ideogram. Chromosomes 25 and 29 are interchanged relative to the original ISCNDA ideogram (Di Bernardino *et al.*, 1990). The basic nomenclature is according to the nomenclature used for the human chromosomes (Paris Conference, 1975). The left side represents the G-band and the right side the R-band nomenclature. Positive G-bands correspond to negative R-bands and vice versa. In some cases a G-band is resolved in several R-bands, which are then indicated as subbands (and vice versa).

chromosomes. The Texas standard restores the original Reading numbering of chromosomes 4 and 6, which was interchanged in Di Berardino *et al.* (1990). Moreover, chromosome 29 has been renamed 25 to accommodate ample evidence that the small fusion partner of the 1/29 fusion is 25 and not 29. This modification of the nomenclature makes it possible to continue naming the

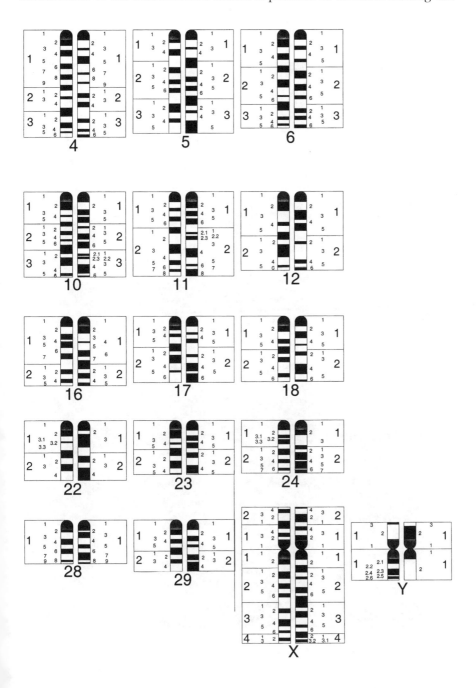

most frequent cattle chromosome aberration '1/29'. However, the nomenclature of chromosomes 25, 27 and 29, relative to the former standards, is still disputed but will hopefully be resolved in the near future. The ISCNDA ideogram, revised according to the Texas standard, is shown in Fig. 10.1. The arrangement of QFQ-banded and RBG-banded chromosomes based on this ideogram is shown in Fig. 10.2.

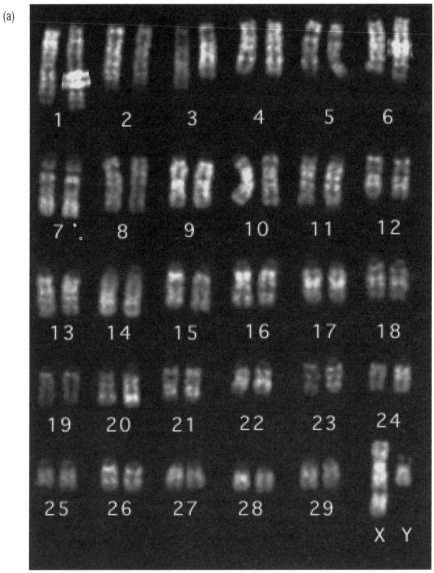

Fig. 10.2 (and opposite). (a) QFQ- and (b) RBG-banded cattle chromosomes arranged according to the revised ISCNDA ideogram.

Location of nucleolar organizer regions

While G-, Q-, R- and C-banding methods produce a characteristic banding of all or most chromosomes, silver staining is used to reveal the regions of active NORs in the telomeric regions of some cattle chromosomes. Revealed by silver staining, NORs have been reported on the telomeres of chromosomes 2, 3, 4,

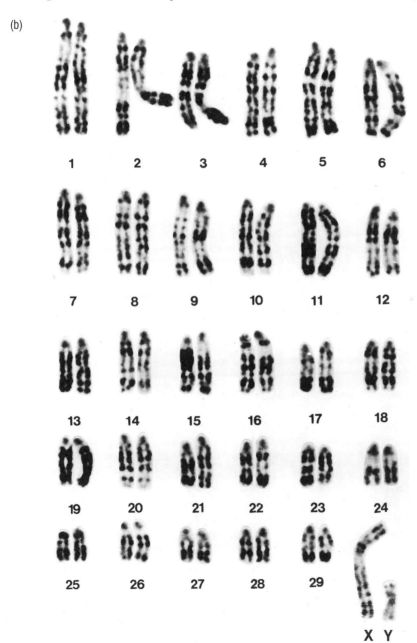

11 and 29 (Di Berardino *et al.*, 1981), of 2, 3, 4, 5 and 28 (Henderson and Bruere, 1979), of 2, 3, 4, 11 (Mayr and Czaker, 1981) and of 2, 3, 6, 11 and 27 (Di Berardino *et al.*, 1985). After fluorescence *in situ* hybridization (FISH) with a human probe for the ribosomal ribonucleic acid (rRNA) gene cluster, strong hybridization signals were observed on the telomeres of bovine chromosomes 2, 3, 6, 11, 28 and 29 (Solinas Toldo *et al.*, 1992). The latter report also provided evidence for the heterozygous presence/absence (or extremely small size) of NORs on chromosome 3, 28 and 29. Chromosomes 2, 3 and 11 have consistently been shown to be carriers of NORs. The discrepancy between 4 and 6 is due to nomenclature problems, which have been solved in the meantime, i.e. chromosome 4 is most probably a carrier of a NOR. Discrepancies in the NOR assignment to the small chromosomes are due to difficulties with chromosome identification and polymorphism phenomena. However, if the Texas standard is used as a basis for chromosome identification, 28 and 25 (29 according to former standards, see above) are most probably occupied by NORs.

The meiotic chromosomes

Conventional meiotic studies, as well as synaptonemal complex analysis by electron microscopy, are very important tools of domestic animal cytogenetics, since they are the basis for understanding the meiotic behaviour of chromosomes and thus for explaining the effects of chromosome abnormalities on reproductive performance. Conventional meiotic studies provide a glimpse of the postsynaptic events, while synaptonemal complex analysis allows detailed observation of the actual synaptic events at a high resolution.

The prophase of the first meiotic division follows the same general course in bulls as in other mammalian species. The leptoptene stage shows thin chromatin material scattered randomly in the nuclear region. In the zygotene, it is possible to observe paired homologous chromosomes jumbled and overlapping each other. In the leptotene, zygotene and early pachytene the heteropyknotic sex vesicle, usually peripherally located, becomes visible. The sex vesicle is uniformly stained in leptoptene and zygotene, but, in the late pachytene and diplotene stage, the structure of the coiled sex bivalent becomes visible inside. The cells in the diakinesis stage show 30 bivalents (Fig. 10.3). An average of one to four chiasmata are observed in each bivalent. The average total number of chiasmata per male meiotic cell was estimated at 53.7 for *B. taurus* (Popescu, 1971) and 58.2 for *B. indicus* (Sharma and Popescu, 1986). Other estimates of the male chiasmata frequency in cattle range from 47.8 to 49.5. Chiasmata counts in female meiosis indicated an average number of 35.7 per cell (Jagiello *et al.*, 1974). In humans, the mean (male) chiasmata frequency amounts to 50.07 (Paris Conference, 1975). The mean chiasmata frequency increases with the chromosome length, but not proportionally. The assumption is that each chiasma is the cytological manifestation of a crossing-over event. Since crossing-over can affect two chromatids within the same

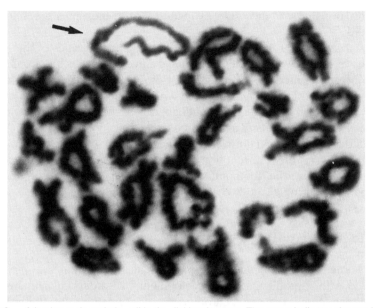

Fig. 10.3. A bovine male meiotic cell in the diakinesis stage. The arrow indicates the sex bivalent. There is a total of 30 bivalents.

chromosome or chromatids of homologous chromosomes, one chiasma represents a genetic length of 0.5 morgans (M) or 50 centimorgans (cM). Consequently, the genetic length of cattle derived from chiasmata counts is in the range of 24–29 M in the male and 18 M in the female. These findings are in conflict with the general observation that meiosis in the heterogametic sex results in shorter genetic maps than meiosis in the homogametic sex and pose the question to what extent chiasmata counts reflect the actual crossing-over frequency. Furthermore, both the male and female linkage maps of cattle are substantially larger than the estimates based on chiasmata counts (see also Chapter 11).

The sex chromosomes are often found as a bivalent, showing end-to-end association in the diplotene and diakinesis stages, as seen in conventional meiotic preparations. In surface-spread spermatocytes, the X–Y bivalent shows a small synaptonemal complex and indicates an association between the long arm of the X chromosome and the short arm of the Y chromosome (Basrur and Switonski, 1990). An unstained gap appeared between this complex and the identifiable remnant of the Y chromosome (Bouvet and Popescu, 1989). These cytological findings suggest the location of the pseudo-autosomal region (PAR) in the telomeric regions of the long arm of the X chromosome and the short arm of the Y chromosome, respectively.

It is well recognized that in female mammals a large proportion of the germ cells degenerate at different stages of meiosis and very few, probably 5%, arrive at the diplotene stage. The mechanism of this degeneration is not known. However, it has been hypothesized that the presence of the active X chromosome is absolutely necessary for a normal meiotic process. The germ cells with unpaired or incompletely paired X chromosomes are selectively eliminated.

Koykul et al. (1997) recently reported pairing irregularities of the X chromosomes in fetal bovine ovaries of all age-groups. However, the incidence of these irregularities increased with age, i.e. from 35.2% in 81–90-day embryos to 69.3% in 121–130-day fetuses. The frequency of the synaptic errors in autosomes was relatively low and did not change with the age of the fetuses. The paternal X chromosome is inactivated before fertilization and the female X is inactivated and reactivated during the differentiation of the cow's ovarian cells. Thus, changes in the transcriptional status of the X chromosomes could be responsible for the temporal increase in the frequency of abnormal synaptonemal pairing of the X chromosomes (Koykul et al., 1997).

Chromosome Abnormalities

Chromosome abnormalities, both numerical and structural, are very often associated with reproductive and developmental disorders. In humans, 0.6% of the living newborns carry chromosome abnormalities. In stillbirths, the incidence is ten times higher (6%) and 100 times (60%) higher in spontaneous abortions (Plachot and Popescu, 1991). In domestic animals, comparatively few data are available on the incidence of chromosome abnormalities, mostly in the context of fertility problems and reproductive failure.

Numerical chromosome abnormalities

There are two types of numerical chromosome aberrations: euploid heteroploidy or polyploidy and aneuploid heteroploidy. Polyploidy is the result of abnormal fertilization (polyandry or polygyny), the suppression of the first cleavage division or the fusion of embryonic cells. Aneuploid heteroploidy arises from non-disjunction of homologous chromosomes during meiosis or mitosis. The most common cases are monosomies and trisomies.

Few monosomies and trisomies have been described in cattle, most probably due to the elimination of embryos carrying such abnormalities prior to implantation. Two well-documented cases of autosomal trisomies for chromosomes 17 (Herzog et al., 1977) and 18 (Herzog et al., 1982), associated with branchygnatia and nanismus, respectively, have been reported. However, several types of sex-chromosome trisomies, namely XXX, XXY and XYY, have been described in cattle (Popescu, 1990). Numerical sex-chromosome abnormalities may lead to sterility or subfertility but are otherwise compatible with a normal development, due to the gene dosage mechanisms inherent in mammalian X chromosomes.

Another type of sex-chromosome aberration in cattle is the freemartin syndrome. According to Ohno (1969), the word freemartin is derived from 'farrow-mart', which meant 'sterile cow' in Scotland and northern England centuries ago. A freemartin is a female born co-twin to a male. Masculinization of the internal genital tract in these animals usually leads to sterility. Due to the fusion of placental membranes of the twins and the exposure of the female twin to anti-Müllerian hormone, the differentiation of the gonads and the Müllerian duct is inhibited between days 50 and 80 of pregnancy, the masculinization of the gonads and the development of the Wolffian duct occurring later, around day 90 (Vigier et al., 1984). Placental anastomosis also leads to the exchange of haematopoietic progenitor cells. This cellular exchange, however, is not responsible for the initiation of the abnormal development of the gonads in the female co-twin (Vigier et al.,1976). From a cytogenetic point of view, the freemartin is a chimerical organism with both XX- and XY-carrying cells in the blood and haematopoietic organs. The ratio of XX to XY karyotypes in cultures of blood lymphocytes varies considerably from individual to individual; it is, however, stable within a postnatal freemartin animal and is not related to the degree of masculinization, and typical freemartin animals are found without apparent chimerism (Greene et al., 1977). The male co-twin to a female freemartin animal is also chimerical, but the genital organs are normal. However, because of poor quality of the semen, some male co-twins to freemartins turn out to be subfertile (Dunn et al., 1979). The possibility of germ-cell chimerism has been discussed and been substantiated by findings of a surplus of female progeny sired by chimerical bulls. However, there is no satisfactory cytological evidence for the presence of XX germ cells in testes of chimerical bulls (Dunn et al., 1979).

The economic consequences of numerical aberrations are limited to the loss or reduction of fertility in the affected animal. Therefore, the spreading of such aberrations and subsequent deleterious effects on a population level do not pose a problem.

Structural chromosome abnormalities

Structural abnormalities occur after chromosome breakage in premeiotic or meiotic stages. The resulting chromosome fragments can be lost (deletion), inverted (inversion) or translocated to another chromosome (translocation, mostly reciprocal). Structural abnormalities are categorized as balanced (without missing or extra chromosomal material) or unbalanced (with missing or extra chromosomal material). Balanced abnormalities are usually phenotypically inert, although the order of genes could be modified. They are often associated with reproductive disorders because of the formation of unbalanced gametes during meiosis. The unbalanced gametes, which are able to participate in fertilization, give rise to unbalanced, not viable zygotes. By way of balanced gametes, the abnormalities are passed to the offspring.

The most common structural chromosome abnormality in cattle is a special type of translocation, the Robertsonian translocation or centric fusion, involving breakage and subsequent fusion of the centromeric regions of two acrocentric chromosomes. Forty-two different types of Robertsonian

Table 10.2. Robertsonian translocations in cattle.

Robertsonian translocation	Breed	Country	Method of chromosome identification	Reference
1;4		Czechoslovakia	M	Lojda et al. (1976)
1;7				Cited by Frank and Robert (1981)
1;21	Holstein Friesian	Japan	G	Miyake et al. (1991)
1;23		Czechoslovakia	M	Lojda et al. (1976)
1;25	Piebald	Germany	G	Stranzinger and Förster (1976)
1;26	Holstein Friesian	Japan	G	Miyake and Kaneda (1987)
1;28		Czechoslovakia	M	Lojda et al. (1976)
1;29	Different breeds	Different countries		Gustavsson and Rockborn (1964)
2;8	Friesian	England	G	Pollock (1974)
2;27				Yu and Xin (1991)
3;4	Limousine	France	R	Popescu (1977)
3;27	Friesian	Romania	M	Samarineanu et al. (1977)
4;4		Czechoslovakia	M	Lojda et al. (1975)
4;8	Chianina	Italy	R	De Giovanni et al. (1988)
4;10	Blonde d'Aquitaine	France	G, R, C	Bahri-Darwich et al. (1993)
5;18	Simmental	Hungary	G	Papp and Kovàcs (1980)
5;21	Japanese Black	Japan		Masuda et al. (1978)
5;22	Polish Red	Poland		Sysa and Slota (1992)
5;23	Brune Roumaine	Romania	M	Samarineanu et al. (1977)
6;16	Dexter	England	G	Logue and Harvey (1978)
6;28		Czechoslovakia	M	Lojda et al. (1976)
7;21	Japanese Black	Japan	G	Hanada et al. (1981)
8;9	Brown Swiss	Switzerland	M	Tschudi et al. (1977)
8;23	Ukrainian Grey	Russia	G, R, C	Biltueva et al. (1994)
9;23	Blonde D'Aquitaine	France	G, R	Cribiu et al. (1989)
11;16	Simmental	Hungary	G	Kovacs (1975)
11;22		Czechoslovakia	M	Lojda et al. (1976)
12;12	Simmental	Germany	G	Herzog and Hohn (1984)
12;15	Holstein Friesian	Argentina	G	Roldan et al. (1984)
13;21	Holstein Friesian	Hungary	M	Kovacs et al. (1973)
13;24	Red and White	Poland		Slota et al. (1988)
14;19	Braunvieh	Switzerland		Stranzinger (1989)
14;20	Simmental	England	G	Logue and Harvey (1978)
14;21	Simmental	Hungary		Kovacs and Szepeshelyi (1977)
14;24	Podolian	Italy	R	Di Berardino et al. (1979)
14;28	Holstein Friesian	USA	G	Ellsworth et al. (1979)
15;25	Barrosa	Portugal	G, G+C/G, R	Iannuzzi et al. (1992)
16;18	Barrosa	Portugal	G, R	Iannuzzi et al. (1993a)
16;19	Marchigiana	Italy	R, C	Malerba (1997)
16;20	Ger. Red Pied × Czech. Red Pied	Czechoslovakia	R, C	Rubes et al. (1996)
16;21	Ger. Red Pied × Czech. Red Pied	Czechoslovakia	R, G	Rubes et al. (1992)
19;21	Holstein Friesian	France	G	Pinton et al. (1997)
20;20	Simmental	Germany	G	Herzog and Hohn (1984)
21;27	Blonde d'Aquitaine	France	R, G, NOR	Berland et al. (1988)
24;27	Holstein hybrid	Egypt	G, C	Mahrous et al. (1994)
25;27	Grey Alpine	Italy	R	De Giovanni et al. (1979)

M, chromosome measurement; C, G, NOR, R: banding method.

translocations have been described (Table 10.2). The fusion event leads to a reduction of the chromosome number by one. In the meiosis of a heterozygous carrier, the two chromosomes involved in the translocation join up with their free homologues to form a trivalent at the diakinesis stage. If the meiotic segregation is normal, the free chromosomes are passed to one cell and the fused chromosomes to the other. The resulting gametes either carry the normal complement or the balanced translocation. In the case of abnormal segregation, one of the two free chromosomes migrates with the fused chromosomes to the daughter cell and leads to the formation of gametes that are disomic or nullisomic for one of the chromosomes involved in the centromeric fusion. After fertilization, such gametes give rise to trisomic or monosomic zygotes, respectively, which are not viable.

The 1/29 translocation

Gustavsson and Rockborn (1964) described for the first time a numerical aberration in cattle, the so-called 1/29 Robertsonian translocation. Since then, this abnormality has been identified in more than 50 breeds distributed over five continents (Plachot and Popescu, 1991). The frequency of this abnormality, which is the most common among the 42 different Robertsonian translocations that have been described, varies considerably from one breed to another. The preferential use of some sires by artificial insemination or natural breeding, along with the isolation of populations, led to a high prevalence of this abnormality in some breeds. In the British White (Eldridge, 1975)and in the Corsican breed (Hari *et al.*, 1984), this abnormality has reached frequencies of up to 60%.

There are two hypotheses for the worldwide distribution of the 1/29 translocation: recurrent mutation or common origin. At present, there is no evidence to support the recurrent mutation hypothesis. Despite the significant number of animals studied and the detected carriers of this anomaly, not a single case of *de novo* appearance of the translocation has been reported. On the other hand, it is well known that chromosomes containing NORs remain attached to the nucleolus during the meiotic prophase and, due to their proximity, they are more prone to translocations. If the two chromosomes implicated in this translocation were carriers of NORs, one could suggest the possibility of recurrence. However, in cattle, neither chromosome 1 nor chromosome 29 (new nomenclature) carry NORs (see above). Two blocks of constitutive heterochromatin in the region of the centromere of the fused chromosome, and therefore termed dicentric, would suggest a recent origin for this fusion, similar to the metacentric chromosomes as found in some strains of mice. However, the fused 1/29 chromosome has always been described as monocentric when centromere-specific banding methods were used, i.e. a single block of constitutive heterochromatin has always been detected at the level of the centromere (Popescu, 1973; Blazak and Eldridge, 1977), favouring the hypothesis of a common and ancient origin of the 1/29 translocation.

The 1/29 translocation causes a 5–10% reduction of fertility due to an increase of the embryonic mortality, which is a consequence of the formation of unbalanced gametes (Dyrendhahl and Gustavsson, 1979). The occurrence of unbalanced gametes has been shown in meiotic studies (Logue, 1977; Logue and Harvey, 1978; Popescu, 1978; Gustavsson, 1979) and by the investigation of embryos sired by a heterozygous parent (King *et al.*, 1980).

In Sweden, the reduction in fertility caused by the 1/29 translocation has been estimated to be responsible for the loss of US$0.5 million year^{-1} for the cattle industry. An eradication programme was implemented in Sweden in 1969, based on karyotype analysis of all bulls before they enter the testing programme and the culling of all carriers (Gustavsson, 1970). Other countries have also adopted more or less strict eradication programmes. Positive effects of the 1/29 translocation on productive traits have not been reported.

The impact of the 1/29 translocation and other chromosome abnormalities

The discovery of the 1/29 translocation had a great impact on the development of domestic animal cytogenetics. More than 230 papers have been published on this abnormality (Popescu and Pech, 1991). Many cytogenetic laboratories have been established all over the world with the aim of searching for this abnormality. As a consequence of these activities, domestic cattle, with probably 50,000 cytogenetically examined animals, represent in terms of cytogenetics one of the most intensively studied mammalian species (besides humans and mouse). Despite these intensive cytogenetic investigations, few cases of other structural abnormalities, such as reciprocal translocations involving interstitial (not centromeric) breakage, tandem translocations or inversions, have been described in cattle (Table 10.3). This may be due to a low occurrence of such abnormalities in cattle or due to difficulties in detecting smaller rearrangements by routine cytogenetics, which usually does not involve banding. In cases where chromosome banding has been applied, quality and resolution are often suboptimal and would not allow detection of subtle chromosome rearrangements.

Approaches to Chromosome Mapping in Cattle

There are different levels of physical genome mapping. They range from the assignment of loci to specific chromosomes to the entire nucleotide sequence of an organism with the ordered position of each gene. The scope of this chapter comprises the chromosome map. The coordinates of this map are the cytogenetic bands as they are listed in the standard karyotype.

Table 10.3. Structural chromosome abnormalities in cattle (other than Robertsonian fusions).

Type of rearrangement	Breed	Number of cases	Chromosomes (bands) involved	Chromosome identification	Phenotype	Fertility	Reference
Tandem translocation	Danish Red	1	Not determined	–	Normal	Reduced by 10%	Hansen (1970)
Reciprocal translocation*	Fleckvieh	1	10/14	G	Normal	NR 30%	Mayr et al. (1979)
	Braunvieh	1	X/autosome	–	Normal	'Repeat breeder'	Gustavsson et al. (1968)
	Braunvieh	1	8/15 (q21;q24)	G	Normal	NR 25%	Mayr et al. (1983)
	Friesian	1	2/20 and 8/27	G	Normal	27% of spermatozoa are abnormal	de Schepper et al. (1982)
	Not known	2	X/23 (p12;q12/13)	Q	Normal/albino	–	Gallagher et al. (1992a)
	Brown Swiss/ Holstein Friesian	2	13/X	G, Q	Normal	'Repeat breeder'	Eldridge (1980)
	Red Holstein	1	8/13 (q11;q24)	G	Normal	Azoospermic	Ansari et al. (1993)
Pericentric inversion	Charolais/Guernsey	1	Not determined	–	Testicular hypoplasia	–	Short et al. (1969)
	Normande	Several	14	R	Normal	NR 58%	Popescu (1972)
Deletion	Simmental	1	X (q22)	G	Hypotrichosis, oligodentia, retarded growth	–	Braun et al. (1988); Stranzinger (1989)

* Translocations are usually considered reciprocal even in the absence of a visible reciprocal exchange.
G, R, Q: banding method; NR: non-return rate.

Rationale of chromosome mapping

In the early days of chromosome mapping, the usefulness of knowing the cytogenetic position of a locus was not obvious. It was thought that knowledge of the meiotic linkage relationship of marker and trait loci would be sufficient, since this knowledge is the basis of marker-assisted selection. In cattle, due to its difficult karyotype, the general feasibility of chromosome mapping was disputed. However, in the meantime, the feasibility of mapping the bovine chromosomes has been amply demonstrated and the strategic function of chromosome mapping is now generally acknowledged. The functions of chromosome mapping can be summarized as follows.

- Aiding the construction of linkage maps (orientation, bridging large gaps).
- Providing information for accessing chromosome regions of interest by microdissection.
- Verifying the position of candidate genes and markers in positional cloning projects.
- Providing information on the physical order and distances of loci in high-resolution genetic mapping and association studies.
- Facilitating the establishment of contigs (i.e. the arrays of overlapping clones).
- Testing large-insert clones for chimerism.
- Information on the location of loci relative to evolutionarily conserved regions.

Somatic-cell hybrid mapping

The first gene map of cattle consisted of groups of syntenic loci, i.e. loci that were identified to be located on the same chromosome based on the analysis of a panel of somatic-cell hybrids (cattle × hamster or cattle × mouse) (Womack and Moll, 1986). Due to the uniform morphology of the bovine chromosomes, chromosome identification in hybrid cells was not possible. However, by 1993 syntenic groups were established for all 29 autosomes and the X chromosome. They were designated U1 to U29 (for unassigned synteny groups) and X (Fries *et al.*, 1993) (see also Table 10.1).

Isotopic and fluorescence in situ hybridization

The assignment of the syntenic groups to specific chromosomes was gradually accomplished first by radioactive and later by non-isotopic *in situ* hybridization of representative probes for the syntenic groups. Chromosome 27 was the last to become 'mapped territory' (Masabanda *et al.*, 1996). Radioactive *in situ* hybridization is a statistical approach that might require chromosome identification and analysis of the autoradiographic silver-grain distribution in up to

100 metaphase spreads for the assignment of a locus, a task that is not trivial in cattle. Therefore, only a few laboratories reported assignments by radioactive *in situ* hybridization. The breakthrough in bovine chromosome mapping came with the advent of non-isotopic approaches to *in situ* hybridization – FISH, using biotinylated cosmid probes and fluorochrome-coupled avidin for signal detection (Pinkel *et al.*, 1986; Lichter *et al.*, 1990). The FISH signals appear as twin spots, resulting from hybridization at both chromatids. Background, if any, appears as much less brilliant single spots. Five to ten metaphase spreads are usually sufficient for an assignment. The identification of the chromosomes carrying the signal is based either on a banding pattern (mainly QFQ) that is photographically or digitally recorded before hybridization or a banding pattern that is induced after *in situ* hybridization (RBA-, QFH-banding). An example of QFQ-based chromosome identification (prior to *in situ* hybridization) is given by Solinas Toldo *et al.* (1993). Examples of RBA- and QFH-banding after *in situ* hybridization are given in Hayes and Petit (1993) and Gallagher *et al.* (1992a), respectively. The limitation of non-radioactive *in situ* hybridization is its relatively low sensitivity. That is, probes smaller than 3–4 kb which are derived from single-copy genes cannot be routinely assigned by non-radioactive *in situ* hybridization. However, localization of large-insert probes, such as cosmids, bacterial artificial chromosomes (BACs) or yeast artificial chromosomes (YACs), can be reliably assigned after suppression of repetitive sequences with unlabelled genomic deoxyribonucleic acid (DNA). The majority of FISH assignments in cattle were based on cosmid probes. Many of these cosmids contain polymorphic microsatellites and their physical as well as meiotic mapping allowed the integration of the two types of maps. A major step toward the integration of the physical and genetic map was achieved through the efforts of the BovMap consortium, an initiative that was financed by the European Commission (Ferretti *et al.*, 1997).

The usefulness of any chromosome map depends on its resolution. The resolution of a cytogenetic map is proportional to the degree of condensation of the chromatin. Metaphase chromosomes are routinely used as hybridization substrate for FISH. Standard metaphase FISH provides a mapping resolution of about 2 Mb. The use of less condensed mitotic prometaphase or meiotic pachytene chromosomes enhances the resolution to 1–2 Mb (Gallagher *et al.*, 1998). Ordering of clones by FISH with conventional chromosome preparations is also facilitated by the use of multiple fluorochromes, called multiple-colour FISH. However, on a submegabase level, the resolution of mitotic and meiotic metaphase or prometaphase chromosomes obtained by standard preparation procedures is no longer sufficient for the ordering of large-insert clones. Mechanically stretched chromosomes (MSCs) allow for a 10–20 times higher resolution than conventional chromosomes (Laan *et al.*, 1995). The fact that MSCs are still intact often allows one to obtain information on the centromere–telomere orientation of the clones. This is a distinct advantage over FISH with interphase nuclei, which provides similar resolution but only over a few hundreds of kilobases (Trask *et al.*, 1989). Maximum resolution, i.e. at the 1 kb level, is achieved by FISH with extended single DNA fibres, called fibre FISH (Parra and Windle, 1993; Fidlerová *et al.*, 1994; Houseal *et al.*, 1994). Measurement of (human) fibre-FISH signals from cosmids of known length indicated an average condensation of 0.33 ± 0.03 µm kb^{-1}, which is very close to the 0.34 µm kb^{-1} derived from the Watson and Crick DNA model (Florijn *et al.*, 1995). Thus, FISH provides mapping tools at all levels of resolution and it will therefore meet most physical mapping requirements in the context of positional cloning in cattle.

Together with the availability of gridded or pooled large-insert libraries (YAC and BAC), FISH provides a straightforward approach to test positional candidate-gene hypotheses. Using human cloned genes or polymerase chain reaction (PCR)-generated probes, large-insert clones specific for the genes of interest are quickly identified in the gridded YAC or BAC library by colony hybridization or by PCR from pools of YAC and BAC clones. The resulting clones are assigned by FISH. If they do not map to a region previously identified to contain the locus of interest, they are either false positives or the specific candidate-gene hypothesis must be rejected. If large-insert clones containing the candidate-gene sequences do map to the region of interest, the candidate sequences are searched for causal variants. This approach recently led to the successful identification of deletions in the growth differentiation factor 8 (*GDF8*) gene that are responsible for the muscular hypertrophy phenotype in cattle (Grobet *et al.*, 1997; Smith *et al.*, 1997).

Radiation-hybrid panels

Physical mapping by FISH is complemented by the analysis of radiation-hybrid panels. Such panels consist of cell lines that result from the fusion of irradiated bovine cells with mouse cells. A part of the bovine chromosome complement is retained in each hybrid cell line in fragmented form as a consequence of radiation. The individual hybrid lines are screened by PCR to identify those hybrids that have retained a given locus. Nearby loci tend to show similar retention patterns. Womack *et al.* (1997) have recently established a panel of 101 radiation-hybrid lines for bovine genome mapping. The resolving power of mapping using this radiation-hybrid panel was demonstrated by separating two loci that could not be resolved by meiotic linkage mapping. The distance between these two loci was 17 cR$_{5000}$ (1 cR$_{5000}$ is equal to 1% probability of breakage between two markers at a radiation dose of 5000 rads of gamma ray). This example, as well as the results of extensive radiation hybrid mapping of the human genome, indicates a submegabase resolution power of this mapping approach (e.g. Hudson *et al.*, 1995). The general application of radiation-hybrid mapping is dependent on the availability of a high-density framework of markers, a requirement that is met in cattle by the dense microsatellite marker maps (see Chapter 11). Radiation-hybrid mapping is a powerful tool for comparative genome mapping, since chromosomal order can be established for expressed loci, which are usually conserved between species but often not amenable to linkage mapping for a lack of allelic variation.

Comparative mapping

Comparative maps will allow access to the wealth of data resulting from the human genome programme, which should result not only in the entire nucleotide sequence of the human genome but also a complete map of all 50,000–100,000 genes. Such maps are the basis for 'comparative positional candidate cloning' of trait loci. Comparative maps are based on the mapping of conserved loci in different species. Linkage mapping added little to comparative mapping, since conserved loci are often not amenable to linkage mapping for a lack of allelic variation. Comparative mapping was therefore largely based on physical methods. Mapping of conserved loci, using a somatic-hybrid panel, provided evidence of extensive synteny conservation of cattle with mouse and humans in the beginning of bovine genome mapping (Womack and Moll, 1986). However, synteny conservation does not imply linkage conservation, i.e. the conservation of locus order within a conserved segment. It becomes increasingly evident that rearrangements in the locus order within conserved syntenies are rather frequent. Radiation-hybrid panel mapping is potentially a powerful tool for the definition of such rearrangements and work is in progress to establish a high-resolution human/bovine comparative map based on this mapping (J.E. Womack, San Diego, 1998, personal communication).

A more direct approach to the delineation of conserved synteny segments than the mapping of individual loci is Zoo-FISH painting. This approach is based on FISH of human chromosome-specific probes on cattle chromosomes. Hayes (1995), Solinas Toldo *et al.* (1995a) and Chowdhary *et al.* (1996) have established the boundaries of the conserved synteny segments for most bovine chromosome regions using Zoo-FISH. The paints resulting from the three Zoo-FISH efforts are remarkably similar (Fig. 10.4). The minor differences are likely to be the consequence of a lack of precision in assigning the FISH signals to the cytogenetic bands and the use of differing probe sets by the three groups.

Chromosome sorting and microdissection

Chromosomes represent a natural portioning of the genome. The sorting of chromosomes allows the physical separation of these portions. The first step of chromosome sorting is the establishment of a flow karyotype. Despite the uniform appearance of the cattle chromosomes, Schmitz *et al.* (1994) were able to produce a bivariate flow karyotype resolving up to 36 peaks, although the anticipated number was 31. This was explained by the resolution of chromosomal pairs into two populations due to heteromorphisms in the centromeric heterochromatin. They were able to sort 22 chromosome types as pure fractions. The identity of these fractions was assessed using a PCR-based amplification procedure, the priming authorizing random mismatches (PARM)-PCR (Milan *et al.*, 1993), and subsequent FISH (see also Chapter 6).

A more direct access not only to entire chromosomes but also to specific regions is provided by microdissection, i.e. the microsurgical removal of specific chromatin. This technique was originally applied to dissect polytene chromosomes in *Drosophila*. Since the polytene chromosomes represent a 1000-fold lateral reduplication of euchromatin, the dissection of a single polytene chromosome provides sufficient material to construct a representative library. To obtain such a library in a vertebrate species, it is necessary to dissect more than 100 chromosomes, which is not practicable in most cases. However, the possibility of cloning the microdissected material in microbial vectors and subsequent PCR with vector-specific primers (Lüdecke *et al.*, 1989) or direct PCR using degenerate oligonucleotide priming (DOP) (Guan *et al.*, 1992) made microdissection a useful tool for vertebrate genome analysis. The cloning-based approach still requires several chromosome fragments to be microdissected, while the DOP-PCR approach has been successfully applied to amplify individual fragments. The option of working from a single chromosome fragment when using DOP-PCR is a distinct advantage in cattle, where chromosome identification poses special problems. The specificity of the microdissected material is assessed by FISH of an aliquot of the material with metaphase spreads. Ponce de Leon *et al.* (1996) used the cloning-based approach to dissect the cattle X chromosome. The resulting material was used in a targeted marker development for the X chromosome and as Zoo-FISH probe in a comparative study of the bovid X chromosomes. Goldammer *et al.* (1996) used microdissection and DOP-PCR for the isolation of region-specific sequences on chromosome 6, which were then used for the development of microsatellite markers in regions of chromosome 6 that were shown to contain lactation-relevant quantitative trait loci (Weikard *et al.*, 1997). (Additional aspects of microdissection are covered in Chapter 6 of this book.)

The Chromosome Map of Cattle

The present chromosome map is shown in Fig. 10.4. It is a cytogenetic map, since the positions of the loci are indicated in cytogenetic coordinates, i.e. individual bands, a region representing several bands or entire chromosomes. The map positions are based on direct physical mapping by *in situ* hybridization or indirect physical mapping using somatic-cell hybrid panels or are derived from linkage of a locus to physically mapped loci. However, only coding loci (type I) are included if a regional assignment is not available. Loci without regional assignments are listed in boxes next to each chromosome. An important function of chromosome maps is to visualize the comparative relationship of the chromosomal position of conserved loci. For this reason, the Zoo-FISH results with human chromosome probes obtained by three different groups are included in the form of bars next to each chromosome representing the human/bovine chromosome homologies. All locus symbols shown on the chromosome map (Fig. 10.4) are alphabetically listed in Table 10.4, together with the locus names and the chromosomal assignments. This table also

Continued on p. 302

Table 10.4. List of coding gene loci and physically mapped D-loci.

Locus symbol	Locus name	Chromosomal assignment	Reference
A2M	Alpha-2-macroglobulin	5	Threadgill and Womack (1990b); Barendse et al. (1992); Echard et al. (1994)
ABL1	Abelson murine leukemia viral (v-abl) oncogene homologue 1	11	Threadgill and Womack (1990a)
ABL2	Abelson murine leukemia viral (v-abl) oncogene homologue 2 (arg, Abelson-related gene)	16	Threadgill and Womack (1990a)
ACADM (MCAD)	Acyl-coenzyme A dehydrogenase, C-4 to C-12 straight chain	3	Beever et al. (1997a, b)
ACLY	ATP citrate lyase	19q17	Comincini et al. (1997)
ACO1	Aconitase, soluble	8	Womack and Moll (1986)
ACO2	Aconitase 2, mitochondrial	5q35	Womack et al. (1991); Ryan and Womack (1994); Comincini et al. (1997)
ACR	Acrosin	5	Friedl et al. (1994a)
ACTA1	Alpha skeletal actin	28	Threadgill et al. (1994); Riggs et al. (1997)
ACTG2	Actin, gamma 2, smooth muscle, enteric	11	Barendse et al. (1997)
ACVR2 (D2S50)	Activin A receptor, type II	2q23–q24	Flavin et al. (1996); Monteagudo et al. (1996); Barendse et al. (1997)
ACY1	Aminoacylase-1	22	Womack and Moll (1986)
ADA	Adenosine deaminase	13	Womack and Moll (1986)
ADCY2	Adenylate cyclase 2 (brain)	15	Bishop et al. (1994); Kappes et al. (1997)
ADCYAP1 (PACAP)	Adenylate cyclase-activating polypeptide 1 (pituitary)	24	Larsen et al. (1996); Agaba et al. (1997a); Barendse et al. (1997)
ADH2	Alcohol dehydrogenase (class I), beta polypeptide	6	Zhang et al. (1992)
ADORA1	Adenosine A1 receptor	16	Barendse et al. (1997)
ADRA2	Adrenergic, alpha-2, receptor	26	Threadgill and Womack (1991a)
ADRB2	Adrenergic, beta-2-, receptor, surface	7	Womack et al. (1991)
ADRBK2	Adrenergic, beta, receptor kinase 2	17	Barendse et al. (1997)
AK1	Adenylate kinase-1, soluble	11	Heuertz and Hors-Cayla (1981)
AK2	Adenylate kinase 2	2	Barendse et al. (1997)
AKAP	Anchor protein regulatory subunit (AKAP75)	10	Barendse et al. (1997)
ALB	Albumin	6	Bouquet et al. (1986); Bishop et al. (1994); Barendse et al. (1997); Kappes et al. (1997)

Table 10.4. *Continued.*

Locus symbol	Locus name	Chromosomal assignment	Reference
ALDH1	Aldehyde dehydrogenase 1, soluble	8	Threadgill and Womack (1990a)
ALDH2	Aldehyde dehydrogenase 2, mitochondrial	17	Threadgill and Womack (1990b); Barendse *et al.* (1994, 1997)
ALDOB	Aldolase B, fructose bisphosphate	8	Threadgill and Womack (1990a); Barendse *et al.* (1994, 1997)
ALPI	Alkaline phosphatase, intestinal	2	Kappes *et al.* (1997)
ALPL	Alkaline phosphatase, liver/bone/kidney	2	Beever *et al.* (1992); Ma *et al.* (1996)
AMH	Anti-Mullerian hormone	7	Womack *et al.* (1989); Rogers *et al.* (1991); Barendse *et al.* (1997); Gao and Womack (1997)
AMY1A	Alpha 1A amylase	3	Threadgill *et al.* (1994); Elduque *et al.* (1996); Beever *et al.* (1997b)
ANT1	Adenine nucleotide translocator 1 (skeletal muscle)	27	Lei and Womack (1997)
ANT2	Adenine nucleotide translocator 2 (fibroblast)	X	Lei and Womack (1997)
ANT3	Adenine nucleotide translocator 3 (liver)	X	Lei and Womack (1997)
ANX5 (ENX2)	Annexin V (endonexin II)	6	Lord *et al.* (1996)
APOE	Apolipoprotein E	18q24	Brzozowska *et al.* (1993)
APP	Amyloid beta (A4) precursor protein (protease nexin-II)	1	Threadgill *et al.* (1991)
APT1 (FAS)	Apoptosis (APO-1) antigen 1	26	Yoo *et al.* (1996a); Kappes *et al.* (1997)
ARR1 (ARRB1)	Arrestin, beta 1	15q25	Barendse *et al.* (1997); Comincini *et al.* (1997)
ASL	Argininosuccinate lyase	25	Womack *et al.* (1989)
ASS	Argininosuccinate synthetase (ASS deficiency, citrullinaemia)	11	Dennis *et al.* (1989); Threadgill and Womack (1990a)
AT3	Antithrombin III	16	Womack *et al.* (1989); Threadgill *et al.* (1994); Beever *et al.* (1997a, b)
ATP2B4 (PMCA4, ATP2B2)	ATPase, Ca^{2+}-transporting, plasma membrane 4	16	Barendse *et al.* (1997)
AVP	Arginine vasopressin (neurophysin II)	13	Dietz *et al.* (1992a); Barendse *et al.* (1994, 1997); Schläpfer *et al.* (1997a)
BF	Properdin factor B	23	Teutsch *et al.* (1989)
BLVR	Bovine leukaemia virus receptor	7q15	Popescu *et al.* (1995)
BMP3	Bone morphogenetic protein 3 (osteogenic)	6	Lord *et al.* (1996)

Table 10.4. *Continued.*

Locus symbol	Locus name	Chromosomal assignment	Reference
BOLA-A	Major histocompatibility complex, class I	23q22	Fries *et al.* (1986); Lindberg and Andersson (1988); Skow *et al.* (1988); Brown *et al.* (1989); Toye *et al.* (1990); Gallagher *et al.* (1992a, b); Kappes *et al.* (1997)
BOLA-B	Major histocompatibility complex, class I	23	Ennis *et al.* (1988); Toye *et al.* (1990)
BOLA-D@	Major histocompatibility complex, class II, cluster	23	
BOLA-DIB	Major histocompatibility complex, class II, DI beta	23	Stone and Muggli-Cockett (1990); Kappes *et al.* (1997)
BOLA-DNA	Major histocompatibility complex, class II, DN alpha	23	Andersson *et al.* (1988)
BOLA-DOB	Major histocompatibility complex, class II, DO beta	23	Andersson *et al.* (1988)
BOLA-DQA	Major histocompatibility complex, class II, DQ alpha	23	Andersson *et al.* (1986, 1988); Sigurdardóttir *et al.* (1988); Groenen *et al.* (1989); Georges *et al.* (1990); van der Poel *et al.* (1990)
BOLA-DQB	Major histocompatibility complex, class II, DQ beta	23	Andersson *et al.* (1986, 1988); Sigurdardóttir *et al.* (1988); Groenen *et al.* (1989, 1990); Georges *et al.* (1990)
BOLA-DRA	Major histocompatibility complex, class II, DR alpha	23	Andersson *et al.* (1986, 1988); Sigurdardóttir *et al.* (1988); Groenen *et al.* (1989); van der Poel *et al.* (1990)
BOLA-DRB1	Major histocompatibility complex, class II, DR beta 1	23	Andersson *et al.* (1988); Sigurdardóttir *et al.* (1988); Kappes *et al.* (1997)
BOLA-DRB2	Major histocompatibility complex, class II, DR beta 2	23	Muggli-Cockett and Stone (1989); Kappes *et al.* (1997)
BOLA-DRB3	Major histocompatibility complex, class II, DR beta 3	23	Burke *et al.* (1991); Groenen *et al.* (1990); Lewin *et al.* (1992); van Eijk *et al.* (1992); Ma *et al.* (1996)
BOLA-DRBP1	Major histocompatibility complex, class II, DR beta, pseudogene	23	Muggli-Cockett and Stone (1988); Groenen *et al.* (1990); Creighton *et al.* (1992)
BOLA-DYA	Major histocompatibility complex, class II, DY alpha	23q12/q13	Andersson *et al.* (1988); Georges *et al.* (1990); van der Poel *et al.* (1990); van Eijk *et al.* (1992); Ma *et al.* (1996); Skow and Nall (1996); Skow *et al.* (1996); Barendse *et al.* (1997)
BOLA-DYB	Major histocompatibility complex, class II, DY beta	23	Andersson *et al.* (1988); Georges *et al.* (1990)

Table 10.4. *Continued.*

Locus symbol	Locus name	Chromosomal assignment	Reference
BRN (RNS, RNASE)	Brain ribonuclease	10	Barendse *et al.* (1994, 1997); Bishop *et al.* (1994); Kappes *et al.* (1997)
BTN	Butyrophilin	23q21–q23	Ashwell *et al.* (1996); Brunner *et al.* (1996); Taylor *et al.* (1996); Karall *et al.* (1997)
BTY1	BTY1 expressed sequence	Yq	Comincini *et al.* (1997)
C4	Complement component 4	23	Andersson *et al.* (1988); Lindberg and Andersson (1988)
C5	Complement component 5	8	Threadgill and Womack (1990a); Barendse *et al.* (1997)
C9	Complement component 9	1	Zhang and Womack (1992)
CA2	Carbonic anhydrase II	14	Threadgill *et al.* (1990); Barendse *et al.* (1991); Kappes *et al.* (1997)
CALB1	Calbindin 1 (28 kD)	14	Broad *et al.* (1995a)
CALD1	Caldesmon 1	4	Ryan *et al.* (1997)
CALR	Calrectulin	7	Gao and Womack (1996)
CAST	Calpastatin	7	Bishop *et al.* (1994); Kappes *et al.* (1997)
CAT	Catalase	15	Womack and Moll (1986); Womack *et al.* (1989)
CATHL@	Cathelicidin gene family	22q24	Castiglioni *et al.* (1996b)
CATHL1	Cathelicidin 1 (cyclic dodecapeptide, bactenin 1)	22q24	Castiglioni *et al.* (1996b)
CATHL2	Cathelicidin 2 (bactenin 5)	22q24	Castiglioni *et al.* (1996b)
CATHL3	Cathelicidin 3 (bactenin 7)	22q24	Castiglioni *et al.* (1996b)
CATHL4	Cathelicidin 4 (indolicidin)	22q24	Castiglioni *et al.* (1996b)
CATHL5	Cathelicidin 5	22q24	Castiglioni *et al.* (1996b)
CATHL6	Cathelicidin 6	22q24	Castiglioni *et al.* (1996b)
CBS	Cystathionine-beta-synthase	1	Threadgill *et al.* (1991)
CCN1, CCNA	Cyclin A	6	Lord *et al.* (1996)
CD3D	Antigen CD3D, delta polypeptide (TiT3 complex)	15	Li *et al.* (1992); Agaba *et al.* (1997b)
CD3E	Antigen CD3E, epsilon polypeptide (TiT3 complex)	15	Li *et al.* (1992)
CD3Z	CD3Z antigen, zeta polypeptide (TiT3 complex)	3q11q14	Ansari *et al.* (1994); Agaba *et al.* (1997b)
CD5 (LEU1)	CD5 antigen (p56–62)	29	Barendse *et al.* (1997)
CD8A	Antigen CD8A (p32)	11	Li *et al.* (1992); Barendse *et al.* (1997)
CDH2 (NCAD)	Cadherin 2 (*N*-cadherin 1), *N*-cadherin	24	Larsen *et al.* (1996); Agaba *et al.* (1997a); Barendse *et al.* (1997)
CEA	Cacinoembryonic antigen	18	Gao and Womack (1996)
CENPB	Centromere protein B (80 kDa)	13	Burkin *et al.* (1996)
CENPC	Centromere protein B (140 kDa)	6	Burkin *et al.* (1996)

Table 10.4. Continued.

Locus symbol	Locus name	Chromosomal assignment	Reference
CFTR	Cystic fibrosis transmembrane conductance regulator	4(q23–q25)	Wallis and Womack (1994); Tebbutt et al. (1996)
CGA	Glycoprotein hormones, alpha polypeptide	9	Dietz et al. (1992a); Barendse et al. (1997)
CGN1	Conglutinin	28q18	Gallagher et al. (1993c)
CHGA	Chromogranin A, parathyroid secretory protein 1	21	Dietz et al. (1992a)
CHGB (SCG1)	Chromogranin B (secretogranin 1)	13	Barendse et al. (1997)
CHRNB1	Cholinergic receptor, nicotinic, delta polypeptide	19q14–q22	Echard et al. (1994); Yang and Womack (1995, 1996); Yang et al. (1998)
CHRND	Cholinergic receptor, nicotinic, delta polypeptide	2q42–q45	Echard et al. (1994)
CLAPB1	Clathrin-associated/assembly/adaptor protein, large, beta 1	19	Yang et al. (1998)
CLCN1	Chloride channel 1, skeletal muscle (Thomsen disease, autosomal dominant)	4	Ryan et al. (1997)
CLTLA2	Clathrin, light chain A2	8	Gallagher et al. (1991)
CLTLB	Clathrin, light chain B	7	Gallagher et al. (1991)
CNCG	Cyclic nucleotide gated channel (photoreceptor), cGMP gated	6	Gallagher et al. (1992d)
CNR1	Cannaboid receptor 1 (brain)	9q22	Pfister-Genskow et al. (1997)
CNTFR	Ciliary neurotrophic factor receptor	8q21	Wigger et al. (1998)
COL2A1	Collagen, type II, alpha 1 (primary osteoarthritis, spondyloepiphyseal dysplasia, congenital)	5	Barendse et al. (1997)
COL3A1	Collagen, type III, alpha1	2q22	Echard et al. (1994); Fisher et al. (1997a)
COL4A1	Collagen, type IV, alpha 1	12	Sun et al. (1997)
COL6A1	Collagen, type VI, alpha 1	1q12–q14	Threadgill et al. (1991); Schmutz et al. (1994)
COL6A2	Collagen, type VI, alpha 2	1	Threadgill et al. (1991)
COX8H	Cytochrome c oxidase subunit VIII, heart	29	Lomax et al. (1995)
COXP1	Cytochrome c oxidase, pseudogene	5	Dietz et al. (1992a)
CP	Ceruloplasmin	1	Larsen (1977); Threadgill and Womack (1991c)
CR2	Complement receptor 2	16	Threadgill et al. (1994)

Table 10.4. *Continued.*

Locus symbol	Locus name	Chromosomal assignment	Reference
CRH	Corticotrophin-releasing hormone	14	Barendse *et al.* (1991, 1997); Broad *et al.* (1995a)
CRYA1	Crystallin, alpha A	1	Skow *et al.* (1988); Barendse *et al.* (1997)
CRYA2	Crystallin, alpha polypeptide 2	15	Womack *et al.* (1989)
CRYB1 (CRYBA1)	Crystallin, beta A1	19	Yang and Womack (1995, 1996); Yang *et al.* (1998)
CRYG1	Crystallin, gamma polypeptide 1	2	Adkison *et al.* (1988b)
CRYGS (CRYG8)	Crystallin, gamma S	1	Barendse *et al.* (1994, 1997); Ma *et al.* (1996)
CSF1R	Colony-stimulating factor 1 receptor	7	Zhang and Womack (1992)
CSF2	Colony-stimulating factor 2 (granulocyte–macrophage)	7	Zhang and Womack (1992); Barendse *et al.* (1997); Gao and Womack (1997)
CSN1S1	Casein, alpha-S1	6q24–q33	Grosclaude *et al.* (1965); Larsen and Thymann (1966); Hines *et al.* (1969); Ferretti *et al.* (1990); Threadgill and Womack (1990c); Prinzenberg *et al.* (1997)
CSN1S2	Casein, alpha-S2	6q24–q33	Grosclaude *et al.* (1978); Ferretti *et al.* (1990); Threadgill and Womack (1990c)
CSN2	Casein, beta	6q24–q33	Grosclaude *et al.* (1965); Larsen and Thymann (1966); Hines *et al.* (1969); Ferretti *et al.* (1990); Threadgill and Womack (1990c); Fisher *et al.* (1997b)
CSN3	Casein, kappa	6q24–q33	Grosclaude *et al.* (1965); Larsen and Thymann (1966); Hines *et al.* (1969); Ferretti *et al.* (1990); Threadgill and Womack (1990c); Monteagudo *et al.* (1992)
CSNK2A1	Casein kinase 2, alpha 1 polypeptide	13	Gao *et al.* (1997); Schläpfer *et al.* (1997a)
CSNK2A2	Casein kinase 2, alpha prime polypeptide	5	Gao *et al.* (1997)
CSNK2B	Casein kinase 2, beta polypeptide	23	Gao *et al.* (1997)
CTSB (CPNB)	Cathepsin B	8	Barendse *et al.* (1997)
CYB5	Cytochrome b-5	24	Agaba *et al.* (1997a); Barendse *et al.* (1997)
CYM	Prochymosin	3	Dietz *et al.* (1992a)
CYP11B1	Cytochrome P450, subfamily XIB (steroid 11-beta-hydroxylase), polypeptide 1	14	Broad *et al.* (1995a)

Table 10.4. *Continued.*

Locus symbol	Locus name	Chromosomal assignment	Reference
CYP17	Cytochrome P450, subfamily XVII (steroid 17-alpha-hydoxylase)	26	Echard et al. (1994)
CYP19	Cytochrome P450, subfamily XIX (aromatization of androgens)	10q26	Goldammer et al. (1994)
CYP19P1	Cytochrome P450, subfamily XIX (aromatization of androgens), pseudogene	10q26	Goldammer et al. (1994)
CYP21	Cytochrome P450, subfamily XXI (steroid 21-hydroxylase)	23	Skow et al. (1988); Creighton et al. (1992); Ma et al. (1996); Barendse et al. (1997); Kappes et al. (1997)
CYP2D@	Cytochrome P450, subfamily IID	5	Barendse et al. (1997)
D1S1	DNA segment (CSRD1613)	1q36–q46	Drinkwater et al. (1992a)
D1S12	DNA segment (IOBL26)	1q45–q46	Solinas Toldo et al. (1995b)
D1S33	DNA segment (BL28)	1q42	Bishop et al. (1994); Grosz et al. (1997)
D1S40	DNA segment (JAB6)	1q41	Williams et al. (1996a); Ferretti et al. (1997)
D1S42	DNA segment (INRA212)	1q12	Ferretti et al. (1997)
D1S48	DNA segment (INRA226, cosA14)	1q45	Eggen et al. (1998)
D1S49	DNA segment (clOBT992)	1q31	Prakash et al. (unpublished)
D1S50	DNA segment (clOBT1449)	1q45	Prakash et al. (unpublished)
D1S51	DNA segment (clOBT539)	1q45	Prakash et al. (unpublished)
D1S58	DNA segment (INRA227, CosAE14)	1q45	Eggen et al. (1998)
D1S63	DNA segment (HAUT33, GT9)	1q31	Ferretti et al. (1997); Thieven et al. (1997)
D1S108	DNA segment (clOBT1441)	1q41	Prakash et al. (1998)
D1S109	DNA segment (clOBT650)	1q43	Prakash et al. (1998)
D2S1	DNA segment (GMBT28)	2q12–q21	Georges et al. (1991)
D2S7	DNA segment (IDVGA2)	2q45	Solinas Toldo et al. (1995b); Williams et al. (1996b); Ferretti et al. (1997)
D2S19	DNA segment (IDVGA37)	2q45	Williams et al. (1996b); Ferretti et al. (1997)
D2S25	DNA segment (IDVGA64)	2q44	Mezzelani et al. (1995)
D2S32	DNA segment (clOBT918, RI918)	2q14–q21	Anastassiadis et al. (1996); Ferretti et al. (1997); Williams et al. (1997b)
D2S33	DNA segment (INRAZARA232, cosA21)	2q45	Eggen et al. (1998)
D2S34	DNA segment (INRAZARA233, cosA33)	2q45	Eggen et al. (1998)

Table 10.4. *Continued.*

Locus symbol	Locus name	Chromosomal assignment	Reference
D2S36	DNA segment (IDVGA72)	2q44	Mezzelani *et al.* (1995)
D2S46	DNA segment (IDVGA80)	2q42	Nijman *et al.* (1996b)
D2S78	DNA segment (cIOBT428)	2q12	Prakash *et al.* (unpublished)
D3S1	DNA segment (BMC24A)	3q37	Bishop *et al.* (1994)
D3S21	DNA segment (BL41)	3q21	Bishop *et al.* (1994); Grosz *et al.* (1997)
D3S24	DNA segment (IDVGA27)	3q37	Williams *et al.* (1996b); Ferretti *et al.* (1997)
D3S25	DNA segment (IDVGA35)	3q35	Williams *et al.* (1996b); Ferretti *et al.* (1997)
D3S26	DNA segment (INRA197)	3q36	Ferretti *et al.* (1997); Grosz *et al.* (1997)
D3S29	DNA segment (IDVGA53)	3q21	Williams *et al.* (1996b); Ferretti *et al.* (1997)
D3S30	DNA segment (IDVGA56)	3q23	Mezzelani *et al.* (1995)
D3S36	DNA segment (INRA236, cosA19)	3q14	Eggen *et al.* (1998)
D3S41	DNA segment (HAUT31, GT8)	3q36	Ferretti *et al.* (1997); Thieven *et al.* (1997)
D3S47	DNA segment (TEXAN23)	3q12	Cai *et al.* (1995)
D3S48	DNA segment (TEXAN27)	3q12	Cai *et al.* (1995)
D4S10	DNA segment (BMC1410)	4q13–q14	Bishop *et al.* (1994)
D4S32	DNA segment (IDVGA51)	4q31	Ferretti *et al.* (1997)
D4S33	DNA segment (IDVGA61S1) 2 MS	4q32	Mezzelani *et al.* (1995)
D4S62	DNA segment (IDVGA61S5)	4q32	Mezzelani *et al.* (1995)
D4S63	DNA segment (cIOBT264)	4q21	Prakash *et al.* (unpublished)
D4S64	DNA segment (cIOBT1447)	4q32	Prakash *et al.* (unpublished)
D5S2	DNA segment (ETH2)	5q35	Barendse *et al.* (1994); Grosz *et al.* (1997)
D5S3	DNA segment (ETH10)	5q25	Solinas Toldo *et al.* (1993)
D5S15	DNA segment (BMC1009)	5q23	Bishop *et al.* (1994); Hawkins *et al.* (1995); Grosz *et al.* (1997)
D5S17	DNA segment (BL37)	5q22	Bishop *et al.* (1994); Grosz *et al.* (1997)
D5S27	DNA segment (IDVGA9)	5q35	Williams *et al.* (1996b); Ferretti *et al.* (1997)
D5S30	DNA segment (JAB2)	5q33/q34	Williams *et al.* (1996a); Ferretti *et al.* (1997)
D5S34	DNA segment (INRA235, cosA10)	5q21	Eggen *et al.* (1998)
D5S35	DNA segment (INRA240, cosA35)	5q25	Eggen *et al.* (1998)
D5S40	DNA segment (cIOBT568)	5q22	Prakash *et al.* (unpublished)
D5S48	DNA segment (IDVGA25), chimeric	5q13	Redaelli *et al.* (1996)
D5S49	DNA segment (HAUT29)	5q33–q34	Thieven *et al.* (1997)
D5S50	DNA segment (cIOBT995)	5q35	Prakash *et al.* (unpublished)
D5S77	DNA segment (cIOBT738)	5q35	Prakash *et al.* (unpublished)
D5S78	DNA segment (cIOBT772)	5q12–q14	Prakash *et al.* (unpublished)
D6Z1	DNA segment (cIOBT33)	6q12–q15	Solinas Toldo *et al.* (1995b)

Table 10.4. *Continued.*

Locus symbol	Locus name	Chromosomal assignment	Reference
D6S3	DNA segment (ETH8)	6q35	Solinas Toldo et al. (1993)
D6S24	DNA segment (clOBT475, FBN3)	6q12–q14	Barendse et al. (1997); Ferretti et al. (1997)
D6S29	DNA segment (IDVGA65S4)	6q22	Harlizius et al. (1996); Ferretti et al. (1997)
D6S31	DNA segment (clOBT450), chimeric	6q21	Nijman et al. (1996b)
D6S61	DNA segment (clOBT962)	6q36	Prakash et al. (unpublished)
D6S62	DNA segment (clOBT705)	6q15	Prakash et al. (unpublished)
D7S20	DNA segment (clOBT930)	7q22–q24	Anastassiadis et al. (1996); Ferretti et al. (1997)
D7S21	DNA segment (IDVGA62A)	7q15	Zhang et al. (1995); Williams et al. (1996b); Ferretti et al. (1997)
D7S24	DNA segment (clOBT1476)	7q12	Prakash et al. (1998)
D7S30	DNA segment (IDVGA90)	7q14/15	Zhang et al. (1995); Williams et al. (1996b); Ferretti et al. (1997)
D7S34	DNA segment (BL5)	7q14	Bishop et al. (1994); Grosz et al. (1997)
D7S35	DNA segment (BL1043)	7q28	Bishop et al. (1994); Grosz et al. (1997)
D7S37	DNA segment (BOBT24)	7q15–q21	Buitkamp et al. (1996)
D8S1	DNA segment (ETH13)	8q21	Solinas Toldo et al. (1993)
D8S21	DNA segment (IDVGA52)	8q23	Williams et al. (1996b); Ferretti et al. (1997)
D8S26	DNA segment (INRA234, cosA40)	8q22–q23	Eggen et al. (1998)
D8S30	DNA segment (clOBT250, chimeric)	8q12	Olsaker et al. (1996b); Barendse et al. (1997); Prakash et al. (1998)
D8S46	DNA segment (UW63)	8q18	Larsen et al. (1997)
D9S8	DNA segment (BMC701)	9q22	Bishop et al. (1994); Hawkins et al. (1995)
D9S14	DNA segment (INRA144)	9q25	Ferretti et al. (1997)
D9S16	DNA segment (CSSM25)	9q17–q21	Johnson et al. (1995)
D10S1	DNA segment (GMBT19)	10q13–q24	Georges et al. (1991); Barendse et al. (1994)
D10S36	DNA segment (JAB10)	10q15	Williams et al. (1996a); Ferretti et al. (1997)
D11S5	DNA segment (ETH9)	11q27	Solinas Toldo et al. (1993)
D11S17	DNA segment (INRA177)	11q16	Ferretti et al. (1997)
D11S27	DNA segment (IDVGA3)	11q23	Williams et al. (1996b); Ferretti et al. (1997)
D11S34	DNA segment (INRA195)	11q25	Ferretti et al. (1997)
D11S37	DNA segment (IDVGA63)	11q24/q25	Mezzelani et al. (1995)
D11S38	DNA segment (JAB7)	11q26	Williams et al. (1996a); Ferretti et al. (1997)
D11S44	DNA segment (INRA241, cosA24)	11q28	Eggen et al. (1998)
D11S45	DNA segment (clOBT911)	11q28	Prakash et al. (unpublished)

Table 10.4. *Continued.*

Locus symbol	Locus name	Chromosomal assignment	Reference
D11S52	DNA segment (IDVGA8), chimeric	11q12/13	Ferretti *et al.* (1994); Mezzelani *et al.* (1995)
D11S53	DNA segment (IDVGA10), chimeric	11q13/14	Ferretti *et al.* (1994); Mezzelani *et al.* (1995)
D11S56	DNA segment (HAUT30)	11q12–q13	Thieven *et al.* (1997)
D11S99	DNA segment (clOBT1459)	11q12	Prakash *et al.* (unpublished)
D11S100	DNA segment (clOBT739)	11q14	Prakash *et al.* (unpublished)
D11S101	DNA segment (clOBT742)	11q16	Prakash *et al.* (unpublished)
D11S102	DNA segment (clOBT782)	11q21	Prakash *et al.* (unpublished)
D12S16	DNA segment (IDVGA41)	12q15	Mezzelani *et al.* (1994); Williams *et al.* (1996b)
D12S17	DNA segment (INRA209/clOBT361	12q26	Bahri-Darwich *et al.* (1994); Ferretti *et al.* (1997)
D12S21	DNA segment (IDVGA57D) 2 MS	12q13	Mezzelani *et al.* (1995)
D12S22	DNA segment (HAUT1)	12q25	Ferretti *et al.* (1997); Barendse *et al.* (1997); Thieven *et al.* (1997)
D12S25	DNA segment (clOBT323)	12q24	Jørgensen *et al.* (1995)
D12S26	DNA segment (clOBT985)	12q11–q12	Ferretti *et al.* (1997)
D12S29	DNA segment (IDVGA57A) 2 MS	12q13	Mezzelani *et al.* (1995); Ferretti *et al.* (1997)
D12S34 (FLT1)	DNA segment (UW56, FLT1(2))	12q15	Sun *et al.* (1997)
D12S35 (FLT1)	DNA segment (UW57, FLT1(3))	12q15	Sun *et al.* (1997)
D12S37 (RB1)	DNA segment (UW59, RB1(2))	12q13	Sun *et al.* (1997)
D12S39 (F10)	DNA segment (UW61, F10(1))	12q23–qter	Sun *et al.* (1997)
D12S40 (F10)	DNA segment (UW62, F10(2))	12q23–qter	Sun *et al.* (1997)
D13S1	DNA segment (ETH7)	13q21–q22	Grosz *et al.* (1997)
D13S11	DNA segment (BL42)	13q22	Bishop *et al.* (1994); Grosz *et al.* (1997)
D13S13	DNA segment (BMC1222)	13q12–q13	Bishop *et al.* (1994); Hawkins *et al.* (1995); Grosz *et al.* (1997)
D13S16	DNA segment (JAB3)	13q23	Williams *et al.* (1996a); Ferretti *et al.* (1997)
D13S20	DNA segment (INRA225, cosA18)	13q13	Eggen *et al.* (1998)
D13S21	DNA segment (INRA196, CosAE27)	13q24	Eggen *et al.* (1998)
D13S57	DNA segment (clOBT1437)	13q12	Prakash *et al.* (unpublished)
D14S1	DNA segment (GMBT6)	14q11–q16	Georges *et al.* (1991)
D14S34	DNA segment (INRA237, cosA28)	14q12	Eggen *et al.* (1998)
D14S35	DNA segment (IDVGA76)	14q14	Williams *et al.* (1996b, c); Ferretti *et al.* (1997)
D15S5	DNA segment (INRA046)	15q25	Vaiman *et al.* (1993)

Table 10.4. *Continued.*

Locus symbol	Locus name	Chromosomal assignment	Reference
D15S13	DNA segment (IDVGA10), chimeric	15q25	Williams *et al.* (1996b); Ferretti *et al.* (1997)
D15S14	DNA segment (IDVGA28)	15q25	Mezzelani *et al.* (1995)
D15S16	DNA segment (IDVGA32)	15q25	Williams *et al.* (1996b)
D15S17	DNA segment (IDVGA60)	15q21	Mezzelani *et al.* (1995)
D15S18	DNA segment (JAB1)	15q14	Williams *et al.* (1995, 1996a); Ferretti *et al.* (1997)
D15S19	DNA segment (JAB4)	15q21/q22	Williams *et al.* (1995, 1996a)
D15S20	DNA segment (JAB8)	15q21	Williams *et al.* (1995, 1996a); Ferretti *et al.* (1997)
D15S22	DNA segment (INRA224, cosA26)	15q21	Eggen *et al.* (1998)
D15S23	DNA segment (cIOBT395)	15q27–28	Nijman *et al.* (1996a); Ferretti *et al.* (1997)
D15S27	DNA segment (IDVGA23)	15q27	Redaelli *et al.* (1996)
D15S28	DNA segment (IDVGA11), chimeric	15q25	Ferretti *et al.* (1994); Mezzelani *et al.* (1995)
D15S29	DNA segment (HAUT23, GT1)	15q27	Thieven *et al.* (1997)
D16S5	DNA segment (ETH11)	16q21	Solinas Toldo *et al.* (1993); Barendse *et al.* (1994)
D16S21	DNA segment (IDVGA49)	16q17	Mezzelani *et al.* (1995)
D16S23	DNA segment (IDVGA68)	16q16	Tammen and Ferretti (1996); Ferretti *et al.* (1997)
D16S24	DNA segment (IDVGA69)	16q21	Mezzelani *et al.* (1995)
D16S25	DNA segment (JAB9)	16q17	Williams *et al.* (1996a)
D16S26	DNA segment (INRA218, cosA15)	16q16	Eggen *et al.* (1998)
D16S28	DNA segment (cIOBT450), chimeric	16q21–q22	Nijman *et al.* (1996b)
D16S30	DNA segment (IDVGA26)	16q21	Mezzelani *et al.* (1995)
D16S32	DNA segment (IDVGA25C), chimeric	16q15	Redaelli *et al.* (1996)
D16S33	DNA segment (IDVGA25D), chimeric	16q15	Redaelli *et al.* (1996)
D16S34	DNA segment (IDVGA66B)	16q12	Williams *et al.* (1996c); Ferretti *et al.* (1997)
D17S2	DNA segment (ETH4)	17q26	Solinas Toldo *et al.* (1993)
D17S23	DNA segment (cIOBT1482)	17q12–q14	Monteagudo *et al.* (1994)
D17S24	DNA segment (IDVGA40)	17q23	Williams *et al.* (1996b); Ferretti *et al.* (1997)
D17S29	DNA segment (cIOBT975, IOZARA975)	17q26	Ferretti *et al.* (1997)
D17S39	DNA segment (HAUT40, GT12)	17q15	Thieven *et al.* (1997)
D18S10	DNA segment (IDVGA31)	18q12/q13	Williams *et al.* (1996b); Ferretti *et al.* (1997)
D18S16	DNA segment (IDVGA55)	18q24	Ferretti *et al.* (1997)
D18S17	DNA segment (HAUT14)	18q21	Barendse *et al.* (1997); Ferretti *et al.* (1997); Thieven *et al.* (1997)

Table 10.4. *Continued.*

Locus symbol	Locus name	Chromosomal assignment	Reference
D18S20	DNA segment (INRA228, cosA16)	18q15–q21	Eggen *et al.* (1998)
D19S1	DNA segment (GMBT22)	19q17–q23	Georges *et al.* (1991)
D19S2	DNA segment (ETH3)	19q23	Barendse *et al.* (1994); Yang and Womack (1996)
D19S3	DNA segment (ETH12)	19q17	Solinas Toldo *et al.* (1993)
D19S4	DNA segment (cIOBT34)	19q22	Olsaker *et al.* (1996a); Ferretti *et al.* (1997)
D19S14	DNA segment (BMC4213)	19q13	Bishop *et al.* (1994)
D19S15	DNA segment (BMC1012)	19q22	Bishop *et al.* (1994)
D19S16	DNA segment (IDVGA42)	19q15	Mezzelani *et al.* (1995)
D19S17	DNA segment (IDVGA44)	19q22	Williams *et al.* (1996b); Ferretti *et al.* (1997)
D19S18	DNA segment (IDVGA46)	19q16	Williams *et al.* (1996b); Ferretti *et al.* (1997)
D19S19	DNA segment (IDVGA47)	19q17	Mezzelani *et al.* (1995)
D19S21	DNA segment (BMC6059)	19q	Bishop *et al.* (1994)
D19S24	DNA segment (IDVGA48B) 2 MS	19q21	Mezzelani *et al.* (1995)
D19S25	DNA segment (IDVGA54)	19q17	Mezzelani *et al.* (1995)
D19S31	DNA segment (INRA238, cosA44)	19q15	Eggen *et al.* (1998)
D19S32	DNA segment (cIOBT965)	19q17–q21	Anastassiadis *et al.* (1996); Barendse *et al.* (1997)
D19S48	DNA segment (IDVGA48E) 2 MS	19q21	Mezzelani *et al.* (1995)
D19S70	DNA segment (cIOBT1457)	19q17	Prakash *et al.* (unpublished)
D21S1	DNA segment (GMBT16)	21q21–q24	Georges *et al.* (1991); Barendse *et al.* (1994)
D21S3	DNA segment (GMBT15)	21q22–q24	Georges *et al.* (1991)
D21S9	DNA segment (ETH1)	21q17	Solinas Toldo *et al.* (1993)
D21S23	DNA segment (IDVGA30)	21q23	Williams *et al.* (1996b); Ferretti *et al.* (1997)
D21S24	DNA segment (IDVGA39)	21q23 (proximal)	Williams *et al.* (1996b); Ferretti *et al.* (1997)
D21S25	DNA segment (IDVGA45.1) 2 MS	21q15	Mezzelani *et al.* (1995); Williams *et al.* (1996b); Ferretti *et al.* (1997)
D21S39	DNA segment (IDVGA45.2) 2 MS	21q15	Mezzelani *et al.* (1995)
D21S46	DNA segment (HAUT28)	21q15/q16	Ferretti *et al.* (1997); Thieven *et al.* (1997)
D22S19	DNA segment (INRA194)	22q13	Barendse *et al.* (1997); Ferretti *et al.* (1997)
D22S43	DNA segment (cIOBT754)	22q24	Prakash *et al.* (unpublished)
D23S27	DNA segment (cIOBT528)	23q14–q15	Barendse *et al.* (1997); Ferretti *et al.* (1997)
D23S28	DNA segment (INRA145, CosAE11)	23q22–q23	Eggen *et al.* (1998)

Table 10.4. *Continued.*

Locus symbol	Locus name	Chromosomal assignment	Reference
D23S38	DNA segment (clOBT1479, RI1479)	23q25	Ferretti *et al.* (1997); Prakash *et al.* (unpublished)
D24S1	DNA segment (GMBT5)	24q12–q24	Georges *et al.* (1991); Barendse *et al.* (1994)
D24S9	DNA segment (BMC1404)	24q12–q13.1	Hawkins *et al.* (1995)
D24S11	DNA segment (BL6)	24q12	Grosz *et al.* (1997)
D24S13	DNA segment (clOBT1401)	24q21–q23	Barendse *et al.* (1997); Ferretti *et al.* (1997)
D24S16	DNA segment (INRA223, cosA7)	24q23	Eggen *et al.* (1998)
D25S1	DNA segment (BMC4216)	25q13	Hawkins *et al.* (1995)
D25S6	DNA segment (clOBT355, INRA206)	25q13	Bahri-Darwich *et al.* (1994); Barendse *et al.* (1997)
D25S11	DNA segment (cosA20, INRA222)	25q19	Ferretti *et al.* (1997)
D25S12	DNA segment (IDVGA71)	25q12/13	Ferretti *et al.* (1997); Zhang *et al.* (1995)
D25S19	DNA segment (HAUT39, GT7)	25q17	Ferretti *et al.* (1997); Thieven *et al.* (1997)
D26S1	DNA segment (GMBT11)	26q11–q21	Georges *et al.* (1991); Georges and Massey (1992)
D26S14	DNA segment (IDVGA59)	26q22	Ferretti *et al.* (1997); Williams *et al.* (1996b)
D26S23	DNA segment (clOBT730)	26q22	Olsaker *et al.* (1997b); Prakash *et al.* (unpublished)
D27S1	DNA segment (BM203)	27q23	Bishop *et al.* (1994)
D27S14	DNA segment (clOBT313)	27q23	Barendse *et al.* (1997); Ferretti *et al.* (1997); Olsaker *et al.* (1997a); Prakash *et al.* (unpublished)
D28S10	DNA segment (IDVGA8) chimeric	28q18–q19	Williams *et al.* (1996b); Ferretti *et al.* (1997)
D28S11	DNA segment (BMC1002)	28q14	Hawkins *et al.* (1995)
D28S12	DNA segment (IDVGA29)	28q13	Williams *et al.* (1996b); Ferretti *et al.* (1997)
D28S17	DNA segment (INRA201)	28q12	Ferretti *et al.* (1997)
D28S18	DNA segment (IDVGA43)	28q17	Williams *et al.* (1996b); Ferretti *et al.* (1997)
D28S21	DNA segment (IDVGA88)	28q16	Zhang *et al.* (1995)
D29S1	DNA segment (BMC8012)	29q15	Bishop *et al.* (1994)
D29S2	DNA segment (IDVGA7)	29q24	Mezzelani *et al.* (1994); Williams *et al.* (1996b); Ferretti *et al.* (1997)
D29S6	DNA segment (BMC3224)	29q24	Bishop *et al.* (1994)
D29S8	DNA segment (IDVGA2)	29q15	Solinas Toldo *et al.* (1995b)
D29S10	DNA segment (BMC1206)	29q24	Bishop *et al.* (1994)
D29S12	DNA segment (JAB5)	29q12	Williams *et al.* (1996a); Ferretti *et al.* (1997)
D29S23	DNA segment (INRA142, CosAE4)	29q12	Eggen *et al.* (1998)

Table 10.4. *Continued*.

Locus symbol	Locus name	Chromosomal assignment	Reference
D29S24	DNA segment (INRA211, CosAE17)	29q21	Eggen *et al.* (1998)
D29S25	DNA segment (INRA176, CosAE2)	29q21	Eggen *et al.* (1998)
DXS12	DNA segment (BM4604, BAC204)	Xq26–q31	Bishop *et al.* (1994); Yeh *et al.* (1996)
DXS19	DNA segment (clOBT949)	Xq26–q31	Friedl *et al.* (1994c)
DXS22	DNA segment (clOBT945)	Xq26-q31	Prakash *et al.* (1996, 1997)
DXS30	DNA segment (IDVGA82)	Xq34	Zhang *et al.* (1995)
DXYS4	DNA segment (clOBT1489, IOZARA1489)	Xq43	Martín-Burriel *et al.* (1996); Prakash *et al.* (1996, 1997); Barendse *et al.* (1997); Ferretti *et al.* (1997)
DXYS5	DNA segment (INRA030, BAC57)	Xq42–q43, Yp13–pter	Yeh *et al.* (1996)
DBH	Dopamine beta-hydroxylase (dopamine beta-monooxygenase)	11	Barendse *et al.* (1994, 1997)
DCT (TYRP2)	Tyrosinase-related protein-2	12q23	Hawkins *et al.* (1996); Sun *et al.* (1996a, 1997)
DDC	Dopa decarboxylase (aromatic L-amino acid decarboxylase)	4	Barendse *et al.* (1997)
DEFB@	Beta-defensin	27q12	Gallagher *et al.* (1995)
DIA4	Diaphorase 4	18	Womack *et al.* (1989)
DMD	Muscular dystrophy	X	Livingston (1988)
DNMT	DNA methyltransferase	7q15	Comincini *et al.* (1997)
DNTT (TDT)	Deoxynucleotidyl-transferase, terminal	26	Barendse *et al.* (1997)
DSC1	Desmocollin 1	24q21/q22	Solinas Toldo *et al.* (1995c)
DSC2	Desmocollin 2	24	Solinas Toldo *et al.* (1995c); Agaba *et al.* (1997a)
DSC3	Desmocollin 3	24	Solinas Toldo *et al.* (1995c); Barendse *et al.* (1997)
EAA	A blood group (erythrocyte antigen A)	15	Larsen (1965); Ma *et al.* (1996); Kappes *et al.* (1997)
EAB	B blood group (erythrocyte antigen B)	12	Georges *et al.* (1990); Grosclaude *et al.* (1990); Hines and Larsen (1990); Bishop *et al.* (1994); Ma *et al.* (1996); Kappes *et al.* (1997)
EAC	C blood group (erythrocyte antigen C)	18	Bishop *et al.* (1994); Ma *et al.* (1996); Kappes *et al.* (1997)
EAF	F blood group (erythrocyte antigen F)		Georges *et al.* (1990); Grosclaude *et al.* (1990); Hines and Larsen (1990)
EAJ	J blood group (erythrocyte antigen J)	11	Hines *et al.* (1969); Larsen (1970); Ma *et al.* (1996)
EAL	L blood group (erythrocyte antigen L)	3	Bishop *et al.* (1994); Ma *et al.* (1996); Beever *et al.* (1997b); Kappes *et al.* (1997)

Table 10.4. *Continued.*

Locus symbol	Locus name	Chromosomal assignment	Reference
EAM	M blood group (erythrocyte antigen M)	23	Leveziel and Hines (1984); Lindberg and Andersson (1988); Kappes *et al.* (1997)
EAR'	R blood group (erythrocyte antigen R')	16	Bishop *et al.* (1994); Ma *et al.* (1996); Kappes *et al.* (1997)
EAS	S blood group (erythrocyte antigen S)	21	Georges *et al.* (1987, 1990); Grosclaude *et al.* (1990); Hines and Larsen (1990); Bishop *et al.* (1994); Ma *et al.* (1996); Kappes *et al.* (1997)
EAT'	T' blood group (erythrocyte antigen T')	19	Bishop *et al.* (1994); Ma *et al.* (1996); Kappes *et al.* (1997)
EAZ	Z blood group (erythrocyte antigen Z)	10	Georges *et al.* (1990); Grosclaude *et al.* (1990); Hines and Larsen (1990); Larsen *et al.* (1992); Bishop *et al.* (1994); Ma *et al.* (1996); Taylor *et al.* (1997)
EDNRB	Endothelin receptor type B	12q22	Schläpfer *et al.* (1997b)
EEF2 (EF2)	Eucaryotic translation elongation factor 2	7q15	Gao and Womack (1996)
EGF	Epidermal growth factor	6	Lanneluc *et al.* (1996); Barendse *et al.* (1997)
ELN	Elastin	25	Dietz (1992); Echard *et al.* (1994)
ENO1	Enolase 1	16	Heuertz and Hors-Cayla (1981); Echard *et al.* (1984); Womack and Moll (1986)
EPO	Erythropoietin	25	Barendse *et al.* (1997)
EPOR	Erythropoietin receptor	7	Gao and Womack (1996)
ESD	Esterase D/ formylglutathione hydrolase	12	Sun *et al.* (1996b)
ESR	Oestrogen receptor	9q25–q27	Womack *et al.* (1991); Ansari *et al.* (1994)
ETS2	Avian erythroblastosis virus E26 (v-ets) oncogene homologue 2	1	Threadgill *et al.* (1991)
EVI1	Ecotropic viral integration site 1	1	Broad *et al.* (1994)
F10	Coagulation factor X	12	Dietz *et al.* (1992a); Pearce *et al.* (1995); Sun *et al.* (1997)
F11	Coagulation factor 11 (plasma thromboplastin antecedent)	17	Zhang *et al.* (1992)
F13A1	Coagulation factor XIII, A1 polypeptide	23	Park *et al.* (1994); Ma *et al.* (1996)
F9	Coagulation factor IX (plasma thromboplastic component)	Xq33–q34	Livingston (1988)

Table 10.4. *Continued.*

Locus symbol	Locus name	Chromosomal assignment	Reference
FABP3L	Fatty acid-binding protein 3, muscle- and heart-like	6	Barendse *et al.* (1997)
FBN1	Fibrillin 1	10q26	Tilstra *et al.* (1994); Masabanda *et al.* (1998)
FCG2R (FCAR?)	Fragment crystallizable (Fc) gamma 2 receptor	18	Klungland *et al.* (1997a)
FCGR1	Fc fragment of IgG, high affinity, receptor for (CD64)	3	Bishop *et al.* (1994); Barendse *et al.* (1997)
FCGR2	Fc fragment of IgG, low affinity, receptor for (CD32)	3	Klungland *et al.* (1997b)
FCGR3	Fc fragment of IgG, high affinity, receptor for (CD16)	3	Klungland *et al.* (1997b)
FECH	DNA segment (ferrochelatase (protoporphyria)	24	Straka *et al.* (1991); Agaba *et al.* (1997a); Barendse *et al.* (1997)
FES	Feline sarcoma viral (v-fes) oncogene; v-fps oncogene homologue	21	Tobin-Janzen and Womack (1992)
FGB	Fibrinogen, B beta polypeptide	17	Zhang *et al.* (1992)
FGF1	Fibroblast growth factor, acidic (endothelial growth factor)	7	Zhang and Womack (1992); Barendse *et al.* (1997); Gao and Womack (1997)
FGF2	Fibroblast growth factor, basic	17q23–q25	Pearce *et al.* (1995); Barendse *et al.* (1997); Fisher *et al.* (1997b)
FGF5	Fibroblast growth factor 5	6	Lanneluc *et al.* (1996)
FGFR3	Fibroblast growth factor receptor 3	6	Teres *et al.* (1996); Barendse *et al.* (1997); Usha *et al.* (1997)
FGG	Fibrinogen, gamma polypeptide	17q12–q23	Zhang *et al.* (1992); Johnson *et al.* (1993); Barendse *et al.* (1997)
FGR	Gardner–Rasheed feline sarcoma viral (v-fgr) oncogene homologue	2	Threadgill (1990); Beever *et al.* (1992); Threadgill *et al.* (1994); Ma *et al.* (1996)
FH	Fumarate hydratase	16	Threadgill (1990); Barendse *et al.* (1997)
FIM3	Friend–murine leukaemia virus integration site 3 homologue	1	Threadgill and Womack (1991c)
FLT1	Fms-related tyrosine kinase (vascular endothelial growth factor/vascular permeability factor receptor)	12q15	Sun *et al.* (1996b, 1997)
FN1	Fibronectin 1	2	Adkison *et al.* (1988b); Ma *et al.* (1996); Barendse *et al.* (1997)

Table 10.4. *Continued.*

Locus symbol	Locus name	Chromosomal assignment	Reference
FOS	Murine FBJ osteosarcoma viral (v-fos) oncogene homologue	10	Miller *et al.* (1991); Tobin-Janzen and Womack (1992)
FSHB	Follicle-stimulating hormone, beta polypeptide	15q24–q29	Fries (1989); Barendse *et al.* (1991, 1997); Hediger *et al.* (1991); Kappes *et al.* (1997)
FTZF1 (FTZ1, SF1, SHR)	Fushi tarazu factor (*Drosophila*) homologue 1, steroid hormone receptor	11	Barendse *et al.* (1997)
FUCA1	Fucosidase, alpha-L-1, tissue	2	Beever *et al.* (1992); Threadgill *et al.* (1994); Ma *et al.* (1996)
FUT3	Fucosyltransferase 3 (galactoside 3(4)-L-fucosyltransferase, Lewis blood group included)	7q15	Oulmouden *et al.* (1995)
G6PD	Glucose-6-phosphate dehydrogenase	X	Heuertz and Hors-Cayla (1978); Förster *et al.* (1980); Shimizu *et al.* (1981); Womack and Moll (1986)
GAA	Glucosidase, alpha; acid (glycogen storage disease type II, Pompe's disease)	19	Barendse *et al.* (1991, 1994); Drinkwater *et al.* (1992b)
GABRA2	Gamma-aminobutyric acid (GABA) A receptor, alpha 2	6	Barendse *et al.* (1994, 1997)
GAP43	Growth-associated protein 43	1	Threadgill *et al.* (1991)
GAPD	Glyceraldehyde-3-phosphate dehydrogenase	5	Womack and Moll (1986)
GART	Phosphoribosylglycinamide formyltransferase, phosphoribosylglycinamide synthetase, phosphoribosylamino imidazole synthetase	1	McAvin *et al.* (1988)
GC	Group-specific component (vitamin D-binding protein)	6	Bouquet *et al.* (1986); Bishop *et al.* (1994); Kappes *et al.* (1997)
GCG	Glucagon	2q21–q22	Lopez-Corrales *et al.* (1997)
GCSFR	Colony-stimulating factor 3 receptor (granulocyte)	3	Yoo *et al.* (1995a); Kappes *et al.* (1997)
GDF8 (MSTN, MH)	Growth differentiation factor 8 (myostatin, muscular hypertrophy)	2q14–q15	Grobet *et al.* (1997)
GDH	Glucose dehydrogenase	5	Womack and Moll (1986); Monteagudo *et al.* (1992)
GFAP	Glial fibrillary acidic protein	19	Bishop *et al.* (1994); Barendse *et al.* (1997); Kappes *et al.* (1997)

Table 10.4. *Continued.*

Locus symbol	Locus name	Chromosomal assignment	Reference
GGTA1 (GGTA, GLYT2)	Glycoprotein, alpha-galactosyltransferase 1	11q26	Hayes et al. (1996); Barendse et al. (1997)
GGTB2	Glycoprotein-4-beta-galactosyltransferase 2	8	Threadgill and Womack (1990a); Barendse et al. (1997)
GH	Growth hormone	19q17–q23	Hediger et al. (1990); Barendse et al. (1991, 1997); Miller et al. (1991); Yang and Womack (1996); Kappes et al. (1997); Yang et al. (1998)
GHR	Growth hormone receptor	20	Moody et al. (1995c); Barendse et al. (1997)
GHRH	Growth hormone-releasing hormone	13	Moody et al. (1995a); Barendse et al. (1997)
GJA1 (CX43)	Gap-junction protein, alpha 1, 43 kDa (connexin 43)	9q15/q16	Castiglioni et al. (1996a); Barendse et al. (1997)
GJB1 (CX32)	Gap-junction protein, beta 1, 32 kDa (connexin 32)	Xq22	Castiglioni et al. (1996a)
GLA	Galactosidase alpha	X	Heuertz and Hors-Cayla (1978); Shimizu et al. (1981)
GLI	Glioma-associated oncogene homologue (zinc finger protein)	5	Threadgill and Womack (1990b)
GLNS	Glutamate-ammonia ligase (glutamine synthase)	10q33	Masabanda et al. (1997)
GLNSP1	Glutamate-ammonia ligase (glutamine synthase), pseudogene 1	16q21	Masabanda et al. (1997)
GLO1	Glyoxalase I	23	Womack and Moll (1986)
GNAS1	Guanine nucleotide-binding protein (G protein), alpha-stimulating activity polypeptide 1	13	Gao et al. (1997); Schläpfer et al. (1997a)
GNAZ	Guanine nucleotide-binding protein (G protein), alpha z polypeptide	22	Aleyasin and Barendse (1997b)
GNB1	G-protein beta 1 subunit	16	Threadgill et al. (1994)
GNRHR (LHRHR)	Gonadotrophin-releasing hormone receptor (luteinizing-releasing hormone receptor)	6	Lord et al. (1996)
GOT1 (ASAT)	Glutamic-oxaloacetic transaminase 1, soluble (aspartate aminotransferase 1)	26	Barendse et al. (1997)
GPI	Glucosephosphate isomerase	18q22–q24	Echard et al. (1984); Womack and Moll (1986); Chowdhary et al. (1991)

Table 10.4. *Continued*.

Locus symbol	Locus name	Chromosomal assignment	Reference
GPX1	Glutathione peroxidase 1	22	Threadgill and Womack (1991c); Barendse *et al.* (1997)
GRO1 (MGSA)	GRO1 oncogene (melanoma growth-stimulating activity)	6	Lord *et al.* (1996)
GRP	Gastrin-releasing peptide	24	Larsen *et al.* (1996)
GRP78	Glucose-regulated protein (78 kDa)	11	Threadgill and Womack (1990a)
GSN	Gelsolin	8q28	Threadgill and Womack (1990a); Broad *et al.* (1995b)
GSNL1	Gelsolin-like	5	Barendse *et al.* (1992)
GSR	Glutathione reductase	27	Womack and Moll (1986)
GSTP1	Glutathione S-transferase pi	29	Barendse *et al.* (1997)
GUC2C (GUCY2C)	Guanylate cyclase 2C (heat stable enterotoxin receptor)	17	Barendse *et al.* (1997)
GUSB	Beta-glucuronidase	25	Womack *et al.* (1989)
HBB	Haemoglobin, beta	15q22–q27	Larsen (1965); Fries *et al.* (1988); Foreman and Womack (1989); Barendse *et al.* (1991, 1997); Kappes *et al.* (1997)
HCK	Haemopoietic cell kinase	13	Womack *et al.* (1991); Schläpfer *et al.* (1997a)
HEXA	Hexosaminidase A (alpha polypeptide)	10	Womack *et al.* (1989)
HEXB	Hexosaminidase B (beta polypeptide)	20	Barendse *et al.* (1994, 1997)
HF1 (HF)	H factor 1 (complement)	16	Williams *et al.* (1997a)
HK1	Hexokinase 1	28	Threadgill and Womack (1991a)
HMGCR	3-Hydroxy-3-methylglutaryl-coenzyme A reductase	10	Zhang and Womack (1992)
HOXA@	Homoeobox region 1	4	Womack *et al.* (1989)
HOXB@	Homoeobox region 2	19q17–q23	Womack *et al.* (1989); Barendse *et al.* (1991); Miller *et al.* (1991); Gunawardana and Fries (1992)
HOXC@	Homoeobox region 3	5q13–q22	Threadgill and Womack (1990b); Barendse *et al.* (1992); Gunawardana and Fries (1992)
HP	Haptoglobin	18	Womack *et al.* (1991)
HPRT	Hypoxanthine-guanine phosphoribosyltransferase	X	Heuertz and Hors-Cayla (1978); Förster *et al.* (1980); Shimizu *et al.* (1981)
HRH1	Histamine receptor, H1	22	Bishop *et al.* (1994); Barendse *et al.* (1997); Kappes *et al.* (1997)
HSD3B	Hydroxy-delta-5-steroid dehydrogenase, 3 beta- and steroid delta-isomerase	3q12	Cai *et al.* (1995)
HSP70-1 (HSPA1)	Heat-shock 70 kDa protein 1	23q21–q22	Gallagher *et al.* (1992b, 1993b); Grosz *et al.* (1992); Kappes *et al.* (1997)

Table 10.4. *Continued.*

Locus symbol	Locus name	Chromosomal assignment	Reference
HSP70-2	Heat-shock 70 kDa protein 2	23q21–q22	Gallagher *et al.* (1992b, 1993b); Grosz *et al.* (1992)
HSP70-3	Heat-shock 70 kDa protein 3 (HSPA2 ?, protein 2)	10q34	Grosz *et al.* (1992); Gallagher *et al.* (1993b)
HSP70-4 (HSPA5, HSPA6 or HSPA7)	Heat-shock 70 kDa protein 4 (human: HSPA6 or HSPA7)	3q13	Grosz *et al.* (1992)
HTR1A	5-Hydroxytryptamine (serotonin) receptor	20	Barendse *et al.* (1994, 1997)
HTR2A	5-Hydroxytryptamine (serotonin) receptor 2A	12	Sun *et al.* (1996b)
IDH1	Isocitrate dehydrogenase, soluble	2	Heuertz and Hors-Cayla (1981); Womack and Moll (1986)
IF	Complement component 1	6	Zhang *et al.* (1992); Barendse *et al.* (1997)
IFNA	Interferon, alpha, leucocyte	8q15	Adkison *et al.* (1988a); Ryan *et al.* (1992a, b)
IFNAR	Interferon, alpha; receptor	1	Langer *et al.* (1992); Barendse *et al.* (1997)
IFNB1	Interferon, beta 1, fibroblast	8q15	Adkison *et al.* (1988a); Ryan *et al.* (1992a,b)
IFNG	Interferon, gamma or immune type	5q22–q24	Threadgill and Womack (1990b); Chaudhary *et al.* (1993); Barendse *et al.* (1997)
IFNT	Interferon, trophoblast	8q15	Adkison *et al.* (1988a); Iannuzzi *et al.* (1993b); Ryan *et al.* (1993a)
IFNW	Interferon omega	8q15	Adkison *et al.* (1988a); Iannuzzi *et al.* (1993); Ryan *et al.* (1993a)
IGF1	Insulin-like growth factor I (somatomedin C)	5	Bishop *et al.* (1991); Miller *et al.* (1991); Barendse *et al.* (1997); Ge *et al.* (1997); Kappes *et al.* (1997)
IGF1R	Insulin-like growth factor 1 receptor	21	Womack *et al.* (1991); Barendse *et al.* (1997)
IGF2	Insulin-like growth factor 2	29q21–q24	Womack *et al.* (1991); Ansari *et al.* (1994); Schmutz *et al.* (1996a)
IGF2R	Insulin-like growth factor II receptor	9q27–q28	Friedl and Rottmann (1994a)
IGFBP3	Insulin-like growth factor-binding protein 3	4	Maciulla *et al.* (1997)
IGH@	Immunoglobulin heavy-chain gene cluster	21q23–q24	Miller *et al.* (1991); Gu *et al.* (1992); Ponce de Leon *et al.* (1992); Tobin-Janzen and Womack (1992)
IGHA	Immunoglobulin alpha	21q24	Ponce de Leon *et al.* (1992)
IGHG1	Immunoglobulin gamma 1	21q24	Gu *et al.* (1992); Tobin-Janzen and Womack (1992)

Table 10.4. *Continued.*

Locus symbol	Locus name	Chromosomal assignment	Reference
IGHG2	Immunoglobulin gamma 2	21q23–q24	Miller et al. (1991); Gu et al. (1992); Ponce de Leon et al. (1992)
IGHG3	Immunoglobulin gamma 3	21q24	Gu et al. (1992); Ponce de Leon et al. (1992)
IGHM	Immunoglobulin mu	21	Tobin-Janzen and Womack (1992)
IGHML1	Immunoglobulin mu-like 1	11q23	Tobin-Janzen and Womack (1992)
IGJ	Immunoglobulin J polypeptide	6	Zhang et al. (1992); Fisher et al. (1997b)
IGL@	Immunoglobulin light chain, lambda gene cluster	17	Womack et al. (1989)
IL10	Interleukin 10	16	Beever et al. (1997a, b)
IL1RA	Interleukin 1 receptor, alpha; type I receptor	11	Broom and Hill (1994); Kappes et al. (1997)
IL1RB	Interleukin 1 receptor, beta; type II receptor	11	Kappes et al. (1997)
IL2	Interleukin 2	17q22–q23	Zhang et al. (1992); Chowdhary et al. (1994); Fisher et al. (1997b)
IL2RA	Interleukin 2 receptor, alpha	13q13–q14	Threadgill and Womack (1991a); Yoo et al. (1995b); Schläpfer et al. (1997a)
IL2RG	Interleukin 2 receptor, gamma	Xq22–23	Yoo et al. (1996b)
IL3	Interleukin 3 (colony-stimulating factor, multiple)	7q15/q21	Hawken et al. (1996)
IL4	Interleukin 4	7q15/q21	Buitkamp et al. (1995); Barendse et al. (1997); Kappes et al. (1997)
IL5	Interleukin 5 (colony-stimulating factor, eosinophil)	7q15/q21	Hawken et al. (1996)
IL6 (IFNB2)	Interleukin 6 (interferon, beta 2)	4	Barendse et al. (1997)
IL7	Interleukin 7	14	Barendse et al. (1997)
INHA	Inhibin, alpha	2q36–q42	Barendse et al. (1994, 1997); Brunner et al. (1995b)
INHBA	Inhibin, beta A	4q26	Womack et al. (1991); Neibergs et al. (1993)
INHBB	Inhibin, beta B (activin AB beta polypeptide)	2q31–q33	Goldammer et al. (1995)
INSR	Insulin receptor	7	Womack et al. (1991)
ITGB2 (CD18, BLAD)	Integrin, beta 2 (antigen CD18 (p95), lymphocyte function-associated antigen 1; macrophage antigen 1 (mac-1) beta subunit), leucocyte adhesion deficiency	1	Threadgill et al. (1991); Shuster et al. (1992); Czarnik and Kamiński (1997)

Table 10.4. *Continued.*

Locus symbol	Locus name	Chromosomal assignment	Reference
ITIL	Inter-alpha-trypsin inhibitor (protein HC), light	8	Threadgill and Womack (1990a)
ITPA	Inosine triphosphatase A	13	Womack and Moll (1986)
IVL	Involucrin	1q41–q46	Schmutz *et al.* (1998)
JUNB	Jun B proto-oncogene	7	Gao and Womack (1996)
JUND	Jun D proto-oncogene	7	Gao and Womack (1996)
KAP8	Keratin-associated protein 8	1	Barendse *et al.* (1997)
KCNA4	Potassium voltage-gated channel, shaker-related subfamily, member 4	15	Byrne *et al.* (1996); Barendse *et al.* (1997); Kappes *et al.* (1997)
KIT	Hardy–Zuckerman 4 feline sarcoma viral (v-kit) oncogene homologue	6	Zhang *et al.* (1992); Barendse *et al.* (1997)
KNG	Kininogen	1	Aleyasin and Barendse (1997a); Barendse *et al.* (1997)
KRAS2	Kirsten rat sarcoma 2 viral (v-Ki-ras2) oncogene homologue	5	Threadgill and Womack (1990b); Barendse *et al.* (1992)
KRT1	Cytokeratin Ia (human cytokeratin 1)	5q14–q23	Fries *et al.* (1991)
KRT10	Cytokeratin VI (human cytokeratin 10)	19q16–q23	Fries *et al.* (1991); Kappes *et al.* (1997); Yang *et al.* (1998)
KRT10L1	Cytokeratin 51-like (human cytokeratin 10-like)	18q12–q15	Biltueva *et al.* (1996)
KRT19	Cytokeratin 22 (human cytokeratin 19)	19q16–q23	Fries *et al.* (1991)
KRT1L1	Cytokeratin Ia-like 1 (human cytokeratin 1-like)	5q25–q33	Fries *et al.* (1991)
KRT3	Cytokeratin Ib (human cytokeratin 3)	5q14–q23	Fries *et al.* (1991)
KRT5	Cytokeratin III (human cytokeratin 5)	5q14–q23	Fries *et al.* (1991)
KRT5L1	Cytokeratin III-like 1 (human cytokeratin 5-like)	5q25–q33	Fries *et al.* (1991)
KRT8	Cytokeratin 8 (human cytokeratin 8)	5q14–q23	Fries *et al.* (1991)
KRT8L1	Cytokeratin 8-like 1 (human cytokeratin 8-like)	10q31–q36	Fries *et al.* (1991)
KRTA@	Cytokeratin class I gene cluster (KRT10, KRT19)	19q16–q23	Fries *et al.* (1991)
KRTB@	Cytokeratin class II gene cluster (KRT1, KRT3, KRT5, KRT8)	5q14–q23	Fries *et al.* (1991)
LALBA	Lactalbumin, alpha	5q21	Threadgill and Womack (1990c); Hayes *et al.* (1993); Osta *et al.* (1995); Martín-Burriel *et al.* (1997a); Prinzenberg and Erhardt (1997)
LAMC1	Laminin, gamma	16q22–q24	Pearce *et al.* (1995)

Table 10.4. *Continued.*

Locus symbol	Locus name	Chromosomal assignment	Reference
LBCL1 (LFC)	Murine lfc oncogene homologue	3q21	Comincini *et al.* (1997)
LDHA	Lactate dehydrogenase A	29q21–q24	Dain *et al.* (1984); Womack and Moll (1986); Schmutz *et al.* (1996a)
LDHB	Lactate dehydrogenase B	5	Heuertz and Hors-Cayla (1981); Echard *et al.* (1984); Womack and Moll (1986)
LDLR	Low-density lipoprotein receptor	7	Dietz *et al.* (1992a); Barendse *et al.* (1997)
LEP (OB, OBS)	Obesity (murine homologue, leptin)	4q32	Pfister-Genskow *et al.* (1996a); Stone *et al.* (1996); Kappes *et al.* (1997); Lien *et al.* (1997); Pomp *et al.* (1997); Wilkins and Davey (1997)
LEPR	Leptin receptor	3q33	Pfister-Genskow *et al.* (1996b)
LGB	Lactoglobulin, beta	11q28	Hines *et al.* (1969); Larsen (1970); Threadgill and Womack (1990c); Monteagudo *et al.* (1992); Hayes and Petit (1993); Barendse *et al.* (1997); Kappes *et al.* (1997)
LHB	Luteinizing hormone beta polypeptide	18	Womack *et al.* (1991)
LIF	Leukaemia inhibitory factor (cholinergic differentiation factor)	17	Piedrahita *et al.* (1997)
LPL	Lipoprotein lipase	8	Threadgill and Womack (1991b); Lien *et al.* (1995); Barendse *et al.* (1997)
LPO	Lactoperoxidase	19q13	Hayes *et al.* (1993)
LTF	Lactotransferrin	22q24	Schwerin *et al.* (1994); Martín-Burriel *et al.* (1997b)
LYZ@	Lysozyme, gene cluster (LYZ1, LYZ2, LYZ3)	5q23	Sigurdardóttir *et al.* (1990); Gallagher *et al.* (1993a); Weikard *et al.* (1996); Barendse *et al.* (1997)
LYZ1	Lysozyme 1	5q23	Sigurdardóttir *et al.* (1990); Gallagher *et al.* (1993a); Brunner *et al.* (1995a)
LYZ2	Lysozyme 2	5q23	Sigurdardóttir *et al.* (1990); Gallagher *et al.* (1993c); Brunner *et al.* (1995a)
LYZ3	Lysozyme 3	5q23	Sigurdardóttir *et al.* (1990); Gallagher *et al.* (1993a); Brunner *et al.* (1995a)
LYZL	Lysozyme-like	7q23	Gallagher *et al.* (1993a)
MANBA	Beta-mannosidase	6q31–q36	Schmutz *et al.* (1996b)

Table 10.4. *Continued.*

Locus symbol	Locus name	Chromosomal assignment	Reference
MAOA	Monoamine oxidase A	X	Barendse *et al.* (1997)
MAP1B	Microtubule-associated protein	20q14	Eggen *et al.* (1998a); Nonneman *et al.* (1996); Barendse *et al.* (1997)
MAPT	Microtubule-associated protein tau	19	Barendse *et al.* (1994, 1997); Ma *et al.* (1996); Yang and Womack (1996); Kappes *et al.* (1997)
MB	Myglobin	5	Barendse *et al.* (1997)
MBP	Myelin basic protein	24	Ansari *et al.* (1994)
MC1R	Melanocortin 1 receptor (alpha melanocyte-stimulating hormone receptor)	18	Werth *et al.* (1996); Barendse *et al.* (1997)
MC2R (ACTHR)	Melanocortin 2 receptor (adrenocorticotrophic hormone receptor)	24	Larsen *et al.* (1996)
MCAM	Melanoma adhesion molecule	15	Barendse *et al.* (1997); Moore and Byrne (1997)
MDH2	Malate dehydrogenase, mitochondrial	25	Womack and Moll (1986)
ME1	Malic enzyme, cytoplasmic	9	Heuertz and Hors-Cayla (1981); Echard *et al.* (1984); Womack and Moll (1986)
MGF (SCF)	Mast-cell growth factor (white heifer disease)	5	Charlier *et al.* (1996a); Tisdall *et al.* (1996)
MOG	Myelin oligodendrocyte glycoprotein	23	Barendse *et al.* (1997)
MOS	Moloney murine sarcoma viral (v-mos) oncogene homologue	14	Threadgill *et al.* (1990); Barendse *et al.* (1991)
MPI	Mannosephosphate isomerase	21	Womack and Moll (1986)
MPO	Myelin peroxidase	19	Yang and Womack (1995)
MT2A (MT2)	Metallothionein 2A	18	Barendse *et al.* (1997)
MTNR1A	Melatonin receptor 1A	27	Messer *et al.* (1997)
MTTP	tRNA proline	6	Fisher *et al.* (1997b)
MX1	Myxovirus (influenza) resistance 1, homologue of murine (interferon-inducible protein p78)	1	Broad *et al.* (1994)
MYC	Avian myelocytomatosis viral (v-myc) oncogene homologue	14	Threadgill *et al.* (1990); Barendse *et al.* (1991)
MYF5	Myogenic factor 5	5	Bishop *et al.* (1994); Kappes *et al.* (1997); Ryan *et al.* (1997)
MYF6	Myogenic factor 6	5	Ryan *et al.* (1997)
MYL4	Myosin, light polypeptide 4, alkali, atrial, embryonic	19	Yang and Womack (1995, 1996); Yang *et al.* (1998)

Table 10.4. *Continued.*

Locus symbol	Locus name	Chromosomal assignment	Reference
MYOD1	Myogenic factor 3	15q23	Ryan *et al.* (1997)
MYOG	Myogenin (myogenic factor 4)	16	Beever *et al.* (1997a, b); Ryan *et al.* (1997)
NCAM	Neural-cell adhesion molecule (CD56)	15	Barendse *et al.* (1994, 1997); Ma *et al.* (1996)
NDUFV2	NADH dehydrogenase (ubiquinone) flavoprotein 2 (24 kDa)	5	Agaba *et al.* (1997a)
NEB	Nebulin	2q23–q24	Lopez-Corrales *et al.* (1997)
NEFL	Neurofilament, light polypeptide	8	Threadgill and Womack (1991b)
NEFM	Neurofilament, medium polypeptide	8	Threadgill and Womack (1991b)
NF1	Neurofibromin 1 (neurofibromatosis, von Recklinghausen disease, Watson disease)	19	Yang and Womack (1995, 1996); Yang *et al.* (1998)
NGFB	Nerve growth factor, beta polypeptide	3q23	Elduque and Womack (1992, 1997); Ansari *et al.* (1994); Elduque *et al.* (1997a, b)
NKNB	Neurokinin B	5	Threadgill and Womack (1990b)
NP	Nucleoside phosphorylase	10	Womack and Moll (1986); Miller *et al.* (1991)
NPPA (PND, ANP)	Natriuretic peptide precursor A	16	Beever *et al.* (1997b)
NPR3 (ANPRC, ANP1)	Natriuretic peptide receptor C	20	Bishop *et al.* (1994); Barendse *et al.* (1997); Kappes *et al.* (1997)
NRAMP (LSH)	Natural resistance-associated macrophage protein (might include leishmaniasis)	2q41–q42	Pitel *et al.* (1994); Feng *et al.* (1996); Riggs *et al.* (1997)
NRAS	Neuroblastoma RAS viral (v-ras) oncogene homologue	3	Threadgill *et al.* (1994)
NTS	Neurotensin	5q13	Echard *et al.* (1994)
OAT	Ornithine aminotransferase	26	Threadgill and Womack (1991a)
ODF	Outer dense fibre of sperm tails 1	14q23–q24	Friedl *et al.* (1994b)
OGCP	2-Oxoglutarate carrier protein	19	Yang *et al.* (1998)
OPCM, OCAM	Opioid-binding protein and cell adhesion molecule	29	Dietz (1992); Kappes *et al.* (1997)
OPSN	Opsin	22	Dietz (1992)
OSG	Oviduct-specific glycoprotein	3	Barendse *et al.* (1997)
OXT	Prepro-oxytocin (neurophysin I)	13	Dietz *et al.* (1992a); Schläpfer *et al.* (1997a)

Table 10.4. *Continued.*

Locus symbol	Locus name	Chromosomal assignment	Reference
P4HB	Procollagen-proline, 2-oxygluturate 4-dioxygenase (proline 4-hydroxylase), beta polypeptide (protein disulphide isomerase; thyroid hormone-binding protein p55)	19	Yang and Womack (1995, 1996); Yang *et al.* (1998)
PAG1B	Pregnancy-associated glycoprotein 1	29	Martín-Burriel *et al.* (1998)
PAH	Phenylalanine hydroxylase	5	Threadgill and Womack (1990b); Barendse *et al.* (1992)
PAI1 (PLANH1)	Plasminogen-activator inhibitor, type I	25	Barendse *et al.* (1997)
PAI2 (PLANH2)	Plasminogen-activator inhibitor, type II (arginine–serpine)	24	Larsen *et al.* (1996); Agaba *et al.* (1997a); Barendse *et al.* (1997); Kappes *et al.* (1997)
PCCB	Propionyl coenzyme A carboxylase, beta polypeptide	1q31–q36	Schmutz *et al.* (1998)
PDE1 (PDE1A)	Phosphodiesterase 1A, calmodulin-dependent, 3',5' cyclic nucleotide phosphodiesterase,	2	Barendse *et al.* (1997)
PDE6A (PDEA)	Phosphodiesterase 6A, cGMP-specific, rod, alpha	2	Barendse *et al.* (1997)
PDEA	Phosphodiesterase, cGMP (rod receptor), alpha polypeptide	7	J.E. Womack (College Station, 1993, personal communication); Pearce *et al.* (1995
PDEA2	Phosphodiesterase, cGMP (cone receptor)	26	Gallagher *et al.* (1992d)
PDEB (PDE6B)	Phosphodiesterase 6B, cGMP-specific, rod, beta	6q33–q36	Gallagher *et al.* (1992d); Lord *et al.* (1996); Barendse *et al.* (1997)
PDEG	Phosphodiesterase, cGMP (rod receptor), gamma polypeptide	19	Gallagher *et al.* (1992d)
PDGFRA	Platelet-derived growth factor receptor, alpha polypeptide	6	Lord *et al.* (1996)
PDGFRB	Platelet-derived growth factor receptor, beta polypeptide	7	Zhang and Womack (1992)
PDHA2	Pyruvate dehydrogenase (lipoamide) alpha 2	6	Lanneluc *et al.* (1996)
PDME	Progressive degenerative myeloencephalopathy ('weaver')	4	Stuart and Leipold (1985); Georges *et al.* (1993)

Table 10.4. *Continued.*

Locus symbol	Locus name	Chromosomal assignment	Reference
PEPB	Peptidase B	5	Heuertz and Hors-Cayla (1981); Dain et al. (1984); Echard et al. (1984); Womack and Moll (1986)
PEPC	Peptidase C	2	Dain et al. (1984); Threadgill (1990)
PFKL	Phosphofructokinase, liver type	1	Threadgill et al. (1991)
PFKM	Phosphofructokinase muscle type	5	Threadgill et al. (1994)
PGD	6-Phosphogluconate dehydrogenase	16	Heuertz and Hors-Cayla (1981); Echard et al. (1984); Beever et al. (1997b); Womack and Moll (1986)
PGK1	Phosphoglycerate kinase-1	Xq21–q22	Heuertz and Hors-Cayla (1978); Shimizu et al. (1981); Livingston (1988)
PGM1	Phosphoglucomutase-1	3	Echard et al. (1984); Womack and Moll (1986)
PGM2	Phosphoglucomutase-2	6	J.E. Womack (College Station, 1993, personal communication)
PGM3	Phosphoglucomutase-3	9	Heuertz and Hors-Cayla (1981); Womack and Moll (1986)
PGY3	P glycoprotein 3/ multiple drug resistance 3	4	Womack et al. (1991)
PI2 (PI?)	Protease inhibitor-2 (anti-elastase, alpha-1-antitrypsin)	21	Georges et al. (1987, 1990)
PIGR	Polymeric immunoglobulin receptor	16q13	Kulseth et al. (1992, 1994); Barendse et al. (1997)
PIM1	Pim-1 oncogene	18	Barendse et al. (1997)
PIT1	Pituitary-specific transcription factor (growth-hormone factor 1)	1q21–q25	Harlizius et al. (1995); Moody et al. (1995b); Schmutz et al. (1998)
PKM2	Pyruvate kinase, muscle	10	Heuertz and Hors-Cayla (1981); Womack and Moll (1986)
PL1	Placental lactogen	23	Dietz et al. (1992b)
PLAT	Plasminogen activator, tissue	27	Threadgill and Womack (1991b)
PLAU	Plasminogen activator, urokinase	28	Threadgill and Womack (1991a)
PLCG1	Phospholipase C, gamma 1 (formerly subtype 148)	13	Schläpfer et al. (1997a, c)
POMC	Pro-opiomelanocortin (adrenocorticotrophin/beta-lipotrophin)	11	Womack et al. (1991)
PPARA	Peroxisome proliferative activated receptor, alpha	5	Sundvold et al. (1997)
PRKCA	Protein kinase, C alpha	19	Barendse et al. (1991)

Table 10.4. *Continued.*

Locus symbol	Locus name	Chromosomal assignment	Reference
PRKCB1	Protein kinase C, beta 1 polypeptide	25	Womack *et al.* (1991)
PRKCG (PKCG)	Protein kinase C, gamma	18	Gao and Womack (1996)
PRL	Prolactin	23	Hallerman *et al.* (1988); Creighton *et al.* (1992); Lewin *et al.* (1992); van Eijk *et al.* (1992); Ma *et al.* (1996); Barendse *et al.* (1997); Kappes *et al.* (1997)
PRLR	Prolactin receptor	20q17	Hayes *et al.* (1996); Barendse *et al.* (1997)
PRM1	Protamine 1	25q12–q13	Threadgill *et al.* (1991); Friedl and Rottmann (1994c); Barendse *et al.* (1997)
PRM2	Protamine 2	25q12–q13	Friedl and Rottmann (1994c)
PRNP	Prion protein	13	Womack *et al.* (1991); Ryan and Womack (1993); Schläpfer *et al.* (1997a)
PRP@	Prolactin-related protein, gene cluster (PRP1, PRP3, PRP6, PRP10)	23	Dietz *et al.* (1992b)
PRP1	Prolactin-related protein 1	23	Dietz *et al.* (1992b)
PRP10	Prolactin-related protein 10	23	Dietz *et al.* (1992b)
PRP3	Prolactin-related protein 3	23	Dietz *et al.* (1992b)
PRP6	Prolactin-related protein 6	23	Dietz *et al.* (1992b)
PSAP (SAP2)	Prosaposin (sphingolipid activator protein 1 and 2)	28	Elduque and Womack (1992); Barendse *et al.* (1997); Elduque *et al.* (1997c)
PSMB9 (LMP2, RING12)	Proteasome (prosome, macropain) subunit, beta type, 9 (large multi-functional protease 2)	23	Ma *et al.* (1996)
PTH	Parathyroid hormone	15q22–q27	Fries *et al.* (1988); Foreman and Womack (1989); Barendse *et al.* (1991, 1997); Kappes *et al.* (1997)
QDPR	Quinoid dihydropteridine reductase	6	Zhang *et al.* (1992); Barendse *et al.* (1997)
RAF1	Murine leukaemia viral (v-raf-1) oncogene homologue 1	22	Threadgill and Womack (1991c)
RASA	RAS p21 protein activator (GTP-ase activating protein)	7q24–q28	Eggen *et al.* (1992); Ma *et al.* (1996); Barendse *et al.* (1997); Kappes *et al.* (1997)
RB1	Retinoblastoma	12q13	Sun *et al.* (1997)
RBP3	Retinol-binding protein 3, interstitial	28	Threadgill and Womack (1991a); Barendse *et al.* (1997); Kappes *et al.* (1997)
REN	Renin	16	Threadgill *et al.* (1994)

Table 10.4. *Continued.*

Locus symbol	Locus name	Chromosomal assignment	Reference
RHCE	Rhesus blood group, CcEe antigens	2q45	Méténier-Delisse *et al.* (1997)
RHD (RH)	Rhesus blood group, D antigen	2q45	Méténier-Delisse *et al.* (1997)
RHO	Rhodopsin	22q23–q24	Hallerman *et al.* (1988); Womack *et al.* (1989); Ansari *et al.* (1994)
RNS1	Ribonuclease A (pancreatic)	27	Monteagudo *et al.* (1992)
RPS14	Ribosomal protein S14	7	Zhang and Womack (1992)
RYR1	Ryanodine receptor 1 (skeletal; calcium release channel)	18q23–q24	Miller *et al.* (1991); Chowdhary *et al.* (1992)
S100B	S100 protein, beta polypeptide	1	Threadgill *et al.* (1991)
SCYA2 (MCP1)	Small inducible cytokine A2 (monocyte chemotactic protein 1, homologous to mouse Sig-je)	19	Yang *et al.* (1998)
SI	Sucrase-isomaltase	1	Threadgill and Womack (1991c)
SKI	Avian sarcoma viral v-ski oncogene homologue	27	Ryan and Womack (1995)
SLC2A2 (GLUT2)	Solute carrier family 2 (facilitated glucose transporter), member 2	1	Broad *et al.* (1994)
SMPP	Spasmolytic polypeptide	1	Threadgill and Womack (1991d)
SOD1	Superoxide dismutase, soluble	1q12–q14	Heuertz and Hors-Cayla (1981); Womack and Moll (1986); Schmutz *et al.* (1996c); Band *et al.* (1997); Barendse *et al.* (1997); Marquess *et al.* (1997)
SOD1L1	Superoxide dismutase, soluble-like	13	Gallagher *et al.* (1992c); Schläpfer *et al.* (1997a)
SOD2	Superoxide dismutase-2, mitochondrial	9	Heuertz and Hors-Cayla (1981); Echard *et al.* (1984); Womack and Moll (1986); Barendse *et al.* (1997)
SOD3	Superoxide dismutase 3, extracellular	6	Gallagher *et al.* (1992c)
SOX2	SRY (sex-determining region Y) – box 2	1q33	Payen *et al.* (1995); Hayes *et al.* (1996)
SPARC2	Secreted protein, acidic, cysteine-rich (osteonectin)	7	Womack *et al.* (1989); Rogers *et al.* (1991); Barendse *et al.* (1997); Gao and Womack (1997)
SPP1	Secreted phosphoprotein 1 (osteopontin, bone sialoprotein I, early T-lymphocyte activation 1)	6	Lanneluc *et al.* (1996); Barendse *et al.* (1997)
SRN (RNS, RNASE)	Seminal ribonuclease	10	Barendse *et al.* (1997)

Table 10.4. *Continued.*

Locus symbol	Locus name	Chromosomal assignment	Reference
SRY	Sex-determining region Y	Yq12.5	Cui *et al.* (1995)
SSBP	Single-stranded DNA-binding protein	2	Barendse *et al.* (1997)
SST	Somatostatin	1q23–q25	Dietz and Womack (1989); Barendse *et al.* (1997)
STAT5A	Signal transducer and activator of transcription 5A	19q17	Goldammer *et al.* (1997)
SY	Syndactyly	15	Charlier *et al.* (1996b)
SYT1	Synaptotagmin 1	5	Barendse *et al.* (1997)
TCP1	T-complex 1	23	Andersson (1988); Andersson *et al.* (1988)
TCRA	T-cell receptor, alpha	10q15	Li *et al.* (1992); Agaba *et al.* (1997b); Barendse *et al.* (1997)
TCRB	T-cell receptor, beta cluster	4	Li *et al.* (1992); Pearce *et al.* (1995); Agaba *et al.* (1997b); Barendse *et al.* (1997)
TCRD	T-cell receptor, delta	10	Li *et al.* (1992)
TCRG	T-cell receptor, gamma cluster	4	Li *et al.* (1992); Agaba *et al.* (1997b); Barendse *et al.* (1997)
TF	Transferrin	1q41-46	Larsen (1977); Threadgill *et al.* (1991); Broad *et al.* (1994); Elduque *et al.* (1995); Kappes *et al.* (1997); Schmutz *et al.* (1998)
TG	Thyroglobulin (hereditary goitre)	14q12-q16	Ricketts *et al.* (1987); Threadgill *et al.* (1990); Barendse *et al.* (1991, 1997); Daskalchuk and Schmutz (1997); Kappes *et al.* (1997)
TGFBR1	Transforming growth factor, beta receptor I (activin A receptor type II-like kinase, 53 kDa)	8	Roelen *et al.* (1998)
THBD	Thrombomodulin	13	Schläpfer *et al.* (1997a, c)
THBS (SCO)	Thrombospondin, SCO-spondin	4q36	Popescu *et al.* (1997)
THH	Trichohyalin	1q41–q46	Schmutz *et al.* (1998)
THRA1	Thyroid hormone receptor, alpha 1 (avian erythroblastic leukaemia viral (v-erb-a) oncogene homologue 1, formerly ERBA1)	19	Yang and Womack (1995, 1996); Yang *et al.* (1998)
TK1	Thymidine kinase 1, soluble	19	Yang *et al.* (1998)
TNFA	Tumour necrosis factor, alpha (cachectin)	23	Lester *et al.* (1996); Barendse *et al.* (1997); Dietz *et al.* (1997)
TNP1	Transition protein 1 (during histone to protamine replacement)	2	Friedl and Rottmann (1994c); Heriz *et al.* (1994)

Table 10.4. *Continued.*

Locus symbol	Locus name	Chromosomal assignment	Reference
TNP2	Transition protein 2 (during histone to protamine replacement)	25q12–q13	Friedl and Rottmann (1994c)
TP53	Tumour protein p53 (Li–Fraumeni syndrome)	19q15	Coggins *et al.* (1995); Yang and Womack (1995, 1996); Yang *et al.* (1998)
TPI1	Triosephosphate isomerase	5	Echard *et al.* (1984); Womack and Moll (1986)
TSHB	Thyroid-stimulating hormone, beta	3	Beever *et al.* (1997a, b)
TSPY@	Testis-specific protein, Y-linked, gene family	Yp13–q12.6	Vogel *et al.* (1997a, b)
TTR (PALB)	Transthyretin (prealbumin, amyloidosis type I)	24	Larsen *et al.* (1996); Agaba *et al.* (1997a)
TYMS (TS)	Tymidylate synthetase	24	Larsen *et al.* (1996)
TYR	Tyrosinase	29	Womack *et al.* (1989)
UMPS	Uridine monophosphate synthetase (uridine monophosphate synthetase deficiency, DUMPS)	1q31	Schwenger and Schöber (1991); Schwenger *et al.* (1991); Friedl and Rottmann (1994b); Ryan *et al.* (1994); Barendse *et al.* (1997)
UPK1A	Uroplakin IA	18	Ryan *et al.* (1993b)
UPK1B	Uroplakin IB	1	Ryan *et al.* (1993b)
UPK2	Uuroplakin II	15	Ryan *et al.* (1993b)
UPK3A	Uroplakin IIIA	6	Ryan *et al.* (1993b)
UPK3B	Uroplakin IIIB	5	Ryan *et al.* (1993b)
VEGF	Vascular endothelial growth factor	23	Barendse *et al.* (1994, 1997); Kappes *et al.* (1997)
VIL	Villin	2q34	Mathiason *et al.* (1995)
VIM	Vimentin	13	Threadgill and Womack (1991a); Barendse *et al.* (1997); Schläpfer *et al.* (1997a)
VWF	Von Willebrand factor	5q35	Janel *et al.* (1996)
WNT1	Murine mammary tumour virus integration site (v-int-1)	5	Threadgill and Womack (1990b); Barendse *et al.* (1992)
WT1	Wilms tumour 1	15	Payen *et al.* (1995)
YES1	Yamaguchi sarcoma viral (v-yes-1) oncogene homologue 1	24	Womack *et al.* (1991)
ZFX	Zinc finger protein, X-linked	Xq34	Xiao *et al.* (1998)
ZFY	Zinc finger protein, Y-linked	Yp13	Xiao *et al.* (1998)
ZNF146	Zinc finger protein 146	18q24	Hayes *et al.* (1996)
ZNF164	Zinc finger protein 164	17q24	Le Chalony *et al.* (1995); Hayes *et al.* (1996)

ATP, adenosine triphosphate; ATPase, adenosine triphosphatase; cGMP, cyclic guanosine monophosphate; DNA, deoxyribonucleic acid; tRNA, transfer ribonucleic acid; NADH, nicotinamide adenine dinucleotide; GTPase, guanosine triphosphatase.
@ indicates a cluster of loci.

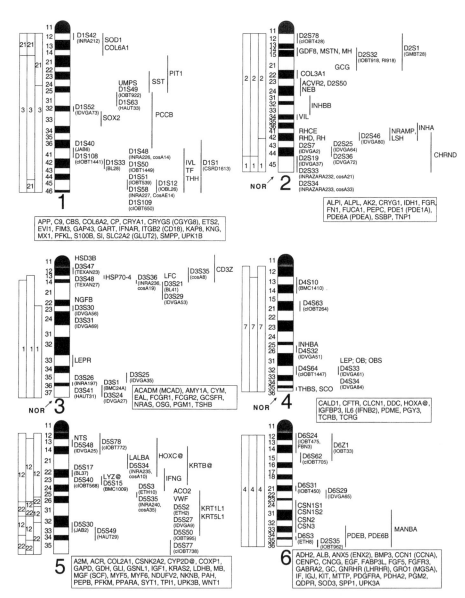

Fig. 10.4. The chromosome map of cattle. The ideogram corresponds to the revised G-band ideogram (see Fig. 10.3). The vertical lines on the right side of each chromosome delineate the location of the loci, which are indicated by the locus symbols. The locus names for each symbol and the mapping references are given in Table 10.4. Locus designations are, whenever possible, chosen according to the corresponding entries in the genome database for human mapping data (http://gdbwww.gdb.org/gdb). Anonymous loci are designated by D-numbers, e.g. *D1S42* stands for the 42nd anonymous locus (D-segment) on chromosome 1. Designations which were given by the laboratories that identified the loci are included in parentheses. The D-loci are in most cases microsatellites derived from large-insert clones mapped by fluorescence *in situ* hybridization. In some cases, two microsatellites were identified in the same large-insert clone (indicated by two different

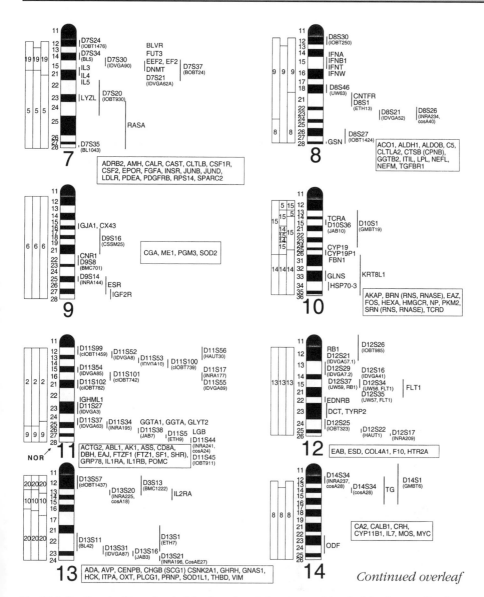

Fig. 10.4. *Continued* D-numbers). Other large-insert clones were chimerical, i.e. they produced signals at two or more sites within a chromosome or on different chromosomes. In the beginning of physical gene mapping, the regional assignments were based on radioactive *in situ* hybridization, with subsequent analysis of the autoradiographic grain clusters. These assignments usually produced large mapping intervals (long vertical lines). Later, the signals resulting from fluorescence *in situ* hybridization allowed for narrower mapping intervals (shorter vertical lines). On the left side of each chromosome, the numbers in each open bar indicate the homologous human chromosomes as identified by Zoo-FISH by three different groups. The bars next to each chromosome summarize the results obtained by Solinas Toldo *et al.* (1995a), the middle bars those by Hayes (1995) and the leftmost bars those by Chowdhary *et al.* (1996). **PAR** indicates the pseudo-autosomal region on the X- and Y-chromosomes, **NOR** the presumptive locations of nucleolar organizer regions.

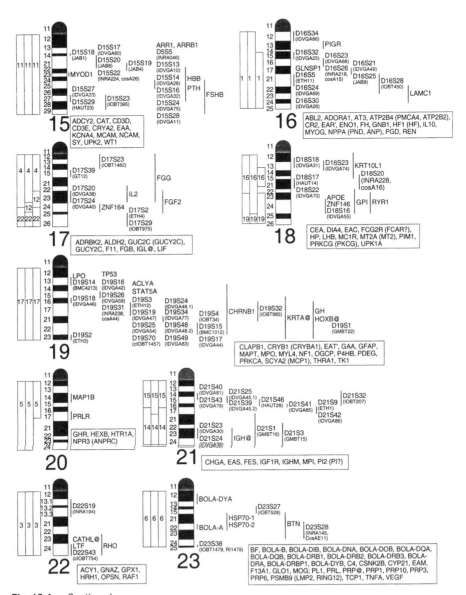

Fig. 10.4. *Continued*

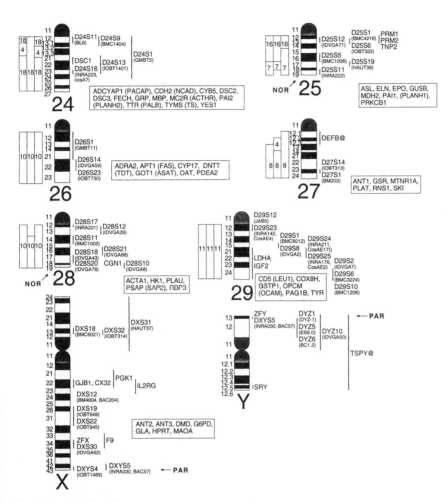

Fig. 10.4. Continued

provides the literature references to the mapping of each locus. The chromosome map contains a total of 516 coding loci and 214 physically mapped D loci. Most coding loci are identified by markers at the DNA levels, with the exception of blood-group and serum-protein loci, as well as the loci underlying the genetic disorders weaver (progressive degenerative myeloencephalopathy (PDME)) and syndactyly (SY).

Acknowledgement

The authors are grateful to Don Gallagher for critical reading of the manuscript and useful comments.

References

Adkison, L.R., Leung, D.W. and Womack, J.E. (1988a) Somatic cell mapping and restriction fragment analysis of bovine alpha and beta interferon gene families. *Cytogenetics and Cell Genetics* 47, 62–65.

Adkison, L.R., Skow, L.C., Thomas, T.L., Petrash, M. and Womack, J.E. (1988b) Somatic cell mapping and restriction fragment analysis of bovine genes for fibronectin and gamma crystallin. *Cytogenetics and Cell Genetics* 47, 155–159.

Agaba, M.K., Kemp, S.J., Barendse, W. and Teale, A. (1997a) Comparative mapping in cattle of genes located on human chromosome 18. *Mammalian Genome* 8, 530–532.

Agaba, M., Kemp, S.J., Barendse, W. and Teale, A. (1997b) Genetic mapping of bovine T-cell receptor complex loci. *Animal Genetics* 28, 235–237.

Aleyasin, A. and Barendse, W. (1997a) Kininogen (*KNG*) is linked to loci on cattle chromosome 1 and extends the syntenic conservation with human chromosome 3. *Mammalian Genome* 8, 78–79.

Aleyasin, A. and Barendse, W. (1997b) Novel conserved synteny between human chromosome 22 and cattle chromosome 22 established by linkage mapping of transducin alpha-1 subunit (GNAZ). *Mammalian Genome* 8, 458–459.

Anastassiadis, C., Leyhe, B., Olsaker, I., Friedl, R., Rottmann, O., Hiendleder, S. and Erhardt, G. (1996) Three polymorphic microsatellites for bovine chromosomes 7, 12 and 19. *Animal Genetics* 27, 125–126.

Andersson, L. (1988) Genetic polymorphism of a bovine t-complex gene (TCP1): linkage to major histocompatibility genes. *Journal of Heredity* 79, 1–5.

Andersson, L., Böhme, J., Peterson, P.A. and Rask, L. (1986) Genomic hybridization of bovine class II major histocompatibility genes: 2. Polymorphism of DR genes and linkage disequilibrium in the DQ–DR region. *Animal Genetics* 17, 295–304.

Andersson, L., Lunden, A., Sigurdardottir, S., Davies, C.J. and Rask, L. (1988) Linkage relationships in the bovine MHC region: high recombination frequency between class II subregions. *Immunogenetics* 27, 273–280.

Ansari, H.A., Jung, H.R., Hediger, R., Fries, R., König, H. and Stranzinger, G. (1993) A balanced autosomal reciprocal translocation in an azoospermic bull. *Cytogenetics and Cell Genetics* 62, 117–123.

Ansari, H.A., Pearce, P.D., Maher, D.W. and Broad, T.E. (1994) Regional assignment of conserved reference loci anchors unassigned linkage and syntenic groups to ovine chromosomes. *Genomics* 24, 451–455.

Ashwell, M.S., Ogg, S.L. and Mather, I.H. (1996) The bovine butyrophilin gene maps to chromosome 23. *Animal Genetics* 27, 171–173.

Bahri-Darwich, I., Cribiu, E.P., Berland, H.M. and Darre, R. (1993) A new Robertsonian translocation in Blonde d'Aquitaine cattle, rob(4;10). *Genetics, Selection, Evolution* 25, 413–419.

Bahri-Darwich, I., Vaiman, D., Olsaker, I., Oustry, A. and Cribiu, E.P. (1994) Assignment of bovine synteny groups U27 and U8 to R-banded chromosome 12 and 27, respectively. *Hereditas* 120, 261–265.

Band, M., Eggen, A., Bishop, M.D. and Ron, M. (1997) Isolation of microsatellites from a bovine YAC clone harbouring the SOD1 gene. *Animal Genetics* 28, 363–366.

Barendse, W., Armitage, S.M., Womack, J.E. and Hetzel, D.J. (1991) Substantial conservation of gene order of three bovine chromosomal segments when compared to humans. *Cytogenetics and Cell Genetics* 58, 2123.

Barendse, W., Armitage, S.M., Womack, J.E. and Hetzel, J. (1992) Linkage relations between A2M, HOX3, INT1, KRAS2, and PAH on bovine chromosome 5. *Genomics* 14, 38–42.

Barendse, W., Armitage, S.M., Kossarek, L.M., Shalom, A., Kirkpatrick, B.W., Ryan, A.M., Clayton, D., Li, L., Neibergs, H.L., Zhang, N., Grosse, W.M., Weiss, J., Creighton, P., McCarthy, F., Ron, M., Teal, A.J., Fries, R., McGraw, R.A., Moore, S.S., Georges, M., Soller, M., Womack, J.E. and Hetzel, D.J.S. (1994) A genetic linkage map of the bovine genome. *Nature Genetics* 6, 227–235.

Barendse, W., Vaiman, D., Kemp, S.J., Sugimoto, Y., Armitage, S., Williams, J., Sun, H., Eggen, A., Agaba, M., Aleyasin, A., Band, M., Bishop, M., Buitkamp, J., Byrne, K., Collins, F., Cooper, L., Coppetiers, W., Denys, B., Drinkwater, R., Easterday, K., Elduque, C., Ennis, S., Erhardt, G., Ferretti, L., Flavin, N., Gao, Q., Georges, M., Gurung, R., Harlizius, B., Hawkins, G., Hetzel, J., Hirano, T., Hulme, D., Jorgensen, C., Kessler, M., Kirkpatrick, B., Konfortov, B., Kostia, S., Kuhn, C., Lenstra, J., Leveziel, H., Lewin, H., Leyhe, B., Li, L., Martin Burriel, I., McGraw, R., Miller, R., Moody, D.E., Moore, S.S., Nakane, S., Nijman, I., Olsaker, I., Pomp, D., Rando, A., Ron, M., Shalom, A., Teale, A.J., Thieven, U., Urquhart, B.G.D., Vage, D.-I., Van de Weghe, A., Varvio, S., Velmala, R., Vilkki, J., Weikard, R., Woodside, C., Womack, J.E., Zanotti, M. and Zaragoza, P. (1997) A medium-density genetic linkage map of the bovine genome. *Mammalian Genome* 8, 21–28.

Basrur, P.K. and Switonski, M. (1990) Rearranging genetic interpretations through meiotic studies on domestic animals. *Revista Brasilera de Reproduçao Animal* 2 (Suppl.) 2, 20–31.

Beever, J.E., Hines, H.C. and Lewin, H.A. (1992) Linkage relationships between bovine FGR, FUCA1, ALPL, PGD and the B and C blood group loci. *Animal Genetics* 23(Suppl. 1), 78.

Beever, J.E., Fisher, S.R. and Lewin, H.A. (1997a) Polymorphism identification in the ACADM, AT3, IL10, MYOG and TSHB genes of cattle. *Animal Genetics* 28, 373–374.

Beever, J.E., Fisher, S.R., Guerin, G. and Lewin, H.A. (1997b) Mapping of eight human chromosome 1 orthologs to cattle chromosomes 3 and 16. *Mammalian Genome* 8, 533–536.

Berland, H.M., Sharma, A., Cribiu, E.P., Darre, R., Boscher, J. and Popescu, C.P. (1988) A new case of Robertsonian translocation in cattle. *Journal of Heredity* 79, 33–36.

Biltueva, L., Sharshova, S., Sharshov, A., Ladygina, T., Borodin, P. and Graphodatski, A. (1994) A new Robertsonian translocation, 8;23, in cattle. *Genetics, Selection, Evolution* 26, 159–165.

Biltueva, L.S., Sablina, O.V., Beklemisheva, V.R., Shvets, Y., Tkachenko, A., Dukhanina, O., Lushnikova, T.P., Vorobieva, N.V., Graphodatsky, A.S. and Kisselev, L.L. (1996) Localization of rat K51 keratin-like locus (KRT10L) to human and animal chromosomes by *in situ* hybridization. *Cytogenetics and Cell Genetics* 73, 209–213.

Bishop, M.D., Tavakkol, A., Threadgill, D.W., Simmen, F.A., Simmen, R.C.M., Davis, M.E. and Womack, J.E. (1991) Somatic cell mapping and restriction fragment length polymorphism analysis of bovine insulin-like growth factor I (IGF-I). *Journal of Animal Science* 69, 4306–4311.

Bishop, M.D., Kappes, S.A., Keele, J.W., Stone, R.T., Sunden, S.L.F., Hawkins, G.A., Solinas Toldo, S., Fries, R., Grosz, M.D., Yoo, J. and Beattie, C.W. (1994) A genetic linkage map for cattle. *Genetics* 136, 619–639.

Blazak, W.F. and Eldridge, F.E. (1977) A Robertsonian translocation and its effect upon fertility in Brown Swiss cattle. *Journal of Dairy Science* 60, 1133–1142.

Bouquet, Y., van de Weghe, A., van Zeveren, A. and Varewyck, H. (1986) Evolutionary conservation of the linkage between the structural loci for serum albumin and vitamin D binding protein (Gc) in cattle. *Animal Genetics* 17, 175–182.

Bouvet, A. and Popescu, C.P. (1989) Synaptonemal complexes analyis in a bull carrying a 4:8 Robertsonian translocation. *Annales de Génétique* 32, 193–199.

Braun, U., Ansari, H.A., Hediger, R., Süss, U. and Ehrensperger, F. (1988) Hypotrichose und Oligodentie, verbunden mit einer Xq-Deletion, bei einem Kalb der Schweizerischen Fleckviehrasse. *Tierärztliche Praxis* 16, 39–44.

Broad, T.E., Burkin, D.J., Cambridge, L.M., Maher, D.W., Lewis, P.E., Ansari, H.A., Pearce, P.D. and Jones, C. (1994) Seven loci on human chromosome 4 map onto sheep chromosome 6: a proposal to restore the original nomenclature of this sheep chromosome. *Mammalian Genome* 5, 429–433.

Broad, T.E., Burkin, D.J., Cambridge, L.M., Maher, D.W., Lewis, P.E., Ansari, H.A., Pearce, P.D. and Jones, C. (1995a) Assignment of five loci from human chromosome 8q onto sheep chromosome 9. *Cytogenetics and Cell Genetics* 68, 102–106.

Broad, T.E., Lewis, P.E., Burkin, D.J., Gleeson, A.J., Carpenter, M.A., Jones, C., Pearce, P.D., Maher, D.W. and Ansari, H.A. (1995b) Thirteen loci physically assigned to sheep chromosome 2 by cell hybrid analysis and *in situ* hybridization. *Mammalian Genome* 6, 862–866.

Broom, M.F. and Hill, D.F. (1994) Construction of a large-insert yeast artificial chromosome library from sheep DNA. *Mammalian Genome* 5, 817–819.

Brown, P., Spooner, R.L. and Clark, A.J. (1989) Cloning and characterization of a BoLA class I cDNA clone. *Immunogenetics* 29, 58–60.

Brunner, R.M., Henke, M., Guérin, G., Goldammer, T., Seyfert, H.-M. and Schwerin, M. (1995a) The macrophage expressed variant of the bovine lysozyme-encoding gene maps to chromosome 5q23. *Mammalian Genome* 5, 833.

Brunner, R.M., Goldhammer, T., Hiendleder, S., Jäger, C. and Schwerin, M. (1995b) Comparative mapping of the gene coding for inhibin-alpha (INHA) to chromosome 2 in sheep and cattle. *Mammalian Genome* 6, 309.

Brunner, R.M., Guérin, G., Goldammer, T., Seyfert, H.-M. and Schwerin, M. (1996) The bovine butyrophilin encoding gene (BTN) maps to chromosome 23. *Mammalian Genome* 7, 635–636.

Brzozowska, A., Fries, R., Womack, J., Grimholt, U., Myklebost, O. and Rogne, S. (1993) Isolation, sequencing and expression analyis of a bovine apolipoprotein E (APOE) cDNA and chromosomal localization of the APOE locus. *Mammalian Genome* 4, 53–57.

Buitkamp, J., Schwaiger, F.-W., Solinas Toldo, S., Fries, R. and Eppeln, J.T. (1995) The bovine interleukin-4 gene: genomic organization, localization and evolution. *Mammalian Genome* 6, 350–356.

Buitkamp, J., Obexerruff, G., Kessler, M. and Epplen, J.T. (1996) A microsatellite (BOBT24) located between the bovine IL4 and IL13 loci is polymorphic in cattle and goat. *Animal Genetics* 27, 212–213.

Burke, M.G., Stone, R.T. and Muggli-Cockett, N.E. (1991) Nucleotide sequence and Northern analysis of a bovine major histocompatibility class II DRbeta-like cDNA. *Animal Genetics* 22, 343–352.

Burkin, D.J., Jones, C., Burkin, H.R., McGrew, J.A. and Broad, T.E. (1996) Sheep CENPB and CENPC genes show a high level of sequence similarity and conserved synteny with their human homologs. *Cytogenetics and Cell Genetics* 74, 86–89.

Byrne, K., Barendse, W. and Moore, S.S. (1996) Dinucleotide polymorphism at the bovine potassium voltage gated channel locus (KCNA4). *Animal Genetics* 27, 437.

Cai, L., Taylor, J.F., Wing, R.A., Gallagher, D.S., Woo, S.S. and Davis, S.K. (1995) Construction and characterization of a bovine bacterial artificial chromosome library. *Genomics* 29, 413–425.

Castiglioni, B., Ferretti, L., Tenchini, M.L., Mezzelani, A., Simonic, T. and Duga, S. (1996a) Physical mapping of connexin 32 (GJB1) and 43 (GJA1) genes to bovine chromosomes Xq22 and 9q15/16 by fluorescence *in situ* hybridization. *Mammalian Genome* 7, 634–635.

Castiglioni, B., Scocchi, M., Zanetti, M. and Ferretti, L. (1996b) Six antimicrobial peptide genes of the cathelicidin family map to bovine chromosome 22q24 by fluorescence *in situ* hybridization. *Cytogenetics and Cell Genetics* 75, 240–242.

Charlier, C., Denys, B., Belanche, J.I., Coppieters, W., Grobet, L., Mni, M., Womack, J., Hanset, R. and Georges, M. (1996a) Microsatellite mapping of the bovine roan locus – a major determinant of white heifer disease. *Mammalian Genome* 7, 138–142.

Charlier, C., Farnir, F., Berzi, P., Vanmanshoven, P., Brouwers, B., Vromans, H. and Georges, M. (1996b) Identity-by-descent mapping of recessive traits in livestock – application to map the bovine syndactyly locus to chromosome 15. *Genome Research* 6, 580–589.

Chaudhary, R., Chowdhary, B.P., Johansson, M. and Gustavsson, I. (1993) The gene for the bovine interferon gamma (IFNG) maps to the q22–q24 bands of chromosome 5 in cattle. *Hereditas* 119, 7–10.

Chowdhary, B.P., Harbitz, I., Davies, W. and Gustavsson, I. (1991) Chromosomal localization of the glucose phosphate isomerase (GPI) gene in cattle, sheep and goat by *in situ* hybridization – chromosomal banding homology versus molecular conservation in Bovidae. *Hereditas* 114, 161–170.

Chowdhary, B.P., Harbitz, I., Davies, W. and Gustavsson, I. (1992) Chromosomal localization of the calcium release channel (CRC) gene in cattle and horse by *in situ* hybridization – evidence of a conserved synteny with glucose phosphate isomerase. *Animal Genetics* 23, 43–50.

Chowdhary, B.P., Hassanane, M.S. and Gustavsson, I. (1994) Regional localization of the bovine interleukin-2 (IL2) gene to chromosome 17q22–q23 by in situ hybridization. *Cytogenetics and Cell Genetics* 65, 166–168.

Chowdhary, B.P., Frönicke, L., Gustavsson, I. and Scherthan, H. (1996) Comparative analysis of the cattle and human genomes: detection of ZOO-FISH and gene mapping-based chromosomal homologies. *Mammalian Genome* 7, 297–302.

Coggins, L.W., Scobie, L., Jackson, M.E. and Campo, M.S. (1995) Assignment of the bovine p53 gene (TP53) to chromosome 19q15 by fluorescence *in situ* hybridization. *Mammalian Genome* 6, 687–688.

Comincini, S., Drisaldi, B. and Ferretti, L. (1997) Isolation of coding sequences from bovine cosmids containing microsatellite markers by means of exon-trapping. *Mammalian Gemome* 8, 486–490.

Creighton, P., Eggen, A., Fries, R., Jordan, S.A., Hetzel, J., Cunningham, E.P. and Humphries, P. (1992) Mapping of bovine markers CYP21, PRL and BOLA DRBP1 by genetic linkage analysis in reference pedigrees. *Genomics* 14, 526–528.

Cribiu, E.P., Matejka, M., Darre, R., Durand, V., Berland, H.M. and Bouvet, A. (1989) Identification of chromosomes involved in a Robertsonian translocation in cattle. *Annales de Génétique et Sélection Animale* 21, 555–560.

Cui, X., Kato, Y., Sato, S. and Sutou, S. (1995) Mapping of bovine Sry gene on the distal tip of the long arm and murine Sry on the short arm of the Y chromosome by the method of fluorescence *in situ* hybridization (FISH). *Animal Science and Technology (Japan)* 66, 441–444.

Czarnik, U. and Kamiński, S. (1997) Detection of intronic sequence in the bovine ITGB2 gene. *Animal Genetics* 28, 320–321.

Dain, A.R., Tucker, E.M., Donker, R.A. and Clarke, S.W. (1984) Chromosome mapping in cattle using mouse myeloma/calf lymph node cell hybridomas. *Biochemical Genetics* 22, 429–439.

Daskalchuk, T.E. and Schmutz, S.M. (1997) Genetic mapping of thyroglobulin on bovine chromosome 14. *Mammalian Genome* 8, 74–76.

De Giovanni, A.M., Succi, G., Molteni, L. and Castiglioni, M. (1979) A new autosomal translocation in 'Alpine grey cattle'. *Annales de Génétique et Sélection Animale* 11, 115–120.

De Giovanni, A.M., Molteni, L., Succi, G., Galliani, C., Boscher, J. and Popescu, C.P. (1988) A new type of Robertsonian translocation in cattle. In: *Proceedings of the 8th European Colloquium on Cytogenetic of Domestic Animals, Bristol, UK*, pp. 53–59.

Dennis, J.A., Healy, P.J., Beaudet, A.L. and O'Brien, W.E. (1989) Molecular definition of bovine arginiosuccinate synthetase deficiency. *Proceedings of the National Academy of Sciences of the USA* 86, 7947–7951.

de Schepper, G.C., Aalbers, J.G. and Te Brake, J.H.A. (1982) Double reciprocal translocation heterozygosity in a bull. *Veterinary Record* 110, 197–199.

Di Berardino, D., Iannuzzi, L., Ferrara, L. and Matassino, D. (1979) A new case of Robertsonian translocation in cattle. *Journal of Heredity* 70, 436–438.

Di Berardino, D., Iannuzzi, L., Bettini, T.M. and Matassion, D. (1981) Ag-NORs variation and banding homologies in two species of Bovidae: *Bubalus bubalis* and *Bos taurus*. *Canadian Journal of Genetics and Cytology* 23, 89–99.

Di Berardino, D., Lioi, M.B. and Iannuzzi, L. (1985) Identification of nucleolus organizer chromosomes in cattle (*Bos taurus*, L.) by sequential silver staining + RBA banding. *Caryologia* 38, 95–102.

Di Berardino, D., Hayes, H., Fries, R. and Long, S.E. (1990) International System for Cytogenetic Nomenclature of Domestic Animals (ISCNDA 1989). *Cytogenetics and Cell Genetics* 53, 65–79.

Dietz, A.B. (1992) Analysis of the prolactin related proteins and placental lactogen family and the development of a sequence tagged site based bovine gene map. Dissertation, Texas A&M Universtiy.

Dietz, A.B. and Womack, J.E. (1989) Somatic cell mapping of the bovine somatostatin gene. *Journal of Heredity* 80, 410–412.

Dietz, A.B., Georges, M., Threadgill, D.W., Womack, J.E. and Schuler, L.A. (1992a) Somatic cell mapping, polymorphism, and linkage analysis of bovine prolactin-related proteins and placental lactogen. *Genomics* 14, 137–143.

Dietz, A.B., Neibergs, H.L. and Womack, J.E. (1992b) Assignment of eight loci to bovine syntenic groups by use of PCR: extension of a comparative gene map. *Mammalian Genome* 3, 106–111.

Dietz, A.B., Neibergs, H.L., Womack, J.E. and Kehrli, M.E., Jr (1997) Rapid communication: single strand conformational polymorphism (SSCP) of bovine tumor necrosis factor alpha. *Journal of Animal Science* 75, 2567.

Drinkwater, R., Johnson, S.E., Aspden, W., Harrison, B., Womack, J.E. and Hetzel, D.J. (1992a) Assignment of bovine syntenic group U10 to chromosome 1 of cattle using a locus specific minisatellite probe. In: *Proceedings, Third Australasian Gene Mapping Workshop, Brisbane, Australia*, p.17.

Drinkwater, R., Aspden, W., Harrison, B., Wisselaar, H.A., Hermans, M.M.P., Kroos, M.A., Reuser, A.J.J. and Hetzel, D.J.S. (1992b) Pompe's disease (glycogenosis type II) in Australian Brahman cattle. *Animal Genetics* 23(Suppl. 1), 104.

Dunn, H.O., McEntee, K., Hall, C.E., Johnson, R.H., Jr and Stone, W.H. (1979) Cytogenetic and reproductive studies of bulls born co-twin with freemartins. *Journal of Reproduction and Fertility* 57, 21–30.

Dyrendhahl, I. and Gustavsson, I. (1979) Sexual functions, semen characteristics and fertility of bulls carrying the 1/29 chromosome translocation. *Hereditas* 90, 281–289.

Echard, G., Gellin, J., Benne, F. and Gillois, M. (1984) Progress in gene mapping of cattle and pigs using somatic cell hybridization. *Cytogenetics and Cell Genetics* 37, 458–459.

Echard, G., Broad, T.E., Hill, D. and Pearce, P. (1994) Present status of the ovine gene map (*Ovis aries*) – comparison with the bovine m161ap (*Bos taurus*). *Mammalian Genome* 5, 324–332.

Eggen, A., Solinas-Toldo, S., Dietz, A.B., Womack, J., Stranzinger, G. and Fries, R. (1992) RASA contains a polymorphic microsatellite and maps to chromosome 7q2.4–qter and bovine syntenic group U22. *Mammalian Genome* 3, 559–563.

Eggen, A., Solinas Toldo, S. and Fries, R. (1998a) A cosmid specific for sequences encoding a microtubule associated protein, MAP1B, contains a polymorphic microsatellite and maps to bovine chromosome 20q14. *Journal of Heredity* 89, 359–363.

Eggen, A., Bahri-Darwich, I., Elduque, C., Petit, E., Oustry, A., Cribiu, E.P. and Levéziel, H. (1998b) Physical mapping of new cosmids containing microsatellite for mapping the bovine genome. *Animal Genetics* (in press).

Eldridge, F.E. (1975) High frequency of a Robertsonian translocation in a herd of British White Cattle. *Veterinary Records* 96, 71–73.

Eldridge, F.E. (1980) X-autosome translocation in cattle. In: *Proceedings of the 4th European Colloquium on Cytogenetics of Domestic Animals, Uppsala, Sweden*, pp. 23–30.

Elduque, C. and Womack, J.E. (1992) Somatic cell mapping and restriction fragment analysis of the beta-nerve growth factor and the sphingolipid activator protein-2 genes in cattle. *Animal Genetics* 23(Suppl. 1), 80.

Elduque, C. and Womack, J.E. (1997) Localization of the beta-nerve growth factor gene (*NGFB*) to bovine chromosome 3. *Mammalian Genome* 8, 73–74.

Elduque, C., Rodellar, C. and Zaragoza, P. (1995) A MspI polymorphism at the transferrin (TF) locus in cattle. *Animal Genetics* 26, 280.

Elduque, C., Rodellar, C. and Zaragosa, P. (1996) RFLPs at the amylase 1 locus in cattle. *Animal Genetics* 27, 213.

Elduque, C., Laurent, P., Hayes, H., Rodellar, C., Levéziel, H. and Zaragoza, P. (1997a) Assignment of the beta-nerve growth factor (NGFB) to bovine chromosome 3 band q23 by *in situ* hybridization. *Cytogenetics and Cell Genetics* 77, 306–307.

Elduque, C., Li, L., Barendse, W. and Womack, J.E. (1997b) Physical and genetic mapping of the prosaposin gene (*PSAP*) to bovine chromosome 28. *Mammalian Genome* 8, 73.

Elduque, C., Zaragoza, P. and Womack, J.E. (1997c) An EcoRI polymorphism at the beta-nerve growth factor (NGFB) locus in cattle. *Animal Genetics* 28, 322.

Ellsworth, S.M., Paul, S.R. and Bunch, T.D. (1979) A 14/28 dicentric robertsonian translocation in a Holstein cow. *Theriogenology* 11, 165–170.

Ennis, P.D., Jackson, A.P. and Parham, P. (1988) Molecular cloning of bovine class I MHC cDNA. *Journal of Immunology* 141, 642–651.

Feng, J.W., Li, Y.J., Hashad, M., Schurr, E., Gros, P., Adams, L.G. and Templeton, J.W. (1996) Bovine natural resistance associated macrophage protein 1 (Nramp1) gene. *Genome Research* 6, 956–964.

Ferretti, L., Leone, P. and Sgaramella, V. (1990) Long range restriction analysis of the bovine casein genes. *Nucleic Acids Research* 18, 6829–6833.

Ferretti, L., Leone, P., Pilla, F., Zhang, Y., Nocart, M. and Guérin, G. (1994) Direct characterization of bovine microsatellites from cosmids: polymorphism and synteny mapping. *Animal Genetics* 25, 209–214.

Ferretti, L., Urquhart, B.G.D., Eggen, A., Olsaker, I., Harlizius, B., Castiglioni, B., Mezzelani, A., Solinas Toldo, S., Thieven, U., Zhang, Y., Morgan, A.L.G., Teres, V.M., Schwerin, M., Martin-Buriel, B.P., Chowdhary, B.P., Erhardt, G., Nijman, I.J., Cribiu, E.P., Barendse, W., Leveziel, H., Fries, R. and Williams, J.L. (1997) Cosmid-derived markers anchoring the bovine genetic map to the physical map. *Mammalian Genome* 8, 29–36.

Fidlerová, H., Senger, G., Kost, M., Sanseau, P. and Sheer, D. (1994) Two simple procedures for releasing chromatin from routinely fixed cells for fluorescence in situ hybridization. *Cytogenetics and Cell Genetics* 65, 203–205.

Fisher, S.R., Beever, J.E. and Lewin, H.A. (1997a) Genetic mapping of *COL3A1* to bovine chromosome 2. *Mammalian Genome* 8, 76–77.

Fisher, S.R., Beever, J.E. and Lewin, H.A. (1997b) Genetic mapping of five human chromosome 4 orthologues to bovine chromosomes 6 and 17. *Animal Genetics* 28, 253–257.

Flavin, N., Heriz, A., Monteagudo, L.V., Ennis, S., Martin, F., Barendse, W., Arruga, M.V. and Rogers, M. (1996) Cloning of the bovine activin receptor type II gene (ACVR2) and mapping to chromosome 2 (BTA2). *Cytogenetics and Cell Genetics* 75, 25–29.

Florijn, R.J., Bonden, L.A.J., Vrolijk, H., Wiegant, J., Vaandrager, J.-W., Baas, F., den Dunnen, J.T., Tanke, H.J., van Ommen, G.-J.B. and Raap, A.K. (1995) High-resolution DNA fiber-FISH for genomic DNA mapping and colour bar-coding of large genes. *Human Molecular Genetics* 4, 831–836.

Ford, C.E., Pollok, D.L. and Gustavsson, I. (1980) Proceedings of the First International Conference for the Standardization of Banded Karyotypes of Domestic Animals, Reading, England, 1976. *Hereditas* 92, 145–162.

Foreman, M.E. and Womack, J.E. (1989) Genetic and synteny mapping of parathyroid hormone and beta hemoglobin in cattle. *Biochemical Genetics* 27, 541–550.

Förster, M., Stranzinger, G. and Hellkuhl, B. (1980) X-chromosome gene assignment of swine and cattle. *Naturwissenschaften* 67, 48.

Frank, M. and Robert, J.M. (1981) La pathologie chromosomique. Etude chez *Bos taurus*. *Revue de Sciences Vétérinaires* 132, 405–411.

Friedl, R. and Rottmann, O. (1994a) Assignment of the cation independent mannose 6-phosphate/insulin-like growth factor II receptor to bovine chromosome 9q27-28 by fluorescent *in situ* hybridization. *Animal Genetics* 25, 191–193.

Friedl, R. and Rottmann, O.J. (1994b) Assignment of the bovine uridine monophosphate synthase gene to the bovine chromosome region 1q34–36 by FISH. *Mammalian Genome* 5, 38–40.

Friedl, R. and Rottmann, O.J. (1994c) Bovine protamine 1, protamine 2 and transition protein 2 map to BTA 29q12–13. *Mammalian Genome* 5, 523.

Friedl, R., Adham, I.M. and Rottmann, O.J. (1994a) Mapping of the gene encoding bovine preproacrosin (ACR) to chromosome BTA5 region q35. *Mammalian Genome* 5, 830–831.

Friedl, R., Adham, I.M. and Rottmann, O.J. (1994b) Physical mapping of bovine outer dense fiber gene (ODF) to chromosome BTA14q23–24 in cattle. *Mammalian Genome* 5, 833–834.

Friedl, R., Anastassiadis, C., Olsaker, I. and Rottmann, O. (1994c) Characterization and localization of bovine derived cosmids containing microsatellites. *Animal Genetics* 25(Suppl. 2), 47 (abstract).

Fries, R. (1989) The gene for the β-subunit of the follicle-stimulating hormone maps to bovine chromosome 15. *Journal of Heredity* 80, 401–403.

Fries, R., Hediger, R. and Stranzinger, G. (1986) Tentative chromosomal localization of the bovine major histocompatibility complex by *in situ* hybridization. *Animal Genetics* 17, 287–294.

Fries, R., Hediger, R. and Stranzinger, G. (1988) The loci for parathyroid hormone and beta-globin are closely linked and map to chromosome 15 in cattle. *Genomics* 3, 302–307.

Fries, R., Threadgill, D.W., Hediger, R., Gunawardana, A., Blessing, M., Jorcano, J.L., Stranzinger, G. and Womack, J.E. (1991) Mapping of bovine cytokeratin sequences to four different sites on three chromosomes. *Cytogenetics and Cell Genetics* 57, 135–141.

Fries, R., Eggen, A. and Womack, J.E. (1993) The bovine genome map. *Mammalian Genome* 4, 405–428.

Gallagher, D.S., Threadgill, D.W., Jackson, A.P., Parham, P. and Womack, J.E. (1991) Somatic cell mapping of bovine clathrin light chain genes: identificiation of a new bovine syntenic group. *Cytogenetics and Cell Genetics* 56, 154–156.

Gallagher, D.S.J., Basrur, P.K. and Womack, J.E. (1992a) Identification of an autosome to X chromosome translocation in the domestic cow. *Journal of Heredity* 83, 451–452.

Gallagher, D.S., Grosz, M., Basrur, P.K., Skow, L. and Womack, J.E. (1992b) Chromosomal localization in cattle of BoLA and HSP70 genes by fluorescent in situ hybridization and confirmation of the identity of an autosome to X chromosome translocation. *Animal Genetics* 23(Suppl. 1), 81.

Gallagher, D.S., Gibbs, L.S., Shafer, J.B. and Womack, J.E. (1992c) Somatic cell mapping of bovine EC-SOD and SOD1L loci. *Genomics* 12, 610–612.

Gallagher, D.S., Womack, J.E., Baehr, W. and Pittler, S.J. (1992d) Syntenic mapping of visual transduction genes in cattle. *Genomics* 14, 699–706.

Gallagher, D.S.J., Threadgill, D.W., Ryan, A.M., Womack, J.E. and Irwin, D.M. (1993a) Physical mapping of the lysozyme gene family in cattle. *Mammalian Genome* 4, 368–373.

Gallagher, D.S., Grosz, M.D., Womack, J.E. and Skow, L.C. (1993b) Chromosomal localization of HSP70 genes in cattle. *Mammalian Genome* 4, 388–390.

Gallagher, D.S., Ryan, A.M., Liou, L.S., Sastry, K.N. and Womack, J.E. (1993c) Somatic cell mapping of conglutinin (CGN1) to cattle syntenic group U29 and fluorescence in situ localization to chromosome 28. *Mammalian Genome* 4, 716–719.

Gallagher, D.S., Ryan, A.M., Diamond, G., Bevins, C.L. and Womack, J.E. (1995) Somatic cell mapping of beta-defensin genes to cattle syntenic group U25 and fluorescence in situ localization to chromosome 27. *Mammalian Genome* 6, 554–556.

Gallagher, D.S.J., Yang, Y.-P., Burzlaff, J.D., Womack, J.E., Stelly, D.M., Davis, S.K. and Taylor, J.F. (1998) Physical assignment of six type I anchor loci to bovine chromosome 19 by fluorescence *in situ* hybridization. *Animal Genetics* 29, 130–134.

Gao, Q. and Womack, J.E. (1996) Comparative mapping of anchor loci from HSA19 to cattle chromosomes 7 and 18. *Journal of Heredity* 88, 524–527.

Gao, Q. and Womack, J.E. (1997) A genetic map of bovine chromosome 7 with an interspecific hybrid backcross panel. *Mammalian Genome* 8, 258–261.

Gao, Q., Li, L. and Womack, J.E. (1997) Assignment of the casein kinase II gene family to cattle chromosomes. *Animal Genetics* 28, 146–149.

Ge, W., Davis, M.E. and Hines, H.C. (1997) Two SSCP alleles detected in the 5′-flanking region of bovine IGF1 gene. *Animal Genetics* 28, 155–156.

Georges, M. and Massey, J. (1992) Polymorphic DNA markers in Bovidae. Patent application WO 92/13102.

Georges, M., Swillens, S., Bouquet, Y., Lequarré, A.S. and Hanset, R. (1987) Genetic linkage between the bovine plasma protease inhibitor 2 (Pi-2) and S blood group loci. *Animal Genetics* 18, 311–316.

Georges, M., Lathrop, M., Bouquet, Y., Hilbert, P., Marcotte, A., Schwers, A., Roupain, J., Vassart, G. and Hanset, R. (1990) Linkage relationships among 20 genetic markers in cattle: evidence for linkage between two pairs of blood group systems: B–Z and S–F/V respectively. *Animal Genetics* 21, 95–105.

Georges, M., Gunawardana, A., Threadgill, D., Lathrop, M., Olsaker, I., Mishra, A., Sargeant, L., Schoeberlein, A., Steele, M., Terry, C., Threadgill, D.S., Zhao, X., Holm, T., Fries, R. and Womack, J. (1991) Characterization of a set of variable number of tandem repeat markers conserved in Bovidae. *Genomics* 11, 24–32.

Georges, M., Lathrop, M., Dietz, A.B., Libert, F., Mishra, A., Nielsen, D., Sargeant, L.S., Steele, M.R., Zhao, X., Leipold, H. and Womack, J.E. (1993) Microsatellite mapping of the gene causing Weaver disease in cattle will allow the study of an associated QTL. *Proceedings of the National Academy of Science of the USA* 90, 1058–1062.

Goldammer, T., Guerin, G., Brunner, R.M., Vanselow, J., Furbaß, R. and Schwerin, M. (1994) Chromosomal mapping of the bovine aromatase gene (CYP19) and an aromatase pseudogene to chromosome 10 and syntenic group U5. *Mammalian Genome* 5, 822–823.

Goldammer, T., Brunner, R.M., Hiendleder, S. and Schwerin, M. (1995) Comparative mapping of sheep inhibin subunit beta(b) to chromosome 2 in sheep and cattle by fluorescence *in situ* hybridization. *Animal Genetics* 26, 199–200.

Goldammer, T., Weikard, R., Brunner, R.M. and Schwerin, M. (1996) Generation of chromosome fragment specific bovine DNA sequences by microdissection and DOP-PCR. *Mammalian Genome* 7, 291–296.

Goldammer, T., Meyer, L., Seyfert, H.-M., Brunner, R.M. and Schwerin, M. (1997) *STAT5A* encoding gene maps to chromosome 19 in cattle and goat and to chromosome 11 in sheep. *Mammalian Genome* 8, 705–706.

Greene, W.A., Dunn, H.O. and Foote, R.H. (1977) Sex-chromosome ratios in cattle and their relationship to reproductive development in freemartins. *Cytogenetics and Cell Genetics* 18, 97–105.

Grobet, L., Royo Martin, L.J., Poncelet, D., Pirottin, D., Brouwers, B., Riquet, J., Schoeberlein, A., Dunner, S., Ménissier, F., Masabanda, J., Fries, R., Hanset, R. and Georges, M. (1997) A deletion in the bovine myostatin gene causes the double-muscled phenotype in cattle. *Nature Genetics* 17, 71–74.

Groenen, M.A.M., van der Poel, J.J., Dijkhof, R.J.M. and Giphart, M.J. (1989) Cloning of the bovine major histocompatibility complex class II genes. *Animal Genetics* 20, 267–278.

Groenen, M.A.M., van der Poel, J.J., Dijkhof, R.J.M. and Giphart, M.J. (1990) The nucleotide sequence of bovine MHC class II DQB and DRB genes. *Immunogenetics* 31, 37–44.

Grosclaude, F., Puyolle, J., Garnier, J. and Ribadeau-Dumas, B. (1965) Déterminisme génétique des caséines K du lait de vache: étroite liaison du locus K-Cn avec alphaS-Cn et beta-Cn. *Comptes Rendus de l'Académie des Sciences* 261, 5229.

Grosclaude, F., Joudrier, P., Mahé, M.-F. and Hazé, G. (1978) Polymorphisme de la caséine alpha-s2 bovine: étroite liaison du locus alpha-s2-Cn avec les loci alpha-s1-Cn, beta-Cn et kappa-Cn; mise en évidence d'une délétion dans le variant alpha-s2-Cn D. *Annales de Génétique et Selection Animale* 10, 313–327.

Grosclaude, F., François, D. and Wimitzky, M. (1990) Evidence for absence of linkage between the B and Z as well as between the F and S systems of cattle blood groups. *Animal Genetics* 21, 427–429.

Grosz, M.D., Womack, J.E. and Skow, L.C. (1992) Syntenic conservation of HSP70 genes in cattle and humans. *Genomics* 14, 863–868.

Grosz, M.D., Solinas-Toldo, S., Stone, R.T., Kappes, S.M., Fries, R. and Beattie, C.W. (1997) Chromosomal localization of six bovine microsatellite markers. *Animal Genetics* 28, 39–40.

Gu, F., Chowdhary, B.P., Andersson, L., Harbitz, I. and Gustavsson, I. (1992) Assignment of the bovine immunoglobulin gamma heavy chain (IGHG) gene to chromosome 21q24 by *in situ* hybridization. *Hereditas* 117, 237–240.

Guan, X.-Y., Meltzer, P.S., Cao, J. and Trent, J.M. (1992) Rapid generation of region-specific genomic clones by chromosome microdissection: isolation of DNA from a region frequently deleted in malignant melanoms. *Genomics* 14, 680–684.

Gunawardana, A. and Fries, R. (1992) Assignment of the HOX2 and HOX3 gene clusters to the bovine chromosome regions 19q17–q23 and 5q13–22. *Animal Genetics* 23, 195–199.

Gustavsson, I. (1970) Economic importance of a translocation in Swedish cattle. In: *1. Europäisches Kolloquium über Zytogenetik (Chromosomenpathologie) in Veterinärmedizin und Säugetierkunde, Giessen*, pp. 108–111.

Gustavsson, I. (1979) Distribution and effects of the 1/29 Robertsonian translocation in cattle. *Journal of Dairy Science* 62, 825–835.

Gustavsson, I. and Rockborn, G. (1964) Chromosome abnormality in three cases of lymphatic leukaemia in cattle. *Nature* 203, 990.

Gustavsson, I., Fraccaro, M., Tiepolo, L. and Lindsten, J. (1968) Presumptive X-autosome translocation in a cow: preferential inactivation of the normal X chromosome. *Nature* 218, 183–184.

Hallerman, E.M., Theilman, J.L., Beckmann, J.S., Soller, M. and Womack, J.E. (1988) Mapping of bovine prolactin and rhodopsin genes in hybrid somatic cells. *Animal Genetics* 19, 123–131.

Hanada, M., Muramatsu, S., Abe, T. and Fukushima, T. (1981) Robertsonian chromosome polymorphism found in a local herd of the Japanese Black cattle. *Annales de Génétique et Sélection Animale* 13, 205–211.

Hansen, K.M. (1970) Tandem Fusion, Translokation und Unfruchtbarkeit beim Rind. In: *1. Europäisches Kolloquium über Zytogenetik (Chromosomenpathologie) in Veterinärmedizin und Säugetierkunde, Giessen, Germany*, pp. 115–118.

Hari, J.J., Franceschi, P., Casabianca, F., Boscher, J. and Popescu, C.P. (1984) Etude cytogénétique d'une population de bovins corses. *Compte Rendu de l'Academie d'Agriculture de France* 70, 191–199.

Harlizius, B., Hetzel, J. and Barendse, W. (1995) Comparative mapping of the proximal part of bovine chromosome 1. *Mammalian Genome* 6, 481–483.

Harlizius, B., Guérin, G. and Ferretti, L. (1996) IDVGA65 (D6S29), a SSCP marker assigned to BTA6 by means of FISH, genetic and synteny mapping. *Animal Genetics* 27, 379.

Hawken, R.J., Broom, M.F., van Stijn, T.C., Lumsden, J.M., Broad, T.E. and Maddox, J.F. (1996) Mapping the ovine genes encoding IL3, IL4, IL5, and CSF2 to sheep chromosome 5q13–q15 by FISH. *Mammalian Genome* 7, 858–859.

Hawkins, G.A., Solinas Toldo, S., Bishop, M.D., Kappes, S.M., Fries, R. and Beattie, C.W. (1995) Physical and linkage mapping of the bovine genome with cosmids. *Mammalian Genome* 6, 249–254.

Hawkins, G.A., Eggen, A., Hayes, H., Elduque, C. and Bishop, M.D. (1996) Tyrosinase-related protein-2 (DCT TYRP2) maps to bovine chromosome 12. *Mammalian Genome* 7, 474–475.

Hayes, H. (1995) Chromosome painting with human chromosome-specific DNA libraries reveals the extent and distribution of conserved segments in bovine chromosomes. *Cytogenetics and Cell Genetics* 71, 168–174.

Hayes, H. and Petit, E. (1993) Mapping of the beta-lactoglobulin gene and of an immunoglobulin M heavy chain-like sequence to homoeologous cattle, sheep, and goat chromosomes. *Mammalian Genome* 4, 207–210.

Hayes, H., Popescu, P. and Dutrillaux, B. (1993) Comparative gene mapping of lactoperoxidase, retinoblastoma, and alpha-lactalbumin genes in cattle sheep, and goats. *Mammalian Genome* 4, 593–597.

Hayes, H., Le Chalony, C., Goubin, G., Mercier, D., Payen, E., Bignon, C. and Kohno, K. (1996) Localization of ZNF164, ZNF146, GGTA1, SOX2, PRLR and EEF2 on homoeologous cattle, sheep and goat chromosomes by fluorescent *in situ* hybridization and comparison with the human gene map. *Cytogenetics and Cell Genetics* 72, 342–346.

Hediger, R., Johnson, S.E., Barendse, W., Drinkwater, R.D., Moore, S.S. and Hetzel, J. (1990) Assignment of the growth hormone gene locus to 19q26–qter in cattle and to 11q25–qter in sheep by *in situ* hybridization. *Genomics* 8, 171–174.

Hediger, R., Johnson, S.E. and Hetzel, D.J.S. (1991) Localization of the beta-subunit of follicle stimulating hormone in cattle and sheep by *in situ* hybridization. *Animal Genetics* 22, 237–244.

Henderson, L.M. and Bruere, A.N. (1979) Conservation of nucleolus organizer regions during evolution in sheep, goat, cattle and aoudad. *Canadian Journal of Genetics and Cytology* 21, 1–8.

Heriz, A., Arruga, M.V., Monteagudo, L.V., Tejedor, M.T., Pitel, F. and Echard, G. (1994) Assignment of the transition protein 1 (TNP1) gene to U17 bovine synteny group by PCR. *Mammalian Genome* 5, 742.

Herzog, A. and Hohn, H. (1984) Two new translocation type trisomies in calves, 60,XX,t(12;12),+12 and 60,XX,t(20;20),+20. In: *Proceedings of 6th European Colloquium on Cytogenetics of Domestic Animals, Zurich, Switzerland*, pp. 313–317.

Herzog, A., Hohn, H. and Rieck, G.V. (1977) Survey of recent situation of chromosome pathology in different breeds of German cattle. *Annales de Génétique et Sélection Animale* 9, 471–491.

Herzog, A., Hohn, H. and Olyschlager, F. (1982) Autosomal trisomy in calves with dwarfism. *Deutsche Tierärztliche Wochenschrift.* 89, 400–403.

Heuertz, S. and Hors-Cayla, M.-C. (1978) Carte génétique des bovins par la technique d'hybridation cellulaire. Localisation sur le chromosome X de la glucose-6-phosphate déshydrogénase, la phosphoglycérate kinase, l'alpha-galactosidase A et l'hypoxanthine guanine phosphorybosyl transférase. *Annales de Génétique* 21, 197–202.

Heuertz, S. and Hors-Cayla, C. (1981) Cattle gene mapping by somatic cell hybridization study of 17 enzyme markers. *Cytogenetics and Cell Genetics* 30, 137–145.

Hines, H.C. and Larsen, B. (1990) Evidence against linkage between two pairs of bovine blood group loci: B–Z and F–S. *Animal Genetics* 21, 431–432.

Hines, H.C., Kiddy, C.A., Brum, E.W. and Arawe, C.W. (1969) Linkage among cattle blood and milk polymorphisms. *Genetics* 62, 401–412.

Houseal, T.W., Dackowski, W.R., Landes, G.M. and Klinger, K.W. (1994) High resolution mapping of overlapping cosmids by fluorescence *in situ* hybridization. *Cytometry* 15, 193–198.

Hudson, T.J., Stein, L.D., Gerety, S.S., Ma, J., Castle, A.B., Silva, J., Slonim, D.K. et al. (1995) An STS-based map of the human genome. *Science* 270, 1945–1954.

Iannuzzi, L., Rangel-Figueiredo, T., Di Meo, G.P. and Ferrara, L. (1992) A new Robertsonian translocation in cattle, rob(15;25). *Cytogenetics and Cell Genetics* 59, 280–283.

Iannuzzi, L., Di Meo, G.P., Rangel-Figueiredo, T. and Ferrara, L. (1993a) A new case of centric fusion translocation in cattle, rob(16;18). In: *Proceedings of the 8th North American Colloquium on Domestic Animal Cytogenetics and Gene Mapping, Guelph, Canada*, pp. 127–128.

Iannuzzi, L., Gallagher, D.S., Ryan, A.M., Di Meo, G.P. and Womack, J.E. (1993b) Chromosomal localization of omega and trophoblast interferon genes in cattle and river buffalo by sequential R-banding and fluorescent *in situ* hybridization. *Cytogenetics and Cell Genetics* 62, 224–227.

Jagiello, G.M., Miller, W.A., Ducayen, M.B. and Lin, J.S. (1974) Chiasma frequency and disjunctional behavior of ewe and cow oocytes matured *in vitro*. *Biology of Reproduction* 10, 354–363.

Janel, N., Schibler, L., Oustry, A., Kerbiriou-Nabias, D., Cribiu, E.P. and Vaiman, D. (1996) The localization of the von Willebrand factor gene on cattle, sheep and goat chromosomes illustrates karyotype evolution in mammals. *Mammalian Genome* 7, 633–634.

Johnson, S.E., Moore, S.S., MacKinnon, R., Hetzel, D.J.S. and Barendse, W. (1995) The cosmid CSSM25 assigns syntenic group U2 to bovine chromosome 9 and is localized to ovine chromosome 8. *Mammalian Genome* 6, 529–531.

Johnson, S.E., Barendse, W. and Hetzel, D.J.S. (1993) The gamma-fibrinogen gene (FGG) maps to bovine chromosome 17 in both cattle and sheep. *Cytogenetics and Cell Genetics* 62, 176–180.

Jørgensen, C.B., Olsaker, I. and Thomsen, P.D. (1995) A polymorphic bovine dinucleotide repeat D12S25 (IOBT323) at chromosome 12q24. *Animal Genetics* 26, 447.

Kappes, S.M., Keele, J.W., Stone, R.T., McGraw, R.A., Sonstegard, T.S., Smith, T.P., Lopez-Corrales, N.L. and Beattie, C.W. (1997) A second-generation linkage map of the bovine genome. *Genome Research* 7, 235–249.

Karall, C., Looft, C., Barendse, W. and Kalm, E. (1997) Detection and mapping of a point mutation in the bovine butyrophilin gene using F-SSCP-analysis. *Animal Genetics* 28, 66.

King, W.A., Linares, T., Gustavsson, I. and Bane, A. (1980) Presumptive translocation type trisomy in embryos sired by bulls heterozygous for the 1/29 translocation. *Hereditas* 92, 167–169.

Klungland, H., Vage, D.I. and Lien, S. (1997a) Linkage mapping of the *Fcgamma2* receptor gene to bovine chromosome 18. *Mammalian Genome* 8, 300–301.

Klungland, H., Gomez-Raya, L., Howard, C.J., Collins, R.A., Rogne, S. and Lien, S. (1997b) Mapping of bovine FcgammaR (*FCGR*) genes by sperm typing allows extended use of human map information. *Mammalian Genome* 8, 573–577.

Kovacs, A. (1975) Über eine neue autosomale Translokation beim ungarischen Fleckviehrind. In: *Proceedings of the 2nd European Colloquium on Cytogenetics of Domestic Animals, Giessen, Germany*, pp. 162–167.

Kovacs, A. and Szepeshelyi, F. (1977) Chromosomal screening of breeding bulls in Hungary. *Journal of Dairy Science* 70, 236.

Kovacs, A., Meszaros, I., Sellyei, M. and Vass, I. (1973) Mosaic centromeric fusion in a Holstein-Friesian bull. *Acta Biologica Academia Sciencia Hungaria* 24, 215–220.

Koykul, W., Switonski, M. and Basrur, P.K. (1997) The X bivalent in fetal bovine oocytes. *Hereditas* 126, 59–65.

Kulseth, M.A., Womack, J.E. and Rogne, S. (1992) The sequencing and chromosomal assignment of bovine polymeric immunoglobulin receptor cDNA. *Animal Genetics* 23(Suppl. 1), 46–47.

Kulseth, M.A., Solinas Toldo, S., Fries, R., Womack, J.E., Lien, S. and Rogne, S. (1994) Chromosomal localization and detection of DNA polymorphisms in the bovine polymeric immunoglobulin receptor gene. *Animal Genetics* 25, 113–117.

Laan, M., Kallioniemi, O.P., Hellsten, E., Alitalo, K., Peltonen, L. and Palotie, A. (1995) Mechanically stretched chromosomes as targets for high-resolution FISH mapping. *Genome Research* 5, 13–20.

Langer, J.A., Puvanakrishnan, R. and Womack, J.E. (1992) Somatic cell mapping of the bovine interferon-alpha receptor. *Mammalian Genome* 3, 237–240.

Lanneluc, I., Mulsant, P., Saidi-Mehtar, N. and Elsen, J.M. (1996) Synteny conservation between parts of human chromosome 4q and bovine and ovine chromosomes 6. *Cytogenetics and Cell Genetics* 72, 212–214.

Larsen, B. (1965) Test for linkage of the genes for haemoglobin types and the loci controlling blood and transferrin types in cattle. *Aarsberetning fra Institut for Sterilitetsforskning, KVI* 1965, 183–191.

Larsen, B. (1970) Linkage relations of blood group and polymorphic protein systems in cattle. *Aarsberetning fra Institut for Sterilitetsforskning, KVI* 13, 165–194.

Larsen, B. (1977) On linkage relations of ceruloplasmin polymorphism (Cp) in cattle. *Animal Blood Groups and Biochemical Genetics* 8, 111–113.

Larsen, B. and Thymann, M. (1966) Studies on milk protein polymorphism in Danish cattle and the interaction of the controlling genes. *Acta Veterinaria Scandinavica* 7, 189–205.

Larsen, B., Christensen, K. and Agerholm, J.S. (1992) A possible location of the bovine Z blood group system on chromosome 8. *Animal Genetics* 23(Suppl. 1), 77–78.

Larsen, N.J., Womack, J.E. and Kirkpatrick, B.W. (1996) Seven genes from human chromosome 18 map to chromosome 24 in the bovine. *Cytogenetics and Cell Genetics* 73, 184–186.

Larsen, N.J., Eggen, A., Hayes, H., Byla, B., Doud, L., Bishop, M.D. and Kirkpatrick, B.W. (1997) UW54 and UW63, two polymorphic bovine microsatellites on chromosomes 6 and 8q18, respectively. *Animal Genetics* 28, 377–378.

Le Chalony, C., Pibouin, L., Hayes, L., Apiou, F., Dutrillaux, B. and Goubin, G. (1995) Partial nucleotide sequence and chromosomal localization of a bovine zinc finger gene ZNF164. *Cytogenetics and Cell Genetics* 70, 192–194.

Lei, L. and Womack, J.E. (1997) Somatic cell mapping of the adenine nucleotide translocator gene family in cattle. *Mammalian Genome* 8, 773–774.

Lester, D.H., Russell, G.C., Barendse, W. and Williams, J.L. (1996) The use of denaturing gradient gel electrophoresis in mapping the bovine tumor necrosis factor alpha gene locus. *Mammalian Genome* 7, 250–252.

Leveziel, H. and Hines, H.C. (1984) Linkage in cattle between the major histocompatibility complex (BoLA) and the M blood group system. *Genetics, Selection, Evolution* 16, 405–416.

Lewin, H.A., Schmitt, K., Hubert, R., van Eijk, M.J.T. and Arnheim, N. (1992) Close linkage between bovine prolactin and BoLa-DRB3 genes: genetic mapping by single sperm typing. *Genomics* 13, 44–48.

Li, L., Teale, A., Bensaid, A., Dunlap, S., Dietz, A.B. and Womack, J.E. (1992) Somatic cell mapping of T-cell receptor CD3 complex and CD8 genes in cattle. *Immunogenetics* 36, 224–229.

Lichter, P., Tang, C.-J.C., Call, K., Hermanson, G., Evans, G.A., Housman, D. and Ward, D.C. (1990) High resolution mapping of human chromosome 11 by *in situ* hybridization with cosmid clones. *Science* 247, 64–69.

Lien, S., Gomezraya, L. and Vage, D.I. (1995) A BsmAI polymorphism in the bovine lipoprotein lipase gene. *Animal Genetics* 26, 283–284.

Lien, S., Sundvold, H., Klungland, H. and Våge, D.I. (1997) Two novel polymorphisms in the bovine obesity gene (OBS). *Animal Genetics* 28, 245.

Lindberg, P.-G. and Andersson, L. (1988) Close association between DNA polymorphism of bovine major histocompatibility complex class I genes and serological BoLA-A specificities. *Animal Genetics* 19, 245–255.

Livingston, R.J. (1988) The mapping of phosphoglycerate kinase, Duchenne muscular dystrophy, and clotting factor IX to the bovine X chromosome by *in situ* hybridization. Dissertation, Texas A&M Universtiy.

Logue, D.N. (1977) Meiosis in the domestic ruminants with particular reference to Robertsonian translocations. *Annales des Génétique et Séléction Animale* 9, 493–507.

Logue, D.N. and Harvey, M.J.A. (1978) Meiosis and spermatogenesis in bulls heterozygous for a presumptive 1/29 Robertsonian translocation. *Journal of Reproduction and Fertility* 54, 159–165.

Lojda, L., Mikulas, L. and Rubes, L. (1975) Einige Ergebnisse der Chromosomenuntersuchungen im Rahmen der staatlichen Erbgesundheitskontrolle beim

Rind. In: *Proceedings of the 2nd European Colloquium of Cytogenetics of Domestic Animals, Giessen, Germany*, pp. 269–276.

Lojda, L., Rubes, J., Staiksova, M. and Havrandsova, J. (1976) Chromosomal findings in some reproductive disorders in bulls. In: *Proceedings of the 8th International Congress of Animal Reproduction and Artificial Insemination, Krakow, Poland*, pp. 1–158.

Lomax, M.I., Riggs, P.K. and Womack, J.E. (1995) Structure and chromosomal location of the bovine gene for the heart muscle isoform of cytochrome c oxidase subunit VIII. *Mammalian Genome* 6, 118–122.

Lopez-Corrales, N.L., Sonstegard, T.S., Smith, T.P.L. and Beattie, C.W. (1997) Physical assignment of glucagon and nebulin in cattle, sheep, and goat. *Mammalian Genome* 8, 428–429.

Lord, E.A., Lumsden, J.M., Dodds, K.G., Henry, H.M., Crawford, A.M., Ansari, H.A., Pearce, P.D., Maher, D.W., Stone, R.T., Kappes, S.M., Beattie, C.W. and Montgomery, G.W. (1996) The linkage map of sheep chromosome 6 compared with orthologous regions in other species. *Mammalian Genome* 7, 373–376.

Lüdecke, H.-J., Senger, G., Claussen, U., and Horsthemke, B. (1989) Cloning defined regions of the human genome by microdissection of banded chromosomes and enzymatic amplification. *Nature* 338, 348–350.

Ma, R.Z., Beever, J.E., Da, Y., Green, C.A., Russ, I., Park, C., Heyen, D.W., Everts, R.E., Fisher, S.R., Overton, K.M., Teale, A.J., Kemp, S.J., Hines, H.C., Guérin, G. and Lewin, H.A. (1996) A male linkage map of the cattle (*Bos taurus*) genome. *Journal of Heredity* 87, 261–271.

McAvin, J.C., Patterson, D. and Womack, J.E. (1988) Mapping of bovine PGRS and PAIS genes in hybrid somatic cells: syntenic conservation with human chromosome 21. *Biochemical Genetics* 26, 9–18.

Maciulla, J.H., Zhang, H.M. and DeNise, S.K. (1997) A novel polymorphism in the bovine insulin-like growth factor binding protein-3 (IGFBP3) gene. *Animal Genetics* 28, 375.

Mahrous, K.F., Hassanane, M.S. and El-Kholy, A.F. (1994) Robertsonian translocation and freemartin cases in hybrid Friesian cows raised in Egypt. *Egyptian Journal of Animal Production* 31, 213–220.

Malerba, F. (1997) Individuazione di una nuova translocazione Robertsoniana in un torello di razza Marchigiana: sua determinazione e studio dei suoi effetti sulla fertilità. Corso di Laurea in Scienze Agrarie, Università degli Studi di Milano, Milano, Italy.

Marquess, F.L., Brenneman, R.A., Schmutz, S.M., Taylor, J.F. and Davis, S.K. (1997) A highly polymorphic bovine dinucleotide repeat SOD1MICRO2. *Animal Genetics* 28, 70.

Martín-Burriel, I., Chowdhary, B.P., Prakash, B., Zaragoza, P. and Olsaker, I. (1996) A polymorphic bovine dinucleotide repeat DXYS4 (IOZARA1489) at the pseudo-autosomal region of the sex chromosomes. *Animal Genetics* 27, 287.

Martín-Burriel, I., Osta, R., Barendse, W. and Zaragoza, P. (1997a) Linkage mapping of a bovine alpha-lactalbumin pseudogene (LALBAps) using a LINE-associated polymorphism. *Animal Genetics* 28, 316.

Martín-Burriel, I., Osta, R., Barendse, W. and Zaragoza, P. (1997b) New polymorphism and linkage mapping of the bovine lactotransferrin gene. *Mammalian Genome* 8, 704–705.

Martín-Burriel, I., Elduque, C., Osta, R., Laurent, P., Barendse, W. and Zaragoza, P. (1998) SINEVA polymorphism and mapping of the bovine pregnancy-associated glycoprotein 1 gene. *Mammalian Genome* 9, 179–180.

Masabanda, J., Kappes, S.M., Smith, T.P.L., Beattie, C.W. and Fries, R. (1996) Mapping of a linkage group to the last bovine chromosome (BTA27) without an assignment. *Mammalian Genome* 7, 229–230.

Masabanda, J., Wigger, G., Eggen, A., Stranzinger, G. and Fries, R. (1997) The bovine glutamine synthase gene (*GLUL*) maps to 10q33 and a pseudogene (*GLULP*) to 16q21. *Mammalian Genome* 8, 794–795.

Masabanda, J., Ewald, D., Buitkamp, J., Potter, K. and Fries, R. (1998) Molecular markers for the bovine fibrillin 1 gene (FBN1) map to 10q26. *Animal Genetics* 29, 460–461.

Masuda, H., Takahaschi, T., Soejima, A. and Waido, Y. (1978) Centric fusion of chromosome in a Japanese Black bull and its offspring. *Japanese Journal of Zootechnological Sciences* 49, 853–858.

Mathiason, K.J., Honeycutt, D.A., Burns, B., Taylor, J.F. and Skow, L.C. (1995) Identification and mapping of dinuecleotide repeat near the bovine villin locus. In: *Proceedings of the 9th North American Colloquium on Domestic Animal Cytogenetics and Gene Mapping, College Station, Texas, USA*, p. 53.

Mayr, B. and Czaker, R. (1981) Variable positions of nucleolus organizer regions in Bovidae. *Experientia* 37, 564–565.

Mayr, B., Themsessel, H., Wochl, F. and Schleger, W. (1979) Reziproke Translokation 60,XY,t(10;14) beim Rind. *Journal of Animal Breeding and Genetics* 96, 44–47.

Mayr, B., Krutzler, H., Auer, H. and Schleger, W. (1983) Reciprocal translocation 60,XY,t(10;14)(21;24) in cattle. *Journal of Reproduction and Fertility* 69, 629–630.

Messer, L.A., Wang, L., Tuggle, C.K., Yerle, M., Chardon, P., Pomp, D., Womack, J.E., Barendse, W., Crawford, A.M., Notter, D.R. and Rothschild, M.F. (1997) Mapping of the melatonin receptor 1a (MTNR1A) gene in pigs, sheep, and cattle. *Mammalian Genome* 8, 368–370.

Méténier-Delisse, L., Hayes, H., Leroux, C., Giraud-Delville, C., Levéziel, H., Guérin, G., Martin, P. and Grosclaude, F. (1997) Isolation and molecular characterization of bovine Rhesus-like transcripts and chromosome mapping of the relevant locus. *Animal Genetics* 28, 202–209.

Mezzelani, A., Solinas Toldo, S., Nocart, M., Guérin, G., Ferretti, L. and Fries, R. (1994) Mapping of syntenic groups U7 and U27 to bovine chromosomes 25 and 12, respectively. *Mammalian Genome* 5, 574–576.

Mezzelani, A., Zhang, Y., Redaelli, L., Castiglioni, B., Leone, P., Williams, J., Solinas Toldo, S., Wigger, G., Fries, R. and Ferretti, L. (1995) Chromosomal localization and molecular characterization of 53 cosmid-derived bovine microsatellites. *Mammalian Genome* 6, 629–635.

Milan, D., Yerle, M., Schmitz, A., Chaput, B., Vaiman, M., Frelat, G. and Gellin, J. (1993) A PCR-based method to amplify DNA with random primers: determining the chromosomal content of procine flow-karyotype peaks by chromosome painting. *Cytogenetics and Cell Genetics* 62, 139–141.

Miller, J.R., Thomsen, P.D., Dixon, S.C., Tucker, E.M., Konfortov, B.A. and Harbitz, I. (1991) Synteny mapping of the bovine IGHG2, CRC and IGF1 genes. *Animal Genetics* 23, 51–58.

Miyake, Y.I. and Kaneda, Y. (1987) A new type of Robertsonian translocation (1/26) in a bull with unilateral cryptorchidism, probably occurring *de novo*. *Japanese Journal of Veterinary Science* 49, 1015–1019.

Miyake, Y.I., Murakami, R.K. and Kaneda, Y. (1991) Inheritance of the Robertsonian translocation (1/21) in the Holstein-Friesian Cattle I. Chromosome analysis. *Journal of Veterinary Medical Science* 53, 113–116.

Monteagudo, L.V., Arruga, M.V., Tejedor, M.T., Savva, D. and Skidmore, C.J. (1992) Mapping of six genes in cattle using a panel of interspecific hamster × cattle hybrid clones. *Animal Genetics* 23(Suppl. 1), 82.

Monteagudo, L.V., Yerle, M., Olsaker, I., Mazhar, K., Skidmore, C.J., Savva, D. and Arruga, M.V. (1994) Mapping of cosmid clones carrying microsatellites to bovine chromosomes using FISH. *Animal Genetics* 25(Suppl. 2), 53 (abstract).

Monteagudo, L.V., Heriz, A., Flavin, N., Rogers, M., Ennis, S. and Arruga, M.V. (1996) Fluorescent *in situ* localization of the bovine activin receptor type IIA locus on chromosome 2 (2q2.3-2.4). *Mammalian Genome* 7, 869.

Moody, D.E., Pomp, D. and Barendse, W. (1995a) Rapid communication: restriction fragment length polymorphism in amplification products of the bovine growth hormone-releasing hormone gene. *Journal of Animal Science* 73, 3789.

Moody, D.E., Pomp, D. and Barendse, W. (1995b) Restriction fragment length polymorphism in amplification products of the bovine pit1 gene and assignment of pit1 to bovine chromosome. *Animal Genetics* 26, 45–47.

Moody, D.E., Pomp, D., Barendse, W. and Womack, J.E. (1995c) Assignment of the growth hormone receptor gene to bovine chromosome 20 using linkage analysis and somatic cell mapping. *Animal Genetics* 26, 341–343.

Moore, S.S. and Byrne, K. (1997) A SSCP at the bovine melanoma adhesion molecule (MCAM) locus. *Animal Genetics* 28, 243–244.

Muggli-Cockett, N.E. and Stone, R.T. (1988) Identification of genetic variation in the bovine major histocompatibility complex DRβ-like genes using sequenced bovine genomic probes. *Animal Genetics* 19, 213–225.

Muggli-Cockett, N.E. and Stone, R.T. (1989) Partial nucleotide sequence of a bovine major histocompatibility class II DR-beta-like gene. *Animal Genetics* 20, 361–370.

Neibergs, H.L., Gallagher, D.S., Georges, M., Sargeant, L.S., Dietz, A.B. and Womack, J.E. (1993) Physical mapping of inhibin beta-A in domestic cattle. *Mammalian Genome* 4, 328–332.

Nijman, I.J., Lenstra, J.A., Schwerin, M. and Olsaker, I. (1996a) Polymorphisms and physical locations of three bovine microsatellite loci – IOBT395, IOBT528, IOBT1401. *Animal Genetics* 27, 221–222.

Nijman, I.J., Lenstra, J.A., Schwerin, M., Guérin, G., Castiglioni, B., Ferretti, L. and Olsaker, I. (1996b) Physical and genetic mapping of two polymorphic bovine dinucleotide repeats – IOBT450 (D6S31) and IDVGA80 (D2S46). *Animal Genetics* 27, 377–378.

Nonneman, D., Shibuya, H. and Johnson, G.S. (1996) A *Bsm*AI PCR/RFLP in the bovine microtubule-associated protein-5 (MAP1B) gene. *Animal Genetics* 27, 288–289.

Ohno, S. (1969) The problem of the bovine freemartin. *Journal of Reproduction and Fertility* Suppl. 7, 53–61.

Olsaker, I., Solinas Toldo, S. and Fries, R. (1996a) A highly polymorphic bovine dinucleotide repeat D19S4 (IOBT 34) at chromosome 19q21. *Animal Genetics* 27, 58.

Olsaker, I., Chowdhary, B.P. and Guerin, G. (1996b) A highly polymorphic bovine dinucleotide repeat D3S40 (IOBT 250) derived from a chimeric cosmid clone. *Animal Genetics* 27, 59.

Olsaker, I., Prakash, B., Guttersrud, O.A. and Chowdhary, B.P. (1997a) A genetic and physical marker for bovine chromosome 27: IOBT 313 (D27S14). *Animal Genetics* 28, 240.

Olsaker, I., Gundersen, L. and Chowdhary, B.P. (1997b) A highly polymorphic bovine dinucleotide repeat at chromosome 26q22: IOBT 730 (D26S23). *Animal Genetics* 28, 242.

Osta, R., García-Muro, E. and Zarazaga, I. (1995) A MspI polymorphism at the bovine alpha-lactalbumin gene. *Animal Genetics* 26, 204–205.

Oulmouden, A., Vaiman, D., Oustry, A., Cribiu, E.P. and Julien, R. (1995) Localization of fucosyl-transferase gene to bovine and goat chromosome 7 and sheep chromosome 5. *Mammalian Genome* 6, 760–761.

Papp, M. and Kovàcs, A. (1980) 5/18 dicentric Robertsonian translocation in a Simmental bull. In: *Proceedings of the 4th European Colloquium on Cytogenetics of Domestic Animals, Uppsala, Sweden*, pp. 51–54.

Paris Conference (1975) Supplement (1975): standardization in human cytogenetics. *Cytogenetics and Cell Genetics* 15, 201–238.

Park, C., Russ, I., Da, Y. and Lewin, H.A. (1994) Mapping of F13A to BTA23: gene order and recombination differences between bulls detected by sperm typing. *Animal Genetics*, 25, 50.

Parra, I. and Windle, B. (1993) High resolution visual mapping of stretched DNA by fluorescent hybridization. *Nature Genetics* 5, 17–21.

Payen, E., Saidi-Mehtar, N., Pailhoux, E. and Cotinot, C. (1995) Sheep gene mapping – assignment of ALDOB, CYP19, WT and SOX2 by somatic cell hybrid analysis. *Animal Genetics* 26, 331–333.

Pearce, P.D., Ansari, H.A., Maher, D.W. and Broad, T.E. (1995) Five regional localizations to the sheep genome – first assignments to chromosomes 5 and 12. *Animal Genetics* 26, 171–176.

Pfister-Genskow, M., Hayes, H., Eggen, A. and Bishop, M.D. (1996a) Chromosomal localization of the bovine obesity (OBS) gene. *Mammalian Genome* 7, 398–399.

Pfister-Genskow, M., Hayes, H., Eggen, A. and Bishop, M.D. (1996b) The leptin receptor (*LRPR*) gene maps to bovine chromosome 3q33. *Mammalian Genome* 8, 227.

Pfister-Genskow, M., Weesner, G.D., Hayes, H., Eggen, A. and Bishop, M.D. (1997) Physical and genetic localization of the bovine cannaboid receptor (*CNR1*) gene to bovine chromosome 9. *Mammalian Genome* 8, 301–302.

Piedrahita, J.A., Weaks, R., Petrescu, A., Shrode, T.W., Derr, J.N. and Womack, J.E. (1997) Genetic characterization of the bovine leukaemia inhibitory factor (LIF) gene: isolation and sequencing, chromosome assignment and microsatellite analysis. *Animal Genetics* 28, 14–20.

Pinkel, D., Straume, T. and Gray, J.W. (1986) Cytogenetic analysis using quantitative, high-sensitivity fluorescence hybridization. *Proceedings of the National Academy of Sciences of the USA* 83, 2934–2938.

Pinton, A., Ducos, A., Berland, H.M., Séguéla, A., Blanc, M.F., Darré, A., Mimar, S. and Darré, R. (1997) A new Robertsonian translocation in Holstein-Friesian cattle. *Genetics, Selection, Evolution* 29, 523–526.

Pitel, F., Lantier, I., Riquet, J., Lanneluc, I., Tabet-Aoul, K., Saïdi-Mehtar, N., Lantier, F. and Gellin, J. (1994) Cloning, sequencing, and localization of an ovine fragment of the NRAMP gene, a candidate for the ITY/LSH/BCG gene. *Mammalian Genome* 5, 834–835.

Plachot, M. and Popescu, C.P. (1991) Les anomalies chromosomiques et géniques. Leurs conséquences sur le développement et la reproduction. In: Thibault, E.C.

and Levasseur, M.C. (eds) *Reproduction chez les Mammifères*. Ellipses, Paris, pp. 687–712.

Pollock, D.L. (1974) Chromosome studies in artificial insemination sires in Great Britain. *Veterinary Records* 95, 266–277.

Pomp, D., Zou, T., Clutter, A.C. and Barendse, W. (1997) Rapid communication: mapping of leptin to bovine chromosome 4 by linkage analysis of a PCR-based polymorphism. *Journal of Animal Science* 75, 1427.

Ponce de Leon, F.A., Osborne, B.A. and Goldsby, R.A. (1992) Chromosomal localization of the bovine IgG2, IgG3, and IgA immunoglobulin heavy chain genes. In: *Proceedings, 10th European Colloquium on Cytogenetics and Cell Genetics of Domestic Animals, Utrecht, The Netherlands*, p. 15.

Ponce de Leon, F.A., Ambady, S., Hawkins, G.A., Kappes, S.M., Bishop, M.D., Robl, J.M. and Beattie, C.W. (1996) Development of a bovine X chromosome linkage group and painting probes to assess cattle, sheep, and goat X chromosome segment homologies. *Proceedings of the National Academy of Sciences of the USA* 93, 3450–3454.

Popescu, C.P. (1971) Les chromosomes méiotiques du boeuf. *Annales de Génétique et Sélection Animale* 3, 125–143.

Popescu, C.P. (1972) Un cas possible d'inversion péricentrique chez les bovins. *Annales de Génétique* 15, 197–200.

Popescu, C.P. (1973) L'hétérochromatine constitutive dans le caryotype bovin normal et anormal. *Annales de Génétique* 16, 183–188.

Popescu, C.P. (1977) A new type of Robertsonian translocation in cattle. *Journal of Heredity* 68, 139–142.

Popescu, C.P. (1978) A study of meiotic chromosomes in bulls carrying the 1/29 translocation. *Annales de Biologie Animale, Biochimie, Biophysique* 18, 383–389.

Popescu, C.P. (1990) Chromosomes of the cow and bull. In: McFeely, R.A. (ed.) *Advances in Veterinary Science and Comparative Medicine, Domestic Animal Cytogenetics*. Vol. 34, Academic Press, San Diego, pp. 41–71.

Popescu, C.P. and Pech, A. (1991) Une bibliographie sur la translocation 1/29 de bovins dans le monde (1964-1990). *Annales de Zootechnie* 40, 271–305.

Popescu, C.P., Boscher, J., Hayes, H.C., Ban, J. and Kettmann, R. (1995) Chromosomal localization of the BLV receptor candidate gene in cattle, sheep, and goat. *Cytogenetics and Cell Genetics* 69, 50–52.

Popescu, C.P., Long, S., Riggs, P., Womack, J., Schmutz, S., Fries, R. and Gallagher, D.S. (1996) Standardization of cattle karyotype nomenclature: report of the committee for the standardization of the cattle karyotype. *Cytogenetics and Cell Genetics* 74, 259–261.

Popescu, C.P., Hayes, H., Meiniel, R., Creveaux, I. and Meiniel, A. (1997) Localization of the SCO-spondin gene to cattle chromosome 4. *Chromosome Research* 5, 276–277.

Prakash, B., Kuosku, V., Olsaker, I., Gustavsson, I. and Chowdhary, B.P. (1996) Comparative FISH mapping of bovine cosmids to reindeer chromosomes demonstrates conservation of the X-chromosome. *Chromosome Research* 4, 214–217.

Prakash, B., Olsaker, I., Gustavsson, I. and Chowdhary, B.P. (1997) FISH mapping of three bovine cosmids to cattle, goat, sheep and buffalo X-chromosomes. *Hereditas* 126, 115–119.

Prinzenberg, E.M. and Erhardt, G. (1997) Bovine beta-lactoglobulin I is identified by amplification created restriction site using SmaI. *Animal Genetics* 28, 379.

Prinzenberg, E.M., Hiendleder, S. and Erhardt, G. (1997) A trinucleotide repeat polymorphism is present in bovine CSN1S1. *Animal Genetics* 28, 379–380.

Redaelli, L., Zhang, Y., Castiglioni, B., Mezzelani, A., Comincini, S., Guerin, G. and Ferretti, L. (1996) Characterization and mapping of three bovine polymorphic microsatellite loci. *Animal Genetics* 27, 121.

Ricketts, M.H., Simons, M.J., Parma, J., Mercken, L., Dong, Q. and Vassart, G. (1987) A nonsense mutation causes hereditary goitre in the Afrikander cattle and unmasks alternative splicing of thyroglobulin transcripts. *Proceedings of the National Academy of Sciences of the USA* 84, 3181–3184.

Riggs, P.K., Owens, K.E., Rexroad, C.E., Jr, Amaral, M.E. and Womack, J.E. (1997) Development and initial characterization of a *Bos taurus* × *B. gaurus* interspecific hybrid backcross panel. *Journal of Heredity* 88, 373–379.

Roelen, B.A., Van Eijk, M.J., Van Rooijen, M.A., Bevers, M.M., Larson, J.H., Lewin, H.A. and Mummery, C.L. (1998) Molecular cloning, genetic mapping, and developmental expression of a bovine transforming growth factor beta (TGF-beta) type I receptor. *Molecular Reproduction and Development* 49, 1–9.

Rogers, D.S., Gallagher, D.S. and Womack, J.E. (1991) Somatic cell mapping of the genes for Anti-Müllerian hormone and osteonectin in cattle: identification of a new bovine syntenic group. *Genomics* 9, 298–300.

Roldan, E.R.S., Merani, M.S. and Von Lawzewitsch, I. (1984) Two abnormal chromosomes found in one cell line of a mosaic cow with low fertility. *Genetics, Selection, Evolution* 16, 135–142.

Rubes, J., Borkovec, L., Borkovcova, Z. and Urbanova, J. (1992) A new robertsonian translocation in cattle, rob(16;21). In: *Proceedings of 10th European Colloquium on Cytogenetics of Domestic Animals, Utrecht, The Netherlands*, pp. 201–205.

Rubes, J., Musilova, P., Borkovec, L., Borkovcova, Z., Svecova, D. and Urbanova, J. (1996) A new Robertsonian translocation in cattle, rob(16;20). *Hereditas* 124, 275–279.

Ryan, A.M. and Womack, J.E. (1993) Somatic cell mapping of the bovine prion protein gene and restriction fragment length polymorphism studies in cattle and sheep. *Animal Genetics* 24, 23–26.

Ryan, A.M. and Womack, J.E. (1994) Somatic cell mapping of the mitochondrial aconitase gene (ACO2) to bovine chromosome 5. *Animal Genetics* 25, 123.

Ryan, A.M. and Womack, J.E. (1995) Somatic cell mapping of the ski proto-oncogene to bovine syntenic group U25 (Bta 27). *Mammalian Genome* 6, 560–561.

Ryan, A.M., Gallagher, D.S. and Womack, J.E. (1992a) Syntenic mapping and chromosomal localization of bovine alpha and beta interferon genes. *Mammalian Genome* 3, 575–578.

Ryan, A.M., Gallagher, D.S. and Womack, J.E. (1992b) Synteny mapping, fluorescent *in situ* localization and allele frequencies for type I interferon genes in cattle. *Animal Genetics* 23(Suppl. 1), 81.

Ryan, A.M., Gallagher, D.S. and Womack, J.E. (1993a) Somatic cell mapping of omega and trophoblast interferon genes to bovine syntenic group U18 and *in situ* localization to chromosome 8. *Cytogenetics and Cell Genetics* 63, 6–10.

Ryan, A.M., Womack, J.E., Yu, J., Lin, J.-H., Wu, X.-R., Sun, T.-T., Clarke, V. and D'Eustachio, P. (1993b) Chromosomal localization of uroplakin genes of cattle and mice. *Mammalian Genome* 4, 656–661.

Ryan, A.M., Gallagher, D.S., Schober, S., Schwenger, B. and Womack, J.E. (1994) Somatic cell mapping and *in situ* localization of the bovine uridine monophosphate synthase gene (UMPS). *Mammalian Genome* 5, 46–47.

Ryan, A.M., Schelling, C.P., Womack, J.E. and Gallagher, D.S., Jr (1997) Chromosomal assignment of six muscle-specific genes in cattle. *Animal Genetics* 28, 84–87.

Samarineanu, N.E., Livescu, B. and Granciu, I. (1977) Identification of Robertsonian translocation in some breeds of cattle. *Lucrarile Stiintifce ale Institutului de Cercetari si Cresterea Taurinelor Corbeanca* 3, 53–60.

Schläpfer, J., Yang, Y., Rexroad, C., 3rd and Womack, J.E. (1997a) A radiation hybrid framework map of bovine chromosome 13. *Chromosome Research* 5, 511–519.

Schläpfer, J., Gallagher, D.S., Burzlaff, J.D., Davis, S.K., Taylor, J.F. and Womack, J.E. (1997b) Physical mapping of the endothelin receptor type B to bovine chromosome 12. *Mammalian Genome* 8, 380–381.

Schläpfer, J., Kata, S.R., Amarante, M.R. and Womack, J.E. (1997c) Syntenic assignment of thrombomodulin (THBD) and phosphatidylinositol-specific phospholipase C (PLC-II) to bovine chromosome 13. *Animal Genetics* 28, 308–309.

Schmitz, A., Oustry, A., Chaput, B., Bahri-Darwich, I., Yerle, M., Milan, D., Frelat, G. and Cribiu, E.P. (1994) The bovine bivariate flow karyotype and peak identification by chromosome painting and PCR generated probes. *Mammalian Genome* 6, 415–420.

Schmutz, S.M., Berryere, T.G., Moker, J.S., Thue, T.D. and Winkelman, D.C. (1994) Gene mapping from a bovine 1;29 DNA library prepared using chromosome microdissection. *Mammalian Genome* 5, 138–141.

Schmutz, S.M., Moker, J.S., Gallagher, D.S., Kappes, S.M. and Womack, J.E. (1996a) *In situ* hybridization mapping of LDHA and IGF2 to cattle chromosome 29. *Mammalian Genome* 7, 473.

Schmutz, S.M., Moker, J.S., Leipprandt, J.R. and Friderici, K.H. (1996b) Beta-mannosidase maps to cattle chromosome 6. *Mammalian Genome* 7, 474.

Schmutz, S.M., Cornwell, D., Moker, J.S. and Troyer, D.L. (1996c) Physical mapping of SOD1 to bovine chromosome 1. *Cytogenetics and Cell Genetics* 72, 37–39.

Schmutz, S.M., Moker, J.S. and Berryere, T.G. (1998) *In situ* hybridization of five loci to cattle chromosome 1. *Cytogenetics and Cell Genetics* 8, 51–53.

Schwenger, B. and Schöber, S. (1991) Development of a DNA test for DUMPS diagnosis: characterization of the bovine UMP-synthase gene. *Animal Genetics* 23(Suppl. 1), 57–58.

Schwenger, B., Schöber, S. and Simon, D. (1991) A TaqI polymorphism at the bovine locus for uridinemonophosphate synthase (UMPS). *Animal Genetics* 23, 82.

Schwerin, M., Solinas Toldo, S., Eggen, A., Brunner, R., Seyfert, H.M. and Fries, R. (1994) The bovine lactoferrin gene (LTF) maps to chromosome 22 and syntenic group U12. *Mammalian Genome* 5, 486–489.

Sharma, S. and Popescu, C.P. (1986) The meiotic chromosomes of Creole cattle from Guadeloupe. *Revue d'Elevage et de Médecine Vétérinaire des Pays Tropicaux* 38, 353–357.

Shimizu, N., Shimizu, Y., Koudo, I., Woods, C. and Wegner, T. (1981) The bovine genes for phosphoglycerate kinase, glucose-6-phosphate dehydrogenase, alpha-galactosidase, and hypoxanthine phosphoribosyltransferase are linked to the X chromosome in cattle–mouse cell hybrids. *Cytogenetics and Cell Genetics* 29, 26–31.

Short, R.V., Smith, J., Mann, T., Evans, E.P., Hallet, J., Fryer, A. and Hamerton, J.L. (1969) Cytogenetic and endocrine studies of a freemartin heifer and its bull co-twin. *Cytogenetics* 8, 369–388.

Shuster, D.E., Kehrli, M.E., Jr, Ackermann, M.R. and Gilbert, R.O. (1992) A prevalent mutation responsible for leukocyte adhesion deficiency in Holstein cattle. *Proceedings of the National Academy of Sciences of the USA* 89, 9925–9929.

Sigurdardóttir, S., Lunden, A. and Andersson, L. (1988) Restriction fragment length polymorphism of DQ and DR class II genes of the bovine major histocompatibility complex. *Animal Genetics* 19, 133–150.

Sigurdardóttir, S., Lundén, A. and Andersson, L. (1990) Restriction fragment length polymorphism of bovine lysozyme genes. *Animal Genetics* 21, 259–265.

Skow, L.C. and Nall, C.A. (1996) A second polymorphism in exon 2 of the BOLA-DYA gene. *Animal Genetics* 27, 216–217.

Skow, L.C., Womack, J.E., Petrash, J.M. and Miller, W.L. (1988) Synteny mapping of the genes for 21 steroid hydroxylase, alpha A crystallin, and class I bovine leukocyte antigen in cattle. *DNA* 7, 143–149.

Skow, L.C., Snaples, S.N., Davis, S.K., Taylor, J.F., Huang, B. and Gallagher, D.H. (1996) Localization of bovine lymphocyte antigen (BOLA) DYA and class I loci to different regions of chromosome 23. *Mammalian Genome* 7, 388–389.

Slota, E., Danielak, B. and Kozubska-Sobocinska, A. (1988) The Robertsonian translocation in the cattle quintuplet. In: *Proceedings of the 8th European Colloquium on Cytogenetics of Domestic Animals, Bristol, UK*, pp. 122–123.

Smith, T.P., Lopez-Corrales, N.L., Kappes, S.M. and Sonstegard, T.S. (1997) Myostatin maps to the interval containing the bovine mh locus. *Mammalian Genome* 8, 742–744.

Solinas Toldo, S., Pienkowska, A., Fries, R. and Switonski, M. (1992) Localization of nucleolar organizer regions in farm animals by *in situ* hybridization method with a probe from a human rRNA gene. In: *Proceedings of the 10th European Colloquium on Cytogentics of Domestic Animals, Utrecht, The Netherlands*, pp. 232–236.

Solinas Toldo, S., Fries, R., Steffen, P., Neibergs, H.L., Barendsc, W., Womack, J.E., Hetzel, D.J.S. and Stranzinger, G. (1993) Physically mapped, cosmid–derived microsatellite markers as anchor loci on bovine chromosomes. *Mammalian Genome* 4, 720–727.

Solinas Toldo, S., Lengauer, C. and Fries, R. (1995a) Comparative genome map of human and cattle. *Genomics* 27, 489–496.

Solinas Toldo, S., Mezzelani, A., Hawkins, G.A., Bishop, M.D., Olsaker, I., Mackinlay, A.G., Ferretti, L. and Fries, R. (1995b) Combined Q-banding and fluorescence *in situ* hybridization for the identification of bovine chromosomes 1 to 7. *Cytogenetics and Cell Genetics* 69, 1–6.

Solinas Toldo, S., Troyanovski, R., Weitz, S., Lichter, P., Franke, W.W. and Fries, R. (1995c) Bovine desmocollin genes (DSC1, DSC2, DSC3) cluster on chromosome 24q21/q22. *Mammalian Genome* 6, 484–486.

Stone, R.T. and Muggli-Cockett, N.E. (1990) Partial nucleotide sequence of a novel bovine major histocompatibility complex class II beta-chain gene, BoLA-DIB. *Animal Genetics* 21, 353–360.

Stone, R.T., Kappes, S.M. and Beattie, C.W. (1996) The bovine homolog of the obese gene maps to chromosome 4. *Mammalian Genome* 7, 399–400.

Straka, J.G., Hill, H.D., Krikava, J.M., Kools, A.M. and Bloomer, J.R. (1991) Immunochemical studies of ferrochelatase protein: characterization of the normal and mutant protein in bovine and human protoporphyria. *American Journal of Human Genetics* 48, 72–78.

Stranzinger, G. (1989) Zytogenetische Kontrolluntersuchungen an Nutztieren. *Landwirtschaft Schweiz* 2, 355–362.

Stranzinger, G.F. and Förster, M. (1976) Autosomal chromosome translocation of piebald cattle and brown cattle. *Experientia* 32, 24–27.

Stuart, L.D. and Leipold, H.W. (1985) Lesions in bovine progressive degenerative myeloencephalopathy ('weaver') of Brown Swiss cattle. *Veterinary Pathology* 22, 13–23.

Sun, H.S., Womack, J.E. and Kirkpatrick, B.W. (1996a) Syntenic assignment of dopamine tautomerase (DCT) to bovine chromosome 12. *Animal Genetics* 27, 421–422.

Sun, H.S., Womack, J.E. and Kirkpatrick, B.W. (1996b) Syntenic assignment of human serotonin receptor subtype 2 (HTR2), esterase D (ESD), and fms-related tyrosine kinase (FLT) homologs to bovine chromosome 12. *Mammalian Genome* 7, 518–519.

Sun, H.S., Cai, L., Davis, S.K., Taylor, J.F., Doud, L.K., Bishop, M.D., Hayes, H., Barendse, W., Vaiman, D., McGraw, R.A., Hirano, T., Sugimoto, Y. and Kirkpatrick, B.W. (1997) Comparative linkage mapping of human chromosome 13 and bovine chromosome 12. *Genomics* 39, 47–54.

Sundvold, H., Olsaker, I., Gomez-Raya, L. and Lien, S. (1997) The gene encoding the peroxisome proliferator-activated receptor (PPARA) maps to chromosome 5 in cattle. *Animal Genetics* 28, 374.

Sysa, P.S. and Slota, E. (1992) The investigation of Karyotype in cattle within the new system of cytogenetic control of bulls in Polland. In: *Proceedings of the 10th European Colloquium on Cytogenetics of Domestic Animals, Utrecht, The Netherlands*, pp. 248–249.

Tammen, I. and Ferretti, L. (1996) A bovine polymorphic microsatellite locus IDVGA68 (D16S23) assigned to BTA 16. *Animal Genetics* 27, 433.

Taylor, C., Everest, M. and Smith, C. (1996) Restriction fragment length polymorphism in amplification products of the bovine butyrophilin gene – assignment of bovine butyrophilin to bovine chromosome 23. *Animal Genetics* 27, 183–185.

Taylor, J.F., Lutaaya, E., Sanders, J.O., Turner, J.W. and Davis, S.K. (1997) A medium density microsatellite map of BTA10: reassignment of INRA69. *Animal Genetics* 28, 360–362.

Tebbutt, S.J., Broom, M.F., van Stijn, T.C., Montgomery, G.W. and Hill, D.F. (1996) Genetic and physical mapping of the ovine cystic fibrosis gene. *Cytogenetics and Cell Genetics* 74, 245–247.

Teres, V., Lester, D.H. and Williams, J.L. (1996) Mapping of bovine fibroblast growth factor receptor 3 (FGFR3) to the telomeric region of chromosome 6 by SSCP analysis. *Animal Genetics* 27, 371.

Teutsch, M.R., Beever, J.E., Stewart, J.A., Schook, L.B. and Lewin, H.A. (1989) Linkage of complement factor B gene to the bovine major histocompatibility complex. *Animal Genetics* 20, 427.

Thieven, U., Solinas-Toldo, S., Fries, R., Barendse, W., Simon, D. and Harlizius, B. (1997) Polymorphic CA-microsatellites for the integration of the bovine genetic and physical map. *Mammalian Genome* 8, 52–55.

Threadgill, D.S. (1990) Comparative mapping of homologous loci from four human chromosomes in the bovine. Dissertation, Texas A&M Universtiy.

Threadgill, D.W. and Womack, J.E. (1990a) Syntenic conservation between humans and cattle. I. Human chromosome 9. *Genomics* 8, 22–28.

Threadgill, D.W. and Womack, J.E. (1990b) Syntenic conservation between humans and cattle. II. Human chromosome 12. *Genomics* 8, 29–34.

Threadgill, D.W. and Womack, J.E. (1990c) Genomic analysis of the major bovine milk protein genes. *Nucleic Acids Research* 18, 6935–6942.

Threadgill, D.W. and Womack, J.E. (1991a) Synteny mapping of human chromosome 8 loci in cattle. *Animal Genetics* 22, 117–122.

Threadgill, D.S. and Womack, J.E. (1991b) Mapping HSA10 homologous loci in the bovine. *Cytogenetics and Cell Genetics* 57, 123–126.

Threadgill, D.S. and Womack, J.E. (1991c) Mapping HSA 3 loci in cattle: additional support for the ancestral synteny of HSA 3 and 21. *Genomics* 11, 1143–1148.

Threadgill, D.S. and Womack, J.E. (1991d) The bovine pancreatic spasmolytic polypeptide gene maps to syntenic group U10: implications for the evolution of BCEI and SP. *Journal of Heredity* 82, 496–498.

Threadgill, D.W., Fries, R., Faber, L.K., Vassart, G., Gunawardana, A., Stranzinger, G. and Womack, J.E. (1990) The thyroglobulin gene is syntenic with the MYC and MOS proto-oncogenes and carbonic anhydrase II and maps to chromosome 14 in cattle. *Cytogenetics and Cell Genetics* 53, 32–36.

Threadgill, D.S., Kraus, J.P., Krawetz, S.A. and Womack, J.E. (1991) Evidence for the evolutionary origin of human chromosome 21 from comparative gene mapping in the bovine and the mouse. *Proceedings of the National Academy of Sciences of the USA* 88, 154–158.

Threadgill, D.S., Threadgill, D.W., Moll, Y.D., Weiss, J.A., Zhang, N., Davey, H.W., Wildeman, A.G. and Womack, J.E. (1994) Syntenic assignment of human chromosome 1 homologous loci in the bovine. *Genomics* 22, 626–630.

Tilstra, D.J., Li, L., Potter, K.A., Womack, J. and Byers, P.H. (1994) Sequence of the coding region of the bovine fibrillin cDNA and localization to bovine chromosome 10. *Genomics* 23, 480–485.

Tisdall, D.J., Quirke, L.D. and Galloway, S.M. (1996) Ovine stem cell factor gene is located within a syntenic group on chromosome 3 conserved across mammalian species. *Mammalian Genome* 7, 472–473.

Tobin-Janzen, T.C. and Womack, J.E. (1992) Comparative mapping of IGHG1, IGHM, FES, and FOS in domestic cattle. *Immunogenetics* 36, 157–165.

Toye, P.G., MacHugh, N.D., Bensaid, A.M., Alberti, S., Teale, A.J. and Morrison, W.I. (1990) Transfection into mouse L cells of genes encoding two serologically and functionally distinct bovine class I MHC molecules from a MHC-homozygous animal: evidence for a second class I locus in cattle. *Immunology* 70, 20–26.

Trask, B., Pinkel, D. and van den Engh, G. (1989) The proximity of DNA sequences in interphase cell nuclei is correlated to genomic distance and permits ordering of cosmids spanning 250 kilobase pairs. *Genomics* 5, 710–717.

Tschudi, P., Zahner, B., Kupfer, U. and Stampfli, G. (1977) Chromosomen untersuchungen an Schweizerischen Rinderrassen. *Schweizerisches Archiv für Tierheilkunde* 119, 329–336.

Usha, A.P., Lester, D.H. and Williams, J.L. (1997) Dwarfism in Dexter cattle is not caused by the mutations in FGFR3 responsible for achondroplasia in humans. *Animal Genetics* 28, 55–57.

Vaiman, D., Bahri-Darwich, I., Mercier, D., Yerle, M., Eggen, A., Leveziel, H., Guérin, G., Gellin, J. and Cribiu, E.P. (1993) Mapping of new bovine microsatellites on cattle chromosome 15 with somatic cell hybrids, linkage analysis, and fluorescence *in situ* hybridization. *Mammalian Genome* 4, 676–679.

van der Poel, J.J., Groenen, M.A.M., Dijkhof, R.J.M., Ruyter, D. and Giphart, M.J. (1990) The nucleotide sequence of the bovine MHC class II alpha genes: DRA, DQA, and DYA. *Immunogenetics* 31, 29–36.

van Eijk, M.J.T., Russ, I. and Lewin, H.A. (1992) The order of bovine DRB3, DYA and PRL determined by single sperm typing: use of primer extension preamplification for multilocus gene mapping. *Animal Genetics* 23(Suppl. 1), 79.

Vigier, B., Locatelli, A., Prepin, J., Mesnil du Buisson, F. and Jost, A. (1976) Les premiéres manifestations du 'freemartinisme' chez le foetus de veau ne dépendant pas du chimérisme chromosomique XX/XY [The first manifestations of freemartinism in the calf fetus do not depend on XX/XY chromosomal chimerism]. *Comptes Rendus Hebdomadaires Séances de l'Académie des Sciences, Série D* 282, 1355–1358.

Vigier, B., Tran, D., Legeai, L., Bezard, J. and Josso, N. (1984) Origin of anti-Mullerian hormone in bovine freemartin fetuses. *Journal of Reproduction and Fertility* 70, 473–479.

Vogel, T., Borgmann, S., Dechend, F., Hecht, W. and Schmidtke, J. (1997a) Conserved Y-chromosomal location of TSPY in Bovidae. *Chromosome Research* 5, 182–185.

Vogel, T., Dechend, F., Manz, E., Jung, C., Jakubiczka, S., Fehr, S., Schmidtke, J. and Schnieders, F. (1997b) Organization and expression of bovine TSPY. *Mammalian Genome* 8, 491–496.

Wallis, D. and Womack, J.E. (1994) Mapping the bovine homolog of the human cystic fibrosis gene. *Journal of Heredity* 85, 490–492.

Weikard, R., Henke, M., Kühn, C., Barendse, W. and Seyfert, H.M. (1996) A polymorphic microsatellite within the immunorelevant bovine lysozyme-encoding gene. *Animal Genetics* 27, 125.

Weikard, R., Goldammer, T., Kühn, C., Barendse, W. and Schwerin, M. (1997) Targeted development of microsatellite markers from the defined region of bovine chromosome 6q21–31. *Mammalian Genome* 8, 836–840.

Werth, L.A., Hawkins, G.A., Eggen, A., Petit, E., Elduque, C., Kreigesmann, B. and Bishop, M.D. (1996) Rapid communication: melanocyte stimulating hormone receptor (MC1R) maps to bovine chromosome 18. *Journal of Animal Science* 74, 262.

Wigger, G., Masabanda, J., Stranzinger, G. and Fries, R. (1998) A molecular marker for the bovine ciliary neurotrophic factor receptor gene (*CNTFR*) maps to 8q21. *Animal Genetics* 29, 64.

Wilkins, R.J. and Davey, H.W. (1997) A polymorphic microsatellite in the bovine leptin gene. *Animal Genetics* 28, 376.

Williams, J.L., Urquhart, B.G. and Barendse, B. (1995) Three bovine chromosome 15 microsatellite markers. *Animal Genetics* 26, 124.

Williams, J.L., Morgan, A.L.G., Guérin, G. and Urquhart, B.G.D. (1996a) Ten new cosmid-derived bovine microsatellite markers. *Animal Genetics* 27, 380.

Williams, J.L., Urquhart, B.G.D., Barendse, W. and Ferretti, L. (1996b) Genetic mapping of 26 cosmid-derived bovine microsatellite markers. *Animal Genetics* 27, 380–381.

Williams, J.L., Urquhart, B.G., Castiglioni, B. and Ferretti, L. (1996c) Mapping two bovine marker loci, IDVGA66 (D16S34) and IDVGA76 (D14S35) by SSCP. *Animal Genetics* 27, 442.

Williams, J.L., Lester, D.H., Teres, V.M., Barendse, W., Sim, R.B. and Soames, C.J. (1997a) Mapping the bovine factor H gene to chromosome 16 by SSCP analysis. *Mammalian Genome* 8, 77–78.

Williams, J.L., Olsaker, I. and Teres, V.M. (1997b) Using SSCP to facilitate mapping microsatellite loci. *Mammalian Genome* 8, 79–80.

Womack, J.E. and Moll, Y.D. (1986) Gene map of the cow: conservation of linkage with mouse and man. *Journal of Heredity* 77, 2–7.

Womack, J.E., Threadgill, D.W., Moll, Y.D., Faber, L.K., Foreman, M.L., Dietz, A.B., Tobin, T.C., Skow, L.C., Zneimer, S.M., Gallagher, D.S. and Rogers, D.S. (1989)

Syntenic mapping of 37 loci in cattle: chromosomal conservation with mouse and man. HGM10. *Cytogenetics and Cell Genetics* 51, 1109.
Womack, J.E., Dietz, A.B., Gallagher, D.S., Li, L., Zhang, N., Neibergs, H.L., Moll, Y.D. and Ryan, A.M. (1991) Assignment of 47 additional comparative anchor loci to the bovine synteny map. *Cytogenetics and Cell Genetics* 58, 2132.
Womack, J.E., Johnson, J.S., Owens, E.K., Rexroad, C.E., III, Schläpfer, J. and Yang, Y.-P. (1997) A whole-genome radiation hybrid panel for bovine genome mapping. *Mammalian Genome* 8, 854–856.
Xiao, C., Tsuchiya, K. and Sutou, S. (1998) Cloning and mapping of bovine ZFX gene to the long arm of the X-chromosome (Xq34) and homologous mapping of ZFY gene to the distal region of the short arm of the bovine (Yp13), ovine (Yp12–p13), and caprine (Yp12-p13) Y chromosome. *Mammalian Genome* 9, 125–130.
Yang, Y.-P. and Womack, J.E. (1995) Human chromosome 17 comparative anchor loci are conserved on bovine chromosome 19. *Genomics* 27, 293–297.
Yang, Y.-P. and Womack, J.E. (1996) Construction of a bovine chromosome 19 linkage map using an interspecies hybrid backcross. *Mammalian Genome* 8, 262–266.
Yang, Y.P., Rexroad, C.E., 3rd, Schläpfer, J. and Womack, J.E. (1998) An integrated radiation hybrid map of bovine chromosome 19 and ordered comparative mapping with human chromosome 17. *Genomics* 48, 93–99.
Yeh, C.C., Taylor, J.F., Gallagher, D.S., Sanders, J.O., Turner, J.W. and Davis, S.K. (1996) Genetic and physical mapping of the bovine X chromosome. *Genomics* 32, 245–252.
Yoo, J., Kappes, S.M., Stone, R.T. and Beattie, C.W. (1995a) Linkage analysis assignment of bovine granulocyte colony stimulating factor receptor (G-CSFR) to chromosome 3. *Mammalian Genome* 6, 686–687.
Yoo, J., Ponce de Leon, F.A., Stone, R.T. and Beattie, C.W. (1995b) Cloning and chromosomal assignment of the bovine interleukin-2 receptor alpha (*IL-2Rα*) gene. *Mammalian Genome* 6, 751–753.
Yoo, J., Stone, R.T., Kappes, S.M., Solinas Toldo, S., Fries, R. and Beattie, C.W. (1996a) Genomic organization and chromosomal mapping of the bovine Fas/APO-1 Gene. *DNA and Cell Biology* 15, 377–385.
Yoo, J., Stone, R.T., Solinas Toldo, S., Fries, R. and Beattie, C.W. (1996b) Cloning and chromosomal mapping of bovine interleukin-2 receptor gamma gene. *DNA and Cell Biology* 15, 453–459.
Yu, R. and Xin, C.L.I. (1991) The 2/27 Robertsonian translocation in wenling lump cattle. *Hereditas (Beijing)* 13, 17–18.
Zhang, N. and Womack, J.E. (1992) Synteny mapping in the bovine: genes from human chromosome 5. *Genomics* 14, 126–130.
Zhang, N., Threadgill, D.W. and Womack, J.E. (1992) Synteny mapping in the bovine: genes from human chromosome 4. *Genomics* 14, 131–136.
Zhang, Y., Redaelli, I., Castiglioni, B., Mezzelani, A. and Ferretti, I. (1995) Five polymorphic bovine microsatellite loci – IDVGA-62A, IDVGA-71, IDVGA-82, IDVGA-88, IDVGA-90. *Animal Genetics* 26, 365–366.

Genetic Linkage Mapping, the Gene Maps of Cattle and the Lists of Loci

W. Barendse[1] and R. Fries[2]

[1]*CSIRO Tropical Agriculture, Level 3 Gehrmann Laboratories, Research Road, University of Queensland, St Lucia 4072, Australia;*
[2]*Lehrstuhl für Tierzucht, Technische Universität München, 85350 Freising-Weihenstephan, Germany*

Introduction to Genetic Linkage	329
Chromosomes and genetic linkage	330
Modern genetic linkage mapping	333
Cattle Families	335
Using cattle in linkage experiments	335
Cattle reference families	337
Polymorphisms	337
Kinds of polymorphism	338
Methods for detecting polymorphism	339
The Linkage Maps of Cattle	343
References	360

Introduction to Genetic Linkage

One of the challenges in biology is the identification of genes that contribute to continuous variation. The most important tool for dissecting that variation into the underlying genetic factors is the genetic linkage map. Most traits that are of evolutionary importance, such as fertility, fecundity, size and shape, are affected by many genes and the environment and so they appear as continuous variation (e.g. Lewontin, 1974; Falconer, 1981). The methods of biometric genetics have been used on continuous traits in predicting the responses of individuals and populations to selection and describing the overall resemblance of relatives. However, the black box of the genome has remained closed. The tools to dissect continuous variation in any species have taken more than a century to develop and these methods have become commonplace only in the last decade. Although these methods have been spectacularly successful in identifying the molecular genetic basis for Mendelian or discontinuous variation (Collins, 1995), we are now only on the threshold of identifying the genetic factors that contribute to a continuous trait. The methods use genetic linkage maps, highly polymorphic or variable deoxyribonucleic acid (DNA) markers, statistical models of the effects of genes on continuous traits and the latest techniques of molecular biology. The rationale for their use is that DNA can be described by gene maps and so the genetic factors affecting continuous variation can also be located on a genetic map. By a process of progressive refinement and educated guesswork, in theory, the locations of these genetic factors can be identified with sufficient accuracy for the identity of the genetic factors to become plain. These identities spawn hypotheses for further genetic testing and formal methods of gene transfer.

Here we shall describe genetic linkage in general, the requirements for linkage studies in cattle, the kinds of variation that are found in genetic linkage maps and finally the current status of linkage maps in cattle.

Chromosomes and genetic linkage

The chromosome is the repository of the nuclear genetic material in a species (Sutton, 1902) and its structure and mode of transmission from one generation to the next constrains the analysis of genetic linkage. Each chromosome is in essence a single DNA molecule wound around protein, which supports, packages and contributes to the control of DNA. In eukaryotes, the rod-like chromosomes appear during cell division, replacing the darkly staining nucleus of the cell. These chromosomes consist of two chromatids attached to each other at the centromere and so each chromosome appears split along its length. There are two chromatids, as the DNA molecule has been copied during the interphase between cell divisions. During cell division, called mitosis, the chromosomes migrate to one plane of the cell and then the chromatids are separated to opposite poles, after which the cell divides and daughter nuclei form. Thus genes are organized in linear arrays, which are inherited as larger groups. In the germ cells, a more complicated process, meiosis, occurs. Chromosomes usually come in homologous pairs, one member derived from each parent. The paired homologues exchange genetic material, so that the chromosomes inherited from the parents of the individual are recombined in the gametes provided for its own offspring. The locations of the interchange often correspond to structures called chiasmata, where it seems the chromatids wrap themselves around one another (Sturtevant, 1913). The homologues separate in a first division, halving the number of chromosomes, and then the recombinant chromatids separate in a second division, with the formation of the gametes. This reduction usually results in the haploid number of chromosomes, and the full complement of chromosomes, the diploid number, is restored upon gametic nuclear fusion or fertilization. Not all chromosomes are paired; in many species the sex chromosomes are not homologues or sex is determined by the number of sex chromosomes. Furthermore, in a small percentage of species, the whole genome has been duplicated and a variety of higher-order ploidies, such as triploid, tetraploid, hexaploid and octaploid, are known, and meiosis is complicated by more than two homologues exchanging genetic material.

These physical movements of chromosomes can be observed in genetic experiments in which the parental chromosomes are marked by genetic variation. At any genetic locus, and in modern molecular terms each DNA base pair is potentially mutable (Benzer, 1961), an individual may be a homozygote or a heterozygote. If it is heterozygous, the DNA strands will differ from each other at the particular locus, while, if it is homozygous, it will have an identical DNA sequence at that locus. It will be hemizygous if it possesses only one chromosome at that locus; for example, male mammals normally possess only one X chromosome. The offspring of a heterozygous parent can be classed into two groups on the basis of which allele or genetic variant they inherited, and so indicate the inheritance of chromosomal segments in offspring. Obviously, to show linkage, the parent must be a heterozygote at both loci under investigation. Such a parent will have a different genetic composition on its homologous chromosomes, since each allele at one locus will be on a separate DNA strand and will be physically linked to a different allele at the second locus. This is the gametic phase it inherited from its own parents. The initial experiments that confirmed genetic linkage (Morgan, 1910) used two physical mutants, miniature wing and white eye, known to be on the *Drosophila melanogaster* X chromosome. The offspring of these crosses mainly represented the parental X chromosomes, flies with either a mutant wing or eye. A small number carried either both mutants or neither mutant, indicating that recombination had occurred between the two loci on the homologous X chromosomes. The proportion of recombinants in the total is the recombination fraction (Sturtevant, 1913). Sturtevant used the recombination fraction to locate six loci along the X chromosome in *D. melanogaster*, choosing the gene order so that the loci with the smallest recombination fractions are adjacent. Distances between loci that are not adjacent are the sums of the distances between the intermediate loci. Interestingly, he foreshadowed the fact that a genetic map and a physical map should have the same order of loci but there is not a uniform correspondence between physical and genetic distance.

A genetic linkage map is the order of the loci along the chromosome, with the genetic distances specified between loci and the total length being the sum of the genetic distances. This simple specification hides several points of complexity (Sturtevant; 1913, Haldane; 1919, Fisher; 1922, Kosambi; 1944). First, even when one parent is a heterozygote for all the loci, the recombination fractions calculated from the offspring do not add up exactly. The recombination fractions between adjacent loci are greater than or equal to the recombination fractions of loci further apart – the distance $AC \leq AB + BC$. A recombination in the interval AB and one in BC in the same individual, a double recombinant, will cause the interval AC to appear shorter than it should be. The further apart three loci are, the greater the impact of this factor as double and triple recombination become more common. Second, the closer the three loci the more likely there is to be interference – that is, further recombination is inhibited in a region due to the occurrence of a recombination event. This reduces the occurrence of double or higher-order recombination events. At scales of 1% recombination, interference is essentially complete, and 1% recombination is equivalent to a 1 centimorgan (cM) genetic distance (Haldane, 1919). Several mapping functions have been derived, some of which allow for the explicit setting of the amount of interference and some are empirical formulations for individual species (e.g. Haldane, 1919; Kosambi, 1944; Rao *et al.*, 1977). Third, there are inconsistencies in the distances between loci in a map from one experiment to another, chiefly due to differences in accuracy, with differences in sample size and smaller distances being measured more accurately than larger distances, but also due to real differences in recombination fractions in different individuals and between the sexes. Maximum likelihood methods (Fisher, 1922) are used to weight recombination fractions for each interval, with the weights derived from the number of recombinants and non-recombinants from each study. This allows combination of data not only for the same interval from different experiments but also where the interval in one experiment formed a part of the interval between more distant loci in another experiment. Unlike more modern maximum likelihood methods, this method maximizes on the odds of the recombination fractions.

Thoday (1961) and his associates were able to localize genetic factors influencing bristle number to chromosomal intervals of the *D. melanogaster* genome, demonstrating that it was possible to use the methods of map localization to dissect a continuous or metric trait. The method is conceptually simple, although in practice the results are less satisfactory than those obtained for discrete mutants or polymorphisms. When discrete polymorphisms are used, the recombinant gametes can be identified without ambiguity in appropriate crosses because each locus can be scored. The polymorphism can then be localized to a point on the map. When a continuous trait is analysed, the phenotypic value for the trait is the sum of the effects of several genes, some linked and some segregating independently, non-linear genetic effects and a smoothing effect of the environment. Since the level of the phenotypic value cannot be ascribed to a particular gene, recombinants cannot be counted directly. However, since the offending gene will lie within a segment bounded by discrete polymorphisms, it can be determined whether recombination has occurred in the interval between the discrete polymorphisms. In practice, then, intervals can be designated as recombinant or parental and these can be correlated to the phenotypes of each individual. The means and variances of the phenotypes for each interval can then be compared and in this way genes affecting continuous traits can be associated with chromosomal intervals. This approach is the basis of most modern methods to dissect continuous traits using the analysis of linkage.

Modern genetic linkage mapping

While it is in principle possible to make a genetic linkage map in any species using these methods, it has become feasible only in recent times through the use of the logarithm of the ratio of the odds (lod)-score methods and the discovery of pervasive, easily sampled DNA polymorphims. As late as 1982, preliminary maps had been constructed for fewer than 70 species, with only a handful, such as the fruit fly, bacteriophage, *Escherichia coli*, *Neurospora*, the nematode, the tomato and the mouse, with many loci mapped (O'Brien, 1983). Most of the polymorphism was due to phenotypic mutants and each mutant had to be described extensively, usually maintained in a reference laboratory, and it required trained observers to distinguish many of the mutants from the wild type.

It was not until Morton (1955) devised a sequential test using lods that information from very small families could be added together in a robust fashion. The lod is an expression of the probability of linkage at a specific recombination fraction compared with linkage at a recombination fraction of 0.5. In the simplest case, where the parental gametic phase is known and recombinants can be identified by deduction, the n recombinants and k non-recombinants are counted and, for the recombination fraction θ, the lod is

$$z = \log_{10} \left[\frac{\theta^n + (1-\theta)^k}{0.5^{(n+k)}} \right]$$

This approach can be used on families where the gametic phase is unknown: in essence, the above equation is divided into two parts, representing the two alternative phases, and half the odds for one potential gametic phase is added to half the odds of the other potential gametic phase. In effect, the information from one offspring is sacrificed in the process. Strictly, only one arbitrary value of θ should be compared with $\theta = 0.5$ but, in practice, a range of values is compared and a maximum is usually calculated by numerical interpolation over all families. Adding the lods from several families implies sequential testing to define points at which linkage is declared or is rejected. Linkage is declared when the lod exceeds a value of 3 and is rejected when it exceeds a value of −2. A value of the lod of 3 represents an odds ratio of 1000 : 1, but this is not equivalent to a probability value of $P = 0.001$. These threshold values represent a type I error rate of less than 0.05 in accepting linkage where none exists, while maximizing the ability of the test to detect linkage when it does occur in species with a genome the size of that of humans. These methods were subsumed in the generalization for pedigree analysis of Elston and Stewart (1972). In complex pedigrees, the previous method was to break these pedigrees into simpler units and then analyse the units as if they were independent. Instead, the odds of observing the genotypes in the most recent generation are calculated and summed across all genotypes and these are then attached by multiplication to the odds for the previous generation. This procedure is followed recursively and a lod score calculated. Ott (1974) applied this method in a computer program for generalized linkage analysis for pairs of loci. The approach has been expanded to multiple loci (Lathrop *et al.*, 1984; Lander and Green, 1987), but in these analyses the odds ratio is maximized, not the odds of recombination fractions, as in the method of Fisher. This multilocus approach has been criticized for losing information and generating biased maps of chromosomes (Morton, 1988), mainly because interference is ignored.

Polymorphisms can now be identified for individual DNA base pairs and these are plentiful, often multiallelic and easy to detect, but it was not always so. Until plentiful DNA polymorphisms were discovered (Meselson and Yuan, 1968; Botstein *et al.*, 1980; Jeffreys *et al.*, 1985; Tautz, 1989; Weber and May, 1989), most polymorphisms were for physical mutants, erythrocytic blood groups, leucocyte antigens and allozymes. Apart from the systems such as leucocyte antigens of the major histocampatibility complex (MHC), most of the blood groups and protein polymorphisms have two or three alleles, often with large differences in allele frequency between the alleles (e.g. Harris and Hopkinson, 1972). The critical limitations with these polymorphisms are that there are few of them, it is often difficult to find individuals that are heterozygous at several loci and only part of the genome is screened, since these polymorphisms were generally in coding regions and some of them showed dominant inheritance, which reduces the number of families that can be analysed simply. The DNA polymorphisms removed these difficulties, since they were plentiful, generally codominant and easy to sample and individuals could be readily obtained who were heterozygous for many loci.

The combination of abundant DNA polymorphisms and the computerization of linkage analysis in the mid-1980s led to the first genetic linkage map of humans (Donis-Keller *et al.*, 1987), largely constructed with anonymous DNA polymorphisms. This was but the first in what would become a major activity of mapping DNA polymorphisms in a large number of plant and animal species that, until then, were considered to be outside the realm of genetic linkage studies for reasons of feasibility or convenience. These included the traditional livestock species, such as cattle, sheep, pigs, goats, horses and chickens, and undomesticated species with no known mutants, such as pine trees, prawns, tuna and oysters.

With the availability of many linkage maps comes the possibility of comparing the gene order for different species. When maps are derived from mutants, there is no guarantee that a mutant of similar aspect in one species is caused by a lesion in the same or homologous gene in another species. It is not until one can identify a protein or a DNA sequence directly that it becomes possible to compare the gene maps of different species. These comparisons provide both valuable and interesting comparisons. The valuable comparisons are when a preliminary genetic map of one species is compared with a comprehensive genetic map of a related species. If sufficient loci are mapped in common, the genetic composition of the preliminary map can be inferred from the genetic composition of the comprehensive map. The interesting comparisons are those that record the passing of evolutionary history and indicate the mechanisms by which chromosomal changes have occurred between species. The comparisons may point to regions of genomes that have remained intact over millions of years of evolutionary change.

Cattle Families

Using cattle in linkage experiments

Cattle are not an intuitive choice as a suitable species for a refined genetic linkage map, partly because of their large size and their slow reproductive rate. Cattle usually produce a single offspring per year and each individual may costs hundreds of dollars to purchase, while linkage mapping is a statistical enterprise requiring large families. Nevertheless, they do have some advantages over other large animals or plants, because a large number of cattle can be obtained and pedigrees can be constructed to suit the geneticist, partly because of their economic value. A large number of phenotypic characters is recorded for a diverse range of uses of the animal. Most of these characters are continuous in distribution. However, the lethal, semilethal and most of the interesting major genes of cattle are generally not treasured as sources of potential information about genes and genome structure, defective cattle are usually selected out of lineages, due to the economic cost to farmers, and there is a strong emphasis among breeders to produce cattle that conform to a particular type.

The effect of these life-history attributes is that cattle families are characteristically half-sib in structure, and often not all the important progenitors are sampled for DNA. Usually, cattle families consist of a sire mated to a large number of dams and so an instant large half-sib family can be generated in one season. With artificial insemination, these families can consist of thousands of offspring, providing all the necessary statistical power to map genes for continuous traits. However, the influence of the dams in these families is ignored, since no informative meioses are derived from them. In these families the gametic phase of the sire is quite often unknown, because his own parents are not sampled, due to the length of the generational interval, and additional analytical problems arise.

In half-sib families, since the dams do not contribute to the linkage analysis in substantive terms, it is most efficient to sample genotypes from the sire, his parents and his offspring. Unless the gametic phase of the dam is known, her single gamete provided to her offspring does not provide information on linkage, so her contribution is merely to help indicate which allele from the sire was inherited by the offspring. Not all offspring require the genotype of the dam for this, only those with the same genotype as the sire, so fewer offspring need the dam's genotype compared with the dams who have to be genotyped. With two alleles at a locus, no more than half the offspring are expected to have the same genotype as a heterozygous sire. The genotype of the dam would be expected to provide information on the sire's contribution to half of those, or the genotype of the dam would be useful for a quarter of the total offspring. But more information might be obtained by genotyping additional offspring. Further, as the number of alleles at the locus increases so the percentage of offspring with the same genotype as the sire decreases. There is no point at which it is more efficient to genotype the dams to get more information from some offspring than to use a larger family in the first place. The only possible exception is where the family size is fixed for other reasons, the dams are available and the family size is only marginal for the detection of linkage.

The lack of some grandparents in the family means that the gametic phase of the parents is unknown and several important analytical advantages are lost. The advantages of known gametic phase are that recombination fractions are not only unbiased but the number of recombinants divided by the total is a maximum likelihood estimate (Ott, 1985), fewer computational resources are required and multipoint maps can be generated, using deduction. The first two are of obvious value but the other two require some discussion. Fewer computational resources are required with three-generation families because the number of possible haplotypes of the parents is small. There are 2^n possible haplotypes for n loci, but if phase is known for each locus there are only two possible haplotypes. Phase will be unknown for some loci, usually where both the sire and the dam have the same genotypes as their own sires and dams. For loci with five alleles, approximately one locus in a thousand (0.001024) will be phase-unknown, and for loci with two alleles approximately one locus in 20 (0.0625), although with half-sib families many more loci will be phase-unknown, since the dams and their parents are not analysed. The total number of possible haplotypes will be $2 + 2^k$ where k is the number of phase-unknown loci. With ten phase-unknown loci, there are more than a thousand possible haplotypes and a large multidimensional array would be used to analyse them. There would be more haplotypes than there are informative offspring in the cattle families generally used to analyse linkage. The array size increases exponentially, which has obvious effects on the increases in computer memory. Logic or the principle of parsimony can be used with phase-known families, since the recombinants can be identified and so loci are ordered by minimizing the number of recombinants. This provides an independent check on the multipoint orders generated by likelihood ratio methods, but it is unavailable if the families are phase-unknown, since recombinants cannot be counted, but are only inferred.

Lack of knowledge on genotypic phase is most noticeable when the genetic distances between loci are small, and this is most important when the density of loci in a map increases. Small distances are usually supported by very few recombinants. If the genotypic error is of the order of 1% and the recombination frequency is of the same close order, then distinguishing between real recombinants and spurious recombinants can be difficult. Error detection usually depends upon knowing the genotypic phase and then searching for those individuals that show recombination events between adjacent or close loci or double recombinants. When phase is unknown, probabilities on the phase are usually derived from the recombination fractions and, when recombination events are rare, there is little information with which to evaluate data integrity.

Cattle reference families

Several cattle families have been described which are used for the construction of genetic linkage maps in cattle. These are the International Bovine Reference Panel (IBRP) family (Barendse *et al.*, 1993, 1994, 1997), the US Department of Agriculture (USDA) Meat Animal Research Center (MARC) Mapping herd (Bishop *et al.*, 1994; Kappes *et al.*, 1997), a large collection of US Holstein families (Georges *et al.*, 1995), the Illinois Reference Resource families (Ma *et al.*, 1996), the Angleton research herd (Yeh *et al.*, 1995) and several smaller groups of cattle families used by individual research groups. The IBRP, MARC and Angleton families use full-sibling families, while the others do not and so represent male-specific genetic maps. The IBRP families do not have a large number of grandparents, so there may be bias in the genetic distances. The MARC families have a relatively small number of offspring in total, so the resolution of gene order is lower than in the other families, although the maps derived from half-sibs use half the potential information from each offspring. None of the studies that use phase-known family structure have analysed their material explicitly to determine the concordance between the likelihood analyses and countable recombination events. No single set of families is ideal, since the Angleton families are derived from a very small number of parents and, since recombination fractions vary between individuals (e.g. Simianer *et al.*, 1997), the genetic distances may be biased. These disadvantages have been overcome by the construction of maps using the raw genotypes from all studies to create consensus maps and this has begun with chromosome 23 (Beever *et al.*, 1996).

Polymorphisms

Polymorphism forms the basis for a linkage map and the usefulness of a polymorphism is usually measured by the polymorphism information coefficient (PIC) value (Botstein *et al.*, 1980). The PIC is an expected heterozygosity derived from allele frequences in a random mating population, and essentially it predicts the backcross and informative intercross frequency, i.e. the frequency with which a heterozygote will be mated to a homozygote plus the frequency with which a heterozygote will be mated to a heterozygote, the latter scaled by a factor of two due to the lower level of information derived from phase-unknown intercrosses. To get information on linkage, at least one parent must be a heterozygote at both loci. The PIC value is for a single locus and so the product of the PIC values gives the probability of getting an informative mating for two loci. In cattle, of course, mating is not at random, nor are herds necessarily random samples of genotypes, so the PIC values are of limited usefulness. The observed heterozygosity, in combination with the number of alleles at the locus, is a more reliable rule of thumb of encountering an informative mating.

Kinds of polymorphism

In a formal sense, there are only two kinds of DNA polymorphism – those due to replacement of DNA bases and those due to insertion or deletion of base pairs. In general, polymorphisms due to replacement of bases – point mutations – usually show two alleles at each locus. Those that are due to insertion/deletion events often show more alleles; those that are due to changes in the number of a particular DNA sequence, such as the nucleotides GT repeated many times, may be hypervariable, with more than five alleles at a locus (e.g. Fries *et al.*, 1990; Moore *et al.*, 1992, 1994; Bishop *et al.*, 1994, Vaiman *et al.*, 1994). In some cases, the DNA sequence that is repeated is not exactly the same every time but there is sufficient similarity to discern the fundamental motif. Depending upon the size of the motif, the element is either a microsatellite (repeat unit of 1–5 base pairs), a minisatellite (repeat unit of tens of base pairs) or a satellite. The hypervariable loci tend to be either microsatellites or minisatellites. However, the hypervariable loci are less common than point mutations, with point mutations being approximately two orders of magnitude more common than microsatellites (e.g. Georges *et al.*, 1987; Steele and Georges, 1991; Steffen *et al.*, 1993; Moore *et al.*, 1994; Vaiman *et al.*, 1994).

The detection of these polymorphisms uses essentially the same technology and it is a rare technique indeed that can be used for only one or the other of these kinds of polymorphism. The techniques either recognize the identity of the sequence or recognize the effects of a change in sequence, and some techniques use a combination of the two. For example, a restriction endonuclease will cut DNA at a specific sequence, and mutations at the sequence will result in DNA of different lengths, corresponding to the presence or absence of the mutation. The lengths can be discriminated once the DNA is separated by size on a gel (Southern, 1975). However, the same system can be used to recognize a polymorphism if there is variation in the length of the fragments, due to repetitive elements, even if there has been no mutation at the restriction endonuclease site (Jeffreys *et al.*, 1985). Then the restriction endonuclease acts merely to cut the DNA to a manageable size for analysis.

Polymorphisms can also be divided into those that are in or extremely near a sequence that is transcribed into ribonucleic acid (RNA), some of which is also translated into polypeptides, and those polymorphisms that are in DNA that is never transcribed into RNA. The first kind, often called type I polymorphisms (O'Brien, 1991), are in sequences that are conserved across species and so are useful for studies of evolutionary history and for the practical matter of interpolating the expected locations of genes between species. The second kind, often called type II polymorphisms (O'Brien, 1991), are in sequences that are usually unique to a species or occur in a narrow range of taxonomic groups. Despite their limitations for some studies, they form the backbone of studies to localize genetic factors to linkage maps, primarily due to their large number. They are often hypervariable.

Methods for detecting polymorphism

The methods of detecting DNA polymorphisms revolve around three major technologies, not all of which are used in every method. These are: (i) gel electrophoresis to separate DNA molecules by either length or volume; (ii) DNA oligonucleotide hybridization, in which oligonucleotides are hybridized to the target DNA; and (iii) DNA visualization, in which the DNA is either stained by silver, intercalated with ethidium bromide, or labelled with radioactive or fluorescent components with detection by various combinations of photography, ultraviolet (UV) fluorescence, autoradiography and laser excitation, with computerized capture of images. These methods have been revolutionized by the polymerase chain reaction (PCR; Mullis *et al.*, 1986), which results in the amplification of specific DNA sequences to quantities of any predetermined amount.

Gel electrophoresis to separate DNA fragments by either length or volume forms the basis for most methods. The DNA has an overall negative charge under physiological conditions, so it migrates in an electric field (e.g. Schwartz and Koval, 1989). By applying this field over a matrix, DNA fragments of different length or volume are separated from each other. The most popular matrices are agarose and polyacrylamide. The first is a refined polysaccharide obtained from seaweeds with a maximum resolution of 8 base pairs (bp) between fragments and a capacity of separating fragments of more than a million bp (Mathew *et al.*, 1988). The second is a plastic of well-defined and predictable characteristics, with a maximum resolution of 1 bp and a capacity of separating fragments up to several thousand bp. In the early methods, the DNA was often transferred from the gel via capillary action to a membrane of nitrocellulose (Southern, 1975) or, later, nylon. Blotting the DNA to a membrane preserves the separation and maintains the resolution of the DNA molecules after agarose electrophoresis. For some uses, agarose and acrylamide are interchangeable, such as for separating native DNA fragments in the size range of 50–2000 bp, in which the difference in length between fragments is greater than 8 bp. Above this size agarose is the medium of choice, while, for applications that involve differences less than 8 bp or denatured or single-stranded DNA, acrylamide is the medium of choice. Native DNA is fractionated by length, but denatured DNA may be fractionated either by length or by volume. If the conditions in the gel maintain the denatured state, the denatured fragments will be separated by length. If the denaturation is relaxed, the single-stranded DNA molecule may anneal to itself and the conformation of the secondary structure will depend upon the ionic composition, the temperature and the degree of residual denaturation (Orita *et al.*, 1989). Differences in the volumes of these secondary conformations can be detected by gel electrophoresis. Gradients in the denaturants can also be used to detect variation (Collins and Myers, 1987; Sheffield *et al.*, 1989). When denaturation is relaxed, DNA may also renature, and mismatches in the pairing of bases caused by differences in the complementary sequences in heterozygous animals will arise (e.g. Zimmerman *et al.*, 1993). The mismatches will cause localized increases in diameter of the double-stranded DNA heteroduplex (HD) molecule, which will slow migration in the gel. Mismatches will only occur in heterozygotes, and the various homozygotes may be indistinguishable from each other on the gel.

Hybridization of DNA oligonucleotides can be used to discriminate the target DNA from other DNA in a mixed pool. The oligonucleotides may have been obtained from a recombinant clone by enzymatic copying (Rigby *et al.*, 1977, Feinberg and Vogelstein, 1983) or they may be synthesized chemically from nucleotide precursors (Wallace *et al.*, 1979, Wood *et al.*, 1985). The single-stranded oligonucleotides are attached to the target DNA after the target has been denatured, and the temperature at which the annealing occurs varies according to the size of the oligonucleotide and the degree of homology between the oligonucleotide and the target sequence. Occasionally, the oligonucleotide can discriminate between alleles at a locus (e.g. Mullis *et al.*, 1986). This occurs where the oligonucleotide is in the range 15–40 bp and the annealing temperature is adjusted so that a mismatch at a single base pair can cause the oligonucleotide to bind with much less affinity to one allele than to the other. With larger oligonucleotides, a single mismatch may be too small to cause a detectable difference in affinity. On other occasions, the temperature of annealing can be reduced and related genes, or the homologous gene in another species, can be identified.

The DNA is visualized by either staining the DNA itself or by the visualization of molecules bound or incorporated into the DNA. Silver salts bind easily to all nucleotides, not just DNA, due to their positive charges. The silver is then deposited, using photographic processes, and bound *in situ* to the gel or other matrix (Bassam *et al.*, 1991). Ethidium bromide is a fluorescent molecule that intercalates between the double strands of DNA, attracted by hydrogen bonds between the nucleotides. Ethidium bromide is fluorescent when excited by UV light, so DNA can be indirectly visualized and photographed (Sharp *et al.*, 1973). The DNA can be copied and, during the copying process, nucleotides labelled with radioactive species or with fluorescent dyes can be incorporated (Rigby *et al.*, 1977; Feinberg and Vogelstein, 1983). The former results in the autoradiography of DNA and the latter may use laser excitation and capture by video of the resultant fluorescence. The nucleotides may be altered to have tags of biotin or digoxygenin and these can be detected using immunological staining, such as with streptavidin (e.g. Kemp *et al.*, 1989). In some cases, a dye is deposited and, in some cases, a fluorescent dye is visualized, using UV light, and recorded by photography.

The detection of polymorphism has been greatly aided by the PCR in which DNA of a specific sequence can be amplified to arbitrarily large amounts (Mullis *et al.*, 1986). The DNA is copied by the DNA–polymerase complex after a DNA olignucleotide primer is bound to the single-stranded template, and the primer is extended by binding the 5′ carbon of the incoming deoxyribose of the nucleoside to the 3′ carbon of the last nucleotide of the primer. So, if two DNA oligonucleotides are designed for the complementary strands of DNA and their 3′ ends face each other, they frame a region of DNA. If the cycle of temperature that causes the denaturation of the target DNA, annealing of the primers to the template and copying of the DNA by the polymerase is repeated several times, DNA will be amplified exponentially in a region that reaches to the 5′ ends of the two DNA primers. After 30 cycles, there will be more than enough DNA to be visualized easily on a gel by staining with ethidium bromide or any of the other methods mentioned above.

These technologies have been combined in a large array of different techniques to detect polymorphisms and these are listed in Table 11.1. Restriction fragment length polymorphism (RFLP), amplification-created restriction sites (ACRS) and variable number of tandem repeats (VNTR) form a group, since a restriction endonuclease is used either to recognize point mutations (RFLP and ACRS) or to cut the DNA to a manageable size for analysis (VNTR), with differences in length resolved using electrophoresis. With the addition of PCR, some VNTR can be analysed without the use of restriction endonucleases, with the PCR acting to provide a manageable size of DNA for analysis (e.g. Boerwinkle

Table 11.1. A matrix of combinations of methods for detecting polymorphisms. Along the vertical axis are the acronyms for each technique and along the horizontal axis are the methods used to detect the polymorphism.

Type	Matrix	DNA oligonucleotides	Labelling	Variation
RFLP	Ag/PAGE	Clone/Synth/PCR	AR/EB/SS/FL	PM/ID
ACRS	Ag/PAGE	Synth/PCR	AR/EB/SS/FL	PM
VNTR	Ag/PAGE	Clone/Synth/PCR	AR/EB/SS/FL	R/PM
SSCP	PAGE	PCR	AR/SS/FL	PM/ID
HD	Ag/PAGE	PCR	AR/EB/SS/FL	PM/ID
DGGE	PAGE	PCR	AR/SS/FL	PM/ID
STRP	Denat.PAGE	PCR/Synth	FL/AR/SS	R/PM
AFLP	Denat.PAGE/Ag	PCR	FL/AR/SS/EB	PM/ID
RAPD	Denat.PAGE/Ag	PCR	AR/EB/FL	PM/ID
ASO	None	Synth/PCR	AR/FL	PM
SNP	None	Synth	FL/AR	PM

RFLP, restriction fragment length polymorphism; ACRS, amplification-created restriction sites; VNTR, variable number of tandem repeats (minisatellites); SSCP, single-strand conformational polymorphism; HD, heteroduplex; DGGE, denaturing gradient gel electrophoresis; STRP, simple tandem repeat polymorphism (microsatellites); AFLP, amplified fragment length polymorphism; RAPD, random amplified polymorphic DNA; ASO, allele-specific oligonucleotide; SNP, single nucleotide polymorphism; Ag, agarose; PAGE, polyacrylamide gel electrophoresis; Clone, oligonucleotides derived from the copying of a clone; Synth, synthesized oligonucleotides; PCR, polymerase chain reaction; AR, autoradiography; EB, ethidium bromide; SS, silver staining; FL, fluorescent labelling; PM, point mutation; ID, insertion/deletion; R, repeat.

et al., 1989). Single-strand conformational polymorphism (SSCP), HD, denaturing gradient gel electrophoresis (DGGE) and simple tandem repeat polymorphism (STRP) form a group since they are dependent upon the PCR to generate the fragments for analysis. Thereafter, the amplified fragments are analysed on different kinds of gel matrices, depending upon the type of polymorphism, length or sequence identity. For SSCP, the differences are detected through perturbations in the secondary structure of the single-stranded conformation of the DNA. The electrophoretic conditions are constant across the gel. Heteroduplexes can usually be seen on an SSCP gel, but they can be run for their own sake. Heteroduplexes can also be visualized by restricting mismatched base pairs with, for example, mung-bean nuclease and then the fragments are separated by electrophoresis. For DGGE, the conditions of the denaturants vary across the gel and the double strands melt at different locations in the gel, depending upon their sequence. Usually STRPs are analysed for repeat length variation on denaturing gels, but they can also be assayed for sequence differences when there is no length variation, using SSCP or DGGE gels. When the length variation contributes to differences in volume, the STRP alleles will correspond to SSCP alleles. Amplified fragment length polymorphism (AFLP) and random amplified polymorphic DNA (RAPD) form a group because they use the power of short DNA oligonucleotides to amplify at random, so a large number of anonymous loci can be analysed simultaneously. Finally, allele-specific oligonucleotide (ASO) and single nucleotide polymorphism (SNP) form a group, with SNP being an extremely developed version of ASO. In ASO, alleles are discriminated through their ability to bind to an oligonucleotide, with either the samples or the oligonucleotide bound to a filter. The positive binding can be detected using autoradiography, fluorescence or simple colour dye in combination. This last version allows yellow for one allele to be combined with blue for the other allele to make green for the heterozygote. With SNP, this is taken several steps further by using massive arrays of DNA oligonucleotides synthesized on a DNA chip (Fodor *et al.*, 1991; Hacia *et al.*, 1996; Marshall, 1997). In combination with the photolithographic process used to make silicon chips for the computer industry, DNA oligonucleotides are built up layer by layer on chips at sufficient accuracy for thousands of oligonucleotides to be placed on a tiny wafer of silicon. Positives are detected with laser-induced fluorescence in a special detector.

Most polymorphisms could be analysed by a great many of the techniques in this table, and the trend in the past 20 years since the invention of the Southern blot has been an improvement in the length of time to collect data on a polymorphism and an increase in the number of loci that can be analysed at the same time. The PCR has shortened the time from a few weeks to an afternoon for an RFLP. Fluorescent labelling and automatic sequencers have shortened the data collection time for STRP analysed by autoradiography from overnight to real-time data collection. While multilocus VNTR has always represented tens of loci, and RAPD and AFLP are polymorphisms for several loci collected at once, different colour fluorescence and the multiple PCR amplification of STRP in the same tube represent the apogee of this technology, since not only can tens of loci be analysed at the same time but also each locus is already uniquely defined. Single nucleotide polymorphism represents turbo-charged ASO, with the simultaneous testing of thousands of single nucleotide polymorphisms in real time.

Due to the history of the development of polymorphism and the timing of the construction of the cattle linkage maps, most of the loci that are used are anonymous STRP. There was a minor role for RFLP early in the development, and SSCP has essentially replaced RFLP as the method of detecting point mutations in or near coding sequences. There has been no use of RAPD in any of the large-scale maps, and AFLP has so far had only a minor role late in the development of these maps. The SNP technique represents a potentially large source of polymorphism, but whether it will be used to make new maps in cattle or whether it will just be used to map traits is not yet known.

The Linkage Maps of Cattle

The genetic linkage maps of cattle are now reaching a mature phase, in which they are useful for the purposes for which they were constructed and the rate of progress, compared with what has been already achieved, is small. They were initially constructed primarily to search for the locations of genes affecting economically important traits and most of these have continuous distributions. As a secondary goal, they would provide information on genomic evolution, but this would only occur if polymorphisms in coding sequences were part of the maps. The initial requirements were for maps that had an average spacing between loci of approximately 20 cM and, in a genome estimated to be between 2500 and 3000 cM, this would require 150 DNA markers. This naïve estimate is clearly insufficient, since not all loci are hypervariable, nor would they be spread relatively evenly over the genome. For loci to be useful in cattle outside a map, they must have many alleles, for the reasons indicated previously. Therefore, in any region, there must be alternative loci that could provide a useful polymorphism with which to detect linkage, and this alone would increase the numbers of loci that are required in a map. Furthermore, the first 150 loci will not provide good coverage of all the chromosomes. This can be seen in the first two genetic maps of cattle (Barendse *et al.*, 1994; Bishop *et al.*, 1994), where 202 and 313 loci, respectively, were genotyped, essentially at random. Some chromosomes were represented by more than one linkage group, some chromosomes had a few unlinked loci assigned to them, the X chromosome was essentially unmapped, few chromosomes had good coverage in all regions, no chromosome had reached anywhere near its final map length, the gene order was disputed in some cases and the distances between some loci have subsequently been shown to be underestimated. Clearly, more loci needed to be mapped to provide a more mature map and one in which alternative loci could be chosen from any region.

The later versions of these two early maps, as well as the additional independent mapping efforts of other laboratories, have resulted in more than

Table 11.2. The anonymous DNA loci mapped by genetic linkage in cattle.

A. The abbreviations for all the loci

CSSM	CSIRO, Tropical Agriculture, Australia
CSFM	CSIRO, Tropical Agriculture, Australia
CSKB	CSIRO, Tropical Agriculture, Australia
CSRD	CSIRO, Tropical Agriculture, Australia
GMBT	Genmark Inc., USA
MGTG	Genmark Inc., USA
TGLA	Genmark Inc., USA
AGLA	Genmark Inc., USA
ETH	ETH, Zurich, Switzerland
HUJ	Hebrew University, Jerusalem, Israel
MAF	AgResearch, New Zealand
OarVH	AgResearch, New Zealand
OarHH	AgResearch, New Zealand
RM	University of Georgia, USA
UWCA	University of Wisconsin, USA
ARO	Agricultural Breeding Unit, Israel
AF	AFRC, Babraham, UK
TXQG	Texas A&M University, Department of Veterinary Pathobiology, USA
HEL	University of Helsinki, Finland
ILSTS	ILRAD, Kenya
HAUT	Veterinary Institute, Hanover
JAB	AFRC, Roslin Institute, UK
RI	AFRC, Roslin Institute, UK
BM	MARC, USDA, USA
BP	MARC, USDA, USA
BR	MARC, USDA, USA
BL	MARC, USDA, USA
BMS	MARC, USDA, USA
BMC	MARC, USDA, USA
INRA	INRA, Jouy-en-Josas, France
INRABERN	INRA, Jouy-en-Josas, France, and University of Berne, Switzerland
MILSTS	University of Milan, Italy
MM	Rijks Universiteit Ghent, Belgium
McM	CSIRO, Division of Animal Health, Australia (ovine-derived)
URB	University of Illinois, USA
IDVGA	IDVGA, Milan, Italy
INRAMTT	MTT, Finland
DIK	Shirakawa Institute of Genetics, Japan
IOBT	Norwegian College of Veterinary Medicine, Oslo
TEXAN	Texas A&M University, Department of Animal Sciences, USA
NOR	Agricultural University of Norway, As
ABS	American Breeders Service, USA
FBN	Dummerstorf, Germany
PZ	Potenza, Italy
INRAZARA	INRA Jouy-en-Josas, France, and Saragossa, Spain
HELMTT	Helsinki University and MTT Finland
IOZARA	Norwegian College of Veterinary Medicine, Oslo, and Saragossa, Spain
DVEPC	INRA, Jouy-en-Josas, France
TUM	Technical University, Munich, Germany
DIAS	National Institute of Animal Science, Foulum, Denmark
UCPC	Universita Cattolica dell Sacro Cuore, Piacenza, Italy
KIEL	Christian-Albrechts-University of Kiel, Germany
ZES	Saragossa, Spain

Table 11.2. *Continued.*

B. The number of anonymous DNA markers mapped to each chromosome

Chromosome	Number of DNA markers
Bta 1	98
Bta 2	68
Bta 3	77
Bta 4	58
Bta 5	69
Bta 6	56
Bta 7	53
Bta 8	56
Bta 9	64
Bta 10	60
Bta 11	91
Bta 12	57
Bta 13	53
Bta 14	40
Bta 15	47
Bta 16	47
Bta 17	50
Bta 18	46
Bta 19	57
Bta 20	39
Bta 21	50
Bta 22	37
Bta 23	37
Bta 24	51
Bta 25	32
Bta 26	31
Bta 27	29
Bta 28	40
Bta 29	40
Bta X	28
Bta Y	8
Bta XY	11
Total	1580

Bta, *Bos taurus*.

1600 anonymous STRP loci located to all regions of the bovine genome, with many alternatives in most genetic regions (Georges *et al.*, 1995; Ma *et al.*, 1996; Barendse *et al.*, 1997; Kappes *et al.*, 1997; summarized in Table 11.2). Anonymous STRPs are clearly the DNA polymorphism of choice in cattle, as they are usually highly polymorphic. They counter the absence of the genotypes of the dams in resource populations. There has been some effort to integrate these STRP markers into systems that can be analysed on automated sequencers, following the lead of the human genome, so that 10–15 loci can be genotyped at the same time (e.g. Heyen *et al.*, 1997). Most of that information, however, is not in the public domain. There is also no common pool of STRP markers organized into panels of different sizes and different fluorescent labels to give the 'stained glass' images of DNA markers, except for the proprietary StockMarks (Perkin Elmer) panel used for parentage testing. This total resource of STRP DNA markers forms the skeleton to which other DNA markers can be pegged. These other loci are usually type I polymorphisms (in or near coding sequences) and are visualized using SSCP, RFLP or DGGE and, in a small number of examples, STRP. More than 250 polymorphisms in or near identified genes have been mapped (Table 11.3). They provide information on genomic evolution as well as candidates for the genetic factors that contribute to the traits of cattle. Polymorphisms such as those assayed using AFLP and SNP do have a role in cattle, since they can provide extremely large numbers of polymorphisms at an extremely reasonable cost and at high speed, but their informativeness will be less than in other species. Their usefulness will increase when they have been mapped to the STRP, so that an association to an AFLP or an SNP will lead to more informative markers.

The major gaps that are left to be filled on the current maps are decreasing in size, although the last few gaps will require unusual effort. The average DNA marker density for the cattle map is 2.5 cM and 3.5 cM for the MARC and IBRP maps, respectively. Nevertheless, there are regions with low densities of DNA markers. The largest gap on an autosome on the genetic maps located at http://spinal.tag.csiro.au/ is the region near the telomere of chromosome 8. Between *HEL9* and *C5* is a distance of 48 cM, in which there are two other loci, *DIK74* and *CSSM47*. On the X chromosome, the region near the centromere has *BM6017*, *HUJ121* and *RMX27* in a space of 36 cM, with a large gap between *HUJ121* and the other two. Other regions of lower density occur near the telomere of chromosome 6, the centromere of chromosome 9, the centromere of chromosome 14, the centromere of chromosome 17 and the subtelomere of chromosome 20. On the maps located at http://sol.marc.usda.gov/ the largest autosomal gap is near the telomere of chromosome 8, where there is a distance of 29 cM between *BM711* and *CSSM47*. The largest X-chromosome gap is between *BMS903* and *BMS811*, a gap of approximately 30 cM, with only *URB10* located in it. Other areas of lower density are the centromere of chromosome 3, the centromere of chromosome 6, the centromere of chromosome 13 and the centromere of chromosome 19. While some of the gaps can be covered by comparing the two maps, clearly the areas of lowest density are similar in the two studies and centromeres seem to have fewer STRP markers than other regions. The other mapping efforts have not provided additional loci to close those gaps. Given that so many loci have already been mapped, it will require a large number of additional loci to fill these last gaps: if prior locations count for anything, the areas with fewer loci could expect to have fewer loci assigned to them in future. It is likely that chromosomal and subchromosomal libraries need to be made to isolate markers for these regions, as has been suggested before (Barendse *et al.*, 1994). Those gaps that are consistent between studies may represent low densities of STRP sequences, and other kinds of polymorphisms may be needed to close the gaps. They could represent areas of higher recombination and this could be tested by physical mapping.

Table 11.3. The coding sequences that have been mapped by genetic linkage in cattle.

Symbol	Location	Name
IFNAR1	Bta 1	Interferon alpha receptor
SOD1	Bta 1	Superoxide dismutase 1
SLC5A3	Bta 1	Solute carrier family 5 (inositol transporter) member 3
KAP8	Bta 1	Keratin-associated protein 8
POU1F1	Bta 1	Pituitary transcription factor 1 [PIT1]
UMPS	Bta 1	Uridine monophosphate synthetase
CRYG8	Bta 1	Crystallin gamma-s
SST	Bta 1	Somatostatin
KNG	Bta 1	Kininogen
CRYAA	Bta 1	Crystallin alpha 1
TF	Bta 1	Transferrin
COL3A1	Bta 2	Collagen 3 alpha 1
NEB	Bta 2	Nebulin
GCG	Bta 2	Glucagon
NRAMP1	Bta 2	Natural resistance-associated macrophage protein 1
PROC	Bta 2	Protein C
ALPI	Bta 2	Alkaline phosphatase (intestinal)
TNP1	Bta 2	Transition protein 1
PDE1	Bta 2	3′, 5′-cyclic nucleotide phosphodiesterase (59/61 kDa)
SSBP	Bta 2	Single-strand binding protein
ACVR2	Bta 2	Activin receptor 2 [UCD2]
PDE6A	Bta 2	cGMP phosphodiesterase alpha subunit
FN1	Bta 2	Fibronectin
INHA	Bta 2	Inhibin alpha
AK2	Bta 2	Adenylate kinase isoenzyme 2
FGR	Bta 2	Gardner–Rasheed feline sarcoma viral (v-fgr) oncogene homologue
FUCA1	Bta 2	Fucosidase, alpha-L-1, tissue
ALPL	Bta 2	Alkaline phosphatase, liver
CD3Z	Bta 3	CD3 antigen zeta
FCGR1	Bta 3	Fcg receptor I, high affinity, CD64
FCGR2	Bta 3	Fcg receptor II, low affinity
FCGR3	Bta 3	Fcg receptor III, low affinity
OSG	Bta 3	Oviduct-specific glycoprotein
ATP1A1	Bta 3	Sodium and potassium ATPase alpha 1
EAL	Bta 3	Erythrocyte antigen L
GCSFR	Bta 3	Colony-stimulating factor receptor
IL6	Bta 4	Interleukin 6
DDC	Bta 4	Dopa decarboxylase [aromatic L-amino acid decarboxylase]
LEP	Bta 4	Leptin [OBS]
TCRB	Bta 4	T-cell receptor beta
TCRG	Bta 4	T-cell receptor gamma
SYT1	Bta 5	Synaptotagmin A
COL2A1	Bta 5	Collagen 2 alpha 1
LALBA	Bta 5	Lactalbumin
LYZ	Bta 5	Lyzozyme
MYF5	Bta 5	Myogenic factor 5
KRT@	Bta 5	Keratin complex [BMC1009]
GLYCAM1	Bta 5	Glycosylated cell-adhesion molecule 1
IFNG	Bta 5	Interferon gamma
IGF1	Bta 5	Insulin-like growth factor 1
MB	Bta 5	Myoglobin
SLC2A3	Bta 5	Glucose transporter 3 [GLUT3] (Solute carrier family 2 member 3)
NUBA	Bta 5	NADH-ubiquinone reductase 24 kDa mitochondrial

Table 11.3. *Continued.*

Symbol	Location	Name
BZRP	Bta 5	Benzodiazepine receptor (peripheral)
DIA1	Bta 5	Diaphorase 1
CYP2D@	Bta 5	Cytochrome P2D
ADSL	Bta 5	Adenylsuccinate lyase
PDGFB	Bta 5	Platelet derived growth factor B
TPI1	Bta 5	Triose phosphate isomerase 1
PPARG	Bta 5	Peroxisome proliferative activated receptor, gamma
ALB	Bta 6	Albumin
CSN1A	Bta 6	Casein alpha s1
CSN2	Bta 6	Casein beta
CSN3	Bta 6	Casein kappa
EGF	Bta 6	Epidermal growth factor
FABP-HL	Bta 6	Fatty acid-binding protein (heart)-like
FGFR3	Bta 6	Fibroblast growth factor receptor 3
GABRA2	Bta 6	Gamma-amino butyric acid receptor alpha 2
IF	Bta 6	Complement component 1
KIT	Bta 6	Hardy–Zuckerman 4 feline sarcoma viral oncogene homologue (v-kit)
PDE6B	Bta 6	Phosphodiesterase B
QDPR	Bta 6	Quininoid dihydropterine reductase
SPP1	Bta 6	Secreted phosphoprotein 1 [Osteopontin]
GC	Bta 6	Vitamin D-binding protein
AMH	Bta 7	Anti-Mullerian hormone
CSF2	Bta 7	Colony-stimulating factor 2
FGF1	Bta 7	Fibroblast growth factor alpha [FGFA]
IL4	Bta 7	Interleukin 4
LDLR	Bta 7	Low-density lipoprotein receptor
MANB	Bta 7	Mannosidase, B
RASA	Bta 7	GAP RAS activator
SPARC	Bta 7	Secreted protein acidic cysteine rich [Osteonectin]
CALP	Bta 7	Calpastatin
ALDOB	Bta 8	Aldolase B
C5	Bta 8	Complement component 5
CPNB	Bta 8	Cathepsin B
GGTB2	Bta 8	*N*-acetylglucosamine galactosyltransferase (beta-1,4-galactosyltransferase)
LPL	Bta 8	Lipoprotein lipase
BMP1	Bta 8	Bone morphogenetic protein 1
CGA	Bta 9	Glycoprotein hormone alpha [GLYAA]
CNR1	Bta 9	Cannabinoid receptor [ABS17]
GJA1	Bta 9	Connexin 43, gap-junction membrane channel protein alpha 1
GNRHL	Bta 9	Gonadotrophin-releasing hormone-like sequence
MYB	Bta 9	Avian myeloblastosis viral v-myb oncogene homologue
SOD2	Bta 9	Superoxide dismutase 2, mitochondrial
AKAP	Bta 10	Anchor protein regulatory subunit (AKAP75) gene
BRN	Bta 10	Brain ribonuclease
CLTA	Bta 10	Clathrin light chain a polymorphism 1
SRN	Bta 10	Seminal ribonuclease
TCRA	Bta 10	T-cell receptor alpha
EAZ	Bta 10	Erythrocytic antigen Z
ACTG2	Bta 11	Actin
CD8A	Bta 11	CD8 antigen
DBH	Bta 11	Dopamine beta hydroxylase
GAPDL	Bta 11	Glyceraldehyde 3 phosphate dehydrogenase-like
GGTA1	Bta 11	Alpha-1-3 galactosyl transferase

Table 11.3. *Continued.*

Symbol	Location	Name
LGB	Bta 11	Lactoglobulin beta
SHR	Bta 11	Steroid hormone receptor
XDH	Bta 11	Xanthine dehydrogenase
ILR1	Bta 11	Interleukin receptor 1
ILR2	Bta 11	Interleukin receptor 2
COL4A1	Bta 12	Collagen 4 alpha 1
DCT	Bta 12	Dopachrome tautomerase
F10	Bta 12	Factor 10
FLT3	Bta 12	fms-related tyrosine kinase
HTR2A	Bta 12	5-hydroxytryptamine (serotonin) receptor 2A
RB1	Bta 12	Retinoblastoma 1
EAB	Bta 12	Erythrocytic antigen B
AVP	Bta 13	Arginine vasopressin [ARVP]
CHGB	Bta 13	Chromogranin B
GHRH	Bta 13	Growth hormone-releasing hormone
OXT	Bta 13	Oxytocin
VIM	Bta 13	Vimentin
CRH	Bta 14	Corticotrophin-releasing hormone
IL7	Bta 14	Interleukin 7
TG	Bta 14	Thyroglobulin
CA2	Bta 14	Carbonic anhydrase 2
ARRB	Bta 15	Arrestin B
CD3D	Bta 15	CD3 antigen delta
FSHB	Bta 15	Follicle-stimulating hormone beta
HBB	Bta 15	Haemoglobin beta
KCNA4	Bta 15	Potassium voltage-gated channel
MCAM	Bta 15	Melanoma-cell adhesion molecule [MUC18 CSSME75]
NCAM	Bta 15	Neural-cell adhesion molecule [ADCYC]
PTH	Bta 15	Parathyroid hormone
EAA	Bta 15	Erythrocytic antigen A
ADORA1	Bta 16	Adenosine A1 receptor
AT3	Bta 16	Antithrombin 3
REN	Bta 16	Renin
ATP2B4	Bta 16	ATP-dependent calcium pump B4 [ATPCP]
FH	Bta 16	Factor H
IL10	Bta 16	Interleukin 10
PIGR	Bta 16	Polyimmunoglobulin receptor
SELP	Bta 16	P selectin
EAR'	Bta 16	Erythrocytic antigen R'
ADRBK2	Bta 17	Beta adrenergic receptor kinase 2
ALDH2	Bta 17	Aldehyde dehydrogenase 2
CRYBA4	Bta 17	Crystallin beta A4
FGF2	Bta 17	Fibroblast growth factor B
FGG	Bta 17	Fibrinogen gamma
GUCY	Bta 17	Guanylate cyclase
LIF	Bta 17	Leukaemia inhibitory factor
PLA2G1B	Bta 17	Phospholipase A2 group 1 B (pancreas) [PLA2]
MC1R	Bta 18	Melanocyte-stimulating hormone receptor
MT2A	Bta 18	Metalothionein 2
PIM1	Bta 18	Pim1 oncogene
FCG2R	Bta 18	Fcg receptor 2
RYR1	Bta 18	Ryanodine receptor 1

Table 11.3. *Continued.*

Symbol	Location	Name
SLC2A3RS	Bta 18	Glucose transporter 3 [GLUT3] (solute carrier family 2 member 3)-related sequence
CEBPA	Bta 18	CCAAT/enhancer binding protein (C/EBP), alpha
EAC	Bta 18	Erythrocytic antigen C
ACACA	Bta 19	Acetyl coenzyme A carboxylase A
CHRNB1	Bta 19	Cholinergic receptor, nicotinic, beta polypeptide 1 (muscle)
CRYBA1	Bta 19	Crystallin beta A1
GFAP	Bta 19	Glial fibrillary acid protein
GH1	Bta 19	Growth hormone beta
IL6R	Bta 19	Interleukin 6 receptor
LPO	Bta 19	Lactoperoxidase
MAPT	Bta 19	Microtubule-associated protein tau
MYL4	Bta 19	Myosin, light polypeptide 4, alkali, atrial
NF1	Bta 19	Neurofibromin 1
P4HB	Bta 19	Procollagen-proline, 2-oxyglutarate 4-dioxygenase beta polypeptide
THRA1	Bta 19	Thyroid hormone receptor, alpha 1
EPOR	Bta 19	Erythropoietin receptor
KRT10	Bta 19	Cytokeratin class I gene cluster
PTF2	Bta 19	Post-transferrin factor 2
EAT′	Bta 19	Erythrocytic antigen T′
ANP1	Bta 20	Atrial natriuretic peptide receptor 1
GHR	Bta 20	Growth hormone receptor
HD5S39	Bta 20	Human D5S39
HEXB	Bta 20	Hexosaminidase B
HTR1A	Bta 20	5-Hydroxytryptamine receptor 1 A
MAP1B	Bta 20	Microtubule-associated protein 1B
PRLR	Bta 20	Prolactin receptor
IGF1R	Bta 21	Insulin-like growth factor 1 receptor
LTF	Bta 21	Lactotransferrin
EAS	Bta 21	Erythrocytic antigen S
GNAZ	Bta 22	Transducin alpha 1 subunit [Guanine nucleotide binding protein, alpha z polypeptide]
GPX1	Bta 22	Glutathione peroxidase
HRH1	Bta 22	Histamine receptor H1
BTN	Bta 23	Butyrophilin
CYP21	Bta 23	Cytochrome p450 21 hydroxylase
DRB3	Bta 23	BoLA DRB3 (MHC class II)
DRBP	Bta 23	BoLA DRBP (pseudogene)
DYA	Bta 23	DYA class II MHC leucocyte antigens
MOG	Bta 23	Myelin/oligodendrocyte glycoprotein
PRL	Bta 23	Prolactin
SMHCC	Bta 23	Ovine MHC class I microsatellite
TNFA	Bta 23	Tumour necrosis factor alpha
VEGF	Bta 23	Vascular endothelial growth factor
EAM	Bta 23	Erythrocytic antigen M
HSP70.1	Bta 23	70 kDa heat-shock protein
DIB	Bta 23	BoLA DIB (MHC class II)
DRB1	Bta 23	BoLA DRB1 (MHC class II)
DRB2	Bta 23	BoLA DRB2 (MHC class II)
F13A	Bta 23	Factor 13 A
LMP2	Bta 23	Large multifunctional protease 2 [PSMB9]
CDH2	Bta 24	*N*-cadherin
CYB5	Bta 24	Cytochrome B5

Table 11.3. *Continued.*

Symbol	Location	Name
DSC3	Bta 24	Desmocollin I type 2
FECH	Bta 24	Ferrochelatase
PACAP	Bta 24	Pituitary adenyl cyclase activating peptide
PAI2	Bta 24	Plasminogen activator inhibitor 2 [PLANH2]
EPO	Bta 25	Erythropoietin
PAI1	Bta 25	Plasminogen activator inhibitor 1 [PLANH1]
PRM1	Bta 25	Protamine P1
GOT1	Bta 26	Glutamic-oxaloacetic transaminase 1, soluble [Aspartate aminotransferase ASAT]
DNTT	Bta 26	Deoxyribonucleotidyl transferase [TDT]
GPRK5	Bta 26	G protein-coupled receptor kinase 5
FAS	Bta 26	Fas/APO-1 oncogene
LAMA1	Bta 27	Laminin alpha 1
MTNR1A	Bta 27	Melatonin receptor 1 A
PSAP	Bta 28	Sphingolipid activator protein 2
RBP3	Bta 28	Interphotoreceptor binding protein (retinol-binding protein 3)
VAV1	Bta 28	Vav1 proto-oncogene
CD5	Bta 29	CD5 antigen
GSTP1	Bta 29	Glutathione S-transferase, class-pi
OPCML	Bta 29	Opioid-binding protein cell adhesion binding molecule-like [OBCAM]
PAG1B	Bta 29	Pregnancy-associated glycoprotein 1 B
MAOA	Bta X	Monoamine oxidase A

Bta, *Bos taurus*; cGMP, cyclic guanosine monophosphate; ATPase, adenosine triphosphatase; NADH, nicotinamide adenine dinucleotide; GAP, guanosine triphosphatase-activating protein; ATP, adenosine triphosphate; BoLA, bovine leucocyte antigen. Alternative names are given in square brackets.

These maps can clearly be used to locate roughly the genetic factors for continuous traits, but it is their performance at fine mapping that is lacking. There are regions of these maps where there are several markers that have been mapped to within 1 cM of each other. Should a genetic factor be mapped to such a region, there would be sufficient genetic resources to begin the fine mapping of that locus. Most regions of the bovine linkage maps do not have such a degree of coverage and investigators would have to generate new markers in most areas of the linkage maps, should they decide to refine the locations of the genetic factors they pursue. In that context, any gap between adjacent loci of more than 3 cM would represent a very large distance, and most of the bovine genome would be gappy. It would require substantial amounts of physical mapping to make a yeast artificial chromosome (Libert *et al.*, 1993) or bacterial artifical chromosome (Cai *et al.*, 1995) contig, and sequencing across such a region would be prohibitively expensive and time-consuming. In any event, the regions of high density are not very well mapped, due to a lack of resolution of the maps based on the existing families, so gene order would need to be confirmed.

The accuracy of genetic maps is ultimately measured in the number of informative meioses for a pair of loci and this is determined by the degree of polymorphism and the number of individuals in families. For most loci, these studies have found that the order of loci separated by approximately 5 or more cM can be specified, but that below this distance it requires a high degree of polymorphism to specify gene order. When data from several studies are merged, it becomes possible to specify the order of loci at lower levels than 5 cM. However, when the gene order reaches 1 cM, to be successful the accuracy of the genotyping must be high, because even 1% genotyping error leads to false recombinants having the same frequency as true recombinants and then it is not possible to discriminate the true gene order.

A high density of DNA markers will be required in those areas where effort is directed to isolating genetic factors for continuous traits. While linkage analysis derived from the segregation of loci in families is the initial stage of most studies of quantitative traits, except in a few species, such as *D. melanogaster*, the most likely location of the polygenic variation is usually resolved to an area of 20–30 cM (e.g. Paterson *et al.*, 1988; Lander and Botstein, 1989; Andersson *et al.*, 1994; Georges *et al.*, 1995). In mammals, this corresponds to approximately 20–30 million base pairs, which is enough space to hide a thousand genes or 10–20 whole bacterial genomes. Quite obviously, the genetic factors have to be located with greater precision before cloning experiments are feasible. In the case of cattle, further breeding and selection experiments would be required if this linkage approach were followed, and the time and money required would be extremely large. An approach familiar to human geneticists circumvents the problems and improves the resolution by an order of magnitude. Resolution can be improved by using many DNA markers in a small genetic region and testing each of them for linkage disequilibrium in randomly drawn samples of the population (Woolf, 1955; Risch, 1983; Copeman *et al.*, 1995). Linkage disequilibrium is usually only detected at scales of around 1–3 cM, although in livestock this value may be slightly larger in some breeds due to the extensive use of a few sires. Determining the peak values for linkage disequilibrium can be done using a multilocus approach, which will indicate the most likely location of the genetic factors. While cloning and sequencing of DNA could be initiated, at this point comparing the genome of cattle with other mammals for likely candidates could result in the rapid identification of some of the genetic factors affecting the trait of interest. So it is of great importance that some of the genes used to test for linkage disequilibrium be genes mapped in other species.

This high density may be required in a substantial part of the genome, since much work this century has indicated that every chromosome is expected to have genetic factors that contribute to a continuous trait (Castle, 1922; Breese and Mather, 1957; Davies, 1971; Shrimpton and Robertson, 1988). Even if not all genetic factors have a large enough effect on the phenotype to make them worth analysing, a wide range of continuous traits is currently being studied in cattle and one could expect their factors to be widely distributed through the genome. It would be ironic if most of them occurred in regions that currently have a low density of DNA markers.

The genetic linkage maps may be replaced by high-resolution physical maps based on whole genome radiation hybrid panels to determine the order

Genetic Linkage Mapping

of markers at high resolution (e.g. Womack *et al.*, 1997). Polymorphism is irrelevant to these physical maps and, while data errors must be guarded against, the maps are built up from data of presence or absence in a somatic-cell hybrid clone. Close distance is not measured by rare events that occur in one in a hundred to one in a thousand meiotic events, so the gene order is less dependent upon the fluctuations associated with rare events. Marker order is the same, irrespective of the mapping technique, so these physical maps will supersede linkage maps for the description of gene order at high resolution. However, linkage maps provide more than the marker order, they provide the genetic distance between loci and there is often not a one-to-one relationship between physical distance and genetic distance. Genetic linkage maps are critical to the estimation of the decay of allelic association (linkage disequilibrium) between alleles at linked loci, the predictions of geneticists during marker assisted selection, genetic risk counselling and other activities that require estimates of the recombination fraction. They are indispensable in the attempts to locate the genetic factors that contribute to continuous variation.

Some of the more spectacular successes of recent time have been the isolation of the genes for discrete traits of cattle using the gene maps. While ultimately the focus of genetic research in cattle is on continuous traits, the first gene affecting a discrete trait has been isolated, using the linkage map as a guide. The trait of muscular hypertrophy or double muscling was located to bovine chromosome 2 (Charlier *et al.*, 1995) and subsequent mapping by that group narrowed the region of interest to a small section. The serendipitous discovery that knockout mutations in the myostatin gene in mice lead to differential growth and that myostatin was located to the same region of bovine chromosome 2 as muscular hypertrophy led to a race to describe mutations at the myostatin gene in cattle, and these were ultimately linked to the double-muscling phenotype. Other discrete traits of cattle have been localized to chromosomes, such as the poll (hornless) gene, the roan coat colour, and the weaver neurological degenerative disease, when the responsible gene is unknown. Maps have not always been necessary for the isolation of genes for traits, since the gene causing the defect can on occasion be guessed, and the genes for traits such as Pompe's disease, bovine leucocyte adhesion disorder, citrullinaemia, deficiency of uridine monophosphate synthetase (DUMPS) and black coat colour have been isolated, while association of chronic posterior spinal paresis with the major histocompatibility complex (MHC) was made because of the similarity to ankylosing spondylosis in humans. The MHC has been shown to be associated with eight infectious and non-infectious diseases or resistance to parasites in cattle.

The maps are now being used to locate the genetic factors that contribute to continuous traits. Before the construction of the linkage maps, many associations were found between markers and traits, and a list of these is maintained by Barendse on http://spinal.tag.csiro.au/. However, the lack of a genetic map has meant that these studies proceeded with whatever polymorphisms were available. Some of these studies may even have been successful in isolating genetic factors affecting the trait, such as the effects of

Continued on p. 360

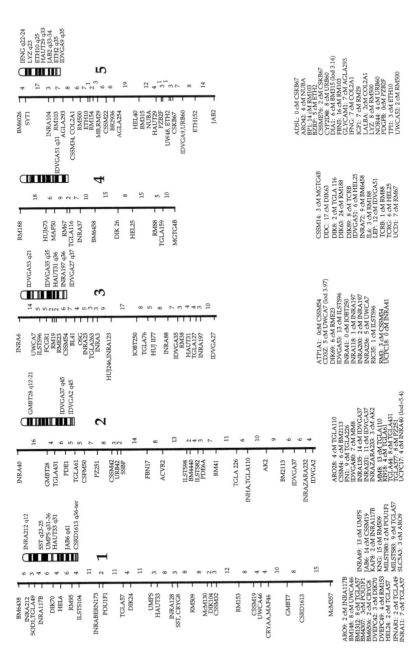

Fig. 11.1. A genetic linkage map of the bovine genome. This map appears on the web site of the Cattle Genome Database in the public domain at /http://spinal.tag.csiro.au/. Chromosome maps of sexes pooled are shown. Distances between loci are given in centimorgans (based on the Kosambi mapping function). Loci included in the stick figures are ordered at greater than or equal to a difference in log odds of 3.00. Loci placed below the stick figures are shown with the distance to the closest ordered marker. This map shows minor differences from that reported in Barendse *et al.* (1997).

Genetic Linkage Mapping

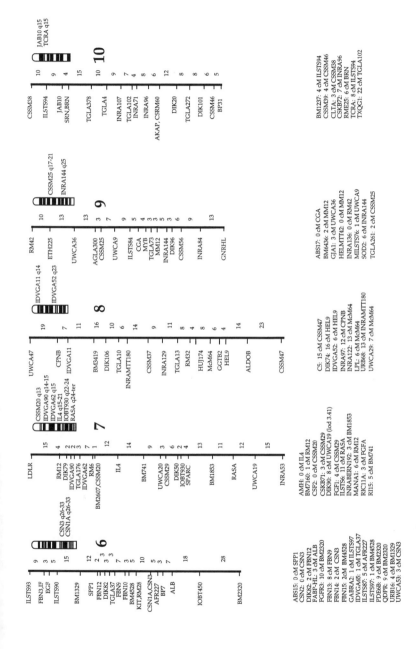

Fig. 11.1. *Continued.*

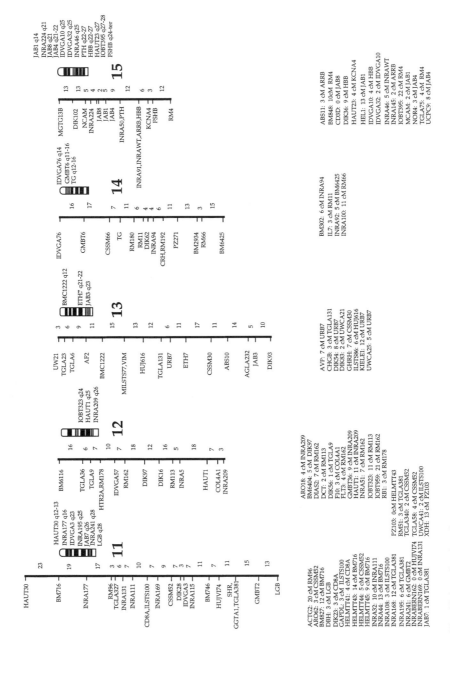

Fig. 11.1. *Continued.*

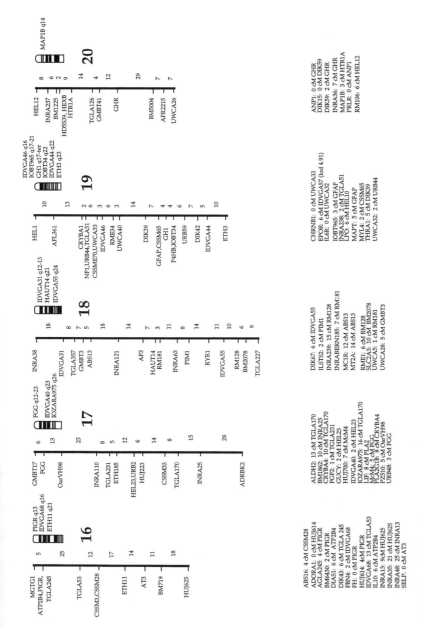

Fig. 11.1. *Continued.*

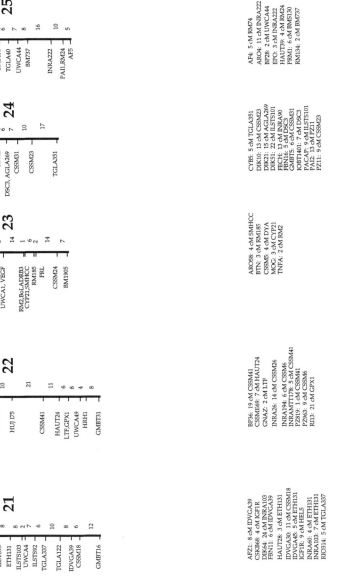

Fig. 11.1. Continued.

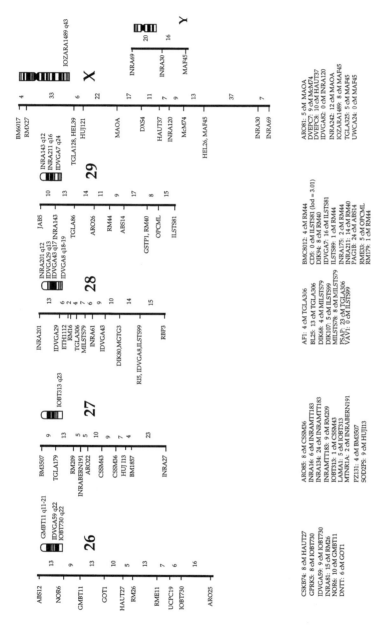

Fig. 11.1. Continued.

the casein protein polymorphism on the characteristics of cheese. However, the lack of large numbers of polymorphisms and the lack of organization of these into linkage maps has meant that positive associations have languished for want of the tools to proceed with the identification of the genetic factors. These tools now exist and the first studies are beginning to appear in which the genetic factors have been located to segments of chromosomes. The resolution is still poor and much further work needs to be done, but the prospect now exists for the identification of the genetic factors that contribute to continuous variation in cattle.

References

Andersson, L., Haley, C.S., Ellegren, H., Knott, S.A., Johansson, M., Andersson, K., Andersson-Eklund, L., Edfors-Lilja, I., Fredholm, M., Hansson, I., Hakansson, J. and Lundstrom, K. (1994) Genetic mapping of quantitative trait loci for growth and fatness in pigs. *Science* 263, 1771–1774.

Barendse, W., Armitage, S.M., Ryan, A.M., Moore, S.S., Clayton, D., Georges, M., Womack, J.E. and Hetzel, J. (1993) A genetic map of DNA loci on bovine chromosome 1. *Genomics* 18, 602–608.

Barendse, W., Armitage, S.M., Kossarek, L.M., Shalom, A., Kirkpatrick, B.W., Ryan, A.M., Clayton, D., Li, L., Neibergs, H.L., Zhang, N., Grosse, W.M., Weiss, J., Creighton, P., McCarthy, F., Ron, M., Teale, A.J., Fries, R., McGraw, R.A., Moore, S.S., Georges, M., Soller, M., Womack, J.E. and Hetzel, D.J.S. (1994) A genetic linkage map of the bovine genome. *Nature Genetics* 6, 227–235.

Barendse, W., Vaiman, D., Kemp, S.J., Sugimoto, Y., Armitage, S.M., Williams, J.L., Sun, H.S., Eggen, A., Agaba, M., Aleyasin, S.A., Band, M., Bishop, M.D., Buitkamp, J., Byrne, K., Collins, F., Cooper, L., Coppetiers, W., Denys, B., Drinkwater, R.D., Easterday, K., Elduque, C., Ennis, S., Erhardt, G., Ferretti, L., Flavin, N., Gao, Q., Georges, M., Gurung, R., Harlizius, B., Hawkins, G., Hetzel, J., Hirano, T., Hulme, D., Jorgensen, C., Kessler, M., Kirkpatrick, B.W., Konfortov, B., Kostia, S., Kuhn, C., Lenstra, J.A., Leveziel, H., Lewin, H.A., Leyhe, B., Li, L., Martin Burriel, I., McGraw, R.A., Miller, J.R., Moody, D.E., Moore, S.S., Nakane, S., Nijman, I.J., Olsaker, I., Pomp, D., Rando, A., Ron, M., Shalom, A., Teale, A.J., Thieven, U., Urquhart, B.G.D., Vage, D-I., Van de Weghe, A., Varvio, S., Velmala, R., Vilkki, J., Weikard, R., Woodside, C., Womack, J.E., Zanotti, M. and Zaragoza, P. (1997) A medium-density linkage map of the bovine genome. *Mammalian Genome* 8, 21–28.

Bassam, B.J., Caetano-Anolles, G. and Gresshoff, P.M. (1991) Fast and sensitive silver staining of DNA in polyacrylamide gels. *Analytical Biochemistry* 196, 80–83.

Beever, J.E., Lewin, H.A., Barendse, W., Andersson, L., Armitage, S.M., Beattie, C.W., Burns, B.M., Davis, S.K., Kappes, S.M., Kirkpatrick, B.W., Ma, R.Z., McGraw, R.A., Stone, R.T. and Taylor, J.F. (1996). Report of the first workshop on the genetic map of bovine chromosome 23. *Animal Genetics* 27, 69–76.

Benzer, S. (1961). On the topography of the genetic fine structure. *Proceedings of the National Academy of Sciences of the USA* 47, 403–415.

Bishop, M.D., Kappes, S.M., Keele, J.W., Stone, R.T., Sunden, S.L.F., Hawkins, G.A., Solinas Toldo, S., Fries, R., Grosz, M.D., Yoo, J. and Beattie, C.W. (1994) A genetic linkage map for cattle. *Genetics* 136, 619–639.

Boerwinkle, E., Xiong, W., Fourest, E. and Chan, L. (1989). Rapid typing of tandemly repeated hypervariable loci by the polymerase chain reaction: application to the apolipoprotein B 3′ hypervariable region. *Proceedings of the National Academy of Sciences of the USA* 86, 212–216.

Botstein, D., White, R.L., Skolnick, M. and Davis, R.W. (1980) Construction of a genetic linkage map in man using restriction fragment length polymorphisms. *American Journal of Human Genetics* 32, 314–331.

Breese, E.L. and Mather, K. (1957) The organisation of polygenic activity within a chromosome in *Drosophila* I. Hair characters. *Heredity* 11, 373–395.

Cai, L., Taylor, J.F., Wing, R.A., Gallagher, D.S., Woo, S.-S. and Davis, S.K. (1995) Construction and characterization of a bovine bacterial artificial chromosome library. *Genomics* 29, 413–425.

Castle, W.E. (1922). *Genetic Studies of Rabbits and Rats*. Publication 320, Carnegie Institution, Washington, DC, pp. 3–55.

Charlier, C., Coppieters, W., Farnir, F., Grobet, L., Leroy, P.L., Michaux, C., Mni, M., Schwers, A., Vanmanshoven, P., Hanset, R. and Georges, M. (1995) The mh gene causing double-muscling in cattle maps to bovine chromosome 2. *Mammalian Genome* 6, 788–792.

Collins, F.S. (1995) Positional cloning moves from perditional to traditional. *Nature Genetics* 9, 347–350.

Collins, M. and Myers, R.M. (1987) Alterations in DNA helix stability due to base modifications can be evaluated using denaturing gradient gel electrophoresis. *Journal of Molecular Biology* 198, 737–744.

Copeman, J.B., Cucca, F., Hearne, C.M., Cornall, R.J., Reed, P.W., Ronningen, K.S., Undlien, D.E., Nistico, L., Buzzetti, R., Tosi, R., Pociot, F., Nerup, J., Cornelis, F., Barnett, A.H., Bain, S.C. and Todd, J.A. (1995) Linkage disequilibrium mapping of a type 1 diabetes susceptibility gene (IDDM7) to chromosome 2q31–q33. *Nature Genetics* 9, 80–85.

Davies, R.W. (1971) The genetic relationship of two quantitative characters in *Drosophila melanogaster*. II. Location of the effects. *Genetics* 69, 363–375.

Donis-Keller, H., Green, P., Helms, C., Cartinhour, S., Weiffenbach, B., Stephens, K., Keith, T.P., Bowden, D.W., Smith, D.R., Lander, E.S., Botstein, D., Akots, G., Rediker, K.S., Gravius, T., Brown, V.A., Rising, M.B., Parker, C., Powers, J.A., Watt, D.E., Kauffman, E.R., Bricker, A., Phipps, P., Muller-Kahe, H., Fulton, T.P., Ng, S., Schumm, J.W., Braman, J.C., Knowlton, R.G., Barker, D.F., Crooks, S.M., Lincoln, S.E., Daly, M.J. and Abrahamson, J. (1987) A genetic linkage map of the human genome. *Cell* 51, 319–337.

Elston, R.C. and Stewart, J. (1972) A general model for the genetic analysis of pedigree data. *Human Heredity* 21, 523–542.

Falconer, D.S. (1981) *Introduction to Quantitative Genetics*, 2nd edn. Longman, London and New York.

Feinberg, A.P. and Vogelstein, B. (1983) A technique for radiolabelling DNA restriction endonuclease fragments to high specific activity. *Analytical Biochemistry* 132, 6–13.

Fisher, R.A. (1922) The systematic location of genes by means of crossover observations. *American Naturalist* 56, 406–411.

Fodor, S.P.A., Read, J.L., Pirrung, M.C., Stryer, L., Lu, A.T. and Solas, D. (1991) Light-directed, spatially addressable parallel chemical synthesis. *Science* 251, 767–773.

Fries, R., Eggen, A. and Stranzinger, G. (1990) The bovine genome contains polymorphic microsatellites. *Genomics* 8, 403–406.

Georges, M., Lequarre, S., Hanset, R. and Vassart, G. (1987) Genetic variation of the bovine thyroglobulin gene studied at the DNA level. *Animal Genetics* 18, 41–50.

Georges, M., Nielsen, D., Mackinnon, M., Mishra, A., Okimoto, R., Pasquino, A.T., Sargeant, L.S., Sorensen, A., Steele, M.R., Zhao, X., Womack, J.E. and Hoeschele, I. (1995) Mapping quantitative trait loci controlling milk production in dairy cattle by exploiting progeny testing. *Genetics* 139, 907–920.

Hacia, J.G., Brody, L.C., Chee, M.S., Fodor, S.P.A. and Collins, F.S. (1996) Detection of heterozygous mutations in BRCA1 using high density oligonucleotide arrays and two-color fluorescence analysis. *Nature Genetics* 14, 441–447.

Haldane, J.B.S. (1919). The combination of linkage values, and the calculation of distances between the loci of linked factors. *Journal of Genetics* 8, 299–309.

Harris, H. and Hopkinson, D.A. (1972) Average heterozygosity per locus in man: an estimate based on the incidence of enzyme polymorphisms. *Annals of Human Genetics* 36, 9–20.

Heyen, D.W., Beever, J.E., Da, Y., Evert, R.E., Green, C., Bates, S.R.E., Ziegle, J.S. and Lewin, H.A. (1997). Exclusion probabilities of 22 bovine microsatellite markers in fluorescent multiplexes for semi-automated parentage testing. *Animal Genetics* 28, 21–27.

Jeffreys, A.J., Wilson, V. and Thien, S.L. (1985). Hypervariable 'minisatellite' regions in human DNA. *Nature* 314, 67–73.

Kappes, S.M., Keele, J.W., Stone, R.T., McGraw, R.A., Sonstegard, T.S., Smith, T.P.L., Lopez-Corrales, N.L. and Beattie, C.W. (1997) A second-generation linkage map of the bovine genome. *Genome Research* 7, 235–249.

Kemp, D.J., Smith, D.B., Foote, S.J., Samaras, N. and Petersen, M.G. (1989) Colorimetric detection of specific DNA segments amplified by polymerase chain reactions. *Proceedings of the National Academy of Sciences of the USA* 86, 2423–2427.

Kosambi, D.D. (1944) The estimation of map distance from recombination values. *Annals of Eugenics (London)* 12, 172–175.

Lander, E.S. and Botstein, D. (1989) Mapping Mendelian factors underlying quantitative traits using RFLP linkage maps. *Genetics* 121, 185–199.

Lander, E.S. and Green, P. (1987) Construction of multilocus genetic linkage maps in humans. *Proceedings of the National Academy of Sciences of the USA* 84, 2363–2367.

Lathrop, G.M., Lalouel, J.M., Julier, C. and Ott, J. (1984). Strategies for multilocus linkage analysis in humans. *Proceedings of the National Academy of Sciences of the USA* 81, 3443–3446.

Lewontin, R.C. (1974). *The Genetic Basis of Evolutionary Change*. Columbia University Press, Irrington, Nebraska, USA.

Libert, F., Lefort, A., Okimoto, R., Womack, J. and Georges, M. (1993) Construction of a bovine genomic library of large yeast artificial chromosomes. *Genomics* 18, 270–276.

Ma, R.Z., Beever, J.E., Da, Y., Green, C.A., Russ, I., Park, C., Heyen, D.W., Everts, R.E., Fisher, S.R., Overton, K.M., Teale, A.J., Kemp, S.J., Hines, H.C., Guerin, G. and Lewin, H.A. (1996) A male linkage map of the cattle (*Bos taurus*) genome. *Journal of Heredity* 87, 261–271.

Maniatis, T., Frisch, E.F. and Sambrook, J. (1982) *Molecular Cloning: a Laboratory Manual*. Cold Spring Harbor Laboratory Press, Cold Spring Harbor, 545 pp.

Marshall, E. (1997). 'Playing chicken' over gene markers. *Science* 278, 2046–2048.

Mathew, M.K., Smith, C.L. and Cantor, C.R. (1988) High-resolution separation and accurate size determination in pulsed-field gel electrophoresis of DNA. 1. DNA size standards and the effect of agarose and temperature. *Biochemistry* 27, 9204–9210.

Meselson, M. and Yuan, R. (1968) DNA restriction enzyme from *E. coli*. *Nature* 217, 1110–1114.

Moore, S.S., Barendse, W., Berger, K.T., Armitage, S.M. and Hetzel, D.J.S. (1992) Bovine and ovine DNA microsatellites from the EMBL and Genbank databases. *Animal Genetics* 24, 463–467.

Moore, S.S., Byrne, K, Berger, K.T., Barendse, W., McCarthy, F., Womack, J.E. and Hetzel, D.J.S. (1994) Characterization of 65 bovine microsatellites. *Mammalian Genome* 5, 84–90.

Morgan, T.H. (1910) The application of the conception of pure lines to sex-limited characters in the same animal. *Proceedings of the Society for Experimental Biology and Medicine* 8, 17.

Morton, N.E. (1955) Sequential tests for the detection of linkage. *American Journal of Human Genetics* 7, 277–318.

Morton, N.E. (1988) Multipoint mapping and the emperor's clothes. *Annals of Human Genetics* 52, 309–318.

Mullis, K., Faloona, F., Scharf, S., Saiki, R., Horn, G. and Erlich, H. (1986) Specific enzymatic amplification of DNA *in vitro*: the polymerase chain reaction. *Cold Spring Harbor Symposia on Quantitative Biology* 51, 263–273.

O'Brien, S.J. (1983) *Genetic Maps*, Vol. 2. *A Compilation of Linkage and Restriction Maps of Genetically Studied Organisms*. Cold Spring Harbor Laboratory, New York.

O'Brien, S.J. (1991) Mammalian genome mapping: lessons and prospects. *Current Opinion in Genetics and Development* 1, 105–111.

Orita, M., Iwahana, H., Kanazawa, H., Hayashi, K. and Sekiya, T. (1989) Detection of polymorphisms of human DNA by gel electrophoresis as single-strand conformation polymorphisms. *Proceedings of the National Academy of Sciences of the USA* 86, 2766–2770.

Ott, J. (1974) Estimation of the recombination fraction in human pedigrees: efficient computation of the likelihood for human linkage studies. *American Journal of Human Genetics* 26, 588–597.

Ott, J. (1985). *Analysis of Human Genetic Linkage*. The Johns Hopkins University Press, Baltimore, Maryland.

Paterson, A.H., Lander, E.S., Hewitt, J.D., Peterson, S., Lincoln, S.E. and Tanksley, S.D. (1988) Resolution of quantitative traits into Mendelian factors by using a complete RFLP linkage map. *Nature* 335, 721–726.

Rao, D.C., Morten, N.E., Lindsten, J., Hulten, M. and Yee, S. (1977) A mapping function for man. *Human Heredity* 27, 99–104.

Rigby, P.W.J., Dieckmann, M., Rhodes, C. and Berg, P. (1977) Labeling deoxyribonucleic acid to high specific activity *in vitro* by nick translation with DNA polymerase I. *Journal of Molecular Biology* 113, 237–251.

Risch, N. (1983). A general model for disease-marker association. *Annals of Human Genetics* 47, 245–252.

Schwartz, D.C. and Koval, M. (1989) Conformational dynamics of individual DNA molecules during gel electrophoresis. *Nature* 338, 520–522.

Sharp, P.A., Sugden, B. and Sambrook, J. 1973. Detection of two restriction endonuclease activities in *Haemophilus parainfluenzae* using analytical agarose–ethidium bromide electrophoresis. *Biochemistry* 12, 3055–3063.

Sheffield, V.C., Cox, D.R., Lerman, L.S. and Myers, R.M. (1989) Attachment of a 40-base-pair G+C-rich sequence (GC-clamp) to genomic DNA fragments by the polymerase chain reaction results in improved detection of single-base changes. *Proceedings of the National Academy of Sciences of the USA.* 86, 232–236.

Shrimpton, A.E. and Robertson, A. (1988) The isolation of polygenic factors controlling bristle score in *Drosophila melanogaster*. I. Allocation of third chromosome sternopleural bristle effects to chromosome sections. *Genetics* 118, 437–443.

Simianer, H., Szyda, J., Ramon, G. and Lien, S. (1997) Evidence for individual and between-family variability of the recombination rate in cattle. *Mammalian Genome* 8, 830–835.

Southern, E.M. (1975) Detection of specific sequences among DNA fragments separated by gel electrophoresis. *Journal of Molecular Biology* 98, 503–517.

Steele, M.R. and Georges, M. (1991). Generation of bovine multisite haplotypes using random cosmid clones. *Genomics* 10, 889–904.

Steffen, P., Eggen, A., Dietz, A.B., Womack, J.E., Stranzinger, G. and Fries, R. (1993). Isolation and mapping of polymorphic microsatellites in cattle. *Animal Genetics* 24, 121–124.

Sturtevant, A.H. (1913) The linear arrangement of six sex-linked factors in *Drosophila*, as shown by their mode of association. *Journal of Experimental Zoology* 14, 43–59.

Sutton, W.S. (1902). On the morphology of the chromosome group in *Brachystola magna*. *Biological Bulletin* 4, 24–39.

Tautz, D. (1989) Hypervariability of simple sequences as a general source of polymorphic DNA markers. *Nucleic Acids Research* 17, 6463–6471.

Thoday, J.M. (1961). Location of polygenes. *Nature* 191, 368–370.

Vaiman, D., Mercier, D., Moazami-Goudarzi, K., Eggen, A., Ciampolini, R., Lepingle, A., Velmala, R., Kaukinen, J., Varvio, S.L., Martin, P., Leveziel, H. and Guerin, G. (1994) A set of 99 cattle microsatellites: characterisation, synteny mapping, and polymorphism. *Mammalian Genome* 5, 288–297.

Wallace, R.B., Shaffer, J., Murphy, R.F., Bonner, J., Hirose, T. and Itakura, K. (1979) Hybridization of synthetic oligodeoxyribonucleotides to φχ 174 DNA: the effect of single base pair mismatch. *Nucleic Acids Research* 6, 3543–3557.

Weber, J. and May, P.E. (1989) Abundant class of human DNA polymorphisms which can be typed using the polymerase chain reaction. *American Journal of Human Genetics* 44, 388–396.

Womack, J.E., Johnson, J.S., Owens, E.K., Rexroad, C.E., III., Schlapfer, J. and Yang, Y.-P. (1997) A whole-genome radiation hybrid panel for bovine gene mapping. *Mammalian Genome* 8, 854–856.

Wood, W.I., Gitschier, J., Lasky, L.A. and Lawn, R.M. (1985) Base composition-independent hybridization in tetramethylammonium chloride: a method for oligonucleotide screening of highly complex gene libraries. *Proceedings of the National Academy of Sciences of the USA* 82, 1585–1588.

Woolf, B. (1955) On estimating the relation between blood group and disease. *Annals of Human Genetics* 19, 251–253.

Yeh, C.C., Taylor, J.F., Gallagher, D.S., Sanders, J.O., Turner, J.W. and Davis, S.K. (1995). Genetic and physical mapping of the bovine X chromosome. *Genomics* 32, 245–252.

Zimmerman, P.A., Carrington, M.N. and Nutman, T.B. (1993). Exploiting structural differences among heteroduplex molecules to simplify genotyping the DQA1 and DQB1 alleles in human lymphocyte typing. *Nucleic Acids Research* 19, 4541–4547.

Genetics of Behaviour in Cattle

12

D. Buchenauer
Institute of Animal Breeding, School of Veterinary Medicine, Bünteweg 170, Hannover D-30559, Germany

Introduction	365
Social Behaviour	367
Group structure	367
Association	369
Aggression	370
Temperament	372
Temperament tests	372
Temperament in different breeds	372
Temperament and sex	374
Temperament, growth and milk yield	374
Sexual Behaviour	375
Male sexual behaviour	375
Female sexual behaviour	379
Maternal Behaviour	380
Feeding Behaviour	381
Grazing selectivity	381
Length of grazing	382
Feeding behaviour in stables	383
Summary	384
Acknowledgements	385
References	385

Introduction

Cattle have been used in human cultures as draught animals, as a source of meat, milk, leather, fertilizer, fuel and by-products and as trusted companions and possessions. But, before cattle were exploited in this way, cattle were needed and kept for religious purposes. Almost all major civilizations were developed in societies with a cattle culture (Albright and Arave, 1997). Selection for behaviour traits has been practised since humans started to tame cattle.

They only kept, reproduced and eventually domesticated animals that suited them best with regard to their behaviour responses. For instance, cattle had to be tame enough to be kept in temple corrals, because they were used to pull carts in temple processions and in sacrificial rites. Religious purposes required of domesticated cattle a quiet disposition (Albright and Arave, 1997).

The genetic changes that occurred during domestication were a result of selection pressures, as well as random processes (inbreeding and genetic drift) (Price, 1997). Differences in behaviour among domestic population groups can be associated with their adaptability to the ecological niches in which they developed or because of different selection goals in different cultures and husbandry systems (Craig, 1994).

The genetics of behaviour involves genetic analysis of behavioural phenotypes. The genetic background, the environment and the interaction of heredity and environment result in the phenotypic expression of a behaviour. Prior experiences are complicating factors, and this has to be considered to minimize confounding from these sources. This requires that the components used in the genetic analysis of behaviour should be recurrently identifiable and classifiable (Siegel, 1975).

Compared with other branches of animal science, there is relatively little information on behavioural genetics available. Faure (1994) gave some reasons for this:

- The importance of environmental influences was often overestimated in the past and ethologists have devoted too much attention to this phenomenon.
- Behaviour is difficult to measure. Measures of behavioural traits are of long duration and therefore it is difficult to obtain enough data for genetic analysis.
- Measures of behavioural traits are frequently not normally distributed and therefore genetic parameters are difficult to estimate.

Another source of problems is when qualitative parameters, rarely displayed traits or behavioural disturbances, which show only a few animals in the observed group, are analysed. In addition, difficulties in studying the genetics of behaviour in cattle compared with other domestic species are based on the long generation interval, few offspring and late maturing of the animals. Another problem is indicated by Siegel (1976): cattle are often kept in herds, so that many behaviour patterns may be acquired by learning from group members and it is difficult or even impossible to distinguish between innate and learned behavioural traits.

In this chapter, particular attention is given to the basic behaviour patterns of social behaviour, temperament, sexual behaviour, maternal behaviour and feeding behaviour. The knowledge of these particular behaviours allows humans to respond appropriately to the needs of the animals and to avoid mistakes that can result in injuries and economic losses. The main objective of this chapter is to review evidence of genetic variation for the above-mentioned behavioural traits in cattle.

Social Behaviour

Group structure

The ancestral form of taurine cattle (*Bos primigenius*) lived in social units and the composition of these changed with the seasons. Like other wild, non-harem-forming species, cows and their offspring formed groups of 20–30 animals, based on maternal relationship. Mature males lived in groups separated from the females during spring and summer, and in September they joined the female groups for mating. During winter, the groups remained together and formed large, sexually mixed herds until spring (Lundberg, 1991). Feral-living domestic cattle show similar social structures. Zeeb (1987) observed a herd of free-range Camargue cattle consisting of 37 animals, which split according to their maternal relationship into three groups of different sizes and some separate individuals, which were newly introduced into the herd. The structure of the groups remained relatively stable. On the other hand, in a semi-wild herd of 120 Scottish Highland cattle, Koene (1997) observed that the social organization consisted of groups of variable size (2–60) whose membership often varied. The groups were more stable from September to December, when at this time only single-sex groups of cows plus calves and bulls existed.

Lazo (1994) found that groups of feral Donana cattle showed two levels of organization in their social structure. At a low level, the animals formed unstable groups (parties) that changed in size and composition depending on ecological factors, e.g. fluctuations in food abundance and distribution, reproductive status, predation risk. At a higher level of social organization subgroups of the cow herds were formed, based on a close female relationship, which were stable social groups. Lazo expressed the opinion that this type of social system is valid for other free-ranging cattle populations.

Dominance

Cattle living in groups establish social hierarchies. The structure of a social hierarchy is either linear, linear-tending or complex (Hafez and Boissou, 1975). The basis of a hierarchy is the mutual recognition of the animals. According to Sambraus (1978), up to 70 animals in a herd may recognize each other. Dominance in cattle depends on height, weight, presence of horns, age, sex and breeds. For example, adult males dominate all members of the group, adult females dominate juveniles except males of about 2.5 years, adult male juveniles from 1.5 years onwards make first attempts to dominate adult females. Male calves aged 6 months dominate females of the same age (Bouissou, 1972).

Beilharz and Zeeb (1982) studied dominance in three dairy cattle herds, each of about 50 animals. These authors observed that there was no single individual animal dominant over all other members of the herd. This means that no cow was completely free of inhibition from at least one other herd member. The social structure varied from herd to herd. The authors concluded that the dominance relationship is a result of learning; once learnt, it persists for a long time.

An advantage of social order is the low incidence of aggression. Dominant animals have been aggressive in the past to obtain their rank position but do not need to be aggressive any more (Reinhardt and Reinhardt, 1975). These authors found significant regressions between age and dominance value as well as between age and body weight. According to them, animals gain in dominance values up to the age of about 9 years in mixed-age herds, where they occupy the highest social ranks of the herd. After the age of about 10 years, cows show progressive decline in dominance values. This observation is in agreement with Beilharz and Zeeb (1981), who found in their study that the top-ranking cows were between 8 and 10 years old. According to Sambraus (1975), the ranking index increased up to an age of 10 years and remained on this level up to the age of 13 years. He found that a decrease in the ranking order of old cows occurred only rarely.

The genetic basis of dominance was studied by Purcell and Arave (1991), using seven female monozygous twins. The pairs were assigned at birth to different rearing systems, isolation or group rearing. After weaning at 60 days, the animals were transferred to a large group of 80 heifers until the dominance rank was assessed at 15–17 months of age. The consistency of ranking between twin pairs was remarkable, with an intraclass correlation of 0.93 between twin pairs, indicating a strong genetic base for dominance.

Heritability for dominance was estimated by Beilharz *et al.* (1966); the behaviour of cows was observed for dominant or submissive behaviours and dominance values were calculated. For this trait, heritability (h^2) was 0.4, indicating that the trait will respond to selection, but the value of selecting animals to change this trait should be assessed carefully.

Dominance in different breeds
In a mixed herd of ten each of Angus, Hereford and Shorthorn cows of the same age, social dominance was studied by Wagnon *et al.* (1966) in 2 successive years. A near-linear dominance was found in the herd. The rankings in social dominance were highly stable. The rank correlation coefficient between the annual dominance rankings in the herd for the 2 years was $r = 0.95$ ($P < 0.01$). A highly significant difference in social dominance among breeds was found. The Angus breed was most dominant and the Hereford the least. Within breeds, dominance rank and weight were positively related; among breeds, this correlation was negative.

In a similar study, Stricklin *et al.* (1980) investigated dominance and spacing behaviour of maternal-lineage families in a mixed-breed herd of Angus and Hereford cows. The results of these authors also showed that Angus cows dominated Herefords, although they were lighter in weight. Another observation was that Angus tended to occupy central positions in a group, whereas Herefords were located on the periphery of the group. For several measures of spacing behaviour, distances from Angus cows to other cows were closer than those from Herefords to their neighbours.

Brakel and Leis (1976) introduced cows into established herds of different breeds to compare breeds for dominance rank. The differences in the dominance value were significant. The breed ranking was Brown Swiss over Holstein over Guernsey over Jersey. In contrast to Wagnon et al. (1966), the breed rankings corresponded with the average body weight of rankings for the breeds.

Oberosler et al. (1982) studied dominance hierarchies in a newly grouped herd, consisting of Holstein cows and cows of the Italian alpine breeds, Bruna Alpina, Grigia delle Alpi and Rendena, which was transferred to alpine pastures. The authors found that Holsteins were dominated by the alpine breeds. The Holstein cows, as a lowland breed, are adapted much less to the change of the environment than the mountain breeds. This might be a reason why they became more passive in agonistic encounters. The ranking of the alpine breeds was Bruna Alpina, Grigia delle Alpi and Rendena.

Rank position and other traits
Relations between rank position of cows and other traits have been studied by different authors. Lundberg (1991) found, in a herd of cows with well-established social relationships, significant correlations between social dominance and marching order to the milking parlour (0.49) or milking order (0.46). In contrast, Reinhardt (1978) did not find relations between the social order and the marching order to the daily grazing area in African cattle. However, the position of the cows within the marching order was very constant.

Koene (1997) calculated correlations between rank position (determined by wins/losses) and aggressive (−0.71), sexual (−0.68) and vocal (−0.61) behaviours. A surprising result of this phenotypic correlation is the strong negative relationship between rank and sexual behaviour. In zebu Rathi cows, dominance ranks were correlated ($P < 0.05$) with first-lactation milk yield, indicating higher social ranks in high-producing cows. Dominance ranks were also highly correlated ($P \leq 0.01$) with heart girth and height at withers, in addition to body weight (Shiv-Prasad et al., 1996).

Association

Associative behaviour means pair bonds between certain animals. Within herds, it is often found that the formation of pairs is a common social strategy. Associative behaviour expresses a conscious choice for company and it must reflect a basic biological need (Fraser and Broom, 1997).

Ewbank (1967) observed four pairs of monozygous twins, one pair of dizygous twins and eight individually born calves paired as artificial twins. The pairs had been reared in double calf pens until being turned out to pasture at 36–72 weeks of age. On pasture, the heifers organized themselves into a herd, in which the members of each pair were usually found close together. The author concluded that the association of pair mates at pasture may be controlled more by the rearing methods used on the twins than by the genetic background of the animals. Similar results were reported by Broom and Leaver (1978). They observed that young Friesian heifers showed more associations when they were reared as calves in the same group than existed between members of different groups. They also found that animals close together in the rank order were more likely to associate with one another.

Lundberg (1991) showed that dairy cows establish associations towards specific activities, with specific partners for lying not being the same as for eating or grooming. He found in herds that 70% of the animals had partners for lying and 46% had partners during eating at the feeding rack.

A common feature of associations is expressed by allo-grooming. In a herd of 20 Holstein cows Sato et al. (1991a, b) showed that social licking was frequently carried out by dominant animals, relatives and familiar cows, e.g. similar age. This behaviour occurred more often in cows that were frequently the nearest neighbours during lying or grazing. The benefits of social licking are seen by the authors as a cleaning function, a psychological and physiological calming effect, and a reduction of behavioural tensions. The authors found significant correlations between social licking and milk yield 0.65 (receiving licking), 0.55 (emitting licking) and dominance values 0.88. But these phenotypic findings do not explain the causal connection of these behaviours.

Aggression

Houpt (1991) described categories of aggression in general but the following are also observed in cattle (Table 12.1).

Aggressive interactions among animals occur after mixing animal groups and while the group is developing its rank order (Syme and Syme, 1979). With the established rank order, there is stability and aggressive encounters are reduced (Fraser and Broom, 1997).

Cattle show aggressive acts if their social distance or their territory is intruded upon by strange conspecifics or persons. Bulls practise a more distinct territorial behaviour than cows, this behaviour increasing with the age of the animals (Albright and Arave, 1997; Fraser and Broom, 1997).

Several environmental conditions can cause aggression among cows. Limited sources for feeding or not enough feeding places, or too many cows

Table 12.1. Categories of aggression (adapted from Houpt, 1991).

Type of aggression	Consequence
Social aggression	To establish rank order
Territorial aggression	To defend home area
Pain-induced aggression	A defence reaction to reduce pain
Maternal aggression	To protect the offspring
Fear-induced aggression	Fear of unknown or particular persons or subjects

for automatic feeders (transponder feeding), as well as not enough cubicles in loose housing, can increase the frequency of aggressive encounters. Limited space in loose-housing stables does not allow the animals to perform all aspects of social behaviour and can result in aggressive interactions (Porzig, 1969; Kempkens, 1989; Empel et al., 1993; Menke et al., 1995). In milking cows indoors a much higher level of gross agonistic behaviour has been observed than at pasture (9.5 vs. 1.1 h^{-1}).

The heritability for aggressive activities was estimated by Baehr (1983), as displacements from an automatic feeding dispenser, $h^2 = 0.28$ and, as displacements from cubicles, $h^2 = 0.48$ in German Holstein Friesians. An interesting result is the higher h^2 for displacements from cubicles than from the feeding dispenser. The occupancy of the cubicles was 120% and the cows were fed silage *ad libitum*, so that these facts might explain the heritability values.

The effect of aggression in two Spanish breeds, Parda and Pirenaica, during the fattening period was studied by Santolaria et al. (1992), who found that the average daily gains were significantly influenced by aggressive behaviour (Table 12.2). The authors used a factorial model to analyse the average daily gains in which the factors breed, aggressive behaviour and sex being included.

The level of aggression also appears to differ among breeds. Sambraus (1994) reported the readiness for fighting in Swiss Eringer. This behaviour is used for cow fights as show attractions. Many cows are kept only for this purpose.

In a herd of various breeds, Brakel and Leis (1976) found differences among breeds in the average encounters in which a cow was involved. The frequency diminished from Ayrshire, Holstein, Jersey and Brown Swiss to Guernsey.

Altmann and Busch (1970) found that the frequency of aggression in a mixed herd of Jersey, German Black Pied, British Friesian and crossbred cows differed among the genetic groups. From all cows involved in displacement activities at a feeding rack, 52.6% were Jersey cows; 36% of them were displacing other animals, 64% were displaced by other animals, 55.2% of the encounters happened within the Jersey breed, and only 8.4% of other genotypes were displaced by Jersey. In another study, crossbred cattle (Brahman × Hereford) were more aggressive and disturbed each other more often during feeding than the purebred Hereford (McPhee et al., 1964).

Table 12.2. Least-square means for average daily gains and social status (from Santolaria et al., 1992).

Social status	Daily gains (g)
Dominant	1467 ± 382
Subdominant	1488 ± 397
Dominated	1268 ± 122

Temperament

Temperament tests

Burrow (1997) defined temperament as the animal's behavioural response to handling by humans. Behavioural responses range from docility to fear, or nervousness, non-responsiveness, escape or withdrawal and aggressive behaviour. Temperament is affected by age or experience, sex, handling of animals, maternal effects, inheritance and breed.

There is a large variety of tests in use to measure temperament (Burrow, 1997). In restrained tests, the movement of the animals is physically restricted. All tests make a subjective assessment of the behaviour in different test situations. The observed behaviour includes the amount of movement, vocalizations, eliminations, tail swishing, kicking, audible respiration, baulking and attempts to escape. This category includes tests such as the bail test, crush test, baulking rating and chute test.

In non-restrained tests, the animals are able to move in a relatively large test area in the presence or absence of an observer. Different behaviours are assessed in the different test situations. Approach/avoidance behaviour, behavioural tests, flight distances, open field test, pound tests and dairy temperament scores are all tests of this type.

Temperament in different breeds

The temperament of Angus, Hereford and Shorthorn steers and heifers was assessed by Tulloh (1961), as the animals moved through yards entering the scales, the crush and the bail, and in the bail. Scores for entering the crush and bail were not related to each other, although mean scores for entering the bail were significantly higher than scores for entering the scales or the crush. Herefords and Angus had lower (better) temperament scores than Shorthorns. Differences between sexes were not significant. Animals with high temperament scores (> 2) had significant lower body weights than those with low temperament scores (< 2).

Hearnshaw and Morris (1984) studied temperament in 8-month-old weaner calves of different genetic origin. Hereford dams were bred with *Bos taurus* sires (Hereford, Simmental and Friesian) or with *Bos indicus* sires (Brahman, Braford and Africander). The scores were obtained when the animals entered the crush and were held in the bail. Seven behavioural responses were scored for 30–60 s while standard management practices were applied to the animals. The heritability of temperament score of *B. taurus* calves was 0.23 ± 0.28, for *B. indicus*-sired calves 0.46 ± 0.37, and for all data combined 0.44 ± 0.25. The high standard errors of the heritability were explained by the design of the experiment, as there were, on average, only 3.77 progeny per sire. The differences between the mean scores of *B. indicus*-sired calves (1.96) and *B. taurus*-sired calves (1.05) were highly significant. The difference was halved to 0.45 points for quarterbred Brahman calves versus *B. taurus* calves, indicating the breed effect, which was additive, with a little heterosis. There were significant differences among calves sired by *B. indicus* bulls, but not among sire lines of *B. taurus* origin.

Morris *et al.* (1994) recorded temperament scores in Angus and Hereford breeds and various crossbred groups when weighing the animals and afterwards when the herd was drafted for natural mating. The authors found significant differences among breeds. The breeds ranked similarly over the various age-groups. Cow differences within breed were repeatable. Heritable effects were generally low to moderate, for example h^2 for the average cow score was 0.22 ± 0.15 (based on 176 sires), for the average yearling this score was 0.32 ± 0.24 (47 sire groups), and for the average calf score it was 0.23 ± 0.12 (53 sire groups). In these results, the high standard errors are remarkable. Traits which are not selected consequently are very heterogeneous and result in high standard errors.

Voisinet *et al.* (1997) compared the temperament of the genetic groups Braford, Simmental × Red Angus, Red Brangus, Simbrah, American Angus and Tarantaise × Angus. The temperament score was based on a numerical scale (1 = calm, no movement, 5 = violent struggling) and recorded during routine weighing and processing. Animals with Brahman inheritance (> 25%) had higher temperament scores (3.45 ± 0.09) and were more 'excitable' than animals without Brahman influence (1.80 ± 0.10, $P < 0.001$). Another comparison of *B. taurus* and *B. indicus* breeds came also to the conclusion that Brahman cattle were usually less docile than European cattle (Fordyce *et al.*, 1985). In all temperament tests, *B. indicus* breeds and their crosses showed poorer results, so that, in general, these genotypes are more difficult to handle than *B. taurus* breeds.

Stricklin *et al.* (1980) scored the temperament of different genetic groups using a restraint test and found that Hereford cattle were more docile than the other British breeds, with Galloway the most excitable. There were highly significant variations among sires within breeds. Heritability estimates from paternal half-sib correlations were 0.48 ± 0.29 (based on the chute scores of 243 purebred bulls) and 0.44 ± 0.18 (based on the chute scores of 388 crossbred calves). Genetic correlations between temperament and various carcass measurements and performance traits were low to moderate and had large standard errors.

Le Neindre *et al.* (1995) carried out docility tests on 906 Limousin heifers descended from 34 sires. The animals were tested individually by a handler, who had 2 min to lead the animal to a corner, keep it there for 30 s, and then stroke it. The observations resulted in a docility score that incorporated several pieces of information. Based on these test results, a second score was defined, the docility criterion, which was a categorical trait with four classes. The tests were performed by seven different handlers. Handlers, birth period, husbandry system and sires had significant effects on test results. Heritability estimates for the docility score and for the docility criterion were 0.22 and 0.18, respectively. The authors concluded that these values may be underestimated;

because the heifers came from many different farms, differences before weaning may have made the docility score more variable. However, the estimates are high enough to allow efficient selection for this trait.

An overview of estimated heritability, given by Burrow (1997), is shown in Table 12.3. The estimates of heritability represented in Table 12.3 show large differences. The results depended on the structure of the data; this in turn was influenced by the experimental design, different tests and scoring systems and previous experiences of the animals, which are confounded with the age of the animals. Another reason for different heritability estimates can be found if animals are culled because of a striking temperament; the variance has changed and heritability estimates are reduced. Burrow (1997) found that heritability estimates are higher in young animals and the variation in temperament scores is larger; in older animals they are reduced, because of additional handling. In general, the estimates of heritability were moderate to high, indicating that this trait will respond to selection.

Temperament and sex

Temperament is significantly influenced by sex. Regardless of observer or temperament ranking system, heifers consistently had higher temperament scores than their male contemporaries (Voisinet *et al.*, 1994). While some investigators confirmed this result and others found similar trends but no significant differences in temperament between the sexes, the authors hypothesized that sex differences may be evident only in certain breeds. Stricklin *et al.* (1980) confirmed that post-weaning sex differences were significant with steers, who were more docile than heifers.

Temperament, growth and milk yield

Voisinet *et al.* (1997) evaluated temperament scores for different breed groups and showed that increased (poorer) temperament scores resulted in decreased daily gains ($P < 0.05$). Cattle that were quieter and calmer during handling had higher average daily gains than cattle that became agitated during routine handling, as seen in Table 12.4.

A relation between temperament and body shape was found by Oikawa *et al.* (1989). Genetic correlations suggested that, in Japanese Black cows, shorter and fatter cows had a better temperament than taller animals, and the h^2 for temperament score was 0.27 ± 0.13. The value may have been reduced by the including of factors such as pregnancy, type of management and quality of stockmanship in the estimation of the error variance.

Burrow (1997) reported that several studies in *B. indicus*-derived cows showed that cows with poor temperament scores were poor in milk yielding and had the poorest milking ability, the longest milking and let-down times and short lactations; docile animals yielded significantly more milk per milking, with the best milking ability in the shortest milking time (Gupta and Mishra, 1978, 1979). Across breeds, Roy and Nagpaul (1984) reported that Karan Fries dairy cows, the most docile breed, had the highest milk yield per day and the best milking abilities, followed by Karan Swiss, while Murrah buffalo cows, the least docile, performed the worst for these traits. But no relationship was found in a study of Khanna and Sharma (1988) between performance and temperament of a group of *B. indicus* × *B. taurus* crossbred cows.

In other studies using *B. taurus* breeds, there was no significant relationship found between temperament and milk yield (for example, Beilharz *et al.*, 1966). In contrast, Lawstuen *et al.* (1988) estimated correlations between temperament and milking speed of 0.36 ± 0.10, between temperament and FCM (full cream milk) of 0.19 ± 0.11 and a final score (sum of 15 traits for type) of 0.36 ± 0.10. As a remarkable result, the authors found higher genetic correlations than phenotypic correlations.

Sexual Behaviour

Male sexual behaviour

Artificial insemination (AI) is the major breeding method in dairy cattle. The goal of the breeding associations is to harvest the highest possible amount of semen from élite bulls. For this reason, maximal stimuli are provided prior to semen collection (Presicce *et al.*, 1993). Environmental conditions, management practices and early experiences may influence the sexual behaviour of bulls. The intensity of sexual activities is considerably reduced during times of physiological stress caused by diseases, insufficient nutrition, climatic extremes, increasing age and injuries caused by slippery surfaces (Hafez and Bouissou, 1975; Houpt, 1991).

Bulls show substantial individual differences in libido. According to Hafez and Bouissou (1975), each bull has a genetically based characteristic level of sexual behaviour, measured as the latency of ejaculation or the numbers of ejaculations in a certain time. These levels are fairly constant for each bull. Investigations in monozygotic twins and triplets, as well as comparisons between sires and their male descendants, showed great similarities of sexual behaviour among closely related animals (Olsen and Petersen 1951; Bane, 1954; Hultnaes, 1959). These early investigations underline the evidence for the genetic control of this trait.

Dual-purpose bulls kept with cow herds copulate 12 times per day on average during the summer season, but bulls are able to copulate up to 30 times per day if they are not mating daily (Sambraus, 1978).

A comparison between Norwegian polled and horned bulls showed that polled bulls had significant shorter penis protrusion and weaker thrust ($P < 0.01$) during semen collection than horned. But there were no significant differences in libido, semen volume and semen quality (Kommisrud and Steine, 1997).

Table 12.3. Estimates of heritability for various measures of temperament (after Burrow, 1997).

Measurement	$h^2 \pm$ S.E.	No. animals	Means ± SD	No. sires	Breed	Sex	Age	Model used	Reference
Non-restrained tests									
Docility score	0.22	904	13.73 units	34	Limousin	F	10–11 months	Sire model	Le Neindre et al. (1995)
Docility criterion	0.18	904	2.13 units	34	Limousin	F	10–11 months	Mixed threshold	Le Neindre et al. (1995)
Flight distance (6 months)	0.40 ± 0.15	485	3.30 m	49	Brahman cross	M	6 months	Paternal half-sib	O'Rourke (1989)
Flight distance (12 months)	0.32 ± 0.14	485	2.78 m	49	Brahman cross	M	12 months	Paternal half-sib	O'Rourke (1989)
Flight distance (24 months)	0.70 ± 0.23	485	2.57 m	49	Brahman cross	M	24 months	Paternal half-sib	O'Rourke (1989)
Flight speed (weaning)	0.54 ± 0.16	561	1.17 s	42	Zebu-derived	M & F	6 months	Paternal half-sib	Burrow et al. (1988)
Flight speed (18 months)	0.26 ± 0.13	558	1.10 s	38	Zebu-derived	M & F	18 months	Paternal half-sib	Burrow et al. (1988)
Restrained tests									
Bail test (movement score)	0.67 ± 0.26	957	3.03 units	na	Various	M & F	10/22 months	Paternal half-sib	Fordyce et al. (1982)
Race test (movement score)	0.17 ± 0.21	957	2.03 units	na	Various	M & F	10/22 months	Paternal half-sib	Fordyce et al. (1982)
Race test (audible respiration)	0.57 ± 0.22	957	1.43 units	na	Various	M & F	10/22 months	Paternal half-sib	Fordyce et al. (1982)
Crush test (movement score)	0.25 ± 0.20	957	1.82 units	na	Various	M & F	10/22 months	Paternal half-sib	Fordyce et al. (1982)
Crush test (audible respiration)	0.20 ± 0.16	957	1.44 units	na	Various	M & F	10/22 months	Paternal half-sib	Fordyce et al. (1982)
Crush test (movement score)	0	1,852	2.08 units	63	Droughtmaster	F	Mature cows	Paternal half-sib	Fordyce and Goddard (1984)
Crush test (audible respiration)	0	1,852	0.31 units	63	Droughtmaster	F	Mature cows	Paternal half-sib	Fordyce and Goddard (1984)
Crush test (movement score)	0.09	1,852	2.08 units	63	Droughtmaster	F	Mature cows	Dam–daughter β	Fordyce and Goddard (1984)
Crush test (audible respiration)	0.05	1,852	0.31 units	63	Droughtmaster	F	Mature cows	Dam–daughter β	Fordyce and Goddard (1984)
Crush test (movement score)	0.10 ± 0.11	485	2.6 units	49	Brahman cross	M	6 months	Paternal half-sib	O'Rourke (1989)
Crush test (movement score)	0.23 ± 0.13	485	2.1 units	49	Brahman cross	M	12 months	Paternal half-sib	O'Rourke (1989)
Crush test (movement score)	0.11 ± 0.11	485	2.2 units	49	Brahman cross	M	24 months	Paternal half-sib	O'Rourke (1989)
Behaviour score	0.40 ± 0.30	144	2.52 units	na	Angus	M & F	12 months	Paternal half-sib	Shrode and Hammack (1971)
Temperament score	0.45	191	na	5	Japanese Black	M & F	Various	Paternal half-sib	Sato (1981)
Temperament score	0.67	191	na	5	Japanese Black	M & F	Various	Dam–offspring β	Sato (1981)
Temperament score (B. taurus)	0.03 ± 0.28	209	1.39 units	~60	Various	M & F	6–9 months	Paternal half-sib	Hearnshaw and Morris (1984)
Temperament score (B. taurus)	0.46 ± 0.37	358	1.39 units	~90	Various	M & F	6–9 months	Paternal half-sib	Hearnshaw and Morris (1984)
Temperament score (combined)	0.44 ± 0.25	567	1.39 units	150	Various	M & F	6–9 months	Paternal half-sib	Hearnshaw and Morris (1984)
Temperament score	0	1,852	2.45 units	63	Droughtmaster	F	Mature cows	Paternal half-sib	Fordyce and Goddard (1984)

Trait	h²	N	mean ± SD	n	Breed	Sex	Age	Method	Reference
Temperament score	0.09	1,852	2.45 units	63	Droughtmaster	F	Mature cows	Dam–daughter β	Fordyce and Goddard (1984)
Temperament score (6 months)	0.14 ± 0.11	485	3.11 units	49	Brahman cross	M	6 months	Paternal half-sib	O'Rourke (1989)
Temperament score (12 months)	0.12 ± 0.11	485	2.42 units	49	Brahman cross	M	12 months	Paternal half-sib	O'Rourke (1989)
Temperament score (24 months)	0.08 ± 0.10	485	2.65 units	49	Brahman cross	M	24 months	Paternal half-sib	O'Rourke (1989)
Dairy temperament score									
Disposition	< 0.15 ± < 0.09	11,106	2.8 ± 0.7 units	na	Holstein	F	Mature cows	Paternal half-sib	Aitchison et al. (1972)
Disposition	0.07 ± 0.02	8,977	2.00 ± 0.2	125	Holstein	F	Mature cows	Sire model	Thompson et al. (1981)
Disposition	< 0.08	5,601	na	187	Holstein	F	Mature cows	Sire model	Agyemang et al. (1982)
Temperament score	0.40 ± 0.09	na	na	na	Various	F	Mature cows	Dam–daughter β	O'Bleness et al. (1960)
Temperament score (young cows)	0.16	1,400	na	133	Holstein	F	< 35 months	Sire model	Van Vleck (1964)
	0	4,080	na	209	Holstein	F	> 35 months	Sire model	Van Vleck (1964)
Temperament score (old cows)	0.53	1,017	1.9 ± 0.8	31	Holstein	F	Mature cows	Paternal half-sib	Dickson et al. (1970)
Temperament score	0.11	4,891	na	157	Friesian	F	Mature cows	Paternal half-sib	Wickham (1979)
Temperament score	0.11	4,171	na	135	Jersey	F	Mature cows	Paternal half-sib	Wickham (1979)
Temperament score	0.19 ± 0.19	319	1.71 ± 0.11	na	Zebu × ex. c.	F	Mature cows	Paternal half-sib	Sharma and Khanna (1980)
Temperament score	0.12 ± 0.02	9,646	28.8 ± 7.6	208	Holstein	F	Mature cows	Sire model	Lawstuen et al. (1988)
Dominance tests									
Dominance value	~0.4	74	na	na	Holstein	F	Mature cows	Twin analysis	Beilharz et al. (1966)
Maternal temperament									
Maternal temperament score	0.32	162	na	na	Hereford	F	Mature cows	na	Brown (1974)
Maternal temperament score	0.17	266	na	na	Angus	F	Mature cows	na	Brown (1974)

na, not available; s.e., standard error; SD, standard deviation; β, regression.

Table 12.4. Least-square means for growth rate and temperament score by breed (after Voisinet et al., 1997).

Breed	Average daily gain (kg day^{-1})	Mean temperament ranking*
Braford	0.95 ± 0.03	3.62 ± 0.15[a]
Red Brangus	0.98 ± 0.04	3.78 ± 0.22[a]
Simbrah	1.10 ± 0.04	2.89 ± 0.22[b]
Angus	1.24 ± 0.06	1.70 ± 0.19[c]
Simmental × Red Angus	1.44 ± 0.02	1.77 ± 0.07[c]
Tarentaise × Angus	1.21 ± 0.09	2.36 ± 0.10[d]

*Steers only. Means with different superscripts differ significantly.

Besides individual differences within breeds, there are differences between breeds in libido. Houpt (1991) reported that European breeds are more likely to mount an inappropriate sexual object used for semen collection than zebu cattle. According to Hafez (1980), bulls of dairy breeds respond more quickly than beef bulls to sexual stimuli. In a breed comparison of the reaction times of bulls being used for semen collection, Holstein bulls had a much shorter reaction time than Angus and Hereford bulls (first mount 1 vs. 13 min, first ejaculation 5 vs. 20 min).

Chenoweth et al. (1996) assessed sexual performances as libido score, number of services, time to first mount and time of sexual inactivity of yearling bulls in Florida during four test periods. Included in these studies were breeds of *B. taurus* (Senepol, Romosinuanu, American Angus and Hereford) and *B. indicus* (Brahman and Nelore × Brahman). The results showed that breeds differed in all sexual behaviour assessments. Besides breed, test period and the interaction had significant effects on these traits. Sexual performance assessments improved with age in *B. taurus* breeds, but not in *B. indicus*. The temperate breeds Angus and Hereford were most sexually active, the tropically adapted *B. taurus* breeds Senepol and Romosinuano were intermediate and the *B. indicus* breeds were least active. Seasonal patterns in sexual performance were not apparent, with breed and year differences occurring. The authors proposed that the failure of *B. indicus* bulls to serve in any tests indicated sexual immaturity or inadequate procedures for testing this breed group. In an earlier study, Chenoweth (1981) reported that Brahman, Africander, Hereford, Brahman crossbred, Africander crossbred and Shorthorn × Hereford bulls in Australia differed in libido and mating ability. In general, Africander and Africander crossbred bulls had higher scores, British breeds and crosses were intermediate, and Brahmans and Brahman crosses were lowest. In Colorado and Wyoming, Chenoweth et al. (1984) studied libido in Angus and Hereford bulls. The breeds did not differ in any of the variables recorded to quantify sex drive, and breed × age interactions also were not important. However, differences in libido were found by Chenoweth et al. (1977) and Berry et al. (1983) in Colorado in bulls from different breeding lines. Lines differed widely, suggesting the existence of genetic variation for this trait.

Blockey et al. (1978) estimated heritability for serving capacity from 157 paternal half-sib bull groups of Hereford and Angus origins on 24 farms in

Australia. With or without body weight as a covariate, h^2 was 0.59 ± 0.16. Hereford bulls had higher mean values for serving capacity than Angus bulls (6.28 vs. 3.98). The authors also noted a close correspondence between monozygotic twin bulls for serving capacity, but large variation between twin sets.

Sexual behaviour among different genotypes was studied by Jezierski et al. (1989). The comparison included Black and White cattle, Hereford and crossbred with 25% European bison. In the period prior to pasturing, each of the genotype groups was kept separately in loose housing, and in summer they were released together on a pasture. The frequency of initiated mounts was highest in Black and White and lowest in crossbred bulls, whereas the frequency of flehmen was lowest directed towards Black and White bulls. Herefords were preferred as partners for mounting by bulls of all genotypes. For aggressive interactions, all bulls preferred partners of their own genotype.

Female sexual behaviour

In all climatic regions, fertility is lower during the winter months and shows a peak during the summer season (Sambraus, 1978). Cows in oestrus increase their walking activity significantly, no matter if they are housed indoors or kept on pasture. The increased activity is associated with reduced time spent lying or feeding. Milk yield and milk-flow rate were not significantly affected by oestrus (Schofield, 1989). Mounting of cows is a regular behaviour during oestrus and occurs more frequently in dairy than in dual-purpose breeds (Sambraus, 1978).

Porzig et al. (1973) summarized that the intensity of oestrus behaviour is a characteristic of breed. The authors noted a more distinctive oestrus behaviour in Black and White, Simmental, Telemark, Swedish Highland, Swedish Red cattle and other red breeds than in white breeds. The oestrus interval is shorter in zebu cattle than in *B. taurus* breeds (Houpt, 1991).

Gwazdauskas et al. (1983) reported that a heritability estimate of oestrus intensity was 0.21, and repeatability within year was 0.29. The intensity of oestrus was positively correlated with conception rate but the correlation was low.

Busch (1972) studied the development of sexual behaviour in 16 heifers from the age of 4 months until 14 months from each of the following genotypes: British Friesian (BF), German Black and White (GBW), Jersey (J), crossbred I (75% GBW + 25% J) and crossbred II (50% GBW + 25% J + 25% BF). All genotypes showed a marked increase in mounting activity between the seventh and eighth month. The highest frequencies in this period were observed in the Jersey and crossbred groups. The highest overall frequency of mounting occurred at 13 months, at which time British Friesian and crossbred II animals were most active in this respect. During the observed developing phase, crossbred II animals showed significantly higher mounting activities compared with all the other genotypes. The order of the other genotypes followed crossbred I > Jersey > British Friesian > Black and White. The order of being mounted by other heifers was Black and White > British Friesian > crossbred I > Jersey > crossbred II. The extremes in mounting and being mounted were negatively related. The development of flehmen showed a first peak in the sixth month of life; this occurred 1 month earlier than mounting. The highest activity of flehmen was observed in crossbred II, followed by Black and White, British Friesian, crossbred I and Jersey.

Maternal Behaviour

Maternal behaviour differences have been observed between beef and dairy cattle, with beef breeds showing more intense maternal behaviour than dairy breeds. Beef cattle in many management systems raise their calves and appropriate maternal behaviour is essential for the calves' survival. Calves from dairy cattle have been reared artificially for many generations and there has been no selection pressure for this behavioural trait.

Maternal behaviour was studied by Selman *et al.* (1970) in beef and dairy cows and dairy heifers during the first 8 h post-partum (p.p.). Beef cows initiated licking of their calves immediately after calving. A number of the dairy cows and dairy heifers were slow or completely failed to initiate this licking of their offspring. The dairy heifers spent significantly less time licking their calves in the initial phase of grooming than the dairy cows or beef cows (11.0, 32.9 and 48.3 min, respectively). The time taken by the dams to stand for their calves teat-seeking advances was not significant. However, the authors emphasized that the beef cows stood quickly to the teat-seeking of their calves, whereas a few dairy heifers and dairy cows remained lying.

Buddenberg *et al.* (1986) studied the maternal behaviour of Hereford, Angus, Charolais and Red Poll. After the calf was born, it was caught for weighing and tattooing. During these activities, the dam was observed for attentiveness to her calf and aggressiveness towards the caretaker. This behaviour was scored a numerical value ranging from 1 to 11, with 1 being most aggressive and 11 being least attentive and exhibiting no maternal instinct. Angus cows scored lowest for maternal behaviour, with a mean rating of 5.3 ± 0.04. Hereford, Charolais and Red Poll cows were similar in their mean maternal rating, with values of 6.2 ± 0.04, 6.0 ± 0.13 and 5.7 ± 0.22, respectively indicating that Angus cows were more attentive to their calves and more aggressive to caretakers. Vigorous defence of the newborn calf was noted for Limousin cows compared with Charolais, Simmental and Gelbvieh, as experienced by the author. According to Buddenberg *et al.* (1986), the estimated heritability for maternal rating was 0.06 ± 0.01. The authors proposed that the low estimate indicated there was only a small genetic component influencing this trait. Repeatability estimates were 0.09 ± 0.02, indicating that there was little similarity in repeated ratings and suggesting the importance of temporary environmental effects.

Bailey and Moore (1980) found no significant differences in maternal behaviour among the following genetic groups: Hereford and Red Poll and their reciprocal crosses, and Angus × Hereford, Brahman × Hereford, Brahman × Angus and Angus × Charolais.

The survival of a newborn calf depends not only on maternal behaviour but also upon an early intake of colostrum after birth. Therefore attention is drawn to the time of the first suckling. The data of Edwards and Broom (1979) show that the lactation number of the cow influenced the time of first suckling. A large number of calves failed to suck within the first 6 h post partum and the numbers increased with the lactation number because older cows have larger udders and teats. Table 12.5 presents a compilation of data on the early suckling activities in calves of different breeds. With the increasing age of the calf, the number of sucklings per day decreased and the duration of the suckling period and the total amount of time spent suckling increased significantly (Lidfors et al., 1993; Alencar et al., 1995).

The suckling behaviour and the general behaviour of calves produced by AI or embryo transfer (ET) were compared by Waltl et al. (1995). The beef crossbred dams raised twins, where one calf was conceived by AI and the other was the product of ET from a donor in the same synchronized oestrus. The results showed no significant differences in the behaviour of the calves, implying that the social affinity between calves and their dams did not depend on the origin of the offspring.

Feeding Behaviour

Grazing selectivity

Cattle are adapted to graze in a wide range of different environments with a large variety of botanical species. General grazing patterns have been described by many authors. Within a 24-h period, cattle graze between 4 and 14 h. Free-range cattle show distinct periods of intense grazing: shortly before sunrise, mid-morning, early afternoon and around dusk. Variations occur due

Table 12.5. Early suckling activities in calves of different breeds.

Breed	First suckling (min p.p.)	No. of sucklings per day	Duration of suckling (min)	Reference
Friesian	259			Langholz et al. (1987)
Simmental	297			
Charolais × Simmental	269			
Simmental × Friesian	246			
Canchim		2.2[a]	8.2[a]	Alencar et al. (1995)
Canchim × Nelore		2.9[b]	7.0[b]	
Beef crossbred				
Early lactation		7.3*	9.2*	Waltl et al. (1995)
Late lactation		5.4	2.9	

*Per 40 h observation. Means with different superscripts differ significantly.

to the availability of botanical resources, palatability, inclement weather conditions, annoyance by insects, etc., and also possibly breeds (Porzig, 1969; Sambraus, 1978; Albright and Arave, 1997; Fraser and Broom, 1997).

Winder *et al.* (1995, 1996) compared plant selectivity of Hereford, Angus and Brangus breeds on a Chihuahuan desert range, using faecal microhistology, and found differences between these breeds during all seasons. The authors proposed that the genetic background of the animals is the important factor in the use of the key plant species in this region. These authors used paternal half-sib analyses to estimate genetic and phenotypic variances and heritability in Brangus cattle. The number of grass species in diets was significantly affected by sire, and heritability estimates for the preference of different plants in different seasons were noted – for example, h^2 values for the number of grass and total species in summer diets were 0.49 and 0.79, respectively. The results seem to be overestimated because the study included only 100 animals.

Cattle select not only between plant species, but also for the nutritive value; Bailey (1995) observed that steers, when given the opportunity to choose between homogeneous or heterogeneous grazing areas developed preferences for the most nutritious patches, as determined by crude protein concentration, in the heterogeneous area and avoided patches with low forage quality.

Length of grazing

Breed differences in the time spent grazing were reported from different environments. Alencar *et al.* (1996) found, under Brazilian conditions, that there were significant differences in the grazing time of Canchim and Nelore breeds (428 and 365 min day^{-1}, respectively). Within the breeds, grazing time was significantly correlated with daily milk yield (0.25) and with milk yield in week 33 of lactation (0.27).

The differences in the grazing time among Senepol, Brahman, American Angus and Hereford cows under tropical summer conditions in Florida were linked to differences in rectal temperature by Hammond and Olson (1994). Senepols and Brahmans grazed significantly longer than the other breeds, due to the higher degree of heat tolerance.

Similar results were found by Langbein and Nichelmann (1993), who compared the grazing behaviour of lactating Holstein Friesian and Siboney de Cuba (5/8 Holstein Friesian × 3/8 Cuban zebu) under the tropical conditions of Cuba. In a first trial, both breeds were kept together on a pasture and no differences occurred in the recorded behaviours. However, significantly higher rectal temperatures for the Holstein Friesian were measured in the afternoon. This indicated a thermal stress for these animals. In a second trial, the breeds were kept on pasture separately; Siboney grazed significantly longer and showed a higher overall activity than Holstein Friesian. The Friesians spent significantly more time in the shade of mango trees.

In a herd of suckler cows consisting of Limousin × Holstein Friesian and Galloway breeds, the crossbred cows grazed significantly longer than Galloway under central European conditions (Barow and Gerken 1997).

In a comparison of heifers from four size–maturity groups (very small, fast-maturing Angus, mature body weight (BW) 387 kg; larger Angus, mature BW 413 kg; Hereford, Polled Hereford and Red Poll, mature BW 468 kg; Charolais and Chianina, mature BW 589 kg), Erlinger et al. (1990) found that the total daily grazing time increased significantly in heifers of larger mature size. At the extremes, the Charolais–Chianina group grazed 70.1 min day^{-1} longer than the group of fast-maturing, small Angus. Diurnal variations in grazing patterns among the groups were also observed. The majority of grazing took place during the early morning and evening hours, but Charolais–Chianina heifers grazed in the early morning for shorter sessions, followed by prolonged periods of lying in the sun. The authors also observed that the Charolais–Chianina group took significantly larger bites than the other groups. In the biting rate, there was no signifcant difference among the groups.

Heritability of grazing time estimated by Macha and Olsarova (1986) was 0.003 ± 0.026 and repeatability was 0.30 ± 0.048, the estimation being based on four groups of half-sisters, each group consisting of 14 animals. The small number of animals cannot result in a reliable heritability estimate.

Morris et al. (1993) found that Friesian, Piedmontese × Friesian and Belgian Blue × Friesian bulls that grazed on ryegrass (*Lolium*) and white clover (*Trifolium repens*) pastures showed a similar intake (kg head^{-1} day^{-1}) of herbage digestible organic matter, whatever the genetic group. There were no differences in the rate of liveweight gain during the intake measurement period of 18 days or in the final liveweight between the genotypes. Belgian Blue × Friesian bulls had lower biting rates and spent more time idling than the other two genetic groups.

Feeding behaviour in stables

A study analysing the temporal patterns of feeding behaviour in dairy cows in order to find regularities in the sequence of duration and frequency of this behaviour was carried out by Stamer et al. (1997). The 46 cows (German Holstein Black and German Holstein Red) were kept under loose housing conditions and were fed roughage *ad libitum* and concentrates according to their milk yield. The cows showed characteristic time patterns of ingestive behaviour. Feed intake occurred in 'meals'. Within these meals, intervals of up to 37 min interrupted the ingestive activities. Intervals longer than 37 min separated the different meals. This objective meal criterion ($T = 37$ min) was not influenced by the number of the lactation, breed, level of concentrates, or different competitive situations. The total daily feeding time was 464.4 ± 114.8 min and the number of meals was 9.0 ± 2.0 per day. The authors concluded that splitting feeding behaviour into meals allows the inclusion of new traits in order to analyse the feed intake.

Baehr (1983) found that repeatability of feeding behaviours was medium to high between observations within lactation, but low to medium between lactations; for example, repeatability of numbers of visits to automatic feed dispersers was 58% and feeding behaviours on the self-fed silage rack was between 57 and 91% within the lactation. The author also estimated heritability for feeding patterns: number of visits to automatic feeding dispenser $h^2 = 0.6124 \pm 0.27$, visits to silage rack $h^2 = 0.27 \pm 0.35$, visiting drinker $h^2 = 0.09 \pm 0.50$, feeding 'quiet' $h^2 = 0.45 \pm 0.40$. The random errors were very high because the heritability estimate was based on 102 cows, descendants from 17 sires, which had at least two offspring in the herd. Also, this study shows only a tendency, because of the small number of animals.

Rumination as a trait of feeding behaviour was investigated by Santha et al. (1988). The 1394 female offspring of 21 bulls were kept in stanchion stables on six dairy farms and fed *ad libitum*. Heritability was estimated for the total daily rumination time within farms $h^2 = 0.42$, for the total sample 0.15, and h^2 for the average duration of rumination periods within farm was 0.25 and for the overall sample 0.20.

The intake of milk-replacer liquid of dairy calves was studied by Mendoza Ordones et al. (1988). A total number of 804 calves of four genetic groups (German Black Pied, Holstein Friesian, Holstein Friesian × German Black Pied and German Black Pied Dairy), kept in three herds, was involved. The time required to drink 1 kg of milk-replacer liquid averaged 22.4 ± 12.9 s, the number of gulps s^{-1} 1.46 ± 0.35 and the amount of liquid consumed per gulp 45.5 ± 27.8 ml. All three traits showed high variation coefficients. Repeatability was > 0.8. The differences between genetic groups and sexes were small and not significant, but there were significant differences between herds in the time required to drink 1 kg. Heritability estimation for the three drinking traits was 0.43, 0.68 and 0.52, respectively. There were significant correlations of body weight on arrival at pasture with drinking time (–0.61), the number of gulps s^{-1} (–0.71) and the amount consumed per gulp (0.74).

Summary

The presented review shows that most investigations concerning the evidence of genetic variation in behavioural traits were based on breed comparisons. In many traits, differences exist between *B. indicus* and *B. taurus*, but there are also differences within the species between the breeds. The reliability of these comparisons depends on the experimental design, the systematic distribution of environmental effects and the interactions between the animals.

In a few studies, monozygous twins were used to study the genetic basis of a behaviour trait, modern microsurgery enabling the multiplication of embryos and the possibility to intensify such studies.

The estimates of genetic parameters, such as heritability and genetic correlations, differed largely, sometimes for the same trait. The majority of heritability estimates were low to moderate, some were relatively high, indicating that these traits were not included in systematic selections so far. Not all authors gave information about the structure of the data, e.g. if they were normally distributed or if a transformation was made. This information is necessessary to assess the reliability of the heritability estimate.

The values of the heritability estimates indicate that most of the traits would respond to selection. Breeding programmes selecting special performance traits could include the effects on behaviour. However, behavioural traits are polygenically inherited and selecting one trait may change other traits. Before behavioural traits are included in breeding programmes, the complete effects for the population should be carefully assessed. Burrow (1997) reported that temperament is incorporated as a trait in a progeny-testing selection index for dairy cattle in a few countries, but a long time period is needed to assess the genetic improvement. Another important parameter is the repeatability of the method. This parameter should be estimated at least with a small part of the data. Some problems imply that correlations and extremely low or extremely high data may influence the results. Such extreme data can lead to misleading or false results or they can differ largely. Therefore the investigator should prove the plausibility of these extreme data.

For the future, clearly defined genetic parameters are needed, which are based on sufficient numbers of animals included in the samples, as other authors have demanded before. With the application of computer programs to analyse video observations, more animals can be observed in a study compared with human observers in order to meet this requirement.

For the future, it can be expected that molecular genetics will come up with interesting new results, identifying special genes for each individual behavioural trait, and a collaboration between ethologists and geneticists will find interesting new aspects in this branch of science.

Acknowledgements

I would like to thank Dr B. Oldigs for his comments on the manuscript, Mr Jörn Wrede for technical support and Mrs Christine Philippi for her language support.

References

Albright, J.L. and Arave, C.W. (1997) *The Behaviour of Cattle*. CAB International, Wallingford, UK.

Alencar, M.M., Tullio, R.R., Correa, L. de A. and da Cruz, G.M. (1995) Suckling traits in Canchim and Canchim × Nelore calves. *Revista da Sociedade Brasiliera de Zootecnica* 24, 706–714. (Abstract cited).

Alencar, M.M. de, Tullio, R.R., da Cruz, G.M. and Correa, L. de A. (1996) Grazing behaviour of beef cows. *Revista da Sociedade Brasiliera de Zootecnica* 25, 13–21. (Abstract cited.)

Altmann, D. and Busch, I. (1970) Ergebnisse von gezielten Verhaltensstudien in einem neuentwickelten Milchvieh-Laufstallsystem. *Tierzucht* 24, 399–370.

Baehr, J. (1983) Verhalten von Milchkühen im Laufstall. Dissertation, University of Kiel, Germany.

Bailey, C.M. and Moore, J.D. (1980) Reproductive performance and birth characters of divergent breeds and crosses of beef cattle. *Journal of Animal Science* 50, 645–652.

Bailey, D.W. (1995) Daily selection of feeding areas by cattle in homogeneous and heterogeneous environments. *Applied Animal Behaviour Science* 45, 183–200.

Bane, A. (1954) Studies on monozygous cattle twins. *Acta Agriculturae Scandinavica* 4, 95–208. (Cited in Hafez, 1975.)

Barow, U. and Gerken, M. (1997) Untersuchungen zur automatisierten Verhaltenserfassung bei Mutterkuehen in ganzjaehriger Aussenhaltung (Study of the automatic registration of behaviour in suckler cows under extensive range conditions). *KTBL-Schrift* 376, 110–119.

Beilharz, R.G. and Zeeb, K. (1982) Social dominance in dairy cattle. *Applied Animal Ethology* 8, 79–97.

Beilharz, R.G., Butcher, D.F. and Freeman, A.E. (1966) Social dominance and milk production in Holsteins. *Journal of Dairy Science* 49, 887–892.

Berry, J.G., Brinks, J.S. and Russel, W.C. (1983) Relationship of seminal vesicle size and measures of libido of beef in yearling bulls. *Theriogenology* 19, 279–284. (Cited in Hohenboken, 1986.)

Blockey, M.A. de B., Straw, W.M. and Jones, L.P. (1978) Heritability of serving capacity and scrotal circumferences in beef bulls. In: *American Society of Animal Science, 70th Annual Meeting*. Abstract 253.

Bouissou, M.-F. (1972) Influence of body weight and presence of horns on social rank in domestic cattle. *Animal Behaviour* 20, 474–477.

Brakel, W.J. and Leis, R.A. (1976) Impact of social disorganization on behaviour, milk yield, and body weight of dairy cows. *Journal of Dairy Science* 59, 716–721. (Cited in Hohenboken, 1986.)

Broom, D.M. and Leaver, J.D. (1978) The effects of group-housing or partial isolation on later social behaviour of calves. *Animal Behaviour* 26, 1255–1263.

Buddenberg, B.J., Brown, C.J., Johnson, Z.B. and Honea, R.S. (1986) Maternal behaviour of beef cows at parturition. *Journal of Animal Science* 62, 42–46.

Burrow, H.M. (1997) Measurement of temperament and their relationship with performance traits of beef cattle. *Animal Breeding Abstracts* 65, 478–495.

Busch, I. (1972) Untersuchungen zum Fortpflanzungsverhalten weiblicher Jungrinder verschiedener Genotypen. Dissertation, Humboldt University Berlin, Germany.

Chenoweth, P.J. (1981) Libido and mating behaviour in bulls, boars and rams: a review. *Theriogenology* 16, 155–177. (Cited in Hohenboken, 1986).

Chenoweth, P.J., Abbitt, B., McInerney, M.J. and Brinks, J.S. (1977) Libido, serving capacity and breeding soundness in beef bulls. In: *Colorado State University General Series* No. 966, pp. 18–21. (Cited in Hohenbroken, 1986.)

Chenoweth, P.J., Farin, P.W., Mateos, E.R., Rupp, G.P. and Pexton, J.E. (1984) Breeding soundness and sex drive by breed and age in beef bulls used for natural mating. *Theriogenology* 22, 341–349. (Cited in Hohenboken, 1986.)

Chenoweth, P.J., Chase, C.C., Larsen, R.E., Thatcher, M.J.D., Bivens, J.F. and Wilcox, C.J. (1996) The assessment of sexual performance in young *Bos taurus* and *Bos indicus* beef bulls. *Applied Animal Behaviour Science* 48, 225–235.

Craig, J.V. (1994) Genetic influences on behaviour associated with well-being and productivity in livestock. In: *Proceedings of the 5th World Congress on Genetics Applied to Livestock Production*, Vol. 20. Guelph, Canada, pp. 150–157.

Edwards, S.A. and Broom, D.M. (1979) The period between birth and first suckling in dairy calves. *Research in Veterinary Science* 26, 255–256.

Empel, W., Jezierski, T., Brzozowski, P., Markiewicz-Grabowska, H., Gimzewska, K. and Kolakowski, T. (1993) Influence of housing type and time elapsing after food supply. *Polish Academy of Sciences, Animal Science Papers and Reports* 114, 301–309 .

Erlinger, L.L., Tolleson, D.R. and Brown, C.J. (1990) Comparison of bite size, biting rate and grazing time of beef heifers from herds distinguished by mature size and rate of maturity. *Journal of Animal Science* 68, 3578–3587.

Ewbank, R. (1967) Behaviour of twin cattle. *Journal of Dairy Science* 50, 1510–1512.

Faure, J.M. (1994) Behavioural genetics: an overview. In: *Proceedings of the 5th World Congress on Genetics Applied to Livestock Production, in Guelph, Canada*, Vol. 20, p.129.

Fraser, A.F. and Broom, D.M. (1997) *Farm Animal Behaviour and Welfare*. CAB International, Wallingford, UK.

Gupta, S.C. and Mishra, R.R. (1978) Temperament and its effect on milking ability of Karan Swiss cows. In: *Proceedings of the XX International Dairy Congress*. (Cited in Burrow, 1997.)

Gupta, S.C. and Mishra, R.R. (1979) Effect of dairy temperament on milking ability of Karan Swiss cows. *Indian Journal of Dairy Science* 32, 32–36. (Cited in Burrow, 1997.)

Gwazdauskas, F.C., Lineweaver, J.A. and Mcgilliard, M.L. (1983) Environmental and management factors affecting estrous activity in dairy cattle. *Journal of Dairy Science* 66, 1510–1514. (Cited in Hohenboken, 1986.)

Hafez, E.S.E. (1980) *Reproduction in Farm Animals*, 4th edn. Lea & Febinger, Philadelphia.

Hafez, E.S.E. and Bouissou, M.F. (1975) The behaviour of cattle. In: Hafez, E.S.E. (ed.) *The Behaviour of Domestic Animals*. Baillière Tindall, London.

Hammond, A.C.and Olson, T.A. (1994) Rectal temperature and grazing time in selected beef cattle breeds under tropical conditions in subtropical Florida. *Tropical Agriculture* 71, 128–134.

Hearnshaw, H. and Morris, C.A. (1984) Genetic and environmental effects on a temperament score in beef cattle. *Australian Journal of Agricultural Research* 35, 723–733.

Hohenboken, W.D. (1986) Inheritance of behavioural characteristics in livestock: a review. *Animal Breeding Abstracts* 54, 623–629.

Houpt, K.A. (1991) *Domestic Animal Behaviour for Veterinarians and Animal Scientists*. Iowa State University Press, Ames.

Hultnaes, C.A. (1959) Studies on variation in mating behaviour and semen picture in young bulls of the Swedish Red-and-White breed and on causes of this variation. *Acta Agriculturae Scandinavica (Suppl.)* 6, 82. (Cited in Hafez, 1975.)

Jezierski, T.A., Koziorowski, M., Goszczynski, J. and Sieradzka, I. (1989) Homosexual and social behaviours of young bulls of different geno- and phenotypes and plasma concentrations of some hormones. *Applied Animal Behaviour Science* 24, 101–113.

Kempkens, K. (1989) Influence of transponder feeding in summer and in winter to the behaviour of dairy cattle in cubicle housing. *KTBL-Schrift* 336, 314–325.

Khanna, A.S. and Sharma, J.S. (1988) Association of dairy temperament score with performance in some Indian breeds and crossbred cattle. *Indian Journal of Animal Sciences* 58, 237–242. (Cited in Burrow, 1997.)

Koene, P. (1997) Social behaviour of semi-wild Scottish Highland cattle and Konik horses in nature-reserves in the Netherlands. In: *Proceedings of the 31st International Congress of International Society of Applied Ethology in Prague*, pp. 85–86.

Kommisrud, E. and Steine, T. (1997) Semen collection, semen production and fertility rates in polled vs. horned bulls in Norway. *Reproduction in Domestic Animals* 32, 221–223.

Langbein, J. and Nichelmann, M. (1993) Weideverhalten von Rindern auf der tropischen Weide als Indikator für eine bestehende Waermebelastung – Probleme der Verhaltensmaskierung in gemischtrassigen Rinderherden (Pasture behaviour of cattle on a tropical pasture as indicator of an existing thermal load – problems of masked behaviour in a mixed breeds herd). *KTBL-Schrift* 356, 78–90.

Langholz, H.J., Schmidt, F.W., Derenbach, J. and Kim, W. (1987) Suckling behaviour, immunoglobulin status and weaning performance in suckler cows. *World Review of Animal Production* 23, 33–38.

Lawstuen, D.A., Hansen, L.B. and Steuernagel, G.R. (1988) Management traits scored linearly by dairy producers. *Journal of Dairy Science* 71, 788–799.

Lazo, A. (1994) Social segregation and the maintenance of social stability in a feral cattle population. *Animal Behaviour* 48, 1133–1141.

Le Neindre, P., Trillat, G., Sapa, J., Menissier, F., Bonnet, J.N. and Chupin, J.M. (1995) Individual differences in docility in Limousin cattle. *Journal of Animal Science* 73, 2249–2253.

Lidfors, L.M., Jensen, P. and Algers, B. (1993) Temporal patterning of suckling bouts in free ranging beef cattle. In: *Proceedings of the International Congress on Applied Ethology, Berlin*, pp. 77–81.

Lundberg, U. (1991) Haltungsbedingte Anpassungsprobleme im Sozialverhalten von Rindern. *KTBL-Schrift* 344, 196–205.

Macha, J.and Olsarova, J. (1986) Variability and heritability of grazing intensity in cattle. *Acta Universitatis Agriculturae Brno, A, Facultas Agronomica* 34, 313–320. (Abstract cited.)

McPhee, C.P., McBride, G. and James, J.W. (1964) Social behaviour of domestic animals. III. Steers in small yards. *Animal Production* 6, 9–12.

Mendoza Ordones, G., Wilke, A. and Seeland, G. (1988) Zur Bedeutung ausgewaehlter Eigenschaften von Milchrindkaelbern waehrend der Nahrungsaufnahme als Hilfsmerkmale der Selektion. *Berichte der Humboldt-Universitaet Berlin* 8, 62–67.

Menke, C., Waiblinger, S. and Foelsch, D.W. (1995) Social behaviour of horned dairy cows in loose housing. *KTBL-Schrift* 370, 107–117.

Morris, S.T., Parker, W.J. and Grant, D.A. (1994) Herbage intake, liveweight gain, and grazing behaviour of Friesian, Piedmontese × Friesian, and Belgian Blue × Friesian bulls. *New Zealand Journal of Agricultural Research* 36, 231–236.

Oberosler, R., Carenzi, C. and Varga, M. (1982) Dominance hierarchies of cows on alpine pastures as related to phenotype. *Applied Animal Ethology* 8, 67–77.

Oikawa, T. Fudo, T. and Kaneji, K. (1989) Estimate of genetic parameters for temperament and body measurements of beef cattle. *Japanese Journal of Zootechnical Science* 60, 894–896. (Abstract cited.)

Olsen, H.H. and Petersen, W.E. (1951) Uniformity of semen production and behaviour in monozygous triplet bulls. *Journal of Dairy Science* 34, 489–490. (Cited in Hafez, 1975.)

Porzig, E. (1969) *Das Verhalten landwirtschaftlicher Nutztiere*. VEB Deutscher Landwirtschaftsverlag, Berlin.
Porzig, E., Laube, R.-B. and Polten, S. (1973) Uberblick über Untersuchungen zu Fragen der Verhaltensgenetik (Survey of investigations on genetics of behaviour). *Archiv für Tierzucht* 16, 175–183.
Presicce, G.A., Brockett, C.C., Cheng, T., Foote, R.H., Rivard, G.F. and Klemm, W.R. (1993) Behavioral responses of bulls kept under artificial breeding conditions to compounds presented for olfaction, taste or with topical nasal application. *Applied Animal Behaviour Science* 37, 273–284 .
Price, E.O. (1997) Behavioural aspects of domestication and feralization. In: *Proceedings of the 31st International Congress of the International Society for Applied Ethology, Prague*, p.47.
Purcell, D. and C.W. Arave (1991) Isolation vs. group rearing in monozygous twins. *Applied Animal Behaviour Science* 31, 147–156.
Reinhardt, V. (1978) Die Marschordnung bei Rindern. *Landwirtschaftliche Zeitschrift* 21, 1366–1367.
Reinhardt, V. and Reinhardt, A. (1975) Dynamics of social hierarchy in a dairy herd. *Zeitschrift für Tierpsychologie* 38, 315–323.
Roy, P.K. and Nagpaul, P.K. (1984) Influence of genetic and non-genetic factors on temperament score and other traits of dairy management. *Indian Journal of Animal Science* 54, 566–568. (Cited in Burrow, 1997.)
Sambraus, H.H. (1975) Beobachtungen und Überlegungen zur Sozialordnung von Rindern. *Züchtungskunde*, 8–14.
Sambraus, H.H. (1978) *Nutztierethologie*. Verlag Paul Parey, Berlin and Munich.
Sambraus, H.H. (1994) *Atlas der Nutztierrassen*. Verlag Eugen, Ulmer.
Santha, T., Prieger, K., Keszthelyi, T. and Czako, J. (1988) Genetic analysis of feeding behaviour of cows. *Allattenyesztes es Takarmanyozas* 37, 501–514. (Abstract cited.)
Santolaria, M.G., Sanudo, C. and Alberti, P. (1992) Effect of sexual and aggressive behaviors on average daily gain and meat quality in feeder young bulls. In: Hagelsoe (ed.) *Animal Genetic Resources for Adaptation to More Extensive Production Systems. Proceedings of a CEC workshop in Foulum, Denmark.*
Sato, S. (1981) Factors associated with temperament of beef cattle. *Japanese Journal of Zootechnical Science* 52, 595–605. (Abstract cited.)
Sato, S., Tarmuizu, K. and Sonoda, T. (1991a) Social, behavioural and physiological functions of allo-grooming in cattle. *Farm Animal Behaviour*, 77–78. (Abstract cited.)
Sato, S., Sako, S. and Maeda, A. (1991b) Social licking patterns in cattle (*Bos taurus*): influence of environmental and social factors. *Applied Animal Behaviour Science* 32, 3–12.
Schofield, S.A. (1989) Oestrus detection methods and oestrus behaviour of dairy cows in different environments. Dissertation, Univerity of Wales, UK.
Selman, I.E., McEwan, E.D. and Fisher, E.W. (1970) Studies on natural suckling in cattle during the first eight hours post partum. I. Behavioural studies. *Animal Behaviour* 18, 276–283.
Shiv-Prasad, Mittal, J.P., Kaushish, S.K. and Prasad, S. (1996) Dominance pattern in free grazing zebu cattle. *Indian Journal of Animal Production and Management* 12, 99–103. (Abstract cited.)
Siegel, P.B. (1975) Behavioural genetics. In: Hafez, E.S.E. (ed.) *The Behaviour of Domestic Animals*, 3rd edn. Baillière Tindall, London.

Stamer, E., Junge, W. and Kalm, E. (1997) Die Zeitstruktur des Futteraufnahmeverhaltens von Michkühen unter Laufstallbedingungen (Temporal pattern of feeding behaviour of dairy cows kept in groups). *Archiv für Tierzucht* 40, 195–214.

Stricklin, W.R., Heisler, C.E. and Wilson, L.L. (1980) Heritability of temperament in beef cattle. *Journal of Animal Science* 5 (Suppl. 1), 109–110.

Syme, G.J. and Syme, L.A. (1979) *Social Structure in Farm Animals*. Elsevier, Amsterdam.

Tulloh, N.M. (1961) Behaviour of cattle in yards. II. A study of temperament. *Animal Behaviour* 9, 25–30.

Voisinet, B.D., Grandin, T., Tatum, J.D., O'Connor, S.F. and Struthers, J.J. (1997) Feedlot cattle with calm temperaments have higher daily gains than cattle with excitable temperaments. *Journal of Animal Science* 75, 892–896.

Wagnon, K.A., Loy, R.G., Rollins, W.C. and Carroll, F.D. (1966) Social dominance in a herd of Angus, Hereford, and Shorthorn cows. *Animal Behaviour* 14, 474–479.

Waltl, B., Appelby, M.C. and Soelkner, J. (1995) Effects of relatedness on the suckling behaviour of calves in a herd of beef cattle rearing twins. *Applied Animal Behaviour Science* 45, 1–9.

Winder, J.A., Walker, D.A. and Bailey, C.C. (1995) Genetic aspects of diet selection in the Chihuahuan desert. *Journal of Range Management* 48, 549–553. (Abstract cited.)

Winder, J.A., Walker, D.A. and Bailey, C.C. (1996) Effect of breed on botanical composition of cattle diets on Chihuahuan desert range. *Journal of Range Management* 49, 209–214. (Abstract cited.)

Zeeb, K. (1987) Das Verhalten freilebender Rinder. *Swiss Vet* 4, 9–18.

Genetics and Biology of Reproduction in Cattle

13

B.W. Kirkpatrick

Department of Animal Sciences, 1675 Observatory Drive, University of Wisconsin, Madison, WI 53706, USA

Introduction	391
Genetics of Twinning and Ovulation Rate	391
Genetics and Biology of Puberty	396
Genetics and Biology of Gestation Length	399
Genetics and Biology of Dystocia	401
Genetics and Biology of Conception Rate	403
Identifying Genes Responsible for Variation in Reproductive Traits	404
References	404

Introduction

This review of the genetics of reproduction in cattle proceeds on a trait-by-trait basis and considers twinning and ovulation rate, age at puberty, dystocia, gestation length and conception rate. In most cases, the pertinent scientific literature has been previously reviewed, and it is the intent of the author of this chapter to summarize the previous reviews and examine the subsequent work in the area.

Genetics of Twinning and Ovulation Rate

The genetics of twinning in cattle has been previously reviewed by Hendy and Bowman (1970), Rutledge (1975), Stolzenburg and Schönmuth (1979), Morris (1984) and Morris and Day (1986). The reviews of Hendy and Bowman (1970), Rutledge (1975) and Stolzenburg and Schönmuth (1979) provide rather extensive lists of studies reporting estimates for frequency of twin birth in various breeds of cattle. Hendy and Bowman report estimates, primarily from dairy cattle, that range from 0.34 to 4.50%. Rutledge provided a more comprehensive list and summarized his findings in part with a contrast of US beef and dairy breed differences. Dairy breed averages (1.3–8.9% twinning) were in general two to three times the level of beef breed averages (0.4–1.1%). Rutledge's survey of beef breeds primarily involved British breeds. As pointed out by Morris (1984), other European breeds, including Simmental, Charolais and Maine-Anjou, have twinning rates considerably higher than the British breeds. While nutritional and other management differences exist between beef and dairy cattle, this summary of breed differences suggests a potential genetic correlation between lactation and twinning rate.

The overwhelming majority of bovine twins are dizygotic, or fraternal, twins as opposed to monozygotic, or identical, twins. Estimates of the frequency of monozygotic twinning are derived by comparing the incidence of like- versus unlike-sex twins. Monozygotic twins are estimated to comprise less than 10% of all twin births when considering typical breed averages for twinning rate (Johansson, 1932; Bonnier, 1946; Johansson and Venge, 1951; Cady and Van Vleck, 1978), and heritability of monozygotic twinning is thought to be nil. As a consequence, twinning rate in cattle is closely related to ovulation rate.

Twinning rate is generally considered a trait with low heritability. Maijala and Syväjärvi (1977) reviewed the literature with regard to estimates of twinning-rate heritability. They report an average heritability estimate of 0.028 ± 0.0004. Subsequent studies with various cattle populations are generally consistent in reporting low heritability estimates for twinning (Table 13.1). Ovulation rate, when considered as a single observation with a binary outcome, likewise has a low heritability. Estimates of heritability for a single observation range from 0.07 to 0.11 (Echternkamp et al., 1990; Van Vleck and Gregory, 1996; Gregory et al., 1997). However, when ovulation rate is evaluated over multiple oestrous cycles, the effective heritability of the average of multiple observations is considerably higher. When considering average ovulation rate over eight cycles, effective heritability rises to approximately 0.38 (Echternkamp et al., 1990; Van Vleck and Gregory, 1996). The genetic correlation between twinning rate and ovulation rate has been estimated to be between 0.75 and 1.00 (Van Vleck and Gregory, 1996; Gregory et al., 1997).

Table 13.1. Recent heritability estimates for twinning rate in cattle, on the observed, binary scale.

Author	Estimate	Population
Cady and Van Vleck, 1978	0.03 to 0.06	US Holstein
Syrstad, 1984	0.006 ± 0.001 to 0.046 ± 0.006	Norwegian dairy cattle
Gregory et al., 1990a	0.02 ± 0.07	USDA-MARC twinning population
Maijala and Osva, 1990	0.005 to 0.007	1st parity Ayrshire and Friesian
	0.023 to 0.051	2nd and later parity
Ron et al., 1990	0.017 to 0.022	Israeli Holstein
Van Vleck and Gregory, 1996	0.03 to 0.08	USDA-MARC twinning population
Gregory et al., 1997	0.09 to 0.10	USDA-MARC twinning population

Given this high genetic correlation, use of ovulation rate as the selection criterion for improving twinning rate makes an excellent case for the potential benefit from indirect selection on a correlated trait.

Several selection experiments aimed at increasing twinning rate are or have been conducted (reviewed by Morris and Day, 1986, 1990b). Efforts in New Zealand (Morris and Day, 1990b) and the USA (Gregory et al., 1997) have used ovulation-rate data as part of the basis for selection. This approach has been particularly successful in the US Department of Agriculture (USDA)-Meat Animal Research Center (USDA-MARC) twinning herd; at the latest report, the twinning rate had reached a level exceeding 30% twin births (Gregory et al., 1997). The twinning rate in this herd is increasing at a rate of approximately 2.5% per year.

Twinning rate is influenced by seasonal effects, although the nature of the seasonal effect (photoperiod, nutritional, etc.) is uncertain. Hendy and Bowman (1970) and Rutledge (1975) reviewed work from eight studies from the 1920s to mid-1960s which examined seasonal effects on twinning rate. Roughly half of these studies identified significant seasonal effects on twinning rate, with peak twinning rate generally corresponding to spring and autumn conception. Studies by Johansson et al. (1974) and Cady and Van Vleck (1978) likewise identified seasonal effects, with highest twinning rates corresponding to spring and autumn conception. Analysis of data from the USDA-MARC twinning population provides a more limited analysis of seasonal effects, given that mating is limited to periods corresponding to spring and autumn calving (conception in mid-June to mid-August and late October to late December, respectively). In this population, twinning rates for the October–December-conceiving cows exceed those for the June–August-bred group, with a difference of 4–6% occurrence of twin births (Gregory et al., 1990a; Van Vleck and Gregory, 1996).

Twinning rate increases with age and parity of dam, with the increase continuing through 8–11 years of age in the work reviewed by Hendy and Bowman (1970) and Rutledge (1975). While this might in part be attributable to improvement in embryo survival with dam age and parity, it probably results in large part from increases in ovulation rate with age (Labhsetwar et al., 1963; Morris et al., 1992). These earlier findings for association of dam age and twinning rate are paralleled by more recent work. Cady and Van Vleck (1978) examined data from US Holsteins and reported twinning frequencies of 1.05, 4.32 and 6.18% for parities 1, 2 and ≥ 3, respectively. Syrstad (1984) examined data from Norwegian dairy cattle and observed twinning frequencies of 0.46, 1.64, 2.31, 3.11 and 3.57% for parities 1–5, respectively. Berry et al. (1994) reported twinning frequencies of 1.3, 6.0 and 9.4% for US Holstein females in parities 1, 2 and ≥ 3, respectively. In studies with the composite USDA-MARC twinning population, Gregory et al. (1990) observed twinning rates of 6, 8–11, 12–13 and 11–14% for cows of 2, 3, 4 or 5 years of age at calving. Morris and Day (1990) analysed data from two milking Shorthorn herds and one Holstein herd; increases in twinning rate with age were observed, with the largest increases occurring over the first five parities and the single biggest increase occurring between parities 1 and 2. Maijala and Osva (1990) examined data from a Finnish Ayrshire and Friesian population and observed increases in twinning rate of roughly 2% and 1% between first and second and second and third parities, respectively.

Twin birth is associated with a number of detriments, including lower perinatal calf survival and poorer cow reproductive performance. These associations have been previously reviewed by Hendy and Bowman (1970) and Cady and Van Vleck (1978). Additional reports of these associations published since these reviews are summarized in Table 13.2. Regarding cow performance, twinning is associated with increased dystocia (due to malpresentation),

Table 13.2. Detrimental association of twinning with calf and cow performance.

Source	Single birth	Twin birth	P value
Perinatal calf mortality (%)			
Cady and Van Vleck, 1978	5.9	22.4	< 0.05
Anderson et al., 1982, heifers	9.1	5.3	NS
Anderson et al., 1982, cows	0	0	NS
Gregory et al., 1990b	5	22	< 0.01
Guerra-Martinez et al., 1990, heifers	8.5	11.9	NS
Guerra-Martinez et al., 1990, cows	4.3	10.8	NS
Day et al., 1995	3.2	15.7	< 0.005
Gregory et al., 1996	3.5	16.5	< 0.01
Retained placenta (%)			
Anderson et al., 1979	0	14	NS
Anderson et al., 1982, heifers	2.3	27.2	< 0.05
Anderson et al., 1982, cows	6.7	33.3	< 0.05
Gregory et al., 1990b	2.8	21.5	< 0.01
Guerra-Martinez et al., 1990, heifers	11.9	35.2	< 0.05
Guerra-Martinez et al., 1990, cows	4.3	24.3	< 0.05
Eddy et al., 1991	2	16	< 0.001
Dystocia (%)			
Anderson et al., 1979	20	21	NS
Anderson et al., 1982, heifers	47.8	21.2	< 0.05
Anderson et al., 1982, cows	6.7	5.6	NS
Gregory et al., 1990b	23	35	< 0.01
Guerra-Martinez et al., 1990, heifers	37.3	21.6	NS
Guerra-Martinez et al., 1990, cows	8.7	10.8	NS
Eddy et al., 1991	5	7	NS
Berry et al., 1994			
lactation 1	30.6	34.5	NS
lactation 2	11.8	30.0	< 0.05
Gregory et al., 1996	20.4	42.2	–
Abortion frequency (%)			
Day et al., 1995	12.0	29.3	< 0.05
Guerra-Martinez et al., 1990, heifers	3.0	4.1	NS
Guerra-Martinez et al., 1990, cows	0	0	NS
Interval, parturition to conception (days)			
Chapin and Van Vleck, 1980	107	132	< 0.05
Gregory et al., 1990b	83	93	< 0.01
Eddy et al., 1991	92	125	< 0.001

NS, not significant.

increased incidence of retained placenta, greater frequency of abortion in twin gestation and longer interval from parturition to first oestrus. Inconsistent results for the incidence of dystocia may reflect two competing dynamics: twinning reduces the incidence of dystocia attributable to large calf size but increases the incidence of dystocia attributable to malpresentation.

After considering costs of twinning on cow and calf performance, the increased calf output is insufficient to make twinning a benefit in typical dairy operations. Recent studies in the Netherlands and the UK suggest lost income of $104 to $108 per cow producing twin calves vs. singles (Eddy *et al.*, 1991; Beerepoot *et al.*, 1992). However, when considering beef production, exploitation of twin birth in an intensive, non-traditional management scheme may provide an opportunity to dramatically increase production efficiency.

The issue of greater frequency of abortion in twin gestations deserves further discussion, as this may impose a limit on the level of twinning that is attainable in a system seeking to exploit high twinning frequency in cattle production. Previous studies of embryo survival in twin pregnancies created by embryo transfer have generally indicated greater embryo survival for bilateral than for unilateral transfers (Hanrahan, 1983). Morris and Day (1990b) report a similar trend in analysis of natural twin pregnancy in the New Zealand twinning herd; in this case, embryo survival for bilateral ovulations (69%) exceeded that for unilateral ovulations (55%). The same authors have suggested that selection based on twinning rate would be selection for increased frequency of bilateral ovulation. Interestingly, there was no difference in embryo survival between unilateral and bilateral pregnancies in the USDA-MARC herd (Echternkamp *et al.*, 1990). In a further analysis of the USDA-MARC data, Echternkamp *et al.* (1990) found the observed proportions of twin pregnancies, single pregnancies and open cows (from those shedding two eggs at ovulation) to fit well with a two-stage model of embryonic loss. In this model, the first stage corresponded to the period prior to placental anastomosis, where loss of one embryo in a twin pregnancy is independent of the other. The latter stage following placental anastomosis is one where the survival of one embryo is dependent on the survival of the second. Results from this study suggest that joint dependency of embryo survival is more the crux of the issue than differences between bilateral and unilateral ovulation. The distribution of twin, single and no pregnancy from double ovulating cows in this study (Echternkamp *et al.*, 1990) was 48%, 29% and 23%, respectively. Twinning frequency as a percentage of pregnant cows would be 62% in this case.

Numerous reports have speculated about the existence of single genes with large effects on twinning rate. Often this speculation follows from the identification of a sire or dam with an exceptional record with regard to twinning rate. Rutledge (1975) and Morris (1984) have listed a number of such reports in the literature. In few, if any, cases have single-gene hypotheses been put to a test that would either build strong support for or refute such a hypothesis, based on segregation of phenotypic classes. Syrstad (1984) and Gregory *et al.* (1990a) examined daughter twinning rates for sons of sires with exceptional twinning rates. The absence of bimodality in progeny twinning rates in both cases was taken as an indication that twinning rate within these exceptional families was not under the control of a single locus. Morris and Day (1990) and Morris and Foulley (1991) categorized sires and dams into high and low groups, based on offspring or own performance. In analysing offspring twinning rates, no evidence of sire group × dam group interaction was observed, suggesting little support for a hypothesis of a single, recessive gene responsible for twinning rate. The advent of highly polymorphic genetic markers and dense marker linkage maps now provides a more powerful tool for dissection of genetic variation in twinning and ovulation rate.

The USDA-MARC twinning population represents a unique resource for identification of ovulation-rate quantitative trait loci (QTL). Initial efforts to map QTL for ovulation rate in the USDA-MARC population have led to the identification of a QTL on bovine chromosome 7 (Blattman *et al.*, 1996). The QTL in question was identified in a Swedish Friesian sire used in the foundation of the USDA-MARC twinning population. Given the sire's ancestry this locus may eventually prove relevant to dairy cattle selection, where the focus would probably be on reducing the incidence of alleles conferring higher a twinning rate.

Identification of an ovulation-rate QTL within the USDA-MARC population may prove useful in cattle selection in two alternative ways. In one case, this information could be used to facilitate introgression of alleles conferring high twinning rate in commercially relevant beef cattle populations. Alternatively, QTL identified in the USDA-MARC herd may help to reveal the location of existing QTL in commercial populations. This information could be used to select for or against alleles conferring higher twinning rates, depending on the population and management system.

Genetics and Biology of Puberty

Martin *et al.* (1992) previously reviewed genetic aspects of puberty in cattle. Much of the information in their review comes from the extensive breed evaluation work conducted by geneticists at the USDA-MARC (Cundiff *et al.*, 1986). Breed differences in age and weight at puberty are in some cases quite dramatic. Information on breed differences previously summarized by Cundiff *et al.* (1986) and Gregory *et al.* (1993) is supplemented in Table 13.3 with additional, recent data from Freetly and Cundiff (1997). There are several relationships that become apparent when examining the data on breed differences. Dairy breeds (Brown Swiss, Holstein, Jersey) attain puberty at younger ages in general than beef breeds, suggesting perhaps that there are genes with pleiotropic effects on lactation and age at puberty. Another general relationship is that breeds with larger mature size in general attain puberty at greater weights and later ages than breeds with lesser mature size. Finally, *Bos indicus* breeds (Boran, Brahman, Sahiwal) and breeds with a major *B. indicus* component (Brangus, Santa Gertrudis) attain puberty at later ages than *Bos taurus* breeds.

Table 13.3. Breed effects on puberty in cattle.

Breed	Weight at puberty (kg)	Age at puberty (days)	Gestation length	% Calf crop born	% Calving difficulty Sire	% Calving difficulty Maternal
Sire breed*						
Angus[†]	344	351	284.6	–	–	–
Belgian Blue	329	347	285.8	–	–	–
Boran	316	396	293.3	–	–	–
Brahman[‡]	323	429	291.7	94	1	10
Brahman[§]	358	426	292.6	–	–	–
Brangus	308	377	285.5	88	12	4
Brown Swiss	279	332	285.0	92	8	8
Charolais	319	384	287.0	88	18	15
Chianina	317	384	287.5	93	12	8
Devon	290	356	284.1	90	4	10
Gelbvieh	284	326	286.3	95	8	11
Hereford[†]	350	355	286.3	–	–	–
Hereford-Angus[‖]	282	357	284.0	91	3	13
Holstein	300	341	282.0	95	5	10
Jersey	235	308	282.9	90	3	7
Limousin	308	384	289.2	89	9	12
Maine Anjou	305	357	285.4	94	20	11
Piedmontese	300	348	289.9	–	–	–
Pinzgauer	277	334	286.0	93	6	13
Red Poll	263	337	285.2	90	4	14
Sahiwal	291	414	294.0	95	2	6
Santa Gertrudis	315	383	286.0	88	4	6
Simmental	302	358	287.3	89	15	17
South Devon	290	350	286.7	88	12	15
Tarentaise	282	349	287.1	91	6	10
Tuli	306	371	291.0	–	–	–

Breed	Weight at puberty (kg)	Age at puberty (days)	Gestation length	% Calf crop born	As a trait of dam Parity 1	As a trait of dam Parity 2
Purebred means						
Angus	317	393	283	81	32	1
Braunvieh	333	350	290	82	74	13
Charolais	370	391	286	81	49	10
Gelbvieh	339	353	287	83	61	8
Hereford	316	411	288	79	49	5
Limousin	338	408	289	75	41	7
Pinzgauer	336	360	287	84	62	16
Red Poll	295	359	288	81	54	3
Simmental	345	363	287	81	49	14

*Offspring sired by bulls of breeds listed with dams of Hereford and Angus breeds.
[†]Offspring from Hereford and Angus sires born after 1981.
[‡]Offspring from Brahman sires born between 1964 and 1975.
[§]Offspring from original Brahman sires (born between 1964 and 1975) and a roughly equal number of sires born between 1984 and 1989.
[‖]Offspring from Hereford and Angus sires born prior to 1970.

Regarding estimates of genetic parameters, Martin et al. (1992) identified nine previous studies in which heritability of age at puberty (female) had been evaluated. The average heritability estimate of 0.40 was relatively high, suggesting that age at puberty is a trait which should be readily amenable to change through selection. While puberty in bulls cannot be determined directly as easily as in heifers, the correlated trait of scrotal circumference can be used as an indicator trait. Martin et al. (1992) cite estimates of genetic correlation between age at puberty in heifers and scrotal circumference in bulls which range between −0.71 and −1.07 (i.e. larger scrotal circumference at a constant age corresponding with earlier attainment of puberty in related females). As with age at puberty in heifers, scrotal circumference in males has a relatively high heritability. Martin et al. (1992) and Brinks (1994) summarized previous estimates of scrotal circumference heritability; the average of the studies listed was 0.47 and 0.49, respectively. There have been few additional reports of heritability estimates since the aforementioned reviews. Gregory et al. (1995) examined various purebred and composite populations and reported a heritability estimate for female age at puberty of 0.31. Keeton et al. (1996) examined field data from the Limousin breed and reported an estimated heritability of 0.46 for scrotal circumference. Both estimates are in reasonable agreement with previous results.

Nutritional and seasonal influences on attainment of puberty in cattle were thoroughly reviewed by Schillo et al. (1992). Age at puberty and nutritional level are inversely related, i.e. heifers fed a higher energy level attain puberty at younger ages than heifers fed a lower energy level. With regard to effects of season, season of birth has a significant effect on age at puberty, with autumn-born heifers reaching puberty at younger ages than spring-born heifers. Studies with controlled environments indicate that much of the seasonal effect can be attributed to the effects of photoperiod. The physiological basis for onset of puberty is thought to be related to increases in frequency of the pulsatile release of luteinizing hormone (LH) from the anterior pituitary. The hypothalamus is fully capable in the prepubertal heifer of producing the gonadotrophin-releasing hormone (GnRH) that causes secretion of LH from the anterior pituitary. However, the hypothalamus and pituitary exhibit a strong negative-feedback response to oestradiol produced in the ovary. It is a decrease in the negative-feedback response that eventually leads to an increase in LH pulse frequency, with a consequent increase in follicular development, increase in oestradiol production by the follicles and a surge of LH that triggers the initial ovulation and oestrous cycle. How nutritional and photoperiodic effects on attainment of puberty are mediated is uncertain.

At least one effort is being made to use selection in an experimental herd to create differences in age at puberty and examine correlated response to selection for pubertal traits. Morris et al. (1993) have created three different selection lines and one control line within an Angus population. One line was selected for increased scrotal circumference at 13 months of age, a second line was selected for reduced age at first oestrus in heifers and increased yearling weight in bulls and a third line was selected for increased age at first oestrus in heifers and increased yearling weight in bulls. The only report from this project to date indicates that both scrotal circumference and age at first oestrus have responded to direct selection after 9 years of selection. However, no correlated response in scrotal circumference was observed in the two lines divergently selected for age at first oestrus while concomitantly selecting for increased yearling weight. Several breeds of beef cattle now include scrotal circumference as a trait for which expected progeny differences (EPDs) are calculated. It is unclear to what extent scrotal circumference is used in practice as a criterion in beef cattle selection.

Genetics and Biology of Gestation Length

Unlike the other traits considered thus far, gestation length is a trait influenced by two genotypes, that of the calf and that of the cow. In most instances, effects of calf and cow genotype are somewhat confounded, given the mother's genetic contribution to the calf. Still, ample evidence exists in the literature that both genotypes are significant contributors to variation in gestation length.

The most extensive information on breed of sire effects on gestation length (through calf genotype) comes from the breed evaluation studies conducted at USDA-MARC (Cundiff *et al.*, 1986; Gregory *et al.*, 1993; Table 13.3). The breed averages listed in the first section of Table 13.3 represent gestation lengths for calves sired by bulls of the listed breeds with mothers of primarily Hereford and Angus breeding. The most pronounced difference evident in these results are the longer gestation lengths for calves sired by bulls of the *B. indicus* breeds (Brahman and Sahiwal). Williamson and Humes (1985) likewise observed gestations of 2–5 days greater length for Brahman-sired calves compared with calves sired by *B. taurus* breeds (all dams were Hereford or Angus). Comparison of Angus-sired and Santa Gertrudis-sired (Gotti *et al.*, 1985) or Brahman-sired (Reynolds *et al.*, 1980) calves also provided evidence of longer gestation lengths for calves sired by the breed (Santa Gertrudis, + 2.0 days; Brahman, + 6.8 days) with *B. indicus* influence. Differences can also be identified between *B. taurus* breeds (Table 13.3), with the most pronounced difference being the shorter gestation lengths associated with dairy sires (Jersey and Holstein). Lawlor *et al.* (1984) compared calves of 25 and 50% Simmental breeding with straightbred Hereford and 50% Angus calves, all produced from Hereford dams. The 50% Simmental group had 2–4-day longer gestation lengths than the Hereford- and Angus-sired groups in this study. The Simmental–Angus difference corresponds well with the breed differences observed in the USDA-MARC study, while the Simmental–Hereford difference is greater. Differences could be attributable to sampling or perhaps heterosis, although other studies suggest heterosis for gestation length is negligible (Sacco *et al.*, 1990; Gregory *et al.*, 1991).

Dam genotype effects on gestation length have also been documented in several cases. When mated to a variety of sire breeds (Brahman, Chianina, Maine Anjou and Simmental), Angus dams had shorter gestation lengths by approximately 2.5 days ($P < 0.001$) than did Hereford dams in a study reported by Williamson and Humes (1985). Reynolds *et al.* (1980) and Gotti *et al.* (1985) compared Santa Gertrudis or Brahman to Angus dams and found Angus dams to have gestation lengths of 4.0 and 4.2 days less, respectively.

Heritability of gestation length has been estimated in several studies since being reviewed by Andersen and Plum (1965). Andersen and Plum identified nine studies estimating heritability of gestation length, with most estimates falling between 0.25 and 0.50. More recent estimates (Table 13.4) generally fall into the same range for estimates of heritability, due to direct additive effects. Estimates of heritability associated with maternal additive effects have been lower. This and the corresponding observation of heritability exceeding estimates of repeatability (Andersen and Plum, 1965) can be taken to indicate that gestation length is under the control more of the fetus than of the dam. Perhaps the most interesting genetic parameter estimate is the negative correlation between growth and gestation length, which has been reported several times (Bourdon and Brinks, 1982; MacNeil *et al.*, 1984; Gregory *et al.*, 1995). Successful selection for reduced gestation length might be beneficial in expanding days available for post-partum reproduction. The negative genetic correlation between growth and gestation length would be favourable to their concomitant improvement.

As indicated in the review by Andersen and Plum (1965) there are non-genetic factors that have repeatedly been associated with differences in gestation length. Dam age or parity is one such factor which has been consistently identified. Mature cows typically have gestation lengths exceeding those of heifers by approximately 1–2 days. Calf gender has also been commonly reported to affect gestation length, with males being carried longer than females, again with a typical difference of 1–2 days.

Table 13.4. Heritability (h^2) estimates for gestation length in cattle.

Author	h^2_{direct}	$h^2_{maternal}$	Population
MacNeil *et al.*, 1984	0.30*	–	USDA-MARC populations
Cundiff *et al.*, 1986	0.64–0.77	–	USDA-MARC populations
Azzam and Nielsen, 1987	0.36–0.45	0.02–0.13	US, multiple breeds
Wray *et al.*, 1987	0.37	0.09	US Simmental
Moore *et al.*, 1990	–	0.01*	Canadian Ayrshire
	–	0.05*	Canadian Holstein
Silva *et al.*, 1992	0.22	–	US Holstein
Gregory *et al.*, 1995	–	0.45*	USDA-MARC populations

*Gestation length considered as trait of the cow.

Genetics and Biology of Dystocia

Dystocia, or calving difficulty, is the leading cause of calf mortality (Patterson *et al.*, 1987; Wittum *et al.*, 1993) and a major source of veterinary expense for cattle producers (Hird *et al.*, 1991; New, 1991; Salman *et al.*, 1991; Wittum *et al.*, 1993). Price and Wiltbank (1978a), Meijering (1984) and Rice (1994) reviewed the literature pertaining to the basis for calving difficulty and potential means for preventing it. It is generally accepted that the primary reason for calving difficulty is incompatibility in size of calf and dam; calf birth weight by itself is the single greatest predictor of calving difficulty. The genetics of dystocia is in large part the genetics of birth weight. Differences in birth weight are moderately heritable, and cattle producers are frequently admonished to choose sires with low birth-weight EPDs for use on immature females. Effects of calf genotype and effects of sire breed on incidence of dystocia have been clearly documented in breed evaluation (see Table 13.3) and other studies. More vexing is the maternal contribution to variation in dystocia.

That there is a maternal genetic component to dystocia is attested to both by breed differences in dystocia (Cundiff *et al.*, 1986; Gregory *et al.*, 1993) and by evidence for heritable differences in calving difficulty. Breed differences in the maternal component of dystocia have been characterized extensively by research conducted at the USDA-MARC (see Table 13.3). F_1 females sired by bulls of various breeds exhibited incidences of calving difficulty that varied significantly from 1% to 17%. Notable among these breed differences are the low incidence of calving difficulty for *B. indicus* breeds (Brahman, Sahiwal) and for Jersey F_1 females, the latter representing the breed cross with smallest skeletal size. The basis for the *B. indicus* superiority for calving difficulty will be addressed later.

Heritability of both direct (i.e. effect of calf genotype) and maternal effects on calving difficulty has been estimated in numerous studies, and those reported since the review of this topic by Meijering (1984) are listed in Table 13.5. In general, heritability of dystocia is higher in heifers than in cows, although heritability of both direct and maternal effects tends to be low. Direct and maternal genetic components of dystocia typically exhibit a negative genetic correlation which works against simultaneous improvement of both aspects of dystocia.

The maternal complement to calf size is pelvic area, and the influence of pelvic area on dystocia has come under considerable study. While the supposition that pelvic area might be a predictor of calving difficulty is intuitively appealing, research studies have provided relatively little support for this idea. As an illustration, one of the most favourable outcomes was reported by Price and Wiltbank (1978b), in which pelvic area accounted for only one quarter of the amount of phenotypic variation in dystocia compared with calf birth weight. After first accounting for the influence of birth weight on dystocia, pelvic area typically accounts for a very minor part of the remaining variation in dystocia. As a consequence, efforts to use pelvic area as a predictor of calving difficulty (Naazie *et al.*, 1989; Basarab *et al.*, 1993; Van Donkersgoed *et al.*,

Table 13.5. Recent genetic parameter estimates for dystocia.

Author	h^2_{direct}	$h^2_{maternal}$	$r_{G\ direct.maternal}$	Population
Meijering, 1984	0.03 to 0.20	0.03 to 0.20	−0.19 to -0.63	Heifers, review article
	0.00 to 0.08	−	−	Cows, review article
Thompson and Rege, 1984	0.24	−	−	US Holstein
Weller et al., 1988	0.03 to 0.06	−	−	Heifers, Israeli Holstein
	0.00 to 0.01	−	−	Cows, Israeli Holstein
Cubas et al., 1991	0.14	0.24	−0.86	US Angus
Naazie et al., 1991	0.28 to 0.37	0.12 to 0.47*	−	Beef and dairy synthetic populations
Manfredi et al., 1991	0.08	0.11	0.15	French Normande
	0.07	0.07	−0.09	French Holstein
Kriese et al., 1994	−	0.09 to 0.11*	−	USDA-MARC purebreds and composites
Gregory et al., 1995	0.18 to 0.42	−	−	USDA-MARC purebreds and composites
	0.04 to 0.29†	−	−	USDA-MARC purebreds and composites

*Calving difficulty score considered a trait of the mother.
†Percentage of difficult births, all others evaluated calving difficulty on a subjective scale.
h^2, heritability; r_G, genetic correlation.

1993) or simulation studies considering the use of pelvic area in animal selection aimed at reducing incidence of dystocia (Cook et al., 1993) have suggested that this is an ineffective strategy.

Pelvic area is only one potential aspect of maternal influence on dystocia. The *B. indicus* advantage in maternal calving ease mentioned above appears to be due to the dam's ability to alter growth of the fetus *in utero*. This is accomplished through reduced placentome weight and reduced uterine blood flow (Ferrell, 1991). Undoubtedly there are other mechanisms which contribute to dam variation in calving ease as well (e.g. variation in relaxin levels or response to relaxin, etc.).

A preponderance of data document the association of two non-genetic factors with dystocia in the dam (Meijering, 1984; Rice, 1994). One of these factors is the age or parity of the dam. Incidence of dystocia is typically three to four times as frequent in females calving for the first time compared with females in second or later parity. The propensity for first-calf heifers to have calves of lighter birth weight is more than offset by the dam's lesser stage of skeletal development relative to mature females (although perhaps other factors play a role in the higher incidence of dystocia as well). The second factor with clearly documented effects on incidence of dystocia is gender of the calf. Male calves typically have greater birth weights than female calves, and there is likewise a greater incidence of dystocia associated with the birth of bull calves compared with heifers (of the order of twofold greater). Clearly, these

Genetics and Biology of Conception Rate

Failure of cows to conceive is the primary factor contributing to loss of reproductive efficiency in beef and dairy cattle production (Wiltbank, 1994). There is some evidence for genetic variation in conception rate, although the proportion of phenotypic variation associated with genetic effects appears to be small.

Fertility data are more commonly recorded for dairy than for beef cattle, owing in large part to the more common use of artificial insemination in dairy cattle. As a consequence, there are numerous reports in the scientific literature of estimates of heritability for fertility traits based on dairy cattle records. The three traits for which heritability estimates are summarized here are conception rate, days open and services per conception. This list is intentionally limited to those traits most closely associated with the cycling cow's ability to conceive. Days open is the furthest removed, given that it also incorporates variation in length of the post-partum anoestrous period. Heritability estimates for these traits are routinely small (Table 13.6), with most being 5% or less. Many

Table 13.6. Recent heritability estimates for conception traits.

Author	Conception rate	Days open	Services per conception	Population
Smith and Legates, 1962	–	0.01	–	US dairy
Everett et al., 1966	–	0.07	–	US Holstein and Guernsey
Janson and Andreasson, 1981	–	0.03*	–	Swedish Red and White
	–	0.04*	–	Swedish Friesian
Seykora and McDaniel, 1983	–	0.05–0.13	–	US dairy
Hansen et al., 1983	–	0.03	–	US Holstein
Taylor et al., 1985	0.08	–	–	US Holstein
Jansen, 1986	0.013	–	–	Swedish Friesian: Parity 0
	0.013	–	–	Parity 1
	0.029	–	–	Parity 2
	0.019	–	–	Parity 3
Jansen et al., 1987	–	0.025	–	Swedish dairy: Parity 1
	–	0.021	–	Parity 2
	–	0.070	–	Parity 3
Hermas et al., 1987	0.03	0.04	–	US Guernsey
Raheja et al., 1989b	–	0.03–0.05	0.03–0.06	Canadian Holstein, cows
Raheja et al., 1989a	–	–	0.04	Canadian Holstein, heifers
Weller, 1989	–	–	0.035†	Israeli Holstein, first parity
	–	–	0.022†	Israeli Holstein, second parity
Moore et al., 1990	–	0.004	0.008	Canadian Ayrshire
	–	0.013	0.013	Canadian Holstein
Hayes et al., 1992	–	0.05	0.03	Canadian Holstein
Buxadera and Dempfle, 1997	0.02	–	0.03	Cuban Holstein

*Calving interval.
†Conception status, similar to the inverse of services per conception.

authors cite the low estimates of heritability in arguing against devoting effort to selection for fertility. However, others have advanced the idea that the low heritability is offset by the appreciable amount of genetic variation existing for fertility traits (Philipsson, 1981; Hermas et al., 1987).

Breed differences and strain differences within breed also provide some evidence for genetic variation in conception rate. Breed evaluation studies at USDA-MARC have examined differences in percentage calf crop at birth and weaning (percentage of cows exposed to a bull that produced a calf at birth or weaning) and found limited evidence for breed differences. The overall variation between breeds (see Table 13.3) is not as great as for other traits, and it is only the most extreme breed differences that were statistically significant in these studies (Cundiff et al., 1986; Gregory et al., 1993). Weller (1989) presented evidence of differences in fertility between strains of Holstein cattle, with Swedish and US Holsteins differing by more than 10% in conception status (roughly the inverse of services per conception).

Given the low heritability for conception rate, the most effective way to genetically exploit this trait is through utilization of heterosis by crossbreeding. Comparisons of straight and crossbreeding mating systems with beef cattle have clearly demonstrated the potential for raising conception rates. Kress et al. (1992) cite an average improvement of 5.5% in calving rate for eight previous studies, while their own and other recent studies show improvements in excess of 10% (Kress et al., 1990; Williams et al., 1990; Brown et al., 1997).

Identifying Genes Responsible for Variation in Reproductive Traits

The development of highly polymorphic genetic markers and extensive linkage maps (Barendse et al., 1997; Kappes et al., 1997) now makes it possible to dissect genetic variation for quantitative traits and identify the chromosomal regions and genes contributing most to variation. As mentioned above, efforts to identify QTL for ovulation rate and twinning rate are already well under way (Blattman et al., 1996). Other traits discussed here should also be considered. Reproductive traits are particularly good candidates for the application of marker or marker-assisted selection, given that their sex-limited phenotypes prohibit the practice of direct phenotypic selection in both genders. The alternative is progeny testing, which is both costly and time-consuming. The primary challenge to studies of this type, in many cases, will be redefining the trait (as in the focus on ovulation rate as a proxy for twinning rate) to effectively raise the heritability of the trait under evaluation.

References

Andersen, H. and Plum, M. (1965) Gestation length and birth weight in cattle and buffaloes: a review. *Journal of Dairy Science* 48, 1224–1235.

Anderson, G.B., Cupps, P.T. and Drost, M. (1979) Induction of twins in cattle with bilateral and unilateral embryo transfer. *Journal of Animal Science* 49, 1037–1042.

Anderson, G.B., BonDurrant, R.H. and Cupps, P.T. (1982) Induction of twins in different breeds of cattle. *Journal of Animal Science* 54, 485–490.

Azzam, S.M. and Nielsen, M.K. (1987) Genetic parameters for gestation length, birth date and first breeding date in beef cattle. *Journal of Animal Science* 64, 348–356.

Barendse, W., Vaiman, D., Kemp, S., Sugimoto, Y., Armitage, S., Williams, J, Sun, H., Eggen, A., Agaba, M., Aleyasin, A., Band, M., Bishop, M., Buitkamp, J., Byrne, K., Collins, F., Cooper, L., Coppettiers, W., Denys, B., Drinkwater, R., Easterday, K., Elduque, C., Ennis, S., Erhardt, G., Ferretti, L., Flavin, N., Gao, Q., Georges, M., Gurung, R., Harlizius, B., Hawkins, G., Hetzel, J., Hirano, T., Hulme, D., Joergensen, C., Kessler, M., Kirkpatrick, B., Konfortov, B., Kostia, S., Kuhn, C., Lenstra, J., Leveziel, H., Lewin, H., Leyhe, B., Li, L., Martin Buriel, I., McGraw, R., Miller, R., Moody, D., Moore, S., Nakane, S., Nijman, I., Olsaker, I., Pomp, D., Rando, A., Ron, M., Shalom, A., Soller, M., Teale, A., Thieven, I., Urquhart, B., Vage, D.I., Van de Weghe, A., Varvio, S., Velmalla, R., Vilkki, J., Weikard, R., Woodside, C., Womack, J., Zanotti, M. and Zaragoza, P. (1997) A medium density genetic linkage map of the bovine genome. *Mammalian Genome* 8, 21–28.

Basarab, J.A., Rutter, L.M. and Day, P.A. (1993) The efficacy of predicting dystocia in yearling beef heifers: I. Using ratios of pelvic area to birth weight or pelvic area to heifer weight. *Journal of Animal Science* 71, 1359–1371.

Beerepoot, G.M.M., Dykhuizen, A.A., Nielen, M. and Schukken, Y.H. (1992) The economics of naturally occurring twinning in dairy cattle. *Journal of Dairy Science* 75, 1044–1051.

Berry, S.L., Ahmadi, A. and Thurmond. M.C. (1994) Periparturient disease on large, dry lot dairies: interrelationships of lactation, dystocia, calf number, calf mortality, and calf sex. *Journal of Dairy Science* 77(Suppl. 1), 379.

Blattman, A.N., Gregory, K.E. and Kirkpatrick, B.W. (1996) A search for quantitative trait loci for ovulation rate in cattle. *Animal Genetics* 27, 157–162.

Bonnier, G. (1946) Studies on monozygous cattle twins. II. Frequency of monozygous twins. *Acta Agriculturae Scandinavica* 1, 147–151.

Bourdon, R.M. and Brinks, J.S. (1982) Genetic, environmental and phenotypic relationships among gestation length, birth weight, growth traits and age at first calving in beef cattle. *Journal of Animal Science* 55(3), 543–553.

Brinks, J.S. (1994) Relationships of scrotal circumference to puberty and subsequent reproductive performance in male and female offspring. In: Fields, M.J. and Sand, R.S. (eds) *Factors Affecting Calf Crop*. CRC Press, Boca Raton, pp. 363–370.

Brown, M.A., Brown, A.H., Jackson, W.G. and Miesner, J.R. (1997) Genotype x environment interactions in Angus, Brahman, and reciprocal cross cows and their calves grazing common bermudagrass and endophyte-infected tall fescue pastures. *Journal of Animal Science* 75, 920–925.

Buxadera, A.M. and Dempfle, L. (1997) Genetic and environmental factors affecting some reproductive traits of Holstein cows in Cuba. *Genetics, Selection, Evolution* 29, 469–482.

Cady, R.A. and Van Vleck, L.D. (1978) Factors affecting twinning and effects of twinning in Holstein dairy cattle. *Journal of Animal Science* 46, 950–956.

Chapin, C.A. and Van Vleck. L.D. (1980) Effects of twinning on lactation and days open in Holsteins. *Journal of Dairy Science* 63, 1881–1886.

Cook, B.R., Tess, M.W. and Kress, D.D. (1993) Effects of selection strategies using heifer pelvic area and sire birth weight expected progeny differences on dystocia in first-calf heifers. *Journal of Animal Science* 71, 602–607.

Cubas, A.C., Berger, P.J. and Healey. M.H. (1991) Genetic parameter estimates for calving ease and survival at birth in Angus field data. *Journal of Animal Science* 69, 3952–3958.

Cundiff, L.V., Gregory, K.E., Koch, R.M. and Dickerson, G.E. (1986) Genetic diversity among cattle breeds and use to increase beef production efficiency in a temperate environment. In: *Proceedings of the 3rd World Congress on Genetics Applied to Livestock Production. IX.* Lincoln, Nebraska, pp. 271–282.

Day, J.D., Weaver, L.D. and Franti, C.E. (1995) Twin pregnancy diagnosis in Holstein cows: discriminatory powers and accuracy of diagnosis by transrectal palpation and outcome of twin pregnancies. *Canadian Veterinary Journal* 36, 93–97.

Echternkamp, S.E., Gregory, K.E., Dickerson, G.E., Cundiff, L.V., Koch, R.M. and Van Vleck, L.D. (1990) Twinning in cattle: II. Genetic and environmental effects on ovulation rate in puberal heifers and postpartum cows and the effects of ovulation rate on embryonic survival. *Journal of Animal Science* 68, 1877–1888.

Eddy, R.G., Davies, O. and David, C. (1991) An economic assessment of twin births in British dairy herds. *Veterinary Record* 129, 526–529.

Everett, R.W., Armstrong, D.V. and Boyd, L.J. (1966) Genetic relationship between production and breeding efficiency. *Journal of Dairy Science* 49, 879–886.

Ferrell, C.L. (1991) Maternal and fetal influences on uterine and conceptus development in the cow: II. Blood flow and nutrient flux. *Journal of Animal Science* 69, 1954–1965.

Freetly, H.C. and Cundiff, L.V. (1997) Postweaning growth and reproduction characteristics of heifers sired by bulls of seven breeds and raised on different levels of nutrition. *Journal of Animal Science* 75, 2841–2851.

Gotti J.E., Benyshek, L.L. and Kiser, T.E. (1985) Reproductive performance in crosses of Angus, Santa Gertrudis and Gelbvieh beef cattle. *Journal of Animal Science* 61, 1017–1022.

Gregory, K.E., Echternkamp, S.E., Dickerson, G.E., Cundiff, L.V., Koch, R.M. and Van Vleck, L.D. (1990a) Twinning in cattle: I. Foundation animals and genetic and environmental effects on twinning rate. *Journal of Animal Science* 68, 1867–1876.

Gregory, K.E., Echternkamp, S.E., Dickerson, G.E., Cundiff, L.V., Koch, R.M. and Van Vleck, L.D. (1990b) Twinning in cattle: III. Effects of twinning on dystocia, reproductive traits, calf survival, calf growth and cow productivity. *Journal of Animal Science* 68, 3133–3144.

Gregory, K.E., Cundiff, L.V. and Koch, R.M. (1991) Breed effects and heterosis in advanced generations of composite populations for preweaning traits of beef cattle. *Journal of Animal Science* 69, 947–960

Gregory, K.E., Cundiff, L.V., Koch, R.M. and Lunstra, D.D. (1993) Differences among parental breeds in germplasm utilization project. In: *Beef Research Progress Report No. 4*. Clay Center, Nebraska, pp. 22–42.

Gregory, K.E., Cundiff, L.V. and Koch, R.M. (1995) Genetic and phenotypic (co)variances for production traits of female populations of purebred and composite beef cattle. *Journal of Animal Science* 73, 2235–2242.

Gregory, K.E., Echternkamp, S.E. and Cundiff, L.V. (1996) Effects of twinning on dystocia, calf survival, calf growth, carcass traits and cow productivity. *Journal of Animal Science* 74, 1223–1233.

Gregory, K.E., Bennett, G.L., Van Vleck, L.D., Echternkamp, S.E. and Cundiff, L.V. (1997) Genetic and environmental parameters for ovulation rate, twinning rate, and weight traits in a cattle population selected for twinning. *Journal of Animal Science* 75, 1213–1222.

Guerra-Martinez, P., Dickerson, G.E., Anderson, G.B. and Green, R.D. (1990) Embryo-transfer twinning and performance efficiency in beef production. *Journal of Animal Science* 68, 4039–4050.

Hanrahan, J.P. (1983) The inter-ovarian distribution of twin ovulations and embryo survival in the bovine. *Theriogenology* 20, 3–11.

Hansen, L.B., Freeman, A.E. and Berger, P.J. (1983) Yield and fertility relationships in dairy cattle. *Journal of Dairy Science* 66, 293–305.

Hayes, J.F., Cue, R.I. and Monardes, H.G. (1992) Estimates of repeatability of reproductive measures in Canadian Holsteins. *Journal of Dairy Science* 75, 1701–1706.

Hendy, C.R.C. and Bowman, J.C. (1970) Twinning in cattle. *Animal Breeding Abstracts* 38, 22–37.

Hermas, S.A., Young, C.W. and Rust, J.W. (1987) Genetic relationships and additive genetic variation of productive and reproductive traits in Guernsey dairy cattle. *Journal of Dairy Science* 70, 1252–1257.

Hird, D.W., Weigler, B.J., Salman, M.D., Danaye-Elmi, C., Palmer, C.W., Holmes, J.C., Utterback, W.W. and Sischo, W.M. (1991) Expenditures for veterinary services and other costs of disease and disease prevention in 57 California beef herds in the National Animal Health Monitoring System (1988–1989). *Journal of the American Veterinary Medical Association* 198, 554–558.

Jansen, J. (1986) Direct and maternal genetic parameters of fertility traits in Friesian cattle. *Livestock Production Science* 15, 153–164.

Jansen, J., Van der Werf, J. and De Boer, W. (1987) Genetic relationships between fertility traits for dairy cows in different parities. *Livestock Production Science* 17, 337–349.

Janson, L. and Andreasson, B. (1981) Studies of fertility traits in Swedish dairy cattle. IV. Genetic and phenotypic correlations between milk yield and fertility. *Acta Agriculturae Scandinavica* 31, 313–322.

Johansson, I. (1932) The sex ratio and multiple births in cattle. *Zeitschrift für Züchtung Reihe B* 24, 183–268.

Johansson, I. and Venge, O. (1951) Studies on the value of various morphological characters for the diagnosis of monozygosity of cattle twins. *Zeitschrift für Tierzüchtung und Züchtungsbiologie* 59, 389–424.

Johansson, I., Lindhé, B. and Pirchner, F. (1974) Causes of variation in the frequency of monozygous and dizygous twinning in various breeds of cattle. *Hereditas* 78, 201–234.

Kappes, S.M., Keele, J.W., Stone, R.T., McGraw, R.A., Sonstegard, T.W., Smith, T.P., Lopez-Corrales, N.L. and Beattie, C.W. (1997) A second-generation linkage map of the bovine genome. *Genome Research* 7, 235–249.

Keeton, L.L., Green, R.D., Golden, B.L. and Anderson, K.J. (1996) Estimation of variance components and prediction of breeding values for scrotal circumference and weaning weight in Limousin cattle. *Journal of Animal Science* 74, 31–36.

Kress, D.D., Doornbos, D.E. and Anderson, D.C. (1990) Performance of crosses among Hereford, Angus and Simmental cattle with different levels of Simmental breeding: IV. Maternal heterosis and calf production by two-year-old dams. *Journal of Animal Science* 68, 54–63.

Kress, D.D., Doornbos, D.E., Anderson, D.C. and Rossi, D. (1992) Performance of crosses among Hereford, Angus, and Simmental cattle with different levels of Simmental breeding: VI. Maternal heterosis of 3- to 8-year-old dams and the dominance model. *Journal of Animal Science* 70, 2682–2687.

Kriese, L.A., Van Vleck, L.D., Gregory, K.E., Boldman, K.G., Cundiff, L.V. and Koch, R.M. (1994) Estimates of genetic parameters for 320-day pelvic measurements of males and females and calving ease of 2-year-old females. *Journal of Animal Science* 72, 1954–1963.

Labhsetwar, A.P., Tyler, W.J. and Casida, L.E. (1963) Analysis of variation in some factors affecting multiple ovulations in Holstein cattle. *Journal of Dairy Science* 46, 840.

Lawlor, T.J., Jr, Kress, D.D., Doornbos, D.E. and Anderson, D.C. (1984) Performance of crosses among Hereford, Angus and Simmental cattle with different levels of Simmental breeding. I. Preweaning growth and survival. *Journal of Animal Science* 58, 1321–1328.

MacNeil, M.D., Cundiff, L.V., Dinkel, C.A. and Koch, R.M. (1984) Genetic correlations among sex-limited traits in beef cattle. *Journal of Animal Science* 58, 1171–1180.

Maijala, K. and Osva, A. (1990) Genetic correlations of twinning frequency with other economic traits in dairy cattle. *Journal of Animal Breeding and Genetics* 107, 7–15.

Maijala, K. and Syväjärvi, J. (1977) On the possibility of developing multiparous cattle by selection. *Zeitschrift für Tierzüchtung und Züchtungsbiologie* 94, 136–150.

Manfredi, E., Ducrocq, V. and Foulley, J.L. (1991) Genetic analysis of dystocia in dairy cattle. *Journal of Dairy Science* 74, 1715–1723.

Martin, L.C., Brinks, J.S., Bourdon, R.M. and Cunliff, L.V. (1992) Genetic effects on beef heifer puberty and subsequent reproduction. *Journal of Animal Science* 70, 4006–4017.

Meijering, A. (1984) Dystocia and stillbirth in cattle – a review of causes, relations and implications. *Livestock Production Science* 11, 143–177.

Moore, R.K., Kennedy, B.W., Schaeffer, L.R. and Moxley, J.E. (1990) Relationships between reproduction traits, age and body weight at calving, and days dry in first lactation Ayrshires and Holsteins. *Journal of Dairy Science* 73, 835–842.

Morris, C.A. (1984) A review of the genetics and reproductive physiology of dizygotic twinning in cattle. *Animal Breeding Abstracts* 52, 803–819.

Morris, C.A. and Day, A.M. (1986) Potential for genetic twinning in cattle. In: *Proceedings of the 3rd World Congress on Genetics Applied to Livestock Production. XI.* Lincoln, Nebraska, pp. 14–29.

Morris, C.A. and Day, A.M. (1990a) Effects of dam and sire group on the propensity for twin calving in cattle. *Animal Production* 51, 481–488.

Morris, C.A. and Day, A.M. (1990b) Genetics and physiology studies of cows in a twin breeding experiment. *Journal of Animal Breeding and Genetics* 107, 2–6.

Morris, C.A. and Foulley, J.L. (1991) A comparison of genetic data from New Zealand and France on twin calving in cattle. *Genetics, Selection, Evolution* 23, 345–350.

Morris, C.A., Day, A.M., Amyes, N.C. and Hurford, A.P. (1992) Ovulation and calving data from a herd selected for twin calving. *New Zealand Journal of Agricultural Research* 35, 379–391.

Morris, C.A., Bennett, G.L. and Johnson, D.L. (1993) Selecting on pubertal traits to increase beef cow reproduction. *Proceedings of the New Zealand Society of Animal Production* 53, 427–432.

Naazie, A., Makarechian, M.M. and Berg, R.T. (1989) Factors influencing calving difficulty in beef heifers. *Journal of Animal Science* 67, 3243–3249.

Naazie, A., Makarechian, M. and Berg, R.T. (1991) Genetic, phenotypic, and environmental parameter estimates of calving difficulty, weight, and measures of pelvic size in beef heifers. *Journal of Animal Science* 69, 4793–4800.

New, J.C., Jr (1991) Costs of veterinary services and vaccines/drugs used for prevention and treatment of diseases in 60 Tennessee cow-calf operations (1987–1988). *Journal of the American Veterinary Medical Association* 198, 1334–1340.

Patterson, D.J., Bellows, R.A., Burfening, P.J. and Carr, J.B. (1987) Occurrence of neonatal and postnatal mortality in range beef cattle. I. Calf loss incidence from birth to weaning, backward and breech presentations and effects of calf loss on subsequent pregnancy rate of dams. *Theriogenology* 28, 557–571.

Philipsson, J. (1981) Genetic aspects of female fertility in dairy cattle. *Livestock Production Science* 8, 307–319.

Price, T.D. and Wiltbank, J.N. (1978a) Dystocia in cattle. a review and implications. *Theriogenology* 9, 195–219.

Price, T.D. and Wiltbank, J.N. (1978b) Predicting dystocia in heifers. *Theriogenology* 9, 221–249.

Raheja, K.L., Burnside, E.B. and Schaeffer, L.R. (1989a) Heifer fertility and its relationship with cow fertility and production traits in Holstein dairy cattle. *Journal of Dairy Science* 72, 2665–2669.

Raheja, K.L., Burnside, E.B. and Schaeffer, L.R. (1989b) Relationships between fertility and production in Holstein dairy cattle in different lactations. *Journal of Dairy Science* 72, 2670–2678.

Reynolds, W.L., DeRouen, T.M., Moin, S. and Koonce, K.L. (1980) Factors influencing gestation length, birth weight and calf survival of Angus, Zebu and Zebu cross beef cattle. *Journal of Animal Science* 51, 860–867.

Rice, L.E. (1994) Dystocia-related risk factors. *Veterinary Clinics of North America – Food Animal Practice* 10, 53–68.

Ron, M., Ezra, E. and Weller, J.I. (1990) Genetics analysis of twinning rate in Israeli Holstein cattle. *Genetics, Selection, Evolution* 22, 349–359.

Rutledge, J.J. (1975) Twinning in cattle. *Journal of Animal Science* 40, 803–815.

Sacco, R.E., Baker, J.F., Cartwright, T.C., Long, C.R. and Sanders, J.O. (1990) Measurements at calving for straightbred and crossbred cows of diverse types. *Journal of Animal Science* 68, 3103–3108.

Salman, M.D., King, M.E., Odde, K.G. and Mortimer, R.G. (1991) Costs of veterinary services and vaccines/drugs used for prevention and treatment of diseases in 86 Colorado cow–calf operations participating in the National Animal Health Monitoring System (1986–1988). *Journal of the American Veterinary Medical Association* 198, 1739–1744.

Schillo, K.K., Hall, J.B. and Hileman, S.M. (1992) Effects of nutrition and season on the onset of puberty in the beef heifer. *Journal of Animal Science* 70, 3994–4005.

Seykora, A.J. and McDaniel, B.T. (1983) Heritabilities and correlations of lactation yields and fertility for Holsteins. *Journal of Dairy Science* 66, 1487–1493.

Silva, H.M., Wilcox, C.J., Thatcher, W.W., Becker, R.B. and Morse, D. (1992) Factors affecting days open, gestation length, and calving interval in Florida dairy cattle. *Journal of Dairy Science* 75, 288–293.

Smith, J.W. and Legates, J.E. (1962) Relationship of days open and days dry to lactation milk and fat yields. *Journal of Dairy Science* 45, 1192–1198.

Stolzenburg, U. and Schönmuth, G. (1979) Genetische Aspekte der Zwillingsträchtigkeit beim Rind. *Akademie der Landwirtschaftswisseenschaften der Deutschen Demokratischen Republik, Berlin* 17, 1–49.

Syrstad, O. (1984) Inheritance of multiple births in cattle. *Livestock Production Science* 11, 373–380.

Taylor, J.F., Everett, R.W. and Bean, B. (1985) Systematic environmental, direct, and service sire effects on conception rate in artificially inseminated Holstein cows. *Journal of Dairy Science* 68, 3004–3022.

Thompson, J.R. and Rege, J.E.O. (1984) Influences of dam on calving difficulty and early mortality. *Journal of Dairy Science* 67, 847–853.

Van Donkersgoed J., Ribble, C.S., Booker, C.W., McCartney, D. and Janzen, E.D. (1993) The predictive value of pelvimetry in beef cattle. *Canadian Journal of Veterinary Research* 57, 170–175.

Van Vleck, L.D. and Gregory, K.E. (1996) Genetic trend and environmental effects in a population of cattle selected for twinning. *Journal of Animal Science* 74, 522–528.

Weller, J.I. (1989) Genetic analysis of fertility traits in Israeli dairy cattle. *Journal of Dairy Science* 72, 2544–2550.

Weller, J.I., Misztal, I. and Gianola, D. (1988) Genetic analysis of dystocia and calf mortality in Israeli-Holsteins by threshold and linear models. *Journal of Dairy Science* 71, 2491–2501.

Williams, A.R., Franke, D.E. and Saxton, A.M. (1990) Genetic effects for reproductive traits in beef cattle and predicted performance. *Journal of Animal Science* 69, 531–542.

Williamson, W.D. and Humes, P.E. (1985) Evaluation of crossbred Brahman and continental European beef cattle in a subtropical environment for birth and weaning traits. *Journal of Animal Science* 61, 1137–1145.

Wiltbank, J.N. (1994) Challenges for improving calf crop. In: Fields, M.J. and Sand, R.S. (eds) *Factors Affecting Calf Crop.* CRC Press, Boca Raton, pp. 1–22.

Wittum, T.E., Salman, M.D., Odde, K.G., Mortimer, R.G. and King, M.E. (1993) Causes and costs of calf mortality in Colorado beef herds participating in the National Animal Health Monitoring System. *Journal of the American Veterinary Medical Association* 203, 232–236.

Wray, N.R., Quaas, R.L. and Pollak, E.J. (1987) Analysis of gestation length in American Simmental cattle. *Journal of Animal Science* 65, 970–974.

Reproductive Technologies and Transgenics

14

N.L. First, M. Mitalipova and M. Kent First
Department of Animal Sciences, 1675 Observatory Drive, University of Wisconsin, Madison, WI 53706, USA

Introduction	411
The *In Vitro* Production of Bovine Embryos	412
Production of Calves of a Predetermined Sex	415
The Production and Uses of Transgenic Livestock	416
History, applications and limitations	416
More efficient methods for making transgenic cattle	417
Cloning Cattle by Nuclear Transfer	421
Overview, biological mechanisms and constraints	421
Use of nuclear transfer to make animals from cultured embryonic stem cells	424
Cloning from differentiated cells	426
Applications of cloning in agriculture	427
References	428

Introduction

Astounding advances have occurred over the past 12 years (1985–1997) in our ability to control, manipulate or change the reproduction and genetics of cattle. These new techniques are the focus of this chapter. The techniques include the following:

1. The ability to produce, culture, study and use commercially *in vitro* produced embryos. The embryos can be biopsied for preimplantation diagnosis of gender and genetic traits. The oocytes can be retrieved in large numbers from valuable cows via biweekly transvaginal follicular retrieval or obtained from an abattoir.

2. The new techniques also include the production of identical twins by splitting embryos or the production of multiple copies of valuable embryos by transferred nuclei from blastomeres, cultured embryonic cells or fetal cells, and in three cases even adult cells, into enucleated oocytes. This technique is being modified in order to improve efficiency.

3. Lastly, to have the ability to introduce deoxyribonucleic acid (DNA) into the genome of animals, including cattle. This technique was developed following the first gene transfers in mice, which resulted in transgenic mice (Gordon *et al.*, 1980; Brinster *et al.*, 1981).

The technologies of *in vitro* embryo production, gene transfer, genetic analysis and diagnosis and embryo cloning have the potential to be used synergistically in cattle breeding and improvement. Both gene transfer and cloning by nuclear transfer (NT) require the ability to culture embryos and preserve them *in vitro* and often the ability to produce the embryos themselves *in vitro*. The efficiency of making or multiplying transgenic animals is improved by the use of NT and subsequent cell-culture techniques. For example, NT allows multiplication of valuable transgenics or the transfer of genes into culture-multiplied cells, of which each has the potential to become an embryo and offspring after use in NT. The use of cultured cells for gene transfer also allows the possibility for gene deletion by DNA homologous recombination transfer (Koller *et al.*, 1989; Joyner, 1991; Melton, 1994; Cibelli, *et al.*, 1998a).

The *In Vitro* Production of Bovine Embryos

The ability to produce embryos of cattle *in vitro* began with studies of Iritani and Niwa (1977) and Brackett *et al.* (1982). Iritani and Niwa (1977) were the first to attempt to fertilize *in vitro*-matured bovine oocytes, while Brackett *et al.* (1982) were the first to produce a live calf from *in vitro* fertilization (IVF). However, this calf was developed from an oocyte matured *in vivo* and an embryo that developed *in vivo*. The first efficient system for maturing bovine oocytes, capacitating sperm, fertilizing *in vitro* and defining the variables was published by Ball *et al.* (1983). Soon thereafter, the first offspring from IVF of *in vitro*-matured oocytes were produced by Critser *et al.* (1986) and by Hanada *et al.* (1986).

In early experiments, such as Brackett *et al.* (1982), embryos resulting from IVF were developed *in vivo* by surgical transfer into the oviducts of cows immediately after fertilization. This *in vivo* culture was replaced by culture in sheep oviducts (Willadsen, 1986; Eyestone and First, 1989), then by co-culture of embryos with oviduct epithelial cells or oviduct-conditioned media (Eyestone and First, 1989) and finally by culture in minimal-composition media supplemented with amino acid(s) (Rosenkrans and First, 1994), media that mimic the composition of the oviduct (Tervit *et al.*, 1972; Takahashi and First, 1992) or media designed to be used in a two-stage sequence to meet embryo nutrient requirements and to reduce cell damage (Gardner *et al.*, 1997; Gardner, 1998). The applications, components, influencing variables, problems and success rates of bovine IVF are the subject of the following discussion.

The *in vitro* production of embryos is a technique that is gaining use in: (i) commercial embryo production from *in vivo*-recovered oocytes of valuable cows (Hasler *et al.*, 1995; van Wagtendonk-de Leeuw *et al.*, 1998); (ii) production of calves from cows not responding to conventional, superovulation and embryo-transfer techniques (Looney *et al.*, 1994); and (iii) commercial production of low-cost embryos from oocytes harvested from abattoir-recovered ovaries, a system called beef production without brood cows (Agca *et al.*, 1998). Collectively, considering all applications, there are at least ten commercial companies producing more than 10,000 calves per year. The process is less than perfect and of lower efficiency than *in vivo* production of embryos (van Wagtendonk-de Leeuw *et al.*, 1998). From some, but not all, laboratories, there are reports of increased frequency of failed or late delivery and larger than normal calves of lower survival rate.

The best success rates are approximately 70–80% of oocytes completing maturation, 70–80% completing fertilization and the first cleavage division, 30–60% of cleaved embryos developing to blastocyst (a stage compatible with non-surgical uterine embryo transfer), 50–60% of these achieving pregnancy and 80% of the pregnancies completing gestation with a live calf. This results in a post-fertilization offspring production efficiency of approximately 20%.

The critical steps in the process are *in vitro* maturation of oocytes, preparation of sperm for participation in fertilization (sperm capacitation and the acrosome reaction), the fertilization process and embryo culture. More and more, it is being recognized that these are not independent events, with later events such as embryo development in culture, highly influenced by variables affecting oocyte maturation and independent of the frequency of oocytes matured. Some of the variables affecting development competence of oocytes (reviewed by Gordon, 1994) are: (i) the age of the females supplying the oocytes; (ii) their health and environmental stress, such as heat stress; (iii) the size and maturity of follicles; (iv) the size of the oocyte (Arlotto *et al.*, 1996); (v) the presence and interaction of cumulus cells with the oocyte; (vi) the conditions of oocyte maturation, such as temperature, pH and gas environment (Lenz *et al.*, 1983); (vii) the number of cohort oocytes; and (viii) the presence of cumulus cells and cell growth factors in the culture media (reviewed by Bavister, 1990; Leese, 1991; Gordon, 1994; Gardner *et al.*, 1997; Leibfried-Rutledge *et al.*, 1997; Gardner, 1998).

Preparation of sperm to bind, penetrate and fertilize oocytes is dependent on preparation of sperm to undergo the acrosome reaction (sperm capacitation) and involves changes in the sperm plasma membrane, allowing the fusion of the outer and inner acrosomal membrane (acrosome reaction). This allows the escape of acrosomal enzymes involved in sperm–oocyte interaction and the passage of sperm through the zona pellucida of the egg (Parrish and First, 1993). The process of sperm capacitation is accomplished at least in part in the oviduct by using heparin sulphate; this compound is the most commonly used agent for causing capacitation of bovine sperm *in vitro* (Parrish *et al.*, 1988, 1989).

The fertilization process and its completion are influenced by conditions of fertilization, such as temperature and media composition (Lenz *et al.*, 1983), by the age of the metaphase II oocyte and by the time kinetics of the sperm–egg interaction (Rose and Bavister, 1992; Dominko and First, 1997). It is also influenced by the sire contributing the sperm (Hillery *et al.*, 1990).

Development of embryos to blastocyst stage, a non-surgical transfer stage, is heavily influenced by conditions of oocyte maturation and the age of the oocyte (Dominko and First, 1997; Leibfried-Rutledge *et al.*, 1997), the sire contributing the sperm (Leibfried-Rutledge *et al.*, 1989; Hillery *et al.*, 1990) and the sex of the embryo (Bredbacka and Bredbacka, 1996). Embryo development is also influenced by conditions of embryo culture, such as media composition, oxidative and peroxidative damage, gas and temperature conditions of incubation and growth factors (Gardner *et al.*, 1997; Leibfried-Rutledge *et al.*, 1997).

The steps in *in vitro* production of embryos are continuously under study. A better understanding of the mechanisms is being achieved and the efficiencies are slowly improving. For greater detail, the reader is referred to several recent reviews (Shamsuddin *et al.*, 1996; Leibfreid-Rutledge *et al.*, 1997; Sirard *et al.*, 1998; Yang *et al.*, 1998).

A modification of the *in vitro* embryo production procedure that has considerably extended its application to oocyte retrieval from live cows and especially cows of high genetic value is transvaginal, ultrasound-guided oocyte recovery (Kruip *et al.*, 1991; Pieterse *et al.*, 1991) as depicted in Fig. 14.1. This

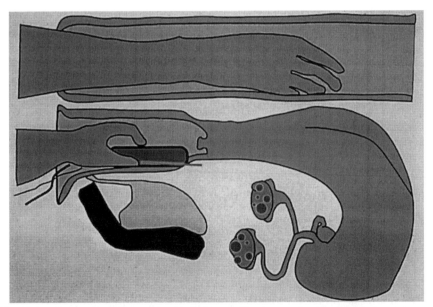

Fig. 14.1. Transvaginal ultrasound-guided retrieval of bovine oocytes. The hand in the rectum places each ovarian follicle on an aspiration needle attached to the ultrasound imaging transducer. The needle penetrates the vaginal wall. Oocytes can be retrieved from unstimulated follicles twice a week. (Redrawn from Kruip *et al.*, 1991.)

is used on a commercial basis to recover primary oocytes from antral follicles, which will be matured, fertilized and cultured to the blastocyst stage using *in vitro* procedures (Looney *et al.*, 1994; Hasler *et al.*, 1995). The procedure requires no superovulation, is minimally invasive and can be done twice per week to recover 15 to 18 oocytes per cow per week (Kruip *et al.*, 1994; Looney *et al.*, 1994; Hasler *et al.*, 1995).

In vitro production of embryos is constantly becoming a more useful tool for maximizing the number of offspring from a valuable cow, producing calves from infertile cows and producing commercial beef cattle in a programme for beef production without brood cows. An important application is the provision of oocyte and embryo culture systems facilitating the practice of gene transfer and animal cloning.

Production of Calves of a Predetermined Sex

The literature is rich in the history of methods attempting to produce offspring exclusively of one sex or the other, as reviewed by Johnson (1994) and Reubinoff and Schenker (1996).

Currently two methods appear feasible for the production of sex-specified cattle and each has limitations concerning how it is applied. The first is the sorting of X and Y chromosome-containing spermatozoa. The only reliable method to date is separation of X and Y by flow-cytometric sorting. This technique is based on determination of the difference in fluorescently labelled DNA of individual spermatozoa in a small flow of medium. Using filtered laser light to illuminate a fluorescent DNA label that spectrophotometrically discriminates between the chromosomes, the sorter gates the spermatozoa according to quantitative fluorescent characteristics of X or Y. The X and Y purity rates range from 85 to 90% (Johnson *et al.*, 1994; Johnson, 1996).

The number of sperm sexed from one ejaculate is adequate for several IVFs but is, however, of quality and quantity insufficient at present for commercial artificial insemination. In spite of this, calves have been born of the predicted sex after artificial insemination with sexed sperm (Seidel *et al.*, 1996). The sexed sperm are of reduced longevity and cannot at present survive freezing. Best success has been with the use of IVF within approximately 6 h after sorting.

The exposure of Hoechst 33342 DNA-labelled sperm to laser light has the potential to cause chromatin damage. Although no deformed or chromatin-damaged offspring have thus far occurred across several species for offspring derived from this sexing procedure, there is a high rate of early embryonic loss, which is suspected to be due in part to chromatin change (Johnson and Schulman, 1994; Reubinoff and Schenker, 1996).

The high-speed flow cytometers in use sort at a rate of 10,000–13,000 sperm s^{-1}. Sorters of four to five times greater speed are becoming available. With higher sort speed and greater sort efficiency, this method may become commercially useful for cattle artificial insemination.

The second reliable method for producing sex-specified offspring is the sexing of cells biopsied from preimplantation embryos, using polymerase chain reaction (PCR) amplification of DNA sequences specific for the Y or for the X chromosome. This method relies on the availability of well-characterized primers which enable amplification of sex-specific regions or single-copy sequence in a multiplex reaction with a positive amplification control. This has been developed and used successfully in several laboratories and with different X or Y chromosome-specific sequences and method variations of the procedure (Avery *et al.*, 1992; Kirkpatrick and Monson, 1993; Hyttinen *et al.*, 1996).

A limitation of the cell biopsy and X–Y PCR method is poor post-freeze survival and subsequent developmental capacity of the previously biopsied embryos.

The Production and Uses of Transgenic Livestock

History, applications and limitations

The ability to make transgenic animals was initially developed in mice. In 1974, Jaenisch and Mintz showed that injection of Simian Virus 40 (SV40) DNA into the blastocoele cavity of mouse embryos resulted in mice carrying the SV40 sequence. Early studies also demonstrated the ability of other viruses to infect their DNA into cells; however, non-replicating viral vectors for introduction of exogenous DNA had not yet been developed when Gordon and colleagues showed that transgenic mice could be developed from mouse eggs microinjected with DNA (Gordon *et al.*, 1980). Within the next few months, this and four other groups reported similar success in stably microinjecting foreign DNA into the genome of the mouse (Brinster *et al.*, 1981; Costantini and Lacy, 1981; Gordon and Ruddle, 1981; Wagner, E.F. *et al.*, 1981; Wagner, T.E. *et al.*, 1981). Some of the foreign genes were expressed in somatic cells (Brinster *et al.*, 1981; Wagner, E.F. *et al.*, 1981; Wagner, T.E. *et al.*, 1981) and in the germ line (Costantini *et al.*, 1981; Gordon and Ruddle, 1981), with offspring of founder mice continuing to express the foreign DNA (Palmiter *et al.*, 1982). Gordon and Ruddle gave the resulting animals expressing the new gene the name 'transgenic'. The experiments most noted by animal scientists were those of Palmiter *et al.* (1982, 1983), wherein a metallothionein-promoter growth-hormone fusion-gene construct was injected into mouse pronuclei, resulting in mice growing much faster and sometimes two times larger than normal non-transgenic mice. Animal scientists quickly attempted to repeat these studies in attempts to greatly increase animal growth rate in pigs, sheep and cattle. Unfortunately, efficiencies of producing transgenics were low in domestic animals (Table 14.1; reviews in Rexroad, 1992; Brem and Muller, 1994; Eyestone, 1994; Colman, 1996; Wall, 1996). Additionally, domestic species with determinant growth characteristics sometimes produced leaner

Table 14.1. Efficiencies of producing transgenic offspring from microinjected zygotes in livestock.

Species	Number of transgenic offspring/number of zygotes injected (%)		Number of injected zygotes required/ transgenic offspring
Cow*	8/5,030	(0.16%)	625
Sheep	45/5,394	(0.80%)	125
Goat	12/985	(1.20%)	83
Pig	182/22,429	(0.80%)	125

In vivo-produced zygotes only, from Eyestone (1994).

carcasses but without the expected improvements in growth rate or growth efficiency (Rexroad, 1992).

More efficient methods for making transgenic cattle

In spite of the inefficiency of pronuclear microinjection of DNA, a few transgenic sheep, goats and cattle of extremely high commercial value have been made for the purpose of production of valuable pharmaceuticals in the milk. Companies devoted to this effort include Pharming, Protein Products Limited (PPL), Genzyme Transgenics (GTC) and Gala Design. Some of the products include blood-clotting factor IX, antitrypsin, tissue plasminogen activator and hepatitis antigen. The value in expressing a gene for a pharmaceutical product in milk appears to be in quantitative yield of the protein, especially with cattle and in the ability of mammary cells to provide authentic and proper post-translational processing, such as glycosylations of the protein to be produced (for review, see Bremel, 1996; Ziomek, 1998). In spite of several years' effort and considerable financial support for the production of new proteins in milk, none of the new milk-produced proteins has yet reached the market. The first two products, recombinant human antithrombin III (GTC) and alpha I antitrypsin (PPL), are in or have completed phase II human clinical trials, respectively. They are headed towards or are in phase III trials, respectively, and may reach the marketplace around the turn of the century (year 2000) (Ziomek, 1998). The majority of the products are being made in goats and sheep, respectively, in which transgenic efficiencies from DNA microinjection are more than four times greater than in cattle. The major advantage for use of cattle is the higher milk and protein yield (approximately 8000 l year^{-1}, and 10–20-fold greater than in sheep or goats). A disadvantage in cattle is the extended period between the microinjection of DNA into the embryo and the time that the cows first lactate (cows lactate beginning at 30–33 months, approximately twice as late as sheep and goats). The long generation interval is also a disadvantage when producing offspring to be derived from a founder animal.

Critical limitations to transgenesis in cattle are inefficiency of embryo DNA microinjection, and long generation interval (Ziomek, 1998). These limitations are overcome for pharmaceutical milk production by transomatic gene transfer, wherein the DNA is injected into the mammary gland of lactation-induced females (Bremel, 1996). This injection can be by recent, efficient, pseudotyped, viral-vector methods of gene transfer (Chan *et al.*, 1998) and by the ability to selectively produce female offspring as described on pp. 415–416 or by DNA transfer into female cells used to make offspring by NT or embryonic cell–embryo chimeras (Cibelli *et al.*, 1998a, b).

While this transgenic application is a growing industry in itself, the range of products possible for transgenic production are severely limited by the costs associated with the process. Indeed, the range of possibilities as the transgenic process becomes more efficient appears to include, in an economic hierarchical order, pharmaceuticals, antigens, nutraceuticals and new food products, such as sweeteners, new cheeses, etc., specific disease-resistant animals, animals of changed meat quality and increased milk or wool production or more efficient growth.

In spite of the present cost limitations of transgenic cattle for use in animal breeding, there are emerging possibilities for vastly improved efficiencies in transgenics. One approach has been the introduction of DNA into oocytes or zygotes by replication-defective viral vectors. Replication-defective retroviral vectors were developed based on the discoveries of reverse transcription and the helper-cell principle by Temin (Temin, 1987; Varmus and Brown, 1989; Kim *et al.*, 1993; Haskell and Bowen, 1995; Chan *et al.*, 1998). Replication-defective ribonucleic acid (RNA) viral vectors are usually restricted to a DNA insert of 10–15 kb. The virions cannot be concentrated and the vector efficiency is dependent on the virion titre achieved.

A particularly useful modification of the retroviral vector is a pseudotyped or hybrid vector, in which the retroviral envelope is replaced with a vesicular stomatitis virus envelope (VSVG), as developed by Burns *et al.* (1993). Because of its VSVG, this vector can be concentrated to high infectivity, with titres of 10^9–10^{11} virions. This vector combines the VSVG with a common Maloney murine leukaemia retroviral vector (MoMLV).

Chan *et al.* (1998) used this vector to introduce DNA into the perivitelline space outside the plasma membrane of bovine metaphase II oocytes. Metaphase II is a stage wherein chromatin is exposed to cytoplasm after nuclear envelope breakdown and wherein membrane–viral vesiculation transfers the viral and gene DNA sequences of interest into the ooplasm.

When genes for neomycin resistance and β-galactosidase were included in the vector construct, 57% of the resulting embryos were positive for β-galactosidase expression. The MoMLV–VSVG pseudotyped vector was also constructed to contain the gene for hepatitis B surface antigen, as well as neomycin resistance (PLSRNL) and infected into oocytes, as above, or into pronuclear zygotes. Of 836 oocytes and 584 zygotes infected, 21% and 33%, respectively, became embryos or blastocysts. Ten blastocysts resulting from oocyte injection and 12 blastocysts from zygote injection were randomly selected for embryo transfer into five or six cows, respectively. Four calves were born from each group. All four of the oocyte-infected calves expressed the hepatitis B antigen and were transgenic, whereas one of the four zygote infected calves was transgenic.

Pseudotyped viral vector infection of DNA into the metaphase II oocyte immediately prior to fertilization appears to be an efficient method for gene transfer. When used to introduce DNA into oocytes, there is an overall efficiency of approximately 8–10% in producing offspring, as compared with approximately 0.02% for conventional pronuclear injection of DNA, as shown in Table 14.1 for cattle. The principal limitation was not in the viral transgenic methodology but in the efficiency of reproductive steps, wherein only 20% of oocytes became blastocysts and 40% of blastocysts transferred into cows became offspring. The same vector is also being used to infect DNA directly into the mammary gland to produce new proteins in milk. This process is called transomatic infection (Bremel, 1996).

Improvements in the *in vitro* production of embryos should enhance the efficiency of the oocyte–pseudotyped viral-vector method. Two limitations of the method are the small size of the DNA construct that can be built into the vectors (approximately 15 kb) and the fact that some viral DNA could be germ-line-transmitted with the gene of interest. Positive aspects are the potential for high-efficiency transfer of genes into oocytes, cultured cells or embryos.

The pseudotyped viral vector infection of DNA into cells can also be an efficient method for the introduction of DNA into cultured embryonic or fetal cells (Chan *et al*., 1998), which are then used in NT to produce offspring (Cibelli *et al*., 1998a, b; Schnieke *et al*., 1997). Other, but less efficient, methods for the introduction of DNA into cultured cells, are ballistic, electroporation and liposome methods. The advantages of transferring DNA into cultured cells are several. Homologous recombination can be used to target a known sequence of the gene construct to the same known sequence in the genome hierarchy, thereby accomplishing site-specific gene transfer or gene deletion. Additionally, embryos can be screened for gene expression when an expressed and selectable gene marker at an embryonic stage is used in the construct. Only the embryos that are expressing the transgene are transferred into cows. The transgenic embryos can also be used to make embryonic cell lines. The cell line can be used either by NT (Cibelli *et al*., 1998a) or by chimerization of cells (Cibelli *et al*., 1998b) into a normal embryo to produce bovine blastocysts and ultimately offspring. If the cells are multiplied to large numbers in culture, gene deletion can be accomplished in the culture by the technique of homologous recombination (Koller *et al*., 1989; Joyner, 1991; Melton, 1994; Cibelli *et al*., 1998a). The production of transgenic fetal fibroblasts, which are then used as nuclear donors in NT to produce transgenic sheep (Schnieke *et al*., 1997) and transgenic cattle (Cibelli *et al*., 1998a, b) has recently been reported (reviewed by Stice *et al*., 1998).

A variation of site-specific gene transfer is to utilize DNA sequences that can enhance homologous recombinations between foreign DNA and the genome (Wall and Seidel, 1992). A version of this technique that has been used and tested in the mouse is the cre-lox system, in which known lox sites are built by transgenesis into a mouse strain and then reduced to a small number of insertion sites by titrated cleavage of lox sites with the Cre protein. This lox site containing recipient strain is then used in gene transfer with constructs designed to target the lox sites (Rucker and Piedrahita, 1997). The negative aspect of this system for site-specific gene transfer in cattle is the high cost of building and maintaining a lox strain of cattle as recipients for gene transfer.

A third and highly controversial approach to the transfer of genes into animals or birds is the introduction of DNA into spermatozoa by co-incubation, transfection, gene gun or liposome, with the expectation that the introduced DNA will be carried to fertilization of an oocyte, integrated into the zygotic genome and later expressed. This method was developed initially by Lavitrano et al. (1989). Transgenic mice resulted from its use. Numerous other laboratories attempted to repeat the process. Most found the DNA bound to spermatozoa, but transgenic offspring or embryos did not result (Brinster et al., 1989). More recently, it has been shown that interaction of sperm and DNA is dependent on a molecular mechanism involving cooperation of specific protein factors (Bird et al., 1992; Lavitrano et al., 1992; Zani et al., 1995). This DNA binding and uptake process is inhibited by a factor in seminal plasma (Lavitrano et al., 1992). Treatment of the DNA with liposomes enhances DNA uptake in DNA liposome–sperm co-culture (Bachiller et al., 1991). It has been suggested that DNA molecules first bind to sperm and are then integrated into the plasma membrane and migrate into the nucleus (Francolini et al., 1993).

Recently, Shemish has reported liposome introduction of a restriction enzyme–green fluorescent protein (GFP) DNA construct into chicken sperm and claims that lymphocytes of 17 of 19 resulting chickens express the GFP, as do lymphocytes of their offspring (M. Shemish, personal communication). Scientists await confirmation of these studies, but, if repeatable, the latter represents a highly efficient method for producing transgenic birds and perhaps cattle.

Another approach to sperm-mediated DNA transfer is the *in vitro* introduction of DNA into the cells of the testes. The transgenic testis cells are then injected into sterilized seminiferous tubules of an animal to repopulate the tubule and become spermatozoa expressing a new gene. This approach was sparked by Brinster and Zimmerman (1994) when they transplanted testis cells into sterilized seminiferous tubules of mice and showed that spermatozoa resulted from the transferred cells. Subsequently, Kim et al. (1997) transferred a liposome Lac-Z DNA preparation into seminiferous tubules of mice and pigs after the native population of spermatocytes had been destroyed by busulphan treatment. The DNA was introduced into a spermatogonial stem-cell population, which eventually became Lac-Z DNA-containing spermatozoa (Kim et al., 1997).

Considering the transgenic chicks (M. Shemish, personal communication) and the DNA-carrying spermatozoa of mice and pigs (Kim et al., 1997), it seems that we may be very near a time when transgenic animals could be efficiently made from spermatozoa carrying exogenous DNA. The promising results with viral introduction of DNA, engineering animals from transgenic cells and perhaps sperm-mediated DNA transfer suggest that animal geneticists and breeders may soon consider transgenic approaches to genetic change to be possible and affordable. The specificity of gene introduction into a population as compared with gene selection techniques could have an impact on the speed and precision with which genetic change is made and excites the intellect with possibilities for the introduction of genes that are foreign to a population. This could be especially useful in meat animals, such as pigs, cattle and poultry, to capitalize on the muscle development potential realized by suppression of the myostatin gene (Mcpherron and Lee, 1997; Grobet *et al.*, *1998*).

Cloning Cattle by Nuclear Transfer

Overview, biological mechanisms and constraints

The cloning of animals received much attention in 1997 and 1998, with the production of lambs and calves by nuclear transplantation of differentiated or immortalized fetal and adult cells into enucleated metaphase II oocytes (Schnieke *et al.*, 1997; Wilmut *et al.*, 1997; Cibelli *et al.*, 1998a, b). Although not yet highly efficient, the ability to clone animals from differentiated cells has challenged our understanding of the irrevocable nature of cell differentiation and germline totipotency, based largely on early amphibian studies. Essentially, cloning from differentiated cells rests in the ability of a cell nucleus to retain and retranscribe the complete array of messages previously turned on and turned off with cell differentiation, as well as the ability of a properly timed and prepared metaphase II oocyte to completely erase the differentiation repertoire of the donor cell. This has to be done in such a way that the introduced nucleus is capable of re-expressing its entire genome. The efficiency of the NT process depends on the completeness with which all donor-cell chromatin is captured within the new zygotic nucleus, the cell-cycle matching of donor cell and oocyte and the ability of the oocyte to demethylate or dedifferentiate all DNA of the donor nucleus. It depends on the appropriate incorporation of oocyte proteins, such as nuclear lamins, into the nucleus of the new cell. It also depends on the re-establishment of normal-length chromosome telomeres and the ability of the first one or two zygotic cell cycles to ensure chromatin normality through exercise of cell-cycle checkpoints and DNA repair mechanisms. Historically, research leading to this point was focused on the use of embryonic or germline lineage cells, expected to be totipotent. The ability to clone animals from differentiated fetal or adult cells has scientific value, in causing re-examination of issues such as the differentiation and totipotency of cells, the ageing of cells and gene expression.

The principal application of cloning cattle appears to be in propagation of valuable transgenic cattle or high-performance cattle. Although the genes of clones are identical, geneticists estimate from twin studies that environmental influences will modulate expressed phenotype, leaving clones approximately 75% alike in performance (G. Shook, personal communication).

Cloning of animals began in the 1950s and resulted in the dogma that offspring could be produced from NT using non-differentiated totipotent cells of the germline but not from differentiated somatic cells (reviewed by Di Bernardino, 1997). The first attempt to clone mammals was in mice, in which Illmensee and Hoppe (1981) produced three mice from surgical transplantation of nuclei from blastocyst inner cell mass into recently fertilized mouse eggs. Other scientists have not been able to repeat these results, and development to young in mice has been primarily from use of donor nuclei no later in development than the four-cell stage (reviewed by Di Bernardino, 1997).

The ability to perform the NT procedure was greatly enhanced when McGrath and Solter (1983) showed that the procedure could be done by fusing rather than injecting a nucleus into an oocyte. This procedure resulted in a superior cell-survival rate compared with microinjection and has been used in most subsequent studies with mammals.

The next milestone in cloning research was the use of electrofusion to introduce 8–16-cell-stage donor blastomeres into enucleated metaphase II oocytes, from which live sheep were produced by Willadsen (1986). Eventually sheep were produced from cell stages as late as the blastocyst inner cell mass (Smith and Wilmut, 1989) and cultured cells from inner cell mass (Campbell et al., 1996; Wells et al., 1997; Wilmut et al., 1997). Goats have also been cloned and recloned from embryonic blastomeres (Yong and Yuqiang, 1998).

In cattle, the first calves from NT involved transfer by electrofusion of donor nuclei from the 2–32-cell stage into enucleated metaphase II bovine oocytes (Prather et al., 1987). A model for cloning cattle and sheep from embryonic cells is shown in Fig. 14.2. Here, as in most pre-1994 cattle and sheep NTs, aged oocytes were used because metaphase II oocytes could not be activated by means other than by spermatozoa until aged (Ware et al., 1989). Recently, the combined use of 6-dimethylaminopurine (6-DMAP) to inhibit oocyte kinases with calcium (Ca^{2+}) ionophores in order to induce activation (Susko-Parrish et al., 1994) has allowed activation of oocytes at any post-metaphase II age. Distinct differences exist between activation of the oocyte early after metaphase II vs. late. With early NT-activated oocytes, nuclear envelope breakdown occurs and donor chromatin is exposed to egg cytoplasm. Chromatin is decondensed and then condensed and incorporated totally, along with oocyte proteins, such as nuclear lamins, into a new nuclear envelope, which proceeds normally through the first cell cycle. Use of such early-activated metaphase II oocytes requires cell-cycle synchrony of donor nucleus and oocytes, usually at G_0–G_1 stages in the cell cycle (Fulka et al., 1996, 1997b, 1998; Poccia and Collas, 1997; Collas and Poccia, 1998). Asynchrony can result in more than one nuclear organizing region and ploidy problems in resulting embryos (Navara et al., 1994), as well as failure to complete DNA replication and passage of cell-cycle checkpoint screening at the first mitotic cell cycle.

When aged oocytes are used, donor-cell nuclear envelope breakdown does not occur. Rather, this oocyte proceeds directly to cleavage and first exposes the donor nucleus to new cytoplasm at the two-cell stage. Aged

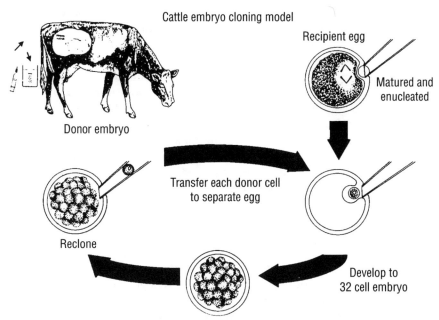

Fig. 14.2. Procedures used to multiply embryos by nuclear transfer. Valuable morula-stage embryos are recovered by non-surgical flush from the uterus of an inseminated cow, the individual cells or blastomeres are removed from the morula and transferred into enucleated oocytes and the new embryos are developed to the morula stage, when the process can be repeated. (Presented originally as a report to W.R. Grace and Company by N.L. First.)

ooplasm is not very competent for dedifferentiation and redifferentiation of differentiated or differentiation-committed nuclei. With aged oocytes, cell-cycle synchrony is not necessary because chromatin is not exposed to cytoplasm until the two-cell stage (First *et al.*, 1992; Leibfried-Rutledge *et al.*, 1992; Campbell *et al.*, 1993; Fulka *et al.*, 1996, 1997b, 1998). Both systems produce offspring when non-differentiated embryonic cells are used as the nuclear donor. However, differentiated cells require long-term exposure of chromatin to ooplasm and synchrony of cell cycles.

Use of oocytes at an age period between early metaphase II and aged oocytes usually results in failed embryo development, because chromatin is often incompletely returned to the nucleus and because ploidy problems occur (Leibfried-Rutledge *et al.*, 1992; Poccia and Collas, 1997).

To date more, than 1000 calves have been produced from NT using embryonic cells as nuclear donors. Recloning has been accomplished (Stice and Keefer, 1993) and frozen donor cells and *in vitro*-matured oocytes (Barnes *et al.*, 1993) have been used. At best, 20–50% of the NTs result in blastocysts for transfer into cows, of which approximately 50% become pregnancies. However, the pregnancy losses are greater than normal, with as many as 20% failing to deliver and requiring induction of labour with larger than normal calves resulting (Wilson *et al.*, 1995). Several aberrations of pregnancy and parturition have been reported. Failure in placental development, with fewer than normal cotyledons, has been observed, with some evidence of a mono- rather than a bilayered chorioallantoic placenta (Stice *et al.*, 1996). The prolonged gestation and large offspring syndrome has been shown to exist, but at a lower frequency for some cultured or *in vitro*-produced embryos of cattle and sheep (Thompson *et al.*, 1995; Walker *et al.*, 1996; Kruip and den Daas, 1997). Even though delivery may be induced at term, some of the calves are larger than normal (Behboodi *et al.*, 1995; Farin and Farin, 1995) and a higher frequency of postnatal deaths occurs (Kruip and den Daas, 1997). It has been reported that some calves derived from IVF or NT, even though not larger than normal initially, suffer from delivery-related metabolic problems such as hypoxia, hypoglycaemia, hypothermia and abnormalities in metabolic hormones (Avery and Greve, 1995; Garry *et al.*, 1996; Kruip and den Daas, 1997). If given excellent postnatal care, such calves survive and appear normal by 2 days after birth (Wilson *et al.*, 1995; Garry *et al.*, 1996). There are at least two schools of thought regarding the higher than normal incidence of pregnancy loss and parturition failures. One hypothesis is that faulty *in vitro* culture conditions allow formation of products of amino acid degradation, which damage the cells, causing failure of embryonic and trophoblast (preplacental) cells. Antioxidant media, which reduce oxidative damage, are available, and this school of thought also advocates elimination of serum proteins from the media. The second school hypothesizes that embryo culture, IVF and NT cause failed or inappropriate gene expression.

Nuclear transfer provides opportunities for failures of many kinds (Fulka *et al.*, 1996, 1997b, 1998), including abnormal chromatin of the donor cells (the oocyte does not provide cell-cycle checkpoint screening), inappropriate or incomplete nuclear or cytoplasmic DNA and protein reprogramming or incorporation within the new nuclear envelope, leading to failed gene expression, damage from the transfer, visualization or culture processes, and probably other cell-damaging factors, including timing of events. It is reasonable to expect multiple causes of pregnancy losses. The answers require time and research. Due to placental loss and problems of failed parturition, attempts to bring NT into use for commercial production of cattle or other domestic species have not reached application. Applications such as combinations of gene transfer and NT to produce high-value founder or speciality cattle are feasible and beginning to be used.

Use of nuclear transfer to make animals from cultured embryonic stem cells

The ability to culture embryonic cells to large numbers or to immortalize the cells could allow massive up-front production of clones, which might be selected before transfer to eliminate those destined for pregnancy loss or parturition failure.

Cultured embryonic or adult cells also provide the ability to study cell lineage, cell differentiation, cell ageing and, when genetically engineered or selected, specifically differentiated cell lines for cell transplants or tumour therapy. For the later uses requiring genetic engineering, the cells are amenable to gene transfer or deletion by use of homologous DNA recombination.

Mice were first produced from embryonic stem (ES) cells derived from the inner cell mass (ICM) of embryonic blastocysts by Evans and Kaufman (1981). Because NT has not been successful for late-stage mouse embryos, the offspring in this and later studies have been obtained by chimerization of the NT product with normal mouse embryonic cells and then selection, after a backcross or sibling mating of the mice that are true-breeding, for a genetic marker. From this early work and the pronuclear transgenics based on microinjection of DNA into the pronuclei of the one-cell zygote, more than 10,000 strains of transgenic mice have now been made, and different cell lines have been used to answer questions concerning genomic control of specific traits, cell lineage, cell differentiation, gene expression and genetic diseases in virtually every major research institution of the world. In some cases gene-deletion 'knockout mice' have been made for every sequenced gene in a few gene families, such as the transforming growth factor β (TgFβ) superfamily. Here the mouse product is used to identify the function of that particular sequence of the gene family.

Initial attempts to culture bovine ICM cells and presumed ES cells, using mouse ES technology, resulted in long-term culture and passage and demonstrated pluripotency, but not totipotency (reviewed by Anderson, 1992; Stice et al., 1994, 1996; Strelchenko and Stice, 1994; Polejaeva et al., 1995; Strelchenko et al., 1995), as tested by NT or chimerism. Offspring have been produced from bovine ICM cells cultured in microdrops for 4 weeks, with approximately four microdrop passages (Sims and First, 1994), from sheep embryonic-disc cells cultured for up to 13 passages (Campbell et al., 1996; Wells et al., 1997; Wilmut et al., 1997), from immortalized primordial germ cells of cattle (American Breeders Service, press release) and from primordial germ cells of pigs (Piedrahita et al., 1998) and transgenic chimeric pigs from cultured embryonic cells (Golueke et al., 1998), as well as chimeric calves from long- term passaged (tenth passage) bovine embryonic stem cells (Cibelli et al., 1998b).

Biologically, ES cells are totipotent cells, capable of differentiating into the precursor cells for any body cell, including functional germ cells. For animal multiplication and gene transfer, these cells must also be totipotent after extensive multiplication and passage in culture, and this may require immortalization of the cell line. Primordial germ cells are the germ cells populating the undifferentiated gonad and are derived from ES cells, one or two cell divisions earlier. The reason why ICM-derived ES cells have been totipotent only after short-term passage is probably due to increased chromatin abnormalities with advanced passages and telomere shortening. This may relate to individual laboratories and their methods of cell derivation and culture or to changes in totipotency after numerous passages. There is evidence in mice that totipotency of ES cells can be reduced after extensive passage or regained by passaged primordial germ cells for a methylation-inactive gene (Stewart et al.,

1994). It has been suggested that one change that could affect totipotency of cultured cells is a progressive demethylation of genes of cultured cells to reduce totipotency or a post-mature demethylation, resulting in return to expression of a methylation-inactivated gene, such as Xist (Stewart et al., 1994). Based primarily on mouse studies (Brandeis et al., 1993; Surani et al., 1993), both sperm and egg enter fertilization with DNA moderately methylated, although sperm are hypomethylated relative to oocytes. The DNA of cleavage-stage embryos is poorly methylated but becomes extensively methylated at late blastocyst, the time at which ES cells are derived from ICM cells (Brandeis et al., 1993; Surani et al., 1993). Cells of long-term passaged cell lines suffer chromatin damage and deletion, most often due to telomere shortening with culture age and passage. This problem is avoided by immortalizing the cell line (Counter, 1996; Bodnar et al., 1998). Theoretically, the precursor cells of the totipotent ICM cells should also be totipotent, unless their genome has not acquired a mature methylation pattern sufficient to allow totipotency.

We have cultured, passaged and made reaggregated blastocysts from cells of pooled embryos from the four-cell, eight-cell and 16-cell stages, as well as from inner morula cells and ICM cells (First et al., 1994; Mitalipova et al., 1997; M. Sims, unpublished).

As seen from the data of Table 14.2, none of the above studies with cattle or sheep ES cells are of efficiencies sufficient for commercial livestock production. The efficiencies are adequate for use in producing and multiplying high-value transgenic animals. For this purpose, the ES-cell transgenic efficiencies are higher than pronuclear DNA microinjection in cattle. A successful recent approach is the use of NT to produce ES cells from transgenic fetal fibroblast nuclei. The resulting ES cells can then be used to make offspring by chimerization into normal embryos or by NT (Cibelli et al., 1998a, b).

Cloning from differentiated cells

For several years and since testing of the Spemann hypothesis in amphibians in the 1950–1960s, scientists have considered that embryonic germline cells, such as embryonic cells, epiblast cells, primordial germ cells and possibly gonocytes, are the totipotent cell lineage and that each is capable of becoming any other cell type. Differentiated cells only occasionally show evidence of totipotency (Di Bernardino, 1997). This idea has been supported by the fact

Table 14.2. Cloning from cultured embryonic cells.

	Cell type	Number of offspring	Year	Authors
Cattle	ES*	4	1993	Sims and First
Cattle	PGC	1	1997	ABS (press release)
Sheep	ES	3	1996	Campbell et al.
Sheep	ES	4	1997	Wilmut et al.

*Not immortalized ES cell line but four- to five-passage cultured cells.
ES, embryonic stem cells; PGC, primordial germ cells.

that the totipotent cells show the presence of the receptor for the cKit oncogene, whereas differentiated cells do not, but differentiated cells, when passed through long-term exposure to oocyte cytoplasm, regain expression of the cKit receptor (Mitalipova et al., 1997). At the time of cell polarization, differentiation-committed (20–30-cell stage) outer and polarized cells of bovine embryos have a low frequency of development to morula or blastocyst stage after use in NT (7%), whereas the inner and non-polarized cells develop after NT to morula or blastocyst at a high frequency (47%) (Navara et al., 1992). The statistically significant difference was interpreted to mean that polarized or differentiation-committed cells had essentially lost totipotency.

The dogma of totipotent cells vs. differentiated cells was challenged for mammals initially when epithelial cells differentiated from ES cells produced offspring after use in NT (Campbell et al., 1996). This challenge became well known with the birth of lambs from transfer of either bovine fetal fibroblast cells, or in the case of the lamb Dolly, cultured mammary epithelial cells into enucleated sheep oocytes (Wilmut et al., 1997). To date, the production of offspring from differentiated fetal epithelial cells has been confirmed (Wells et al., 1997), as has the use of fetal fibroblasts (sheep (Schnieke et al., 1997) and cattle (Cibelli et al., 1998a, b)). The production of offspring from adult differentiated cells has now occurred for the sheep Dolly (Wilmut et al., 1997) and recently mice (Wakayama et al., 1998) and cattle (Kato et al., 1998).

Bovine oocytes, when used in the NT process, have been shown capable of reprogramming nuclei of five species so that they proceed through embryonic development after NT and on a schedule commensurate with the species of the nucleus. These interspecies NTs result in pregnancies with failed early development (Dominko et al. 1998; Mitalipova et al. 1998). Failure may in part be due to a species mismatch in mitochondria.

It is significant to note that the procedures used in successful production of offspring from differentiated cells have differed from earlier failed attempts, both in the preparation of the donor cells for cell and genomic reprogramming and in the timing and activation of the oocyte events of the reprogramming.

Much research now needs to be done to understand why and how differentiated cells can sometimes be reprogrammed to again express their entire genome. The successful production of transgenic lambs (Schnieke et al., 1997) and calves (Cibelli et al., 1998a, b) from culture-multiplied genetically engineered fetal fibroblasts and subsequent NT or embryonic-cell chimerization, as well as production of calves from fetal muscle cells (Vignon et al., 1998), indicates that methods for producing transgenic founder animals or transgenic animals for production of valuable products are available for use.

Applications of cloning in agriculture

Applications of cloning in animal agriculture or biofarming may be arranged in a hierarchy of efficiency tolerance. These applications range from the production of valuable pharmaceuticals in transgenic milk to multiplication of lesser-

Table 14.3. Cloning from fetal cells.

	Cell type	Number of offspring	Year	Transgenic	Authors
Sheep	Epithelial	2	1996	–	Campbell et al.
Sheep	Epithelial	3	1997	–	Wells et al.
Sheep	Fibroblast	3	1997	–	Wilmut et al.
Sheep	Fibroblast	6	1997	3	Schnieke et al.
Cattle	PGC	1	1997	–	ABS (press release)
Cattle	Fibroblast	3	1998a	3	Cibelli et al.
Cattle*	Fibroblast	7	1998b	6	Cibelli et al.

*Calves produced by chimerization of cells into normal embryos – chimeric calves resulted.

value transgenic animals or extremely high-producing animals for traits such as milk production. For use in improving milk production, the pregnancy loss with clones *in utero* must be no greater than normal to get farmer acceptance of the technology. Additionally, clones, while genetically identical, are not phenotypically identical. This is largely due to different *in utero*, postnatal and lactation environments. Dairy-cattle geneticists, using data from identical-twin studies, estimate that milk production of cloned offspring will be 70–74% of that of the high producing animal whose cells were used to make the clones (G. Shook, personal communication).

Developing efficient systems for cloning from adult cells of cattle has improved since the sheep, Dolly, was produced from a mammary cell line. The present hope comes from a more efficient adult cell cloning process using cumulus cells and shown by Wakayama *et al.* (1998) to result in > 30 genetically identical mouse clones. Successful application using ovarian cumulus cells in cattle (Kato *et al.*, 1998) makes possible the production of clones from the highest milk-producing cows.

Also, on the positive side, the ability to make genetically-engineered cell lines and transgenic animals from culture-multiplied and genetically-engineered fetal fibroblasts (Schnieke *et al.*, 1997, Cibelli *et al.*, 1998a, b) provides opportunities for genetically engineering founder animals for traits deemed important in animal breeding or biofarming. Application of the high-efficiency sperm-mediated gene transfer as demonstrated in chickens could make gene transfer feasible for ordinary cattle breeding and challenge our thinking for ways to engineer cattle with improved meat and milk quality and quantity or for improved, specific disease resistance or environmental adaptation. For further review of these applications see Anderson and Seidel (1998), First and Thomson (1998) and Stice *et al.* (1998).

References

Agca, Y., Monson, R.L., Northey, D.L., Abas Mazni, O., Schaefer D.M. and Rutledge, J.J. (1998) Transfer of fresh and cryopreserved IVP bovine embryos: normal calving, birth weight, and gestation lengths. *Theriogenology* 50, 14–162.

Anderson, G.B. (1992) Isolation and use of embryonic stem cells from livestock species. *Animal Biotechnology* 3, 165–175.

Anderson, G.B. and Seidel, G.E. (1998) Development biology – cloning for profit. *Science* 280, 1400–1401.

Arlotta, T., Schwartz, J.L., First, N.L. and Leibfreid-Rutledge, M.L. (1996) Follicle and oocyte stage that affect *in vitro* maturation and development of bovine oocytes. *Theriogenology* 45, 941–956.

Avery, B. and Greve, T. (1995) Development of *in vitro* matured and fertilized bovine embryos, culltured from days 1–5 post insemination in either Menezo-B2 medium or in HECM-6 medium. *Theriogenology* 44(7), 935–945.

Avery, B., Jorgensen, C.B., Madison, V. and Greve, T. (1992) Morphological development and sex of bovine *in vitro*-fertilized embryos. *Molecular Reproduction and Development* 32, 265–270.

Bachiller, D., Schellander, K., Peli, J. and Ruther, U. (1991) Liposome-mediated DNA uptake by sperm cells. *Molecular Reproduction and Development* 30, 194–200.

Ball, G.D., Leibfried, M.L., Lenz, R.W., Ax, R.L., Bavister, B.D. and First, N.L. (1983) Factors affecting successful *in vitro* fetilization of bovine follicular oocytes. *Biology of Reproduction* 28, 717–725.

Barnes, F., Collas, P., Powell, R., King, W., Westhusin, M. and Shepered, D. (1993) Influence of recipient oocyte cell cycle stage on DNA synthesis, nuclear envelope breakdown, chromosome constitution and development in nuclear transplant embryos. *Molecular Reproductive Development* 36, 33–41.

Bavister, B.D. (1990) Tests of sperm-fertilizing ability. In: Asch, R.H., Balmaceda, J.P. and Johnston, I. (eds) *Gamete Physiology*. Serono Symposia USA, Norwell, Massachusetts, pp. 77–105.

Behboodi, E., Anderson, G.B., BonDurant, R.H., Cargill, S.L., Kreuscher, B.R., Medrano, J.F. and Murray, J.D. (1995) Birth of large calves that developed from *in vitro* derived bovine embryos. *Theriogenology* 44(2), 227–232.

Bird, J.M., Powell, R., Horan, R., Cannon, F. and Houghton, J.A. (1992) The binding of exogenous DNA fragments to bovine spermatozoa. *Animal Biotechnology* 3, 181–200.

Bodnar, A.G., Ouellette, M., Frolkis, M., Holt, S.E., Chiu, C.P., Morin, G.B., Harley, C.B., Shay, J.W. and Wright, W.E. (1998) Extension of lifespan by introduction of telomerase in normal human cells. *Science* 279, 349–352.

Brackett, B.G., Bousquet, D., Boice, M.L., Donawick, W.J., Evans, J.F. and Dressel, M.A. (1982) *Biology of Reproduction* 27, 147–158.

Brandeis, M., Ariel, M. and Cedar, H. (1993) Dynamics of DNA methylation during development. *BioEssays* 15, 11709–11713.

Bredbacka, K. and Bredbacka, P. (1996) Glucose controls sex-related growth rate differences of bovine embryos produced *in vitro*. *Journal of Reproduction and Fertility* 106, 169–172.

Brem, G. and Muller, M. (1994) Large transgenic animals. In: Maclean, N. (ed.) *Animals with Novel Genes*. Cambridge University Press, Cambridge, England, pp. 179–244.

Bremel, R.D. (1996) Potential role of transgenesis in dairy production and related areas. *Theriogenology* 45, 51–56.

Brinster, R.L. and Zimmerman, J.W. (1994) Spermatogenesis following male germ-cell transplantation. *Proceedings of the National Academy of Sciences of the USA* 91, 11298–11302.

Brinster, R.L., Chen, H.Y., Trurnabauer, M., Senear, A.W., Warren, R. and Palmiter, R.D. (1981) Somatic expression of herpes thymidine kinase in mice following injection of a fusion gene into eggs. *Cell* 27, 223–231.

Brinster, R.L., Sandgren, E.P., Behringer, R.R. and Palmiter, R.D. (1989) No simple solution for making transgenic mice. *Cell* 59, 239–241.

Burns, J.C., Friedmann, T., Driever, W., Burrascano, M. and Yee, J.K. (1993) Vesicular stomatitis virus G glycoprotein pseudotyped retroviral vectors: concentration to very high titer and efficient gene transfer into mammalian and nonmammalian cells. *Proceedings of the National Academy of Science USA* 90, 8033–8037.

Campbell, K.H., Ritchie, W.A. and Wilmut, I. (1993) Nuclear-cytoplasmic interactions during the first cell cycle of the nuclear transfer reconstructed bovine embryos: implications for deoxyribonucleic acid replication and development. *Biology of Reproduction* 49, 933–942.

Campbell, K.H.S., McWir, J., Ritchie, W.A. and Wilmut, I. (1996) Sheep cloned by nuclear transfer from a cultured cell line. *Nature* 380, 64–66.

Chan, A.W.S., Homan, E.J., Ballou, L.U., Burns, J.C. and Bremel, R.D. (1998) Transgenic cattle produced by reverse transcribed gene transfer in oocytes. *Proceedings of the National Academy of Sciences of the USA*, 95, 14028–14033.

Cibelli, J.B., Stice, S.L., Golueke, P.J., Kane, J.J., Jerry, J., Blackwell, C., Ponce de Leon, A.F. and Robl, J.M. (1998a) Cloned transgenic calves produced from nonquiescent fetal fibroblasts. *Science* 280, 1256–1258.

Cibelli, J.B., Stice, S.L., Golueke, P.J., Kane, J.J., Jerry, J., Blackwell, C., Ponce de Leon, A.F. and Robl, J.M. (1998b) Transgenic bovine chimeric offspring produced from somatic cell-derived stem-like cells. *Nature Biotechnology* 16(7), 642–646.

Collas, P. and Poccia, D. (1998) Remodelling the sperm nucleus into a male pronucleus at fertilization. *Theriogenology* 49, 67–81.

Colman, A. (1996) Production of proteins in the milk of transgenic livestock: problems, solutions and successes. *American Journal of Clinical Nutrition* 63, 639S–645S.

Costantini, F. and Lacy, E. (1981) Introduction of a rabbit globin gene into mouse germline. *Nature* 294, 92–94.

Counter, C.M. (1996) The role of telomeres and telomerase in cell lifespan. *Mutation Research – Reviews in Genetic Toxicology* 366, 45–63.

Critser, E.S., Leibfried-Rutledge, M.L., Eyestone, W.H., Northey, D.L. and First, N.L. (1986) Acquisition of developmental competence during maturation *in vitro*. *Theriogenology* 25, 150.

Di Bernardino, M. (1997) *Genomic Potential of Differentiated Cells*. Columbia University Press, New York, 385 pp.

Dominko, T. and First, N.L. (1997) Timing of meiotic progression in bovine oocytes and its effect on early embryo development. *Molecular Reproduction and Development* 47, 456–467.

Dominko, T., Mitalipova, M., Haley, B., Beyhan, Z., Memili, E. and First, N.L. (1998) Bovine oocytes as a universal recipient cytoplasm in mammalian nuclear transfer. *Theriogenology* 46, 385.

Evans, M.J. and Kaufman, M.H. (1981) Establishment in culture of pluripotential cells from mouse embryos. *Nature* 292, 154–156.

Eyestone, W.H. (1994) Challenges and progress in the production of transgenic cattle. *Reproduction Fertility and Development* 6, 647–652.

Eyestone, W.H. and First, N.L. (1989) Co-culture of early cattle embryos to the blastocyst stage with oviductal tissue or in conditioned medium. *Journal of Reproduction and Fertility* 30, 493–497.

Farin, P.W. and Farin, C.E. (1995) Transfer of bovine embryos produced *in vivo* or *in vitro*: survival and fetal development. *Biology of Reproduction* 52(3), 676–682.

First, N.L. and Thomson, J.A. (1998) From cows stem therapies? *Nature Biotechnology*, 16(7), 620–621.

First, N.L., Leibfried-Rutledge, M.L., Northey, D.L. and Nuttleman, P.R. (1992) Use of *in vitro* matured oocytes 24 hr. of age in bovine nuclear transfer. Paper presented to the IETS Annual Meeting, Denver, Colorado.

First, N.L., Sims, M.M., Park, S.P. and Kent-First, M.J. (1994) Systems for production of calves from cultured bovine embryonic cells. *Journal of Reproduction, Fertility and Development* Suppl. 6, 553–562.

Francolini, M., Lavitrano, M., Lamia, L.L., French, D., Frati, L., Cotelli, F. and Spadafora, C. (1993) Evidence for nuclear internalization of exogenous DNA into mammalian sperm cells. *Molecular Reproductive Development* 34, 133–139.

Fukui, Y., Glew, A.M., Gandolfi, F. and Moor, R.M. (1988) Ram-specific effects on *in vitro* fertilization and cleavage of sheep oocytes matured *in vitro*. *Journal of Reproduction and Fertility* 82, 337–340.

Fulka, J., Jr, First, N.L. and Moor, R.M. (1996) Nuclear transplantation in mammal remodeling of transplanted nuclei under the influence of maturation promoting factor. *BioEssays* 18, 835–840.

Fulka, J., Jr, First, N. L., Lee, C. and Moore, R.M. (1997a) Induction of DNA replication in germinal vesicles and in nuclei formed in maturing mouse oocytes by 6-DMAP treatment. *Zygote* 5, 213–217.

Fulka, J., Jr, Kalab, P., First, N.L. and Moor, R.M. (1997b) Damaged chromatin does not prevent the exit from metaphase I in fused mouse oocytes. *Human Reproduction* 12(11), 2473–2476.

Fulka, J., Jr, First, N.L. and Moor, R.M. (1998) Nuclear and cytoplasmic determinants involved in the regulation of mammalian oocyte maturation. *Molecular Human Reproduction* 4, 41–49.

Gardner, D.K. (1998) Changes in requirements and utilization of nutrients during mammalian preimplantation embryo development and their significance in embryo culture. *Theriogenology* 49, 83–102.

Gardner, D.K., Lane, M., Kouridakis, K. and Schoolcraft, W.B. (1997) Complex physiologically based serum-free culture media increase mammalian embryo development. In: *In Vitro Fertilization and Assisted Reproduction*, pp. 187–191.

Garry, F.B., Adams, R., McCann, J.P. and Odde, K.G. (1996) Postnatal characteristics of calves produced by nuclear transfer cloning. *Theriogenology* 45(1), 141–152.

Golueke, P.G., Cibelli J.B., Balise, J., Morris, J., Kane, J.J., Ponce de Leon, F.A., Jerry, J., Robl, M.J. and Stice, S.L. (1998) Transgenic chimeric pigs derived from cultured embryonic cell lines. *Theriogenology* 49, 238.

Gordon, I. (1994) *Laboratory Production of Cattle Embryos*. CAB International, Wallingford.

Gordon, J.W. and Ruddle, F.H. (1981) Integration and stable germ line transmission of genes injected into mouse pronuclei. *Science* 214, 1244–1246.

Gordon, J.W., Scangos, G.A., Plotkins, D.J., Barbosa, J.A. and Ruddle, F.H. (1980) Genetic transplantation of mouse embryos by microinjection of purified DNA. *Proceedings of the National Academy of Sciences of the USA* 77, 7380–7384.

Grobet, L., Poncelet, D., Royo, L.J., Brouwers, B., Pirottin, D., Michaux, C., Menissier, F., Zanotti, M., Dunner, S. and Georges, M. (1998) Molecular definition of an allelic series of mutations disrupting the myostatin function and causing double muscling in cattle. *Mammalian Genome* 9(3), 210–213.

Gutierrez, A., Cush W.T., Anderson, G.B. and Medrano, J.F. (1997) Ovine-specific Y-chromosome RAPD-SCAR marker for embryo sexing. *Animal Genetics* 28, 135–138.

Hanada, A., Enya, U. and Suzuki, T. (1986) Birth of calves by nonsurgical transfer of *in vitro* fertilized embryos obtained from oocytes matured *in vitro*. *Japanese Journal of Animal Reproduction* 32, 208.

Haskell, R.E. and Bowen, R.A. (1995) Efficient production of transgenic cattle by retroviral infection of early embryos. *Molecular Reproduction and Development* 40, 386–390.

Hasler, J.F., Henderson, W.B., Hurtgen, P.J., Jin, Z.Q., McCauley, A.D., Mower, S.A., Neely, B., Shuey, L.S., Stokes, E. and Trimmer, S.A. (1995) Production, freezing and transfer of bovine IVF embryos and subsequent calving results. *Theriogenology* 32, 727–734.

Hillery, F.L., Parrish, J.J. and First, N.L. (1990) Theriogenology 39, 232 (abstract).

Hyttinen, J.M., Peura, T., Tolvanen, M., Aalto, J. and Janne, J. (1996) Detection of microinjected genes in bovine preimplantation embryos with combined DNA digestion and polymerase chain reaction. *Molecular Reproduction and Development* 43, 150–157.

Illmensee, K. and Hoppe, P.C. (1981) Nuclear transplantation in *Mus musculus*: developmental potential of nuclei from preimplantation embryos. *Cell* 23, 9–18.

Iritani, A. and Niwa, K. (1977) Capacitation of bull spermatozoa and fertilization *in vitro* of cattle follicular oocytes matured in culture. *Journal of Reproduction and Fertilization* 41, 119–121.

Jaenisch, R. and Mintz, B. (1974) Simian virus 40 sequences in DNA of healthy adult mice derived from preimplantation blastocysts injected with viral DNA. *Proceedings of the National Academy of Sciences of the USA* 71, 1250–1254.

Johnson, L.A. (1994) Isolation of X- and Y-bearing sperm for sex preselection. *Oxford Reviews of Reproductive Biology* 16, 303.

Johnson, L.A. (1996) Gender preselection in mammals: an overview. *Deutsche Tierarztliche Wachenschrift* 103, 288–291.

Johnson, L.A. and Schulman, J.D. (1994) The safety of sperm selection by flow-cytometry. *Human Reproduction* 9, 757–759.

Johnson, L.A., Cran, D.G. and Polge, C. (1994) Recent advances in sex preselection of cattle-flow cytometric sorting of X-chromosome and Y-chromosome bearing sperm based on DNA to produce progeny. *Theriogenology* 41, 51–56.

Joyner, A.L. (1991) Gene targeting and gene trap screens using embryonic stem cells; new approaches to mammalian development. *BioEssays* 13, 649–658.

Kato, Y., Tani, T., Sotomaru, Y., Kurokawa, K., Kato, J.Y., Doguchi, H., Yasue, H. and Tsunoda, Y. (1998) Eight calves cloned from somatic cells of a single adult. *Science* 282, 2095–2098.

Kim, J.H., Jung-Ha, H.S., Lee, H.T. and Chung, K.S. (1997) Development of a positive method for male stem-cell-mediated gene transfer in mouse and pig. *Molecular Reproduction and Development* 46, 1–12.

Kim, T., Leibfried-Rutledge, M.L. and First, N.L. (1993) Gene transfer in bovine blastocysts using replication-defective retroviral vectors packaged with Gibbon ape leukemia virus envelope. *Molecular Reproduction and Development* 35, 105–113.

Kirkpatrick, B.W. and Monson, R.L. (1993) Sensitive sex determination assay applicable to bovine embryos derived from IVM and IVF. *Journal of Reproduction and Fertilization* 98, 335.

Koller, B.H., Hagemann, L.J., Doetschman, T., Hagerman, J.R., Huang, S., Williams, P.J., First, N.L., Meada, N. and Smithies, O. (1989) Germ-line transmission of a planned alteration made in a hypoxanthine phosphoribosyltransferase gene by homologous recombination in embryonic stem cells. *Proceedings of the National Academy of Sciences of the USA* 86, 8927–8931.

Kruip, T.A.M. and den Daas, J.H.G. (1997) In vitro produced and cloned embryos: effects on pregnancy, parturition and offspring. *Theriogenology* 47(1), 43–52.

Kruip, T.A.M., Pieterse, M.C., van Beneden, T.H., Vos, P.L.A.M., Wurth, Y.A. and Taverne, M.A.M. (1991) A new method for bovine embryo production – a potential alternative to superovulation. *Veterinary Record* 128, 208–210.

Kruip, T.A.M., Boni, R., Wurth, Y.A., Roelofsen, M.W.M. and Pieterse, M.C. (1994) Potential use of ovum pick-up for embryo production and breeding in cattle. *Theriogenology* 42, 675–684.

Lavitrano, M., Camaioni, A., Fraiti, V.M., Dolci, S., Farace, M.G. and Spadafora, C. (1989) Sperm cells as vectors for introducing foreign DNA into eggs: genetic transformation of mice. *Cell* 57, 717–723.

Lavitrano, M., French, D., Zani, M., Frati, L. and Spadafora, C. (1992) The interaction between exogenous DNA and sperm cells. *Molecular Reproduction and Development* 31, 161–169.

Leese, H.J. (1991) Metabolism of the preimplantation mammalian embryo. *Oxford Reviews of Reproductive Biology* 13, 35–76.

Leibfried-Rutledge, M.L. (1996) Gene expression during early embryonic development. *Biennial Symposium on Reproduction. Journal of Animal Science.*

Leibfried-Rutledge, M.L., Northey, D.L., Nuttleman, P.R. and First, N.L. (1992) Processing of donated nucleus and timing of post-activation events differ between recipient oocytes 24 or 42 hr age. Paper presented to IETS Annual Meeting, Denver, Colorado.

Leibfried-Rutledge, M.L., Dominko, T., Critser, E.S. and Critser, J.K. (1997) Tissue maturation *in vivo* and *in vitro*: gamete and early embryo ontogeny. In: Critser, J.K. (ed.) *Reproductive Tissue Banking.* Academic Press, New York, pp. 23–138.

Leibfried-Rutledge, M.L., Critser, E.S., Parrish, J.J. and First, N.L. (1989) In vitro maturation and fertilization of bovine oocytes. *Theriogenology* 31, 61–74.

Lenz, R.W., Ball, G.D., Leibfried, M.L., Ax, R.L. and First, N.L. (1983) *In vitro* maturation and fertilization of bovine oocytes are temperature-dependent processes. *Biology of Reproduction* 29, 173–179.

Looney, C.R., Lindsay, B.R., Gonseth, C.L. and Johnson, D.L. (1994) Commercial aspects of oocyte retrieval and *in vitro* fertilization (IVF) for embryo production in problem cows. *Theriogenology* 41, 67–72.

McGrath, J. and Solter, D. (1983) Nuclear transplantation in the mouse embryo by microsurgery and cell fusion. *Science* 200, 1300–1320.

Mcpherron, A.C. and Lee, S.J. (1997) Double muscling in cattle due to mutations in the myostatin gene. *Proceedings of the National Academy of Sciences of the USA* 94(23), 12457–12461.

Melton, D.W. (1994) *BioEssays* 16, 633–638.

Mitalipova, M., Chan, A. and First, N. (1997) Pluripotency of bovine embryonic stem cells derived from precompacting bovine embryos. In: *Transgenic Animals in Agriculture,* p. 55.

Mitalipova, M., Dominko, T., Haley, B., Beyhan, Z., Memili, E. and First, N. (1998) Timing and polarization in bovine embryos and developmental potential of polarized blastomeres. *Biology of Reproduction* 46(Suppl. 1), 82 (abstract).

Navara, C.S., Sims, M.M. and First, N.L. (1992) Timing of polarization in bovine embryos and developmental potential of polarized blastomeres. *Biology of Reproduction* 46 (Suppl.1).

Navara C.S., First N.L. and Schatten, G. (1994) Microtubule organization in the cow during fertilization, polyspermy, pathogenesis, and nuclear transfer: the role of the sperm aster. *Developmental Biology* 162(1), 29–40.

Palmiter, R.D., Brinster, R.L., Hammer, R.E., Trumbaur, M.E., Rosenfeld, M.G., Bimberg, N.C. and Evans, R.M. (1982) Dramatic growth of mice that develop from eggs microinjected with metallothionein-growth hormone fusion genes. *Nature* 300, 611–615.

Palmiter, R.D., Norstedt, G., Gelinas, R.E., Hammer, R.E. and Brinster, R.L. (1983) Metallothionein-human GH fusion genes stimulate growth of mice. *Science* 222, 809–814.

Parrish, J.J. amd First, N.L. (1993) Fertilization. In: King, G. (ed.) *Reproduction in Domesticated Animals*. Elsevier Press, New York, pp. 195–228.

Parrish, J.J., Susko-Parrish, J.L., Leibfried-Rutledge, M.L., Crister, E.S., Eyestone, W.H. and First, N.L. (1986) Bovine *in vitro* fertilization with frozen thawed semen. *Theriogenology* 25, 591–600.

Parrish, J.J., Susko-Parrish, J., Winer, M.A. and First, N.L. (1988) Capacitation of bovine sperm by heparin. *Biology of Reproduction* 38, 1171–1180.

Parrish, J.J., Susko-Parrish, J.L. and First, N.L. (1989) Capacitation of bovine sperm by heparin – inhibitory effect of glucose and role of intracellular pH. *Biology of Reproduction* 41, 683–699.

Piedrahita, J.A., Moore, K., Oetama, B., Lee, C.K., Scales, N., Ramsoondar, J., Bazer, F.W. and Ott, T. (1998) Generation of transgenic porcine chimeras using primordial germ cell-derived colonies. *Biology of Reproduction* 58, 321–1329.

Pieterse, M.C., Vos P.L.A.M., Kruip, T.A.M., Wurth, Y.A., van Beneden, T.H., Willemse, A.H. and Taverne, M.A.M. (1991) Transvaginal ultrasound guided follicular aspiration of bovine oocytes. *Theriogenology* 35, 19–24.

Poccia D. and Collas P. (1997) Nuclear envelope dynamics during male pronuclear development. *Development of Growth Differentiation* 39, 541–550.

Polejaeva, I.A., White, K.L., Ellis, L.C. and Reed, W.A. (1995) Isolation and long term culture of mink and bovine embryonic stem cells. *Theriogenology* 43, 300.

Prather, R.S., Barnes, F.L., Sims, M.M., Robl, J.M., Eyestone, W.H. and First, N.L. (1987) Nuclear transplantation in the bovine embryo: assessment of donor nuclei and recipient oocyte. *Biology of Reproduction* 37, 859–866.

Reubinoff, B.E. and Schenker, J.G. (1996) New advances in sex preselection. *Fertility and Sterility* 66, 343–350.

Rexroad, C.E.J. (1992) Transgenic technology in animal agriculture. *Animal Biotechnology* 3, 1–13.

Rose, T.A. and Bavister, B.D. (1992) Effect of oocyte maturation medium on *in vitro* development of *in vitro* fertilized bovine embryos. *Molecular Reproduction Development* 31, 72–77.

Rosenkrans, C.F., Jr and First, N.L. (1994) Effect of free amino acids and vitamins on cleavage and development rate of bovine zygotes *in vitro*. *Journal of Animal Science* 72, 434–437.

Rucker, E.B. and Piedrahita, J.A. (1997) CRE-mediated recombination at the murine whey acidic protein (mWAP) locus. *Molecular Reproduction and Development* 48(3), 324–331.

Schnieke, A.E., Kind, A.J., Ritchie, W.A., Mycock, K., Scott, A.R., Ritchie, M., Wilmut, I., Colman, A. and Campbell, K.H.S. (1997) Human factor IX transgenic sheep produced by transfer of nuclei from transfected fetal fibroblasts. *Science* 278, 2130–2133.

Seidel, G.E., Jr, Johnson, L.A., Allen, C.A. and Welch, G.R. (1996) Artifical insemination with X- and Y-bearing bovine sperm. *Theriogenology* 45, 309.

Shamsuddin, M., Niwa, K., Larsson, B. and Rodriguezmartinez, H. (1996) *In vitro* maturation and fertilization of bovine oocytes. *Reproduction in Domestic Animals* 31(4–5), 613–622.

Sims, M. and First, N.L. (1994) Production of calves by transfer of nuclei from cultured inner cell mall cells. *Proceedings of the National Academy of Sciences of the USA* 91, 6143–6147.

Sirard, M.A., Richard, F. and Mayes, M. (1998) Controlling meiotic resumption in bovine oocytes. *Theriogenology* 49, 483–498.

Smith, L.C. and Wilmut, I. (1989) Influence of nuclear and cytoplasmic activity on the development *in vivo* of sheep embryos. *Biology of Reproduction* 40, 1027–1035.

Stewart, C.L., Gadi, I. and Bhat, H. (1994) Stem cells from primordial germ cells can re-enter the germ line. *Developmental Biology* 161, 626–628.

Stice, S.L. and Keefer, C.L. (1993) Multiple generational bovine embryo cloning. *Biology of Reproduction* 48, 715–719.

Stice, S.L., Keefer, C.L. and Matthews, L. (1994) Bovine nuclear transfer embryos: oocyte activation prior to blastomere fusion. *Molecular Reproduction and Development* 38, 61–68.

Stice, S.L., Strelchenko, N., Keefer, C.L. and Mattews, L. (1996) Pluripotent bovine embryonic cell lines direct embryonic development following nuclear transfer. *Biology of Reproduction* 54, 100–110.

Stice, S.L., Robl, J.M., Ponce de Leon, F.A., Jerry, J., Golueke, P.G., Cibelli, J.B. and Kane, J.J. (1998) Cloning: new breakthroughs leading to commercial opportunities. *Theriogenology* 49, 129–138.

Strelchenko, N.S. and Stice, S.L. (1994) Bovine embryonic pluripotent cell lines derived from morula stage embryos. *Theriogenology* 41, 304.

Strelchenko, N.S., Mitalipova, M.M. and Stice, S. (1995) Further characterization of bovine pluripotent stem cells. *Theriogenology* 43, 327.

Surani, M.A., Sasaki, H., Ferguson-Smith, A.C., Allen, N.D., Barton, S.C., Jones, P.A. and Reik, W. (1993) *Philosophical Transactions of the Royal Society, London, B* 339, 65–172.

Susko-Parrish, J.L., Leibfried-Rutledge, M.L., Northey, D.L., Schutzkus, V. and First, N.L. (1994) Inhibition of protein kinases after an induced calcium transient causes transition of bovine oocytes to embryonic cycles, without meiotic completion. *Developmental Biology* 166, 729–739.

Takahashi, Y. and First, N.L. (1992) *In vitro* development of bovine one-cell embryos: influence of glucose, lactate, pyruvate, amino acids and vitamins. *Theriogenology* 37, 963–978.

Temin, H.M. (1987) Retrovirus vectors for gene transfer. Efficient integration into and expression of exogenous DNA in vertebrate cell genomes. In: Kuchelapti, R. (ed.) *Gene Transfer*. Plenum, New York, pp. 149–187.

Tervit, H.R., Whittingham, D.G. and Rowson, L.E.A. (1972) Successful culture *in vitro* of sheep and cattle ova. *Journal of Reproductive Fertility* 30, 493–497.

Thompson, J.G., Gardner, D.K., Pugh, P.A., McMillan, W.H. and Tervit, H.R. (1995) Lamb birth-weight is affected by culture system utilized during *in-vitro* pre-elongation development of ovine embryo. *Biology of Reproduction*, 53(6), 1385–1391.

van Wagtendonk-de Leeuw, A.M., Aerts, B.J.D. and den Daas, J.H.G. (1998) Abnormal offspring following *in vitro* production of bovine preimplantation embryos: a field study. *Theriogenology* 49, 883–894.

Varmus, H. and Brown, P. (1989) Retroviruses. In: Berg, D.E. and Howe, M.M. (eds) *Mobile DNA*. American Society for Microbiology, Washington, DC, 53–108.

Vignon, X., Chesñe, P., LeBourhis, D., Heyman, Y. and Renard, J. (1998) Developmental potential of bovine embryos reconstructed with somatic nuclei from cultured skin and muscle fetal cells. *Theriogenology* 49, 392.

von Zglinicki, T. and Saretzki, G. (1997) The molecular biology of proliferative senescence. *Zeitschrift für Gerontologie und Geriatrie* 30, 2428.

Wagner, E.F., Stewart, T.A. and Mintz, B. (1981) The human β-globin gene and a functional viral thymidine kinase gene in developing mice. *Proceedings of the National Academy of Sciences of the USA* 78, 5016–5020.

Wagner, T.E., Hoppe, P.C., Jollick, J.D., Scholl, D.R., Hodinka, R. and Gault, J.B. (1981) Micro-injection of a rabbit β-globin gene into zygotes and subsequent expression in adult mice and their offspring. *Proceedings of the National Academy of Sciences of the USA* 78, 6376–6380.

Wakayama, T., Perry, A.C.F., Zuccotti, M., Johnson, K.R. and Yanagimachi, R. (1998) Full-term development of mice from enucleated oocytes injected with cumulus cell nuclei. *Nature* 394, 369–374.

Walker, S.K., Hartwich, K.M. and Seamark, R.F. (1996) The production of unusually large offspring following embryo manipulation: concepts and challenges. *Theriogenology* 45, 111–120.

Wall, R.J. (1996) Transgenic livestock: progress and prospects for the future. *Theriogenology* 45, 57–68.

Wall, R.J. and Seidel, G.E.J. (1992) Transgenic farm animals – a critical analysis. *Theriogenology* 38, 337–357.

Ware, C.B., Bames, F.L., Maiki-Laurilla, M. and First N.L. (1989) Age dependence of bovine oocyte activation. *Gamete Research* 22, 265–275.

Weima, S. and Mummery, C. (1995) *Molecular Reproductive Fertility* 40, 444–454.

Wells, D., Misica, P., Day, T. and Tervit, R. (1997) Production of cloned lambs from an established embryonic cell line: a comparison between *in vivo* and *in vitro* matured cytoplasts. *Biology of Reproduction* 57, 385–393.

Willadsen, S.M. (1986) Nuclear transplantation in sheep embryos. *Nature* 320, 63–65.

Wilmut, I., Schinieko, A.E., Mowhir, J., Kind, A.J. and Campbell, K.H.S. (1997) Viable offspring derived from fetal and adult mammalian cells. *Nature* 395, 810–813.

Wilson, J.M., Williams, J.D., Bondioli, K.R., Looney, C.R., Westhusin, M.E. and McCalla, D.F. (1995) Comparison of birth weight and growth characteristics of bovine calves produced by nuclear transfer (cloning), embryo transfer and natural mating. *Animal Reproductive Sciences* 38, 73–83.

Yang, X., Kubota, C., Suzuki, H., Taneja, M., Bois, P.E.J. and Presice, G.A. (1998) Control of oocyte maturation in cows – biological factors. *Theriogenology* 49, 471–482.

Yong, Z. and Yuqiang, L. (1998) Nuclear–cytoplasmic interaction and development of goat embryos reconstructed by nuclear transplantation: production of goats by serially cloning embryos. *Biology of Reproduction* 58, 266–269.

Zani, M., Lavitrano, M.L., French, D., Lulli, V., Maione, B., Sperandio, S. and Spadafora, C. (1995) The mechanisms of binding exogenous DNA to sperm cells: factors controlling the DNA uptake. *Experimental Cell Research* 217, 57–64.

Ziomek, C.A. (1998) Commercialization of proteins produced in the mammary glands. *Theriogenology* 49, 39–144.

Developmental Genetics 15

A. Ruvinsky[1] and L.J. Spicer[2]

[1]*Animal Science, SRSNR, University of New England, Armidale, NSW 2351, Australia;* [2]*Department of Animal Science, Oklahoma State University, Stillwater, OK 74078, USA*

Introduction	438
Developmental Stages of the Cattle Embryo	438
Genetic Control of Cleavage and Blastocyst Formation	440
Expression of maternal genes	440
Genome activation	441
Embryonic gene expression	442
Trophoblast gene expression	444
Gametic imprinting	445
Implantation and Maternal Recognition of Pregnancy	446
Implantation	446
Cytokines in maternal recognition of pregnancy	447
Placental gene expression	448
Genes Involved in Control of Morphogenesis	449
Gastrulation	449
Notochord formation	450
Hox genes and the development of axial identity	451
Organogenesis: T-box, *Pax*, and other genes	452
Muscle development and gene regulation	453
Developmental effects of coat-colour mutations	455
Sex Determination and Differentiation	456
The major steps in gonad differentiation	456
The *SRY* gene	457
The cycle of the X chromosome	458
Defects of sex determination as a result of interactions with other genes	458
Freemartinism in Cattle	458
Phenomenology	458
Specific features of the bovine placenta	459
XX/XY chimerism	459
Diagnosis of freemartinism and practical aspects	460

Totipotency and Cloning	460
Acknowledgement	461
References	461

Introduction

Recent progress in developmental genetics has been dramatic. This includes extensive studies of mammalian development. However, the vast majority of data concerning mammalian development were generated by murine genetics. This means that, in discussing the genetic aspects of development in cattle, we cannot avoid significant gaps and have to refer to other species, mainly to the mouse.

Despite the high level of similarity in mammalian development, there are numerous contrasts between species, resulting from their morphological differences, placental structure, longevity and schedule of development. These distinctive features of development are certainly based on genetic differences, many of which are still awaiting clarification. Thus, the main objective of this chapter is to collect and pull together all available and relevant data concerning genetic determination of cattle development, and to supplement these data with information from other species in order to achieve better coverage of the topic. It is our hope that future reviews in this area will be able to promote significantly knowledge in developmental genetics of cattle using this initial set of data.

Developmental Stages of the Cattle Embryo

Gamete maturation and fertilization, which comprise the first and very fateful part of each new developmental cycle in mammals, have been considered in the previous chapters. The embryological events and their genetic determination which follow after fertilization are discussed here. Table 15.1 summarizes essential events and the timing of conceptus and fetal development in cattle. Accordingly, three consecutive periods are emphasized: ovum, embryonic and fetal.

The ovum (or preimplantation) period covers the first 15–16 days after fertilization and results in a mature blastocyst ready to implant. This period is characterized by several crucial events, including cleavage, morula formation and compaction, cavitation and blastocyst development. The latter includes hatching (release) from the zona pellucida, blastocyst elongation, gastrulation, somite formation and trophoblast development. The ovum stage in cattle and sheep is longer than in some studied mammalian species (Calarco and McLaren, 1976; Ménézo and Renard, 1993). For the first 5 days, seven cycles of cell division take place and the morula appears. Several hours later, compaction of the morula brings the developing embryo to the next stage – the blastocyst. Tight intercellular junctions develop and these provide a condition

Table 15.1. Essential events and timing of cattle prenatal development (compiled from: Winters et al., 1942; Cruz and Pedersen, 1991; Bazer et al., 1993; Guillomot et al., 1993; Jainudeen and Hafez, 1993; Ménézo and Renard, 1993).

Stage of development	Days after fertilization	Cells/stage
Ovum period	0–16	
Cleavage		
Two-cell	1	2
Eight-cell	2–3	8
Genome activation	3–4	8–16
Entry into the uterus	4	16–32
Morula compaction	5–6	64–128
Blastocyst formation	6–7	128
Hatching	9–10	~200
Blastocyst elongation	11–16	~1000
Trophoblast differentiation	12–18	Thousands
Gastrulation	14–16	Thousands
Notochord	15–16	Thousands
Open neural tube	15–17	Thousands
Differentiation of the first somite	16–17	Thousands
Embryonic period	16–45	
Maternal recognition of pregnancy	16–19	Early somites
Implantation	18–20	Early somites
Head fold	19–20	Somites
Initial placentation	22	Somites
Closed neural tube	22–23	18–19 somites
Optic and otic vesicles, heart	22–23	18–19 somites
Gastrointestinal structures and mesonephros	26	Later somites
Visible forelimb and hindlimb buds	27–28	Vertebrae develop
Fetal period	46–parturition	
Eyelids close	54–59	Growing fetus
More developed form of head and neck	55–60	Growing fetus
Further development of limbs	60–70	Growing fetus
Hair follicles	90 and later	Growing fetus
Horn pits	100	Growing fetus
Hair coat	230	Growing fetus
Birth	276–290 (depending on the breed)	

for the accumulation of fluid within the central cavity (the blastocoele). The majority of cells in the blastocyst, called trophoblast or trophoectoderm, create a layer, which later becomes the chorion and has important trophic functions. The small group of cells located at one pole beneath the trophoblast forms the embryoblast or inner cell mass (ICM). Later, during gastrulation, the ICM differentiates into the three primary germ layers of the embryo (ectoderm, mesoderm and endoderm). During the same time, starting from day 12 of gestation, the trophoblast develops significantly. Further development of the conceptus assembles the essential conditions for implantation. These include the development of the embryo *per se* (notochord and neural tube formation, differentiation of the first somite, etc.), significant development of extra-embryonic tissues and development of maternal recognition of pregnancy.

Starting from this point (about 15–16 days after fertilization), the bovine embryo is entering into a new period of its development – the embryonic stages. The important features of this stage are embryo attachment to the uterine wall and further development of extraembryonic structures and the placenta. This attachment in the cow is superficial and non-invasive (Winters *et al.*, 1942; Bazer *et al.*, 1993). Details are discussed later in this chapter in relation to freemartinism in cattle. During this period, the embryo develops all of its main organ systems and the shape of the embryo changes dramatically. Essential morphogenetic events, such as head, vertebrae and appendage formation, and the development of the nervous system, blood circulation and all other major internal organs occur during this time. By day 45 of gestation, the embryo develops well-recognizable features of the species.

The fetal period continues significantly longer than the two previous periods and covers the last 230–240 days of gestation, when the bovine fetus undergoes extensive growth and final development. Numerous morphological changes, although definite, are not radical. These changes are rather gradual and shape all fetal structures and functions towards requirements for postnatal life.

A comparison of developmental events and gene regulation during embryogenesis in cattle with other farm mammals can be fruitful (Cockett, 1997; Pomp and Geisert, 1998).

Genetic Control of Cleavage and Blastocyst Formation

Expression of maternal genes

The first three cleavage divisions in cattle occur mostly without activation of the embryo genome (Ménézo and Renard, 1993). This stage is covered by information, energy and structural molecules, mainly accumulated during oogenesis. The total ribonucleic acid (RNA) content in the zygote and in early blastomeres in mammals is commonly much higher than in somatic cells. The oocyte and the following early stages of development have the ability to perform polypeptide syntheses in the absence of active transcription. Experiments with a specific inhibitor of RNA polymerase II (α-amanitin) show that cleavage and probably polypeptide synthesis are not significantly affected until the four- to eight-cell embryo, but when added in later stages it completely inhibits further embryonic development (Barnes and First, 1991; Liu and Foote, 1997). Synthesis of heterogeneous RNA (hnRNA) is absent or very slight until the eight-cell embryo (Plante *et al.*, 1994; Viuff *et al.*, 1996; Lavoir *et al.*, 1997). Somatic histone H1, which supposedly regulates critical aspects of chromatin activity during early embryogenesis, assembles on embryonic chromatin during the fourth to sixth cell cycle after fertilization (Smith *et al.*, 1995).

Protein synthesis is not well pronounced until the 16-cell stage in bovine embryos, and embryonic cell ultrastructure supports biochemical observations. Nevertheless, some protein processing evidently does occur (Lavoir *et al.*, 1997). During this period, cellular mass decreases about 20%; however, nuclei increase in size (McLaren, 1974; Bazer *et al.*, 1993).

It is well known that in *Drosophila melanogaster* and *Caenorhabditis elegans* gradients of morphogens in the zygote and early embryo are crucial for establishing positional information (St Johnston and Nüsslein-Volhard, 1992; Nüsslein-Volhard, 1996). These gradients are essentially products of the expression of maternal genes. In what degree similar gradients and elements of cytoskeleton are important during the earliest stage of mammalian development remain to be seen, but it seems unlikely that these factors are not essential, at least in specifying major polarities (Holliday, 1990). Increasing cell polarity was described at the eight-cell stage of mouse and rat development (Reeve, 1981; Gueth-Hallonet and Maro, 1992). Cell fate, controlled by positional information, seems reversible and provides the developing embryo with a certain degree of flexibility. In cattle, cellular polarization occurred in some blastomeres at the nine- to 15-cell stage, but typical distinct polarity was not manifested until after the 16-cell stage, with approximately 40% polar cells per embryo (Koyama *et al.*, 1994).

There are data indicating a low gene expression in the mouse embryo during the first divisions (Davis and Schultz, 1997), but it is rather likely that the replication of deoxyribonucleic acid (DNA) and maintenance of the majority of cellular functions during the first two to three divisions are provided by RNAs and proteins accumulated during oocyte maturation. Oviduct proteins are also involved in this early stage of development. Oestrus-associated glycoprotein (EGP), for instance, is involved in modulation of the cleavage rate (Nancarrow and Hill, 1995).

Genome activation

The latest data suggest that the change from maternal to embryonic control starts as early as the two-cell stage in cattle (Memili *et al.*, 1998). Changes in nucleolus organizer regions (Ag-NORs) and in nucleolus ultrastructure occurring around the eight- to 16-cell stage in bovine embryos suggest a significant transcriptional activation of the ribosomal RNA (rRNA) genes and hnRNA production (King *et al.*; 1988, Lavoir *et al.*, 1997). This is clearly an important step in the activation of protein synthesis machinery. Cytogenetic investigations have also shown that on day 5 the developmental rates were slowest in haploid and polyploid embryos, intermediate in aneuploid embryos and fastest in mixoploid and diploid embryos (Kawarsky *et al.*, 1996). These data indicate active involvement of the embryonic genome at this stage, as well as the importance of diploid balance. In the mouse and probably in other mammalian species, the vast majority of more or less serious deviations from the standard chromosome set are not viable (Dyban and Baranov, 1988).

Physiological studies clearly show significant changes in the metabolic activity of the bovine embryos during early stages of development (Thompson *et al.*, 1996). For instance, adenosine triphosphate (ATP) production and oxygen consumption increase with compaction and blastulation. These data revealed that bovine embryos were dependent on oxidative phosphorylation

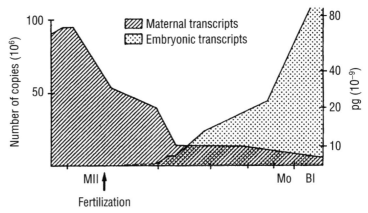

Fig. 15.1. Evolution of polyA+ RNA during the preimplantation development of the mouse embryo (from Ménézo and Renard, 1993, with permission of Ellipses). MII, second meiotic metaphase; Mo, morula; Bl, blastula.

for energy production at all stages of pre-elongation development, with a shift in dependence towards glycolysis in conjunction with compaction. Figure 15.1 shows evolution of the polyA RNA profile during preimplantation development in the mouse embryo. A similar picture is expected for bovine embryos, except with a different time-scale. Expression of various developmentally regulated markers provides valuable information concerning early gene expression in bovine development. It was shown that lamin B appeared as a constitutive component of nuclei at all preimplantation stages but lamins A/C had a stage-related distribution (Shehu *et al.*, 1996). The nuclei from the early cleavage stages contained lamins A/C, which generally disappeared later, with a few possible exceptions in the morula and blastocyst. Several other proteins essential for morphogenetic events appear in developing bovine embryos. This includes several cytoskeletal and cytoskeleton-related components, such as F-actin, α-catenin and E-cadherin. These proteins appear starting from day 6 and their appearance or polarized distribution is relevant to morula compaction (Shehu *et al.*, 1996). Pig data suggest that several more morphogenetically important proteins appear during early cleavage and compaction, such as actin and actin-associated proteins, α-fodrin, vinculin and E-cadherin (Reima *et al.*, 1993). This is a result of activation of the corresponding genes and coincides well with early morphogenetic events. In the pig, these molecules are distributed evenly in blastomeres during early cleavage but are then gradually accumulated in regions of intercellular contacts towards the blastocyst stage (Reima *et al.*, 1993).

Embryonic gene expression

The blastocyst formation creates two totally different cell lineages: non-polarized inner mass cells and polarized trophectoderm or outer cells, having microvilli. It was shown that in mice this process is accompanied by specific redistribution of the actin-associated protein ezrin, which has been proposed to play a role in the formation of microvillous structures (Louvet et al., 1996). These microvilli play an early role in implantation (see section on implantation in this chapter). Before morula compaction, ezrin is located around the cell cortex. However, later, after blastocyst formation, it segregates to outer trophectoderm cells, with microvilli. Two phosphorylated forms of ezrin are present from the ovum period throughout preimplantation development, but the amount continuously decreases. A third non-tyrosine-phosphorylated isoform appears at the eight-cell stage and increases to blastocyst stage (Louvet et al., 1996); the activation of a specific gene would be a reason for this. Several other actin-associated proteins mentioned above (α-fodrin, vinculin and E-cadherin) are involved in cytokeratin bundles, which are not observed until the early blastocyst in both the mouse and the pig (Reima et al., 1993). It was found that the E-cadherin cell-adhesion function is essential in the establishment and maintenance of epithelial-cell morphology during embryogenesis and adulthood. Mouse embryos homozygous for a targeted mutation show severe abnormalities before implantation, because dissociation of adhesive cells of the morula occurs shortly after compaction and their morphological polarization is then destroyed. Maternal E-cadherin is able to initiate compaction, but cannot maintain the process (Riethmacher et al., 1995). Significant defects in the cell-junctional and cytoskeletal organization were found in E-cahedrin-negative mouse embryos (Oshugi et al., 1997).

In spite of the differences found in the mouse and the pig in the timing of events, in both species a close correlation between developmental stage and the organization of the cytoskeleton was observed. In the expanded bovine blastocyst, the distribution of cytoskeletal and cytoskeleton-related proteins looks similar (Shehu et al., 1996). Extracellular fibronectin was first detected in the early blastocyst before differentiation of the primitive endoderm, and at this stage it was localized at the interface between the trophectoderm and extraembryonic endoderm (Shehu et al., 1996). A connection between switching on particular genes, accumulation of proteins and morphogenetic events is visible. Cingulin, the tight-junction peripheral-membrane protein, also contributes to morphological differentiation in early development, and is found in mice. It is likely that other mammals have the same gene. Its synthesis is tissue-specific in blastocysts and is upregulated in the trophectoderm and downregulated in the ICM (Javed et al., 1993).

It is commonly accepted now that proto-oncogenes are deeply involved in numerous processes of embryonic development determining nuclear transcription factors, intracellular signal transducers, growth factors and growth-factor receptors. For instance, activation of the *c-fos* and *c-jun* proto-oncogenes in sheep conceptuses occurs during the period of rapid growth and elongation (Wu, 1996). A similar pattern could be typical for bovine embryos in the time around implantation. These proto-oncogenes are involved in the regulation of gene expression, cell proliferation and differentiation.

To estimate the dynamics of developmental activity of some housekeeping genes in *in vitro* bovine embryos, the following gene products were studied: two mitochondrial transcripts, 12S rRNA and cytochrome b messenger RNA (mRNA); two RNAs involved in the processing of other RNAs, U2 and U3 snRNA; and two nuclear-derived transcripts, β-actin mRNA and histone H3 mRNA (Bilodeau-Goeseels and Schultz, 1997). The RNA levels for the various genes studied remained constant or decreased slightly from the mature oocyte to the six- to eight-cell or morula stage and greatly increased in blastocysts. Increases in gene expression were significant, ranging from two- to sixfold to 110–118-fold.

It was found that at least some regulatory substances secreted by the uterus can act as growth factors. Together with a number of growth factors and their receptors produced by the embryo, they create the medium essential for development. A detailed review of these regulators of mammalian embryonic development is published elsewhere (Schultz and Heyner, 1993). In early bovine embryos, transcripts for insulin-like growth factors (IGF-I, IGF-II) and mRNAs encoding receptors for insulin were detectable at all embryo stages, including the blastocyst. It is suggested that these transcripts are products of both the maternal and embryonic genomes (Schultz *et al.*, 1992). Genes for insulin-like growth factors (*IGF*s), transforming growth factors (*TGF*s), fibroblast growth factor (*FGF*), platelet-derived growth factor (*PDGF*) and also receptors for insulin, *IGF*, *PDGF*, *TGF* α and epidermal growth factor (*EGF*) are expressed by early bovine embryos (Harvey *et al.*, 1995). Harvey and co-workers (1995) reported that successful development of the blastocyst is dependent on the action of EGF and leukaemia inhibitory factor (LIF). The latter may act directly or indirectly, by inducing the expression of other cytokines, to regulate the temporal and spatial production of proteases and protease inhibitors, and to create conditions for implantation at least in the mouse (Harvey *et al.*, 1995).

Many efforts have been contributed toward answering the question whether development of preimplantation embryos depends on internal and/or external factors. Most experiments with rodents show that it is unlikely that preimplantation development is significantly dependent on exogenous factors. Furthermore, none of the known endogenously produced factors and their receptors are essential during preimplantation development to the blastocyst stage (Stewart and Cullinan, 1997). However, later during development, the importance of growth factors rises sharply.

Trophoblast gene expression

Differentiation of trophoblast cells, the first and perhaps the most radical event during mammalian embryonic development, provides an embryonic component for the future fetal–maternal interface during implantation and placentation. A detailed description of current knowledge about the genetic control of trophoblast development and implantation is presented in a special issue of *Developmental Genetics* (Schultz and Edwards, 1997). Clearly, many features of these processes are common for the majority of eutherian mammals and applicable to cattle. At the moment, 44 loci with different functions implicated in preimplantation or peri-implantation events have been identified (Rinkenberger *et al.*, 1997).

A basic helix–loop–helix (bHLH) transcription factor gene, *Hxt*, is expressed in the early trophoblast and in differentiated giant cells of mouse embryos (Cross *et al.*, 1995). The negative HLH regulator, *Id-1*, inhibited rat trophoblast (Rcho-1) differentiation and placental lactogen-I transcription. These data demonstrate a role for HLH factors in regulating trophoblast development, at least in rodents, and indicate a positive role for *Hxt* in promoting the formation of trophoblast giant cells. A separate gene, *Hed*, encodes a related protein, which is expressed in maternal decidium surrounding the implantation site (Cross *et al.*, 1995). One of them is *Mash-2*, a locus homologous to the *Drosophila achaete/scute* complex genes, and this determines the transcription factor. Its expression begins during preimplantation development, but is restricted to trophoblast lineage after the blastocyst stage (Nakayama *et al.*, 1997). This murine locus belongs to the quite rare category of imprinted genes (Guillemot *et al.*, 1995). Mouse embryos that inherit a mutant allele from the mother and normal from the father die after implantation. The cause of death is a lack of placental spongiotrophoblast (McLaughlin *et al.*, 1996). The *MMp9* gene, which is involved in development of giant trophoblast cells in mice, is one more candidate for imprinted genes (Newman-Smith and Werb, 1997). Several more genes are currently under investigation to determine their role in trophoblast development and implantation/placentation events (Schultz and Edwards, 1997).

Genetic determination of integrin trafficking, which regulates adhesion to fibronectin during the differentiation of the mouse peri-implantation blastocyst, has been studied by Schultz *et al.* (1997). The regulation of several metalloproteinase and corresponding genes may also shed additional light on the process of implantation and further trophoblast development (Bass *et al.*, 1997; Das *et al.*, 1997). Because placental development in rodents is very different from that in cattle, further research will be needed to determine if these murine trophoblast genes are relevant to bovine trophoblast development.

Gametic imprinting

Gametic or genomic imprinting is a developmental phenomenon based on differential expression of maternal and paternal alleles in some genes typical for eutherian mammals. Until now, the main bulk of information regarding imprinting comes from the mouse and, to a lesser extent, from humans (Barlow, 1995). The number of imprinted genes known for the mouse is about 20. The genes are essential for regulation of embryonic and placental growth. These include paternally expressed genes – *Igf2*, *Snrpn*, *Ins*, *Znf127* – and maternally expressed genes – *Wt1*, *H19*, *Igf2r*. Imprint acquisition is believed to occur before fertilization and imprint propagation takes place until the morula–blastocyst stage (Ruvinsky and Agulnik, 1990; Shemer et al., 1996). The molecular mechanisms of gametic imprinting are still under intensive investigation. It seems likely, however, that primary gametic signals are not simply copied from the gametes, but rather that a methylation pattern typical for imprinted genes is gradually established during early development (Shemer et al., 1996; Surani, 1998). The developmental function of gametic imprinting is also not absolutely clear, but an explanation proposed by Moore and Haig (1991) is widely accepted. It is based on the idea of genetic conflict arising during pregnancy between maternally and paternally inherited genes. Thus, it is likely that gametic imprinting evolved in mammals to regulate intrauterine growth and to increase safeness of embryonic development. Data on farm animals are still limited and include some indirect and direct evidence of gametic imprinting (Ruvinsky, 1999). In sheep the callipyge gene responsible for pronounced muscle hypertrophy in hindquarters is controlled by a novel form of imprinting, referred to as polar overdominance (Cockett, 1997). The latest experiments confirmed the existence of gametic imprinting in ruminants (Feil et al., 1998).

Lack of maternally or paternally derived alleles in a zygote causes embryonic mortality in several instances and should impose strict requirements on the stability of imprinting signals. A recent report of cloning using somatic cells of adult sheep (Wilmut et al., 1997) is the first indication that differential imprinting signals are maintained in somatic cells long after intrauterine development and are very stable.

It seems possible that knowledge about the influence of the pathway (paternal or maternal) used by an allele to enter the next generation will sooner rather than later be adopted by selection programmes. Selection of modifier genes may significantly change the effect of gametic imprinting and this knowledge should also be taken into consideration.

Implantation and Maternal Recognition of Pregnancy

Implantation

Placental attachment in cattle initially (between week 2 and 3 of pregnancy; see Table 15.1) involves minute papillae of embryonic membranes, which penetrate the vestibule of the uterine glands. These papillae disappear before day 30 of pregnancy (Guillomot and Guay, 1982) and are replaced by interdigitating microvillus connections of the placental allantochorion to maternal caruncular crypts of the uterine endometrium (King et al., 1979). These placenta microvilli grow and become vascularized, as do the associated caruncular areas, which leads to the formation of placentomes, which take on a convex characteristic in cattle. Also, no syncytia form in the intercaruncular epithelium, but abundant giant cells are present during the fourth week of pregnancy and decrease in numbers thereafter (King et al., 1981). Thus implantation leads eventually to formation of extraembryonic membranes and placenta, which are vital for normal embryonic growth and development. Until recently, knowledge about genes involved in implantation was limited and this is still true for cattle. However, in the mouse significant progress has been achieved (Rinkenberger *et al.*, 1997). At least 40 different genes were identified, whose activity is necessary for normal development. A lack of any of them causes peri-implantation or placentation failure. The proposed or known functions of these genes are very diverse, including transcription factors, hormone receptors, cytokines and receptors, DNA-serving enzymes, adhesion molecules, etc. This breakthrough opens opportunities for similar investigations in cattle.

Cytokines in maternal recognition of pregnancy

In order to maintain a successful pregnancy in domestic mammals, the process of luteolysis has to be inhibited, as well as the immune response from the mother to foreign tissue in the uterus (i.e. the embryo). The conceptus is thought to be responsible for this signalling during its first few weeks of life (Roberts *et al.*, 1992; Bazer *et al.*, 1994; Thatcher *et al.*, 1995; Godkin *et al.*, 1997; Martal *et al.*, 1997). Conceptus signalling may be accomplished through the synthesis and secretion of an interferon (IFN)-like polypeptide known as trophoblast protein 1 (TP-1; Roberts *et al.*, 1992; Bazer *et al.*, 1994; Martal *et al.*, 1997), now termed IFN-τ. This protein has been identified in sheep (oTP-1), bovine (bTP-1) and caprine conceptuses (Stewart *et al.*, 1989; Geisert *et al.*, 1992; Bazer *et al.*, 1994). The conceptus's IFN-like protein, which is similar to bovine IFN-αII, has antiviral activity, and is effective against many viruses. Because of their structural resemblance, oTP-1 and IFN-αII compete for the same receptor (Stewart *et al.*, 1989; Roberts *et al.*, 1992). Numerous other cytokines, including interleukin (IL)-1, 2, 4, 6 and 10, colony-stimulating factor and tumour necrosis factor α (TNFα), have been implicated in mediating communications between the conceptus and maternal uterine epithelium, both before implantation and as the placenta develops (Robertson *et al.*, 1994).

Interferon τ prevents luteolysis through the inhibition of endometrial prostaglandin F2α (PGF2α) release in response to oestrogen and oxytocin triggering, and by suppressing inositol phosphate/diacylglycerol second-messenger pathways for PGF2α synthesis (Roberts *et al.*, 1992; Bazer *et al.*, 1994; Godkin *et al.*, 1997). Oestradiol and oxytocin are necessary mediators for the release of PGF2α (Geisert *et al.*, 1992; Bazer *et al.*, 1994; Godkin *et al.*, 1997) required for the initiation of luteolysis. Non-pregnant ewes that have been exposed to uterine infusion of oTP-1 between days 12 and 18 of an oestrous cycle have extended interoestrous intervals through maintenance of the corpus luteum (Vallet *et al.*, 1988). During maternal recognition of pregnancy, ovarian follicular populations are altered in cattle (Thatcher *et al.*, 1989; Spicer and Geisert, 1992). Thus, cytokines may play a role in the suppression of follicular development during this early stage of pregnancy by reducing the secretion of oestradiol, which could otherwise stimulate uterine secretion of PGF2α (Spicer and Geisert, 1992; King and Thatcher, 1993). In support of these suggestions, exogenous IFN-α reduced progesterone concentrations in the blood of heifers (Newton *et al.*, 1990) and *in vitro* studies showed similar negative effects of IFNs on the steroidogenesis of bovine luteal cells (Fairchild and Pate, 1991) and bovine granulosa cells (Alpizar and Spicer, 1994). Furthermore, TNFα but not IL-1β inhibited luteinizing hormone (LH)-stimulated progesterone production by bovine luteal cells (Nothnick and Pate, 1990; Fairchild and Pate, 1991; Benyo and Pate, 1992), and IFN receptors have been identified in ovine corpora lutea (Godkin *et al.*, 1984). Collectively, these results indicate that cytokines have inhibitory effects on follicular and luteal steroidogenesis in cattle. Although TNFα is localized in bovine corpora lutea (Roby and Terranova, 1989) and receptors for TNFα are present in bovine granulosa and thecal cells (Spicer, 1998), as well as in porcine corpora lutea (Richards and Almond, 1994), the presence of factors that regulate TNFα receptors during luteal development in cattle has not been identified to date. Whether these and other cytokines are involved in the maternal recognition of pregnancy remains to be elucidated.

Placental gene expression

As mentioned earlier in the section 'Genetic Control of Cleavage and Blastocyst Formation', the limited evidence in bovine embryos indicates that the embryonic genome is not activated until the eight- to 16-cell stage (Camous *et al.*, 1986; Kopecny *et al.*, 1989; Kopecny and Fakan, 1992). Initiation of embryonic transcription during the eight- to 16-cell stage was further confirmed by studies using bovine embryos produced from *in vitro* matured and fertilized oocytes (Barnes and First, 1991; Rieger *et al.*, 1992). Recent evidence indicates that transcriptional regulation of a small number of specific genes in bovine embryos may occur as early as the four-cell stage (Edwards *et al.*, 1997). Specific genes expressed during preimplantation development were also described earlier (see the section 'Genetic Control of Cleavage and Blastocyst Formation'). Cytokine-related genes, such as bTP-1, are expressed as early as day 18 in bovine conceptuses (Cross and Roberts, 1991), whereas colony-stimulating factor receptor gene expression begins around day 29 in post-attachment bovine trophoblasts (Beauchamp and Croy, 1991).

Bovine placentae express numerous proteins and corresponding mRNAs during development, including prolactin-related protein I, II, III and IV (Schuler and Hurley, 1987; Kessler *et al.*, 1989; Milosavljevic *et al.*, 1989; Yamakawa *et al.*, 1990; Zieler *et al.*, 1990; Tanaka *et al.*, 1991; Anthony *et al.*, 1995; Kessler and Schuler, 1997), placental lactogen (Schuler *et al.*, 1988), prolactin receptor (Schuler *et al.*, 1997), aromatase (Hinshelwood *et al.*, 1995; Fürbass *et al.*, 1997), steroidogenic acute regulatory protein (StAR; Pescador *et al.*, 1996; Pilon *et al.*, 1997), procollagen III (Shang *et al.*, 1997) and an aspartic proteinase, bovine pregnancy-associated glycoprotein 1 (bPAG1; Xie *et al.*,

1994, 1995). The prolactin-related proteins are expressed predominantly in full-term bovine placentae (Yamakawa et al., 1990; Tanaka et al., 1991) and their genes have been assigned to bovine chromosome 21 (Dietz et al., 1992), whereas procollagen III gene expression is detected in the bovine chorioallantois by day 20 of gestation and increases through day 36 (Shang et al., 1997). The latter observation suggests that procollagen III may be involved in the development of the allantois. The temporal pattern of placental expression of placental lactogen, prolactin receptor, aromatase, StAR and bPAG1 has not been characterized in cattle to date.

Genes Involved in Control of Morphogenesis

Gastrulation

The development of the bovine epiblast starting from gastrulation and onwards requires further investigations. The data are very limited, particularly with respect to gene expression activity. It is known that bovine blastocysts hatch at 9–10 days (1.6 mm) and enlarge rapidly thereafter: 3.75 mm on days 11–12, about 10 mm on days 12–14. By day 16, the blastocyst has transformed into an attenuated, bilaminar blastodermic vesicle.

Gastrulation starts about this time, i.e. approximately 3–4 days before implantation (Cruz and Pedersen, 1991). Cell proliferation and rearrangement in the germinal disc are the main events during gastrulation in eutherian embryos. The most obvious feature of commencing gastrulation is the formation of the primitive streak. 'This process begins with the production and proliferation of mesodermal progenitor cells at the proximal (allantoic) end of the primitive streak; this position marks the future caudal end of the fetus. As ectodermal cells migrate through the primitive streak, they move both laterally and distally towards future cranial end of the embryo, extending the primitive streak towards the distal lip' (Wilkins, 1993).

The genes that are responsible for gastrulation in mammals are mainly unknown. The same is true for establishing anterior–posterior orientation. Two genes, the homoeobox gene *goosecoid* (*gsc*) and the winged-helix gene *Hepatic Nuclear Factor-3beta* (*HNF-3beta*) are co-expressed in all three germ layers in the anterior primitive streak and at the rostral end of mouse embryos during gastrulation (Filosa et al., 1997). A member of the *FGF* family, *Ggf-4*, shows expression restriction to the primitive streak and expresses sequentially in developmental pathways such as mesoderm formation and myogenesis, playing a role in specific epithelial–mesenchymal interactions (Niswander and Martin, 1992).

The recently discovered murine *Axin* gene seems to be a crucial regulator in embryonic axis formation in vertebrates. This gene inhibits the so-called Wnt signalling pathway, arranged from several polypeptides and enzymes (Zeng et al., 1997).

The *T* gene, which is required for extension of the posterior axis (Clements *et al.*, 1996) and for several other essential steps in mammalian development, including notochord formation, is discussed below. In the mouse and other mammals, the next step of development is the so-called 'head process', which gives rise to the notochord and contributes to part of the endodermal lining of the gut.

Notochord formation

The notochord is a rod-shaped structure that extends along the embryo and represents the initial axial skeleton, playing an important role in the induction of the neural plate, chondrogenesis and somite formation (Gomercic *et al.*, 1991). Notochord development in the bovine embryo commences 15–16 days after fertilization (a couple of days before implantation), which is different from the mouse and human. It was found that the bovine notochord begins decomposing at the end of the embryonal and the start of the fetal period (45–50 days). After 55–65 days, it no longer presents an unbroken cord of notochord cells. The activity of several oxidative mitochondrial enzymes increases significantly from the early stage of notochord development (Gomercic *et al.*, 1991). Clearly, an activation of some nuclear genes responsible for basic morphogenetic rearrangements is required for notochord formation and development. The *T* gene, which was first described as *Brachyury* mutation in mice 70 years ago, is a very important participant in events required for differentiation of the notochord and the formation of mesoderm during posterior development. The T protein is located in cell nuclei and acts as a tissue-specific transcription factor (Kispert *et al.*, 1995). The cloning and sequencing of the *T* gene led to the discovery of the T-box gene family, which is characterized by a conserved sequence, called the T-box (Bollag *et al.*, 1994). This ancient family of transcription factors underwent duplication around 400 million years ago and is common to vertebrates (Ruvinsky and Silver, 1997). There are indications that several mouse T-box genes play an important role in different mesodermal subpopulations and one in early endoderm during gastrulation (Papaioannou, 1997).

Formation of the notochord leads to several very important events, including induction of the neural tube and development of the gut, heart and brain. A putative morphogen secreted by the floor plate and notochord, sonic hedgehog (Shh), specifies the fate of multiple cell types in the ventral aspect of the vertebrate nervous system. In turn, Shh induces expression of oncogene *Gli-1*, which affects the later development of the dorsal midbrain and hindbrain (Hynes *et al.*, 1997).

Somitogenesis proceeds soon after this (16–17 days) in the opposite, posterior, direction. At this stage (18–20 days), the bovine embryo begins implanting into the uterine wall (see the section 'Implantation and Maternal Recognition of Pregnancy').

Hox *genes and development of axial identity*

The homoeotic genes were first described in *Drosophila* as the primary determinants of segment identity, and they encode transcription factors. These genes share a very similar 180-bp DNA sequence, named the homoeobox. Comparative analysis of the *Drosophila* homoeotic gene complex, called *HOM-C*, and the mammalian homoeobox genes, called the *Hox* complex, showed a very striking case of evolutionary conservation. The *Hox* gene family determines a set of transcription factors crucial for development of axial identity in a very wide range of animal species (Maconochie *et al.*, 1996). Figure 15.2 shows the surprising similarity and collinearity found in the molecular anatomy of the insect and mammalian (vertebrate) complexes. The main difference is the number of the complexes per genome. In insects there is only one, while mammals and other vertebrates have four paralogous sets of genes.

There are no direct data about the structure and function of the bovine *Hox* complex, but it is very likely that the main features of the murine *Hox* complex are typical for it. The *Hox* genes are expressed in segmental fashion in developing somites and the central nervous system. Each *Hox* gene acts from a particular anterior limit in the posterior direction. The anterior and posterior limits are different for different *Hox* genes (Fig. 15.2). The genes located at the 3′ end have the most anterior limit of activity. The transcription of the genes, however, moves in the usual 5′ to 3′ direction. The genes located at the 3′ end express earlier and genes located at the 5′ end work later. The process of segmentation moves along the anterior–posterior (A–P) axis. There are differences in the development of segmentation between the hindbrain and the trunk (Maconochie *et al.*, 1996). Thus, the vertebrate body is at least partially a result of interactions of *Hox* genes, which provide cells with the essential positional and functional information they require to migrate to an appropriate destination and generate the necessary structures. Retinoids can affect the expression of *Hox* genes and there is a 5′ to 3′ gradient in responsiveness of *Hox* genes to retinoids, based on the response elements of some *Hox* genes (Marshall *et al.*, 1996).

A key role of the neural crest as the source for numerous cell lineages, including sensory neurons, glia cells, melanocytes, some bone and cartilage cells, thyroid cells and smooth muscle, is well known (Le Douarin, 1982). Progress during the past few years has been significant in the identification of the genes controlling the development of the neural crest and cell migration (Anderson, 1997). Several growth factors affect the developmental fate of neural-crest cells: the glial growth factor (GGF), TGF-β (promoting smooth-muscle differentiation) and BMP2/4 (involved in bone morphogenesis). Transcription factors are also important in neural-crest lineage determination, among them bHLH transcription factors. The genes *Mash1* and *Mash2* are responsible for the production of these factors (Anderson, 1997).

Organogenesis: T-box, Pax and other genes

Some of T-box gene family are involved in limb morphogenesis and specification of forelimb/hindlimb identity. It has been shown that *Tbx5* and *Tbx4* expression is primarily restricted to developing the fore- and hindlimb buds,

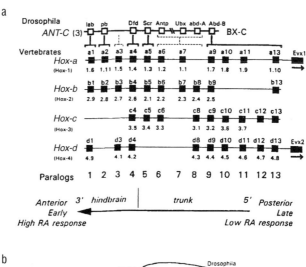

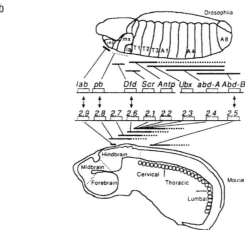

Fig. 15.2. (a) Alignment of the four mouse *Hox* complexes with that of *HOM-C* from *Drosophila*. The vertical shaded boxes indicate related genes. The 13 paralogous groups are noted at the bottom of the alignment. The colinear properties of the *Hox* complexes with respect to timing of expression, anteroposterior level and retinoic acid (RA) response are also noted at the bottom. (From Maconochie et al., 1996, with the authors' permission.) (b) Summary of *HOM-C* and *Hox-2* expression patterns. The upper part of the figure is a diagram of a 10-h *Drosophila* embryo with projections of expression patterns of different genes from *HOM-C* complex to particular body segments. The lower part of the figure is a diagram of a 12-day mouse embryo with projections of expression patterns of different genes from the *Hox-2* complex to particular body segments. (From McGinnis and Krumlauf, 1992; with the authors' permission).

respectively. These two genes have probably been divergently selected in vertebrate evolution to play a role in the differential specification of fore- (pectoral) versus hind- (pelvic) limb identity (Gibson-Brown et al., 1996). Mutations in human TBX3 cause the ulnar–mammary syndrome, characterized by posterior limb deficiencies or duplications, mammary gland dysfunction and genital abnormalities. It was suggested that *TBX3* and *TBX5* evolved from a common ancestral gene and each has acquired specific, yet complementary, roles in patterning the mammalian upper limb (Bamshad et al., 1997).

Pax genes are another family of developmental genes, coding nuclear transcription factors. They contain the paired domain, a conserved amino acid motif with DNA-binding activity. *Pax* genes are key regulators in organogenesis of some tissues and organs, including the kidney, eye, ear, nose, limb muscles, vertebral column and brain. Vertebrate *Pax* genes are involved in pattern formation, possibly by determining the time and place of organ initiation or morphogenesis (Dahl et al., 1997). *Pax-1*, for instance, is a mediator of notochord signals during the dorsoventral specification of vertebrae (Koseki et al., 1993). *Pax-3* may mediate the activation of *MyoD* and *Myf-5*, the myogenic regulatory factors, in response to muscle-inducing signals from either axial tissues or overlying ectoderm, and may act as a regulator of somitic myogenesis (Maroto et al., 1997). Mutations in *Pax-6* result in eye malformation, known as Aniridia in humans and small eye syndrome in mice, demonstrating involvement of this gene in eye formation; the renal coloboma syndrome is a result of mutation in *Pax-2* (Dahl et al., 1997). Another eyes-absent gene (*Eya2*) is involved in eye development in several metazoan phyla and should also play an important role in cattle. Like the *Pax-6* gene family, *Eya2* was probably recruited for visual system formation significantly later than its occurrence (Duncan et al., 1997). Several more genes, such as *Bmp-4*, *Msx-1* and *Msx-2*, coding bone-morphogenetic proteins, which are expressed before neural-tube closure and later on, interact with *Pax-2* and *Pax-3* (Monsoro-Burq et al., 1996).

The study of the dynamics of gene expression and regulation in the bovine photoreceptor system, including rhodopsin, arrestin, the rod alpha-subunit of transduction, interphotoreceptor retinoid-binding protein and the rod alpha-subunit of cyclic guanosine monophosphate (cGMP) phosphodiesterase (Timmers et al., 1993), as well as opsin (DesJardin et al., 1995), shows a complicated picture of gene activity changes and interactions during the final maturation of the system.

Muscle development and gene regulation

The development of muscle tissue is of particular interest for cattle, as this can provide important information from a practical point of view. For instance, it was shown that fetal muscle development differs in cattle that have different postnatal growth patterns by as early as 100 days of gestation, and this is a muscle hyperplasia-dependent phenomenon (Gore et al., 1994). The most recent progress on genetic mechanisms of muscle development was reviewed by Firulli and Olson (1997). Skeletal, cardiac and smooth-muscle cells express overlapping sets of muscle-specific genes; however, there are genes expressed in one particular muscle type. So-called modules or independent *cis*-regulatory regions are required to direct the complete developmental pattern of expression of individual muscle-specific genes, even within a single muscle-cell type. The temporospatial specificity of these myogenic regulatory modules is established by unique combinations of transcription factors (Firulli and Olson, 1997).

At least two generations of cells during fetal development in cattle were observed. One appears at a very early stage and gives rise to adult fast type I fibres. A second generation of cells gives rise to adult fast type IIA and IIB fibres and to type IIC. The beginning of myogenesis is characterized by expression of transitory myosin forms that are not found in adult cattle (Picard *et al.*, 1994). Insulin-like growth factor II modulates myogenesis and, between 60 and 162 days of gestation, the majority of IGF-II was localized to developing muscle cells. Probably, IGF-II acts as an autocrine-acting growth factor during myogenesis (Listrat *et al.*, 1994). Bovine cathepsin B, a lysosomal cysteine proteinase, is involved in fetal muscle development and is encoded by two different transcripts resulting from alternative polyadenylation. These two mRNAs declined similarly from 80 to 250 days of fetal age (Bechet *et al.*, 1996). The bovine fetal skeletal muscle myosin heavy-chain gene expression is influenced by genotype and indirectly by external factors (Gore *et al.*, 1995).

Molecular and developmental studies of double-muscling in cattle caused by the recessive mutation *mh*, located at chromosome 2, shed additional light on skeletal-muscle development (Charlier *et al.*, 1995; Grobet *et al.*, 1997). It was shown that an 11-bp deletion in the coding sequence of the myostatin gene, which belongs to the *TGF*-beta superfamily, caused the muscular hypertrophy (double-muscled phenotype) in Belgian Blue (Grobet *et al.*, 1997; Kambadur *et al.*, 1997). In Piedmontese cattle, a G–A transition in the same region of the gene is responsible for the phenotype (Kambadur *et al.*, 1997). It is likely that myostatin is a negative regulator of muscle growth in cattle and other mammals. It was also found that differentiation of the muscle fibres occurs at a slower rate in double-muscled fetuses, particularly during the first two-thirds of gestation (Picard *et al.*, 1995). Experiments where one fetus was double-muscled and the co-twin was normal show that genetically double-muscled bovine fetuses do not develop their muscles as might be expected. These data strongly support the idea that blood-borne factors regulate muscle hypertrophy in fetal cattle (Gerrard *et al.*, 1995).

A description of bovine fetal growth and development from the 90th to the 255th day of gestation is presented elsewhere (Prior and Laster, 1979). This includes numerous data concerning the dynamics of fetal weight, ash, protein, RNA and DNA content and other parameters. A more detailed study of fetal and placental growth in Herefords and crossbreds has been conducted by Anthony *et al.* (1986).

Developmental effects of coat-colour mutations

Classical coat-colour genetics in mammals has recently acquired a developmental and molecular orientation (Jackson, 1994). Most data were obtained within mouse genetics, but the high homology of mammalian genomes provides a sufficient foundation for extension to cattle (for more information, see Chapter 3).

There is a group of mutations that affect melanocyte differentiation, proliferation and migration and are responsible for development of a different kind of white spots. It was shown that neural-crest melanoblast migration is dependent on a signal encoded by *Steel* and a receptor encoded by the *W* locus in mice. A tyrosine-kinase receptor protein is a product of the proto-oncogene *Kit* (*W* locus), and its ligand, encoded by the *Steel* locus, is a membrane-bound protein. Mutations in one or both loci cause white spots and numerous neurological, haematological and fertility effects, which are in many cases lethal (Fleischman, 1993). The *roan* locus in cattle was mapped to the same interval on chromosome 5 as the candidate gene *Steel* (Charlier *et al.*, 1996). The dominant *R* allele plays the critical role in the determination of white heifer disease, which stems from its pleiotropic effect on fertility. Details are considered below in the paragraph devoted to sex differentiation.

Mutations that affect melanocyte morphology and create dilute colours are common in mammals. It was shown that mutations in a myosin protein (Jackson, 1994), which may be caused by proviral insertion (Jenkins *et al.*, 1981), lead to lack of dendrites by melanocytes and diluted coat colour. Interestingly, neuronal dendrites are not affected by these mutations.

Several other mutations affect melanogenic enzymes and related proteins. Tyrosinase gene mutations lead to albino variants. Brown colours, at least in mice, are the product of mutations in locus coding tyrosinase-related protein (Jackson, 1994). Pleiotropic effects, including decreased viability, are known.

Mutations of two other coat-colour loci, *agouti* and *extension*, affect the regulation of melanogenesis. It was shown that the ratio between black eumelanin and yellow phaeomelanin is regulated by the α-melanocyte-stimulating hormone (αMSH). The product of the *agouti* gene works as an antagonist of αMSH and the *extension* gene encodes the αMSH-receptor. Again, there are several developmental effects of these loci, including obesity and tumour formation, which attract further attention. Several more examples demonstrating connections between coat colour and adaptations in cattle are presented in Chapter 22. Black, brown and red coat colour in cattle are postulated to be produced by the E- and A-loci allele interactions (Adalsteinsson *et al.*, 1995) and corresponding mutations in the MSH receptor were observed (Klungland *et al.*, 1995). Red colour in Holstein cattle is indeed associated with a deletion in the MSHR gene (Joerg *et al.*, 1996).

Thus, molecular analysis of coat-colour mutations provides an excellent opportunity for a better understanding of some basic developmental processes.

Sex Determination and Differentiation

The major steps in gonad differentiation

It was known long ago that in mammals the sex determination mechanism is based on the presence or absence of the Y chromosome. Embryos without the Y chromosome develop as females and those with the Y chromosome develop as males. A recent breakthrough in the molecular understanding of sex determination and differentiation in the mouse and human (Goodfellow and Lovell-Badge, 1993) paved the way for other mammals, including cattle.

In humans and mice, gonadal differentiation starts relatively late in embryonic development and morphological differences in XY embryos appear prior to those in XX embryos. A similar situation is probably true for cattle sex differentiation. In cattle, androstenedione metabolism commences in male embryos at about 25–27 days of gestation and in female embryos several days later, and this occurs at the stage when gonads still look similar (Juarez-Oropeza *et al.*, 1995). The morphologically observable sex differentiation of ovaries takes place in bovine embryos in about 45-day-old female embryos (Kurilo *et al.*, 1987). It is suggested that differentiation of the gonads starts about a week earlier in males than in females. Entry of the oocytes into meiotic prophase occurs in 9-month-old fetuses.

It was shown that testicular development is a key element in establishing mammalian sex. However, it is noteworthy to mention that XY embryos show faster development during the first 8 days than XX embryos, long before the indifferent gonad even appears. These findings suggest that sex-related gene expression affects the development of embryos soon after activation of the embryonic genome (Xu *et al.*, 1992).

The chromosomal sex of germ cells determines their migration pattern and final differentiation into a testis or an ovary. Testicular development is triggered by a gene on the Y chromosome coding for a testis-determining factor (*SRY*). This factor induces, in genetic males, differentiation of Sertoli cells (reviewed by McLaren, 1991) secreting anti-Müllerian hormone (AMH). The latter belongs to the family of TGF-β, causes regression of the Müllerian ducts, promotes the development of Wolffian ducts and promotes differentiation of Leydig cells. The Leydig cells secrete the male steroid hormone, testosterone (Behringer, 1995). Testosterone binds to androgen receptors, which in turn act as transcription factors. More details about AMH and its activity in bovine development are presented elsewhere (Cate and Wilson, 1993). In bovine fetuses, the regression of Müllerian ducts occurs simultaneously in males and freemartins between 50 and 80 days of development (Vigier *et al.*, 1984). A whole chain of developmental events follows, and the phenotype typical for males arises. In females, Müllerian ducts develop, no Leydig cells form in the gonad, no testosterone is produced and development moves steadily towards the female phenotype. The female developmental programme is basic or 'default', while the male programme ultimately requires switching on the *SRY* gene.

The SRY gene

The testis-determining role of the *SRY* gene in mammals is widely accepted after impressive experiments (reviewed by Goodfellow and Lovell-Badge, 1993). The bovine *SRY* gene has been cloned and sequenced. The putative bovine SRY protein consists of 229 amino acid residues, with sequence conservation between species, notably in the region of the so-called high-mobility group (HMG) box (Daneau *et al.*, 1995). This conserved 79 amino acid motif confers DNA-binding ability and probably acts as a transcription factor (Goodfellow and Lovell-Badge, 1993). Apart from the HMG box, the bovine SRY structure shows greater resemblance to human *SRY* than to mouse *Sry* (Daneau *et al.*, 1995).

Expression of the *SRY* gene was confirmed in bovine embryos, both for a short time at the sex-determining stage of development around the period of the primitive undifferentiated gonad and in adult testes (Gutierrez-Adan *et al.*, 1997). Expression of SRY was also reported as early as the four- to eight-cell stage up to the blastocyst stage in bovine embryos. Expression of *SRY* at these early stages and the previous observation that *in vitro*-produced male bovine embryos develop faster in culture than female embryos suggest that sex differences are evident prior to gonadal differentiation and that preimplantation bovine embryos have sexually dimorphic gene expression at least with respect to *SRY* transcripts (Gutierrez-Adan *et al.*, 1997).

Several more genes have been implicated in the process of mammalian sex determination since the cloning of the *SRY* gene (Ramkissoon and Goodfellow, 1996). This set includes the *SRY*-related HMG box (*SOX*) gene family, which displays properties of both classical transcription factors and the architectural components of chromatin (Pevny and Lovell-Badge, 1997). *Sox9* has an essential function in sex determination, possibly downstream of *Sry* in mammals, and is critical for Sertoli-cell differentiation (Morais da Silva *et al.*, 1996).

The *SRY* gene is among the limited number of genes located on the Y chromosome. If this gene is missing from the Y chromosome, XY cattle develop and express gonadal hypoplasia (Kawakura *et al.*, 1997). Mosaicism in a heifer carrying predominantly XY cells seems to be a cause of non-developed gonads, malformation of Müllerian ducts and infertility (Pinheiro *et al.*, 1990). Detection of the bovine fetal SRY sequence from the peripheral blood of pregnant cows by polymerase chain reaction (PCR) analysis is possible and allows the prediction of sex (Yang *et al.*, 1996).

The testis-specific protein, Y-encoded gene (*TSPY*) is one more bovine gene that has been studied recently. It is expressed in fetal bovine testes at about 6–7 months of age and in some adult tissues, shows a differential splicing and is believed to act during spermatogonial proliferation (Vogel *et al.*, 1997).

Both Y and X chromosomes contain transcriptionally active amelogenin genes, which are highly conserved enamel-matrix proteins, which may not have obvious relations with sex determination. An analysis of amelogenin mRNA during bovine tooth development was conducted recently (Yuan *et al.*, 1996).

The cycle of the X chromosome

Since the first publication (Lyon, 1961), it has been commonly accepted that one X chromosome in mammalian females undergoes inactivation during early embryonic development. Numerous investigations shed light on different aspects of X chromosome behaviour, including random inactivation in the ICM, preferential inactivation of the paternal X chromosome in the trophoblast and molecular mechanisms of inactivation, to name a few. These data seem fully applicable to the cycle of the X chromosome in cattle, but experimental evidence remains to be seen. The study of the X bivalent in fetal bovine oocytes allows one to hypothesize that changes occur in the transcriptional status that involve activation, inactivation and reactivation of the X chromosomes during embryonic and ovarian differentiation in the conceptus (Koykul *et al.*, 1997).

Defects of sex determination as a result of interactions with other genes

The *roan* locus in Belgian Blue Cattle gives a good example of such an interaction, being involved in determination of the previously mentioned white heifer disease. This disorder is characterized by a group of anomalies of the female genital tract, which stem from abnormal development of the Müllerian ducts. This includes numerous anatomical lesions in the vagina, cervix and uterus and consequent disturbances in their function. Interestingly, no disorders were found in the ovaries. The *roan* locus was mapped to bovine chromosome 5, and the candidate *Steel* gene coding for the mast-cell growth factor, was mapped in the same interval (Charlier *et al.*, 1996).

Another example of interaction between coat-colour locus and sex differentiation is gonadal hypoplasia, found in white Swedish Highland cattle (Lauvergne, 1970). Additional information can also be obtained from Chapter 3.

Freemartinism in Cattle

Phenomenology

A freemartin is a sterile heifer (genetic female, XX) born co-twin to a bull. The frequency of male–female twin pairs is about 50% of the total number of non-identical twins, as expected. With the exception of bovine freemartinism, intersexuality is rarely reported in domestic animals (Cribiu and Chaffaux, 1990). Freemartinism results from the sexual modification of a female twin by *in utero* exchange of blood with its male co-twin. The dramatic reduction in the size of the uterus and oviducts in freemartin cattle (Khan and Foley, 1994) is due in part to the female fetus exposure to AMH from the male fetus (Vigier *et al.*, 1984). Between 50 and 80 days, Müllerian duct regression occurs simultaneously in males and freemartins and positively correlates with serum AMH concentrations. However, gonadal production of AMH in freemartins was very low (Vigier *et al.*, 1984). Beyond day 70 of fetal life, gubernacular development in freemartins shows male characteristics (van der Schoot *et al.*, 1995), and the ovaries produce abnormally high amounts of testosterone and little or no oestradiol (Shore and Shemesh, 1981). There are indications that the testes of the male co-twin to a freemartin display abnormal steroidogenesis, being, for instance, responsive to LH stimulation (Shore *et al.*, 1984). However, the majority of these males do not differ from normal bulls later. Interestingly, transgenic female murine fetuses expressing the human *AMH* gene have regressed Müllerian ducts and reduced ovarian aromatase activity (Lyet *et al.*, 1995).

Specific features of the bovine placenta

As mentioned earlier, the placenta in cattle is comprised of specialized areas on the fetal chorion, called cotyledons, which are in direct contact with the uterine epithelium of the dam at specialized areas, called caruncles. Oxygen and nutrients pass from the maternal blood to the blood of the fetus, and waste products pass from the fetal blood into the blood of the dam. However, these connections prevent exchange of fetal and maternal blood. Shortly after the implantation of twin pregnancies in cattle, chorionic fusion and vascular anastomosis of the two fetuses usually occur (Mellor, 1969). This allows for the exchange of blood-forming cells between fetuses, the consequence of which, in heterosexual twins, is the formation of a chimerism (60, XX/XY) in peripheral-blood mononuclear leucocytes (Dunn *et al.*, 1968; Basrur *et al.*, 1970).

XX/XY chimerism

Sex-chromosome chimerism in blood leucocytes of bovine freemartins has been well documented (Eldridge and Blazak, 1977; Mayr and Hager, 1978; Summers *et al.*, 1984; Murakami *et al.*, 1989; Cribiu and Chaffaux, 1990). Interestingly, 100% of Friesian–Brahman-cross twins were chimeric, whereas 50% of Jersey–Brahman-cross twins were chimeric (Summers *et al.*, 1984), indicating that breed type may influence the incidence of freemartin chimerism. Most studies indicate that the sex-chromosome chimerism is found in the blood cells of heterosexual twins but not in the germ cells (Vigier *et al.*, 1973; Ford and Evans, 1977).

Diagnosis of freemartinism and practical aspects

Several methods have been described for the detection of freemartinism (van Haeringen and Hradil, 1993), including sex-chromatin karyotyping (Bhatia and Shanker, 1985; Khan and Foley, 1994; Zhang et al., 1994), blood-group serology (Justi et al., 1995) and using PCR (Schellander et al., 1992; Olsaker et al., 1993; Justi et al., 1995). Of these reported methods, the PCR method is rapid and very sensitive and the most suitable for routine testing (Aasen and Medrano, 1990; Horvat et al., 1993; Justi et al., 1995). The PCR method can also be used to detect the sex of bovine preimplantation embryos (Kirkpatrick and Monson, 1993; Ennis and Gallagher, 1994; Hyttinen et al., 1996).

Some breeding programmes aim to reduce the costs of rearing calves. This encourages selection for twins in cattle and eventually increases the number of sterile freemartin heifers (Kastli and Hall, 1978). Introduction of modern diagnostics has suggested that about 17.5% of heifers born with male co-twins may not be freemartins (Zhang et al., 1994), but further work on a larger number of animals is needed to verify this suggestion. Implementation of the diagnostics may preserve thousands of calves with high breeding value and prevent unnecessary economic losses (Zhang et al., 1994). From a production point of view, slaughter-age freemartins do not differ from normal females in growth traits (Hallford et al., 1976; Gregory et al., 1996), but freemartins have higher marbling scores than normal females (Gregory et al., 1996).

Totipotency and Cloning

The recent reports on the cloning of sheep (Wilmut et al., 1997) and later cattle (Kato et al., 1998) using a nuclei from adult cells shake the ground of the current interpretation of totipotency. It has been well known since classical experiments conducted by R. Briggs and T.J. King that nuclear transplantations can be successful, at least in frogs. Later, Gurdon et al. (1975) reported successful transplantation of nuclei from cells derived from adult frogs. However, adult frogs never appeared in these experiments. More recent experiments with frogs have shown that differentiated somatic nuclei transferred to the cytoplasm of oocytes at first meiotic metaphase display enhanced genomic and developmental potential over those transplanted to diplotene oocytes and eggs, at least for the three nuclear cell types tested from the peripheral blood (Di Berardino and Orr, 1992).

The development of nuclear transplantation methods for mammals in the early 1980s resulted in numerous successful experiments (see Chapter 14). The general knowledge obtained from these studies showed that the ability of nuclei to provide normal development decreases significantly from earlier to later stages. This was generally considered as an indication of the progressive loss of totipotency during development and did not contradict data previously obtained with frogs.

Cloning mammals using a nucleus from an adult cell shows that epigenetic changes, which occur in the nucleus during development, can be reversible under certain conditions. Does this mean that the discussion about the loss and restoration of totipotency in consecutive developmental cycles is over? Perhaps not, because in natural conditions nuclei do not jump from one cell to the other, as in transplantation experiments, but stay in differentiated cells. Even if the germ cells are considered, one may assume them to be well-differentiated cells prepared for carrying out a unique function – meiosis. The above notion may be separated into two topics: totipotency of a cell and totipotency of a nucleus artificially transplanted in an egg cell. Meanwhile, significant efforts have been put into the development of totipotent bovine embryonic cell cultures, which may have a great value in cattle breeding (Chapter 14). Thus, regular loss and restoration of totipotency, based on the acquisition and erasing of epigenetic signals, may be a normal feature of the life cycle in many animal species (Ruvinsky, 1997).

Acknowledgement

The authors are thankful to S.E. Echternkamp and R. Geisert for a critical reading of the manuscript and useful comments.

References

Aasen E. and Medrano, J.F. (1990) Amplification of the ZFY and ZFX genes for sex identification in humans, cattle, sheep and goats. *Biotechnology* 8, 1279–1281.

Adalsteinsson, S., Bjarnadottir, S., Våge, D.I. and Jonmundsson, J.V. (1995) Brown coat colour in Icelandic cattle produced by the loci *Extension* and *Agouti*. *Journal of Heredity* 86, 395–398.

Alpizar, E. and Spicer, L.J. (1994) Effects of interleukin-6 on proliferation and follicle-stimulating hormone-induced estradiol production by bovine granulosa cells *in vitro*: dependence on size of follicle. *Biology of Reproduction* 50, 38–43.

Anderson, D.J. (1997) Cellular and molecular biology of neural crest cell lineage determination. *Trends in Genetics* 13, 276–280.

Anthony, R.V., Bellows, R.A., Short, R.E., Staigmiller, R.B., Kaltenbach, C.C. and Dunn, T.G. (1986) Fetal growth of beef calves. II. Effect of sire on parental development of the calf and related placental characteristics. *Journal of Animal Science* 62, 1375–1387.

Anthony, R.V., Liang, R., Kayl E.P. and Pratt, S.L. (1995) The growth hormone/prolactin gene family in ruminant placentae. *Journal of Reproduction and Fertility* Supplement 49, 83–95.

Bamshad, M., Lin, R.C., Law, D.J., Watkins, W.C., Krakowiak, P.A., Moore, M.E., Franceschini, P., Lala, R., Holmes, L.B., Gebuhr, T.C., Bruneau, B.G., Schinzel, A., Seidman, C.E. and Jorde, L.B. (1997) Mutations in human TBX3 alter limb, apocrine and genital development in ulnar-mammary syndrome. *Nature Genetics* 16, 311–315.

Barlow, D.P. (1995) Gametic imprinting in mammals. *Science* 270, 1610–1613.

Barnes, F.L. and First, N.L. (1991) Embryonic transcription in *in vitro* cultured bovine embryos. *Molecular Reproduction and Development* 29, 117–123.

Basrur, P.K., Kosaka, S. and Kanagawa, H. (1970) Blood cell chimerism and freemartinism in heterosexual bovine quadruplets. *Journal of Heredity* 61, 15–18.

Bass, K.E., Li, H., Hawkes, S.P., Howard, E., Bullen, E., Vu, T-K.H., McMaster, M., Janatroup, M. and Fisher, S. (1997) Tissue inhibitor of metalloproteinase-3 expression is upregulated during human cytotrophoblast invasion *in vitro*. *Developmental Genetics* 21, 61–67.

Bazer, F.M., Geisert, R.D. and Zavy, M.T. (1993) Fertilization, cleavage, and implantation. In: Hafez, E.S.E. (ed.) *Reproduction in Farm Animals*, 6th edn. Lea & Febiger, Philadelphia, pp. 188–212.

Bazer, F.W., Ott, T.L. and Spencer, J.L. (1994) Pregnancy recognition in ruminants, pigs and horses: signals from the trophoblast. *Theriogenology* 41, 79–94.

Beauchamp, J.L. and Croy, B.A. (1991) Assessment of expression of the receptor for colony-stimulating factor-1 (fms) in bovine trophoblast. *Biology of Reproduction* 4, 811–817.

Bechet, D.M., Deval, C., Robelin, J., Ferrara, M.J. and Obled, A. (1996) Developmental control of cathepsin B expression in bovine fetal muscles. *Archive of Biochemistry and Biophysics* 334, 362–368.

Behringer, R.R. (1995) The mullerian inhibitor and mammalian sexual development. *Philosophical Transactions of the Royal Society, London B Biological Sciences* 350 (1333), 285–288, discussion 289.

Benyo, D.F. and Pate, J.L. (1992) Tumour necrosis factor-alpha alters bovine luteal cell synthetic capacity and viability. *Endocrinology* 130, 854–860.

Bhatia, S. and Shanker, V. (1985) Sex chromatin as a useful tool for detection of freemartinism in bovine twins. *British Veterinary Journal* 141, 42–48.

Bilodeau-Goeseels, S. and Schultz, G.A. (1997) Changes in the relative abundance of various housekeeping gene transcripst in *in vitro*-produced early bovine embryos. *Molecular Reproduction and Development* 47, 413–420.

Bollag, R.J., Siegfried, Z., Cebra-Thomas, J.A., Davison, E.M. and Silver, L.M. (1994) An ancient family of embryonically expressed mouse genes sharing a conserved protein motif with the *T* locus. *Nature Genetics* 7, 383–389.

Calarco, P.G. and McLaren, A. (1976) Ultrastructural observations of preimplantation stages of the sheep. *Journal of Embryology and Experimental Morphology* 36, 609–622.

Camous, S., Kopechy, V. and Flechon, J.E. (1986) Autoradiographic detection of the earliest stage of [3H]-uridine incorporation into the cow embryo. *Biology of the Cell* 58, 195–200.

Cate, R.L. and Wilson, C. A. (1993) Müllerian-inhibiting substance. In: Gwatkin, R.L. (ed.) *Genes in Mammalian Reproduction*. Wiley-Liss, New York, pp. 185–205.

Charlier, C., Coppieters, W., Farnir, F., Grobet, L., Leroy, P.L., Michaux, C., Mni, M., Schwers, A., Vanmanshoven, P., Hanset, R. and Georges, M. (1995) The *mh* gene causing double-muscling in cattle maps to bovine chromosome 2. *Mammalian Genome* 6, 788–792

Charlier, C., Denys, B., Belanche, J.I., Coppieters, W., Grobet, L., Mni, M., Hanset, R. and Georges, M. (1996) Microsatellite mapping of the bovine roan locus: a major determinant of white heifer desease. *Mammalian Genome* 7, 138–142.

Clements, D., Taylor, H.C., Herrmann, B.G. and Stott, D. (1996) Distinct regulatory control of the *Brachyury* gene in axial and non-axial mesoderm suggests separation of

mesoderm lineages early in mouse gastrulation. *Mechanisms of Development* 56, 139–149.
Cockett, N.E. (1997) Developmental genetics. In: Piper, L. and Ruvinsky, A. (eds) *The Genetics of Sheep*. CAB International, Wallingford, UK, pp. 409–432.
Cribiu, E.P. and Chaffaux, S. (1990) [Intersexuality in domestic mammals]. *Reproduction, Nutrition, Développement Supplément* 1, 51S–61S. [In French.]
Cross, J.C. and Roberts, R.M. (1991) Constitutive and trophoblast-specific expression of a class of bovine interferon genes. *Proceedings of the National Academy of Sciences of the USA* 88, 3817–3821.
Cross, J.C., Flannery, M.L., Blanar, M.A., Steingrimsson, E., Jenkins, N.A., Copeland, N.G., Rutter, W.J. and Werb, Z. (1995) *Hxt* encodes a basic helix–loop–helix transcription factor that regulates trophoblast cell development. *Development* 121, 2513–2523.
Cruz, Y.P. and Pedersen, R.A. (1991) Origin of embryonic and extraembryonic cell lineages in mammalian embryos. In: Pedersen, R.A., McLaren, A. and First, N.L. (eds) *Animal Applications of Research in Mammalian Development*. Current Communications in Cell and Molecular Biology, 4, Cold Spring Harbor Laboratory Press, Cold Spring Harbor, New York, pp. 147–204.
Dahl, E., Koseki, H. and Balling, R. (1997) *Pax* genes and organogenesis. *Bioessays* 19, 755–765.
Daneau, I., Houde, A., Ethier, J.F., Lussier, J.G. and Silversides, D.W. (1995) Bovine SRY gene locus: cloning and testicular expression. *Biology of Reproduction* 52, 591–599.
Das, S.K., Yano, S., Wang, J., Edwards, D.R., Nagase, H. and Dey, S.K. (1997) Expression of matrix metalloproteinases and tissue inhibitors of metalloproteinases in the mouse uterus during the peri-implantation period. *Developmental Genetics* 21, 44–54.
Davis, W., Jr and Schultz, R.M. (1997) Role of the first round of DNA replication in reprogramming gene expression in the preimplantation mouse embryo. *Molecular Reproduction and Development* 47(4), 430–434.
DesJardin, L.E., Lockwood, M.K. and Hauswirth, W.W. (1995) Bovine opsin gene expression exhibits a late fetal to adult regulatory switch. *Journal of Neuroscience Research* 40, 728–736.
Di Berardino, M.A. and Orr, N.H. (1992) Genomic potential of erythroid and leukocytic cells of *Rana pipiens* analyzed by nuclear transfer into diplotene and maturing oocytes. *Differentiation* 50, 1–13.
Dietz, A.B., Georges, M., Threadgill, D.W., Womack J.E. and Schuler, L.A. (1992) Somatic cell mapping, polymorphism, and linkage analysis of bovine prolactin-related proteins and placental lactogen. *Genomics* 14, 137–143.
Duncan, K.M., Kos, L., Jenkins, N.A., Gilbert, D.J., Copeland, N.G. and Tomarev, S.I. (1997) Eyes absent: a gene family found in several metazoan phyla. *Mammalian Genome* 8, 479–485.
Dunn, H.O., Kenney, R.M. and Lein, D.H. (1968) XX–XY chimerism in a bovine true hermaphrodite: an insight into the understanding of freemartinism. *Cytogenetics* 7, 390–402.
Dyban, A.P. and Baranov, V.S. (1988) *Cytogenetics of Mammalian Development*. Oxford University Press, Oxford, UK, 376 pp.
Edwards, J.L., Ealy, A.D., Monterroso, V.H. and Hansen, P.J. (1997) Ontogeny of temperature-regulated heat shock protein 70 synthesis in preimplantation bovine embryos. *Molecular Reproduction and Development* 48, 25–33.

Eldridge, F.E. and Blazak, W.F. (1977) Chromosomal analysis of fertile female heterosexual twins in cattle. *Journal of Dairy Science* 60, 458–463.

Ennis, S. and Gallagher, T.F. (1994) A PCR-based sex-determination assay in cattle based on the bovine amelogenin locus. *Animal Genetics* 25, 425–427.

Fairchild, D.L. and Pate, J.L. (1991) Modulation of bovine luteal cell synthetic capacity by interferon-gamma. *Biology of Reproduction* 44, 357–363.

Feil, R., Khosla, S., Cappai, P. and Loi, P. (1999) Genomic imprinting in ruminants: allele-specific gene expression in parthenogenetic sheep. *Mammalian Genome* 9, 831–834.

Filosa, S., Rivera-Perez, J.A., Gomez, A.P. Gansmuller, A., Sasaki, H., Behringer, R.R. and Ang, S.L. (1997) Goosecoid and HNF-3beta genetically interact to regulate neutral tube patterning during mouse embryogenesis. *Development* 124, 2843–2854.

Firulli, A.B. and Olson, E.N. (1997) Modular regulation of muscle gene transcription; a mechanism for muscle cell diversity. *Trends in Genetics* 13, 364–369.

Fleischman, R.A. (1993) From white spots to stem cells: role of the Kit receptor in mammalian development. *Trends in Genetics* 9, 285–290.

Ford, C.E. and Evans, E.P. (1977) Cytogenetic observations XX/XY chimeras and a reassessment of the evidence of germ cell chimerism in heterosexual twin cattle and marmosets. *Journal of Reproduction Fertility* 49, 25–33.

Fürbass, R., Kalbe, C. and Vanselow, J. (1997) Tissue-specific expression of the bovine aromatase-encoding gene use multiple transcriptional start sites and alternative first exons. *Endocrinology* 138, 2813–2819.

Geisert, R.D., Short, E.C. and Zavy, M.T. (1992) Maternal recognition of pregnancy. *Animal Reproduction Science* 28, 287–298.

Gerrard, D.E., Grant, A.L., Anderson, D.B., Lemenager, R.P. and Judge, M.D. (1995) *In vivo* analysis of serum-borne growth factors in developing co-twinned fetuses. *Journal of Animal Science* 73, 1689–1693.

Gibson-Brown, J.J., Agulnik, S.I., Chapman, D.L., Alexiou, M., Garvey, N., Silver, L.M. and Papaioannou, V.E. (1996) Evidence of a role for T-box genes in the evolution of limb morphogenesis and the specification of forelimb/hindlimb identity. *Mechanisms of Development* 56, 93–101.

Godkin, J.D., Bazer, F.W. and Roberts, R.M. (1984) Ovine trophoblast protein 1, an early secreted blastocyst protein, binds specifically to uterine endometrium and affects protein synthesis. *Endocrinology* 114, 120–130.

Godkin, J.D., Smith, S.E., Johnson, R.D. and Dore, J.J. (1997) The role of trophoblast interferons in the maintenance of early pregnancy in ruminants. *American Journal of Reproductive Immunology* 37, 137–143.

Gomercic, H., Vukovic, S., Gomercic, V. and Skrtic, D. (1991) Histological and histochemical characteristics of the bovine notochord. *International Journal of Developmental Biology* 35, 353–358.

Goodfellow, P.N. and Lovell-Badge, R. (1993) *SRY* and sex determination in mammals. *Annual Review of Genetics* 27, 71–92.

Gore, M.T., Young, R.B., Claeys, M.C., Chromiak, J.A., Rahe, C.H., Marple, D.N., Hough, J.D., Griffin, J.L. and Mulvaney, D.R. (1994) Growth and development of bovine fetuses and neonates representing three genotypes. *Journal of Animal Science* 72, 2307–2318.

Gore, M.T., Young, R.B., Bird, C.R., Rahe, C.H., Marple, D.N, Griffin, J.L. and Mulvaney, D.R. (1995) Myosin heavy chain gene expression in bovine fetuses and neonates representing genotypes with contrasting patterns of growth. *Proceedings of the Society for Experimental Biology and Medicine* 209, 86–91.

Gregory, K.E., Echternkamp, S.E. and Cundiff, L.V. (1996) Effects of twinning on dystocia, calf survival, calf growth, carcass traits, and cow productivity. *Journal of Animal Science* 74, 1223–1233.

Grobet, L., Martin, L.J., Poncelet, D., Pirottin, D., Brouwers, B., Riquet, J., Schoeberlein, A., Dunner, S., Menissier, F., Massabanda, J., Fries, R., Hanset, R. and Georges, M. (1997) A deletion in the bovine myostatin gene causes the double-muscled phenotype in cattle. *Nature Genetics* 17, 71–74

Gueth-Hallonet, C. and Maro, B. (1992) Cell polarity and cell diversification during early mouse embryogenesis. *Trends in Genetics* 8, 274–279.

Guillemot, F., Caspary, T., Tilghman, S.M., Copeland, N.G., Gilbert, D.J., Jenkins, N.A., Anderson, N.A., Joyner, A.L., Rossant, J. and Nagy, A. (1995) Genomic imprinting of Mash-2, a mouse gene required for trophoblast development. *Nature Genetics* 9, 235–242.

Guillomot, M. and Guay, P. (1982) Ultrastructural features of the cell surfaces of uterine and trophoblastic epithelia during embryo attachment in the cow. *Anatomical Record* 204, 315–322.

Guillomot, M., Fléchon, J.-E. and Leroy, F. (1993) Blastocyst development and implantation. In: Thibault, C., Levasseur, M.C. and Hunter, R.H.F. (eds) *Reproduction in Mammals and Man*. Ellipses, Paris, pp. 387–411.

Gurdon, J.B., Laskey, R.A. and Reeves, O.R. (1975) The developmental capacity of nuclei transplanted from keratinized skin cells of adult frogs. *Journal of Embryology and Experimental Morphology* 34, 93–112.

Gutierrez-Adan, A., Behboodi, E., Murray, J.D. and Anderson, G.B. (1997) Early transcription of the *SRY* gene by bovine preimplantation embryos. *Molecular Reproduction and Developments* 48, 246–250.

Hallford, D.M., Turman, E.J., Selk, G.E., Walters, L.E. and Stephens, D.F. (1976) Carcass composition in single and multiple birth cattle. *Journal of Animal Science* 42, 1098–1103.

Harvey, M.B., Leco, K.J., Arcellana-Panlilio, M.Y., Zhang, X., Edwards, D.R. and Schultz, G.A. (1995) Roles of growth factors during peri-implantation development. *Human Reproduction* 10, 712–718.

Hinshelwood, M.M., Liu, Z., Conley, A.J. and Simpson, E.R. (1995) Demonstration of tissue-specific promoters in nonprimate species that express aromatase P450 in placentae. *Biology of Reproduction* 53, 1151–1159.

Holliday, R. (1990) Mechanisms for the control of the activity during development. *Biological Review* 65, 431–471.

Horvat, S., Medrano, J.F., Behboodi, E., Anderson, G.B. and Murray, J.D. (1993) Sexing and detection of gene construct in microinjected bovine blastocysts using the polymerase chain reaction. *Transgenic Research* 2, 134–140.

Hynes, M., Stone, D.M, Dowd, M., Pitts-Meek, S., Goddard, A., Gurney, A. and Rosenthal, A. (1997) Control of cell pattern in the neural tube by the zink finger transcription factor and oncogene *Gli-1*. *Neuron* 19, 15–26.

Hyttinen, J.M., Peura, T., Tolvanen, M., Aalto, J. and Janne, J. (1996) Detection of microinjected genes in bovine preimplantation embryos with combined DNA digestion and polymerase chain reaction. *Molecular Reproduction and Development* 43, 150–157.

Jackson, I.J. (1994) Molecular and developmental genetics of mouse coat colour. *Annual Review of Genetics* 28, 189–217.

Jainudeen, M.R. and Hafez, E.S.E. (1993). Gestation, prenatal physiology, and parturition. In Hafez, E.S.E. (ed.) *Reproduction in Farm Animals*, 6th edn. Lea & Febiger, Philadelphia, pp. 213–236.

Javed, Q., Fleming, T.P., Hay, M. and Citi, S. (1993) Tight junction protein cingulin is expressed by maternal and embryonic genomes during early mouse development. *Development* 117, 1145–1151.

Jenkins, N.A., Copeland, N.G., Taylor, B.A. and Lee, B.K. (1981) Dilute (*d*) coat colour mutation of DBA/2J mice is associated with the site of integration of an ecotropic MuLV genome. *Nature* 293, 370–374.

Joerg, H., Fries, H.R., Meijerink, G.F. and Stranzinger, G.F. (1996) Red colour in Holstein cattle is associated with a deletion in the *MSHR* gene. *Mammalian Genome* 7, 317–318.

Juarez-Oropeza, M.A., Lopez, V., Alvarez-Fernandez, G., Gomez, Y. and Pedernera, E. (1995) Androstenedione metabolism in the indifferent stage of bovine gonad development. *Journal of Experimental Zoology* 271, 373–378.

Justi, A., Hecht, W., Herzog, A. and Speck J. (1995) Comparison of different methods for the diagnosis of freemartinism blood group serology, cytology and polymerase chain reaction. *Deutsche Tierarztliche Wochenschrift* 102, 471–474.

Kambadur, R., Sharma, M., Smith, T.P. and Bass, J.J. (1997) Mutation in myostatin (GDF8) in double-muscled Belgian Blue and Piedmontese cattle. *Genome Research* 7, 910–916.

Kastli, F. and Hall, J.G. (1978) Cattle twins and freemartin diagnosis. *Veterinary Records* 102, 80–83.

Kato, Y., Tani, T., Sotomaru, Y., Kurokawa, K., Kato, J., Doguchi, H., Yasue, H. and Tsunoda, Y. (1998) Eight calves cloned from somatic cells of a single adult. *Science* 282, 2095–2098.

Kawakura, K., Miyake, Y., Murakami, R.K., Kondoh, S., Hirata, T.I. and Kaneda, Y. (1997) Abnormal structure of the Y chromosome detected in bovine gonadal hypoplasia (XY female) by FISH. *Cytogenetics and Cell Genetics* 76, 36–38.

Kawarsky, S.J., Basrur, P.K., Stubbings, R.B., Hansen, P.J. and King, W.A. (1996) Chromosomal abnormalities in bovine embryos and their influence on development. *Biology of Reproduction* 54, 53–59.

Kessler, M.A. and Schuler, L.A. (1997) Purification and properties of placental prolactin-related protein-I. *Placenta* 18, 29–36.

Kessler, M.A., Milosavljevic M., Zieler, C.G. and Schuler, L.A. (1989) A subfamily of bovine prolactin-related transcripts distinct from placental lactogen in the fetal placenta. *Biochemistry* 28, 5154–5161.

Khan, M.Z. and Foley, G.L. (1994) Retrospective studies on the measurements, karyotyping and pathology of reproductive organs of bovine freemartins. *Journal of Comparative Pathology* 110, 25–36.

King, G.J. and Thatcher, W.W. (1993) Pregnancy. In: King, G.L. (ed.) *Reproduction in Domesticated Animals*. Elsevier Science Publications, New York, pp. 229–243.

King, G.J., Atknison, B.A. and Robertson, H.A. (1979) Development of the bovine placentome during the second month of gestation. *Journal of Reproduction and Fertility* 55, 173–180.

King, G.J., Atkinson, B.A. and Robertson, H.A. (1981) Development of the intercaruncular areas during early gestation and establishment of the bovine placenta. *Journal of Reproduction and Fertility* 61, 469–474.

King, W.A., Niar, A., Chartrain, I., Betteridge, K.J. and Guay, P. (1988) Nucleolus organizer regions and nucleoli in preattachment bovine embryos. *Journal of Reproduction and Fertility* 82, 87–95.

Kirkpatrick, B.W. and Monson, R.L. (1993) Sensitive sex determination assay applicable to bovine embryos derived from IVM and IVF. *Journal of Reproduction and Fertility* 98, 335–340.

Kispert, A., Koschorz, B. and Herrmann, B.G. (1995) The protein encoded by *Brachyury* is a tissue-specific transcription factor. *European Molecular Biology Organization Journal* 14, 4763–4772.

Klungland, H., Våge, D.I., Gomez-Raya, L., Adalsteinsson, S. and Lien, S. (1995) The role of melanocyte-stimulating hormone (MSH) receptor in bovine coat colour determination. *Mammalian Genome* 6, 636–639.

Kopecny, V. and Fakan, S. (1992) Extranucleolar genome reactivation: topochemical studies on early bovine embryo. A review. *Acta Histochemica Supplement* 42, 301–309.

Kopecny, V., Fulka, J., Jr, Pivko, J. and Petr, J. (1989) Localization of replicated DNA-containing sites in preimplantation bovine embryo in relation to the onset of RNA synthesis. *Biology of the Cell* 65, 231–238.

Koseki, H., Wallin, J., Wilting, J. Mizutani, Y., Kispert, A., Ebensperger, C., Herrmann, B.G., Christ, B. and Balling, R. (1993) A role for *Pax-1* as a mediator of notochordal signals during the dorsoventral specification of vertebrae. *Development* 119, 649–660.

Koyama, H., Suzuki, H., Yang, X., Jiang, S. and Foote, R.H. (1994) Analysis of polarity of bovine and rabbit embryos by scanning electron microscopy. *Biology of Reproduction* 50, 163–170.

Koykul, W., Switonski, M. and Basrur, P.K. (1997) The X bivalent in fetal bovine oocytes. *Hereditas* 126, 59–65.

Kurilo, L.F., Tepliakova, N.P. and Lavrikova, G.V. (1987) Development of ovaries in bovine fetuses. *Ontogenez* 18, 500–506. [In Russian.]

Lauvergne, J.J. (1970) Gonadal hypoplasia and white coat color in Swedish Highland cattle. *Journal of Heredity* 61, 43–44.

Lavoir, M.-C., Kelk, D., Rumph, N., Barnes, F., Betteridge, K.L. and King, W.A. (1997) Transcription and translation in bovine nuclear transfer embryos. *Biology of Reproduction* 57, 204–213.

Le Douarin, N.M. (1982) *The Neural Crest*. Cambridge University Press, London and New York.

Listrat, A., Gerrard, D.E., Boulle, N., Groyer, A. and Robelin, J. (1994) *In situ* localization of muscle insulin-like growth factor-II m RNA in developing bovine fetuses. *Journal of Endocrinology* 140, 179–187.

Liu, Z. and Foote, R.H. (1997) Effects of amino acids and alpha-amanitin on bovine embryo development in simple protein-free medium. *Molecular Reproduction and Development* 46, 278–285.

Louvet, S., Aghion, J., Santa-Maria, A., Mangeat, P. and Maro, B. (1996) Ezrin becomes restricted to outer cells following assymmetrical division in the preimplantaion mouse embryo. *Developmental Biology* 177, 568–579.

Lyet, L., Louis, F., Forest, M.G., Jasso, N., Behringer, R.R. and Vigier, B. (1995) Ontogeny of reproductive abnormalities induced by deregulation of anti-mullerian hormone expression in transgenic mice. *Biology of Reproduction* 52, 444–454.

Lyon, M.F. (1961) Gene action in the X-chromosome of the mouse (*Mus musculus* L.). *Nature* 190, 372–373.

McGinnis, W. and Krumlauf, R. (1992) Homeobox genes and axial patterning. *Cell* 68, 283–302.
McLaren, A. (1974) Fertilization, cleavage and implantation. In Hafez, E.S.E. (ed.) *Reproduction in Farm Animals*, 3rd edn. Lea & Febiger, Philadelphia, pp. 143–165.
McLaren, A. (1991) Development of the mammalian gonad: the fate of the supporting cell lineage. *BioEssays* 13, 151–156.
McLaughlin, K.J., Szabo, P., Haegel, H. and Mann, J.R. (1996) Mouse embryos with paternal duplication of an imprinted chromosome 7 region die at midgestation and lack placental spongiotrophoblast. *Development* 122, 265–270.
Maconochie, M., Nonchev, S., Morrison, A. and Krumlauf, R. (1996) Paralogous *Hox* genes: function and regulation. *Annual Review of Genetics* 30, 529–556.
Maroto, M., Reshef, R., Munsterbergh, A.E., Koester, S., Goulding, M. and Lassar, A.B. (1997) Ectopic Pax-3 activates MyoD and Myf-5 expression in embryonic mesoderm and neural tissue. *Cell* 89, 139–148.
Marshall, H., Morrison, A., Studer, M., Popperl, H. and Krumlauf, R. (1996) Retinoids and *Hox* genes. *Federation of American Societies for Experimental Biology Journal* 10, 969–978.
Martal, J., Chene, N., Camous, S., Huynh, L., Lantier, F., Hermier, P., L'Haridon, R., Charpigny, G., Charlier, M. and Chaouat, G. (1997) Recent developments and potentialities for reducing embryo mortality in ruminants: the role of IFN-τ and other cytokines in early pregnancy. *Reproduction Fertility and Development* 9, 355–380.
Mayr, B. and Hager, G. (1978) Chimerism in cattle. *Deutsche Tierarztliche Wochenschrift* 85, 170–173.
Mellor, D.J. (1969) Chorioinic fusion and the occurrence of free-martins: a brief review. *British Veterinary Journal* 125, 442–444.
Memili, E., Dominko, T. and First, N.L. (1998) Onset of transcription in bovine oocytes and preimplantation embryos. *Molecular Reproduction and Development* 51, 36–41.
Ménézo, Y. and Renard, J.-P. (1993) The life of the egg before implantation. In: Thibault, C., Levasseur, M.C. and Hunter, R.H.F. (eds) *Reproduction in Mammals and Man*. Ellipses, Paris, pp. 349–367.
Milosavljevic, M., Duello, T.M. and Schuler, L.A. (1989) *In situ* localization of two prolactin-related messenger ribonucleic acids to binucleate cells of bovine placentomes. *Endocrinology* 125, 883–889.
Monsoro-Burq, A.H., Duprez, D., Watanabe, Y., Bontoux, M., Vincent, C., Brickel, P. and Le Dourian, N. (1996) The role of bone morphogenetic proteins in vertebral development. *Development* 122, 3607–3616.
Moore, T. and Haig, D. (1991) Genomic imprinting in mammalian development: a parental tug-of-war. *Trends in Genetics* 7, 45–49.
Morais da Silva, S., Hacker, A., Harley, V., Goodfellow, P., Swain, A. and Lovell-Badge, R. (1996) *Sox9* expression during gonadal development implies a conserved role for the gene in testis differentiation in mammals and birds. *Nature Genetics* 14, 62–68.
Murakami, R., Miyake, Y. and Kaneda, Y. (1989) Cases of XY female, single-birth freemartin and trisomy (61, XX, +20) observed in cytogenetic studies on 18 sterile heifers. *Nippon Juigaku Zasshi* 51, 941–945.
Nakayama, H., Liu,Y., Stifani, S. and Cross, J.C. (1997) Developmental restriction of *Mash-2* expression in trophoblast correlates with potential activation of the Notch-2 pathway. *Developmental Genetics* 21, 21–30.

Nancarrow, C.D. and Hill, J.L. (1995) Oviduct proteins in fertilization and early embryo development. *Journal of Reproduction and Fertility, Supplement* 49, 3–13.

Newton, G.R., Martinod, S., Hansen, P.J., Thatcher, W.W., Siegenthaler, B., Gerber, C. and Voirol, M.J. (1990) Effect of bovine interferon on acute changes in body temperature and serum progesterone concentration in heifers. *Journal of Dairy Science* 73, 3439–3448.

Newman-Smith, E. and Werb, Z. (1997) Functional analysis of trophoblast giant cells in parthenogenetic mouse embryos. *Developmental Genetics* 20, 1–10.

Niswander, L. and Martin, G.R. (1992) Fgf-4 expression during gastrulation, myogenesis, limb and tooth development. *Development* 114, 755–768.

Nothnick, W.B. and Pate, J.L. (1990) Interleukin-1beta is a potent stimulator of prostaglandin synthesis in bovine luteal cell. *Biology of Reproduction* 43, 989–903.

Nüsslein-Volhard, C. (1996) Gradients that organize embryo development. *Scientific American* 275, 54–55.

Oshugi, M., Larue, L., Schwarz, H. and Kemler, R. (1997) Cell-junctional and cytoskeletal organization in mouse blastocysts lacking E-cadherin. *Developmental Biology* 185, 261–271.

Olsaker, I., Jorgensen, C.B., Hellemann, A.L., Thomsen, P.D. and Lie, O. (1993) A fast and highly sensitive method for detecting freemartinism in bovine twins using immunomagnetic beads and Y-specific PCR primers. *Animal Genetics* 24, 311–313.

Papaioannou, V.E. (1997) T-box family reunion. *Trends in Genetics* 13, 212–213.

Pescador, N., Soumano, K., Stocco, D.M., Price, C.A. and Murphy, B.D. (1996) Steroidogenic acute regulatory protein in bovine corpora lutea. *Biology of Reproduction* 55, 485–491.

Pevny, L.H. and Lovell-Badge, R. (1997) Sox genes find their feet. *Current Opinions in Genetics and Development* 7, 338–344.

Picard, B. Robelin, J., Pons, F. and Geay, Y. (1994) Comparison of the foetal development of fibre types in four bovine muscles. *Journal of Muscle Research and Cell Motility* 15, 473–486.

Picard, B., Cagniere, H., Robelin, J. and Geay, Y. (1995) Comparison of the foetal development in normal and double-muscled cattle. *Journal of Muscle Research and Cell Motility* 16, 629–639.

Pilon, N., Daneau, I., Brisson, C., Ethier J.F., Lussier, J.G. and Silversides, D.W. (1997) Porcine and bovine steroidogenic acute regulatory protein (StAR) gene expression during gestation. *Endocrinology* 138, 1085–1091.

Pinheiro, L.E., Mikich, A.B., Bechara,, G.H., Almeida, I.L. and Basrur, P.K. (1990) Isochromosome Y in an infertile heifer. *Genome* 33, 690–695.

Plante, L., Plante, C., Shepherd, D.L. and King, W.A. (1994) Cleavage and 3H-uridine incorporation in bovine embryos of high *in vitro* developmental potential. *Molecular Reproduction and Development* 44, 493–498.

Pomp, D. and Geisert, R. (1998) Developmental genetics. In: Rothschild, M.F. and Ruvinsky, A. (eds) *The Genetics of the Pig*. CAB International, Wallingford, UK, pp. 375–404.

Prior, R.L. and Laster, D.B. (1979) Development of the bovine foetus. *Journal of Animal Science* 48, 1546–1553.

Ramkissoon, Y. and Goodfellow, P. (1996) Early steps in mammalian sex determination. *Current Opinions in Genetics and Development* 6, 316–321.

Reeve, W.J. (1981) Cytoplasmic polarity develops at compaction in rat and mouse embryos. *Journal of Experimental Morphology* 62, 351–367.

Reima, I., Lehtonen, E., Virtanen, I. and Flechon, J.E. (1993) The cytoskeleton and associated proteins during cleavage, compaction and blastocyst differentiation in the pig. *Differentiation* 54, 35–45.

Richards, R.G. and Almond, G.W. (1994) Tumour necrosis factor-alpha differentially alters progesterone and prostaglandin F2 alpha production by porcine luteal cells. *Journal of Endocrinology* 143, 75–83.

Rieger, D., Loskutoff, N.M. and Bettridge, K.J. (1992) Developmentally related changes in the metabolism of glucose and glutamine by cattle embryos produced and co-cultured *in vitro*. *Journal of Reproduction and Fertility* 95, 585–595.

Riethmacher, D., Brinkman, V. and Birchmeier, C. (1995) A targeted mutation in the mouse E-cadherin gene results in defective preimplantation development. *Proceedings of the National Academy of Sciences of the USA* 92, 855–859.

Rikenberger, J.L., Cross, J.C. and Werb, Z. (1997) Molecular genetics of implantation in mouse. *Developmental Genetics* 21, 6–20.

Roberts, R.M., Cross, J.C. and Leaman, D.W. (1992) Interferons as hormones of pregnancy. *Endocrine Reviews* 13, 432–453.

Robertson, S.A., Seamark, R.F., Guilbert, L.J. and Wegmann, T.G. (1994) The role of cytokines in gestation. *Critical Reviews in Immunology* 14, 239–292.

Roby, K.F. and Terranova, P.A. (1989) Localization of tumour necrosis factor (TNF) in rat and bovine ovary using immunocytochemistry and cell blot: evidence for granulosal production. In: Hirshfield, A.N. (ed.) *Growth Factors and the Ovary*. Plenum Press, New York, pp. 273–278.

Ruvinsky, A. (1997) Sex, meiosis and multicellularity. *Acta Biotheoretica* 45, 127–141.

Ruvinsky, A. (1999) Basis of gametic imprinting. *Journal of Animal Science* (in press).

Ruvinsky, A.O. and Agulnik, A.I. (1990) Gametic imprinting and manifestation of the *Fused* gene in the house mouse. *Developmental Genetics* 11, 263–269.

Ruvinsky, I. and Silver, L.M. (1997) Newly identified paralogous groups on mouse chromosomes 5 and 11 reveal the age of a t-box cluster duplication. *Genomics* 40, 262–266.

St Johnston, D. and Nüsslein-Volhard, C. (1992) The origin of pattern and polarity in the Drosophila embryo. *Cell* 68, 201–219.

Schellander, K., Peli, J., Taha, T.A., Kopp, E. and Mayr, B. (1992) Diagnosis of bovine freemartinism by the polymerase chain reaction method. *Animal Genetics* 23, 549–551.

Schuler, L.A. and Hurley, W.L. (1987) Molecular cloning of a prolactin-related mRNA expressed in bovine placenta. *Proceedings of the National Academy of Sciences of the USA* 84, 5650–5654.

Schuler, L.A., Shimomura, K., Kessler, M.A., Zieler, C.G. and Bremel, R.D. (1988) Bovine placental lactogen: molecular cloning and protein structure. *Biochemistry* 27, 8443–8448.

Schuler, L.A., Nagel, R.J., Gao, J., Horseman, N.D. and Kessler, M.A. (1997) Prolactin receptor heterogeneity in bovine fetal and maternal tissues. *Endocrinology* 138, 3187–3194.

Schultz, G.A., and Edwards, D.R. (1997) Biology and genetics of implantation. *Developmental Genetics* 21, 1–5.

Schultz, G.A. and Heyner, S. (1993) Growth factors in preimplantation mammalian embryos. *Oxford Reviews of Reproductive Biology* 15, 43–81.

Schultz, G.A., Hogan, A., Watson, A.J., Smith, R.M. and Heyner, S. (1992) Insulin, insulin-like growth factors and glucose transporters: temporal patterns of gene expression in early murine and bovine embryos. *Reproduction, Fertility and Development* 4, 361–371.

Schultz, J.F., Mayernik, L., Rout, U.K. and Armant, R. (1997) Integrin regulates adhesion to fibronectin during differentiation of mouse peri-implantation blastocysts. *Developmental Genetics* 21, 31–43.

Shang, W., Dore, J.J. and Godkin, J.D. (1997) Developmental gene expression of procollagen III in bovine extraembryonic membranes during early pregnancy. *Molecular Reproduction and Development* 48, 18–24.

Shehu, D., Marsicano, G., Flechon, J.E. and Gali, C. (1996) Developmentally regulated markers of *in vitro*-produced preimplantaion bovine embryos. *Zygote* 4, 109–121.

Shemer, R., Birger, Y., Dean, W.L., Reik, W., Riggs, A.D. and Razin, A. (1996) Dynamic methylation adjustment and counting as part of imprinting mechanisms. *Proceedings of the National Academy of Sciences of the USA* 93, 6371–6376.

Shore, L. and Shemesh, M. (1981) Altered steroidogenesis by the fetal bovine freemartin ovary. *Journal Reproduction and Fertility* 63, 309–314.

Shore, L.S., Shemesh, M. and Mileguir, F. (1984) Foetal testicular steroidogenesis and responsiveness to LH in freemartins and their male co-twins. *International Journal of Andrology* 7, 87–93.

Smith, L.C., Meirelles, F.V., Bustin, M. and Clarke, H.J. (1995) Assembly of somatic histone H1 onto chromatin during bovine early embryogenesis. *Journal of Experimental Zoology* 273, 317–326.

Spicer, L.J. (1998) Tumor necrosis factor-α (TNF-α) inhibits steroidogenesis of bovine ovarian granulosa and thecal cells *in vitro*: involvement of TNF-α receptors. *Endocrine* 8, 109–115.

Spicer, L.J. and Geisert, R.D. (1992) Concentrations of insulin-like growth factor-I, estradiol and progesterone in follicular fluid of ovarian follicles during early pregnancy in cattle. *Theriogenology* 37, 749–760.

Stewart, C.L. and Cullinan, E.B. (1997) Preimplantation development of the mammalian embryo and its regulation by growth factors. *Developmental Genetics* 21, 91–101.

Stewart, H.J., Flint, A.P.F., Lamming, G.E., McCann, S.H.E. and Parkinson, T.J. (1989) Antiluteolytic effect of blastocyst-secreted interferon investigated *in vitro* and *in vivo* in the sheep. *Journal of Reproduction and Fertility* 37, 127–138.

Summers, P.M., Shelton, J.N., Morris, B. and Bell, K. (1984) Interspecific chimerism – the characterization and immunological responsiveness of *Bos taurus–Bos indicus* haemopoietic chimeras produced by embryo transfer. *Australian Journal of Experimental Medical Science* 62, 27–45.

Surani, M.A. (1998) Imprinting and the initiation of gene silencing in the germ line. *Cell* 93, 309–312.

Tanaka, M., Minoura, H., Ushiro, H. and Nakashima, K. (1991) A novel cDNA clone encoding a prolactin-like protein that lacks the two C-terminal cysteine residues isolated from bovine placenta. *Biochimica et Biophysica Acta* 1088, 385–389.

Thatcher, W.W., MacMillan, K.L., Hansen, P.H. and Drost, M. (1989) Concepts for regulation of corpus luteum function by the conceptus and ovarian follicles to improve fertility. *Theriogenology* 31, 149–161.

Thatcher, W.W., Meyer, M.D. and Danet-Desnoyers, G. (1995) Maternal recognition of pregnancy. *Journal of Reproduction and Fertility* 49, 15–28.

Thompson, J.G., Partridge, R.J., Houghton, F.D., Cox, C.I. and Leese, H.J. (1996) Oxygen uptake and carbohydrate metabolism by *in vitro* derived bovine embryos. *Journal of Reproduction and Fertility* 106, 299–306.

Timmers, A.M., Newton, B.R. and Hauswirth, W.W. (1993) Synthesis and stability of retinal photoreceptor m RNA are coordinately regulated during fetal development. *Experimental Eye Research* 56, 257–265.

Vallet, J.L., Bazer, F.W., Fliss, M.F.B. and Thatcher, W.W. (1988) Effect of ovine conceptus secretory proteins and purified ovine trophoblast protein-1 on interoestrous interval and plasma concentrations of prostaglandin F2α in cyclic ewes. *Journal of Reproduction and Fertility* 84, 492–504.

van der Schoot, P., Vigier, B., Prepin, J., Perchellet, J.P. and Gittenberger-de Groot, A. (1995) Development of the gubernaculum and processus vaginalis in freemartinism: further evidence in support of a specific fetal testis hormone governing male-specific gubernacular development. *Anatomical Record* 241, 211–224.

van Haeringen, H. and Hradil, R. (1993) Twins in cattle: freemartin or not? Current aspects. *Tijschrift Voor Diergeneeskunde* 118, 648–649.

Vigier, B., Prepin, J. and Jost, A. (1973) Absence of XX/XY chimerism in somatic tissues of freemartin calf fetuses and their male twins. *Annales de Génétique* 16, 149–155.

Vigier, B., Tran, D., Legeai, L., Bezard, J. and Josso, N. (1984) Origin of anti-Müllerian hormone in bovine freemartin fetuses. *Journal Reproduction and Fertility* 70, 473–479.

Viuff D., Avery, B., Greve, T., King, W.A. and Hyttel, P. (1996) Transcriptional activity in *in vitro* produced bovine two-and four cell embryos. *Molecular Reproduction and Development* 43, 171–179.

Vogel, T., Dechend, F., Manz, E., Jung, C., Jakubiczka, S., Fehr, S., Schidtke, J. and Schnieders, F. (1997) Organization and expression of bovine *TSPY*. *Mammalian Genome* 8, 491–496.

Wilkins, A.S. (1993) *Genetic Analysis of Animal Development*, 2nd edn. Wiley-Liss, New York, 546 pp.

Wilmut, I., Schnieke, A.E., McWhir, J., Kind, A.J. and Campbell, K.H.S. (1997) Viable offspring derived from fetal and adult mammalian cells. *Nature* 385, 810–813.

Winters, L.M., Green, W.W. and Comstock, R.E. (1942) *Prenatal Development of the Bovine*. Technical Bulletin 151, University of Minnesota, Agricultural Experimental Station. Reprinted December 1953.

Wu, B. (1996) Expression of c-foc and c-jun proto-oncogenes by ovine preimplantation embryos. *Zygote* 4, 211–217.

Xie, S., Low, B.G., Nagel, R.J., Beckers, J.F. and Roberts, R.M. (1994) A novel glycoprotein of the aspartic proteinase gene family expressed in bovine placental trophectoderm. *Biology of Reproduction* 51, 1145–1153.

Xie, S., Green, J., Beckers, J.F. and Roberts, R.M. (1995) The gene encoding bovine pregnancy-associated glycoprotein-1, an inactive member of the aspartic proteinase family. *Gene* 159, 193–197.

Xu, K.P., Yadav, B.R., King, W.A. and Betteridge, K.J. (1992) Sex-related differences in developmental rates of bovine embryos produced and cultured *in vitro*. *Molecular Reproduction and Development* 31, 249–252.

Yamakawa, M., Tanaka, M., Koyama, M., Kagesato, Y., Watahiki, M., Yamamoto, M. and Nakashima, K. (1990) Expression of new members of the prolactin growth hormone gene family in bovine placenta: isolation and characterization of two prolactin-like cDNA clones. *Journal of Biological Chemistry* 265, 8915–8920.

Yang, J., Wang, L., Jiang, X., Jiang, Y. and Liu, L. (1996) Detection of bovine fetal Y-specific Sry sequence from maternal blood. *Chinese Journal of Biotechnology* 12, 185–188.

Yuan, Z.A., Collier, P.M., Rosenbloom, J. and Gibson, C.W. (1996) Analysis of amelogenin mRNA during bovine tooth development. *Archive of Oral Biology* 41, 205–213.

Zeng, L., Fagotto, F., Zhang, T., Hsu, W., Vasicek, T.J., Perry, W.L. III, Lee, J.J., Tilghman, S.M., Gumbiner, B.M. and Costantini, F. (1997) The mouse *Fused* locus encodes Axin, an inhibitor of the Wnt signaling pathway that regulates embryonic axis formation. *Cell* 90, 181–192.

Zhang, T., Buoen, L.C., Seguin, B.E., Ruth, G.R. and Weber, A.F. (1994) Diagnosis of freemartinism in cattle: the need for clinical and cytogenic evaluation. *Journal of the American Veterinary Association* 204, 1672–1675.

Zieler, C.G., Kessler, M.A. and Schuler, L.A. (1990) Characterization of a novel prolactin-related protein from bovine fetal placenta. *Endocrinology* 126, 2377–2382.

Genetic Resources and Conservation

16

D.L. Simon
Institute of Animal Breeding and Genetics, Hannover School of Veterinary Medicine, Buenteweg 17p, 30559 Hannover, Germany

Introduction	475
Distribution of Breed Resources of Cattle by World Regions	476
Present Approaches to Conservation	477
Regions with mainly developing countries	477
European approaches to conservation	479
Costs of Conservation	481
Objectives of Conservation, Deduced from Present Practice	482
Conservation Strategies for Different Conservation Objectives	484
Are breeds sufficient expressions of genetic resources and appropriate units of conservation?	484
Which criteria should be used for assessing the status of endangerment of a breed?	484
Criteria for selecting endangered breeds for actual conservation	487
Conservation methods	489
An Alternative Philosophy for Conservation	490
Conclusions	492
Acknowledgements	493
References	493

Introduction

For farm animal species, the term 'genetic resources' is usually understood as synonymous with the term 'breeds'. Breeds have been formed by the activities of humans and by natural selection. Due to the success of animal breeding, highly productive breeds in most farm animal species have been developed during recent decades, which, in principle, have become available throughout the globe. This is particularly the case for breeds of cattle, whose genetic material can easily be moved in the form of semen and embryos.

Since the keeping of animals of highly productive breeds in general is more economic and allows a quicker increase of food supply for a growing human population and may offer a higher prestige value to its owners, less productive breeds tend to be neglected, changed by crossbreeding or replaced (Maijala et al., 1984; Simon, 1984; Engelhardt, 1996). According to the Food and Agriculture Organization's (FAO)'s World Watch List (1995), globally 30% of breeds of the major farm animal species are classified as endangered or critical; for Europe and North America, with generally favourable production conditions, the figure is 43%. In parallel with the growing awareness of the decreasing number of breeds, scientists, non-government organizations (NGOs), governments and grass-roots organizations have expressed concern over this situation (e.g. FAO, 1966, 1981; Bowman, 1974; Maijala, 1974; Alderson, 1981; Maijala et al., 1984; Ollivier, 1996).

One of the main arguments for the conservation of endangered breeds is the concern not to lose genetic diversity, which could become valuable for future breeding options and/or which has not been fully recognized for animal production in adverse environments. Additional arguments are the usefulness of a wider genetic variety for scientific investigations and the conviction that endangered breeds deserve to be conserved as objects of human heritage or for cultural or local reasons.

The fact that decreasing genetic resources in farm animals, particularly in cattle, is the result of positive human activities and that the arguments for conservation are both use- and non-use-orientated makes the topic of conservation of genetic resources a somewhat complex issue. It will be presented here in sections on the following topics: the number of available cattle breeds as an expression of available genetic resources in this species; present approaches to conservation in world regions, including the main conservation methods currently applied; the costs of conservation; the major objectives of conservation; conservation strategies depending on the primary objectives, including the various concepts of risk assessment; selection of breeds for actual conservation and suitable conservation methods; and finally an alternative philosophy for conservation.

Distribution of Breed Resources of Cattle by World Regions

Cattle are the farm animal species with the highest number of animals worldwide (FAO, 1994). Approximately three-quarters of all cattle live in Africa, Asia and the Pacific region, Latin America and the Near East, i.e. broadly speaking in developing countries, and one-quarter in Europe and North America (Table 16.1, adapted from the World Watch List (FAO, 1995)).

However, if we look at the proportion of breeds with sufficient population data, breeds classified as being at risk and breeds with active conservation programmes, the percentage values for the regions of Europe and North America increase to 52%, 74% and 96%, respectively. In other words, although the largest part of the world's cattle population is kept in developing countries,

Table 16.1. Distribution of cattle and of cattle breeds by world regions (adapted from World Watch List, FAO, 1995).

World region	Number of cattle	Number of cattle breeds (in 1000s)			
		On file	With population data	At risk	Maintained*
1 Africa	156,648	120	84	9	0
2 Asia and Pacific Region	426,539	190	118	12	0
3 Latin America and Caribbean Region	321,717	62	39	13	2
4 Near East	60,273	62	38	1	0
5 North America	113,294	48	20	6	1
6 Europe	201,423	305	283	94	52
Subtotal regions 1 to 4	965,177	434	279	35	2
	(75)	(55)	(48)	(26)	(4)
Subtotal regions 5 and 6	314,717	353	303	100	53
	(25)	(45)	(52)	(74)	(96)
World total	1,279,894	787	582	135	55

* With active conservation programmes or maintained by commercial companies or research institutes.
( ), percentage of total.

most of the breeds or genetic resources known to be 'at risk' and even more of the conservation activities are reported from Europe and North America, or, broadly speaking, from developed countries. It has to be assumed that in the developing countries, due to lack of sufficient information, the true number of endangered breeds is higher than that reported so far. Nevertheless, the figures of Table 16.1 make it meaningful in the debate on the conservation of genetic resources to draw attention to possible differences between developing and developed countries.

Present Approaches to Conservation

Regions with mainly developing countries

In Africa, indigenous livestock provide the only practical means of using vast areas of natural grassland, where crop production is impractical. The number of breeds or strains of cattle is of the order of 100–150. Their importance derives from their adaptation to harsh conditions and poor food quality. These abilities, however, are generally difficult to measure. The deficiencies in documentation of the specific genetic qualification of indigenous livestock breeds, in combination with their smaller frame and lack of uniformity compared to exotic breeds, have given rise to a false impression that they are inferior. Therefore Africa's indigenous cattle breeds in general are under threat (Rege and Bester, 1998). On the other hand, zebu × *Bos taurus* crosses have demonstrated heterosis in several economically important traits. This has led to the formation of several zebu-based composite breeds for both milk and beef production, which are important for medium and high production-potential areas.

The formation of composite breeds can be understood as an effective method of sustainable conservation of cattle genetic resources. In addition, the need for the evaluation, improvement and conservation of purebred indigenous breeds is expressed (Rege and Bester, 1998). However, according to FAO's World Watch List (Table 16.1), no active conservation programmes have been reported for this region so far.

According to Sarmiento *et al.* (1998), who report on the situation in Asia and the Pacific region, the general livestock production system of this region can be described so far as small-scale, low-input and well integrated with crop agriculture. There are 159 cattle breeds reported in 12 countries, which in general have become well adapted to different prevalent agroecological conditions. The access to highly productive and high-input genetic material from Western countries, in combination with the increase in demand for food, has led to widespread crossbreeding activities in almost all Asian countries. This has, in fact, threatened the future of many animal genetic resources. In order to stop the erosion of domestic animal diversity, the FAO launched the regional pilot project 'Conservation and Use of Animal Genetic Resources (AnGR) in Asia and the Pacific', covering 12 countries in the region. The programme has three major objectives: to document and monitor livestock breeds, to develop and use genetic material to achieve highly productive sustainable agriculture, and, finally, to conserve unique genetic resources for possible future use. So far no active conservation programmes have been registered in the FAO World Watch List (1995) for this region. Similarly to the situation in Africa, well-planned crossbreeding with high-producing exotic breeds is regarded as an essential tool for the necessary improvement of production efficiency.

The cattle population of Latin America is descended mostly from breeds of the Iberian Peninsula, which became adapted to the large range of environments of the New World. Zebu cattle from Asia have also contributed to the region's cattle genetic resources. Crossbreeding with *B. taurus* from Europe and *Bos indicus* from Asia, which was stimulated by an increased need for animal products, has led to the almost complete absorption of the adapted 'local' breeds (Mariante and Fernandez-Baca, 1998). Conservation activities are under way in several countries of the region, mostly for the criollo breeds of cattle. In Brazil, conservation projects are generally organized as research projects within a national programme on the conservation and utilization of genetic resources. They include the following main elements: (i) identification of the population; (ii) phenotypic and genetic characterization; and (iii) evaluation of production potential (Mariante and Fernandez-Baca, 1998).

It seems typical of developing countries that indigenous and local breeds in general are well adapted to the prevalent unfavourable production conditions. However, their survival as purebreeding populations is increasingly endangered because of a generally high interest in the use of more productive exotic breeds, either by crossbreeding or by breed replacement. Conservation activities in general are in the phase of identification, characterization and evaluation of breeds in relation to their specific environment.

European approaches to conservation

The situation in Europe can be regarded as representative of the more developed countries. In this region the knowledge of the available breeds of cattle is much better than in developing countries. The number of available breeds with population data is high in relation to the total number of cattle (Table 16.1). Production conditions are generally good or can be adapted to the requirements of high-producing breeds. Breeds with lower production potential are rather quickly in danger of becoming replaced or genetically changed by upgrading.

Since 1987, the European Association of Animal Production (EAAP), by means of its Animal Genetic Data Bank in Hannover (EAAP-AGDB), has been active in monitoring information on the breed resources available in Europe. During 1989–1992, this institution served as the EAAP/FAO Global Animal Genetic Data Bank (Ollivier, 1998), became the pacemaker for FAO's Global Domestic Animal Diversity Information System (DAD-IS) and continues to be the largest supplier of information concerning farm animal genetic resources.

By 1997 the EAAP-AGDB had accumulated information on 305 breeds of cattle in some 35 European countries. An analysis of these data revealed that 156 (51%) are classified as being at risk. The number of live animal conservation programmes ($n = 139$) is quite impressive. However, as can be seen from Table 16.2, there seems to be no clear relation between the percentage of conservation programmes and the status of endangerment (rank correlation $r_s = 0.50$, n.s.).

The driving forces for the conservation of endangered breeds of cattle are farmers, NGOs, scientific institutions and national governments. Since 1992, the European Union (EU) has become an important supporter of local breeds which are considered as being in danger of extinction. In 1997, 89 endangered breeds of cattle were supported in ten EU member countries by the EU regulation 2078/92 (D. Dessylas, Brussels, 1997, personal communication).

Conservation is usually carried out in the form of live animals in reproducing herds on private farms. Endangered breeds of cattle are also kept in so-called ark-farm projects and in farm parks. The latter are quite popular in Great Britain, where each year approximately 100,000 visitors are attracted by each farm park (L. Alderson, UK, 1997, personal communication). Farm parks

Table 16.2. Cattle breeds in Europe classified for endangerment, conservation programmes, number and percentage per class of endangerment (data from EAAP-AGDB).

Number of	Total	Class of endangerment					
		Not endangered (1)	Potentially endangered (2)	Minimally endangered (3)	Endangered (4)	Critically endangered (5)	At risk (2–5)
Breeds	305	149	74	27	13	42	156
Conservation programmes	139	51	41	17	6	24	88
		34%	56%	63%	46%	57%	56%

offer visible evidence of endangered breeds to the public and thus contribute to an increased awareness of the need for conservation.

Cryoconservation of semen is used for most cattle breeds; however, it is sometimes difficult to differentiate between storage for commercial use and for conservation. More conclusive is the number of projects for cryoconservation of embryos, especially of projects with a number of involved sires. Only 49 projects of the latter kind have been registered so far by the EAAP-AGDB. This shows that the use of cryopreservation for the purpose of conservation of genetic resources of cattle has been relatively insignificant compared with live animal conservation so far.

As can be concluded from the accumulated information of the EAAP-AGDB, much has already been done in Europe for the conservation of endangered breeds of cattle. Nevertheless, there seems to be no agreement among acting institutions on the main objectives of conservation, on criteria for defining the status of endangerment of breeds and on appropriate requirements for the selection of breeds for actual conservation if many are endangered. Table 16.3 gives an example of conservation programmes which, according to the EAAP-AGDB, are under way for similar cattle breeds in different countries. Obviously, in many cases, decisions to conserve breeds are not only independent of the status of endangerment but also of the existence of conservation programmes for similar breeds in other countries. This results in duplication of efforts.

Before we proceed to deduce the objectives of conservation that seem to be prevalent in developing and developed countries and before we try to point out meaningful conservation strategies for these situations, it seems appropriate to draw attention to the costs of conservation.

Table 16.3. Conservation programmes (CP) for 'similar' breeds (SB) of cattle in Europe, total and in class of endangerment (data from EAAP-AGDB).

Subgroup of similar breeds, formed by EAAP-AGDB	Total number of			Number of SB and CP			
				In class 1 (not endangered)		In classes 2–5 (at risk)	
	Countries	SB	CP	SB	CP	SB	CP
Cattle							
1.2 Original Black Pied	6	8	6	3	1	5	5
3.7 White Lineback	4	5	5	3	2	3	3
5.2 Alpine Brown	4	6	4	3	2	3	2
5.4 Iberian Brown	2	11	9	5	4	6	5
6.2 Grey Mountain	5	7	5	3	2	4	3
Total	21	37	29	16	11	21	18

Costs of Conservation

Information concerning the costs of conservation of genetic resources is sparse. Their definition is difficult, because they depend on the magnitude of the economic disadvantage of keeping animals of a specific endangered breed in comparison with an alternatively available high-producing breed. In addition, they depend on the size of the conserved population and on the kind of conservation method. Brem et al. (1984) and Smith (1984) presented estimates on the costs of conservation of cattle breeds, in German marks and British pounds, by three different methods: (i) live animal conservation in reproducing herds; (ii) cryoconservation of semen; and (iii) cryoconservation of embryos (and semen). The assumptions underlying the computations are somewhat different. Nevertheless, the general conclusion is clear: live animal conservation is expensive, even if a rather low effective population size of $Ne = 25–50$ is assumed. Cryoconservation of semen appears to be cheaper by a factor of 13–30 (Table 16.4).

If not only conservation is of interest but also the later use of conserved material, these results are misleading, because they do not take into account the costs necessary for re-establishing the population from frozen material. If these are considered, cryoconservation of semen turns out to be almost the most expensive method, whereas the combination of live animal conservation and cryoconservation of semen becomes relatively attractive (Lömker and Simon, 1994) (Table 16.5). Even then, conservation remains costly.

Cunningham (1996) used an interesting approach to draw attention to the aspect of costs. Using the method of discounting costs and possible returns, he asks for the required benefits after n years of conservation relative to the required annual costs, if breaking even for the total investment is expected. As

Table 16.4. Estimated costs of conservation by different methods, per breed of cattle (according to Brem et al., 1984; Smith, 1984).

Conservation methods with assumed population size, number of male (m) and female (f) breeding animals, respectively	Costs		Total costs accumulated over years in 1000s and relative to frozen semen (/)	
	For initiation and collection	Per year	20 years	50 years
Brem et al. (1984)[1]				
1 Reproducing population m = 5, f=25	50,000	15,000	350/29	800/30
2 Frozen semen, 500 doses m = 25	2,500	500	12/1	27/1
3 Frozen embryos, n = 100 f = 25 + frozen semen, 500 doses m = 25	42,500	1,000	62/5	92/3
Smith (1984)[2]				
1 Reproducing population m = 10, f = 26	0	5,000	100/8	250/13
2 Frozen semen, 1,250 doses m = 25	9,000	200	13/1	19/1
3 Frozen embryos, n = 625, f = 25	75,000	500	85/6	100/5

[1] Costs in German marks; [2] costs in British pounds.

can be seen from Table 16.6, for a conservation period of 50 years and a discount rate of 0.05, the benefit B in year 50 has to be 229 times larger than the annual costs A. For a rational approach to conservation, it can be concluded that it is essential to minimize the annual costs for conservation, e.g. by a low actual population size (however, with an effective population size of Ne ~85 (see the section on criteria for assessing endangered breed status), and by rapid transfer of the benefit B of year n to a wider population and/or to an extended period of time.

Table 16.5. Capital values (Deutschmark) of the conservation strategies live animals (LA), cryoconservation of semen (CS), cryoconservation of embryos (CE) and combinations, by discounting costs and returns of 50 years of conservation, effective population size of Ne ~50 (adapted from Lömker and Simon, 1994).

Kinds of costs	CS (113,250 doses from 25 bulls, reactivation by upgrading starting year 37)	CE + CS (300 embryos from 90 donors, 2,500 semen doses from 25 bulls, 300 recipients)	LA (64 cows and 16 bulls in 16 reproducing herds)	LA + CS (35 cows and 42,000 semen doses from 20 bulls)
Setting up	114,100	326,400	133,248	88,660
Maintenance	53,919	2,866	A 245,068 B 212,313 C 1,295,444	A 150,940 B 133,049 C 738,142
Reactivation	1,251,902	14,099	0	0
Total	1,419,921	343,365	A 378,316 B 345,601 C 1,428,688	A 239,600 B 221,709 C 826,802

Production level of cows: A, no milking; B and C, 5000 and 3000 kg milk cow^{-1} year^{-1}, respectively.

Table 16.6. Ratio of ultimate benefit after n years of conservation to annual support cost required for investment to break even (adapted from Cunningham, 1996).

	Years of conservation			
Discount rate	25	50	75	100
0.025	74	137	255	472
0.050	68	229	777	2,630
0.075	81	496	3,024	18,441
0.100	106	1,174	12,719	137,806

Objectives of Conservation, Deduced from Present Practice

As can be seen from the various approaches to conservation in different parts of the world, the arguments for conservation of farm animal genetic resources and thus for endangered breeds of cattle are different. They are influenced mainly by two situations.

1. Whether there is a need for a rapid increase of food production, especially of animal protein, for a rapidly growing human population.
2. Whether the natural conditions for animal production are generally unfavourable and can hardly be changed in the foreseeable future, which makes it necessary to use available, well-adapted, local breeds as the basis of the required increased animal production.

In general, both situations are typical of developing countries, but not of the developed countries of Europe or North America.

As a result we have to consider two use-orientated objectives of conservation.

- Conservation of local breeds that are well adapted to unfavourable production conditions but which are nevertheless endangered because of a low to medium production potential, by use and improvement for a better supply of animal products to a growing human population; in other words, 'conservation by sustainable use, now' (CSUN). This reasoning can be regarded as the primary objective of conservation in developing countries; it is not a rational objective for developed countries, where the production conditions in general can be adapted to the specific requirements of high-producing breeds and where efficiency of production is more important than an increase in production.
- Conservation of breeds that are in danger of becoming extinct because they are not competitive any more in favourable production conditions, but which could possess a specific genetic potential – unknown so far – which could become useful in future with possibly changed production conditions and changed requirements of humans; in other words, 'conservation for potential use, later' (CPUL). This reasoning can be regarded as a rational objective for conservation, especially for developed countries.

In addition, two other objectives have to be mentioned which are non-use-orientated, but which are the driving force behind many conservation activities, especially in the developed countries of Europe.

- Conservation of endangered breeds for cultural, historical, ethical and/or local reasons. We know of examples where individual farmers have spent much of their time and money in order to prevent a breed of their liking from getting lost. This objective is beyond a rational reasoning and deserves to be respected. If, however, support from public funds is requested, conditions should be imposed in order to avoid duplication of efforts.
- Conservation because of endangerment. According to the present policy of the EU (Regulation 2078/92, 1992; Working Document VI 5104/92, 1992), a cattle breed has to fulfil two essential requirements in order to be qualified for EU support: it has to be local and it has to be in danger of

becoming extinct. This reasoning can hardly be accepted as an objective for conservation, because it means that, in principle, each endangered breed is qualified for conservation, regardless of the existence of the same or similar breeds elsewhere.

Conservation Strategies for Different Conservation Objectives

Promising conservation strategies require clarification of some major questions.

Are breeds sufficient expressions of genetic resources and appropriate units of conservation?

This can be accepted for the objective 'Conservation for cultural, historical, ethical and/or local reasons', which need not be further explained here.

For the two use-orientated objectives CSUN and CPUL, the situation is different. In essence, the object of conservation is not the endangered breed but its unique genetic potential, such as a known adaptation to a specific harsh environment (in the case of CSUN), or an assumed valuable genetic potential for future breeding options (in the case of CPUL), potentials, that are based on specific genes or gene combinations. However, for an assumed genetic potential, identification of the respective genes and gene combinations is not yet possible and, for a known adaptive potential of an endangered breed, there will hardly be enough time for identification.

In addition it can be argued that conservation of live animals in reproducing herds is much more attractive to the public than conservation of genes. Live animals also offer the possibility of further assessment, mutations and adaptation. These aspects sum up to the conclusion that breeds can indeed be regarded as units of conservation, if one keeps in mind that they are only 'containers' and working units for hopefully available unique genes and gene combinations.

Which criteria should be used for assessing the status of endangerment of a breed?

Several systems for assessing the endangerment of breeds have been proposed (e.g. Maijala *et al.* (1984); DGfZ, (1991); Simon and Buchenauer (1993); Bodo (1994); FAO (1995), European Commission (1992). In addition, some of the NGOs which are active in conservation, such as the Rare Breeds Survival Trust (Alderson, 1978), use their own systems. In Table 16.7, factors are listed which are currently in use by the FAO, the EU and for assessing the breeds of the EAAP-AGDB in Table 16.2.

Table 16.7. Factors for classifying a breed of cattle as being 'not endangered', used for EAAP data in Table 16.2, in FAO's World Watch List (1995) and by the European Union (ECC Reg. 2078/92) (EU, 1992), (European Commission, 1992), simplified.

Criterion	EAAP data	FAO WWL2	European Union
Main factor	F-50 ≤ 10 %	nf > 1000 and nm > 20	nf > 5000
	Numbers are equivalent to an effective population size (Ne)		
	$Ne \geq 84$	$Ne \geq 82$	$Ne \geq 400$
Additional factors	• Decreasing nf and nf < 1000 • % purebreeding • Number of herds < 10 and nf < 500 • Incrossing ≥ 20% of matings	• Decreasing/increasing population size • % purebreeding • Active conservation programme in place	• Decreasing or increasing nf
Classes of endangerment	5	5	2

F-50, assumed accumulated coefficient of inbreeding in 50 years based on present nm and nf; nm, number of male breeding animals; nf, number of female breeding animals. Assumed mating nm : nf, 1 : 40.

First of all endangerment is a function of the number of breeding animals, in addition to conditions which may affect the existence of a breed rather rapidly, such as a downward or upward trend in the number of breeding animals, or a low number of breeding locations. Last but not least, endangerment is dependent on the judgement as to whether genetic changes in the breed have to be considered as threats to a valuable genetic potential or not.

The condition 'trend in the number of female breeding animals' is observed in the three evaluation systems mentioned in Table 16.7. A low number of breeding herds or of breeding locations can increase the risk of rapid disappearance of the breed due to disease hazards, natural disasters or a waning of interest. Therefore, the EAAP system uses the condition 'number of breeding herds < 10 and number of female breeding animals < 500 as an additional factor for risk assessment.

For the minimum number of breeding animals necessary to declare a breed of cattle as being not endangered, numbers ranging from 10,000 (Bodo, 1994) to 750 (Alderson, 1978) have been suggested. Since the status of endangerment forms the basis of the decision as to whether a breed should be conserved or not, and since conservation is costly, a rational approach for setting up minimum requirements of population size seems to be necessary.

Instead of some arbitrary numbers of female animals per breed, the Deutsche Gesellschaft für Züchtungskunde (DGfZ, 1991) in its recommendations on conservation proposed the 'effective population size' Ne as the main factor for the assessment of breeds for endangerment. According to population-genetics theory (Falconer, 1989), Ne is an indicator of the increase of the coefficient of inbreeding per generation, of the amount of random genetic drift and of the decrease of genetic variation within a breed as a group of interbreeding animals. The effective population size Ne is mainly affected by the number of breeding males used for reproduction and is defined as

$$Ne = 4 \times nm \times nf/(nm + nf)$$

with nm, nf = number of male and female breeding animals, respectively (Falconer, 1989). If we want to keep the increase of the coefficient of inbreeding, ΔF, below 1% per generation, an effective population size of $Ne = 1/(2\Delta F) = 1/(2 \times 0.01) = 50$ is required. To achieve $Ne = 50$, the following numbers of breeding males are needed with an increasing number of females: 20 with 35 females, 15 with 80 females or 13 with 325 females. With less than 13 males, even 1000 or more females are not sufficient to ensure $Ne = 50$. It can be concluded that, if it is meaningful to limit the increase of inbreeding in the population, it is more meaningful to ask for a minimum Ne than for a minimum number of female breeding animals.

The formula used to compute Ne assumes unrelated males and random variation in the number of offspring per mating. These assumptions are generally not true, particularly in populations with a decreasing number of males, and they can be difficult to evaluate in developing countries with poor infrastructures. This means that, under the conditions of real life, the minimum Ne should be corrected upward.

Since conservation is a long-term operation, Simon and Buchenauer (1993) set a limit to the assumed accumulated coefficient of inbreeding, F-50, in 50 years of conservation and, considering the species-specific generation interval and the number of required reproduction cycles, deduced the maximum increase of inbreeding per generation and the minimum species-specific Ne. Following this reasoning we fixed the limits of F-50 for five classes of endangerment: 1 = not endangered, 2 = potentially endangered, 3 = minimally endangered, 4 = endangered, 5 = critically endangered to 10%, 20%, 30%, 40% and > 40%, respectively, and deduced, by use of an assumed generation interval of 3.5 years, that for the species of cattle the required corresponding minimum Ne with 1 is ≥ 85, with 2 is 84–55, with 3 is 54–40, with 4 is 39–31 and with 5 is < 31. These marginal values formed the basis for classifying the European cattle breeds in Table 16.2.

Of course, the assumed limits of the accumulated coefficient of inbreeding during conservation, F-50, the assumed length of conservation over n years and the assumed generation interval for the species of cattle are open to discussion. Nevertheless, for a rational approach to conservation, especially in pursuit of the CPUL objective, it appears necessary to take a position concerning these conditions.

The question, whether the condition 'genetic change' of a breed should be considered as an additional factor in the assessment of endangerment again depends on the major objective of conservation.

For the CSUN objective, the aim is not only to conserve local adapted breeds as they are but also to integrate them into a strategy for improved animal production, which will include planned genetic changes by effective selection within breeds and probably by genetic upgrading with high-producing exotic breeds. In other words, for CSUN genetic change is an indispensable element of conservation and cannot to be considered as a factor of endangerment.

For the CPUL objective, the situation is completely different. Here, the aim is to conserve an assumed but unknown genetic potential of a breed for an unknown length of time in order to be able to serve unknown future needs. In this situation genetic changes constitute a real danger for the preservation of the unknown genetic potential and have to be considered as an additional factor for the assessment of endangerment. For the assessment of the European cattle breeds in Table 16.2, incrossing in the order of > 20% of matings was considered as factor for downgrading of breeds into a class of higher endangerment.

For the objective 'Conservation for cultural, historical, ethical and/or local reasons', the question of genetic changes within a breed appears of minor importance for the assessment of endangerment. It probably depends on the judgement of people at the local level as to the degree the external appearance of the breed should be preserved and the extent to which genetic changes should be tolerated if the external appearance remains unchanged.

It can be concluded that the risk definition of breeds is dependent on the objective of conservation. Different approaches in developing countries and developed countries appear appropriate.

Since the status of endangerment of breeds is affected by more than one factor, it appears meaningful to differentiate breeds not only into the two classes 'not endangered' and 'endangered' but to form classes of increased endangerment, such as the systems of the EAAP-AGDB (Simon and Buchenauer, 1993) or the FAO (1995). Classification of breeds by the degree of endangerment can also function as an early warning system and help to counteract the endangerment at an early stage.

Criteria for selecting endangered breeds for actual conservation

Since conservation of genetic resources is costly, it will hardly be possible to conserve all breeds that have been classified as being endangered. Therefore, criteria are needed in order to decide which of the endangered breeds should be conserved and which not. The answer depends again on the primary objective of conservation.

Within the context of conservation for cultural, historical, ethical and/or local reasons, the preference for conservation of a specific breed is usually expressed by the people who actually work with the breed. In this situation, it is probably not adequate to impose criteria from outside, as long as no support from public sources is requested. Nevertheless, the Rare Breeds Survival Trust in the UK requires in its acceptance procedure the following criteria: use of a herd book, breeding true to type and less than 20% of the genetic make-up of the breed being contributed from other breeds (Alderson, 1978).

With regard to CSUN, candidate breeds for the combined objective of conservation and improvement should be the most promising adapted local breeds, preferably evaluated on the basis of reliable data on their adaptive value and of their combining ability with highly productive exotic breeds. Less promising local breeds should be treated as breeds in pursuit of the CPUL objective.

With regard to CPUL, candidates for conservation are endangered breeds which are generally in favourable production conditions but are not competitive any more and which are not needed for present food production. Of these, the breeds should be selected which – although unknown so far – could possess a genetic potential which could become valuable in the future and which cannot be expected to be available in the currently more popular breeds. The main criterion, therefore, should be the degree of genetic uniqueness or the degree of genetic distance in comparison with other breeds, i.e. both with the more popular breeds and with other candidate breeds. Genetic uniqueness has already been asked for as a prerequisite of conservation by the recommendations of the Deutsche Gesellschaft für Züchtungskunde (DGfZ, 1979), as well as Camussi et al. (1985), Weitzman (1993), Barker (1994), Ollivier (1996) and others.

Assessing the genetic uniqueness of a breed can be based on different information, each, however, with inherent limitations (given in parentheses).

- External appearance (inference for the total genome is questionable).
- Breed history (has to be known and has to be true).
- Pedigree analysis (requires reliable data; results are only expectations).
- Quantitative traits (results are affected by environment; requires recording systems).
- Blood groups (inference for the total genome is questionable).

New possibilities arise from the recent developments in molecular genetics, through which polymorphisms on the deoxyribonucleic acid (DNA)-level can be detected and used (Barker, 1994; Ciampolini et al., 1995; Moazami-Goudarzi et al., 1997). Of special interest are so-called microsatellites, in which the underlying loci are highly polymorphic and spread over the whole genome. This makes microsatellites especially suitable for the estimation of genetic distances among breeds. Nei and Takezaki (1994) discussed the statistical techniques available for the measurement of genetic distances between pairs of breeds. For comparisons of estimates among breeds of different countries, agreements are needed on the choice of markers. In an extended concerted action, currently more than 20 European laboratories are genotyping a large set of cattle breeds of several European countries with a common set of microsatellite markers (J.L. Williams, Edinburgh, 1998, personal communication). It is probably meaningful to combine genetic distance estimates at DNA level with information from other sources, such as breed history, performance in quantitative traits and reaction to environmental conditions, in order to come to a final decision which out of several endangered breeds should be selected for conservation.

An interesting approach to the problem 'what to conserve' was presented by Weitzman (1993). His aim is to maximize a 'diversity function' which is dependent on three elements:

- pairwise genetic distances between breeds, based on DNA information;
- extinction probabilities of breeds, based on population size, living conditions and trends;
- costs of altering the extinction probabilities, i.e. the costs of conservation programmes.

This shows that the answer to the question 'what to conserve' is not easy. It requires reliable and comparable data from good recording systems and from genetic analysis and probably more rational thoughts on actual conservation.

Conservation methods

The principal conservation methods for endangered breeds of cattle are: (i) keeping of live animals in reproducing herds (LA); (ii) cryoconservation of semen (CS); and (iii) cryoconservation of embryos (CE) (normally in combination with CS).

The advantage of LA is that animals are permanently available for inspection, testing, research and, if meaningful, crossbreeding. The main disadvantages are the high annual costs of keeping live animals and the need for continuous reproduction of the breed. In each cycle of reproduction, the factors that can disturb the Hardy–Weinberg equilibrium of a population, i.e. mutation, migration, selection, random genetic drift and inbreeding, can cause changes in the frequencies of genes and of genotypes. This is not a problem for CSUN, for which genetic changes to achieve improvements are essential elements, but it interferes with the CPUL objective, i.e. to conserve unknown genes for potential later use. As a consequence, for CPUL, migration and selection should be avoided and the probability of genetic drift and inbreeding should be reduced by ensuring a sufficent Ne in the order of approximately 85.

Cryoconservation of semen and embryos is especially attractive because of the low annual storage costs. The main disadvantage is that animals of the genetic resource are not available for inspection, for further testing and for immediate use. The quantities and genetic relationships of semen and embryos should be planned to enable the reconstruction of a population without significant inbreeding, preferably with $Ne > 85$. This means that 20–25 unrelated bulls (and 25–50 unrelated cows for embryos) should be represented in the stored material (Hodges, 1992).

The advantage of CE over CS is that reconstruction of the breed is possible within 2 years; however, it seems to be difficult to collect a sufficient number of unrelated embryos of a breed that is already approaching endangerment. Although CS is a quick and cheap method of preserving genetic material, reconstruction of the breed from stored semen requires six generations of backcrossing to achieve an expected value of 98% of the original genes. In addition, reconstruction is time-consuming and costly. Stored semen from an 'active semen reserve' can be used to support LA programmes in the effort to minimize inbreeding, either by direct artificial insemination (AI) or by the production of bulls from planned matings for natural service (Simon, 1993).

Weighing up the advantages and disadvantages of the three principal conservation methods that are in use for cattle, one can conclude that LA is essential for CSUN as well as for conservation for cultural, historical, ethical and/or local reasons. Conservation by LA is also meaningful for CPUL, but here with the specific problem of finding a balance between a low actual population size (to minimize costs) and a sufficiently large Ne (to minimize inbreeding). Cryoconservation of semen and embryos is particularly useful if quick actions are necessary to save genetic material which would otherwise be lost; in addition, it is useful as a supplement to LA conservation programmes and as a last reserve in case of a complete loss of a genetic resource which had been expected to be conserved by LA.

The conservation strategies which appear meaningful in the pursuit of the different conservation objectives are summarized in Table 16.8.

An Alternative Philosophy for Conservation

The present approaches to conserving genetic diversity for potential later use are directed toward non-competitive breeds, which, in the case of LA, also form the working units during conservation. In contrast to this, an 'alternative philosophy for livestock breeding' was suggested by Land (1981, 1986). In view of unknown and unpredictable future needs, he suggested the development and maintenance of several strains or lines within a species with divergent biological characteristics, as a supplement to existing breeding policies. These would increase genetic flexibility by purebred or crossbred use, could facilitate a faster response and could be an aid to the rapid improvement of indigenous breeds. 'Such a policy would ensure the availability of appropriate genetic variation in the future and thus provide a positive complement to the passive conservation of rare breeds' (Land, 1981). Land's proposal was supported by Smith (1985), who arrived at the conclusion that, from a national viewpoint, the costs for developing alternative selection stocks are small relative to the possible returns.

In respect of the species of cattle, it can be noted that several of the required specialized lines would be already available, such as Holsteins for milk yield, Charolais for conformation, Simmental for growth rate, Jersey for milk contents, Highlands for harsh environment and N'Dama for trypanotolerance. Several more would be needed, especially some with specific qualifications in stress tolerance, disease resistance, behaviour, product characteristics, etc. If the specialized lines in total cover the entire recognizable diversity of cattle species, it could be expected that the entire unknown genetic potential would also be conserved. This probability might even be

Table 16.8. Conservation strategies for different objectives of conservation of endangered breeds.

Objectives	Specific aspects, meaningful activities
For cultural, historical, ethical and/or local reasons	Local aspects are of specific importance, assessment (and support) from outside is probably not adequate
	Maintenance of external appearance may be more important than strict purebreeding. Selection in order to maintain the breed standard
	Conservation of live animals in reproducing herds
	Avoidance of inbreeding by planned matings and by ensuring an effective population size (Ne) of ~85, i.e. ΔF ~0.6% per generation
For sustainable use, now, i.e. use and improvement of local adapted breeds, mostly in unfavourable production conditions, for a sufficient food supply for a growing human population	The problem is to combine preservation of genetic potentials for adaptation with rapid improvement in production traits
	Characterization and selection of most promising adapted local breeds on reliable data and selection of the most promising exotic breeds on test results for the introduction of genes for high production
	No terminal crossbreeding, but formation of synthetics or composite breeds with the percentage of exotic breeds depending on the possibility to provide appropriate standards in feeding, health and production conditions
	Treatment of the less promising local breeds according to the objective 'conservation for potential use, later'
For potential use, later, i.e. long-term conservation of non-competitive breeds mostly in favourable production conditions, assuming that they possess or may possess a genetic potential that may become useful for future breeding options	The problem is to preserve an unknown potential for unknown future needs
	Search for candidate breeds that represent additional genetic variants on the basis of genetic uniqueness and genetic distances among endangered breeds across national borders
	Keeping of breeds in reproducing herds to allow further assessment, mutation and natural adaptation
	Avoidance of genetic changes, i.e. keeping population in Hardy–Weinberg equilibrium by avoiding incrossing, genetic drift, inbreeding and selection for highly heritable traits. Effective population size (Ne) ~85, i.e. ΔF ~0.6% per generation, planned matings for reproduction
	Preservation of frozen semen for insurance and to supplement matings
	Consideration of 'conservation by specialized lines' as an alternative strategy of conservation
Because of endangerment	In principle, here every endangered breed is qualified for conservation, regardless of the existence of the same or similar breeds elsewhere
	This concept appears to be unsuitable for an effective and long-term conservation policy; it should be abandoned in favour of one of the preceding objectives

higher than the currently favoured CPUL approach, which does not explicitly consider specific qualifications.

Land's proposal, which we may call 'conservation by specialized lines' (CSL), would be in line with the call of the United Nations (UN) Agenda 21 (1992) 'to conserve and maintain genes, species and ecosystems', and not to conserve breeds. The reason why Land's idea has received no response so far in conservation practice can probably be seen in the fact that the development and later use of divergent lines would require effective coordination and cooperation of acting institutions across national borders. However, since effective conservation of genetic resources in future can hardly be achieved without this attitude, CSL could be considered as a true alternative to the conservation of breeds, at least for the species of cattle in the developed countries of Europe.

Conclusions

Although cattle breeds in developing countries are generally not as productive as the popular exotic breeds of developed countries, they have to be used as the basis for the necessary improvement of animal production because they are generally well adapted to the prevalent unfavourable production conditions. Their conservation is appropriate by sustainable use, which should include the improvement of recording systems, within-breed selection and planned utilization of genes of exotic breeds, preferably by development of composite new breeds which are both adapted and highly productive.

In developed countries, the situation is different. Here, the generally favourable production conditions have enabled the development of highly productive breeds by efficient breeding techniques. These breeds are preferred by cattle producers for economic reasons. As a result many breeds in developed countries have become endangered. Their conservation is considered meaningful not for present use but because of the assumption that they could possess an as yet unknown genetic potential which could become useful in the future under changed conditions and requirements.

Since conservation is costly, not all of the endangered breeds can be conserved. In this situation, a CPUL strategy is required. Two approaches appear meaningful.

- To conserve and preferably to keep in Hardy–Weinberg equilibrium those endangered breeds which can be assumed to be genetically unique and which might possess a genetic potential that could become valuable for future breeding options.
- To develop and maintain several strains or lines with divergent biological characteristics which in total cover the entire range of diversity of the species cattle (CSL).

Each of these approaches can only become effective with true coordination and cooperation of acting institutions across national borders. This requires an efficient information system; however, one should keep in mind that the quality, completeness and appropriate use of the accumulated data are more importent than sophisticated information techniques.

Acknowledgements

I would like to thank Dr Barbara Harlizius for collaboration in the section on criteria for selecting endangered breeds and Mrs Hilke Siegel for checking the English manuscript.

References

Alderson, L. (1978) *The Chance to Survive: Rare Breeds in a Changing World.* Cameron & Taylor, London, 192 pp.

Alderson, L. (1981) The conservation of animal genetic resources in the United Kingdom. In: *Report of the FAO/UNEP Technical Consultation on Animal Genetic Resources Conservation and Management, Rome,* 1980. FAO Animal Production Health Paper No. 24, FAO, Rome, pp. 53–76.

Barker, J.S.F. (1994) A global protocol for determining genetic distances among domestic livestock breeds. In: *Proceedings of the 5th World Congress on General Applied Livestock Production,* Vol. 21, Guelph, Canada, pp. 501–508.

Bodo, I.,(1994) Minimum number of individuals in preserved domestic animal populations. In: *Proceedings of the Third Global Conference on Conservation of Domestic Animal Genetic Resources.* Rare Breeds International, Kingston, Canada, pp. 57–65.

Bowman, J.C. (1974) Conservation of rare livestock breeds in the United Kingdom. In: *Proceedings of the 1st World Congress on General Applied Livestock Production, Madrid,* Vol. 2, pp. 23–29.

Brem, G., Graf, F. and Kräusslich, H. (1984) Genetic and economic differences among methods of gene conservation in farm animals. *Livestock Production Science* 11, 65–68.

Camussi, A., Ottaviano, E., Calinski, T. and Kaczmarek, Z. (1985) Genetic distances based on quantitative traits. *Genetics,* 111, 945–962.

Ciampolini, R., Moazami-Goudarz, M., Vaiman, D., Dillman, C., Mazznati, E., Foulley, J.L., Leveziel, H. and Cianic, D. (1995) Individual multilocus genotypes using microsatellite polymorphisms to permit the analysis of genetic variability within and between Italian beef cattle breeds. *Journal of Animal Science* 73, 3259–3268.

Cunningham, P. (1996) Genetic diversity in domestic animals: strategies for conservation and development. In: *Beltsville Symposia in Agricultural Research, XX: Biotechnology's Role in the Genetic Improvement of Farm Animals,* ASAS, Savoy, USA, pp. 13–23.

DGfZ (1979) Stellungnahme zur Bildung von Genreserven in der Tierzuechtung. *Züchtungskunde* 51, 329–331.

DGfZ (1991) Empfehlung zur Kryokonservierung von Sperma, Embryonen und Erbsubstanz in anderer Form zur Erhaltung genetischer Vielfalt bei einheimischen landwirtschftlichen Nutztieren. *Züchtunghskunde* 63, 81–83.

Engelhardt, I. (1996) Inzucht, bedeutende Ahnen und Wahrscheinlichkeit fuer BLAD-Merkmalstraeger in der Deutschen Schwarzbzuntzucht. Dr. med.vet. Dissertation, Hannover School of Veterinary Medicine, Hannover, Germany.

European Commmission (1992) Council Regulation No. 2078/92 of June 1992. *Official Journal of the European Communities* L215, 85–90; with Comité des Structures Agricoles et de Development Rural (STAR) Working Document VI/5404/92.

Falconer, D.S. (1989) *Introduction to Quantitative Genetics*, 3rd edn. Longman, New York, 438 pp.

FAO (1966) *Report of the FAO Study Group on the Evaluation, Utilization and Conservation of Animal Genetic Resources*: FAO, Rome, 33 pp.

FAO (1981) *Animal Genetic Resources Conservation and Management*. FAO Animal Production Health Paper 24, FAO, Rome, 388 pp.

FAO (1994) *FAO Yearbook, 48, Production*. FAO Statistics Series 125, FAO, Rome.

FAO (1995) *World Watch List for Domestic Animal Diversity*, 2nd edn (ed. Scherf, B.D.). FAO, Rome, 769 pp.

Hodges, J. (1992) Recommendations for the preservation of animal genetic diversity in livestock breeds. *EAAP News, Livestock Production Science* 32, 97–99.

Land, R.B. (1981) An alternative philosophy for livestock breeding. *Livestock Production Science* 8, 95–99.

Land, R.B. (1986) Genetic resources requirements under favourable production marketing systems: priority and organisation. In: *Proceedings of the 3rd World Congress on General Applied Livestock Production, Lincoln, USA*, Vol. 12, pp. 486–491.

Lömker, R. and Simon, D.L. (1994) Costs of and inbreeding in conservation strategies for endangered breeds of cattle. In: *Proceedings of the 5th World Congress on General Applied Livestock Production, Guelph, Canada*, Vol. 21, pp. 393–396.

Maijala, K. (1974) Conservation of animal breeds in general. In: *Proceedings of the 1st World Congress on General Applied Livestock Production, Madrid, Spain*, Vol. 2, pp. 37–46.

Maijala, K., Cherekaev, A.V., Devillard, J.M., Reklewski, Z., Rognoni, G., Simon, D.L. and Steane, D.E. (1984) Conservation of animal genetic resources in Europe, final report of an EAAP Working Party. *Livestock Production Science* 11, 3–22.

Mariante, A., da S. and Fernandez-Baca, S. (1998) Animal genetic resources and sustainable development in the Americas. In: *Proceedings of the 6th World Congress on General Applied Livestock Production, Armidale, Australia*, Vol. 28, pp. 27–34.

Moazami-Goudarzi, K., Laloe, D., Furet, J.P. and Grosclaude, F. (1997) Analysis of genetic relationships between 10 cattle breeds with 17 microsatellites. *Animal Genetics* 28, 338–345.

Nei, M. and Takezaki, N. (1994) Estimation of genetic distances and phylogenetic trees from DNA analysis. In: *Proceedings of the 5th World Congress on General Applied Livestock Production, Guelph, Canada*, Vol. 21, pp. 405–412.

Ollivier, L. (1996) The role of domestic animal diversity in the improvement of animal production. *AAA Biotec, Ferrara*, 8–11 October, 1–11.

Ollivier, L. (1998) Guest editorial: history of the EAAP Animal Genetic Data Bank (EAAP-AGDB) and EAAP–FAO collaboration on animal genetic resources. *EAAP News, Livestock Production Science* 54, 67–70.

Rege, J.E.O. and Bester, J. (1998) Livestock resources and sustainable development in Africa. In: *Proceedings of the 6th World Congress on General Applied Livestock Production, Armidale, Australia*, Vol. 28, pp. 19–26.

Sarmiento, J.H., Bouahom, B. and Tshering, L. (1998) Animal genetic resources and sustainable development in Asia and the Pacific. In: *Proceedings of the 6th World Congress on General Applied Livestock Production, Armidale, Australia*, Vol. 28, pp.11–18.

Simon, D.L. (1984) Conservation of animal genetic resources – a review. *Livestock Production Science* 11, 23–36.
Simon, D.L. (1990) The global animal genetic data bank. In: Wiener, G. (ed.) *Animal Genetic Resources: a Global Programme for the Sustainable Development.* Proceedings of an FAO Expert Consultation, Rome, September 1989. FAO Animal Production and Health Paper No. 80, pp. 153–166.
Simon, D.L. (1993) Kryokonservierung zur Unterstuetzung der Lebenderhaltung gefaehrdeter Rinderrassen. *Zuechtungskunde* 65, 91–101.
Simon, D.L. and Buchenauer, D. (1993) *Genetic Diversity of European Livestock Breeds.* EAAP Publication No. 66, Wageningen, 580 pp.
Smith, C. (1984) Estimated costs of genetic conservation of farm animals. In: *Animal Genetic Resources, Conservation by Management, Data Banks and Training.* Animal Production and Health Paper No. 44, FAO, Rome, pp. 21–30.
Smith, C. (1985) Scope for selecting many breeding stocks of possible economic value in the future. *Animal Production* 41, 403–412.
United Nations (1992) *Agenda 21: Programme of Action for Sustainable Development.* United Nations Publication, Sales No. E.93.1.11, New York, 135 pp.
Weitzman, M.L. (1993) What to preserve? An application of diversity theory to crane conservation. *Quarterly Journal of Economics* 157–183.

Marker-assisted Selection 17

M.R. Dentine
*Department of Dairy Science, University of Wisconsin,
1675 Observatory Drive, Madison, WI 53706, USA*

Introduction	497
Theoretical Background to Mendelian Sampling	498
Opportunities for Marker-assisted Selection	500
Adult selection	501
Juvenile selection	501
Multistage selection	502
Utility of Marker Data	503
Parentage and species verification	503
Qualitative traits	503
Quantitative traits	503
Specialized uses	505
Potential Pitfalls	506
Timing of selection	506
Accuracy of estimates of allele substitution	506
Linkage disequilibrium problems	507
Short-term versus long-term results	507
Summary	508
References	508

Introduction

Marker-assisted selection (MAS) has been suggested as an improvement to the use of phenotypic records for genetic change of cattle populations. The appeal of an approach that looks directly at genes rather than inferring genotype from phenotype has always been obvious, but only recently have techniques for low-cost and comprehensive genotyping been available. Efforts to map quantitative trait loci (QTL) have begun and will continue. In this chapter, the assumption will be made that loci with major effects on performance have been mapped to chromosomal location and that accurate tracking of segregation of chromosomal segments containing these loci can be done by using the causative loci themselves or brackets of linked anonymous markers. Use of this information for cattle improvement is the next step.

The first utilization of these techniques was in the detection of recessive genetic diseases, such as bovine leucocyte adhesion deficiency (BLAD) (Kehrli *et al.*, 1992) and deficiency of uridine monophosphate synthetase (DUMPS) Shanks *et al.*, 1987; Schoeber *et al.*, 1993). Other single-gene traits, such as polled (Georges *et al.*, 1993; Harlizius *et al.*, 1997) and coat colour (Klungland *et al.*, 1995), are also being improved, using molecular testing, but selection for improving economic traits, such as muscling, growth, reproduction, meat and milk quality and health, is just beginning (Cowan *et al.*, 1997).

Breeders and scientists are incorporating molecular data in a number of ways, but optimum strategies are still under investigation. The term itself, marker-assisted selection, implies the most likely use of molecular data as additional information on the genetic value of animals. Combinations of phenotypic records and molecular genotypes into an index for selection will be the most likely implementation. The relative weight placed on phenotype or genotype for optimal efficiency will depend on many factors. The most limiting of these is the size of the effect of a major locus on economic merit. Contributions of major loci to genetic variance are likely to be small to moderate (Smith and Simpson, 1986). Even if a major contribution to genetic variance is due to a few loci, these loci may not be accurately mapped or characterized in every case. With some major loci tracked with molecular data and other untracked loci distributed across the genome, the most efficient index will combine all available molecular and phenotypical information.

In general, cattle breeding uses a hierarchical structure, with élite animals of high genetic value concentrated in pedigreed populations and commercial animals benefiting from advances made at the élite level. Improved genetics are distributed using natural breeding and reproductive technologies such as artificial insemination and embryo transfer, with the potential for technologies such as cloning (Wilmut *et al.*, 1992; Bishop *et al.*, 1995). Use of molecular data at various parts of this infrastructure will have different cost–benefit ratios (Beckmann and Soller, 1983).

Previous studies have indicated that genotyping costs could be balanced by improvements in élite populations (Brascamp *et al.*, 1993). Rapid changes in genotyping technologies have dropped the cost on a per-animal basis and high-throughput strategies are likely to accelerate this trend. Even commercial herds might utilize molecular data profitably in some instances. Strategies for MAS depend on the population structure, state of knowledge of genetic associations, traits of interest and comparison with alternative strategies. Selection at different stages in the life cycle may differ in the contributions that could be made by adding markers to existing improvement schemes, since accuracy of selection differs according to the phenotypical records available.

Theoretical Background to Mendelian Sampling

Use of MAS depends on the correlation of molecular genotypes with genetic value for trait differences. Translation of molecular data into additive genetic value in diploid species is based on the additive value of the two alleles of an individual for a given trait. In order to use a consistent approach to assigning value, Falconer (1989) utilized a notation where the additive breeding value of an individual for a given locus was the sum of the additive effects of each of two alleles carried on homologous chromosomes. These additive effects were indicated as α_1 and α_2 for individual alleles on each chromosome of a homologous pair. Subscripts refer to the two chromosomes, which may carry identical or different alleles. Additive genetic value is based on the value of substitution of one allele for another in a given population and may include the average effect of both additive and dominant gene actions. This notation can be used to illustrate the relationship between alleles in parents and offspring, using the following arguments.

Consider an individual offspring with a sire and dam. At a given locus, let the sire have alleles S_1 and S_2, where the allele designated S_1 is the allele transmitted by Mendelian sampling to the offspring and the allele S_2 is the allele not transmitted to this particular offspring. Similarly, the dam carries alleles D_1 (that was transmitted) and D_2 (not transmitted to this offspring). Then the breeding value of the sire is $\alpha_{S1} + \alpha_{S2}$, the breeding value of the dam is $\alpha_{D1} + \alpha_{D2}$ and the breeding value of the offspring is $\alpha_{S1} + \alpha_{D1}$.

An interesting parameterization involves expressing the breeding value (BV) of the offspring in the following way. Starting with:

$$\text{BV of offspring} = \alpha_{S1} + \alpha_{D1} \qquad (1)$$

Adding and subtracting the same quantity in two places (no net change):

$$\text{BV of offspring} = \alpha_{S1} + \left(\frac{\alpha_{S2}}{2} - \frac{\alpha_{S2}}{2}\right) + \alpha_{D1} + \left(\frac{\alpha_{D2}}{2} - \frac{\alpha_{D2}}{2}\right) \qquad (2)$$

Rearranging:

$$\text{BV of offspring} = \underbrace{\frac{\alpha_{S1} + \alpha_{S2}}{2}}_{\text{term 1}} + \underbrace{\frac{\alpha_{S1} - \alpha_{S2}}{2}}_{\text{term 2}} + \underbrace{\frac{\alpha_{D1} + \alpha_{D2}}{2}}_{\text{term 3}} + \underbrace{\frac{\alpha_{D1} - \alpha_{D2}}{2}}_{\text{term 4}} \qquad (3)$$

In this parameterization, the breeding value of an individual is expressed in four terms. The first and third terms represent the breeding values of the sire and dam of the individual. Thus the sum of the first and third terms is the pedigree contribution to any offspring of these two parents. The second and fourth terms are the Mendelian sampling terms, representing the deviation of this individual from the parental average due to the particular set of alleles passed to this offspring.

Note that this parameterization has an interesting property; the additive value of the offspring is expressed as a function of all the parental alleles, not just those passed to this offspring. In particular, Mendelian sampling terms are expressed as differences in the value of the parental alleles. This difference in alleles is termed by Falconer the average effect of gene substitution. Given today's technology, with the opportunity to do exchanges of genetic material even between species, gene substitution could be interpreted in several ways. Perhaps a preferable term for Falconer's concept would be average effect of allele substitution. Mendelian sampling terms are one-half of this effect, representing the deviation of this offspring inheriting one allele from the average of all offspring.

This theory can be extended to more than one locus affecting the trait of interest. In the simplest case of independent additive gene action, loci from separate chromosomes segregate independently and effects can be summed. For loci linked in clusters, the effects are similar but need to be adjusted for physical linkage in phase relationships of alleles and for potential recombinations between loci. Even with an infinitesimal model of many loci with small effect, a single marker can have a substitution effect for an entire linked chromosome segment (Dekkers and Dentine, 1991).

Variances of the four terms are equal (each one-quarter of additive genetic variance) if no covariances exist between additive values of alleles in parents or offspring (for instance, in the absence of inbreeding or assortative mating). In these situations, half of the variance of breeding values in the population is due to the variances of the pedigree values and half to Mendelian segregations (Dekkers and Dentine, 1991).

In using molecular information about alleles present in individuals, prediction of these four terms is affected differently based on stage of life cycle and availability of phenotypical information. For planned matings, the prediction of offspring is limited to the prediction of the pedigree merit (terms 1 and 3), since the segregation terms are not yet determined. After the zygote has been formed, prediction of breeding value could include prediction of the Mendelian sampling terms if molecular data are used for the offspring or if phenotypic data are available on the offspring itself or its progeny. Thus the portion of the genotype that can be inferred with phenotype only may vary. For calves too young for phenotypes or offspring, family records (phenotypes or genotypes) can only address the pedigree terms; for animals with their own records and/or records on progeny, phenotypes and genotypes contribute to estimation of all four terms.

Opportunities for Marker-assisted Selection

For various traits, phenotypical data provide more or less information for estimating breeding value, depending on heritability. Low-heritability traits use family information more heavily and, until extensive records on individual and progeny are available, correlations of actual additive genetic value and estimated breeding values remain low. In these situations, MAS will have more advantages. High-heritability traits leave little to be improved and MAS will provide fewer advances for these traits unless the traits are sex-limited or phenotypes are difficult to obtain. Carcass traits, longevity, disease resistance and reproduction are traits where MAS may have unique advantages (Sellier, 1994; Ruane and Colleau, 1996).

In cattle, most selection schemes utilize phenotypical information on individuals and their relatives in complete animal models, using best linear unbiased prediction (BLUP) or some approximation to this selection-index approach. These methods work most efficiently when accurate and complete pedigree information is known and when large numbers of phenotypical records are available. For some cattle-improvement schemes, pedigrees may be unavailable or uncertain. The utilization of molecular information on individuals has different contributions to make in these various schemes. Clearly, when family information is unavailable and heritability from individual records is low, molecular data can provide additional selection tools. If pedigrees are known, loci with direct effects and loci with known linkages within families can be used. If pedigrees are not known, only those loci with causative effects or those in population disequilibrium with known markers could be used.

Adult selection

For cattle populations using progeny-test schemes with many offspring records, little additional information can be obtained from genotyping of élite sires (Dentine, 1992). Correlations of estimated breeding value with true breeding value are already close to 1, and molecular data cannot improve these estimates for the sires themselves. At the other extreme, animals with no pedigree information and no performance records might benefit most from molecular testing, since MAS would provide the only estimate of breeding value. Selection of grade animals, verification of parentage in pasture-bred herds and identification of homozygotes for favourable qualitative traits, such as polled, would be most enhanced by MAS for adult cattle.

Juvenile selection

Young animals present a unique opportunity for MAS. Estimates of breeding values for parents, regardless of the records used in the estimation technique, can only predict the pedigree terms (terms 1 and 3 in equation 3). Even sib data only contribute to the accuracy of the prediction of pedigree terms. The Mendelian sampling terms are not available, regardless of the extent of phenotypical records on parents or sibs. Thus the maximum correlation of estimated breeding value with true additive genetic merit is limited. The 'effective heritability', the squared correlation between the criteria for selection and actual additive breeding value, is limited to 50% if no records on self or progeny are included.

Molecular data can be obtained on young calves or even embryos to assist in the prediction of the breeding values for terms 2 and 4. Thus MAS has a particular role for juvenile selections where reproductive technologies allow early selections and no other records are available. Molecular data may be the only selection criteria with very rapid generation turnovers (e.g. velogenetics; Georges and Massey, 1991) or may be followed by phenotypical or progeny testing. Early molecular testing has been used to sex embryos (Colleau, 1991), and young dairy bulls are currently screened for a number of genetic defects prior to progeny testing.

Multistage selection

Artificial insemination bulls

Selection of bulls for artificial insemination is a very large contributor to decisions affecting genetic progress in current dairy-cattle schemes (Van Vleck, 1977). Investments in bulls due to testing costs can be high and the opportunity to improve the accuracy of selection is very attractive. In addition to using the molecular data to predict the value of young bull calves, molecular data may also be an additional tool used to evaluate the genetic superiority of the parents. Genotyping of élite progeny-tested sires may not be efficient, due to the high accuracy with which their breeding value is estimated, but questions about the superiority of bull dams could be addressed. Bull dams are highly selected, but may have lower accuracy of selection than bull sires. Bull dams are often young to shorten the generation interval and have limited phenotypical records. In addition, preferential treatment of élite cows may not allow for dependable estimates of their additive breeding values.

Young bulls

For young bulls, MAS provides early information on Mendelian sampling (Kashi *et al.*, 1990). In particular, molecular data provide the only data on inheritance of major qualitative loci, such as genetic defects or exact genotypes for desired qualitative traits. For quantitative traits, MAS applied within families allows some discrimination between full-sibs with identical pedigrees or between half-sibs with similar dams.

Another strategy that might benefit from MAS is the sequential selection decisions based on traits expressed at different times throughout life. If traits are expressed late in life, early molecular testing might allow prescreening of individuals at an earlier age to avoid the expenses of testing for all individuals. Lifetime reproductive performance, longevity or freedom from arthritis might be traits that could benefit from knowledge of loci associated with these late performance traits. Given the expense of progeny testing, a prescreening of candidates to identify those with more potential will be an important use of molecular data.

Utility of Marker Data

Parentage and species verification

The ability to track segregation of alleles from parent to offspring and the uniqueness of genotypes for individuals allow use of molecular data directly in cattle improvement. Even if loci are not linked to genetic disease or major trait effects, knowledge of a deoxyribonucleic acid (DNA) pattern, or fingerprint, can be used to improve cattle. Animals and carcasses can be positively identified to connect animals to phenotypes such as meat quality and other post-mortem measurements. Identification of species or an individual can prevent fraud in meats (Meyer *et al.*, 1995) or be used in potential theft cases (Wagner *et al.*, 1994). Parentage can be verified for cases of pasture breeding by multiple bulls or to improve accuracy of progeny testing.

Qualitative traits

A number of qualitative traits can also be managed more easily with identification of heterozygotes as carriers of genetic disease or to distinguish homozygotes from heterozygotes for favourable single alleles, such as polled. In addition to checking for undesirable alleles in young bulls entering progeny testing, homozygous normal heifers or embryos might bring a premium price. Coat colour has economic value in some markets and matings could be made based on knowledge of genotype to ensure that offspring would have the most desirable phenotypes. Sexing of embryos would also add value to embryos and would contribute efficiencies to herd replacement operations (Colleau, 1991).

Although molecular data will help eliminate undesirable alleles and increase favourable alleles, there are costs associated with testing. At very low frequencies of undesirable alleles, the cost of testing would not be balanced by the potential gains from knowing genotype. Thus, undesirable alleles would probably not be eliminated completely by testing. If testing is stopped, based on the small potential gain, but some undesirable alleles rise again in frequency to the point where the impact rises to previous levels, molecular testing could be resumed.

Quantitative traits

Additive variation
Use of marker data to improve selection for continuous quantitative traits has been of great interest. Although most scientists agree that a combination of marker information and phenotypic records is superior to either alone, the weighting of the two sources of data is still under investigation. Clearly the maximum use of additive genetics for the subsequent generation would result

from estimation of combined breeding value for major genes and other genetics in a single value. Methods for this estimation have been proposed by Fernando and Grossman (1989) but have not been utilized, due to computational and logistic difficulties. With only a few individuals genotyped, reluctance of genotyping organizations to make data public and reliance on linked markers for genotyping, the complete approach is unlikely to be utilized. Various other indices of merit have been proposed and used in simulations to investigate the likely results of utilizing these methods (Larzul et al., 1997). A few general conclusions can be drawn as a result of these studies.

1. Modest improvements in genetic progress are likely under selection on a combined index of molecular and phenotypical data, as contrasted with the use of predictions from phenotypical records alone.

2. Putting too much emphasis on a few major loci can result in lower overall progress than using phenotypical data alone.

3. Greater progress is possible under low initial frequency of favourable alleles, more accurate estimates of major gene effects, larger effects from fewer loci, shorter time horizons, traits of lower heritability, traits that are not measured on every individual (sex-limited, carcass traits, etc.) and instances where family relationships are unknown, incomplete or misidentified.

4. Although increases in selection accuracy may be modest, changes in generation interval may also occur if marker information is used for earlier decisions (Edwards and Page, 1994). Most studies have assumed selection decisions occurring at the same time in two alternative schemes. With genotypic information available much earlier than most phenotypical records, strategies that shorten the generations may accelerate genetic progress primarily through increasing the turnover of generations.

5. Advantages of MAS for a single major locus are realized fairly quickly (Fournet et al., 1997) and continued progress depends on continued discovery of QTL with major trait effects (Meuwissen and Goddard, 1996).

Several other strategies could potentially provide genetic improvement through MAS faster than in traditional individual or family-index selection. In the case of multiple trait selection, MAS could utilize individual loci with pleiotropic effects on several traits. Genetic correlations that unfavourably retard desired genetic progress are composed of the joint effects of all loci involved in both traits. But individual loci may not have pleiotropic effects consistent with the average genetic correlations. Individual loci with favourable, or at least less undesirable, joint effects could be used in MAS to make faster progress toward the desired goals.

Non-additive variation
Non-additive variation is also a target for MAS utilizing specific loci. Many QTL detection schemes allow estimation of dominance or epistasis for a set of loci. Schemes that deploy genotypes across a variety of environments could also be used to estimate genotype × environment interactions. Quantitative trait loci that exhibit such interactions could be used to specifically target assignment of

individuals to environments, using marker genotypes. Maintenance of genetic variability and avoidance of inbreeding could also be objectives of MAS (Schoen and Brown, 1993). Mating strategies could incorporate these data to take advantage of non-additive merit for individuals or breeding populations. Finally, selection for heterosis at single locus or as a general index across loci would be possible using molecular data for individual selection.

Specialized uses

Transgenics

Several special cases involve the tracking of introduced alleles into populations. Use of molecular tags on genes introduced by gene insertion or site-directed mutagenesis can track the inheritance of alleles in the offspring of the original (usually hemizygous) founder animals. The insertion of pharmaceutical genes into embryos that originated from ova of cull cows has been suggested. If the original transgenic animal was created from embryos of lesser overall merit relative to the population, the successful transgene insertion must be transferred to a better background genotype to make the resulting transgenic line economically valuable (Cundiff *et al.*, 1993; Hillel *et al.*, 1993).

Introgression

Similarly, superior alleles not currently present in improved breeds can be introduced into populations by conventional breeding and their introgression accelerated if molecular data can track the introduced alleles (Hospital *et al.*, 1992). An additional advantage comes from the ability to discriminate against the other parts of the introduced genome by selection against other introduced (and presumably inferior) alleles (for instance, Markel *et al.*, 1997). Various schemes have been proposed to save the cost of genotyping every individual and the conclusions of several authors favour use of molecular data in early generations and use of phenotypical data on most loci, with genotyping only at the introduced locus.

Marketing

Income from sale of improved genetics is not entirely based on objective measures of genetic merit. Market demand can influence price, based on the reputation of the provider of genetics, popular perceptions of the competitiveness of organizations and enthusiasm for new approaches. In the case of MAS, customer preference for higher accuracies of selection and the ability to market an organization as forward-thinking may create temporary advantages that increase the attractiveness of genotype data. Buyers of cattle, embryos and semen may be willing to pay premium prices for genotyped animals in excess of the true genetic value of the information.

Potential Pitfalls

Timing of selection

Although marker-assisted selection has considerable potential for genetic improvement, some additional considerations must be included. The first is the cost of testing and analysis to determine desirable genotypes. The logistics of getting DNA samples and phenotypical records on appropriate individuals will continue to be difficult in dispersed cattle populations. Original estimates of costs of molecular testing were based on Southern hybridization techniques (Beckmann and Soller, 1983) and are overestimates under current polymerase chain reaction technology. More advanced laboratory techniques, such as the DNA chip technologies, clever statistical designs and improved software for analyses are likely to decrease costs even further. Cost–benefit analyses will still need to be included in choice of strategies for employing MAS but may not be limiting factors.

A more biological difficulty is the timing of detection of QTL and utilization. In general, detection of QTL occurs by observing the performance of progeny (or grand-progeny; Weller *et al.*, 1990) and relating this performance to alleles inherited from parents or grandparents. Estimates are relevant to the previous population allele frequencies. With the long generation intervals in cattle, changes in allele frequency or traits of importance may have occurred that change the economic value of the loci. Utilization is most advantageous in young animals and is most likely to occur in an even later generation than those involved in the detection. Thus the long time span between the genetic basis for MAS and the utility of that knowledge will continue to decrease the impact of MAS.

Additionally, designs for detection are most powerful at intermediate frequencies of favourable alleles, but the advantages of MAS are greatest for quickly increasing favourable alleles at low frequency. No doubt some favourable alleles will be detected that are near fixation; in these instances, MAS has little to offer. Some advantages to knowing that such loci have significant effects on traits may still be useful in cases where favourable alleles can be introgressed into other populations or where searches of the locus may discover previously undetected alleles superior to the currently prevalent ones. These loci will also be used in basic biological studies to determine what gene actions are involved in allele superiority.

Accuracy of estimates of allele substitution

Most simulations that have addressed the use of molecular data have presumed that estimates of allelic effects would be known without error, although some have considered losses from recombination. Inaccuracies of estimation, particularly overestimates, will put undue emphasis on molecular data. Uncertainty about size of allelic effects will undoubtedly lower genetic progress

(Spelman and van Arendonk, 1997). The power of detection in most schemes is not able to detect smaller QTL, and efforts to avoid high false-positive rates can bias estimates. Some simulations have looked at risk in MAS strategies and have concluded that risks were not increased by the use of markers (Meuwissen and van Arendonk, 1992). Continued verification of allele substitution effects associated with markers may avoid some of the problems with false-positive identification or inaccurate estimation of size or location (Gimelfarb and Lande, 1994)

Linkage disequilibrium problems

Most uses of marker data will be in populations already under selection for one or more traits. In these populations, some linkage disequilibrium between loci will already exist and further changes to phase disequilibrium will occur with MAS. In some instances, existing unfavourable linkages can be more efficiently handled with marker data. An example is the apparent close linkage of a genetic defect in Brown Swiss and favourable alleles for production traits at one or more loci (Hoeschele and Meinert, 1990). If a cluster of QTL is present in populations, MAS provides a mechanism to select favourable recombinant haplotypes that might otherwise increase in frequency very slowly.

Unlinked QTL can also be in phase disequilibrium from mutation, migration, selection or drift (Hospital and Chevalet, 1996). These relationships may complicate estimates of QTL allele substitution effects in segregating populations (Mackinnon and Georges, 1992). Negative covariances between effects at loci that influence traits could slow genetic progress and reduce genetic variance (Bulmer, 1971). This effect occurs regardless of the method of selection used, but some strategies for MAS could modify the relationships in positive or negative directions.

Short-term versus long-term results

One particular demonstration of this disequilibrium effect on genetic progress is seen in the comparison of short-term and long-term response. Gibson (1994) showed that some strategies for MAS could produce a greater short-term response but lower longer-term gains. This phenomenon also occurs with other methods of selection, such as family index, and is attributable to decreases in genetic variance due to linkage disequilibrium. Dekkers and van Arendonk (1998) showed that MAS could be modified to optimize response at a given planning horizon, with weights that varied with allele frequencies.

Summary

One unifying principle in genetic selection is the importance of using all known information in the selection process. With increasing knowledge of the position and effects of major loci for quantitative variation, modifications of traditional selection procedures based only on phenotypes will be needed. Complications of the use of MAS will require customized strategies to maximize benefits and avoid problems. Although MAS may not be as simple as first proposed, some combinations of molecular and phenotypical data are likely to be used profitably in cattle selection programmes.

References

Beckmann, J.S. and Soller, M. (1983) Restriction fragment length polymorphisms in genetic improvement: methodologies, mapping and costs. *Theoretical and Applied Genetics* 67, 35–43.

Bishop, M.D., Hawkins, G.A. and Keefer, C.L. (1995) Use of DNA markers in animal selection. *Theriogenology* 43, 61–70.

Brascamp, E.W., van Arendonk, J.A.M. and Groen, A.F. (1993) Economic appraisal of the utilization of genetic markers in dairy cattle breeding. *Journal of Dairy Science* 76, 1204–1213.

Bulmer, M.G. (1971) The effect of selection on genetic variability. *American Naturalist* 105, 201–211.

Colleau, J.J. (1991) Using embryo sexing within closed mixed multiple ovulation and embryo transfer schemes for selection on dairy cattle. *Journal of Dairy Science* 74(11), 3973–3984.

Cowan, C.M., Meland, O.M., Funk, D.C. and Erf, D.F. (1997) Realized genetic gain following marker-assisted selection of progeny test dairy bulls. In: *Plant and Animal Genome V. Proceedings*, San Diego, California, Poster 296.

Cundiff, L.V., Bishop, M.D. and Johnson, R.K. (1993) Challenges and opportunities for integrating genetically modified animals into traditional animal breeding plans. *Journal of Animal Science* 71(Suppl. 3), 20–25.

Dekkers, J.C.M. and Dentine, M.R. (1991) Quantitative genetic variance associated with chromosomal markers in segregating populations. *Theoretical and Applied Genetics* 81, 212–220.

Dekkers, J.C.M. and van Arendonk, J.A.M. (1998) Optimum selection on identified quantitative trait loci. In: *Proceedings of 6th World Congress on Genetics Applied to Livestock, Armidale, Australia*, Vol. 26, pp. 361–364.

Dentine, M.R. (1992) Marker-assisted selection in cattle. *Animal Biotechnology* 3(1), 81–93.

Edwards, M.D. and Page, N.J. (1994) Evaluation of marker-assisted selection through computer simulation. *Theoretical and Applied Genetics* 88, 376–382.

Falconer, D.S. (1989) *Introduction to Quantitative Genetics*, 3rd edn. Roland Press, New York, pp. 112–128.

Fernando, R.L. and Grossman, M. (1989) Marker assisted selection using best linear unbiased prediction. *Genetics, Selection and Evolution* 21, 467–477.

Fournet, F., Elsen, J.M., Barbieri, M.E. and Manfredi, E. (1997) Effect of including major gene information in mass selection: a stochastic simulation in a small population. *Genetics, Selection and Evolution* 29, 35–56.

Georges, M. and Massey, J.M. (1991) Velogenetics, or the synergistic use of marker-assisted selection and germ-line manipulation. *Theriogenology* 35, 151–158.

Georges, M., Drinkwater, R., King, T., Mishra, A., Moore, S.S., Nielsen, D., Sargeant, L.S., Sorensen, A., Steele, M.R., Zhao, X., Womack, J.E. and Hetzel, J. (1993) Microsatellite mapping of a gene affecting horn development in *Bos taurus*. *Nature Genetics* 4(2), 206–210.

Gibson, J.P. (1994) Short term gain at the expense of long-term response with selection of identified loci. In: *Proceedings of 5th World Congress on Genetics Applied to Livestock Production, Guelph, Canada*, Vol. 21, pp. 201–204.

Gimelfarb, A. and Lande, R. (1994) Simulation of marker assisted selection in hybrid populations. *Genetical Research Cambridge* 63, 39–47.

Harlizius B., Tammen, I., Eichler, K., Eggen, A. and Hetzel, D.J. (1997) New markers on bovine chromosome 1 are closely linked to the polled gene in Simmental and Pinzgauer cattle. *Mammalian Genome* 8(4), 255–257.

Hillel, J., Gibbins, A.M.V., Etches, R.J. and Shaver D.McQ. (1993) Strategies for the rapid introgression of a specific gene modification into a commercial poultry flock from a single carrier. *Poultry Science* 72, 1197–1211.

Hoeschele, I. and Meinert, T.R. (1990) Association of genetic defects with yield and type traits: the weaver locus effect on yield. *Journal of Dairy Science* 73, 2503–2515.

Hospital, F. and Chevalet, C. (1996) Interactions of selection, linkage and drift in the dynamics of polygenic characters. *Genetical Research* 67(1), 77–87.

Hospital, F., Chevalet, C. and Mulsant, P. (1992) Using markers in gene introgression breeding programs. *Genetics* 132, 1199–1210.

Kashi, Y., Hallerman, E. and Soller M. (1990) Marker-assisted selection of candidate bulls for progeny testing programmes. *Animal Production* 51, 63–74.

Kehrli, M.E., Jr, Ackermann, M.R., Shuster, D.E., van der Maaten, M.J., Schmalstieg, F.C., Anderson, D.C. and Hughes B.J (1992) Bovine leukocyte adhesion deficiency: beta 2 integrin deficiency in young Holstein cattle. *American Journal of Pathobiology* 140(6), 1489–1492.

Klungland, H., Vage, D.I., Gomez-Raya, L., Adalsteinsson, S. and Lien, S. (1995) The role of melanocyte-stimulating hormone (MSH) receptor in bovine coat color determination. *Mammalian Genome* 6(9), 636–639.

Larzul, C., Manfredi, E. and Elsen, J.M. (1997) Potential gain from including major gene information in breeding value estimation. *Genetics, Selection and Evolution* 29, 161–184.

Mackinnon, M.J. and Georges, M.A. (1992) The effects of selection on linkage analysis for quantitative traits. *Genetics* 132(4), 1177–1185.

Markel, P., Shu, P., Ebeling, C., Carlson, G.A., Nagle, D.L., Smutko, J.S. and Moore, K.J. (1997) Theoretical and empirical issues for marker-assisted breeding of congenic mouse strains. *Nature Genetics* 17, 280–284.

Meuwissen, T.H.E. and Goddard, M.E. (1996) The use of marker haplotypes in animal breeding schemes. *Genetics, Selection and Evolution* 28, 161–176.

Meuwissen, T.H.E. and van Arendonk, J.A. (1992) Potential improvements in rate of genetic gain from marker-assisted selection in dairy cattle breeding schemes. *Journal of Dairy Science* 75(6), 1651–1659.

Meyer, R., Hofelein, C., Luthy, J. and Candrian, U. (1995) Polymerase chain reaction–restriction fragment polymorphism analysis: a simple method for species identification in food. *Journal of AOAC, International* 78(6), 1542–1551.

Ruane, J. and Colleau, J.J. (1996) Marker-assisted selection for a sex-limited character in a nucleus breeding population. *Journal of Dairy Science* 79, 1666–1678.

Schoeber, S., Simon, D. and Schwenger, B. (1993) Sequence of the cDNA encoding bovine uridine monophosphate synthase. *Gene* 124(2), 307–308.

Schoen, D.J. and Brown, A.H.D. (1993) Conservation of allelic richness in wild crop relatives is aided by assessment of genetic markers. *Proceedings of the National Academy of Sciences of the USA* 90, 10623–10627.

Sellier, P. (1994) The future role of molecular genetics in the control of meat production and meat quality. *Meat Science* 36, 29–44.

Shanks, R.D., Bragg, D.St. A. and Robinson, J.L. (1987) Incidence and inheritance of deficiency for uridine monophosphate synthase in Holstein bulls. *Journal of Dairy Science* 70, 1893–1897.

Smith, C. and Simpson, S.P. (1986) The use of genetic polymorphisms in livestock improvement. *Journal of Animal Breeding and Genetics* 103, 205–217.

Spelman, R. and van Arendonk, J.A. (1997) Effect of inaccurate parameter estimates on genetic response to marker-assisted selection in an outbred population. *Journal of Dairy Science* 80(12), 3399–3410.

Van Vleck, L.D. (1977) Theoretical and actual progress in dairy cattle. In: *International Conference on Quantitative Genetics*. Iowa State University Press, Ames, Iowa, pp. 543–568.

Wagner, V., Schild, T.A. and Geldermann, H. (1994) Application of polymorphic DNA sequences to differentiate the origin of decomposed bovine meat. *Forensic Science International* 64, 89–95.

Weller, J.I., Kashi, Y. and Soller, M. (1990) Power of daughter and granddaughter designs for determining linkage between marker loci and quantitative trait loci in dairy cattle. *Journal of Dairy Science* 73(9), 2525–2537.

Wilmut, I., Haley, C.S. and Wooliams, J.A. (1992) Impact of biotechnology on animal breeding. *Animal Reproductive Science* 28, 149–162.

Genetic Improvement of Dairy Cattle

18

M.E. Goddard[1,3] **and G.R. Wiggans**[2]
[1]*Animal Genetics and Breeding Unit, University of New England, Armidale, New South Wales 2351, Australia ;* [2]*Animal Improvement Programs Laboratory, Agricultural Research Service, USDA, Beltsville, MD 20705-2350, USA;* [3]*Institute of Land and Food Resources, University of Melbourne, Parkville, Australia*

Introduction	512
Breeding Objectives	512
Profit functions	512
Fitness traits	514
Genetic Variation	515
Breed differences	515
Within-breed variation	516
Inbreeding and heterosis	520
Genotype × environment interaction	522
Individual genes affecting milk production	522
Genetic Evaluation	525
Evaluation models	526
National evaluations	526
International evaluation	527
Traits evaluated	527
Genetic trend for yield	527
Economic indices	528
Future enhancements	528
Design of Breeding Programmes	528
Artificial insemination	528
Multiple ovulation and embryo transfer	530
Minimizing inbreeding	531
Cloning	531
Marker-assisted selection	532
Conclusions	532
References	533

Introduction

Dairy cattle are one of the most highly studied, genetically, of all domesticated mammalian species. Genetic improvement of dairy cattle involves determining which improvements are desirable, which traits provide information on the goal, how heritable those traits are and how to evaluate them, and how to design a breeding programme to achieve the goals. This chapter describes how to determine which goals should be established in order to emphasize profit or efficiency as the ultimate goal of the dairy enterprise. The traits typically measured are listed, along with how they are related and the genetic parameters utilized in the selection process. Evaluation procedures used to establish genetic rankings are derived from observations on related animals and are reviewed. Scientific innovations, such as artificial insemination (AI), marker assisted selection and cloning, are reviewed, as well as their effect on the design of breeding programmes.

Breeding Objectives

The first task in the design of breeding programmes is to define the breeding objective. The usual purpose of breeding programmes is assumed to be an economic one, i.e. to increase the profitability of dairy farming. Therefore, the objective is defined by a profit function, which shows how a change in each trait influences profit. This profit function is based on a bioeconomic model of the farm and obviously depends on the prices the farmer receives for milk and other products and the prices he/she pays for inputs. The methodology for defining profit functions is reviewed by Gomez *et al.* (1997) and Goddard (1998) and an example of its application to pasture-based dairy farming is given by Visscher *et al.* (1994).

Profit functions

The profit function can be non-linear if the effect of a trait on profit is curvilinear. For instance, the effect on profit of increasing fertility might decrease as the mean fertility of the herd increases. However, the profit function can usually be approximated by a linear function:

profit = $\Sigma\, a_i\, \mathrm{bv}_i$

where bv_i is the breeding value for the ith trait and a_i is the economic weight for the ith trait. Thus a_i is the effect on profit of a 1-unit increase in trait i when all other traits are held constant.

The economic weights for a particular trait depend on the other traits that are included in the profit function. For instance, if feed intake is included in the profit function, the economic weight for cow body weight is positive, because increasing body weight increases income from the sale of cull cows. However, if feed intake is not included in the profit function, the economic weight of body weight may be negative, because larger cows have greater feed requirements for maintenance. Generally, including all traits that directly affect income and costs in the profit function is best, but a common practice has been to leave feed intake out of the profit function and to adjust the economic weights of other traits to reflect the change in intake caused by a change in each of the other traits. Thus, the economic weights can be thought of as partial regression coefficients of profit on the breeding value for each trait.

It is important to distinguish between the traits which form the objective and the traits upon which selection is based (selection criteria). For instance, long herd life and low incidence of mastitis may be goals, but selection for those traits is inconvenient because herd life is only known late in life and mastitis is not necessarily recorded. However, conformation traits, such as udder depth, may be genetically correlated with herd life and mastitis incidence and consequently may be useful selection criteria. In this situation, udder depth would not be part of the breeding objective.

Profit can be viewed from the perspective of the individual farmer, the industry or the community and expressed per litre of milk, per cow or per farm. If, when all costs are included, mean profit is close to zero, and market signals are passed along the chain from consumer to cattle breeder, then the relative economic weights are the same from all perspectives and for all units of expression. Mean profit is expected to be close to zero when returns to capital, management and labour are included as costs; otherwise, investment capital would flow into the industry until profitability declined to that of alternative investments.

However, if market signals are not passed along the marketing chain, economic weights can be severely distorted. Quotas are an example of artificial prices leading to distorted market signals and distorted breeding objectives. Gibson (1989) shows how quotas decrease the economic weight for the product under quota. In consequence, farmers may point genetic improvement in a direction that does not maximize the economic benefit to the community as a whole.

The yields of milk, fat and protein are the major determinants of income to dairy farmers and the most important traits in the objective. Their relative economic weights depend on the pricing formula by which farmers are paid. If feed intake is not included in the profit function, the economic weights for milk, fat and protein need to include the extra feed cost associated with extra yield. If the milk is used for manufacturing, the protein is most valuable and the fat is of some value, but the volume is of negative value because it must be transported from farm to factory and evaporated to make some products. Combining the prices received with the feed costs for products, Visscher *et al.* (1994) derived economic weights for Australian dairying of protein ($3.51 kg^{-1}), fat ($1.10 kg^{-1}) and volume (−$0.04 l^{-1}). When the relative economic weights are expressed per genetic standard deviation, they are 1.0 for protein, 0.4 for fat and −0.4 for volume. Although milk pricing and feed cost vary from country to country and over time, these relative economic weights are not atypical.

Fitness traits

Other traits commonly included in breeding objectives are health, fertility, calving ease, body weight, feed intake, milking speed, temperament and length of herd life. More detailed consideration of their economic weights is given by Gomez *et al.* (1997).

Among health traits, the incidence of mastitis is the most important, because mastitis causes milk loss, treatment costs and reduced milk quality. In Scandinavia, mastitis is recorded and the bulls are progeny-tested for incidence of mastitis among their daughters. However, in most countries only somatic cell count (SCC) is recorded. This trait is genetically correlated with mastitis incidence and hence is a selection criterion, but it also has an economic value of its own if milk price is reduced for milk with high SCC. The economic weight of mastitis per genetic standard deviation is approximately one-quarter to one-half that of milk protein yield.

Cow fertility influences AI and veterinary costs, the interval between calvings and hence the pattern and yield from later lactations. In Europe, the economic weight per genetic standard deviation of cow fertility is estimated to be approximately half that of milk protein yield (Philipsson *et al.*, 1994). However, where dairy calves are of much lower value and where farmers can manage cows with long calving intervals so that those cows have long persistent lactations, the economic weight of cow fertility may be much less.

Calving ease is valuable because dystocia results in veterinary costs, extra labour costs, lost calves and cows, reduced milk yield and infertility. The economic weight depends heavily on the incidence of dystocia, which is usually only high in heifers. Calving ease is affected by the genetic merit of the calf and the cow; therefore, selection needs to consider calving ease as a trait of the cow and of the calf.

Feed for cows costs money, so the economic weight for feed intake (when milk yield and other traits are held constant) is negative. The size of this economic weight depends on the proportion of all costs that are proportional to the feed requirement of the herd. In grazing systems, most costs are related to farm size and this in turn determines total feed available. However, in environments where cows must be housed, the housing cost is large and therefore the proportion of all costs due to feed is reduced, as is the economic weight for feed intake.

It has been suggested that the economic weight of feed intake should be positive because cows with high intakes would have better health and fertility and could be fed a less energy-dense, less expensive ration. Including health and fertility in the profit function directly would be logical, so they do not contribute to the economic weight of feed intake. However, when considering selection criteria, the genetic correlations between intake, disease and fertility would need to be considered. The ability of cows to sustain milk yield when fed a less expensive diet is a trait separate from feed intake, which should be investigated further.

Milking speed is of economic value because slow milkers increase the labour cost of milking. In some milking systems, the variability of milking speed is important, because one slow cow can delay the whole shed. In fact, the objective might be to reduce the total labour needed for milking, but to do this we would need to identify the traits of the cow which determine total labour needed and which show genetic variability. Good temperament, while it may be difficult to assign a monetary value to it, is valued highly by dairy farmers in Australia and New Zealand who milk large numbers of cows and want to avoid the disruption and danger caused by wild cows.

A long mean herd life increases profitability, because it decreases replacement costs and increases the proportion of the herd in the most productive, mature age-groups. However, culling of cows is a management decision of the farmer and is done to minimize the economic loss caused by cows of low production, fertility or health. Thus, when these traits are included in the profit function, culling for these reasons should not be included in the definition of herd life. The correct procedure is to include the trait 'reduced herd-life due to traits not in the profit function' (Goddard, 1998). If all traits causing culling were included directly, it would not be necessary to include herd life in the breeding objective.

Economic weights differ between countries and individual farms and are likely to change in future. Possible changes might be caused by environmental effects of dairy farming or automated milking machines that demand cows with a consistent udder anatomy.

Genetic Variation

Genetic parameters indicate the rate of genetic change that is possible and are required for estimation of genetic merit. Of these parameters, heritability describes what portion of the variation (variance) in a trait is of genetic origin and correlations among these traits indicate how genetic change in one trait can affect others. When multiple traits are evaluated, covariances indicate to what degree the information from one trait affects others. If an animal has more than one observation for a trait, repeatability describes the expected similarity among those observations. Other genetic parameters include the effects of dominance, individual genes, breed, inbreeding, heterosis (crossbreeding) and the interaction of genetics with the environment.

Breed differences

The world dairy cattle population is classified into breeds, most of which originally arose in Europe. Registry organizations maintain pedigree records, which enable animals to be traced to the origin of the breed, or importation. With globalization, selection goals have become more similar, and the technology to support high yields is available around the world, particularly in temperate regions. In this environment, the Holstein breed has become dominant, because of its high yield. The Jersey has emerged as the primary alternative breed, because of high component yields and smaller size, along with the collection of Red breeds. Other breeds have regional importance. Crossbreeding programmes have been proposed as a way of upgrading indigenous cattle to a high-producing breed or as a way of obtaining the benefits of hybrid vigour. Table 18.1 displays the differences in yields for the five most common dairy breeds in the USA.

Within-breed variation

Yield traits

Milk yield is usually defined as production during the 305 days following calving, with milk produced after this period not included in genetic evaluations. Individual lactations (parities) of a cow are generally regarded as repeated measurements of the same genetic trait, although some countries have implemented multitrait systems that allow for correlations of < 1.00 among parities. Estimates of heritabilities for milk, fat and protein yields are quite similar across countries (Table 18.2), with heritability estimates for percentage fat and protein content usually much higher than for total yield. Dominance variation has been found to be of minor importance for yield traits, as Misztal et al.

Table 18.1. Standardized lactation averages by breed for 1,861,284 cows with records used in genetic evaluations and calving in 1996 in the USA.

Breed	% of cows	Milk	Fat %	Protein %
Ayrshire	0.5	7102	3.9	3.3
Brown Swiss	0.9	8088	4.0	3.5
Guernsey	0.7	6431	4.5	3.5
Holstein	92.4	9962	3.6	3.1
Jersey	5.5	6848	4.6	3.8

Table 18.2. Genetic parameters used for national evaluation of Holstein yield traits by ten countries that provide bull evaluations for the International Bull Evaluation Service (Interbull).

	Heritability					
Country	Milk yield*	Fat yield*	Fat content†	Protein yield*	Protein content†	Repeatability†
Australia	0.25	0.25	0.45	0.25	0.60	0.50
Canada	0.33	0.33	–	0.33	–	–
Denmark	0.29	0.27	–	0.30	–	–
France	0.30	0.30	0.50	0.30	0.50	–
Germany	0.30–0.36	0.25–0.35	0.40	0.26–0.34	0.25	–
Italy	0.30	0.30	–	0.30	–	0.50
New Zealand	0.35	0.28	–	0.31	–	0.60
The Netherlands	0.35	0.35	–	0.35	–	–
United Kingdom	0.35	0.35	–	0.35	–	0.55
USA	0.30	0.30	–	0.30	–	0.55

*Source: Interbull (1997).
†Source: Interbull (1992).

Table 18.3. Estimated sire standard deviations (diagonal) and genetic correlations (above diagonal) considered in the Interbull evaluation for dairy production traits of February 1998; sire standard deviation estimates reflect the scale for record preadjusting in various countries and are expressed in kg (lbs in the USA).

	CAN	DEU	DNK	FRA	ITA	NLD	SWE	USA	GBR	NZL	AUS
CAN	11.75	0.91	0.91	0.95	0.93	0.93	0.91	0.96	0.93	0.80	0.84
DEU		7.50	0.91	0.92	0.90	0.93	0.90	0.89	0.90	0.77	0.80
DNK			7.29	0.91	0.90	0.94	0.94	0.91	0.93	0.76	0.81
FRA				9.73	0.95	0.93	0.91	0.95	0.93	0.78	0.82
ITA					8.65	0.91	0.90	0.95	0.91	0.77	0.81
NLD						7.70	0.94	0.93	0.93	0.80	0.83
SWE							8.58	0.90	0.91	0.76	0.80
USA								21.66	0.92	0.77	0.81
GBR									6.50	0.81	0.84
NZL										4.72	0.90
AUS											4.08

CAN, Canada; DEU, Germany; DNK, Denmark; FRA, France; ITA, Italy; NLD, Netherlands; SWE, Sweden; USA, United States of America; GBR, United Kingdom; NZL, New Zealand; AUS, Australia.

(1998) estimated additive and dominance effects to be 41–44% and 5–7%, respectively, of phenotypical variance for Holstein milk, fat and protein yields in the USA. Dominance variation is due to interactions among genes at a specific locus. When all gene action is additive, each gene adds its influence to the expressed merit of the animal. When there is dominance, one member of the allele pair masks the expression of the other. Dominance variation measures the size of this influence. Genetic correlations were reported to be 0.69 between milk and fat yields, 0.90 between milk and protein yields, and 0.78 between fat and protein yields for registered US Holsteins (Misztal et al., 1992).

Genetic standard deviations and correlations among countries for protein yield are in Table 18.3. The correlation is highest between the USA and Canada (0.96) and lowest between New Zealand and all other countries (0.76 to 0.81) except Australia (0.90). These differences are related to the management systems predominant within the country (grazing in Australia and New Zealand vs. confinement feeding in North America and Europe). The lower correlations for Australia and New Zealand indicate an interaction between genotype and environment; that is, somewhat different genes are required for high performance in North America and Europe from those required in Australia and New Zealand.

The genetic correlations among the individual parities provide an indication of the appropriateness of the assumption that later lactations are repeated observations of the lactation trait. Table 18.4 reports results from Spain (Garcia-Cortez et al., 1995) showing declining correlations as lactations are more distant. One reason for lactations not having a correlation of 1 is because cows reach their mature production level at different rates. Genetic differences between merit for individual lactations can be due to this factor.

Lactation production is measured by sampling approximately 1 day of production per month. The correlations among the individual daily productions

Table 18.4. Heritabilities, additive and residual correlations between the first four lactations for protein yield obtained in a four-trait analysis.*

Lactation	First	Second	Third	Fourth
First	**0.24**	0.89	0.78	0.69
Second	0.39	**0.25**	0.86	0.66
Third	0.32	0.42	**0.26**	0.65
Fourth	0.23	0.35	0.46	**0.19**

*Heritabilities on diagonals, additive genetic correlations above diagonals and residual correlations below diagonals.

Table 18.5. Heritabilities (on diagonal and bold), genetic correlations (above diagonal), and phenotypic correlations (below diagonal) for protein yields.

Lactation stage*	First	Second	Third	Fourth
First	**0.15**	0.92	0.83	0.75
Second	0.47	**0.15**	0.97	0.92
Third	0.40	0.56	**0.18**	0.97
Fourth	0.32	0.45	0.58	**0.18**

*Lactation stage: first, test day nearest to 43 between 6 and 80 days; second, test day nearest to 118 between 81 and 155 days; third, test day nearest to 193 between 156 and 230 days; fourth, test day nearest to 268 between 231 and 305 days.

indicate how the stages of lactation are related. These correlations for first-lactation Holsteins were estimated by Gengler *et al.* (1997). Table 18.5 provides these correlations. As with lactation yields, correlations decline as test days become more distant.

Conformation traits
Visual appraisals of cows for conformation (type) traits have been collected for many years. In many countries, conformation traits are scored on a linear scale and include udder, locomotion and other body traits. Heritability estimates for conformation traits are given in Table 18.6 for ten major dairy countries that participate in Interbull. Additive and dominance effects were estimated to be 45 and 7%, respectively, of phenotypical variance for stature, 28 and 8% for strength, 34 and 10% for body depth, 23 and 5% for dairy form, and 24 and 5% for fore udder attachment in US Holsteins (Misztal *et al.*, 1998).

Reproduction traits
To reduce losses from difficult calvings, calving ease (performance) is often considered when breeding heifers. Heritability estimates for calving ease range from 0.05 (Australian Holsteins) to 0.15 (US Holsteins) (Interbull, 1996). A genetic correlation of −0.27 between daughter and dam calving performance is assumed for Canadian dairy cattle (Interbull, 1996).

Workability traits
Workability traits include milking speed, temperament and likeability. Heritability estimates for milking speed range from 0.21 (Canadian dairy

Table 18.6. Heritability estimates used for national evaluation of conformation traits by ten countries that provide bull evaluations for Interbull (from Interbull, 1996).

Country	Udder*	Locomotion†	Other body traits‡
Australia	0.17–0.33	0.10–0.20	0.17–0.45
Canada	0.08–0.24	0.07–0.20	0.18–0.40
Denmark	0.17–0.43	0.09–0.30	0.16–0.63
France	0.30	0.30	0.50
Germany	0.30–0.36	0.25–0.35	0.40
Italy	0.30	0.30	–
New Zealand	0.35	0.28	–
The Netherlands	0.35	0.35	–
United Kingdom	0.35	0.35	–
United States	0.30	0.30	–

*Includes texture; depth; fore attachment; rear attachment height and width; support; suspensory ligament; cleft; and teat length, placement and diameter (thickness).
†Includes rear leg set and view (side and rear), hock and bone quality, and foot angle.
‡Includes size, stature, strength, capacity, body length and depth, top line; rump length, width, and angle; chest width and floor, thurl (pin) width and set, loin, bone quality, angularity, dairy character (form), muzzle width.

breeds) to 0.25 (Australian Holsteins). Australia also reports heritability estimates of 0.16 for temperament and 0.20 for likeability.

Health traits

The health trait of most concern to dairy producers is resistance to mastitis. Milk samples collected to determine fat and protein content are also evaluated for SCC, an indicator of udder health. High cell counts are associated with mastitis and depressed milk yield. Because SCC data are positively skewed and have markedly heterogeneous variances among groups, they are usually transformed to log, base 2, equivalents (somatic cell scores (SCS)). Somatic cell scores have a more normal distribution and a higher heritability than SCC (Ali and Shook, 1980), though a lower genetic correlation with clinical mastitis (Shook, 1988).

Longevity

Longevity is an overall measure of a cow's fertility and disease resistance and is often referred to as survival, stayability or productive life. In countries including Canada, France, Germany and the Netherlands, longevity is adjusted to reflect the effect of culling for low milk yield, whereas countries such as New Zealand, Australia and the USA report overall longevity. Because of culling for low yield, milk yield has a moderate correlation with overall longevity and contributes to a higher heritability for overall longevity than is found for the adjusted measure. Regardless of whether culling for low yield is considered, heritability estimates for longevity are < 0.10 for all major dairy countries (Interbull, 1996).

Inbreeding and heterosis

An animal is inbred if its parents are related. More technically, the inbreeding coefficient is the probability that an animal receives the same gene from both parents. A simple way to detect inbreeding is to determine if the same ancestor appears in the pedigree of both the sire and dam, creating the possibility of passing on the exact same gene to offspring through both parents. A consequence of industry-wide intense selection of bulls is the increase in inbreeding within the population. The use of an animal model for evaluation tends to further increase inbreeding, because families of animals tend to be selected. The consideration of all relationships tends to make the evaluations of family members similar.

Calculation of inbreeding is computationally intensive and without special techniques would require a matrix of the order of the size of the population. VanRaden (1992) proposed a method that constructs the relationship matrix of one animal at a time, thus greatly reducing memory requirements. He expressed inbreeding relative to a base population that is assumed unrelated and non-inbred. With this base, inbreeding is a measure of increase in homozygosity since that base. Inbreeding levels for the Holstein population in the USA are given in Fig. 18.1. The base population was animals born before 1960. For the 20 years until 1980, inbreeding increased slowly at about 0.044% $year^{-1}$. More recently, during the period from 1988, the rate of increase has been 0.275% $year^{-1}$.

A consequence of receiving the same genes from both ancestors is that the likelihood of undesirable recessives increases. This leads to decreased productivity called inbreeding depression. For Holstein cows in the USA Wiggans *et al.* (1995) found the values given in Table 18.7.

Heterosis can be viewed as the opposite of inbreeding and describes an increase in heterozygosity, reducing the likelihood of deleterious homozygous recessives. Heterosis measures the degree that offspring exceed the average of

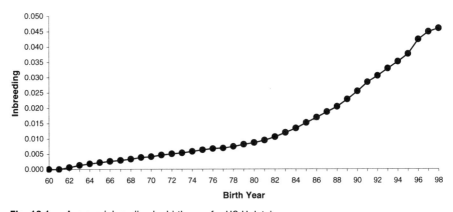

Fig. 18.1. Average inbreeding by birth year for US Holsteins.

Table 18.7. Estimates of inbreeding depression for a 1% increase in inbreeding for US Holsteins.

Trait	Inbreeding depression
Milk (kg)*	−29.6
Fat (kg)*	−1.08
Protein (kg)*	−0.97
Days of productive life (days)†	−13.07
First calving interval (days)†	+0.26
Somatic cell score (scores)†	−0.004

*Wiggans et al. (1994).
†Smith et al. (unpublished observations).

Table 18.8. Effect of heterosis on lactation yield for Holstein–Jersey crosses in New Zealand (from Harris et al., 1996).

Trait	Effect of heterosis
Fat (kg)	+6.8
Protein (kg)	+5.0
Live weight (kg)	+7.2
Survival (%)*	+4.7

*First to second lactation.

the performance of their parents, the magnitude of which depends on the genetic distance between the parents. Heterosis is usually a result of breed crosses. If the parental breeds are quite different in the trait, the benefit of heterosis is unlikely to make the progeny competitive with the higher producing parental breed. Heterosis may contribute a significant advantage in fitness. In New Zealand, where most milk is used in manufacturing, the Jersey breed (less milk, but high in fat) is perceived as competitive with Holstein (more milk, but less fat) and the progeny of crossing these breeds are highly regarded (Table 18.8).

Heterosis has also been a concern in Europe, with the introduction of semen and embryos from North America. In this situation, the crossbred progeny were backcrossed to the North American bulls, so the performance of generations past the F_1 is important. If epistatic gene combinations in the parent breeds have a positive effect on yield, often breaking these combinations up in subsequent generations will reduce yield. The loss of these epistatic effects is called recombination loss. Table 18.9 shows estimates of recombinant loss, which are negative, meaning that the segregating generations perform worse than expected from the performance of the parent breeds and the F_1. Although epistasis is a possible cause of these results, there are also other possible explanations, i.e. preferential treatment of the F_1.

Table 18.9. Estimates of heterosis and recombination for several European populations (from Harbers, 1997).

Country	Breed	Trait	Heterosis (kg)	Recombination (kg)
NL	HF × DF	Milk	120.0	−100.0
		Fat	6.0	−1.3
		Protein	4.4	−3.5
D	HF × RD	Milk	210	−2382
		Fat	10	−77
		Protein	7	83
UK	HF × F	Milk	100	−156
		Fat	4.5	−2.1
		Protein	3.6	−3.8
NL	HF × MRY	Milk	140	−295
		Fat	6.0	−11.1
		Protein	5.3	−8.2

NL, Netherlands; D, Denmark; UK, United Kingdom; HF, Holstein Friesian; DF, Dutch Friesian; RD, Red Danish; F, Friesian; MRY, Dutch MRY.

Genotype × environment interaction

An interaction between genetics and environment exists when the effect of genes is different in different environments. This can lead to a reranking of animals. Even if the ranking does not change, a smaller response in one environment is still indicative of an interaction. Procedures to account for heterogeneous variance and scale effects may eliminate this interaction.

An interaction reduces the value of information from other environments. Interbull evaluations incorporate correlations between countries of less than 1, effectively assuming some genotype × environment (G × E) interaction. In the extreme, G × E interaction is a concern in introducing high-producing cattle into marginal environments to upgrade indigenous cattle. It may be that the native cattle are well adapted to the harsh conditions and will survive in environments where the improved cattle do not. Thus the native cattle are superior in that environment.

Individual genes affecting milk production

Yields of milk and its components are classical quantitative traits affected by many genes, as well as environmental factors. However, this does not exclude the possibility that there are some individual genes which have a moderate effect. If these genes could be identified, it would increase our understanding of the genetics of milk production and be of practical use when selecting for increased production.

Originally, attention focused on those few genes whose inheritance could be easily followed, for instance blood groups, blood protein and milk protein polymorphisms (Ng-Kwai-Hang *et al.*, 1984, 1986, 1987; Kroeker *et al.*, 1985;

Gonyon *et al.*, 1987; Aleandri *et al.*, 1990; Van Eenennaam and Medrano, 1991a, b; Bovenhuis *et al.*, 1992; Bovenhuis and Weller, 1994; Famula and Medrano, 1994; Ehrmann *et al.*, 1997). Only in the last case (milk proteins) was there an obvious reason to expect that these genes would affect milk production. Many experiments found significant effects of single genes on milk production traits, but generally they have not been large enough or consistent enough to have been used in commercial breeding programmes. To understand the inconsistency of the results, it is necessary to understand the reasons that could cause an association between an animal's genotype at a specific locus and its milk production.

1. The gene could have a direct effect on milk production.
2. The gene could be in linkage disequilibrium with a gene affecting milk production.
3. The gene could be linked to a gene for milk production in linkage equilibrium in the population but in linkage disequilibrium in the sample of families studied.
4. The effect could be totally spurious and due to the statistical analysis ignoring the family structure of the animal samples. (For instance, a particular allele may occur largely in the descendants of one exceptional ancestor, causing the appearance that this allele increases milk production.)
5. False-positive results are expected in one in 20 tests and many significant tests are carried out because there are several markers multiplied by several traits.

Some, but not all, experiments have attempted to distinguish between these five causes of association. Hines (1990) reviewed work in his own laboratory and elsewhere on blood groups, transferrin and milk proteins in Holstein and Guernsey cattle. Among the most consistent effects of blood groups and proteins were an effect of the B blood group on fat % and transferrin type on milk yield, both in Holsteins. Although these appear to be direct effects, rather than linked markers, they could be due to linkage disequilibrium between the marker and a quantitative trait locus (QTL) for milk production.

The milk protein genes are more logical candidates for a direct effect on milk production and do show somewhat more consistent results, although there are still clear differences between experiments. These have been reviewed by Bovenhuis and Weller (1994), as well as Hines (1990).

At the β-lactoglobulin gene, the AA genotype tends to have the highest protein %, with an advantage of about 0.03% over the BB genotype. This occurs because the AA genotype increases the concentration of β-lactoglobulin itself (Ng-Kwai-Hang *et al.*, 1987; Ehrmann *et al.*, 1997), and hence whey. There may be some compensatory decrease in casein % (Ng-Kwai-Hang *et al.*, 1986) but not enough to eliminate the increase in total protein %. There is a tendency, usually non-significant, for the AA genotype to increase milk yield by up to 100 l. This may explain part of the decrease in casein %. The AA genotype also decreases fat % by about 0.05% compared with the BB genotype. The consistency of this effect suggests it is due to the β-lactoglobulin gene itself, but Bovenhuis and Weller (1994) also found evidence for a gene linked to β-lactoglobulin affecting fat %.

The κ-casein gene also affects the concentration of its own protein, with the BB genotype causing the highest concentration. In AB heterozygotes, the concentration of B protein in the milk is almost twice that of A protein (Van Eenennaam and Medrano, 1991b). This effect is most easily explained by a polymorphism in a regulatory region of the gene which is in linkage disequilibrium with the polymorphism in the coding region. This explanation is supported by the absence of an effect of κ-casein genotype in breeds other than Holstein. Due to this increase in κ-casein, the total protein % is about 0.05% higher in BB than AA genotypes. The yield of protein (kg) in most studies is also highest in BB genotypes, but there is no consistent effect on milk volume. There is no consistent effect on fat % or yield, although Hines (1990) suggested a QTL for fat % linked to κ-casein.

The effect of the β-casein genotype is less clear-cut than that of β-lactoglobulin and κ-casein, but it appears that, in Holsteins, the B allele increases the production of β-casein (Kroeker *et al.*, 1985; Ng-Kwai-Hang *et al.*, 1987; Ehrmann *et al.*, 1997) and hence increases casein % (Ng-Kwai-Hang *et al.*, 1986) and protein %. The advantage of the B allele does not lead to an increase in protein yield (kg), due to a possible depressing effect on milk volume. There have been some studies with significant effects of β-casein genotype on fat %, with the A_1 and B alleles associated with highest fat % (Ng-Kwai-Hang *et al.*, 1984, 1986; Bovenhuis *et al.*, 1992; Bovenhuis and Weller, 1994).

At the α_{s1}-casein locus, the C allele increases protein %, possibly because it increases the proportion of α-casein itself (Ng-Kwai-Hang *et al.*, 1987; Ehrmann *et al.*, 1997). However, the C allele does not cause a consistent increase in protein yield and in fact the B allele appears to be more often associated with the highest milk yield.

At the β-lactoglobulin and κ-casein genes, the BB genotypes have the highest ratio of casein to whey proteins and they also have greater cheese yield and shorter renneting times than the AA genotypes (Graham *et al.*, 1984; Marziali and Ng-Kwai-Hang, 1986; Aleandri *et al.*, 1990).

In summary, evidence is building that the genotype at milk protein loci has its primary effect on the synthesis of its own protein. Poorly understood regulatory mechanisms may then lead to changes in the synthesis of other proteins and perhaps even milk volume and fat. In addition, there is evidence for genes affecting fat % which are linked to the casein complex and β-lactoglobulin (Ron *et al.*, 1994).

With the advent of deoxyribonucleic acid (DNA) technology other candidate genes have been investigated, and prolactin (Cowan *et al.*, 1990) and growth hormone (Hoj *et al.*, 1993; Falaki *et al.*, 1996) are reported to affect milk production.

A more systematic approach to finding genes for milk production is to map QTL, using a linkage study and a collection of genetic markers that cover all 30 chromosomes. Using a granddaughter design (Weller *et al.*, 1990),

Georges *et al.* (1995) found five chromosome regions that affected milk, fat or protein yield. One QTL mapped to chromosome 6, which includes the casein loci, but the QTL mapped to a different part of the chromosome.

When performing significance tests with 159 genetic markers, as Georges *et al.* (1995) did, there is a high probability of false positives. To guard against this, Georges *et al.* used a very stringent significance level. This reduces the likelihood of false positives but means that some QTL that were segregating in their families are likely to have missed detection. Ashwell *et al.* (1997) performed a similar but smaller experiment, with only 16 markers and a less stringent significance test. They found many markers that were significant for some trait in some family, but a number of these are expected to be false positives.

Other studies vary in the stringency of their significance tests and hence their trade-off between too many false positives and failing to detect QTL that are segregating. However, cases of agreement between studies confirm the presence of QTLs.

Boichard and Bishop (1997) confirmed the existence of a QTL affecting protein yield and milk yield near the beginning of chromosome 1, found by Georges *et al.* (1995). Both Mosig (1998) and Ron (1998) reported a QTL for protein % on chromosome 3 at about 50 centimorgans (cM). On chromosome 6, there appear to be at least two QTL (Kuhn, 1996), one around 30–50 cM (Georges *et al.*, 1995; Boichard and Bishop, 1997; Gonaz-Raya, 1998; Mosig *et al.*, 1998) and one around 80 cM (Mosig *et al.*, 1998), close to the casein genes. The first QTL increases milk volume without changing fat and protein yield and so decreases fat % and protein % (Georges *et al.*, 1995) or at least protein % (Spelman *et al.*, 1996). On chromosome 9, Georges *et al.* (1995) found a QTL affecting fat and protein yield at about 60 cM; in a similar location Mosig (1988) detected a QTL affecting protein % (they did not measure the effect on other traits); and Vilkki *et al.* (1997), in a different breed, found evidence, although not significant, for a QTL affecting milk and protein yield.

In contrast, on chromosome 10, Georges *et al.* (1995) found a QTL at about 20 cM affecting fat %, Mosig (1998) a QTL at about 40 cM affecting protein % and Ron (1998) a QTL at about 80 cM affecting protein %. On chromosome 23, Boichard and Bishop (1997) and Ashwell *et al.* (1997) found suggestive evidence for a QTL affecting fat yield. The prolactin gene also maps to chromosome 23.

It seems likely that, with further research, many QTLs affecting milk production will be mapped and hopefully these genes will eventually be identified. Some will probably turn out to be known genes, such as the caseins, and some previously unknown genes.

Genetic Evaluation

The goal of a genetic evaluation system is to produce rankings of animals that will enable progress in attaining a breeding objective when selection decisions are made based on the rankings. Both national and international genetic evaluation systems have been developed over the last 60 years. As statistical techniques and computing power have advanced, evaluation systems have become more accurate in estimating genetic differences.

Evaluation models

One of the earliest methods for genetic evaluation of dairy bulls was a daughter–dam comparison. This method assumed that the difference in yield between a bull's daughter and its dam resulted from the genetics of the bull; that is, the effect of environment was assumed to be constant. The daughter–dam comparison was followed by the herdmate comparison, which accounted for the effect of environment by comparing animals that produced in the same herd and calved during the same season. However, the daughter–dam comparison did not account for genetic differences between herds or consider the genetic contribution from parents. Relationships among cows through their sires (and later maternal grandsires) were able to be accounted for by using best linear unbiased prediction procedures with a sire model, thereby joining the genetic considerations of the daughter–dam comparison and the environmental considerations of the herdmate comparison.

Currently, an animal model is used by nearly all major dairy countries. An animal model allows consideration of all relationships among animals and results in simultaneous evaluation of cows and bulls. A particular animal's evaluation is a function of the evaluations of its parents and its progeny, as well as its own records. However, because the system is simultaneous, information from one animal can affect the evaluations of others.

In Australia, New Zealand and the north-eastern USA, the lactation measure is calculated from yield deviations on individual test days. This test-day model allows more accurate accounting for environment, because effects of specific test days are estimated. The test-day model is an animal model that uses the test-day yields instead of the lactation yields as input. One advance in this model is to allow for genetic differences by test day. Jamrozik *et al.* (1997) of Canada have proposed fitting a lactation curve for each cow and lactation as a random genetic effect. An alternative proposed by Wiggans and Goddard (1997) is to define test-day yields as separate effects which are correlated and to analyse them in a multitrait analysis. Both of these approaches support analysis of persistency and should reduce the fluctuation in evaluations of bulls when many second-crop daughters' partial lactation records are added. This fluctuation may be caused by daughters whose lactation curves differ from the norm.

National evaluations

Each country has adapted its evaluation system to model the structure of its data. Some of the ways in which systems differ include calculation of lactation records, parameter estimates, accounting for age, definition of environmental groups, definition of unknown-parent groups, accounting for inbreeding and heterosis, and reporting scale for evaluations.

International evaluation

The extensive marketing of bull semen and embryos internationally has generated an interest in international comparison of bulls. The Interbull Centre in Uppsala, Sweden, combines bull evaluations from participating countries to generate rankings that include the bulls from all countries, but reported on each country's evaluation scale. This multitrait, across-country evaluation (MACE) (Schaeffer, 1985) recognizes correlations of less than one between performance in different countries, so the rankings may differ. The MACE procedure was first used in 1994. Previously, conversion equations were used. The Interbull Centre also conducts research to improve international ranking and ways to extend the procedure to other traits.

Traits evaluated

Milk yield is the fundamental trait analysed, although milk fat percentage determination has been part of most milk recording systems since the beginning. With the growing importance of manufacturing and the improvements in laboratory equipment, determination of protein content has become almost universal in major dairy producing countries. Somatic cell count collection is also widespread and is used as an indicator of udder health and the presence of subclinical mastitis. Most countries also have a programme to collect conformation trait data. Recent research effort has focused on using these traits to select for increased profitability through prolonged herd life or greater disease resistance or as an indicator of maintenance cost.

Genetic trend for yield

Average breeding value of cows by birth year is a common measure of genetic trend and an indication of the success of a national breeding programme. However, the evaluation model and the adjustments for age effects can affect trend estimates. Table 18.10 shows the trend in breeding values for yield traits

Table 18.10. Genetic trend in yield traits of US Holstein cows born in 1994.

Trait	Trend (kg)	% of mean production
Milk	131	1.3
Fat	3.9	1.1
Protein	4.1	1.3

of US Holstein cows. Because trend has been increasing in recent years, the values are specific for cows born in 1994.

Economic indices

With the large number of traits analysed, it is necessary to define a breeding goal and develop an index that weights individual traits according to their contribution to that goal. In many countries, there is a negative weight on milk volume, because of the cost of production, hauling and removal. Protein yield receives heavy weight in most indices. Scandinavian countries have an extensive system for collection of health data and include health traits in their index.

Future enhancements

With the rising cost of labour, most milk recording systems are attempting to find less expensive ways to collect data. This has led to the popularity of a.m.–p.m. plans where only one milking per month is recorded, at alternating times (a.m. or p.m.). Large herds with electronic meters able to collect daily weights find that collection of samples is a considerable inconvenience and so collect samples only quarterly. A test day model is well suited to accommodate a wide range of testing plans.

With the advances in DNA technology, genes that influence yield of milk and components are likely to be discovered. Evaluation systems can be adapted to estimate the effect of various alleles and improve accuracy of evaluations by using that information.

Design of Breeding Programmes

Estimated breeding values (EBVs) provide cattle breeders with a tool for identifying the best bulls and cows for breeding. Genetic progress is also affected by which specific matings are made. The design of breeding programmes specifies how these cattle are mated. In dairy cattle, research into the design of breeding programmes has focused on obtaining maximum benefit from new technology, especially reproductive technology, such as AI and multiple ovulation and embryo transfer (MOET).

Artificial insemination

The availability of AI led to breeding programmes based on progeny testing. A group of young bulls are progeny-tested by producing a number of daughters each. After the progeny test, bulls selected for widespread use are mated to produce replacement heifers and a new generation of young bulls. These young bulls are bred from the best cows available. Selection of cows to produce replacement heifers is also practised, but is of limited value because the low reproductive rate of cows means that nearly all cows are needed to maintain the herd size. Thus there are four types of selection decisions: bulls to breed bulls, bulls to breed cows, cows to breed bulls and cows to breed cows, but the selection intensity on the fourth pathway is low.

The design parameters which have attracted most attention are the proportion of cows to be mated to young bulls, the number of young bulls' progeny tested per year and the number of daughters per young bull. The optimum values of these parameters vary widely between studies, from 15% to almost 100% of cows mated to young bulls and 20 to 400 daughters per young bull. For a given-size population of cows, these two parameters determine the number of young bulls to be progeny-tested (Skjervold and Langholz, 1964; Van Vleck, 1964; Lindhe, 1968; Hinks, 1970; Hunt et al., 1972; Brascamp, 1973; Oltenacu and Young, 1974; Petersen et al., 1974; Stitchbury and Goddard, 1985; Dekkers et al., 1996).

Several factors explain the different optima found. If the rate of genetic progress is the objective, the optimum number of daughters per bull is low and the optimum proportion of cows mated to young bulls is high (Skjervold and Langholz, 1964). However, if the economic benefit from the programme is the objective, the optimum number of daughters per bull is higher and the proportion of mating to young bulls is reduced (Van Vleck, 1964; Lindhe, 1968; Hinks, 1970; Brascamp, 1973; Petersen et al., 1974; Dekkers et al., 1996). Because it is expensive to purchase and raise new bulls, it costs less to produce an extra daughter from a bull already being tested than the first daughter from a new bull (Meuwissen and Goddard, 1997). Also, discounting future benefits favours using proved bulls instead of young bulls, so that the benefits of selection are achieved more quickly.

The breeding objective also affects the optimum design. If the objective is for dual-purpose cattle, bulls can be selected for progeny-testing based on their own performance (i.e. growth rate). This increases the optimum proportion of matings to young bulls and increases the number of daughters per young bull. However, if the objective includes traits of low heritability (e.g. mastitis resistance and fertility), the optimum number of daughters per young bull increases (Skjervold and Langholz, 1964; Meuwissen and Woolliams, 1993).

As the population size increases, the optimum proportion of matings to young bulls decreases, the number of daughters per bull increases, the number of bulls' progeny tested increases and the rate of genetic gain increases (Skjervold and Langholz, 1964; Stitchbury and Goddard, 1985). In practice, the advantages of large population size are being exploited by the use of a 'global' breeding programme in which bulls to breed bulls are selected from a worldwide pool of bulls. These optima are comparatively flat, so there is little cost in departing slightly from the optimum value of a parameter. When economic benefit in specialized dairy cattle is the objective, 100–200 daughters per young bull is close to the optimum and in fact major dairy breeding programmes are using such designs (Van Vleck, 1964; Lindhe, 1968; Hinks, 1970; Brascamp, 1973; Petersen *et al.*, 1974; Stitchbury and Goddard, 1985; Dekkers *et al.*, 1996).

Multiple ovulation and embryo transfer

At first, the availability of MOET did not appear to be of much value to dairy breeding programmes, because it was too expensive to use on the 'cows to breed cows' pathway and the cows to breed bulls were already highly selected. New designs were needed to gain benefit from MOET. Nicholas and Smith (1989) proposed nucleus breeding herds, with selection of bulls based on the performance of their sibs and older relatives. In the so-called 'adult' scheme, bulls and cows were selected at 3½ years of age, when the cows have a partial lactation record. In the 'juvenile' scheme, bulls and cows were selected at 15 months of age, based on their parents' EBVs. With MOET, both nucleus schemes had faster predicted genetic gain than a traditional progeny-testing programme. More recent calculation of the predicted rates of genetic gain have reduced the superiority of the nucleus MOET schemes, but not eliminated it (Lohuis *et al.*, 1993; Lohuis, 1995). These designs have a reduced generation interval but less accurate selection than progeny testing. In practice, the very short generation interval is hard to achieve.

The nucleus and progeny-testing designs could be combined by opening the nucleus to élite cows from the general population and by using progeny-tested bulls as sires within the nucleus ('hybrid' schemes). Meuwissen (1991) showed that MOET increased genetic gain by 13% in these hybrid schemes. Estimated breeding values, which can compare animals across age-groups, provide a logical way to select between bulls and cows of different ages, and inside and outside the nucleus. In this way, some young bulls and some proved bulls can be used as sires within the nucleus.

Selection on EBV maximizes the genetic merit of the next generation, but it is not necessarily the policy which maximizes the merit of future generations. There may be an advantage in selecting animals with a lower EBV but of lower reliability, because this improves the opportunity for selection in later generations (Goddard and Howarth, 1994). For instance, Meuwissen (1991) found that selecting cows outside the nucleus on EBV could actually decrease the rate of genetic gain. Similarly, it might be worthwhile to use young bulls to breed bulls even if they have a lower EBV than the best proved bulls. However, the improvement in genetic gain by doing this is usually small.

With aspiration of oocytes from the ovaries, followed by *in vitro* maturation and fertilization (*in vitro* embryo production (IVEP)), it is possible to increase the reproductive rate of cows above that possible with MOET and to achieve a further small increase in genetic gain (Kinghorn *et al.*, 1991; de Boer and Van Arendonk, 1994; Leitch *et al.*, 1995; Lohuis, 1995).

Minimizing inbreeding

Intense selection implies a small number of parents for the next generation and in time this causes inbreeding. Small effective population size and inbreeding cause inbreeding depression, increased incidence of recessive abnormalities, reduced genetic variation and random fluctuations in the mean of the population. Consequently, it is desirable to minimize inbreeding and maximize genetic gain, which, because they are conflicting objectives, implies some trade-off between them (Leitch *et al.*, 1994). In traditional progeny-testing schemes, the number of bulls used in the nucleus or used to breed bulls largely determines the rate of inbreeding. Most studies have found the optimum balance between genetic gain and inbreeding was to use two new bulls each year (Skjervold and Langholz, 1964; Hunt *et al.*, 1974; Petersen *et al.*, 1974; Stitchbury and Goddard, 1985). This high intensity of selection may not be appropriate for large populations, such as the global black and white cow population (Goddard, 1990). Fortunately, slight differences in breeding objectives between countries can lead to different bulls being selected in different countries, and consequently the total number of bulls used is increased (Goddard, 1990).

Nucleus schemes using MOET, with their short generation intervals, have higher rates of inbreeding than traditional designs. When minimizing inbreeding or variance of the mean is part of the objective, this causes the optimum design to move toward use of proved bulls and an open nucleus. There is also an advantage to factorial mating schemes in which each cow is mated to several bulls (Leitch *et al.*, 1994; Meuwissen and Woolliams, 1994; Luo *et al.*, 1995).

As the reproductive rate of cows increases (e.g. by using IVEP), the number of cows needed as parents decreases and this further increases inbreeding. Optimum designs may then use as many bulls as cows in a square factorial mating system or even more bulls than cows, because the accuracy of selection of bulls is less than that of cows at 4 years of age (de Boer and Van Arendonk, 1994).

Cloning

The technology to produce many genetically identical cows may soon be available (Seidel, 1996). Cloning would not greatly increase the rate of genetic gain in the nucleus (de Boer and Van Arendonk, 1994). However, it could dramatically reduce the amount by which the average commercial cow lags behind the nucleus. In this scenario many clones would be produced and distributed as embryos to commercial dairy farmers. This would provide a once-only lift in the genetic merit of the commercial cow population, but it would be necessary to maintain existing breeding programmes to generate ongoing genetic improvement.

Marker-assisted selection

Some genes that cause variation in milk yield or other important traits have been identified or mapped by linkage to genetic markers, and the number of these genes will undoubtedly increase (Ng-Kwai-Hang et al., 1984, 1986, 1987; Kroeker et al., 1985; Gonyon et al., 1987; Aleandri et al., 1990; Van Eenennaam and Medrano, 1991; Bovenhuis et al., 1992; Bovenhuis and Weller, 1994; Famula and Medrano, 1994; Ehrmann et al., 1997). Deoxyribonucleic acid tests for genotype at these loci provide additional information on the genetic value of bulls and cows and so could lead to more accurate selection. The genotype information is most useful for traits which are otherwise difficult to select for or for genes which show non-additive gene action (Larzul et al., 1997).

Computer simulations of MOET nucleus herds show that tests for markers linked to QTL could increase the rate of genetic gain by up to 20% (Ruane and Colleau, 1996; Meuwissen and Goddard, 1997). However, in traditional progeny testing programmes the benefit is less (Spelman and Garrick, 1997). It takes some years before the maximum benefit from the markers is achieved because the initial data are used to establish linkage phase. In the long term, the benefit from markers declines, because the QTL to which they are linked becomes fixed and so the markers are no longer useful. In the very long term, selection using the markers may even achieve less progress than selection ignoring the markers (Gibson, 1994). If a test existed for the QTL itself, instead of the markers linked to it, this would allow faster genetic progress and be easier to implement, because data to establish linkage phase would not be needed.

A DNA test is available for one mutation at the melanocyte-stimulating hormone (MSH) locus, which causes red coat colour (Klungland et al., 1995), and for some disease genes, such as citrullinaemia (Dennis et al., 1989). These tests help to identify carriers of undesirable genes but the economic benefit from this is usually small compared with an increase in traits such as milk protein yield.

Conclusions

Rapid progress in genetic improvement of dairy cattle has been achieved in recent years. This progress has resulted from a focus on yields of milk and components, the traits of primary economic importance. The investment of producers in milk recording and AI organizations in testing a large number of young bulls each year has been an important contribution to this success. Data collection is somewhat easier with dairy cattle than with some other farm species, because of the intensive nature of production and the relatively high value of the individual animals. This situation has led to a highly developed system of data collection, genetic evaluation and young sire development. Future developments in marker-assisted selection, evaluation methods and breeding plans hold promise for further increases in the rate of genetic improvement.

References

Aleandri, R., Buttazzoni, L.G., Schneider, J.C., Caroli, A. and Davoli, R. (1990) The effects of milk protein polymorphisms on milk components and cheese-producing ability. *Journal of Dairy Science* 73, 241–255.

Ali, A.K.A. and Shook, G.E. (1980) An optimum transformation for somatic cell concentration in milk. *Journal of Dairy Science* 63, 487–490.

Ashwell, M.S., Rexroad, C.E. Jr, Miller, R.H., VanRaden, P.M. and Da, Y. (1997) Detection of loci affecting milk production and health traits in an elite US Holstein population using microsatellite markers. *Animal Genetics* 28, 216–222.

Boichard, D. and Bishop, M.D. (1997) Detection of QTLs influencing milk production and mastitis resistance with a granddaughter design in Holstein cattle. In: *48th Annual Meeting of the EAAP Commission on Animal Genetics*. Vienna, Austria, pp. 1–5.

Bonczek, R.R. and Young, C.W. (1980) Comparison of production and reproduction traits of two inbred lines of Holstein cattle with attention to the effect of inbreeding. *Journal of Dairy Science* 63(Suppl. 1); 107 (abstract).

Bovenhuis, H. and Weller, J.I. (1994) Mapping and analysis of dairy cattle quantitative trait loci by maximum likelihood methodology using milk protein genes as genetic markers. *Genetics* 137, 267–280.

Bovenhuis, H., Van Arendonk, J.A.M. and Korver, S. (1992) Associations between milk protein polymorphisms and milk production traits. *Journal of Dairy Science* 75, 2549–2559.

Brascamp, E.W. (1973) Model calculations concerning economic optimization of AI breeding with cattle. *Zeitschrift für Tierzüchtung und Züchtungsbiologie* 90, 126.

Cowan, C.M., Dentine, M.R., Ax, R.L. and Schuler, L.A. (1990) Structural variation around prolactin gene linked to quantitative traits in an elite Holstein sire family. *Theoretical and Applied Genetics* 79, 577–582.

de Boer, I. and Van Arendonk, J. (1994) Additive response to selection adjusted for effects of inbreeding in a closed dairy cattle nucleus assuming a large number of gametes per female. *Animal Production* 58, 173–180.

Dekkers, J.C.M, Vandervoort, G.E. and Burnside, E.B. (1996) Optimal size of progeny groups for progeny-testing programs by artificial insemination firms. *Journal of Dairy Science* 79, 2056–2070.

Dennis, J.A., Healy, P.J., Beaudet, P.J. and O'Brien, W.E. (1989) Molecular definition of bovine arginosuccinate synthetase deficiency. *Proceedings of the National Academy of Science, USA* 86, 7947–7951.

Ehrmann, S., Bartenschlager, H. and Geldermann, H. (1997) Quantification of gene effects on single milk proteins in selected groups of dairy cows. *Journal of Animal Breeding and Genetics* 114, 121–132.

Falaki, M., Sneyers, M., Prandi, A., Massart, S., Corradini, C., Formigoni, A., Burny, A., Portetelle, D. and Renaville, R. (1996) *Taq* I growth hormone gene polymorphism and milk production traits in Holstein–Friesian cattle. *Animal Science* 63, 175–181.

Famula, T.R. and Medrano, J.F. (1994) Estimation of genotype effects for milk proteins with animal and sire transmitting ability models. *Journal of Dairy Science* 77, 3153–3162.

Garcia-Cortez, L.A., Moreno, C., Varona, L., Rico, M. and Altarriba, J. (1995) (Co)variance component estimation of yield traits between different lactations using an animal model. *Livestock Production Science* 43, 111–117.

Gengler, N., Tijani, A., Wiggans, G.R., Van Tassell, C.P. and Philpot, J.C. (1997) Estimation of (co)variances of test day yields for first lactation Holsteins in the United States. *Journal of Dairy Science* 82, 225.

Georges, M., Nielsen, D., Mackinnon, M., Mishra, A., Okimoto, R., Pasquino, A., Sargeant, L., Sorensen, A., Steele, M., Zhao, X., Womack, J. and Hoeschele, I. (1995) Mapping quantitative trait loci controlling milk production in dairy cattle by exploiting progeny testing. *Genetics* 139, 907–920.

Gibson, J.P. (1989) Economic weights and index selection of milk production traits when multiple production quotas apply. *Animal Production* 49, 171–181.

Gibson, J.P. (1994) Short-term gain at the expense of long-term response with selection of identified loci. In: *Proceedings of the 5th World Congress on Genetics Applied to Livestock Production*, Vol. 21. Guelph, Ontario, Canada, pp. 201–204.

Goddard, M.E. (1990) Optimal effective population size for the global population of black and white dairy cattle. *Journal of Dairy Science* 75, 2902–2911.

Goddard, M.E. (1998) Consensus and debate in the definition of breeding objectives. *Journal of Dairy Science* 2, 6–18.

Goddard, M.E. and Howarth, J.M. (1994) Dynamic selection rules and selection of males to progeny test. In: *Proceedings of the 5th World Congress on Genetics Applied to Livestock Production*, Vol. 18. Guelph, Ontario, Canada, pp. 306–309.

Gomez, A., Steine, T., Colleau, J., Pedersen, J., Pribyl, J. and Reinsch, N. (1997) Economic values in dairy cattle breeding, with special reference to functional traits – report of an EAAP working group (review). *Livestock Production Science* 49, 1–21.

Gomez-Raya, L. (1998) Mapping QTL for milk production traits in Norwegian cattle. In: *Proceedings of the 6th World Congress on Genetics Applied to Livestock Production*, Vol. 26. Armidale, New South Wales, Australia, pp. 429–432.

Gonyon, D.S., Mather, R.E., Hines, H.C., Haenlein, G.F.W., Arave, C.W. and Gaunt, S.N. (1987) Associations of bovine blood and milk polymorphism with lactation traits: Holsteins. *Journal of Dairy Science* 70, 2585–2598.

Graham, E.R.B., McLean, D.M. and Zviedrans, P. (1984) The effect of milk protein genotypes on the cheesemaking properties of milk and on the yield of cheese. In: *Fourth Conference of the Australian Association of Animal Breeding and Genetics*. Adelaide, South Australia, pp. 136–137.

Harbers, A.G.F. (1997) The usage of heterosis correction in a multiple breed genetic evaluation. In: *Proceedings of 1997 Interbull Meeting*, Vol. 16. Vienna, Austria, pp. 89–93.

Harris, B.L., Clark, J.M. and Jackson, R.G. (1996) Across breed evaluation of dairy cattle. In: *Proceedings of the New Zealand Society of Animal Production*. New Zealand Society of Animal Production, Waikato University, pp. 12–15.

Hines, H. (1990) Genetic markers for quantitative trait loci in dairy cattle. In: *Proceedings of the 4th World Congress on Genetics Applied to Livestock Production*, Vol. 13. Edinburgh, Scotland, pp. 121–124.

Hinks, C.J.M. (1970) The selection of dairy bulls for artificial insemination. *Animal Production* 12, 569–576.

Hoj, S., Fredholm, M., Larsen, N.J. and Nielsen, V.H. (1993) Growth hormone gene polymorphism associated with selection for milk fat production in lines of cattle. *Animal Genetics* 24, 91–96.

Hunt, M.W., Burnside, E.B., Freeman, M.G. and Wilton, J.W. (1972) Impact of selection, testing, and operational procedures on genetic progress in a progeny testing artificial insemination stud. *Journal of Dairy Science* 55, 829.

Hunt, M.S., Burnside, E.B., Freeman, M.G. and Wilton, J.W. (1974) Genetic gain when sire sampling and proving programs vary in different artificial insemination population sizes. *Journal of Dairy Science* 57, 251–257.

Interbull (1992) *Sire Evaluation Procedures for Dairy Production Traits Practised in Various Countries, 1992*. Bulletin No. 5. Department of Animal Breeding and Genetics, SLU, Uppsala, Sweden, 84 pp.

Interbull (1996) *Sire Evaluation Procedures for Non-dairy-production and Growth and Beef Production Traits Practised in Various Countries, 1996*. Bulletin No. 13. Department of Animal Breeding and Genetics, SLU, Uppsala, Sweden, 201 pp.

Interbull (1997) Interbull routine genetic evaluation for dairy production traits, August 1997. http://www-Interbull.slu.se/lastev/lastev1.html. Accessed 20 January, 1998.

Jamrozik, J., Schaeffer, L.R. and Dekkers, J.C.M. (1997) Genetic evaluation of dairy cattle using test day yields and random regression model. *Journal of Dairy Science* 80, 1217–1226.

Kinghorn, B.P., Smith, C. and Dekkers, J.C.M. (1991) Potential genetic gains in dairy cattle with gamete harvesting and *in vitro* fertilization. *Journal of Dairy Science* 74, 611–622.

Klungland, H., Vage, D.I., Gomez-Raya, L., Adalsteinsson, S. and Lein, S. (1995) The role of melanocyte-stimulating hormone (MSH) receptor in bovine coat color determination. *Mammalian Genome* 6, 636–639.

Kroeker, E.M., Ng-Kwai-Hang, K.F., Hayes, J.F. and Moxley, J.E. (1985) Effects of environmental factors and milk protein polymorphism on composition of casein fraction in bovine milk. *Journal of Dairy Science* 68, 1752–1757.

Kuhn, C.H. (1996) Isolation and application of chromosome 6 specific microsatellite markers for detection of QTL for milk-production traits in cattle. *Journal of Animal Breeding and Genetics* 113, 355–362.

Larzul, C., Manfredi, E. and Elsen, J. (1997) Potential gain from including major gene information in breeding value estimation. *Genetics Selection Evolution* 29, 161–184.

Leitch, H., Smith, C., Burnside, E. and Quinton, M. (1994) Genetic response and inbreeding with different selection methods and mating designs for nucleus breeding programs of dairy cattle. *Journal of Dairy Science* 77, 1702–1718.

Leitch, H., Smith, C., Burnside, E. and Quinton, M. (1995) Effects of female reproductive rate and mating design on genetic response and inbreeding in closed nucleus dairy herds. *Animal Production* 60, 389–400.

Lindhe, B. (1968) Model simulation of AI breeding within a dual purpose breed of cattle. *Acta Agriculture Scandinavica* 18, 33–39.

Lohuis, M. (1995) Potential benefits of bovine embryo-manipulation technologies to genetic improvement programs. *Theriogenology* 43, 51–60.

Lohuis, M., Smith, C. and Dekkers, J. (1993) MOET results from a dispersed hybrid nucleus programme in dairy cattle. *Animal Production* 57, 369–378.

Luo, A., Woolliams, J. and Thompson, R. (1995) Controlling inbreeding in dairy MOET nucleus schemes. *Animal Production* 60, 379–387.

Marziali, A.S. and Ng-Kwai-Hang, K.F. (1986) Relationships between milk protein polymorphisms and cheese yielding capacity. *Journal of Dairy Science* 69, 1193–1201.

Meuwissen, T.H.E. (1991) Expectation and variance of genetic gain in open and closed nucleus and progeny testing schemes. *Animal Production* 53, 133–141.

Meuwissen, T.H.E. and Goddard, M.E. (1997) Optimization of progeny tests with prior information on young bulls. *Livestock Production Science* 52, 57–68.

Meuwissen, T.H.E. and Woolliams, J.A. (1993) Responses of multi-trait selection in open nucleus schemes for dairy cattle breeding. *Animal Production* 56, 293–299.

Meuwissen, T.H.E. and Woolliams, J.A. (1994) Maximizing genetic response in breeding schemes of dairy cattle with constraints on variance of response. *Journal of Dairy Science* 77, 1905–1916.

Misztal, I., Lawlor, T.J., Short, T.H. and VanRaden, P.M. (1992) Multiple-trait estimation of variance components of yield and type traits using an animal model. *Journal of Dairy Science* 75, 544–551.

Misztal, I., Varona, L., Culbertson, M., Gengler, N., Bertrand, J.K., Mabry, J., Lawlor, T.J. and Van Tassell, C.P. (1998) Studies on the value of incorporating effect of dominance in genetic evaluations of dairy cattle, beef cattle, and swine. In: *Proceedings of the 6th World Congress on Genetics Applied to Livestock Production*, Vol. 25. Armidale, New South Wales, Australia, pp. 513–516.

Mosig, M.O. (1998) Mapping QTL affecting milk protein in Israel Holstein dairy cattle by selective DNA pooling with dinucleotide microsatellite markers. In: *Proceedings of the 6th World Congress on Genetics Applied to Livestock Production*, Vol. 26. Armidale, New South Wales, Australia, pp. 253–256.

Ng-Kwai-Hang, K.F., Hayes, J.F., Moxley, J.E. and Monardes, H.G. (1984) Association of genetic variants of casein and milk serum proteins with milk, fat and protein production by dairy cattle. *Journal of Dairy Science* 67, 835–840.

Ng-Kwai-Hang, K.F., Hayes, J.F., Moxley, J.E and Monardes, H.G. (1986) Relationships between milk protein polymorphisms and major milk constituents in Holstein–Friesian cows. *Journal of Dairy Science* 69, 22–26.

Ng-Kwai-Hang, K.F., Hayes, J.F., Moxley, J.E. and Monardes, H.G. (1987) Variation in milk protein concentrations associated with genetic polymorphism and environmental factors. *Journal of Dairy Science* 70, 563–570.

Nicholas, F.W. and Smith, C. (1989) Increased rates of genetic change in dairy cattle by embryo transfer and splitting. *Animal Production* 36, 341–353.

Oltenacu, P.A. and Young, C.W. (1974) Genetic optimization of a young bull sampling program in dairy cattle. *Journal of Dairy Science* 57, 894–897.

Petersen, P.H., Gjøl Christenson, L., Bech Andersen, B. and Ovesen, E. (1974) Economic optimization of breeding structure within a dual purpose cattle population. *Acta Agriculturae Scandinavica* 24, 247–258.

Philipsson, J., Banos, G. and Arnason, T. (1994) Present and future uses of selection index methodology in dairy cattle. *Journal of Dairy Science* 77, 3252–3261.

Plowman, R.D. and McDaniel, B.T. (1968) Changes in USDA Sire Summary procedures. *Journal of Dairy Science* 51, 306–311.

Ron, M. (1998) A new approach to the problem of multiple comparisons for detection of quantitative trait loci. In: *Proceedings of the 6th World Congress on Genetics Applied to Livestock Production*, Vol. 26. Armidale, New South Wales, Australia, pp. 229–232.

Ron, M., Yoffe, O., Ezra, E., Medrano, J.F. and Weller, J.L. (1994) Determination of effects of milk protein genotype on production traits of Israeli Holsteins. *Journal of Dairy Science* 77, 1106–1113.

Ruane, J. and Colleau, J.J. (1996) Marker-assisted selection for a sex-limited character in a nucleus breeding population. *Journal of Dairy Science* 79, 1666–1678.

Schaeffer, L.R. (1994) Multiple-country comparison of dairy sires. *Journal of Dairy Science* 77, 2671–2678.

Seidel, G.E. (1995) Sexing, bisection and cloning embryos. In: Enne, G., Greppi, G.F. and Lauria, A. (eds) *Reproduction and Animal Breeding: Advances and Strategy*. Elsevier, Amsterdam, pp. 147–154.

Shook, G.E. (1988) Selection for disease resistance. *Journal of Dairy Science* 72, 1349–1362.

Skjervold, H. and Langholz, H. (1964) Factors affecting the optimum structure of AI breeding in dairy cattle. *Zeitschrift für Tierzüchtung und Züchtungsbiologie* 80, 25.

Spelman, R. and Garrick, D. (1997) Utilisation of marker assisted selection in a commercial dairy cow population. *Livestock Production Science* 47, 139–147.

Stitchbury, J.W. and Goddard, M.E. (1985) Genetic progress – what is possible and how to get there? In: *Proceedings of the Challenge: Efficient Dairy Production*. Albury–Wodonga, Australia, pp. 432–454.

US Department of Agriculture (1962) New DHIA sire record. *Dairy Herd Improvement Letter* 38(4), ARS-44-116, 6 pp.

Van Eenennaam, A.L. and Medrano, J.F. (1991a) Differences in allelic protein expression in the milk of heterozygous κ-casein cows. *Journal of Dairy Science* 74, 1491–1496.

Van Eenennaam, A.L. and Medrano, J.F. (1991b) Milk protein polymorphisms in California dairy cattle. *Journal of Dairy Science* 74, 1730–1742.

VanRaden, P.M. (1992) Accounting for inbreeding and cross breeding in genetic evaluation of large populations. *Journal of Dairy Science* 75, 3136.

Van Vleck, L.D. (1964) Sampling the young sire in artificial insemination. *Journal of Dairy Science* 47, 441–446.

Vilkki, H.J. (1997) Genetics and breeding, multiple marker mapping of quantitative trait loci of Finnish dairy cattle by regression. *Journal of Dairy Science* 80, 198–204.

Visscher, P.M., Bowman, P.J. and Goddard, M.E. (1994) Breeding objectives for pasture based dairy production systems. *Livestock Production Science* 40, 123–137.

Weller, J.I., Kashi, Y. and Soller, M. (1990) Power of daughter and granddaughter designs for determining linkage between marker loci and quantitative trait loci in dairy cattle. *Journal of Dairy Science* 73, 2525.

Wiggans, G.R. and Goddard, M.E. (1996) A computationally feasible test day model with separate first and later lactation genetic effect. In: *Proceedings of the New Zealand Society of Animal Production*. New Zealand Society of Animal Production, Waikato University, pp. 19–21.

Wiggans, G.R. and Goddard, M.E. (1997) A computationally feasible test day model for genetic evaluation of yield traits in the United States. *Journal of Dairy Science* 80, 1795–1800.

Wiggans, G.R., VanRaden, P.M. and Zuurbier, J. (1995) Calculation and use of inbreeding coefficients for genetic evaluation of United States dairy cattle. *Journal of Dairy Science* 78, 1584–1590.

Molecular Genetics of Milk Production

19

W.S. Bawden and K.R. Nicholas
*Victorian Institute of Animal Science, Lactation Department,
475 Mickleham Road, Attwood, Victoria 3049, Australia*

Introduction	540
Bovine Milk Composition	541
The Caseins and Their Genes	541
The casein locus	541
Evolution of the casein genes	543
The Major Whey Proteins and Their Genes	544
Alpha-lactalbumin	544
Beta-lactoglobulin	545
Lactoferrin	547
Endocrine, Autocrine and Paracrine Regulation of Milk Protein Gene Expression	547
Endocrine regulation	547
Local control of milk composition and milk production	548
Regulation by the extracellular matrix	549
Transcriptional Regulation of Milk-protein Gene Expression	550
Prolactin-mediated activation of transcription	551
The glucocorticoid receptor	554
Pregnancy-specific mammary nuclear factor – a progesterone-responsive factor	555
Regulation by single-stranded DNA-binding proteins	556
Factors mediating mammary-specific expression	556
Transcriptional regulation by the extracellular matrix	558
Transgenic Studies Utilizing Bovine Milk-protein Gene Promoters	559
Expression studies	559
The alteration of milk composition via transgenesis	561
Conclusion	563
Acknowledgements	563
Editorial Note	564
References	564

©CAB International 1999. *The Genetics of Cattle* (eds R. Fries and A. Ruvinsky)

Introduction

Increasing international competition for export and import replacement of primary produce has resulted in the adoption of a wide range of technologies to improve productivity in the dairy industry. Improved milk production and, more particularly, improved milk composition for value-added processing are major goals of the industry. Protein is the most valuable component of milk and the industry would benefit considerably from cattle with increased protein yield without an additional increase in volume (Karatzas and Turner, 1997). Therefore, an understanding of the molecular genetics of the milk proteins and of mammary gland function is central to these goals.

Appropriate nutrition is essential for maximal milk production and to limit seasonal variation in milk composition, but significant additional economic gains in improved milk protein content are not likely by manipulation of diet alone (see Haresign, 1979). Breeding programmes have used genetic selection based on progeny testing to target superior genetics, and the result has been a small, but cumulative, improvement in milk production and composition. Conventional selection based on production traits can be enhanced by the use of gene marker-assisted selection, which requires identifying chromosomal markers linked to traits that improve production (see Chapters 17 and 18, this volume). These markers can be used to identify superior animals for breeding and, in the longer term, may direct researchers to candidate genes, which themselves produce the observed result. Identifying genes that enhance milk quality is important, as it is now possible to transfer additional copies of genes to an animal and to direct their expression to a specific tissue. These transgenic animals may, for example, express introduced genes in the mammary gland and secrete the product in the milk to enhance its nutritional or manufacturing quality, or express a gene in a tissue which permits secretion of the protein into the peripheral circulation, where it may have a systemic effect on lactational performance. The technology to produce transgenic mice was developed in the early 1980s and was followed by the production of transgenic livestock, including pigs, sheep and, more recently, cattle (Wall *et al.*, 1997; Eyestone *et al.*, 1998). New technology is now emerging which allows bovine and ovine embryos to be cloned from the cells of a single embryo in early stages of development or from the inner mass cells obtained later in embryological development. More recently, it has been reported that sheep and cattle may be cloned from differentiated mammary cells and embryo fibroblasts (Campbell *et al.*, 1996; see also Stice *et al.*, 1998) and this technology is now available to clone transgenic livestock (Cibelli *et al.*, 1998).

A necessary prerequisite to successfully achieving these advances in improved milk composition in cattle is a substantial understanding of the regulation of milk-protein genes. A number of reviews have been published recently which discuss the molecular structure and regulation of expression of milk-protein genes, particularly in laboratory species (Mercier and Vilotte, 1993; Groenen and van der Poel, 1994; Rosen *et al.*, 1996a). This chapter reviews the current status of our knowledge of the structure, function and control of the milk-protein genes in cattle.

Bovine Milk Composition

Bovine milk consists of a complex mixture of water (87.3%), lactose (4.8%), fat (3.7%), protein (3.5%) and a number of other minor components (Jenness, 1974). Approximately 80% of the protein content of milk consists of the caseins, α_{S1}-, α_{S2}-, β- and κ-casein, which comprise a group of acidic phosphoproteins that precipitate from skim milk at pH 4.6 and 20°C (Jollès, 1975). The whey proteins, including β-lactoglobulin (BLG; Palmer, 1934), α-lactalbumin (Jenness, 1974), lactoferrin (Groves, 1960) and various enzymes constitute the remaining fraction. Approximately 95% of the milk proteins are synthesized by the mammary gland. The remaining 5% originate from the blood, including the immunoglobulins IgG and IgM, as well as transferrin and serum albumin (Jenness, 1974).

The protein content of milk can range from 10 to 200 g kg^{-1} between various mammalian species (Murphy and O'Mara, 1993). In addition, the relative proportion of individual milk proteins can vary markedly from that found in the cow. For example, whey acidic protein (WAP), the major whey protein of mice and rats (Hennighausen and Sippel, 1982; Hobbs and Rosen, 1982), is not present in the milk of cattle or humans (Jenness, 1979). The protein content in different cattle breeds can also vary significantly (Murphy and O'Mara, 1993).

The Caseins and Their Genes

The caseins are predominantly present as a colloidal aggregation complexed with calcium phosphate (7% dry weight) to form casein micelles (Waugh, 1971; McMahon and Brown, 1984). The calcium-sensitive caseins, α_{S1}-, α_{S2}- and β-casein, contain clusters of phosphoseryl residues and, when isolated, will precipitate out of solution at low calcium concentration (Waugh, 1971). In contrast, κ-casein contains only one or two phosphoseryl residues and is able to remain in solution over a broad range of calcium concentrations (Waugh and von Hippel, 1956).

The casein locus

The genomic sequences for all four casein genes have been determined (Alexander *et al.*, 1988; Bonsing *et al.*, 1988; Gorodetskii *et al.*, 1988; Koczan *et al.*, 1991; Groenen *et al.*, 1993). In addition, all of these genes have been mapped to chromosome 6 (6q31) in cattle (Womack *et al.*, 1989; Gallagher *et al.*, 1994) within a 250 kb locus arranged in the order α_{S1}–β–α_{S2}–κ (Ferretti *et al.*, 1990; Threadgill and Womack, 1990; Rijnkels *et al.*, 1997; Figure 19.1).

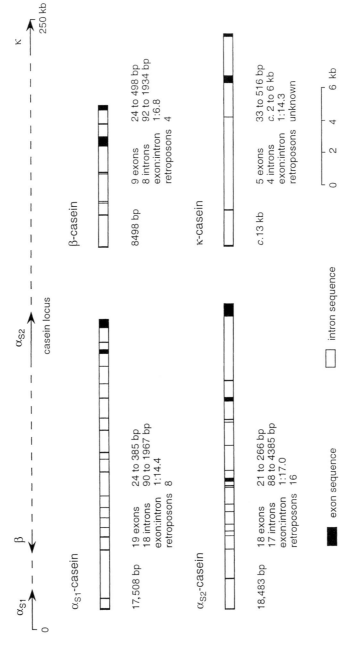

Fig. 19.1. The bovine casein genes. Schematic representation of the bovine casein genes depicting their exon structure (including number and size range of exons and introns, exon to intron ratio and number of transposons), as well as their organization within the casein locus. (Derived from sequence data reported by Alexander *et al.* (1988), Bonsing *et al.* (1988), Koczan *et al.* (1991), Groenen *et al.* (1993) and gene mapping performed by Rijnkels *et al.* (1997).)

The calcium-sensitive caseins possess similar gene structure, containing many small exons and a low exon to intron ratio. Both α_{S1}- and α_{S2}-casein possess relatively large transcriptional units, 17.5 and 18.5 kilobase pairs (kb), respectively, with a similar number of exons (Koczan et al., 1991; Groenen et al., 1993). Beta-casein is approximately half the size of the other calcium-sensitive caseins and contains half the number of exons (Bonsing et al., 1988). Fine mapping of the casein locus has also demonstrated that the β-casein gene is divergently transcribed with respect to the other casein genes (Rijnkels et al., 1997). These genes also include multiple copies of repeat elements, many of which are *Alu*-like Artiodactyla retroposons (Watanabe et al., 1982; Skowronski et al., 1984). The calcium-sensitive caseins are clustered within a 140 kb region of the locus, while κ-casein is positioned 95–120 kb downstream of the α_{S2}-casein transcriptional unit (Rijnkels et al., 1997).

Evolution of the casein genes

Sequence alignments of casein genes from various species reveal evidence of a high mutation rate, including both major insertions and deletions, in addition to sequence rearrangements. Thus the caseins appear to be a rapidly evolving gene family, presumably due to the minimal functional and structural constraints on their amino acid sequence (Stewart et al., 1984; Bonsing and Mackinlay, 1987). This is demonstrated by the fact that silent and replacement mutations are found with similar frequency within the α_S-casein complementary deoxyribonucleic acids (cDNAs). Homologous proteins that possess more defined structural requirements for function, such as enzymes, tend to possess a greater abundance of silent mutations. This lack of conservation of the majority of the coding region is consistent with the perceived loose structure/function relationship of the caseins in forming stable micelles (Bonsing and Mackinlay, 1987).

Despite their highly divergent nature, there are regions of strong similarity between the calcium-sensitive casein genes. For example, the first exon always consists of the 5′ non-coding sequence, which is highly conserved, presumably due to important secondary structure interactions necessary for post-transcriptional regulation (Blackburn et al., 1982; Stewart et al., 1984). In addition, exon 2 is uniformly 63 base pairs (bp) in length and encodes the remaining 5′ non-coding region, the entire signal peptide and two additional amino acid residues of the mature casein. Another interesting feature is that all the protein-coding exons end in a complete codon, so that none of the codons are interrupted by splice sites (Bonsing and Mackinlay, 1987; Mercier and Vilotte, 1993). These similarities have led to the belief that the calcium-sensitive caseins constitute a multigene family, which is thought to have arisen from an ancestral gene via intra- and intergenic duplication (Yu-Lee et al., 1986; Bonsing and Mackinlay, 1987). Evidence for this theory can be seen within the α_{S1}-casein gene, where a 154 bp region spanning exon 10 differs by only four bases from a similar region comprising exon 13 (Koczan et al., 1991). Exons 7–11 and 12–16 within the α_{S2}-casein may also have resulted via internal duplication (Stewart et al., 1987). Although α_{S1}- and α_{S2}-casein appear more closely related to each other on the basis of gene size and number of exons, Groenen et al. (1993) propose that analysis of exon lengths indicates that the α_{S2}-casein gene is more closely related to the β-casein gene.

Although the κ-casein gene is physically linked to the calcium-sensitive caseins, both its amino acid and nucleotide sequence suggest it is more closely related to the fibrinogen gene family (Jollès et al., 1978; Alexander et al., 1988). The bovine gene consists of five exons separated by very large introns. The first three exons are quite small, 65 bp or less, with the majority of the protein-coding sequence contained within exon 4 (Alexander et al., 1988). Inter-species comparisons demonstrate that κ-casein possesses the highest degree of conservation of all the casein genes, which may be related to its essential function of stabilizing the casein micelle (Alexander et al., 1988).

The Major Whey Proteins and Their Genes

The whey proteins comprise a diverse group of globular polypeptides and, in contrast to the caseins, many of their structures have been determined (reviewed in Wong et al., 1996). The casein and whey proteins differ not only in their physical structure and properties, but also in their genomic organization and postulated evolutionary origins (Bonsing and Mackinlay, 1987; Bawden et al., 1994). Unlike the caseins, the whey-protein genes are dispersed throughout the genome and a number of pseudogenes have also been characterized (Soulier et al., 1989; Vilotte et al., 1993; Passey and Mackinlay, 1995).

Alpha-lactalbumin

Alpha-lactalbumin is a calcium metalloprotein which, in combination with β-1, 4-galactosyltransferase, forms the lactose synthase complex situated in the *trans*-Golgi membrane of mammary epithelial cells (MECs; Ebner and Schanbacher, 1974). The formation of this complex is necessary for the synthesis of lactose, the major carbohydrate in milk. Though functionally dissimilar, the high degree of homology between the amino acid sequences of α-lactalbumin and the *c*-type lysozymes suggests they are derived from a common ancestral gene (Brew et al., 1970).

The genomic sequence of bovine α-lactalbumin comprises a transcriptional unit approximately 2 kb in length containing four exons (Vilotte et al., 1987). Comparison with published α-lactalbumin and *c*-type lysozyme sequences shows that the exon organization of these genes has been highly conserved (Hall et al., 1987). The similarity of these sequences has been dramatically demonstrated by the transfer of lysozyme activity to goat

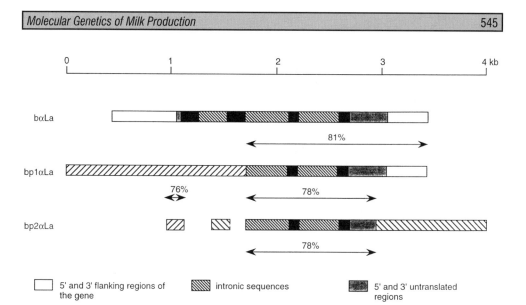

Fig. 19.2. Bovine α-lactalbumin and pseudogenes. Schematic representation of the bovine α-lactalbumin gene (bαLa) and comparison with two pseudogenes (bp1αLa and bp2αLa). Percentage similarity between regions is represented by the arrows (the arrow below bp2αLa corresponds to the comparison between bp2αLa and bαLa). (Reproduced from Vilotte *et al.* (1993), used by permission of Academic Press, Orlando, USA.)

α-lactalbumin following exchange of exon 2 with the same exon from hen lysozyme (Kumagai *et al.*, 1992).

Two pseudogenes for α-lactalbumin have been isolated and sequenced from the bovine genome (Soulier *et al.*, 1989; Vilotte *et al.*, 1993). Both demonstrate approximately 80% homology with the authentic gene and stretch from intron 2 to the 3′ untranslated region (UTR) of exon 4. The 5′ ends of the pseudogenes are also similar to each other and yet have no homology with the native sequence (Fig. 19.2). The α-lactalbumin pseudogenes are present at the same chromosomal location as the authentic gene on chromosome 5 (5q21), supporting the suggestion that they arose via gene duplication (Soulier *et al.*, 1989; Hayes *et al.*, 1993).

Beta-lactoglobulin

Beta-lactoglobulin is the major whey protein of ruminants and is present in bovine milk at a concentration of 3.1 mg ml^{-1} (Dalgleish, 1992). The structure of bovine BLG has been determined by X-ray crystallography (Papiz *et al.*, 1986), which revealed it as a member of the lipocalin superfamily (North, 1991). The ability of BLG from a number of species to bind retinol *in vitro* and its structural homology to retinol-binding protein implicated BLG in the transport of retinol to the suckling infant (Fugate and Song, 1980; Godovac-Zimmermann *et al.*, 1985). However, this view needs to be reassessed in light of data presented by Neuteboom *et al.* (1992), who failed to detect BLG–retinol complexes in the bovine mammary gland microsome, using a combination of HPLC and spectroscopic methods. Thus, despite being isolated over 60 years ago, the biological role of BLG is still largely unknown (Pérez and Calvo, 1995).

The bovine BLG gene consists of a transcriptional unit 4724 bp in length, containing seven exons (Alexander *et al.*, 1993). The BLG gene has been mapped to the short arm of chromosome 3 (3p28) in sheep and to the homologous chromosome 11 (11q28) in both cattle and goats (Hayes and Petit, 1993). Recently, a BLG-like pseudogene (ψ-BLG) was discovered 14 kb upstream from the authentic bovine BLG gene (Passey and Mackinlay, 1995). The pseudogene is of similar length, contains the same number of exons and is in the same orientation as the authentic BLG gene. Most of the introns are highly divergent, while exon similarities range from 60 to 92.5%. It is proposed that both genes originated from a common ancestor via gene duplication. Exon 5 of ψ-BLG contains an in-frame stop codon, but no ψ-BLG product has been detected in the lactating mammary glands of cattle. Interestingly, the milk of several species, including the horse and pig, contains major and minor BLG fractions, designated BLG I and BLG II, respectively. Amino acid analysis suggests that both these proteins are in fact encoded by separate genes (Conti *et al.*, 1984; Halliday *et al.*, 1991, 1993). The inferred translation product for ψ-BLG shows greater similarity to the BLG II sequences from the horse and cat than the published bovine sequence (Fig. 19.3). Due to the highly conserved exon sequences, the authors suggest that ψ-BLG may have been expressed until relatively recently. The genomes of both sheep and goats have also been reported to contain BLG pseudogenes which are present at the same chromosomal location as the authentic genes (Folch *et al.*, 1996).

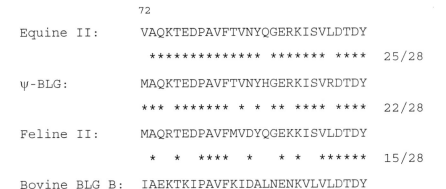

Fig. 19.3. Comparison of a 28-amino acid sequence spanning exon III and exon IV of equine BLG II, feline BLG II and bovine ψ-BLG and BLG. Asterisks represent identities between adjacent sequences (the fraction of conserved residues is shown at the end of the line). The bovine BLG B sequence is that from Alexander *et al.* (1993), while the equine and feline sequences are reported by Halliday *et al.* (1991, 1993). (Reproduced from Passey and Mackinlay (1995), with permission.)

Lactoferrin

Lactoferrin is an iron-binding glycoprotein belonging to the transferrin gene family which also possesses bactericidal properties (Oram and Reiter, 1968; Arnold *et al.*, 1977). The tertiary structure of human lactoferrin has been determined and consists of two globular lobes, each lobe containing a single iron-binding site, which are connected by a hinge region (Anderson *et al.*, 1989). Both lobes possess significant internal homology and are thought to have arisen from an internal duplication of an ancestral gene (Park *et al.*, 1985). The bovine lactoferrin gene is the largest of the characterized milk-protein genes, containing 17 exons within a 34.5 kb transcriptional unit (Seyfert *et al.*, 1994).

Endocrine, Autocrine and Paracrine Regulation of Milk Protein Gene Expression

Mammary gland function is controlled by a number of different mechanisms, all of which act concurrently to regulate the synthesis and secretion of milk. Endocrine factors (hormones and growth factors) stimulate growth of the mammary gland, which, together with paracrine influences (e.g. the extracellular matrix (ECM)), is essential for the expression of the milk protein genes (Topper and Freeman, 1980; Streuli *et al.*, 1995). Autocrine factors (regulatory molecules that are produced by the mammary gland and subsequently 'feed back' to stimulate or inhibit mammary function) have also been shown to modulate the effects of endocrine-stimulated milk-protein synthesis and secretion (Wilde *et al.*, 1995, 1997).

Endocrine regulation

Early studies (Hartmann and Shelton, 1971; Hartmann, 1973) showed that lactogenesis in dairy cattle proceeds in two stages. Stage one occurs prior to parturition and the secretion is a high-protein colostrum. The decline of progesterone at parturition signals stage two lactogenesis and the copious secretion of milk. This role of progesterone in lactogenesis has been confirmed by *in vitro* studies showing that this hormone inhibits the synthesis and secretion of milk proteins in mammary explants from late pregnant and lactating cows (Gertler *et al.*, 1982; Goodman *et al.*, 1983; Shamay *et al.*, 1987). The hormonal regulation of milk-protein gene expression varies among mammals, but generally requires prolactin, adrenal glucocorticoids (usually cortisol) and insulin (Topper and Freeman, 1980; Vonderhaar and Ziska, 1989). Recent studies by Sheehy *et al.* (1997) have reported experiments using a mammary explant culture system with tissue obtained by biopsy from late pregnant cows. Their experiments showed that the casein and whey-protein genes are expressed at elevated levels from at least 30 days prior to calving, but, when this tissue (in phase-one lactogenesis) was cultured for 4 days in media containing insulin and cortisol, the level of messenger ribonucleic acid (mRNA) for the κ-casein, β-casein and BLG genes declined to almost undetectable levels. The subsequent inclusion of prolactin in the culture media did not induce milk-protein gene expression. However, when explants were prepared from mammary tissue biopsied from cows more than 30 days before calving, it was possible to demonstrate induction of these milk-protein genes in the presence of insulin, cortisol and prolactin. These results confirm those of a number of studies which have used mammary-gland explants from pregnant and lactating cows (Gertler *et al.*, 1982; Goodman *et al.*, 1983), mammary acini, primary cell cultures of bovine MECs and bovine cell lines (MacKenzie *et al.*, 1985; Baumrucker *et al.*, 1988; Choi *et al.*, 1988; Talhouk *et al.*, 1990; Gibson *et al.*, 1991; Jung-Youb *et al.*, 1995) to demonstrate that the casein and whey-protein genes are induced in response to insulin, cortisol and prolactin. However, there are no definitive studies reporting the minimal requirement and physiological concentrations of hormones required *in vitro* for the expression of the individual milk-protein genes.

It has been demonstrated that insulin is essential for casein gene expression in mammary explants from the mouse (Bolander *et al.*, 1981) and rat (Kulski *et al.*, 1983). Whereas other growth factors and serum can supplant the role of insulin in maintaining tissue viability, insulin (together with prolactin and cortisol) is required for transcription of the casein genes (Chomczynski *et al.*, 1984). Both prolactin and cortisol have an additional role in stabilizing casein mRNA (Chomczynski *et al.*, 1984, 1986). A role for insulin in the regulation of the bovine milk-protein genes has yet to be examined, and, in general, the intracellular mechanism by which insulin signals milk-protein gene expression has not yet been addressed.

Local control of milk composition and milk production

The rate of milk secretion in dairy cattle is regulated by the frequency and completeness with which milk is removed from the mammary glands. More frequent milking applied bilaterally to two bovine mammary glands increases milk production in those glands (see Wilde *et al.*, 1997). Therefore, an autocrine mechanism is thought to match the supply of milk to the demands of either the nutritional requirements of the suckled offspring or the milking timetable in dairy animals. Bovine milk constituents have been screened for the presence of a chemical inhibitor of milk secretion in a mammary tissue explant bioassay and results indicate that biological activity resides in a small, acidic protein, which has been termed feedback inhibitor of lactation (FIL; Wilde *et al.*, 1995, 1997).

Autocrine modulation of endocrine control may partially explain the developmental responses of the tissue to sustained alterations in milking frequency. In the cow, these developmental responses are incompletely characterized. However, in cattle, as in goats (Wilde *et al.*, 1987), frequency of milking regulates the degree of mammary cell differentiation (Hillerton *et al.*, 1990). Therefore, it is conceivable that autocrine signalling of MECs may result in changes in milk composition, in addition to milk secretion. Recent studies have shown that the down regulation of the expression of two protease inhibitor genes in the tammar wallaby (*Macropus eugenii*) is matched to changes in the sucking pattern of the developing pouch young (Nicholas *et al.*, 1997). It is interesting that the activity of bovine trypsin inhibitor is similarly down regulated at parturition in cattle and correlates with the onset of the sucking stimulus. However, the potential for autocrine regulation of this protein, and the gene which codes for it, remains to be studied.

Regulation by the extracellular matrix

It is now well established that hormonal cues alone are not sufficient to elicit the expression of milk-protein genes within the mammary gland; interactions with the ECM are also required (Lin and Bissell, 1993; Roskelley *et al.*, 1995). This ECM-mediated signal transduction across the cell membrane, and finally to the nucleus, is facilitated by the integrin family of transmembrane proteins. The first step in this pathway requires the binding of laminin-1 to β_1-integrin, which is then thought to interact with the cytoskeleton (Streuli *et al.*, 1991, 1995). The MECs obtained from mid-pregnant mice and cultured on plastic plates are unable to maintain a differentiated state and rapidly cease to express β-casein, even in the presence of lactogenic hormones. However, when cultured on floating collagen gels, these cells are able to synthesize their own basement membrane and the ability to express β-casein is restored (Emerman and Pitelka, 1977; Emerman *et al.*, 1977). In contrast, the WAP gene is not expressed in MECs cultured under the same conditions. Mammary cells cultured on an Engelbreth–Holm–Swarm tumour biomatrix, which is rich in laminin, are able to form spherical alveolar-like structures, termed 'mammospheres'. These cells are polarized and secrete both WAP and β-casein preferentially into the lumen of these structures (Li *et al.*, 1987; Barcellos-Hoff *et al.*, 1989). Thus higher-order mammary differentiation is required for the expression and secretion of WAP, a protein expressed later in pregnancy than β-casein. However, this system does not meet the requirements for the synthesis of α-lactalbumin (Schmidhauser *et al.*, 1995).

Whereas the presence of the ECM in culture induces MECs to adopt a spherical morphology, which is absolutely required for prolactin-induced β-casein gene expression (Roskelley *et al.*, 1994), the synthesis of lactoferrin is dependent on the presence of insulin and cell rounding alone (Roskelley *et al.*, 1994; Close *et al.*, 1997). Studies have shown that lactoferrin gene expression is repressed at both the transcriptional and post-transcriptional level in flattened cells. The induction of cell rounding by the addition of cytochalasin D, which disrupts the cytoskeleton, was able to relieve this repression of expression (Close *et al.*, 1997). The mechanism by which changes in the cytoskeleton, which are correlated with changes in cell shape, regulate gene expression remains elusive. One theory proposes that changes in the organization of the cytoskeleton could alter the three-dimensional organization of the nuclear matrix and thus influence gene expression (Getzenberg, 1994). At the post-transcriptional level, it has been proposed that alterations in the cytoskeleton may affect the stability of mRNA transcripts within the mammary gland (Bissell and Hall, 1987; Blum *et al.*, 1989). Indeed, more than 90% of β-casein mRNA isolated from mouse MECs is found to be associated with the cytoskeleton (Bissell and Hall, 1987). Roskelley *et al.* (1995) propose that cell rounding represents the lowest level in the hierarchy of ECM-mediated signalling required for milk-gene expression, with higher levels of sophistication approaching the *in vivo* environment being required for β-casein, WAP and α-lactalbumin expression, respectively.

Transcriptional Regulation of Milk-protein Gene Expression

The production of milk during lactation requires the transcriptional activation of milk-protein genes, which is facilitated by the binding of nuclear factors to their target sequences, usually located upstream of the coding region. Once bound, these factors can then interact with RNA polymerase in either a negative or a positive manner and thus regulate transcription (Struhl, 1989). Much of the information available regarding the nuclear factors involved in the regulation of milk-protein gene expression comes from the introduction of both murine and rat β-casein–chloramphenicol acetyltransferase (CAT) reporter constructs into the HC11 murine MEC line, in association with promoter deletion analysis and transcription factor binding studies (Doppler *et al.*, 1989; Lee and Oka, 1992; Kanai *et al.*, 1993). Unfortunately, little corresponding information is available for dairy cattle, primarily due to the absence of suitable cell lines to perform these analyses (Watson *et al.*, 1991; Sheehy *et al.*, 1997). However, the similarities between many of the rodent genes discussed below with the corresponding sequences in cattle indicates that similar mechanisms may be functioning within the bovine system. Indeed, many of the transcription factors and their specific DNA-binding motifs originally characterized in rodents have subsequently been detected in cattle (Fig. 19.4).

A common observation that has emerged from these studies is that a number of the nuclear factors involved in the regulation of milk-gene expression are themselves regulated by hormonal cues and thus act as the link between the hormonal status of the animal and milk production. Many factors also demonstrate altered expression profiles and binding activities within the virgin animal and throughout pregnancy, lactation and involution. Thus the occupancy of various transcription factor-binding sites alters with the stage of lactation, as shown by the 'footprint' left by these various proteins within the promoter. The close proximity of many of these binding sites suggests that synergistic and cooperative effects, in addition to competition for the occupancy of various sites, may be occurring among these factors, which in turn contributes to either the activation or the repression of expression. In recent years, the

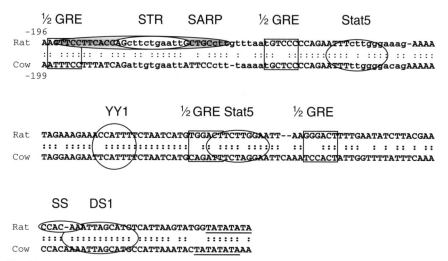

Fig. 19.4. Alignment of the rat and bovine β-casein genes and transcription factor-binding sites. Colons represent identities between aligned sequences, gaps have been introduced to facilitate the alignment, which was performed using the FastA program (Pearson and Lipman, 1988; Pearson, 1990). The rat β-casein sequence is that published by Jones et al. (1985), while the bovine sequence is that reported by Bonsing et al. (1988). Transcription factor-binding sites are those determined for the rat β-casein promoter, as reported by Schmitt-Ney et al. (1991), Altiok and Groner (1993, 1996), Welte et al. (1993), Meier and Groner (1994), Raught et al. (1994, 1995), Doppler et al. (1995) and Lechner et al. (1997a) except for SS and DS1 which were detected within the murine promoter (Saito and Oka, 1996). C/EPB-binding sites are in lower case and the TATA boxes are underlined.

number of transcription factors implicated in the regulation of milk-gene expression has risen dramatically, with a corresponding appreciation of the complexity of the interactions involved (Rosen et al., 1996a, b). A simplified model for the regulation of rat β-casein transcription is presented in Fig. 19.5.

Prolactin-mediated activation of transcription

Promoter deletion analysis has located the hormone response elements necessary for the transcription of the rat and murine β-casein genes within the region 221 bp and 258 bp upstream of the transcription initiation site, respectively (Altiok and Groner, 1993; Kanai et al., 1993). Footprinting analysis has revealed that as many as seven different proteins bind to the −258 to +7 portion of the murine promoter (Kanai et al., 1993). One of these proteins, originally termed the mammary gland-specific nuclear factor (MGF), is critical in mediating the positive response of the β-casein gene to prolactin (Schmitt-Ney et al., 1991). Recently, MGF has been assigned as a member of the signal transducers and activators of transcription (Stat) family of transcription factors and designated Stat5 (Wakao et al., 1994). Once prolactin has bound to the

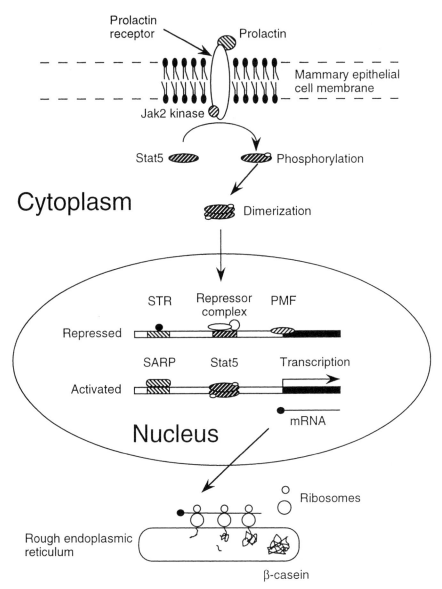

Fig. 19.5. Model of the activation of β-casein transcription. Schematic representation of the repression and activation of β-casein transcription. Transcription is repressed by the binding of the pregnancy-specific mammary nuclear factor (PMF), spanning the start point of transcription, a repressor complex and the single-stranded DNA-binding transcriptional regulator (STR). At parturition serum levels of progesterone decrease as does the binding of PMF. Increased prolactin levels result in the activation of signal transducer and activator of transcription (Stat5), allowing it to displace the repressor complex, which includes Yin Yang 1 (YY1), and bind to its target sequence. The synthesis and transport of β-casein mRNA sequesters STR to the cytoplasm, allowing the single-stranded DNA activator region-binding protein (SARP) to bind to the promoter. This model is based on data from Lee and Oka (1992), Gouilleux *et al.* (1994), Meier and Groner (1994), Raught *et al.* (1994), Welte *et al.* (1994) and Altiok and Groner (1996).

extracellular-binding domain of the long form of the prolactin receptor, cytosolic Stat5 is phosphorylated by Jak2 kinase. Subsequently, Stat5 dimerizes and is then translocated to the nucleus, where it binds to its target sequence to activate transcription of the β-casein gene (Gouilleux *et al.*, 1994; Welte *et al.*, 1994; see Fig. 19.5).

Binding sites for Stat5 have been located between nucleotides −80 to −100 and −130 to −150 of the rat β-casein gene to sites which correspond to the imperfect palindromic sequence 5′-TTCNNNGAA-3′ (Schmitt-Ney *et al.*, 1991; Wakao *et al.*, 1994; see Fig. 19.4). Mutations introduced within the proximal Stat5-binding site abolished the hormonal induction of the β-casein promoter, confirming its essential role in the hormonal responsiveness of this gene (Schmitt-Ney *et al.*, 1991). This transcription factor has also been detected in the mammary gland of lactating cattle and is able to bind to regions within the bovine α-lactalbumin (Kuys *et al.*, 1996) and κ-casein promoters (Adachi *et al.*, 1996), in addition to oligonucleotides containing portions of the bovine $α_{S1}$- and β-casein promoters (Wakao *et al.*, 1992). Consensus binding sites for Stat5 are absent from the bovine lactoferrin promoter (Seyfert *et al.*, 1994), which is in agreement with the observation that the lactoferrin gene can be transcribed in murine MEC culture in the absence of prolactin (Close *et al.*, 1997).

While Stat5 DNA-binding activity has been detected in mammary tissue obtained from lactating cattle, it is absent in non-lactating, pregnant and involuting cows and liver extracts (Kuys *et al.*, 1996), which agrees with the findings of Schmitt-Ney *et al.* (1992) in mice. During lactation, frequent suckling appears necessary to maintain Stat5 activity in both mice (Schmitt-Ney *et al.*, 1992) and cattle (Kuys *et al.*, 1996). Yang *et al.* (1997) have recently established that there is also a positive correlation between Stat5 activity and protein concentration in the milk of cattle.

Murine Stat5 exists as two homologous isoforms, Stat5a and Stat5b, which differ only in their carboxyl terminus (Liu *et al.*, 1995). Interestingly, whereas both WAP and β-casein genes contain Stat5-binding sites in their promoter, only WAP expression is reduced in Stat5a-deficient mice. This result suggests that β-casein gene expression is primarily controlled by other transcription factors (Liu *et al.*, 1997; Udy *et al.*, 1997).

Site-directed mutagenesis of a negative control region within the rat β-casein promoter, which is immediately adjacent to the proximal Stat5-binding site, led to an increased constitutive transcription of a CAT reporter construct in culture (Schmitt-Ney *et al.*, 1991; Altiok and Groner, 1993). These data suggested that the regulation of β-casein gene expression included a relief of repression mechanism. Further investigation of one of the proteins which bound within this region revealed it to be identical to the nuclear protein Yin Yang-1 (YY1; Meier and Groner, 1994; Raught *et al.*, 1994; see Fig. 19.4). Band-shift assays revealed that Stat5 bound to this negative control region with greater affinity than YY1, suggesting that, once Stat5 is activated via Jak2 kinase in response to increased prolactin levels, it is able to displace YY1 from the promoter, thus relieving its repressive effects and activating transcription (Meier and Groner, 1994; Raught *et al.*, 1994; see Fig. 19.5). Murine YY1 is also able to recognize its corresponding binding site within the bovine β-casein promoter, suggesting that a similar hormone-mediated relief of repression mechanism may be operating for this gene (Meier and Groner, 1994).

The glucocorticoid receptor

Studies of β-casein expression have shown that the kinetics of induction by prolactin is more rapid than that for glucocorticoids, indicating that these hormones act by different signal transduction pathways (Doppler *et al.*, 1990). Once complexed with their cytosolic receptors, glucocorticoids are translocated to the nucleus, where they bind to glucocorticoid-response elements (GRE), possessing the pseudo-palindromic consensus sequence GGTACA*nnn*TGTTCT, and function as a transcriptional activator (Beato, 1989). Activated glucocorticoid receptor (GR) is also able to alter the chromatin structure within the promoter and facilitate access of other transcription factors to their recognition sequences (reviewed in Tsai and O'Malley, 1994), which appears to be the case for nuclear factor I (NF-I) within the rat WAP promoter (Li and Rosen, 1994).

Classical GREs have not been detected within the promoters of milk-protein genes; however, multiple half-sites containing the TGT(T/C)CT motif are present (Yu-Lee *et al.*, 1986; Welte *et al.*, 1993; see Fig. 19.4). The functional significance of these half-sites has been demonstrated by their ability to specifically bind purified GR from rat liver at many locations within both the rat β-casein and WAP promoters (Welte *et al.*, 1993). In addition, three separate CCAAT/enhancer-binding proteins (C/EBPs) bind within −180 to −132 of the rat β-casein upstream region, quite close to the GR half-sites and the distal Stat5 element (see Fig. 19.4). Mutations introduced into either of the C/EBP-binding sites located between −180 and −157 impaired the response of the rat β-casein promoter to lactogenic hormones (Doppler *et al.*, 1995). The expression of various C/EBP isoforms differs throughout the development of the mammary gland. A dominant negative isoform of C/EBPβ has a repressive effect, which is negatively regulated by the action of glucocorticoids, which may account for the delayed response to this hormone (Raught *et al.*, 1995). It has been proposed that the increased level of glucocorticoid downregulates the repressive form of C/EBPβ, allowing the active form of C/EBPβ to positively interact with the GR and the β-casein promoter (Raught *et al.*, 1995; Rosen *et al.*, 1996b). The proximity of the GR–C/EBP complex to that of YY1 may also facilitate its dissociation, allowing the binding of activated Stat5. Such a model provides an explanation for the delayed glucocorticoid response and the synergistic effect between prolactin and glucocorticoids for the induction of β-casein expression (Rosen *et al.*, 1996b). Interactions with C/EBPs may be important in stabilizing the binding of the GR to the non-palindromic half-sites (Lechner *et al.*, 1997a). Also, C/EBP proteins have been detected bound within the bovine β- and κ-casein promoters (Korobko *et al.*, 1995), while putative C/EBP recognition sites are present within the promoters of all the major bovine casein and whey-protein genes (Popov, 1996). The importance of C/EBPβ for ductal morphogenesis, lobuloalveolar development and functional differentiation has recently been demonstrated by the analysis of C/EBPβ knockout mice (Seagroves *et al.*, 1998; Robinson *et al.*, 1998).

Mutational inactivation of the proximal Stat5 site within the rat β-casein promoter was found to interrupt not only the prolactin response but also that of glucocorticoids (Schmitt-Ney *et al.*, 1992). Two GR half-sites overlap with the proximal Stat5 element, indicating that a close cooperation may occur between these two transcription factors (see Fig. 19.4). Recent investigations into the extent of this interaction has produced some conflicting data. Lechner *et al.* (1997a, b) have demonstrated that the action of the GR is DNA-template-dependent, as mutations introduced into the three proximal binding sites within the β-casein promoter abolished the augmentation of expression by glucocorticoids in MECs. In contrast, Stoecklin *et al.* (1997) have proposed that activated GR need not bind to DNA at all, but can function as a transcriptional coactivator of Stat5 exclusively via protein–protein interactions. These contrary results were obtained in COS-7 cells expressing various combinations of the prolactin receptor, GR and Stat5. The advantage of this system is the ability to introduce the pathways necessary for milk-protein gene expression within non-MECs to investigate the individual roles of various receptors and transcription factors. However, a limitation of this system is that these factors may be expressed at unphysiologically high levels and therefore may not accurately reflect the *in vivo* environment within MECs. For example, the construct containing the mutated C/EBP sites described above, which were unresponsive to hormones in HC11 cells, displayed hormonal induction in COS-7 cells (Lechner *et al.*, 1997a). It is possible, however, that both DNA-template-dependent and independent interactions between the GR and Stat5 are involved in the regulation of β-casein expression (Lechner *et al.*, 1997b). Whatever the actual mode of interaction between the GR and Stat5, they may be important for ensuring the restriction of milk-protein gene expression to conditions where both prolactin- and glucocorticoid-dependent signalling pathways are functional (Lechner *et al.*, 1997a).

Pregnancy-specific mammary nuclear factor – a progesterone-responsive factor

As mentioned above, casein expression is repressed until parturition by the action of progesterone. A 65-kDa pregnancy-specific mammary nuclear factor (PMF) has been shown to bind at two separate sites possessing the palindromic sequence 5′-TGAT/ATCA-3′ within the murine β-casein promoter (Lee and Oka, 1992). Mutation of these sites leads to the loss of inhibition by progesterone, indicating that PMF is involved in the progesterone-mediated repression of expression. One of these PMF sites straddles the start point of transcription, suggesting that its repressive effect is achieved by inhibiting the formation of the transcription initiation complex (Lee and Oka, 1992; see Fig. 19.5). Putative PMF sites have also been located within the promoters of all four bovine casein genes (Schild *et al.*, 1994; Popov, 1996).

Regulation by single-stranded DNA-binding proteins

Promoter deletion analysis revealed the presence of a regulatory element between nucleotides −170 and −221 of the rat β-casein gene, which possesses overlapping negative and positive regulatory elements (Altiok and Groner, 1993, 1994). A single-stranded DNA-binding transcriptional regulator (STR) is able to repress transcription by binding to the upper strand of the promoter between nucleotides −194 and −163. The activity of STR in the murine mammary gland is high during the latter part of pregnancy and postlactation, but is downregulated during lactation and in HC11 cells cultured in the presence of lactogenic hormones (Altiok and Groner, 1993). The overlapping positive regulatory element between nucleotides −183 and −170 interacts with a single-stranded DNA activator region-binding protein (SARP), which possesses high binding activity during lactation and competes with STR for occupancy of its binding site (Altiok and Groner, 1996). Also, SARP activity is positively regulated by signalling mechanisms from the ECM and suckling stimuli. High SARP- and low STR-binding activity have also been detected in mammary extracts prepared from lactating cattle and sheep (Altiok and Groner, 1996). An additional STR site is also present within the 5′ UTR of β-casein mRNA. It is proposed that, after the initial induction via lactogenic hormones, the STR site within the resulting transcript is able to sequester STR from the promoter and relieve its repressive effects within the nucleus and thus augment β-casein transcription (see Fig. 19.5). The binding of STR to the 5′ UTR of β-casein mRNA may also have a stabilizing effect on the transcript, perhaps by protecting it from degradation via 5′ exonucleases (Altiok and Groner, 1994).

An additional single-stranded DNA-binding complex (SS) has been detected between −62 and −42 (termed block C) of the murine β-casein promoter. This region has significant sequence similarity to the corresponding position within a number of caseins of various species, including bovine α_{S1}- and β-casein (Saito and Oka, 1996). Mutations introduced within this region reduced the hormonal responsiveness of the murine promoter by as much as 84%. A double-stranded DNA-binding complex (DS1) was found to bind adjacent to SS, the binding activity of both complexes being positively regulated by lactogenic hormones (Saito and Oka, 1996; see Fig. 19.4).

Factors mediating mammary-specific expression

Milk-protein gene expression does not occur in the liver, although receptors for prolactin, glucocorticoid and insulin are present in the liver and mammary gland (Topper and Freeman, 1980). This indicates that mammary cell-specific factors, in addition to the action of lactogenic hormones, are important for milk-protein gene expression. Tissue-specific expression may be mediated by tissue-specific factors or via the synergistic effect of ubiquitous factors, in such a way as to restrict the expression of specific genes to certain tissues.

Originally Stat5 was reported as a mammary gland-specific factor; however, it was later found to be expressed in many different tissues (Wakao et al., 1994). Furthermore, mutational inactivation of all three Stat5 sites within the ovine BLG promoter led to a five-fold reduction in the level of BLG expression in transgenic mice, and yet expression remained tissue-specific (Burdon et al., 1994).

The question of which factors mediate the mammary-specific expression of milk-protein genes still remains to be elucidated. However, it appears that there are multiple mechanisms involved, since evidence suggests that the casein and whey-protein genes may respond to different signalling pathways to ensure tissue specificity (Mink et al., 1992; Liu et al., 1997).

The NF-I family of transcription factors has been implicated in the tissue-specific expression of a number of genes (Paonessa et al., 1988). Mutational inactivation of two NF-I sites within the distal promoter of rat WAP abolished its mammary-specific expression in transgenic mice (Li and Rosen, 1995). It has been proposed that a novel mammary-specific member of the NF-I family may exist and elicit the tissue-specific expression of milk-protein genes. Recently, Furlong et al. (1996) have demonstrated that various NF-I isoforms exhibit distinct expression profiles during the development of the murine mammary gland. These include three lactation-associated proteins (46, 68 and 114 kDa), which were replaced by a specific 74 kDa isoform during involution. Interestingly, if day 2 involuting mammary glands are resuckled, the involution-specific isoform disappears and the 114 kDa protein returns (F. Martin, Dublin, 1997, personal communication). Differential expression of NF-I isoforms has also been observed within the mammary gland of cattle, where an increase in the binding activity of NF-I complexes can be correlated with their altered subunit composition during lactation. Footprint analysis has shown that this complex binds to the region within the bovine β-casein promoter between −264 and −239 and two sites within intron VI (Ivanov et al., 1990). The NF-I site within the promoter is also in close proximity to a C/EBP site, with which it may interact (Korobko et al., 1995).

In conjunction with another nuclear factor, termed mammary cell-activating factor (MAF), NF-I was demonstrated to mediate the mammary-specific expression of mouse mammary tumour virus (Mink et al., 1992). The introduction of mutations into the MAF site between −120 and −100 within the mouse WAP promoter was also found to alter the hormone-independent mammary cell-specific transcription of reporter constructs in cell culture (Welte et al., 1994). Further characterization of the factors that bind within this region showed that they were related to the extensive Ets family of transcription factors. Some Ets proteins are expressed during specific stages of development and are restricted to certain tissues, thus being candidates for the regulation of tissue-specific gene expression (Wasylyk et al., 1993). A number of Ets-related proteins are expressed within MECs during lactation, the most abundant of which appears to be GABPα, which is able to bind to the MAF site within the mouse WAP promoter (W. Doppler, Innsbruck, 1997, personal communication). Elements similar to the MAF site have been identified within the regulatory regions of other whey-protein genes but are absent from casein promoters (Mink *et al.*, 1992).

Transcriptional regulation by the extracellular matrix

Promoter deletion analysis and transfection experiments performed in murine MEC culture revealed the presence of an enhancer region between −1517 and −1677 upstream of the bovine β-casein gene, designated β-casein element-1 (BCE-1), which is able to confer both prolactin and ECM responsiveness to an inactive β-casein promoter (Schmidhauser *et al.*, 1990, 1992). This regulation was absent when constructs were introduced into Chinese hamster ovary cells and Mabin–Darby canine kidney cells, suggesting that the enhancer may be mammary-specific (Leliévre *et al.*, 1996). Elements similar to BCE-1 may not be restricted to the β-casein gene, as an alignment with the bovine BLG gene revealed a sequence possessing 53% similarity within the region −741 to −619 (Fig. 19.6). Bleck and Bremel (1993) have also identified a 69 bp region approximately 1.5 kb upstream of the bovine α-lactalbumin gene which contains 75% sequence similarity with the 3′ end of BCE-1.

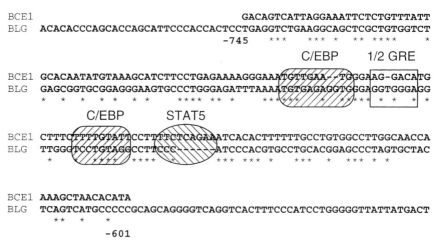

Fig. 19.6. Alignment of BCE-1 and a portion of the bovine β-lactoglobulin gene. Asterisks represent identities between aligned sequences, gaps have been introduced to facilitate the alignment, which was performed using the Clustal W program (Thompson *et al.*, 1994). The BCE-1 element is that published by Schmidhauser *et al.* (1992), while the bovine BLG sequence is that reported by Alexander *et al.* (1993). Putative transcription factor-binding sites within BCE-1 are as published by Raught *et al.* (1995).

Sequence analysis has revealed the presence of a putative Stat5-binding site, a half glucocorticoid response element and two C/EBP-binding sites within BCE-1 (Raught *et al.*, 1995; see Fig. 19.6). In addition, it can be demonstrated that both the ECM- and prolactin-mediated induction of β-casein expression are kinetically and functionally separable (Roskelley *et al.*, 1994). The specific factors required to integrate both ECM and prolactin signals still remain to be positively identified, although mutations introduced at either end of BCE-1 were able to greatly reduce the efficacy of the enhancer (Schmidhauser *et al.*, 1995). These regions correspond to the putative Stat5- and C/EBP-binding site, which both appear to be necessary, but not sufficient, for ECM-dependent activation (Roskelley *et al.*, 1995).

How does the ECM regulate expression at the transcriptional level? Recent investigations have revealed that both Stat5 and SARP DNA-binding activity, crucial for the expression of the β-casein gene, are positively regulated by the ECM (Streuli *et al.*, 1995; Altiok and Groner, 1996; Edwards *et al.*, 1996). The DNA-binding activity of Stat5 isolated from MEC culture was found to be present in cells cultured in the presence of basement membrane or laminin-1, but missing from those cultured on plastic or collagen (Streuli *et al.*, 1995). Further studies revealed that the ECM is able to regulate milk-gene expression by regulating the activity of Jak2 kinase (Edwards *et al.*, 1996). Thus both the ECM and prolactin signalling pathways converge at this junction, which is crucial in the phosphorylation cascade required to activate Stat5. In addition, SARP activity is upregulated in confluent cultures of HC11 cells (Altiok and Groner, 1996). Expression of β-casein cannot be induced in proliferating cultures of HC11 cells, but requires confluent cultures, which seem to be able to deposit their own basement membrane (Ball *et al.*, 1988; Chammas *et al.*, 1994). Deletion of the SARP-binding site within the β-casein promoter removes this repressive effect and permits the expression of β-casein in exponentially growing cells (Altiok and Groner, 1996). Thus the regulation of SARP-binding activity may be another means by which the extracellular environment regulates milk gene expression at the transcriptional level. Further research is required to elucidate the complexity of interactions required to integrate signals from the ECM and endocrine factors necessary to elicit milk-protein gene expression.

Transgenic Studies Utilizing Bovine Milk-protein Gene Promoters

Expression studies

One of the advantages of studying transgene expression in transgenic animals is the opportunity to identify elements required for *in vivo* expression, especially tissue-specific and developmental regulation (Soulier *et al.*, 1992; Burdon *et al.*, 1994; Li and Rosen, 1995). Many of the transgenic experiments performed with milk-protein genes have been reviewed by Bawden *et al.* (1994) and Wall *et al.* (1997). Here we shall give a brief overview of the transgenic studies performed with bovine mammary-specific promoters, many of which

Table 19.1. Transgenic studies utilizing bovine milk protein promoters in various mammalian species (adapted and updated from Bawden et al., 1994, with permission).

Transgene construct	Flanking sequence*	Expression per ml milk	Reference
Mice			
α_{S1}-casein-h urokinase	21; 2 kb	1–2 mg	Meade et al. (1990)
α_{S1}-casein-ht-PA (cDNA)	1.6 kb; SV40	50 µg	Riego et al. (1993)
α_{S1}-casein-h lactoferrin (cDNA)	6.2; 8 kb	0.1–36 µg	Platenburg et al. (1994)
α_{S1}-casein (minigene)	1.35; 1.5 kb	< 0.1% endog.[†]	Clarke et al. (1994)
α_{S1}-casein-h lysozyme (cDNA)	20; 2 kb	To 0.78 mg	Maga et al. (1995)
β-casein	4.3; 1.3 kb	< 1 µg–3 mg	Bawden et al. (unpublished results)
β-casein	16; 8 kb	< 0.1–20 mg	Rijnkels et al. (1995)
α_{S2}-casein	8; 1.5 kb	< 0.1% endog.[†]	Rijnkels et al. (1995)
κ-casein	5; 19 kb	< 0.03% endog.[†]	Rijnkels et al. (1995)
κ-casein-bGH	4.4 kb; bGH	0.1–7.5 ng	Popov et al. (1995)
α-lactalbumin	750; 336 bp	2.5 µg–0.45 mg	Vilotte et al. (1989)
α-lactalbumin	2; 2.7 kb	100 ng–1.5 mg	Bleck and Bremel (1994)
α-lactalbumin-oTP (cDNA)	740; 450 bp	1 µg	Stinnakre et al. (1991)
α-lactalbumin-b β-casein	2 kb; b β-casein	1–12 mg	Bleck et al. (1995)
α-lactalbumin-b β-casein[‡]	2 kb; b β-casein[‡]	2–3 mg	Choi et al. (1996)
β-lactoglobulin	1.2; 2.7 kb	1–6 mg	Bawden et al. (unpublished results)
β-lactoglobulin	8; 25 kb	2 mg	T. Webster (Melbourne, 1997, personal communication)
β-lactoglobulin-hEPO	2.8; 1.9 kb	0.4 µg–0.3 mg	Korhonen et al. (1997)
Rats			
α_{S1}-casein-hGH	671 bp; hGH	87 µg–6.5 mg	Ninomiya et al. (1994)
β-casein-hGH	1.7 kb; hGH	1 µg–10.9 mg	Ninomiya et al. (1994)
α-lactalbumin	0.8; 0.5 kb	0.2 µg–2.4 mg	Hochi et al. (1992)
α-lactalbumin-hGH	738 bp; hGH	1 µg–4.4 mg	Ninomiya et al. (1994)
Rabbits			
α_{S1}-casein-ht-PA (cDNA)	1.6 kb; SV40	8–50 ng	Riego et al. (1993)
α_{S1}-casein-hIGF-1 (cDNA)	2.9; 3.5 kb	0.1–1 mg	Brem et al. (1994)
β-lactoglobulin-hEPO	2.8; 1.9 kb	0.35–0.54 mg	Korhonen et al. (1997)

*Flanking sequence around the designated transcriptional unit (size of 5′ flanking; 3′ flanking sequence or 3′ gene sequence utilized).
[†]As percentage of endogenous bovine expression level.
[‡]Modified protein coding sequence.
h, human; b, bovine; t-PA, tissue plasminogen activator; GH, growth hormone; oTP ovine trophoblast interferon; EPO, erythropoietin; IGF-1, insulin-like growth factor-1.

are summarized in Table 19.1. However, it should be kept in mind that many studies utilize chimeric constructs, containing only portions of the bovine gene, which can make direct comparisons difficult.

Most bovine milk-protein promoters are able to confer mammary-specific expression on native or heterologous protein coding sequences within transgenic organisms (Ninomiya et al., 1994; Rijnkels et al., 1995). The elements responsible for tissue-specific and developmental regulation appear to be localized less than 1 kb from the transcription start site, while those regions required to direct high-level expression (1 mg ml^{-1} or greater) tend to be further upstream (Bawden *et al.*, 1994). For example, 750 bp of upstream region is sufficient to drive the tissue-specific expression of bovine α-lactalbumin (at a maximum of 0.45 mg ml^{-1}) in the milk of transgenic mice (Vilotte *et al.*, 1989). Promoter deletion studies identified the elements required for the tissue-specific expression of this transgene to be located in the region −477 to −220 of this promoter (Soulier *et al.*, 1992).

Whey promoters tend to require less upstream region to direct high-level expression in transgenic mice than do the casein genes (Bawden *et al.*, 1994). High-level expression of α-lactalbumin was achieved with 2 kb of upstream region (Bleck and Bremel, 1994) and yet more than twice this amount is insufficient for the expression of κ-casein (Rijnkels *et al.*, 1995). For this reason, the α-lactalbumin promoter has been utilized to express bovine casein genes in transgenic mice (Bleck *et al.*, 1995; Choi *et al.*, 1996). However, Bleck *et al.* (1995) reported premature involution with their transgenic mice expressing bovine β-casein under the regulation of the α-lactalbumin promoter. It is unlikely that this was due solely to high-level expression of bovine β-casein in mouse milk, as we ourselves and others (Rijnkels *et al.*, 1995) have expressed this protein at similar levels without any deleterious effects. The construct utilized by Bleck *et al.* (1995) actually contains the β-casein promoter cloned downstream of the α-lactalbumin transcription start point and therefore contains two different promoters. However, the reason for premature involution in these lines of mice remains unclear.

Interestingly, high-level expression has been achieved for bovine α_{S1}- and β-casein transgenes, utilizing 21 and 16 kb of upstream sequence, respectively (Meade *et al.*, 1990; Rijnkels *et al.*, 1995), and yet very poor results have been obtained for the α_{S2}- and κ-casein genes (Popov *et al.*, 1995; Rijnkels *et al.*, 1995). As these genes are present in a locus within cattle in the order α_{S1}–β–α_{S2}–κ-casein, it has been proposed that major regulatory elements required for the coordinated high-level expression of these genes during lactation may reside upstream of the locus (Rijnkels *et al.*, 1995), as is the case for the human β-globin locus (Grosveld *et al.*, 1987). This may explain why all α_{S2}- and κ-casein transgenics to date exhibit low expression levels. Alternatively, the regulatory elements required for high-level expression of α_{S2}- and κ-casein may simply be further upstream of the protein coding regions than those fragments already introduced into mice. For example, 21 kb of upstream region is required for the high-level expression of α_{S1}-casein (Meade *et al.*, 1990), and yet no fragments containing greater than 5 kb of κ-casein upstream region have been examined in transgenic mice.

The alteration of milk composition via transgenesis

Targeting transgene expression to the mammary gland in transgenic animals provides an opportunity to modify the composition of all the major milk constituents and subsequently improve the nutritional and manufacturing quality of milk (Maga and Murray, 1995; Karatzas and Turner, 1997; Wall *et al.*, 1997). For example, the introduction of additional casein genes to the bovine genome may increase the total casein content of milk, thereby increasing its manufacturing value for the production of cheese (Richardson, 1985; Kang *et al.*, 1986). In addition, the expression of less favourable genes, such as those implicated in allergic responses, could be downregulated with the use of antisense, ribozyme or gene knockout technology (Stinnakre *et al.*, 1994; Stacey *et al.*, 1995; Sokol and Murray, 1996). Protein engineering has also been performed on milk-protein genes, such as bovine κ-casein (Oh and Richardson, 1991a, b), β-casein (Simons *et al.*, 1993; Choi *et al.*, 1996) and BLG (Cho *et al.*, 1994; Lee *et al.*, 1994), to improve their dairy-processing characteristics. In recent years, a number of studies have been performed with milk-protein genes which have been specifically designed to determine the effects of altered milk-protein expression levels on milk production, utilizing transgenic mice as a model system. These include experiments designed to alter lactose levels (Stinnakre *et al.*, 1994; Stacey *et al.*, 1995; L'Huillier *et al.*, 1996), the introduction of antibacterial proteins (Maga *et al.*, 1994; Platenburg *et al.*, 1994) and the alteration of casein expression levels (Kumar *et al.*, 1994; Gutiérrez-Adán *et al.*, 1996).

Alpha-lactalbumin
Much of the world's population is lactose-intolerant; thus it would be advantageous to reduce lactose levels in bovine milk (Mercier, 1986). Homozygous null mutants for murine α-lactalbumin have been generated via gene targeting in embryonic stem cells by two different laboratories (Stinnakre *et al.*, 1994; Stacey *et al.*, 1995). As predicted, both α-lactalbumin and thus lactose were absent from the milk of these mice. In addition, the milk contained approximately 60% more fat and an 88% increase in protein with respect to wild type milk (Stacey *et al.*, 1995). Unfortunately, in both experiments the milk obtained from these mice was so viscous that very little could be ejected from the teat (Stinnakre *et al.*, 1994; Stacey *et al.*, 1995). These results confirm the importance of lactose in the regulation of milk volume within the mammary gland.

A ribozyme-mediated approach to the downregulation of α-lactalbumin has been pursued by L'Huillier *et al.* (1996). Mice expressing a hammerhead ribozyme, designed to interact with the 3′ UTR of the bovine α-lactalbumin mRNA, were crossed with transgenic mice that expressed bovine α-lactalbumin at a high level. Double-transgenic mice exhibited a 78–50% reduction in bovine α-lactalbumin mRNA expression, while the expression of endogenous α-lactalbumin was unaffected.

Lactoferrin and lysozyme
Udder disease caused by bacterial infection not only causes discomfort in the animal but also leads to loss of milk production and quality (Miles *et al.*, 1992). Milk proteins such as lysozyme and lactoferrin have been demonstrated to possess antimicrobial activity (Fleming, 1922; Arnold *et al.*, 1977) and yet are

expressed at low levels in bovine milk with respect to that of other mammals (Chandan *et al.*, 1968; Masson and Heremans, 1971). Both human lysozyme and lactoferrin have been expressed in the milk of transgenic mice (Maga *et al.*, 1994; Platenburg *et al.*, 1994). Milk samples from transgenic mice secreting human lysozyme have recently been demonstrated to be bacteriostatic against a mastitis-causing strain of *Staphylococcus aureus*, in addition to certain cold-spoilage organisms (Maga *et al.*, 1998). Thus, the high-level expression of antibacterial proteins in bovine milk could confer mastitis resistance on these cows.

Kappa-casein

The primary role of κ-casein in milk is to stabilize the structure of the casein micelle (Waugh and von Hippel, 1956). The chymosin-mediated cleavage of κ-casein leads to the destabilization of the micelle and the precipitation of the caseins important in cheesemaking (Mercier *et al.*, 1973). McGann *et al.* (1980) demonstrated that the total amount of κ-casein present in micelles was proportional to the ratio of surface area to volume. Therefore, it has been proposed that enhanced secretion of κ-casein in milk could reduce micelle size, thereby increasing its surface area and resulting in reduced rennet clotting time (Pearse *et al.*, 1986). Transgenic mice expressing bovine κ-casein have been generated utilizing the caprine β-casein promoter (Gutiérrez *et al.*, 1996); however, whereas this milk possessed micelles of a reduced diameter, there was no difference in rennet clotting time compared with milk from non-transgenic mice, although gel strength did increase (Gutiérrez-Adán *et al.*, 1996).

Conclusion

The cost of producing transgenic cattle (Wall *et al.*, 1997) has been a considerable deterrent, limiting the use of this species for the production of pharmaceutical proteins and impeding genetic engineering of dairy cattle with improved production traits. However, with our increased understanding of the regulation of bovine milk-protein gene expression and the development of technology for nuclear transfer and animal cloning, the efficiency of producing transgenic livestock will improve. This will provide new opportunities for the dairy industry to modify and add proteins to milk for improved processing to value-added products, to enhance milk composition for consumption by humans and to engineer cattle for the production of therapeutic proteins (Lönnerdal, 1996; Karatzas and Turner, 1997; Wall *et al.*, 1997).

Acknowledgements

The authors would like to thank A. Cato, W. Doppler, S. Gorodetskii, B. Groner, J. Kennelly, F. Martin, T. Oka, J. Rosen and J. Yang for supplying updates on their research for inclusion in this chapter. In addition, we are indebted to Maria Constanzo-Servais for her assistance in researching this review, and also J. Martyn, T. Webster and J. Rosen for their helpful comments on the manuscript.

Editorial Note

The bovine casein gene nomenclature has recently been changed to bring it into line with the human nomenclature. The relationships are:

Gene	Old symbol	New symbol
α_{s1}-casein	CASA1	CSN1S1
α_{s2}-casein	CASA2	CSN1S2
β-casein	CASB	CSN2
κ-casein	CASK	CSN3

There was no time to implement the use of new symbols in this chapter due to time constraints.

References

Adachi, T., Ahn, J.Y., Yamamoto, K., Aoki, N., Nakamura, R. and Matsuda, T. (1996) Characterization of the bovine κ-casein gene promoter. *Biosciences, Biotechnology and Biochemistry* 60, 1937–1940.

Alexander, L.J., Stewart, A.F., Mackinlay, A.G., Kapelinskaya, T.V., Tkach, T.M. and Gorodetskii, S.I. (1988) Isolation and characterisation of the bovine κ-casein gene. *European Journal of Biochemistry* 178, 395–401.

Alexander, L.J., Hayes, G., Bawden, W., Stewart, A.F. and Mackinlay, A.G. (1993) Complete nucleotide sequence of the bovine β-lactoglobulin gene. *Animal Biotechnology* 4, 1–10.

Altiok, S. and Groner, B. (1993) Interaction of two sequence-specific single-stranded DNA-binding proteins with an essential region of the β-casein gene promoter is regulated by lactogenic hormones. *Molecular and Cellular Biology* 13, 7303–7310.

Altiok, S. and Groner, B. (1994) β-casein mRNA sequesters a single-stranded nucleic acid binding protein which negatively regulates the β-casein gene promoter. *Molecular and Cellular Biology* 14, 6004–6012.

Altiok, S. and Groner, B. (1996) Regulation of gene expression in mammary epithelial cells by cellular confluence and sequence-specific DNA binding factors. *Biochemical Society Symposia* 63, 115–131.

Anderson, B.F., Baker, H.M., Norris, G.E., Rice, D.W. and Baker, E.N. (1989) Structure of human lactoferrin: crystallographic structure analysis and refinement to 2.8 Å resolution. *Journal of Molecular Biology* 209, 711–734.

Arnold, R.R., Cole, M.F. and McGhee, J.R. (1977) A bacteriocidal effect for human lactoferrin. *Science* 197, 263–265.

Ball, R.K., Friis, R.R., Schoenenberger, C.A., Doppler, W. and Groner, B. (1988) Prolactin regulation of beta-casein gene expression and of a cytosolic 120-kd protein in a cloned mouse mammary epithelial cell line. *EMBO Journal* 7, 2089–2095.

Barcellos-Hoff, M.H., Aggeler, J., Ram, T.G. and Bissell, M.J. (1989) Functional differentiation and alveolar morphogenesis of primary mammary cultures on reconstituted basement membrane. *Development* 105, 223–235.

Baumrucker, C.R., Deemer, K.P., Walsh, R., Riss, T.L. and Akers, R.M. (1988) Primary culture of bovine mammary acini on a collagen matrix. *Tissue and Cell* 20, 541–554.

Bawden, W.S., Passey, R.J. and Mackinlay, A.G. (1994) The genes encoding the major milk specific proteins and their use in transgenic studies and protein engineering. *Biotechnology and Genetic Engineering Reviews* 12, 89–137.

Beato, M. (1989) Gene regulation by steroid hormones. *Cell* 56, 335–344.

Bissell, M.J. and Hall, G.H. (1987) Form and function in the mammary gland, the role of the extracellular matrix. In: Neville, M.C. and Daniel, C.W. (eds) *The Mammary Gland: Development, Regulation and Function.* Plenum Press, New York, pp. 355–373.

Blackburn, D.E., Hobbs, A.A. and Rosen, J.M. (1982) Rat β-casein cDNA: sequence analysis and evolutionary comparisons. *Nucleic Acids Research* 10, 2295–2307.

Bleck, G.T. and Bremel, R.D. (1993) Sequence and single-base polymorphisms of the bovine α-lactalbumin 5′-flanking region. *Gene* 126, 213–218.

Bleck, G.T. and Bremel, R.D. (1994) Variation in the expression of a bovine α-lactalbumin transgene in milk of transgenic mice. *Journal of Dairy Science* 77, 1897–1904.

Bleck, G.T., Jiménez-Flores, R. and Bremel, R.D. (1995) Abnormal properties of milk from transgenic mice expressing bovine β-casein under control of the bovine α-lactalbumin 5′ flanking region. *International Dairy Journal* 5, 619–632.

Blum, J.L., Zeigler, M.E. and Wicha, M.S. (1989) Regulation of mammary differentiation by the extracellular matrix. *Environmental Health Perspectives* 80, 71–83.

Bolander, F.F., Nicholas, K.R., Van Wyk, J. J. and Topper, Y.J (1981) Insulin is essential for accumulation of casein mRNA in mouse mammary epithelial cells. *Proceedings of the National Academy of Sciences of the USA* 78, 5682–5684.

Bonsing, J. and Mackinlay, A.G. (1987) Recent studies on nucleotide sequences encoding the caseins. *Journal of Dairy Research* 54, 447–461.

Bonsing, J., Ring, J.M., Stewart, A.F. and Mackinlay, A.G. (1988) Complete nucleotide sequence of the bovine β-casein gene. *Australian Journal of Biological Sciences* 41, 527–537.

Brem, G., Hartl, P., Besenfelder, U., Wolf, E., Zinovieva, N. and Pfaller, R. (1994) Expression of synthetic cDNA sequences encoding human insulin-like growth factor-1 (IGF-1) in the mammary gland of transgenic rabbits. *Gene* 149, 351–355.

Brew, K., Castellino, F.J. Vanaman, T.C. and Hill, R.L. (1970) The complete amino acid sequence of bovine α-lactalbumin. *Journal of Biological Chemistry* 245, 4570–4582.

Burdon, T.G., Maitland, K.A., Clark, A.J., Wallice, R. and Watson, C.J. (1994) Regulation of the sheep β-lactoglobulin gene by lactogenic hormones is mediated by a transcription factor that binds an interferon-γ activation site-related element. *Molecular Endocrinology* 8, 1528–1536.

Campbell, K.H.S., McWhir, J., Ritchie, W.A. and Wilmut, I. (1996) Sheep cloned by nuclear transfer from a cultured cell line. *Nature* 380, 64–66.

Chammas, R., Taverna, D., Cella, N., Santos, C. and Hynes, N.E. (1994) Laminin and tenascin assembly and expression regulate HC11 mouse mammary cell differentiation. *Journal of Cell Science* 107, 1031–1040.

Chandan, R.C., Parry, R.M. and Shahani, K.M. (1968) Lysozyme, lipase, and ribonuclease in the milk of various species. *Journal of Dairy Science* 51, 606–607.

Cho, Y., Gu, W., Watkins, S., Lee, S.P., Kim, T.R., Brady, J.W. and Batt, C.A. (1994) Thermostable variants of bovine β-lactoglobulin. *Protein Engineering* 7, 263–270.

Choi, B.K., Bleck, G.T., Wheeler, M.B. and Jiménez-Flores, R. (1996) Genetic modification of bovine β-casein and its expression in the milk of transgenic mice. *Journal of Agricultural and Food Chemistry* 44, 953–960.

Choi, Y.J., Keller, W.L., Berg, I.E., Park, C.S. and Mackinlay, A.G. (1988) Casein gene expression in bovine mammary gland. *Journal of Dairy Science* 71, 2898–2903.

Chomczynski, P., Qasba, P. and Topper, Y.J. (1984) Essential role of insulin in transcription of the rat 25,000 molecular weight casein gene. *Science* 226, 1326–1328.

Chomczynski, P., Qasba, P. and Topper, Y.J. (1986) Transcriptional and post-transcriptional roles of glucocorticoid in the expression of the rat 25,000 molecular weight casein gene. *Biochemical and Biophysical Research Communications* 114, 380–387.

Cibelli, J.B., Stice, S.L., Golueke, P.J., Kane, J.J., Jerry, J., Blackwell, C., Ponce de León, F.A. and Robl, J.M. (1998) Cloned transgenic calves produced from nonquiescent fetal fibroblasts. *Science* 280, 1256–1258.

Clarke, R.A, Sokol, D., Rigby, N., Ward, K., Murray, J.D. and Mackinlay, A.G. (1994) Mammary gland-specific expression of bovine α_{S1}-casein derived transgenes in mice. *Transgenics* 1, 313–319.

Close, M.J., Howlett, A.R., Roskelley, C.D., Desprez, P.Y., Bailey, N., Rowning, B., Teng, C.T., Stampfer, M.R. and Yaswen, P. (1997) Lactoferrin expression in mammary epithelial cells is mediated by changes in cell shape and actin cytoskeleton. *Journal of Cell Science* 110, 2861–2871.

Conti, A., Godovac-Zimmermann, J., Liberatori, J. and Braunitzer, G. (1984) The primary structure of monomeric beta-lactoglobulin I from horse colostrum (*Equus caballus*, Perissodactyla). *Hoppe-Seyler's Zeitschrift für Physiologische Chemie* 365, 1393–1401.

Dalgleish, D.G. (1992) Bovine milk protein properties and the manufacturing quality of milk. *Livestock Production Science* 35, 75–95.

Doppler, W., Groner, B. and Ball, R.K. (1989) Prolactin and glucocorticoid hormones synergistically induce expression of transfected rat β-casein gene promoter constructs in a mammary epithelial cell line. *Proceedings of the National Academy of Sciences of the USA* 86, 104–108.

Doppler, W., Höck, W., Hofer, P., Groner, B. and Ball, R.K. (1990) Prolactin and glucocorticoid hormones control transcription of the β-casein gene by kinetically distinct mechanisms. *Molecular Endocrinology* 4, 912–919.

Doppler, W., Welte, T. and Philipp, S. (1995) CCAAT/enhancer-binding protein isoforms β and δ are expressed in mammary epithelial cells and bind to multiple sites in the β-casein gene promoter. *Journal of Biological Chemistry* 270, 17962–17969.

Ebner, K.E. and Schanbacher, F.L. (1974) Biochemistry of lactose and related carbohydrates. In: Larson, B.L. and Smith, V.R. (eds) *Lactation: a Comprehensive Treatise*, Vol. II. Academic Press, New York, pp. 77–113.

Edwards, G.M., Wilford, F.H. and Streuli, C.H. (1996) Extracellular matrix controls the prolactin signalling pathway in mammary epithelial cells. *Biochemical Society Transactions* 24, 345S.

Emerman, J.T and Pitelka, D.R. (1977) Maintenance and identification of morphological differentiation in dissociated mammary epithelium on floating collagen membranes. *In Vitro* 13, 316–328.

Emerman, J.T., Enami, J., Pitelka, D.R. and Nandi, S. (1977) Hormonal effects on intracellular and secreted casein in cultures of mouse mammary epithelial cells on floating collagen membranes. *Proceedings of the National Academy of Sciences of the USA* 74, 4466–4470.

Eyestone, W.H., Gowallis, M., Monohan, J., Sink, T., Ball, S.F. and Cooper, J.D. (1998) Cloning: new breakthroughs leading to commercial opportunities. *Theriogenology* 49, 386.

Ferretti, L., Leone, P. and Sgaramella, V. (1990) Long range restriction analysis of the bovine casein genes. *Nucleic Acids Research* 18, 6829–6833.

Fleming, A. (1922) On a remarkable bacteriolytic element found in tissues and secretions. *Proceedings of the Royal Society of London B* 93, 306–317.

Folch, J.M., Coll, A., Hayes, H.C. and Sanchez, A. (1996) Characterization of a caprine beta-lactoglobulin pseudogene, identification and chromosomal localization by *in situ* hybridisation in goat, sheep and cow. *Gene* 177, 87–91.

Fugate, R.D. and Song, P.S. (1980) Spectroscopic characterization of β-lactoglobulin-retinol complex. *Biochimica et Biophysica Acta* 625, 28–42.

Furlong, E.E.M., Keon, N.K., Thorton, F.D., Rein, T. and Martin, F. (1996) Expression of a 74-kDa nuclear factor 1 (NF1) protein induced in mouse mammary gland during involution. *Journal of Biological Chemistry* 271, 29688–29697.

Gallagher, D.S., Schelling, C.P., Groenen, M.M.A. and Womack, J.E. (1994) Confirmation that the casein gene cluster resides on cattle chromosome 6. *Mammalian Genome* 5, 524.

Gertler, A., Weil, A. and Cohen, N. (1982) Hormonal control of casein synthesis in organ culture of the bovine lactating mammary gland. *Journal of Dairy Research* 49, 387–398.

Getzenberg, R.H. (1994) Nuclear matrix and the regulation of gene expression: tissue specificity. *Journal of Cellular Biochemistry* 55, 22–31.

Gibson, C.A., Vega, J.R., Baumrucker, C.R., Oakley, C.S. and Welsch, C.W. (1991) Establishment and characterisation of bovine mammary epithelial cell lines. *In Vitro Cellular and Developmental Biology* 27A, 585–594.

Godovac-Zimmermann, J., Conti, A., Liberatori, J. and Braunitzer, G. (1985) The amino acid sequence of beta-lactoglobulin II from horse colostrum (*Equus caballus*, Perissodactyla): beta-lactoglobulins are retinol binding proteins. *Biological Chemistry Hoppe-Seyler* 366, 601–608.

Goodman, G.T., Akers, R.M., Friderici, K.H. and Tucker, H.A. (1983) Hormonal regulation of α-lactalbumin secretion from bovine mammary tissue cultured *in vitro*. *Endocrinology* 112, 1324–1330.

Gorodetskii, S.I., Tkach, T.M. and Kapelinskaya, T.V. (1988) Isolation and characterisation of the *Bos taurus* β-casein gene. *Gene* 66, 87–96.

Gouilleux, F., Wakao, H., Mundt, M. and Groner, B. (1994) Prolactin induces phosphorylation of Tyr694 of Stat5 (MGF), a prerequisite for DNA binding and induction of transcription. *EMBO Journal* 13, 4361–4369.

Groenen, M.A.M. and van der Poel, J.J. (1994) Regulation of expression of milk protein genes: a review. *Livestock Production Science* 38, 61–78.

Groenen, M.A.M., Dijkhof, R.J., Verstege, A.J. and van der Poel, J.J. (1993) The complete sequence of the gene encoding bovine alpha S2-casein. *Gene* 123, 187–193.

Grosveld, F., van Assendelft, G.B., Greaves, D.R. and Kollias, G. (1987) Position-independent, high-level expression of human β-globin gene in transgenic mice. *Cell* 51, 975–985.

Groves, M.L. (1960) The isolation of a red protein from milk. *Journal of the American Chemical Society* 82, 3345–3350.

Gutiérrez, A., Meade, H.M., Jiménez-Flores, R., Anderson, G.B., Murray, J.D. and Medrano, J.F. (1996) Expression of a bovine kappa-CN cDNA in the mammary gland of transgenic mice utilising a genomic milk protein gene as an expression cassette. *Transgenic Research* 5, 271–279.

Gutiérrez-Adán, A., Maga, E.A., Meade, H.M., Shoemaker, C.F., Medrano, J.F., Anderson, G.B. and Murray, J.D. (1996) Alteration of the physical characteristics of milk from transgenic mice producing bovine κ-casein. *Journal of Dairy Science* 79, 791–799.

Hall, L., Emery, D.C., Davies, M.S., Parker, D. and Craig, R.K. (1987) Organisation and sequence of the human α-lactalbumin gene. *Biochemical Journal* 242, 735–742.

Halliday, J.A., Bell, K. and Shaw, D.C. (1991) The complete amino acid sequence of feline β-lactoglobulin II and a partial revision of the equine β-lactoglobulin II sequence. *Biochimica et Biophysica Acta* 1077, 25–30.

Halliday, J.A., Bell, K., McAndrew, K. and Shaw, D.C. (1993) Feline β-lactoglobulins I, II and III, and canine β-lactoglobulins I and II: amino acid sequences provide evidence for the existence of more than one gene for β-lactoglobulin in the cat and dog. *Protein Sequence Data Analysis* 5, 201–205.

Haresign, W. (1979) Body condition, milk yield and reproduction in cattle. In: Haresign, W. and Lewis, D. (eds) *Recent Advances in Animal Nutrition*. Butterworths, London, pp. 107–122.

Hartmann, P.E. (1973) Changes in the composition and yield of mammary secretion of cows during the initiation of lactation. *Journal of Endocrinology* 59, 231–247.

Hartmann, P.E. and Shelton, J.N. (1971) Progesterone and the initiation of milk synthesis in the cow and the ewe. *Proceedings of the Australian Physiological and Pharmacological Society* 2, 50.

Hayes, H.C and Petit, E.J. (1993) Mapping the β-lactoglobulin gene and of an immunoglobulin M heavy chain-like sequence to homologous cattle, sheep, and goat chromosomes. *Mammalian Genome* 4, 207–210.

Hayes, H.C., Popescu, P. and Dutrillaux, B. (1993) Comparative gene mapping of lactoperoxidase, retinoblastoma, and α-lactalbumin genes in cattle, sheep, and goats. *Mammalian Genome* 4, 593–597.

Hennighausen, L.G. and Sippel, A.E. (1982) Mouse whey acidic protein is a novel member of the 'four-disulphide core' proteins. *Nucleic Acids Research* 10, 2677–2684.

Hillerton, J.E., Knight, C.H., Turvey, A., Wheatley, S.D. and Wilde, C.J. (1990) Milk yield and mammary function in dairy cows milked four times daily. *Journal of Dairy Research* 57, 285–294.

Hobbs, A.A. and Rosen, J.M. (1982) Sequence of the rat α- and γ-casein mRNA's: evolutionary comparison of the calcium dependent rat casein multigene family. *Nucleic Acids Research* 10, 8079–8098.

Hochi, S.I., Ninomiya, T., Waga-Homma, M., Sagara, J. and Yuki, A. (1992) Secretion of bovine α-lactalbumin into the milk of transgenic rats. *Molecular Reproduction and Development* 33, 160–164.

Ivanov, V.N., Kabishev, A.A., Gorodetskii, S.I. and Gribanovskii, V.A. (1990) Activation of the *trans*-activating transcription factor NF1 in lactating mammary gland. *Molekulyarnaya Biologiyacular* 24, 1605–1615.

Jenness, R. (1974) Caseins and caseinate micelles of various species. *Netherlands Milk and Dairy Journal* 27, 251–257.

Jenness, R. (1979) Comparative aspects of milk proteins. *Journal of Dairy Research* 46, 197–210.

Jollès, P. (1975) Structural aspects of the milk clotting process. Comparative features with the blood clotting process. *Molecular and Cellular Biochemistry* 7, 73–85.

Jollès, P., Loucheaux-Lefebvre, M.H. and Henschen, A. (1978) Structural relatedness of kappa-casein and fibrinogen gamma-chain. *Journal of Molecular Evolution* 11, 271–277.

Jones, W.K., Tu-Lee, L.Y., Clift, S.M., Brown, T.L. and Rosen, J.M. (1985) The rat casein multigene family: fine structure and evolution of the β-casein gene. *Journal of Biological Chemistry* 260, 7042–7050.

Jung-Youb, A., Aoki, N., Adachi, T., Mizuno, Y., Nakamura, R. and Matsuda, T. (1995) Isolation and culture of bovine mammary epithelial cells and establishment of gene transfection conditions in the cells. *Biosciences, Biotechnology and Biochemistry* 59, 59–64.

Kanai, A., Nonomura, N., Yoshimura, M. and Oka, T. (1993) DNA-binding proteins and their *cis*-acting sites controlling hormonal induction of a mouse β-casein::CAT fusion protein in mammary epithelial cells. *Gene* 126, 195–201.

Kang, T., Jiménez-Flores, R. and Richardson, T. (1986) Casein genes and genetic engineering of the caseins. In: Evens, J.W. and Hollaender, A. (eds) *Genetic Engineering of Animals: an Agricultural Perspective.* Plenum Press, New York, pp. 95–111.

Karatzas, C.N. and Turner, J.D. (1997) Toward altering milk composition by genetic manipulation: current status and challenges. *Journal of Dairy Science* 80, 2225–2232.

Koczan, D., Hobom, G. and Seyfert, H.M. (1991) Genomic organisation of the bovine α_{S1}-casein gene. *Nucleic Acids Research* 19, 5591–5596.

Korhonen, V.P., Tolvanen, M., Hyttinen, J.M., Ousi-Oukari, M., Sinervirta, R., Alhonen, L., Jauhiainene, M., Janne, O.A. and Janne, J. (1997) Expression of bovine beta-lactoglobulin/human erythropoietin fusion protein in the milk of transgenic mice and rabbits. *European Journal of Biochemistry* 245, 482–489.

Korobko, I.V., Kazanskii, A.V., Grinenko, N.F. and Gorodetskii, S.I. (1995) Identification of binding sites for c/EBP nuclear factor in the sequences of bovine casein genes. *Doklady Akademii Nauk* 340, 108–110.

Kulski, J.K., Nicholas, K.R., Topper, Y.J. and Qasba, P. (1983) Essentiality of insulin and prolactin for accumulation of rat casein mRNAs. *Biochemical and Biophysical Research Communications* 116, 994–999.

Kumagai, I., Takeda, S. and Miura, K.I. (1992) Functional conversion of the homologous proteins α-lactalbumin and lysozyme by exon exchange. *Proceedings of the National Academy of Sciences of the USA* 89, 5887–5891.

Kumar, S., Clarke, A.R., Hooper, M.L., Horne, D.S., Law, A.J.R., Leaver, J., Springbett, A., Stevenson, E. and Simons, J.P. (1994) Milk composition and lactation of β-casein-deficient mice. *Proceedings of the National Academy of Sciences of the USA* 91, 6138–6142.

Kuys, Y.M., Snell, R.G. and Wheeler, T.T. (1996) Binding of nuclear proteins to the bovine α-lactalbumin gene promoter. *Proceedings of the New Zealand Society of Animal Production* 56, 68–70.

Lechner, J., Welte, T. and Doppler, W. (1997a) Mechanism of interaction between the glucocorticoid receptor and Stat5: role of DNA-binding. *Immunobiology* 197, 175–186.

Lechner, J., Welte, T., Tomasi, J.K., Bruno, P., Cairns, C., Gustafsson, J.A. and Doppler, W. (1997b) Promoter-dependent synergy between glucocorticoid receptor and

Stat5 in the activation of β-casein gene transcription. *Journal of Biological Chemistry* 272, 20954–20960.

Lee, C.S. and Oka, T. (1992) A pregnancy-specific mammary nuclear factor involved in the repression of the mouse β-casein gene transcription by progesterone. *Journal of Biological Chemistry* 267, 5797–5801.

Lee, S.P., Kim, D.S., Watkins, S. and Batt, C.A. (1994) Reducing whey syneresis in yogurt by the addition of a thermolabile variant of β-lactoglobulin. *Biosciences, Biotechnology and Biochemistry* 58, 309–313.

Leliévre, S., Weaver, V.M. and Bissell, M.J. (1996) Extracellular matrix signalling from the cellular membrane skeleton to the nuclear skeleton: a model for gene regulation. *Recent Progress in Hormone Research* 51, 417–432.

L'Huillier, P.J., Soulier, S., Stinnakre, M.G., Lepourry, L., Davis, S.R., Mercier, J.C. and Vilotte, J.L. (1996) Efficient and specific ribozyme-mediated reduction of bovine α-lactalbumin expression in double transgenic mice. *Proceedings of the National Academy of Sciences of the USA* 93, 6698–6703.

Li, M.L., Aggeler, J., Farson, D.A., Hatier, C., Hassell, J. and Bissell, M.J. (1987) Influence of a reconstituted basement membrane and its components on casein gene expression and secretion in mouse mammary epithelial cells. *Proceedings of the National Academy of Sciences of the USA* 84, 136–140.

Li, S. and Rosen, J.M. (1994) Glucocorticoid regulation of rat whey acidic protein gene expression involves hormone-induced alterations of chromatin structure in the distal promoter region. *Molecular Endocrinology* 8, 1328–1335.

Li, S. and Rosen, J.M. (1995) Nuclear factor I and mammary gland factor (STAT5) play a critical role in regulating rat whey acidic protein gene expression in transgenic mice. *Molecular and Cellular Biology* 15, 2063–2070.

Lin, C.Q. and Bissell, M.J. (1993) Multi-faceted regulation of cell differentiation by extracellular matrix. *FASEB Journal* 7, 737–743.

Liu, X., Robinson, G.W., Gouilleux, F., Groner, B. and Hennighausen, L. (1995) Cloning and expression of Stat5 and an additional homologue (Stat5b) involved in prolactin signal transduction in mouse mammary tissue. *Proceedings of the National Academy of Sciences of the USA* 92, 8831–8835.

Liu, X., Robinson, G.W., Wagner, K.U., Garrett, L., Wynshaw-Boris, A. and Hennighausen, L. (1997) Stat5a is mandatory for adult mammary gland development and lactogenesis. *Genes and Development* 11, 179–186.

Lönnerdal, B. (1996) Recombinant human milk proteins – an opportunity and a challenge. *American Journal of Clinical Nutrition* 63, 622S–626S.

McGann, T.C.A., Donnelly, W.J., Kearny, R.D. and Buchheim, W. (1980) Composition and size distribution of bovine casein micelles. *Biochimica et Biophysica Acta* 630, 261–270.

MacKenzie, D.D.S., Brooker, B.E. and Forsyth, I. (1985) Ultrastructural features of bovine mammary epithelial cells grown on collagen gels. *Tissue and Cell* 17, 39–51.

McMahon, D.J. and Brown, R.J. (1984) Composition, structure and integrity of casein micelles: a review. *Journal of Dairy Science* 67, 499–512.

Maga, E.A. and Murray, J.D. (1995) Mammary gland expression of transgenes and the potential for altering the properties of milk. *Bio/Technology* 13, 1452–1457.

Maga, E.A., Anderson, G.B., Huang, M.C. and Murray, J.D. (1994) Expression of human lysozyme mRNA in the mammary gland of transgenic mice. *Transgenic Research* 3, 36–42.

Maga, E.A., Anderson, G.B., Cullor, J.S., Smith, W. and Murray, J.D. (1998) Antimicrobial properties of human lysozyme transgenic mouse milk. *Journal of Food Protection* 61, 52–56.

Masson, P.L. and Heremans, J.F. (1971) Lactoferrin in the milk of different species. *Comparative Biochemistry and Physiology*, 39B, 143–147.

Meade, H., Gates, L., Lacy, E. and Lonberg, N. (1990) Bovine α_{S1}-casein gene sequences direct high-level expression of active human urokinase in mouse milk. *Bio/Technology* 8, 443–446.

Meier, V.S. and Groner, B. (1994) The nuclear factor YY1 participates in repression of the β-casein gene promoter in mammary epithelial cells and is counteracted by mammary gland factor during lactogenic hormone induction. *Molecular and Cellular Biology* 14, 128–137.

Mercier, J.C. (1986) Genetic engineering applied to milk producing animals: some expectations. In: Smith, C., King, J.W.B. and McKay, J.C. (eds) *Exploiting New Technologies in Animal Breeding*. Oxford University Press, London, pp. 122–131.

Mercier, J.C and Vilotte, J.L. (1993) Structure and function of milk protein genes. *Journal of Dairy Science* 76, 3079–3098.

Mercier, J.C., Brignon, G. and Ribadeau-Dumas, B. (1973) Primary structure of bovine kappa B casein complete sequence. *European Journal of Biochemistry* 35, 222–235.

Miles, H., Lesser, W. and Sears, P. (1992) The economic implications of bioengineered mastitis control. *Journal of Dairy Science* 75, 596–605.

Mink, S., Härtig, E., Jennewein, P., Doppler, W. and Cato, A.C.B. (1992) A mammary cell-specific enhancer in mouse mammary tumor virus DNA is composed of multiple regulatory elements including binding sites for CTF/NFI and a novel transcription factor, mammary cell activating factor. *Molecular and Cellular Biology* 12, 4906–4918.

Murphy, J.J. and O'Mara, F. (1993) Nutritional manipulation of milk protein concentration and its impact on the dairy industry. *Livestock Production Science* 35, 117–134.

Neuteboom, B., Giuffrida, M.G. and Conti, A. (1992) Isolation of a new ligand-carrying casein fragment from bovine mammary gland microsomes. *FEBS Letters* 305, 189–191.

Nicholas, K.R., Simpson, K., Wilson, M., Trott, J. and Shaw, D.C. (1997) The tammar wallaby: a model to study putative autocrine-induced changes in milk composition. *Journal of Mammary Gland Biology and Neoplasia* 2, 299–310.

Ninomiya, T., Hirabayashi, M., Sagara, J. and Yuki, A. (1994) Functions of milk protein gene 5′ flanking regions on human growth hormone gene. *Molecular Reproduction and Development* 37, 276–283.

North, A.C.T. (1991) Structural homology in ligand-specific transport proteins. *Biochemical Society Symposia* 57, 35–48.

Oh, S. and Richardson, T. (1991a) Genetic engineering of bovine κ-casein to enhance proteolysis by chymosin. In: Parris, N. and Barford, R. (eds) *Interactions of Food Proteins*. American Chemical Society, Washington, DC, pp. 195–211.

Oh, S. and Richardson, T. (1991b) Genetic engineering of bovine κ-casein to improve its nutritional quality. *Journal of Agriculture and Food Chemistry* 39, 422–427.

Oram, J.D. and Reiter, B. (1968) Inhibition of bacteria by lactoferrin and other iron-chelating agents. *Biochimica et Biophysica Acta* 170, 351–365.

Palmer, A.H. (1934) The preparation of a crystalline globulin from the albumin fraction of cow's milk. *Journal of Biological Chemistry* 104, 359–372.

Paonessa, G., Gounari, F., Frank, R. and Cortese, R. (1988) Purification of a NF1-like DNA-binding protein from rat liver and cloning the corresponding cDNA. *EMBO Journal* 7, 3115–3123.

Papiz, M.Z., Sawyer, L., Eliopoulos, E.E., North, A.C.T., Findlay, J.B.C., Sivaprasadarao, R., Jones, T.A., Newcomer, M.E. and Kraulis, P.J. (1986) The structure of β-lactoglobulin and its similarity to plasma retinol-binding protein. *Nature* 324, 383–385.

Park, I., Schaeffer, E., Sidoli, A., Baralle, F.E., Cohen, G.N. and Zakin, M.M. (1985) Organisation of the human transferrin gene: direct evidence that it originated by gene duplication. *Proceedings of the National Academy of Sciences of the USA* 82, 3149–3153.

Passey, R.J and Mackinlay, A.G. (1995) Characterisation of a second, apparently inactive, copy of the bovine β-lactoglobulin gene. *European Journal of Biochemistry* 233, 736–743.

Pearse, M.J., Linklater, P.M., Hall, R.J. and Mackinlay, A.G. (1986) The effect of casein composition and casein dephosphorylation on the coagulation and syneresis of artificial micelle milk. *Journal of Dairy Science* 53, 381–390.

Pearson, W.R. (1990) Rapid and sensitive sequence comparison with FASTP and FASTA. *Methods in Enzymology* 183, 63–98.

Pearson, W.R. and Lipman D. J. (1988) Improved tools for biological sequence analysis. *Proceedings of the National Academy of Sciences of the USA* 85, 2444–2448.

Pérez, M.D. and Calvo, M. (1995) Interaction of β-lactoglobulin with retinol and fatty acids and its role as a possible biological function for this protein: a review. *Journal of Dairy Science* 78, 978–988.

Platenburg, G.J., Kootwijk, E.P.A., Kooiman, P.M., Woloshuk, S.L., Nuijens, J.H., Krimpenfort, P.J.A., Pieper, F.R., de Boer, H.A. and Strijker, R. (1994) Expression of human lactoferrin in milk of transgenic mice. *Transgenic Research* 3, 99–108.

Popov, L.S. (1996) Some aspects of structure and expression of milk protein genes (a review). *Molecular Biology* 30, 742–753.

Popov, L.S., Korobko, I.V., Andreeva, L.E., Dvoryanchikov, G.A., Kaledin, A.S. and Gorodetskii, S.I. (1995) On the functional significance of the 5′-region of the bovine κ-casein gene. *Doklady Akademii Nauk* 340, 111–113.

Raught, B., Khursheed, B., Kazansky, A. and Rosen, J. (1994) YY1 represses β-casein gene expression by preventing the formation of a lactation-associated complex. *Molecular and Cellular Biology* 14, 1752–1763.

Raught, B., Liao, W.S.L. and Rosen, J.M. (1995) Developmentally and hormonally regulated CCAAT/enhancer-binding protein isoforms influence β-casein gene expression. *Molecular Endocrinology* 9, 1223–1232.

Richardson, T. (1985) Chemical modification and genetic engineering of food proteins. *Journal of Dairy Science* 68, 2753–2762.

Riego, E., Limonta, J., Aguilar, A., Pérez, A., de Armas, R., Solano, R., Ramos, B., Castro, F.O. and de la Fuente, J. (1993) Production of transgenic mice and rabbits that carry and express the human tissue plasminogen activator cDNA under the control of a bovine alpha S1 promoter. *Theriogenology* 39, 1173–1185.

Rijnkels, M., Kooiman, P.M., de Boer, H.A. and Pieper, F.R. (1997) Organization of the bovine casein gene locus. *Mammalian Genome* 8, 148–152.

Rijnkels, M., Kooiman, P.M., Krimpenfort, P.J.A., de Boer, H.A. and Pieper, F.R. (1995) Expression analysis of the individual bovine β-, α_{s2}- and κ-casein genes in transgenic mice. *Biochemical Journal* 311, 929–937.

Robinson, G.W., Johnson, P.F., Hennighausen, L. and Sterneck, K. (1998) The C/EBP beta transcription factor regulates epithelial cell proliferation and differentiation in the mammary gland. *Genes and Development* 12, 1907–1916.

Rosen, J.M., Li, S., Raught, B. and Hadsell, D. (1996a) The mammary gland as a bioreactor: factors regulating the efficient expression of milk protein-based transgenes. *American Journal of Clinical Nutrition* 63, 627S–632S.

Rosen, J.M., Zahnow, C., Kazansky, A. and Raught, B. (1996b) Composite response elements mediate hormonal and developmental regulation of milk protein expression. *Biochemical Society Symposia* 63, 101–113.

Roskelley, C.D., Desprez, P.Y. and Bissell, M.J. (1994) Extracellular matrix-dependent tissue-specific gene expression in mammary epithelial cells requires both physical and biochemical signal transduction. *Proceedings of the National Academy of Sciences of the USA* 91, 12378–12382.

Roskelley, C.D., Srebrow, A. and Bissell, M.J. (1995) A hierarchy of ECM-mediated signalling regulates tissue-specific gene expression. *Current Opinion in Cell Biology* 7, 736–747.

Saito, H. and Oka, T. (1996) Hormonally regulated double- and single-stranded DNA-binding complexes involved in mouse β-casein gene transcription. *Journal of Biological Chemistry* 271, 8911–8918.

Schild, T.A., Wagner, V. and Geldermann, H. (1994) Variants within the 5′-flanking regions of the bovine milk protein genes: I. κ-casein-encoding gene. *Theoretical and Applied Genetics* 89, 116–120.

Schmidhauser, C., Bissell, M., Myers, C.A. and Casperson, G.F. (1990) Extracellular matrix and hormones transcriptionally regulate bovine β-casein 5′ sequences in stably transfected mouse mammary cells. *Proceedings of the National Academy of Sciences of the USA* 87, 9118–9122.

Schmidhauser, C., Casperson, G.F., Myers, C.A., Sanzo, K.T., Bolten, S. and Bissell, M.J. (1992) A novel transcriptional enhancer is involved in the prolactin- and extracellular matrix-dependent regulation of β-casein gene expression. *Molecular Biology of the Cell* 3, 699–709.

Schmidhauser, C., Myers, C.A., Mossi, R., Casperson, G.F., Sanzo, K.T., Bolten, S. and Bissell, M.J. (1995) Extracellular matrix dependent gene regulation in mammary epithelial cells. In: Wilde, C.J., Peaker, M. and Knight, C.H. (eds) *Intercellular Signalling in the Mammary Gland*. Plenum Press, New York, pp. 107–119.

Schmitt-Ney, M., Doppler, W., Ball, R.K. and Groner, B. (1991) β-casein gene promoter activity is regulated by the hormone-mediated relief of transcriptional repression and a mammary-gland-specific nuclear factor. *Molecular and Cellular Biology* 11, 3745–3755.

Schmitt-Ney, M., Happ, B., Ball, R.K. and Groner, B. (1992) Developmental and environmental regulation of a mammary gland-specific nuclear factor essential for transcription of the gene encoding β-casein. *Proceedings of the National Academy of Sciences of the USA* 89, 3130–3134.

Seagroves, T.N., Krnacik, S., Raught, B., Gay, J., Burgess-Beuse, B., Darlington, G.J. and Rosen, J.M. (1998) C/EBPbeta, but not C/EBPalpha, is essential for ductal morphogenesis, lobuloalveolar proliferation, and functional differentiation in the mouse mammary gland. *Genes and Development* 12, 1917–1928.

Seyfert, H.M., Tuckoricz, A., Interthal, H., Koczan, D. and Hobom, G. (1994) Structure of the bovine lactoferrin-encoding gene and its promoter. *Gene* 143, 265–269.

Shamay, A., Zeelon, E., Ghez, Z., Cohen, N., Mackinlay, A.G. and Gertler, A. (1987) Inhibition of casein and fat synthesis and α-lactalbumin secretion by progesterone in explants from bovine lactating mammary glands. *Journal of Endocrinology* 113, 81–88.

Sheehy, P.A., Nicholas, K.R., Gooden, J.M. and Wynn, P.C. (1997) Identification of the major constraints to milk protein synthesis. In: Corbett, J.L., Choct, M., Nolan, J.V. and Rowe, J.B. (eds) *Recent Advances in Animal Nutrition*. University of New England Press, Armidale, pp. 176–183.

Simons, G., van den Heuvel, W., Reynen, T., Frijters, A., Rutten, G., Slangen, C.J., Groenen, M., de Vos, W.M. and Siezen, R.J. (1993) Overproduction of bovine β-casein in *Escherichia coli* and engineering of its main chymosin cleavage site. *Protein Engineering* 6, 763–770.

Skowronski, J., Plucienniczak, A., Bednarek, A. and Jaworski, J. (1984) Bovine 1.709 satellite. Recombination hotspots and dispersed repeated sequences. *Journal of Molecular Biology* 177, 399–416.

Sokol, D.L. and Murray, J.D. (1996) Antisense and ribozyme constructs in transgenic animals. *Transgenic Research* 5, 363–371.

Soulier, S., Mercier, J.C., Vilotte, J.L., Anderson, J., Clark, A.J. and Provot, C. (1989) The bovine and ovine genomes contain multiple sequences homologous to the α-lactalbumin-encoding gene. *Gene* 83, 331–338.

Soulier, S., Vilotte, J.L., Stinnakre, M.G. and Mercier, J.C. (1992) Expression analysis of ruminant α-lactalbumin in transgenic mice: developmental regulation and general location of important *cis*-regulatory elements. *FEBS Letters* 297, 13–18.

Stacey, A., Schnieke, A., Kerr, M., Scott, A., McKee, C., Cottingham, I., Binas, B., Wilde, C. and Colman, A. (1995) Lactation is disrupted by α-lactalbumin deficiency and can be restored by human α-lactalbumin gene replacement in mice. *Proceedings of the National Academy of Sciences of the USA* 92, 2835–2839.

Stewart, A.F., Willis, I.M. and Mackinlay, A.G. (1984) Nucleotide sequences of bovine α_{S1}- and κ-casein cDNAs. *Nucleic Acids Research* 12, 3895–3907.

Stewart, A.F., Bonsing, J., Beattie, C.W., Shah, F., Willis, I.M. and Mackinlay, A.G. (1987) Complete nucleotide sequences of bovine α_{S2}- and β-casein cDNA's: comparisons with related sequences in other species. *Molecular Biology and Evolution* 4, 231–241.

Stice, S.L., Robl, J.M., Ponce de León, F.A., Jerry, J., Golueke, P.G., Cibelli, J.B. and Kane, J.J. (1998) Cloning: new breakthroughs leading to commercial opportunities. *Theriogenology* 49, 129–138.

Stinnakre, M.G., Vilotte, J.L., Soulier, S., L'Haridon, R., Charlier, M., Gaye, P. and Mercier, J.C. (1991) The bovine α-lactalbumin promoter directs expression of ovine trophoblast interferon in the mammary gland of transgenic mice. *FEBS Letters* 284, 19–22.

Stinnakre, M.G., Vilotte, J.L., Soulier, S. and Mercier, J.C. (1994) Creation and phenotypic analysis of α-lactalbumin-deficient mice. *Proceedings of the National Academy of Sciences of the USA* 91, 6544–6548.

Stoecklin, E., Wissler, M., Moriggl, R. and Groner, B. (1997) Specific DNA binding of Stat5, but not of glucocorticoid receptor, is required for their functional cooperation in the regulation of gene transcription. *Molecular and Cellular Biology* 17, 6708–6716.

Streuli, C., Bailey, N. and Bissell, M.J. (1991) Control of mammary epithelial differentiation: basement membrane induces tissue-specific gene expression in the absence of cell–cell interactions and morphological polarity. *Journal of Cell Biology* 115, 1383–1395.

Streuli, C.H., Edwards, G.W., Delcommenne, M., Whitelaw, C.B.A., Burdon, T.G., Schindler, C. and Watson, C.J. (1995) Stat5 as a target for regulation by extracellular matrix. *Journal of Biological Chemistry* 270, 21639–21644.

Struhl, K. (1989) Molecular mechanisms of transcriptional regulation in yeast. *Annual Review of Biochemistry* 58, 1051–1077.

Talhouk, R.S., Neiswander, R.L. and Schanbacher, F.L. (1990) *In vitro* culture of cryopreserved bovine mammary cells on collagen gels: synthesis and secretion of casein and lactoferrin. *Tissue and Cell* 22, 583–599.

Thompson, J.D., Higgins, D.G. and Gibson, T.J. (1994) CLUSTAL W: improving the sensitivity of progressive multiple sequence alignment through sequence weighting, position-specific gap penalties and weight matrix choice. *Nucleic Acids Research* 22, 4673–4680.

Threadgill, D.W. and Womack, J.E. (1990) Genomic analysis of the major bovine milk protein genes. *Nucleic Acids Research* 18, 6935–6942.

Topper, Y.J. and Freeman, C.S. (1980) Multiple hormonal interactions in the developmental biology of the mammary gland. *Physiological Reviews* 60, 1049–1106.

Tsai, M.-J. and O'Malley, B.W. (1994) Molecular mechanisms of action of steroid/thyroid receptor superfamily members. *Annual Review of Biochemistry* 63, 451–486.

Udy, G.B., Towers, R.P., Snell, R.G., Wilkins, R.J., Park, S.H., Ram, P.A., Waxman, D.J. and Davey, H.W. (1997) Requirement of STAT5b for sexual dimorphism of body growth rates and liver gene expression. *Proceedings of the National Academy of Sciences of the USA* 94, 7239–7244.

Vilotte, J.L., Soulier, S., Mercier, J.C., Gaye, P., Hue-Delahaie, D. and Furet, J.P. (1987) Complete nucleotide sequence of bovine α-lactalbumin gene: comparison with its rat counterpart. *Biochimie (Paris)* 69, 609–620.

Vilotte, J.L., Soulier, S., Stinnakre, M.G., Massoud, M. and Mercier, J.C. (1989) Efficient tissue-specific expression of bovine α-lactalbumin in transgenic mice. *European Journal of Biochemistry* 186, 43–48.

Vilotte, J.L., Soulier, S. and Mercier, J.C. (1993) Complete sequence of a bovine α-lactalbumin pseudogene: the region homologous to the gene is flanked by two directly repeated LINE sequences. *Genomics* 16, 529–532.

Vonderhaar, B.K. and Ziska, S.E. (1989) Hormonal regulation of milk gene expression. *Annual Review of Physiology* 51, 641–652.

Wakao, H., Schmitt-Ney, M. and Groner, B. (1992) Mammary gland-specific nuclear factor is present in lactating rodent and bovine mammary tissue and composed of a single polypeptide of 89 kDa. *Journal of Biological Chemistry* 267, 16365–16370.

Wakao, H., Gouilleux, F. and Groner, B. (1994) Mammary gland factor (MGF) is a novel member of the cytokine regulated transcription factor gene family and confers the prolactin response. *EMBO Journal* 13, 2182–2191.

Wall, R.J., Kerr, D.E. and Bodioli, K.R. (1997) Transgenic dairy cattle: genetic engineering on the large scale. *Journal of Dairy Science* 80, 2213–2224.

Wasylyk, B., Hahn, S.L. and Giovane, A. (1993) The Ets family of transcription factors. *European Journal of Biochemistry* 211, 7–18.

Watanabe, Y., Tsukada, T., Notake, M., Nakanishi, S. and Numa, L. (1982) Structural analysis of repetitive DNA sequences in the bovine corticotropin–β-lipotropin precursor gene region. *Nucleic Acids Research* 10, 1459–1469.

Watson, C.J., Gordon, K.E., Robertson, M. and Clark, A.J. (1991) Interaction of DNA-binding proteins with a milk protein gene promoter *in vitro*: identification of a mammary gland specific factor. *Nucleic Acids Research* 19, 6603–6610.

Waugh, D.F. (1971) Formation and structure of casein micelles. In: McKenzie, H.A. (ed.) *Milk Proteins: Chemistry and Molecular Biology*, Vol. II. Academic Press, New York, pp. 3–85.

Waugh, D.F. and von Hippel, P.H. (1956) κ-casein and the stabilization of casein micelles. *Journal of the American Chemical Society* 78, 4576–4582.

Welte, T., Philipp, S., Cairns, C., Gustafsson, J.A. and Doppler, W. (1993) Glucocorticoid receptor binding sites in the promoter region of milk protein genes. *Journal of Steroid Biochemistry and Molecular Biology* 47, 75–81.

Welte, T., Garimouth, K., Philipp, S. Jennewein, P., Huck, C. Cato, A.C.B. and Doppler, W. (1994) Involvement of Ets-related proteins in hormone-independent mammary cell specific gene expression. *European Journal of Biochemistry* 223, 997–1006.

Wilde, C. J., Addey, C.V.P., Boddy-Finch, L.M. and Peaker, M. (1995) Autocrine control of milk secretion: from concept to application. In: Wilde, C., Knight, C. and Peaker, M. (eds) *Intercellular Signalling in the Mammary Gland*. Plenum Press, New York, pp. 227–237.

Wilde, C.J., Knight, C.H. and Peaker, M. (1997) Autocrine regulation of milk secretion. In: Phillips, C.J.C. (ed.) *Progress in Dairy Science*. CAB International, New York, pp. 311–332.

Womack, J.E., Threadgill, D.W., Moll, Y.D., Faber, L.K., Koreman, M.L., Dietz, A.B., Tobin, T.C., Skow, L.C., Zneimer, S.M., Gallagher, D.S. and Rodgers, D.S. (1989) Syntenic mapping of 37 loci in cattle: chromosomal conservation with mouse and man. *Cytogenetics and Cell Genetics* 51, 1109 (abstract).

Wong, D.W.S., Camirand, W.M. and Pavlath, A.E. (1996) Structures and functionalities of milk proteins. *Critical Reviews in Food Science and Nutrition* 36, 807–844.

Yang, J., Baracos, V.E. and Kennelly, J.J. (1997) Role of Stat5 in controlling milk protein synthesis in Holstein cows. *Journal of Dairy Science* 80, 204.

Yu-Lee, L.-Y., Richter-Mann, L., Couch, C.H., Stewart, A.F., Mackinlay, A.G. and Rosen, J.M. (1986) Evolution of the casein multigene family: conserved sequences in the 5′ flanking and exon regions. *Nucleic Acids Research* 14, 1883–1902.

Genetic Improvement of Beef Cattle

20

B.P. Kinghorn[1] and G. Simm[2]

[1]Animal Science, SRSNR, University of New England, NSW 2351, Australia; [2]Genetics and Reproduction Department, Animal Biology Division, Scottish Agricultural College, Bush Estate, Penicuik EH26 0PH, UK

Introduction	577
Breeding Goals	578
Formulation of breeding objectives	579
Breeding goals for beef production systems	581
Breed Resources and Crossbreeding	584
Evaluating breed resources	584
An overview of crossbreeding theory	584
Breeds and crosses used in beef production	587
Selection Within Breeds	588
Evaluating individuals	588
Systems of testing	592
Traits recorded	593
Evaluations across herds, breeds and countries	594
Indices of overall economic merit	595
Evidence of genetic improvement and its value	596
Molecular and Reproductive Methods	597
Detection and use of quantitative trait loci	597
Multiple ovulation and embryo transfer nucleus schemes	598
Sexing and cloning	599
References	600

Introduction

The aim of this chapter is to discuss the breeding goals, genetic resources and methods of genetic improvement used in beef cattle breeding. The emphasis is on genetic improvement in temperate production systems. Food and Agriculture Organizatin (FAO) statistics on the world production of beef and veal by continent show that North and Central America have the highest production at about 30% of the world total, and Europe follows with about 21% of global production (FAO, 1996; Simm, 1998).

Beef cattle breeding in temperate countries is fairly heterogeneous, so it is worth setting the scene a bit further. In most European countries, over 50% of beef production is from pure dairy or dual-purpose breeds, either from cull cows, from male calves or from surplus female calves not required as dairy herd replacements. Traditionally, beef breeding goals and criteria were usually considered in dairy and dual-purpose breeds. However, there is often little or no emphasis on beef traits today, in dairy breeds. In addition to this direct contribution from dairy herds, there is an indirect contribution to beef production through crossing of dairy cows to beef bulls. This produces beef × dairy calves for slaughter and, in some countries, beef × dairy suckler cows (i.e. cows kept for rearing beef calves).

Many European countries have only a small specialized beef cattle breeding industry – in many cases, this comprises purebred terminal sire breeds to supply beef bulls for crossing in dairy or dual-purpose herds. In contrast, pure beef breeds account for a high proportion of total production in France and, to a lesser extent, in Italy and Spain. In Britain and Ireland, suckler herds of beef × dairy cows, derived as a by-product of the use of beef bulls in dairy herds, make an important contribution to total output. In other major temperate beef-producing countries, such as the USA, Canada, parts of South America, New Zealand and parts of Australia, beef production is based on extensively grazed or ranched cows, mainly of pure British beef breeds, like the Hereford, Aberdeen Angus and Shorthorn, or crosses among them. In some of these countries, such as the USA, Canada and parts of Australia, this extensive preweaning regime is usually followed by a more intensive finishing period in feedlots. The extensive nature of many production systems, and the widespread use of crossbred animals in the commercial sector of most beef industries means that performance recording and genetic improvement are usually concentrated in a relatively small sector of the population.

Bos indicus and Sanga beef breeds have been widely used in tropical areas in developing countries. Their use in tropical and subtropical regions of developed countries, such as Australia, has increased markedly in the last few decades, but more emphasis is now being placed on their crosses with *Bos taurus* breeds in an attempt to increase productivity and product quality.

Breeding Goals

From the brief introduction above, it is apparent that there are two broad categories of beef production in many countries: (i) beef production from dairy and dual-purpose herds; and (ii) beef production from specialized beef herds. Within the specialized beef sector, there is further differentiation into terminal sire and maternal breeds, crosses or lines. Terminal sire breeds also get used in dairy and dual-purpose herds. Each of these categories of use requires a distinct set of beef breeding goals, or at least different priorities, and these are discussed below. However, some more general issues of formulation of breeding goals are described first.

Formulation of breeding objectives

Breeding objectives stipulate the animal characteristics to be improved and the desired direction for genetic change. They should be constructed in a manner that allows them to play an appropriate role, together with parameters such as heritability and correlations, as part of a genetic evaluation system, in order to facilitate ranking of animals on genetic merit and implementation of an effective breeding programme design.

To this end, breeding objectives are generally expressed as economic weightings, which describe the economic impact of a unit change in each trait of commercial importance. These economic weightings can be used directly to help evaluate different breeds and crosses, or, more commonly, they can be used in conjunction with genetic parameters and knowledge of population structure to rank animals on an index of genetic merit in monetary units.

The breeding objective traits are not necessarily the same as the selection criterion traits that are measured and used to make selection decisions. For example, lean percentage may be a breeding objective and ultrasonically measured backfat thickness a selection criterion. Knowing the genetic relationship between these two traits permits selection index methods to target the former, using data on the latter.

There are two approaches to calculating these weightings – the economically rational approach and the 'desired gains' approach.

The economically rational approach
The classic approach to calculating economic weightings is economically rational – it takes no account of genetic parameters. This makes sense in that the value of making a unit change in a given trait should not be influenced by how difficult it is to generate this change. These difficulties can be handled appropriately at the genetic evaluation phase. In this setting, breeding objectives should reflect the costs and returns involved in a production system, and should not consider costs and gains generated in a breeding programme.

Help from biological modelling   The economically rational approach assumes that we know the genetic parameters (heritabilities, genetic correlations and phenotypic correlations) for all traits that are measured and/or of economic importance. However, this is often not the case in beef production systems, where it is extremely difficult to measure many of the traits of true importance, such as mature size, the shape of the growth and feeding curves and the patterns of tissue deposition. Such traits are often ignored when developing breeding objectives and yet their direct or indirect effect on profit can be large. In particular, the effect of mature size on production efficiency is such that selecting for efficiency as measured between fixed ages or fixed weights can be quite misleading (Kinghorn, 1997).

The approach generally assumes that the biological interactions among traits are linear in nature. However, this is often not the case in meat production systems, where relationships can be complex, such as the effect of fatness on maternal ability and juvenile survival in heterogeneous environments. It is possible that relationships are neutral at the current levels of trait expression, but that with genetic change in selected traits, thresholds are passed and/or relationships develop.

Biological modelling of production systems can be used to predict such changes. This modelling usually involves a mixture of mechanistic and empirical features (Ball *et al.*, 1998). Mechanistic features give powers of extrapolation beyond what we get through use of empirically derived parameters, such as heritabilities and linear correlations. However, biological modelling cannot be used to reliably separate predictions of genetic relationships and phenotypic relationships, and this casts doubt on its power to help set breeding objectives. In practice, it seems that biological modelling can play a quality-control role, to predict any deleterious effects of breeding objectives set through use of an economically rational approach.

Units of expression All economic weightings in a breeding objective should have the same basis for units of expression, such as 'dollars per head'. Choice of this basis can have an important influence of the consequences of using the breeding objective. A simple basis for unit of expression, such as 'dollars per head', can be used for situations in which all traits are directly related to economic costs or returns, and thus excludes reproductive traits, whose effect is at least partly manifested through progeny. A less simple basis is 'dollars per breeding cow per year', which accommodates both production and reproduction traits. In all cases, each trait should use this same basis. Delays in returns due to expression in progeny can be accommodated by considering the pattern of flow of genes through the population, and discounting future returns to give current values (McClintock and Cunningham, 1974).

Economic weights calculated on a 'dollars per head' or 'dollars per breeding cow per year' basis suffer a potentially important drawback. They relate to dollars per livestock unit, rather than dollars per resource unit, such as 'dollars per hectare'. As an example, consider two breeds of beef cattle:

Breed	Value of weight at slaughter	Value of food consumed	Profit per head	Dollar efficiency
Small	$1000	$500	$500	2 : 1
Large	$1800	$1000	$800	1.8 : 1

The large breed would be targeted by a breeding objective based on 'dollars per head'. However, a breeding objective based on 'dollars per hectare' would target the small breed. A breeding objective based on dollars per resource unit will usually be more appropriate, as long as proper account is made of any fixed costs per head.

Economic values can be calculated from several different perspectives, e.g. with the aim of maximizing the profitability of an enterprise for an individual producer, or with the aim of improving the efficiency of a national livestock industry. Amer (1994) and Weller (1994) discuss these different approaches and the attempts to unify them. In the former category, increasingly sophisticated models have been proposed for deriving economic values, including enterprise models which reoptimize management following genetic improvement (e.g. Amer *et al.*, 1996, 1997).

The 'desired gains' approach
An alternative approach to developing breeding objectives, the 'desired gains' approach, involves declaration of the relative magnitudes of genetic gain desired in the traits of importance. The breeding objective calculations still result in relative economic weights, but these are now influenced by genetic parameters, with generally greater economic weightings for traits that are more difficult to change. A simple subset of this approach is the restricted index, in which the objective is set up to give a predicted zero genetic change in one or more nominated traits. Examples are restrictions for no change in backfat or birth weight.

Brascamp (1984) describes methods that can be used for both restriction and desired gains. He also shows how to use a mixture of the economically rational and the 'desired gains' approaches, with some traits constrained to prechosen levels of response and others influenced just by production economics. In all cases, relative economic weights are calculated, which is useful for demonstrating the 'effective economic weights' that nominated desired gains or restrictions imply.

Breeding goals for beef production systems

Beef breeding goals in dairy and dual-purpose breeds
At first sight, it seems efficient to breed for both milk and meat production from the same type of animal. However, most of the evidence suggests that there is an unfavourable genetic correlation between milk production and growth or carcass characteristics (e.g. Pirchner, 1986).

Some breeds or strains, such as the Simmental strains in several continental European countries, have achieved fairly high productivity in both milk and beef traits, as a result of many generations of selection. Even for these strains, it is difficult to compete nationally and internationally with both specialized milk and specialized beef breeds. As a result, there is a general trend towards milk production from more specialized dairy cattle breeds and strains. In some countries, there is still an attempt to limit the expected deterioration in beef merit by performance-testing dairy bulls for growth and conformation, and preselecting bulls on these traits prior to progeny-testing for milk production. In other countries, the deterioration in beef merit of the specialized dairy strains is compensated for, at least partially, by crossing those females not required to breed replacement dairy heifers to specialized beef breeds. So, in temperate dairying countries with large-scale specialized industries, breeding goals in dairy breeds have little or no emphasis on beef traits. Even in dual-purpose breeds, the emphasis on beef traits is likely to be secondary to that on milk traits (Simm, 1998).

Terminal sires for use in dairy herds and specialized beef herds
Terminal sire beef breeds (i.e. those specially selected to sire the slaughter generation of animals) are used in dairy herds, for two main purposes. The first is to mate to dairy heifers to reduce the risk of calving difficulties, compared with that following matings to a dairy sire. The second is to mate to mature dairy cows that are not required to breed replacement dairy heifers.

Difficult calvings are costly, both directly and because they delay rebreeding, depress milk production and compromise both cow and calf survival and welfare. Hence, dairy heifers have often been mated to bulls from one of the easier-calving beef breeds, such as the Hereford, Aberdeen Angus and Limousin. However, mating dairy heifers to a beef bull is becoming less common as more dairy producers recognize that their heifers are often the highest genetic-merit animals in the herd, and hence valuable as dams of replacements. Also, the wider availability of calving-ease evaluations in dairy breeds means that it is easier to select a dairy sire suitable for mating to heifers.

As the incidence of calving difficulties is lower in mature cows than in heifers, there is more scope to select beef bulls for other attributes to maximize returns from calf sales. Many beef-cross calves born on dairy farms are sold at a young age. So increasing calf weight and conformation (muscularity or shape) is an important breeding goal for dairy farmers choosing a beef breed, or an individual beef sire – although increasing weight and conformation tends to conflict with the aim of reducing calving difficulties.

The performance of beef-cross calves in later life is of little direct concern to most dairy farmers, although, in theory, sire breeds or individual sires with high genetic merit for later performance ought to result in higher rewards in the marketplace. These market signals work reasonably well at the level of sire breed. There is less widespread discrimination among sires within a breed, although in some countries artificial insemination (AI) companies, beef breed societies or recording agencies have schemes to identify and promote beef sires for use in dairy herds which combine acceptable calving ease with good growth and carcass characteristics.

In many of the specialized beef production systems in temperate countries, there is widespread use of crossbreeding. Often this is to achieve complementary use of breeds. Usually small or medium-sized breeds or crosses are used as dam lines, and larger breeds are used as terminal sires. Larger breeds are valuable as terminal sires as they usually have a faster growth rate and produce leaner carcasses at a given weight than smaller breeds. Although ease of calving is still important when terminal sire breeds are used in specialized beef breeding herds, their main role is to improve the growth and carcass characteristics of their crossbred offspring.

The definition of carcass merit depends to some extent on whether commercial animals are sold at live auctions or directly to abattoirs, but it usually encompasses some measure of weight, fatness and conformation. (Breed and sex may also modify the price.) In theory, good communication between sectors of the industry should mean that breeding goals are similar, whether animals are marketed dead or alive. However, in practice they often differ.

In many North American and East Asian markets, a premium is paid for high marbling – that is, high levels of visible intramuscular fat in the eye muscle. Particularly in North America, this premium for marbling is based on its value as an indicator of good eating quality. Recently, interest in marbling in several exporting countries has been fuelled by its importance in the lucrative Japanese beef market.

Meat eating quality is becoming an increasingly important issue with consumers and the meat industry in richer countries. The post-slaughter treatment of carcasses, especially chilling rate, ageing and method of hanging, are known to have important effects on eating quality (Dikeman, 1990; Cuthbertson, 1994). However, there is less information on preslaughter effects on beef eating quality, such as breed, breeding value within breed or production system. The information that is available suggests that there are breed differences in indirect measures of meat quality, especially marbling, colour and fibre type. There are differences in tenderness between breed types: double-muscled breeds generally have the most tender meat, followed by other *B. taurus* breeds, with *B. indicus* breeds ranking lowest. There are less consistent differences in tenderness between the non-double-muscled *B. taurus* breeds, or between any of the breed types, in juiciness and flavour. Despite this, there are consistent reports of substantial within-breed genetic variation in both indirect and direct measures of eating quality (Kemp, 1994). This indicates that there is scope for improvement through within-breed selection, although, in the absence of good live-animal predictors of eating quality, this is difficult to achieve without progeny-testing. In future, molecular markers of eating quality may allow more efficient selection programmes.

Breeding replacement females for specialized beef herds
The main breeding goals for cows in specialized beef herds, in addition to adequate growth and carcass merit, are good fertility, ease of calving, good maternal ability (which includes adequate milk production and good mothering ability) and low or intermediate mature size, to reduce cow maintenance requirements. These individual goals are sometimes aggregated into measures like weight of calf weaned per cow per annum, or weight of calf weaned per kg cow mature weight per annum.

The ability of animals to withstand extreme climates and to tolerate low-quality feed and periods of feed shortage is also important in some areas, and there is often concern about possible genotype × environment interactions for these 'adaptation' traits. These traits are often difficult to define, and the most practical route for within-breed improvement is often simply to record and select on performance in the harsh environment concerned (Simm *et al.*,

1996). The emphasis on each of these traits will vary depending on the production system and breed or crossbreed type of cow used. In some cases, the traits of importance will be best improved by selection; in others, they will be best improved by crossbreeding. For instance, the fertility of crossbred cows is usually high, as a result of heterosis, and so is of somewhat less concern in selection within the component breeds (Simm, 1998).

Breed Resources and Crossbreeding

Evaluating breed resources

Genetic evaluation of breed resources is relatively simple wherever good estimates of mean performance are available for the environment and production systems of interest. This is because the effects involved can be measured with high accuracy from much data, and can be treated as fixed effects. These effects constitute an inventory of genetic resources, and the economic value of each breed genotype can be estimated by simply multiplying predicted performance for each trait by its corresponding economic weight, and summing across traits. In contrast, when we come to evaluating the genetic merit of individual animals, there are many fewer data available per estimate, and the random nature of breeding values makes the process more much more difficult, especially for traits that are difficult to measure, such as feed conversion efficiency and disease resistance.

Breed evaluations can be extended to evaluation of different crossbreeding systems, with breeding objectives being calculated according to the specific role of each component breed or cross. For example, the breeding objective for a terminal sire breed would involve little or no pressure on female fertility traits, as these will only be important within that breed, which will constitute only a small part of the total system.

An overview of crossbreeding theory

The value of crossbreeding
The key reasons for crossbreeding are listed here.

- The averaging of breed effects. For example, to get an animal of intermediate size to fit a particular pasture cycle or market demand. This may involve either regular systems of crossing or the creation of composite breeds (e.g. Cundiff *et al.*, 1986).
- Direct heterosis. Crossbred individuals often exhibit heterosis. Heterosis is measured as the extra performance of the crossbreds over the weighted average of their parent breeds. The percentage increase in performance ranges from about 0 to 10% for growth traits and 5 to 25% for fertility traits (e.g. Gregory *et al.*, 1991). The effect of heterosis on the total production

system can be even more than this, as effects accumulate over traits (e.g. Cundiff *et al.*, 1986).
- Maternal heterosis. Crossbred cows can exhibit considerable heterosis in their ability to raise fast-growing, viable offspring.
- Sire–dam complementation. A good crossbreeding system aims to use breeding cows that are of small or intermediate mature size (but not so small that dystocia is a problem), as well as fertile. When a large terminal-sire breed is used, the proportion of feed directed to growing animals is increased and the production system benefits accordingly.
- Possibly cheap source of breeding animals. This is evident in some crossing systems – for example, in the British and Irish beef industries, where many suckler cows have come from matings between beef bulls and dairy cows.

The genetic basis of heterosis
We need to know the genetic basis of heterosis in order to predict the value of untested genotypes. There are two genetic mechanisms postulated as causing heterosis effects.

- Dominance: where the individual's parents come from two different breeds, the individual will carry a wider range of alleles, sampled from two breeds rather than just one. It is thought that this equips the individual better to perform well, especially under a varying or stressful environment. We would thus expect dominance to be a positive effect, and there is much evidence to support this.
- Epistasis: when we cross breeds, alleles have to interact or 'cooperate' with alleles at other loci which they are 'not used to'. The crossbred animal may thus be out of harmony with itself, and we predict that epistasis, if important, is a negative effect.

The dominance model of heterosis is widely assumed and used, and so this model will be taken here. It should be borne in mind that epistatic loss could cause errors in prediction based on the dominance model alone.

Breed dominance is greatest when all loci consist of two alleles derived from different breeds, as in a first cross (F_1). Other crosses show a proportion of this heterosis, equal to the proportion of loci that are heterozygous with respect to breed of origin. This can be seen in the column D_d (dominance for the direct subtrait) in Table 20.1.

Table 20.1 shows how to predict the merit of untested crossbred genotypes given estimates of crossbreeding effects. These are additive (A) or 'purebreed' effects for each of the three example breeds, and dominance (D) or heterotic effects, here assumed equal for each pair of breeds. Subscripts denote the direct subtrait (d) and the maternal subtrait (m) – both of these being of some importance for weaning weight in cattle. A least-squares analysis of the form $\hat{\beta} = (X'X)^{-1} X'Y$ can be used to estimate the crossbreeding effects (in vector $\hat{\beta}$) from merit (Y), where X is the matrix formed by the body of Table 20.1. The section 'Evaluating breed resources' outlines the simple

Table 20.1. Example of the prediction of merit of weaning weight from estimated crossbreeding parameters. Multiply the coefficients shown in the body of the table by the values of the corresponding effects (see text). Adding the products gives the prediction of weaning weight, merit, in the last column.

Effects:	Mean	A_{d1}	A_{d2}	A_{d3}	A_{m1}	A_{m2}	A_{m3}	D_d	D_m	
Values (kg):	280	+20	0	−20	−6	−1	+7	20	10	Merit
Breed 1	1	1	0	0	1	0	0	0	0	294.0
Breed 2	1	0	1	0	0	1	0	0	0	279.0
Breed 3	1	0	0	1	0	0	1	0	0	267.0
Best F_1 (1 × 2)	1	0.5	0.5	0	0	1	0	1	0	309.0
Best 3 Breed-X (1 × 23)	1	0.5	0.25	0.25	0	0.5	0.5	1	1	318.0
Best Backcross (1 × 12)	1	0.75	0.25	0	0.5	0.5	0	0.5	1	311.5
Balanced (1, 2)	1	0.5	0.5	0	0.5	0.5	0	0.5	0.5	301.5
Synthetics (1, 2, 3)	1	0.33	0.33	0.33	0.33	0.33	0.33	0.67	0.67	300.0
Optimum (1, 2)	1	0.63	0.37	0	0.63	0.37	0	0.47	0.47	302.4
Synthetics (1, 2, 3)	1	0.57	0.31	0.12	0.57	0.31	0.12	0.56	0.56	303.0
Rotations (1, 2)	1	0.5	0.5	0	0.5	0.5	0	0.67	0.67	306.5
(1,2,3)	1	0.33	0.33	0.33	0.33	0.33	0.33	0.86	0.86	305.7

Synthetics at equilibrium; rotations at equilibrium and averaged over years.

Table 20.2. General recommendations on use of crossbreeding.

Purebreed	When no cross is better
F_1 cross	When direct heterosis is important
3-breed cross	When both direct and maternal heterosis are important
4-breed cross	When paternal heterosis is important as well
Backcross	When only two good parental breeds are available and/or when direct heterosis is not important
Rotational cross	When females are too expensive either to buy in or to produce in the same enterprise
Open or closed synthetic	When both males and females are too expensive. A few initial well-judged importations establish the synthetic, and it can then either be closed (which helps to establish a breed 'type') or left open to occasional well-judged importations

approach that can be used to consider all breeding objective traits to help predict the economic merit of different crosses.

Choice of crossing system

Gregory and Cundiff (1980) reported maternal and direct dominance effects between *B. taurus* breeds at 14.8% and 8.5%, respectively, for weight of calf weaned per cow exposed. This indicates the importance of crossbred cows in the production system, even though maternal dominance is generally reduced at older ages, for example at slaughter age. Gregory and Cundiff used these figures to estimate the genetic merit of a wide range of crossing systems in beef cattle.

The best crossing system to use depends to a large extent on the value of the breeds available, as well as the amount of heterosis expressed in crossbred animals. This is illustrated in Table 20.2 by describing the conditions under which each crossbred genotype is worthy of choice.

Of course, care should be taken to consider factors other than the predicted genetic merit of candidate crosses for the traits of importance. The key factor here is the cost of maintaining structured crossing systems, where separate breeding units are required to give an ongoing supply of purebred and/or crossbred parents. These costs often outweigh the genetic benefits of more structured crosses, especially in low-fecundity species such as cattle, where the parental breeding units must be relatively large to supply the final cross.

Breeds and crosses used in beef production

Clearly the predominance of black and white strains in the dairy industry means that they are major contributors to beef output, both directly through surplus calves and cull cows and, in some countries, indirectly through their contribution to the genetic make-up of suckler cows. However, the increasing specialization for milk production in black and white strains means that their predominance is often seen as a disadvantage in beef production. Because of the economic incentive towards specialization for milk production in most temperate countries, the biggest opportunity to improve beef output from dairy breeds is through crossing surplus females to specialized beef breeds.

Of the specialized beef breeds in Europe, the French breeds, particularly the Charolais and Limousin, and to a lesser extent the British breeds, particularly the Hereford and Angus, are most common (Simm, 1998). The popularity of the French breeds is probably due to their high growth rates or high lean meat yield, while the popularity of the British breeds is probably due to their relatively low incidence of calving difficulties (Liboriussen, 1982; Thiessen *et al.*, 1984; Cundiff *et al.*, 1986; Gregory *et al.*, 1991; Amer *et al.*, 1992). Also, the traditional British breeds, especially the Aberdeen Angus, have had something of a renaissance recently, because of perceived benefits in eating quality.

The increased use of the specialized French breeds as terminal sires in Europe, often at the expense of the traditional British breeds, is mirrored in many other temperate beef-producing countries. However, the British breeds remain important in breeding herds, either as purebreds or as components of crossbred maternal lines, in many of these countries (e.g. the USA, Canada, Australia, New Zealand).

Although statistics on numbers of animals are useful, several less numerous breeds have a disproportionate influence through the use of AI, especially in dairy herds. For example, in the UK there are relatively small numbers of purebred Belgian Blue cattle, but this breed was responsible for the second largest number of beef inseminations made by the main AI organizations in 1993/94. The growth in importance of this breed is due to its ability to sire high-conformation crossbred calves, with acceptable levels of calving ease, when mated to dairy cows.

In several major beef producing countries (e.g. the USA, Australia), there is growing interest in the use of composite breeds, especially as maternal lines. The use of these animals is efficient when rotational crossing is impractical or when several breeds have important contributions to overall merit. The growth in interest in composite breeds is in part based on the results of very extensive research progress at the US Meat Animal Research Center in Nebraska over the last few decades (e.g. Gregory *et al.*, 1991).

Bos indicus and Sanga breeds have been increasingly used in crossing systems in tropical beef production regions. There is a general trend to keep the proportion of genes from these breeds low in order to avoid deleterious effect on meat quality. However, research suggests that more variation in meat quality is caused by management and processing factors than by proportion of *B. indicus* genes (Hearnshaw *et al.*, 1998).

Selection Within Breeds

Evaluating individuals

What causes an exceptional animal to be so much better than its contemporaries? There are two basic reasons.

1. The gene variants (alleles) it has inherited are more favourable and/or they are present in more favourable combinations, making the animal genetically superior.

2. It has probably experienced a better 'environment', through good management or good luck.

In seeking genetic change, we are not really interested in how much 'environmental advantage' an animal has had – because that source of superiority cannot be transmitted to the next generation. Moreover, in selection programmes, we are generally not interested in the combination of alleles, as, in general, these combinations cannot be transmitted to the next generation (in the case of intralocus dominance) or are only weakly transmitted (in the case of interlocus epistasis).

We want to be able to choose the animals with alleles that will have the most beneficial effect on progeny, and we do this by selecting animals on the basis of their estimated breeding values. Breeding value (denoted by A, signifying additivity of effect) is a description of the value of an animal's alleles to its progeny. In general, we do not know which alleles an animal carries, so we can never fully know what an animal's breeding value is. However, we can estimate it from a wide range of information sources.

The simplest estimate of an animal's breeding value is that based on just its phenotypic superiority (P, phenotype as a deviation from the contemporary mean):

$$\hat{A} = \frac{V_A}{V_P} P = h^2 P$$

where $\hat{A}$ is estimated breeding value (EBV) and ^ denotes 'estimate', and $h^2 = \frac{V_A}{V_P}$ is heritability of the trait concerned. In conceptual terms, the phenotypic superiority of the animal, P, is regressed or shrunk according to the proportion of phenotypic variation in the trait concerned which is due to effects that cannot be transmitted between generations.

Selection on phenotype gives a percentage response that depends on:

- selection intensity – the smaller the proportion retained for breeding, the higher the response;
- generation interval – the younger the average age of parents, the faster the rate of response;
- heritability – the higher the heritability the higher the response;
- coefficient of variation (CV) – the higher the CV, the higher the response.

The last two factors generally differ between traits. Table 20.3 gives estimates of these for a number of traits in beef cattle.

Use of information from relatives – best linear unbiased prediction

In selecting animals to act as parents, we are interested in choosing those with the most favourable alleles. An animal's own performance gives an indication of the value of its alleles to its progeny. However, some of this animal's alleles are also carried by each of its relatives, and so the performance of an animal's relatives can be used to give a more accurate assessment of the alleles it carries.

Thus progressive breeding programmes make use of information from all known relatives. This is of most value when heritability is low – when an

Table 20.3. Coefficient of variation (CV, phenotypic standard deviation divided by mean) heritability estimates for a range of traits in beef cattle (condensed from Simm, 1998, after Koots, 1994a, b).

Trait	CV (%)	Heritability (%)
Age at first calving	5.7	6
Conception rate – cows*	61.8	17
Perinatal mortality – direct*	674.1	10
Scrotal circumference	8.0	48
Birth weight – direct	12.3	31
Birth weight – maternal	12.6	14
Weaning weight – direct	12.3	24
Weaning weight – maternal	13.6	13
Postweaning gain	13.7	31
Mature cow weight	12.1	50
Gross food conversion ratio	11.0	32
Backfat depth at constant age	24.5	44
Dressing percentage	3.2	39
Marbling score, constant age	34.1	38
Eye muscle area, constant age	10.1	42
Tenderness	18.2	29

*These traits are binomially distributed with a high mean, making CV figures less meaningful.

animal's own performance is a poor indicator of breeding value. As heritability increases, there is a diminishing proportional value of information from relatives, until, at a heritability of unity, an animal's own performance is a perfect indicator of its breeding value, with no room for improvement due to relatives' information.

Traditionally, information from different classes of relatives is combined, after correction for environmental effects, using selection indices. However, today, the method of choice for predicting breeding values, which is an extension of selection index methods, is best linear unbiased prediction (BLUP). Kennedy (1981) and Van Vleck *et al.* (1989) give digestible descriptions of BLUP techniques and Kinghorn (1997) gives a small example analysis. This section will not review these, but the following list describes the key properties of BLUP EBVs.

- Estimated breeding values are generally additive. For example, if a bull has an EBV of $\hat{A}$ = + 20.0 kg and a cow has $\hat{A}$ = + 10.0 kg for 400-day weight, then the prediction is that progeny will have a 400-day weight superiority of (20.0 + 10.0) / 2 = 15.0 kg. This is actually a prediction of progeny genetic value, but, as progeny dominance deviation and environmental deviation are unknown and thus have 'expectations' of zero, it is also a prediction of progeny breeding value and phenotype. Note also that the proportion of parental superiority in EBV that is transmitted to progeny is unity, after accounting for halving due to meiosis. Thus the heritability of EBVs is unity, as they have been pre-regressed.
- Best linear unbiased prediction makes full use of information from all relatives. It does this by use of the numerator relationship matrix, which describes the predicted number of alleles per locus shared by descent between each pair of animals. It is not necessary for BLUP to give separate attention to sib testing, progeny testing, own performance, etc. Use of information from all relatives (even those long dead) is simultaneously handled. This gives greater flexibility, more accurate EBVs and more selection response.
- Best linear unbiased prediction predicts breeding values and accounts for fixed environmental effects simultaneously (management group, herd, season, year, etc.). This means that animals can be compared across groups, giving wider scope for selection. For example, comparing across age-groups means that older animals have to prove their competitiveness at every round of selection. This property of BLUP usually accounts for most of its advantage over less powerful methods.
- Best linear unbiased prediction gives genetic trends. The ability to compare the EBVs of animals born and measured in different years means that year mean EBVs can be calculated and genetic trends reported.
- Best linear unbiased prediction can cater for non-random mating – such that bulls can be compared via their progeny even if some were allocated better cows. This can only be done where the cows were allocated on the

basis of their recorded performance, such that BLUP can account for their EBVs when evaluating the bulls concerned.
- Best linear unbiased prediction can account for selection bias. For example, consider ranking bulls on the weaning weights of their daughters at their first two calvings. The worse bulls, who had worse daughters, will have benefited more from culling of daughters on first weaning performance. However, BLUP accounts for this, given that the information used to make selection decisions (first weaning results in this case) is included in the data set.

BLUP analyses are generally provided as a bureau service in association with organized recording schemes.

Outputs from a BLUP analysis includes EBVs (or $\hat{A}$ values) for each of the traits fitted – which can include both measured criterion traits and breeding objective traits, even if there is missing information on the latter. The breeder only needs to weight EBVs for the objective traits by their economic weights to provide a selection index which s/he can select on: Index = $a_1\hat{A}_1 + a_2\hat{A}_2 + a_3\hat{A}_3 + \ldots$. The selection index is itself an EBV for economic merit.

Some traits are mediated through the maternal environment. For example, weaning weight is influenced genetically not only by the genes in the calf, but also by the genes in its mother, mediated through the maternal environment (e.g. milk supply). Thus the numerator relationship matrix for maternal effects on weaning weight is determined by relationships among the dams of the calves measured. This means that a single set of observations on weaning weight can give rise to both direct EBVs and maternal EBVs. If a breeder is selecting a terminal sire, s/he should ignore the maternal EBV, as this source of genetic merit will never be expressed. However, in order to maximize the weaning weight of the selected bull's grandprogeny via daughters, selection should be based on $\frac{1}{2}\text{EBV}_{\text{maternal}} + \frac{1}{4}\text{EBV}_{\text{growth}}$. This is actually a prediction of the performance of these grandprogeny, and the coefficients result from the fact that the grandprogeny benefit on average from one-quarter of the bull's genes for direct effects, and one-half of the bull's genes, in their mothers, for maternal effects.

Estimated breeding values across breeds

There is an increasing interest in genetic evaluations using information from crossbred animals and genetic evaluations on crossbred animals. Pollak and Quaas (1998) give the technical basis of this and a description of example cases. As a simple concept, analysis can be done to estimate all breed and heterosis effects and to simultaneously fit breeding values in a BLUP analysis. This leads to the prediction of progeny merit from any mating pair, based on the breed constitution of the progeny and the EBVs (free of breed and heterosis effects) of the parents.

However, without very good width of data, it is very difficult to get a reliable splitting of breed direct and maternal effects. Moreover, the genetic correlation between breeding values over different breeds of mate may be significantly less than unity – such that, for example, the EBV ranking of a group of Angus bulls might depend on what breed of cow they are to be mated to.

One problem with implementation is the general need to rank breeds and crosses on the breeding objective traits. There is much room for argument over the publishing and use of such values. This is one reason why genetic evaluations across breeds may take place more readily behind the closed doors of large breeding corporations.

Systems of testing

Most beef cattle genetic improvement programmes are based on performance testing or progeny testing. Both of these depend on performance recording. Essentially, this involves recording the identity, pedigree, birth date, sex and performance (e.g. live weights) of individual animals, plus any major management groupings or treatments likely to influence performance.

Performance testing

Since many of the traits of interest in beef cattle can be recorded in both sexes and prior to sexual maturity, there is a fairly long history of performance recording and performance testing in beef breeding. This dates from the 1940s and 1950s in the USA and slightly later in many other countries. Today performance testing is usually the responsibility of breed associations (e.g. in the USA), government departments or agencies receiving some government support (e.g. in many European countries) or private agencies, either alone or in partnership with each other.

Compared with the situation in dairy cattle breeding, a relatively low proportion of beef cattle are performance-recorded. This is partly because of the greater distinction between commercial and breeding herds than in the dairy industry – especially in countries where crossbreeding is widespread. For example, performance-recorded animals comprise less than 2% of the total beef cattle population in the USA (Middleton and Gibb, 1991), Australia and the UK. However, even within the purebred beef sector, there is usually a much lower proportion of recording than in the dairy industry.

Most performance testing schemes involve recording the preweaning performance of all animals on-farm. In some countries, postweaning performance continues to be measured on farm. In others, central performance testing is used. Central testing of beef cattle has been quite widely used worldwide since the 1950s, especially in the USA, Canada and Europe. It involves submitting some animals, especially higher-performing bulls, from the breeders' own farms to a central station, where they are compared with bulls from other herds in a uniform environment. Despite the potential benefits of this, the correlations between the performance of bulls in central stations and the subsequent performance of their progeny is often lower than expected. This is often attributed to large pretest environmental effects.

Progeny testing

In many countries there is a deliberate strategy of first performance-testing and then progeny-testing bulls, with selection at each stage. As with performance testing, progeny-testing schemes either operate on-farm or at central testing stations.

Sequential testing is particularly common in the specialized beef breeds in France. Large numbers of purebred animals are performance-recorded on farm for weights at birth, 120 and 210 days, and for muscular and skeletal development at weaning (Ménissier, 1988; Bonnett et al., 1994). The best males from on-farm recording are brought to central testing stations after weaning, and tested further from 8 to 14 months of age. About 35 of the best of these bulls go on to be progeny-tested to assess their daughters' maternal ability, in central progeny test stations.

Progeny testing causes an increase in generation interval, with potentially negative effects on overall selection response. An appropriate breeding programme design is thus needed to balance the effects on increased selection accuracy and increased generation interval. In some cases, the high accuracies generated by progeny testing are themselves of commercial value in the seedstock marketplace, and this should also be taken into account.

Cooperative breeding schemes

Although most breeding schemes revolve around performance testing or progeny testing, as outlined above, there are some variations that deserve special mention. The first of these are cooperative breeding schemes, such as group breeding schemes and sire referencing schemes. Group breeding schemes usually involve formation of a central nucleus herd, formed from élite cows from cooperating members' herds. When the nucleus is larger than the members' herds, or when recording and selection are more effective, genetic progress can be accelerated. Perhaps because of the relatively high legal and financial commitment required, and the growth in uptake of national across-herd genetic evaluation procedures, there seems to have been a decline in interest in cattle group breeding schemes over the last decade or so. However, formal or informal sire referencing schemes have been established in several breeds in France, Denmark, Britain, the USA and elsewhere, either before or during this period. These schemes involve the use of an agreed panel of sires on a proportion of the cows in each member's herd, usually by AI. In some cases, these schemes have been formed specifically to create or strengthen genetic links between herds to allow more accurate across-herd or across-test genetic evaluations.

Traits recorded

Generally, on-farm performance recording schemes around the world have concentrated on measuring live weights at regular intervals (or growth rates between these), together with visual scores of muscularity and measurements or scores of height or skeletal development. The development of mobile, reasonably accurate ultrasonic scanners in the 1970s and 1980s allowed measurements of fat and muscle depths or areas to be included in some on-farm recording schemes. Typically these measurements are taken on or over the eye muscle at one of the last ribs, or in the loin region of animals at about a year or 400 days of age. At least in theory, one of the benefits of central testing is that it permits more frequent and more comprehensive measurements to be made. For example, it is rarely practical to measure feed intake of individual animals on farms, but it is fairly common in central performance test stations. Similarly, progeny testing allows actual carcass measurements to be obtained.

Terminal-sire characteristics have generally dominated beef breeding schemes in Europe. With the exception of some breeding schemes in France, few maternal characteristics, such as fertility, have been recorded. As a result, what little objective selection there has been for maternal characteristics has been on traits like calving ease, birth weight and 200-day weight, which are of importance in both terminal sire and maternal lines. However, until recently, methods of separating direct and maternal genetic influences on these traits have not been in widespread use. Maternal traits have received more attention in North America, Australia and New Zealand, where specialized beef herds account for a far higher proportion of beef output. Genetic evaluations for scrotal size (which is an indicator of both male and female fertility and age at puberty) and female fertility (measured as days from the start of the mating period to calving) have been introduced recently for some breeds in Australia and New Zealand. Evaluations for scrotal size and mature cow weight have been introduced for some breeds in the USA.

Many of the traits concerned with reproduction have fairly low heritabilities. However, many are economically important, and there is substantial variation in them, so there is both the incentive and scope for genetic improvement.

Direct heritabilities of growth traits tend to be moderately high, while maternal heritabilities tend to be slightly lower (Table 20.3). The heritabilities of carcass traits tend to be even higher than those for growth traits. However, carcass traits have to be assessed either indirectly on live candidates for selection (e.g. by ultrasonic measurements), or directly on progeny or other relatives of the candidates for selection, so they are not as easy to improve as it seems at first sight. For more details of genetic parameters, see Koots *et al.* (1994a, b).

Evaluations across herds, breeds and countries

To be able to compare BLUP EBVs fairly across contemporary groups and years, genetic links are needed between groups and years. In dairy herds, strong links occur automatically because of the very widespread use of AI. In some countries, there is little use of AI in specialized beef breeds, and this has limited the introduction of national across-herd genetic evaluations. However, AI use is higher in other countries. For example, between 20 and 50% of births in pedigree herds of the major beef breeds in Britain are the result of AI. Also, the recent introduction of foreign breeds to a country or the popularity of imported strains within a breed tends to increase the use of AI. In such cases, there will often be strong enough genetic links between herds and years to make reliable comparisons of EBVs across herds and years.

A major technical limitation to performing evaluations across breeds is that animals of different breeds are rarely kept as contemporaries under similar management and feeding systems. However, as indicated above, across-breed evaluations are becoming feasible using information from crossbred animals or from designed breed comparisons, together with estimates of genetic trends in each of the purebred populations since the breed comparison was made (Amer *et al.*, 1992; Benyshek *et al.*, 1994).

Compared with the situation in dairy cattle, there has been less effort to date in developing international conversions of EBVs or expected progeny differences (EPDs) for beef cattle or performing international genetic evaluations. However, there is growing interest in this area. For example, international conversions have been produced for some beef breeds in use in Canada and the USA. Also, across-country evaluations are being investigated or performed routinely for several breeds in the USA and Canada, France and Luxemburg, and Australia and New Zealand (Benyshek *et al.*, 1994; Graser *et al.*, 1995; Journaux *et al.*, 1996).

These across-flock, breed and country genetic evaluations are starting to have an important impact. They give credible objective comparison between seedstock sources, which in other industries has led to altered buying patterns and a shake-out in the seedstock sector.

Indices of overall economic merit

As noted previously, the selection index provides a means of maximizing response in the breeding objective. Briefly, the selection index apportions selection emphasis in the most appropriate way, based on the relative economic importance of traits in the breeding goal, and on the strength of genetic associations between measured traits and breeding goal traits. Until recently, the emphasis in beef cattle breeding in North America has been on using sophisticated methods to produce individual-trait EBVs. In contrast, in Europe, while less sophisticated methods of evaluation were used until recently, selection indices have been quite widely used in both specialized beef breeds and in dairy and dual purpose breeds.

Much of the emphasis in Europe has been on producing indices for terminal-sire characteristics. For example, a terminal-sire index was introduced in Britain in the mid-1980s and used in most breeds until 1997. The selection objective of this index was to maximize the margin between saleable meat yield and feed costs, taking into account the costs of difficult calvings (Allen and Steane, 1985). Index scores were calculated from the animal's own records of calving difficulty score, 200- and 400-day weight and a visual muscling score. If they were recorded, additional measurements of birth weight, feed intake and ultrasonically measured fat thickness were included, to increase the accuracy of the index.

In the late 1990s, new indices are being introduced for Signet performance-recorded beef herds in Britain. These are more closely linked to market returns (i.e. using associations with carcass weight, fat class and conformation class rather than with saleable meat yield). Also separate indices are being introduced for calving performance and for growth and carcass performance of terminal sires. The calving value ranks animals on genetic merit for calving ease, based mainly on records of birth weight, calving ease and gestation length, while the new beef value ranks them on genetic merit for growth and carcass traits, based mainly on records for weights, fat depth, muscle depth and muscle score. These two indices can be added together to rank animals on overall merit for calving ease and production together. The contributions which the calving value and beef value make to overall merit vary, depending on the importance of calving ease and on variation in the component traits in the breed concerned. However, typically, calving value accounts for about 16% of the variation in overall merit (Amer *et al.*, 1998).

Indices combining BLUP EBVs for reproduction, growth and carcass traits have been developed in Australia. An important feature of these indices is that the economic values applied can be tailored or customized to individual breeders' requirements. This is achieved via a computer software package, which uses data on returns and costs of beef production for individual producers or production systems (Barwick *et al.*, 1994).

Evidence of genetic improvement and its value

Estimates of genetic change achievable

In theory, changes of at least 1% of the mean per annum are possible following selection for weight or growth traits in beef cattle. However, in practice, rates of change are often lower than this. For example, a review of several beef cattle selection experiments showed that average changes of 0.6% and 0.8% per annum were achieved with selection for weaning and yearling weight, respectively (Mrode, 1988).

The increased uptake of across-herd BLUP genetic evaluations over the last decade has permitted more widespread estimation of genetic trends in industry breeding schemes. For example, Crump *et al.* (1997) show estimated genetic trends in birth weight, 200- and 400-day growth since 1980, for the most numerous performance-recorded beef breeds in Britain. The changes in 200- and 400-day weights ranged from 0.15 to 0.5% of the breed mean per annum for the different breeds.

Trends similar to or lower than these have been reported in several breeds in Canada and Australia (Graser *et al.*, 1984; de Rose and Wilton, 1988). Slightly higher trends in weaning weight have been reported in the US Angus and Hereford breeds (Benyshek *et al.*, 1994). This may be explained partly by the earlier availability of BLUP methods in the US beef industry. It is probably also partly due to the higher herd and population sizes for these breeds in the USA. Similar trends in weaning weight (from about 0.2 to 1.1 kg per annum) and positive trends in muscularity have been reported for the major French breeds between 1991 and 1995 (Journaux *et al.*, 1996).

In most of these studies of industry trends, the rates of change achieved are well below those theoretically possible and below those actually achieved in selection experiments. The apparently low rate of change is partly explained by the fact that selection has not been solely for weight traits. However, it is also partly due to the relatively low use of objective methods of selection and the fact that, in at least some of the countries mentioned, only within-herd comparisons could be made for most of the period concerned.

The economic value of genetic improvement

There have been relatively few studies of the value of genetic improvement in beef cattle, although these do show favourable estimates of cost : benefit (Barlow and Cunningham, 1984). A recent study of the costs and benefits of implementation of across-herd BLUP and index selection in the terminal-sire sector of the British beef industry showed that estimated discounted returns exceeded the costs of implementation, including research, within a few years of introduction. Estimated annual discounted returns are expected to reach about £18 million per annum and to exceed annual costs of implementation by a factor of 30 : 1, about 20 years after introduction of these technologies (Simm *et al.*, 1998).

Molecular and Reproductive Methods

Detection and use of quantitative trait loci

Some single loci of major effect, known as major genes or quantitative trait loci (QTL), have been identified and exploited directly. In cattle, these include the double-muscling gene (Georges *et al.*, 1998) and various coat-colour genes. Most such QTL have been detected by inspection of data. However, systematic methods for computer screening have been developed. More importantly, with the recent development of genetic maps for cattle (Barendse *et al.*, 1997; see Chapter 11), genetic marker plus trait performance data can be analysed to detect and locate other QTL of commercial value (reviewed by Kinghorn *et al.*, 1994).

Where QTL have been cloned and deoxyribonucleic acid (DNA) tests developed to determine genotype for individual animals, genetic evaluation at QTL is relatively simple. The QTL genotypes can be treated as fixed effects, and these effects can be estimated very accurately, just as fixed effects of breed and cross means can be estimated accurately. The QTL effects may differ between genetic backgrounds (e.g. breeds) and between environments or production systems – and the power to estimate the range of effects involved constitutes a major advantage over evaluation of polygenes using, for example, BLUP. This advantage extends to the ability to market specific QTL-genotyped seedstock with a performance and product image which is much more tangible than for competing 'high polygenic merit' seedstock.

However, such direct DNA tests may not be as reliable as implied above. Georges *et al.* (1998) located the double-muscled cattle *mh* gene at the myostatin locus. Of the 11 DNA sequence polymorphisms identified at this locus, five would be predicted to disrupt the function of the protein. This means that a DNA test to identify just one of the defective alleles would not be reliable in industry – and so caution is required for any gene locus.

Where QTL alleles can be inferred with imperfect accuracy through use of linked markers, marker assisted selection (MAS) can be used (see Chapter 17 for more detail). As with direct DNA tests for QTL, the value of MAS depends on a number of factors.

- Where heritability is low, the value of information on individual QTL tends to be higher.
- Where the trait(s) of interest cannot be measured on one sex, marker information gives a basis to rank animals of that sex.
- If the trait is not measurable before sexual maturity, marker information can be used to select at a juvenile stage.
- If a trait is difficult to measure, is sex-linked or is measured post-slaughter, marker information can be used instead.

Marker-assisted selection is handicapped by the fact that, unless there is considerable linkage disequilibrium, no one marker allele is consistently associated with a favourable QTL allele, due to recombination events. This means that linkage phase in parents needs to be inferred – something which can be done readily with very large half-sib families, as in dairy cattle. However, beef cattle population structures lead to the need for MAS analysis methods which can operate on general pedigrees, and several appropriate methods have been put forward (for example, van Arendonk *et al.*, 1994).

Multiple ovulation and embryo transfer nucleus schemes

The potential value of multiple ovulation and embryo transfer (MOET) in accelerating response to selection was first reported for beef cattle by Land and Hill in 1975. They estimated that responses to selection for growth rate could be doubled by the use of MOET, albeit with higher rates of inbreeding. As in dairy cattle, these original estimates of the benefits of MOET are now believed to be on the high side. Recent estimates suggest that 30% extra progress is possible, compared with a conventional scheme of similar size and with the same rate of inbreeding (Villanueva *et al.*, 1995).

While MOET has been used widely in beef cattle as a means of importing and exporting genetic material, and to multiply newly imported breeds or valuable individuals more rapidly than possible with natural reproduction, it has not been used widely in structured breed improvement programmes to date.

Sexing and cloning

One of the earliest intended uses of in *in vitro*-produced embryos was to improve the beef merit of calves from dairy or suckler cows, by creating a supply of beef embryos. Initially, the main source of eggs was the ovaries of slaughtered beef heifers. Eggs were collected from beef heifers with a high proportion of continental beef breeds in their genetic make-up, and embryos produced from these by maturing them and then fertilizing them with semen from high-merit proved bulls. These embryos were then marketed for transfer into beef suckler cows or dairy cows. Transfers were made either singly or to create twins, either by transferring an *in vitro*-produced embryo into cows already carrying a natural embryo or by transferring two *in vitro*-produced embryos. Despite a ready supply of ovaries from slaughtered heifers, early techniques produced few transferable embryos per ovary. Also, some *in vitro* culture techniques are implicated in the birth of very large calves, generally with associated calving difficulties (Kruip and den Daas, 1997).

More recently, techniques have been developed to allow the recovery of unfertilized eggs directly from the ovaries of live cows (see Chapter 14 for a review of these and related techniques). These techniques involve collection of eggs through an ultrasonically guided needle inserted into the ovary, usually via the vagina (Kruip, 1994). This type of recovery is called *in vivo* aspiration of oocytes or ovum pick-up (OPU). It has several potential advantages compared with recovery of eggs from slaughtered cows or with conventional embryo recovery techniques. In particular purebred animals of high genetic merit can be used as donors, so the technique is of potential benefit in genetic improvement and not just in dissemination. Moreover, eggs can be collected from donors on a weekly basis, allowing tens or potentially hundreds of embryos to be produced from the same donor. The resulting *in vitro* fertilization allows for cross-classified mating of males and females, which gives a useful boost in selection accuracy under juvenile breeding schemes, in which young animals are selected before measurement, on the basis of their parents' EBVs (Kinghorn *et al.*, 1991).

In most circumstances, sexing of either semen or embryos is probably of little value in accelerating genetic improvement. However, the development of a cheap, reliable technique for sexing semen in large enough quantities for conventional AI could lead to major improvements in the dissemination of genetic improvement and in the efficiency of beef production. Semen or embryo sexing on a smaller scale could still allow more effective dissemination if it is coupled with *in vitro* production of embryos (Cran *et al.*, 1993).

In genetic improvement programmes, cloning could be used to produce many animals of the same genotype in order to improve the accuracy of evaluation, or to allow evaluation of traits normally measured post-slaughter on some members of the cloned group. This would involve implanting some embryos from each cloned line to produce animals for testing, and freezing others to allow subsequent use (or further cloning) of the best-tested cloned lines in breeding or dissemination programmes. One factor to consider here is that clone testing can give accurate estimates of an individual's genetic value (value of alleles to self), but accuracy of EBV (value of alleles to progeny) from clone testing is limited to $\sqrt{V_A / V_G}$, where V_A is variance due to breeding values and V_G is variance due to genetic values. Moreover, if cloning is considered only in the context of closed breeding schemes, with fixed numbers of animals tested, then the expected benefits generally diminish or disappear, as keeping more identical animals means that fewer different families can be kept and so selection intensities will be reduced (Villanueva and Simm, 1994).

While the benefits of cloning in genetic improvement may be limited, the potential of the technique to accelerate dissemination of genetic improvement to commercial herds or flocks is great, especially for cloning from adult material. For this potential to be realized, reliable and cost-effective methods for cloning will be required. Also, improved and cost-effective techniques for delivery will be needed, including reliable methods for freezing cloned embryos and subsequent non-surgical transfer.

In many countries, there is public concern over the application of new technologies in animal production. Most people accept the use of animals for a range of purposes, including food production, providing that the animals are treated humanely. However, it is often difficult to decide whether or not a particular treatment is humane. For discussion of these issues with respect to new reproductive technologies, see report of the Ministry of Agriculture (MAFF, 1995).

References

Allen, D.M. and Steane, D.E. (1985) Beef selection indices. *British Cattle Breeders Club Digest* 40, 63–70.

Amer, P.R. (1994) Economic theory and breeding objectives. In: *Proceedings of the 5th World Congress on Genetics Applied to Livestock Production, Guelph, Canada*, Vol. 18, pp. 197–204.

Amer, P.R., Kemp, R.A. and Smith, C. (1992) Genetic differences among the predominant beef cattle breeds in Canada: an analysis of published results. *Canadian Journal of Animal Science* 72, 759–771.

Amer, P.R., Lowman, B.G. and Simm, G. (1996) Economic values for reproduction traits in beef suckler herds based on a calving distribution model. *Livestock Production Science* 46, 85–96.

Amer, P.R., Emmans, G.C. and Simm, G. (1997) Economic values for carcase traits in UK commercial beef cattle. *Livestock Production Science* 51, 267–281.

Amer, P.R., Crump, R.E. and Simm, G. (1998) A terminal sire selection index for UK beef cattle. *Animal Science* (in press).

Ball, A.J., Thompson, J.M. and Kinghorn, B.P. (1998) Breeding objectives for meat animals: use of biological modelling. *Australian Society of Animal Production* 22, 94–97.

Barendse, W., Vaiman, D., Kemp, S.J. *et al.* (1997) A medium-density genetic linkage map of the bovine genome. *Mammalian Genome* 8, 21–28.

Barlow, R. and Cunningham, E.P. (1984) Benefit–cost analyses of breed improvement programmes for beef and sheep in Ireland. In: *Proceedings of the Second World Congress on Sheep and Beef Cattle Breeding, Madrid*, Vol II, paper P31.

Barwick, S.A., Henzell, A.L. and Graser, H.-U. (1994) Developments in the construction and use of selection indices for genetic evaluation of beef cattle in Australia. In: *Proceedings of the 5th World Congress on Genetics Applied to Livestock Production*, Vol. 18, University of Guelph, Guelph, Canada, pp. 227–230.

Benyshek, L.L., Herring, W.O. and Bertrand, J.K. (1994) Genetic evaluation across breeds and countries: prospects and implications. In: *Proceedings of the 5th World Congress on Genetics Applied to Livestock Production*, Vol. 17, University of Guelph, Guelph, Canada, pp. 153–160.

Bonnett, J.N., Journaux, L., Mocquot, J.C. and Rehben, E. (1994) Breeding cattle for the next millennium. *British Cattle Breeders Club Digest* 49, 10–18.

Brascamp, E.W. (1984) Selection indices with constraints. *Animal Breeding Abstracts* 52, 645–654.

Cran, D.G., Johnson, L.A., Miller, N.G.A., Cochrane, D. and Polge, C. (1993) Production of bovine calves following separation of X- and Y-chromosome bearing sperm and *in vitro* fertilization. *Veterinary Record* 132, 40–41.

Crump, R.E., Simm, G., Nicholson, D., Findlay, R.H., Bryan, J.G.E. and Thompson, R. (1997) Results of multivariate individual animal model genetic evaluations of British pedigree beef cattle. *Animal Science* 65, 199–207.

Cundiff, L.V., Gregory, K.E., Koch, R.M. and Dickerson, G.E. (1986). Genetic diversity among cattle breeds and its use to increase beef production efficiency in a temperate environment. In: *Proceedings of the 3rd World Congress on Genetics Applied to Livestock Production, Lincoln, Nebraska*, Vol. IX, pp. 271–282.

Cuthbertson, A. (1994) Enhancing beef eating quality. *British Cattle Breeders Club Digest* 49, 33–37.

de Rose, F.P. and Wilton, J.W. (1988) Estimation of genetic trends for Canadian station-tested beef bulls. *Canadian Journal of Animal Science* 68, 49–56.

Dikeman, M.E. (1990) Genetic effects on the quality of meat from cattle. In: *Proceedings of the 4th World Congress on Genetics Applied to Livestock Production, Edinburgh*, Vol. XV, pp. 521–530.

Food and Agriculture Organization of the United Nations (FAO) (1996) *FAO Production Yearbook 1995*, Vol. 49. Food and Agriculture Organization of the United Nations, Rome, Italy.

Georges, M., Grobet, L., Poncelet, D., Royo, L.J., Pirottin, D. and Brouwers, B. (1998) Positional candidate cloning of the bovine mh locus identifies an allelic series of mutations disrupting the myostatin function and causing double-muscling in cattle. In: *6th World Congress on Genetics Applied to Livestock Production, Armidale, 11–16 January 1998*, Vol. 26, pp. 195–204.

Graser, H.-U., Hammond, K. and McClintock, A.E. (1984) Genetic trends in Australian Simmental. *Proceedings of the Australian Association of Animal Breeding and Genetics* 4, 86–87.

Graser, H.-U., Goddard, M.E. and Allen, J. (1995) Better genetic technology for the beef industry. *Proceedings of the Australian Association of Animal Breeding and Genetics* 11, 56–64.

Gregory, K.E. and Cundiff, L.V. (1980) Crossbreeding in beef cattle: evaluation of systems. *Journal of Animal Science* 51, 1224–1242.

Gregory, K.E., Cundiff, L.V. and Koch, R.M. (1991) Breed effects and heterosis in advanced generations of composite populations for preweaning traits of beef cattle. *Journal of Animal Science* 69, 947–960. (Also, see other papers by the same authors in Volumes 69 and 70.)

Hearnshaw, H., Gursansky, B.G. Gogel, B., Thompson, J.M., Fell, L.R., Stephenson, P.D., Arthur, P.F., Egan, A.F., Hoffman, W.D. and Perry, D. (1998) Meat quality in cattle of varying Brahman content: the effect of post-slaughter processing, growth rate and animal behaviour on tenderness. In: *44th International Congress of Meat Science and Technology, Barcelona, Spain*, pp. 1048–1049.

Journaux, L., Rehben, E., Laloë, D. and Ménissier, F. (1996) *Main Results of the Genetic Evaluation IBOVAL96 for the Beef Cattle Sires. Edition 96/1.* Institut de l'Elevage, Département Génétique Identification et Contrôle de Performances, Paris, France, and Institut National de Recherche Agronomique, Station de Génétique Quantitative et Appliquée, Jouy-en-Josas, France.

Kemp, R.A. (1994) Genetics of meat quality in cattle. In: *Proceedings of the 5th World Congress on Genetics Applied to Livestock Production*, Vol. 19, University of Guelph, Guelph, Canada, pp. 439–445.

Kennedy, B.W. (1981) Variance component estimations and prediction of breeding values. *Canadian Journal of Genetics and Cytology* 23, 565–578.

Kinghorn, B.P. (1997) Genetic improvement of sheep. In: Ruvinsky, A. and Piper, L. (eds) *The Genetics of Sheep*. CAB International, Wallingford, Oxon., pp. 565–591.

Kinghorn, B.P., Smith, C. and Dekkers, J.C.M. (1991) Potential genetic gains with gamete harvesting and *in vitro* fertilization in dairy cattle. *Journal of Dairy Science* 74, 611–622.

Kinghorn, B.P., van Arendonk, J.A.M. and Hetzel, D.J.S. (1994) Detection and use of major genes in animal breeding. *AgBiotech News and Information* 6, 297N–302N.

Koots, K.R., Gibson, J.P., Smith, C. and Wilton, J.W. (1994a) Analyses of published genetic parameter estimates for beef production traits. 1. Heritability. *Animal Breeding Abstracts* 62, 309–338.

Koots, K.R., Gibson, J.P. and Wilton, J.W. (1994b) Analyses of published genetic parameter estimates for beef production traits. 2. Phenotypic and genetic correlations. *Animal Breeding Abstracts* 62, 825–853.

Kruip, T.A.M. (1994) Oocyte retrieval and embryo production *in vitro* for cattle breeding. In: *Proceedings of the 5th World Congress on Genetics Applied to Livestock Production*, Vol. 20, University of Guelph, Guelph, Canada, pp. 172–179.

Kruip, T.A.M. and den Daas, J.H.G. (1997) *In vitro* produced and cloned embryos: effects on pregnancy, parturition and offspring. *Theriogenology* 47, 43–52.

Land, R.B. and Hill, W.G. (1975) The possible use of superovulation and embryo transfer in cattle to increase response to selection. *Animal Production* 21, 1–12.

Liboriussen, T. (1982) Comparison of paternal strains used in crossing and their interest for increasing production in dairy herds. In: *Proceedings of the 2nd World Congress on Genetics Applied to Livestock Production*, Vol. V, pp. 469–481.

McClintock, A.E. and Cunningham, E.P. (1974) Selection in dual purpose cattle populations: defining the breeding objective. *Animal Production* 18, 237–248.

Ménissier, F. (1988) La sélection des races bovines à viande spécialisées en France. In: *Proceedings of the 3rd World Congress on Sheep and Beef Cattle Breeding, Lincoln, Nebraska*, Vol. 2, pp. 215–236.

Middleton, B.K. and Gibb, J.B. (1991) An overview of beef cattle improvement programs in the United States. *Journal of Animal Science* 69, 3861–3871.

Ministry of Agriculture, Fisheries and Food (MAFF) (1995) *Report of the Committee to Consider the Ethical Implications of Emerging Technologies in the Breeding of Farm Animals*. HMSO, London.

Mrode, R.A. (1988) Selection experiments in beef cattle. Part 2. A review of responses and correlated responses. *Animal Breeding Abstracts* 56, 155–167.

Pirchner, F. (1986) Evaluation of industry breeding programs for dairy cattle milk and meat production. In: *Proceedings of the 3rd World Congress on Genetics Applied to Livestock Production, Lincoln, Nebraska*, Vol. IX, pp. 153–164.

Pollak, E.J. and Quaas, R.L. (1998) Multibreed genetic evaluations of beef cattle. In: *6th World Congress on Genetics Applied to Livestock Production, Armidale, 11–16 January 1998*, Vol. 23, pp. 81–88.

Simm, G. (1998) *Genetic Improvement of Cattle and Sheep*. Farming Press, Ipswich.

Simm, G., Conington, J., Bishop, S.C., Dwyer, C.M. and Pattinson, S. (1996) Genetic selection for extensive conditions. *Applied Animal Behaviour Science* 49, 47–59.

Simm, G., Amer, P.R. and Pryce, J.E. (1998) Returns from genetic improvement of sheep and beef cattle in Britain. In: *SAC Animal Sciences Research Report 1997*. Scottish Agricultural College, Edinburgh, pp. 12–16.

Thiessen, R.B., Hnizdo, E., Maxwell, D.A.G., Gibson, D. and Taylor, St.C.S. (1984) Multibreed comparisons of British cattle: variation in body weight, growth rate and food intake. *Animal Production* 38, 323–340. (Also, see other papers by the same authors in later volumes of *Animal Production*.)

van Arendonk, J.A.M., Tier, B. and Kinghorn, B.P. (1994) Use of multiple genetic markers in prediction of breeding values. *Genetics* 137(1), 319–329.

Van Vleck, L.D., Pollack, E.J. and Oltenacu, E.A.B. (1989) *Genetics for the Animal Sciences*. W.H. Freeman, New York.

Villanueva, B. and Simm, G. (1994) The use and value of embryo manipulation techniques in animal breeding. In: *Proceedings of the 5th World Congress on Genetics Applied to Livestock Production, Armidale, Australia*, Vol. 20, pp. 200–207.

Villanueva, B., Simm, G. and Woolliams, J.A. (1995) Genetic progress and inbreeding for alternative nucleus breeding schemes for beef cattle. *Animal Science* 61, 231–239.

Weller, J.I. (1994) *Economic Aspects of Animal Breeding*. Chapman & Hall, London.

Genetics of Meat Quality

21

D.M. Marshall
Department of Animal and Range Sciences, South Dakota State University, Brookings, SD 57007, USA

Introduction	605
Genetic Parameters and Selection Implications	606
Heritability of meat traits	606
Genetic correlations with meat traits	609
Live-animal ultrasound evaluation	614
Heterosis Effects in Breed Crosses	615
Body composition	615
Technological and sensory quality	615
Major Genes and Quantitative Trait Loci	616
Breed Variation	617
Body composition	617
Technological quality	619
Shear force and sensory quality	619
Genetic Effects on Attributes of Fat	622
Cholesterol content	622
Fatty acid profile	623
Fat colour	625
Prospects for Genetic Improvement in Meat Traits	625
Consumer demand for beef	625
Consistency in eating quality	626
Genetic evaluation and improvement	627
Conclusions	628
References	628

Introduction

Historically, cattle production systems and breeding programmes have often been dictated by cow herd constraints (e.g. dairy production, climate, availability of production resources) rather than consumer preferences for beef product quality. Meat traits were not routinely included in industry genetic evaluation programmes, because of the difficulty in obtaining measurements and the assumption that meat traits were of less importance to economic efficiency than growth or reproduction. More recently, the cattle industries of many countries are under increasing pressure to improve consumer qualities of beef products with regard to meat palatability, diet/health concerns and product convenience. At the same time, increasing competition from the pork and poultry industries have forced cattle producers to seek new methods of improving the cost efficiency of meat production. Consequently, there is currently much interest in the potential for genetic manipulation of such traits and the associated impact on other production parameters.

In this chapter, meat quality is broadly defined to include body composition traits, technological (chemical and physical) attributes and sensory characteristics (visual appeal and eating quality). Cattle breeders have long assumed that body composition traits are generally quite heritable, whereas genetic control of technological and sensory attributes of beef has not been thoroughly studied until recently. In order to design appropriate breeding programmes to produce desirable beef products at a competitive cost, cattle breeders need genetic information, such as heritability, genetic correlations and breed differences. The aim of this chapter is to summarize current knowledge of genetic aspects of meat quality and to discuss potential implications of genetic change for the cattle industry.

Genetic Parameters and Selection Implications

Heritability of meat traits

A survey of the research literature indicated that scientists, like the cattle industry, have recently placed increased emphasis on eating qualities of beef. Previous reviews of genetic parameters in cattle meat traits (Renand, 1988; Koots *et al.*, 1994a, b; Marshall, 1994) included many estimates for carcass composition traits, but relatively few estimates for technological or sensory quality traits. Several very recent studies have included genetic parameters for traits more directly related to the physical appearance and eating quality of beef.

Carcass composition is often measured at or adjusted to constant age, weight or fatness, and each trait–end-point combination could be considered a separate trait. Heritability estimates for commonly measured composition traits are presented in Table 21.1 separately for different end-points when end-point could be determined from the research publication. Included are values averaged across studies from the review of Koots *et al.* (1994a) and from studies reported after that review. Heritability estimates for objective measures of technological quality and for subjective evaluations of sensory traits are shown in Tables 21.2 and 21.3, respectively.

Carcass composition traits are sufficiently heritable for improvement through genetic selection to be relatively effective in many cattle populations.

Table 21.1. Heritability (h^2) estimates of cattle carcass composition traits.

Trait*	Review estimates[†] (mean h^2)	Recent estimates[‡] Mean h^2	(n)[§]	Range
Lean yield (A)	0.47	0.50	(11)	0.26–0.76
Lean yield (W)	0.48			
Lean yield (F)		0.76	(1)	
Carcass weight (A)	0.23	0.35	(11)	0.15–0.59
Carcass weight (F)	0.36	0.10	(1)	
Dressing % (A)	0.39	0.26	(6)	0.06–0.40
Dressing % (W)	0.38			
Dressing % (F)		0.21	(1)	
Fat thickness (A)	0.44	0.39	(11)	0.25–0.56
Fat thickness (W)	0.46			
Kidney or kidney, pelvic and heart (KPH) fat (A)		0.34	(5)	0.28–0.43
Lean %	0.55	0.71	(1)	
Longissimus muscle area (A)	0.42	0.37	(12)	0.06–0.65
Longissimus muscle area (W)	0.41			
Longissimus muscle area (F)		0.45	(2)	0.38–0.52
Fat trim weight (A)		0.32	(1)	
Fat trim % (A)		0.49	(3)	0.35–0.59
Veal carcass fleshiness		0.31	(1)	
Retail product weight (A)		0.41	(3)	0.28–0.50
Bone weight (A)		0.39	(1)	
Bone % (A)		0.37	(3)	0.21–0.47
Rib thickness (A)		0.32	(3)	0.26–0.41
Lean/bone ratio (A)	0.63			
Marbling score (A)	0.38	0.49	(15)	0.19–0.79
Marbling score (F)	0.65	0.32	(3)	0.18–0.52
Marbling score (W)	0.36			

*Letter in parentheses indicates that the trait was evaluated at a constant age or days in feedlot (A), animal or carcass weight (W) or fat thickness (F).
[†]Source: review of Koots et al., 1994a.
[‡]Sources: Van Veldhuizen et al., 1991; Johnston et al., 1992a; Mukai et al., 1993; Robinson et al., 1993; Gregory et al., 1994b, 1995; Renand et al., 1994; Shackelford et al., 1994b; Mukai et al., 1995; Aass, 1996; Hirooka et al., 1996; Wheeler et al., 1996, 1997; Wulf et al., 1996; AAA, 1997; Anderson, 1997; ASA, 1997; O'Connor et al., 1997; Kim et al., 1998; Lee et al., 1998; Splan et al., 1998.
[§]Number of estimates included in the mean.

Intramuscular fat content (usually evaluated in the longissimus dorsi muscle) is often subjectively evaluated by visual inspection of a cross-section of the muscle (i.e. marbling score, Table 21.1), and in some studies has been measured objectively by chemical analysis (i.e. intramuscular lipid percentage, Table 21.2). Both measures indicate that intramuscular fat content is highly heritable. Shear force and myofibrillar fragmentation index are physical and biochemical measures of tenderness, respectively, whereas calpastatin is an inhibitor of the calcium-dependent proteases involved in the enzymatic degradation of myofibrillar proteins during post-mortem storage (ageing). Each of these

Table 21.2. Heritability (h^2) estimates of technological quality traits of beef.

Trait	Review estimates* (mean h^2)	Recent estimates[†] Mean h^2	(n)[‡]	Range
Intramuscular lipid %	0.26	0.54	(6)	0.26–0.93
Shear force	0.30	0.25	(10)	0.02–0.53
Calpastatin activity		0.43	(4)	0.15–0.65
Myofibrillar fragmentation		0.39	(3)	0.17–0.58
Lean colour reflectance	0.26			
L*a*b lightness		0.29	(1)	0.27–0.30
L*a*b redness		0.17	(1)	0.16–0.17
L*a*b yellowness		0.11	(1)	0.08–0.13
Ultimate pH	0.26	0.15	(3)	0.10–0.19
Water loss	0.24			

*Adapted from the reviews of Renand, 1988; Koots et al., 1994a.
[†]Sources: Gregory et al., 1994b, 1995; Renand et al., 1994; Shackelford et al., 1994b; Aass, 1996; Barkhouse et al., 1996; Wheeler et al., 1996; Wulf et al., 1996; O'Connor et al., 1997.
[‡]Number of studies. In studies reporting more than one estimate from the same animals, all estimates were used in the range, but only the within-study mean was used to calculate the across-study mean.

Table 21.3. Heritability (h^2) estimates of subjective sensory traits.*

Trait	Mean h^2	(n)[†]	Range
Lean firmness	0.30	(2)	0.29–0.30
Lean texture	0.14	(1)	
Lean texture and firmness	0.28	(1)	
Lean colour	0.16	(2)	0.12–0.19
Lean colour and gloss	0.24	(1)	
Veal colour	0.16	(1)	
Fat colour	0.00	(1)	
Fat colour, gloss, and quality	0.16	(1)	
Tenderness	0.22	(12)	0.03–0.50
Juiciness	0.14	(7)	0.00–0.26
Flavour intensity	0.10	(9)	0.00–0.43
Flavour desirability	0.01	(1)	
Overall acceptability	0.04	(1)	

*Sources: Dinkel and Busch, 1973; Wilson et al., 1976; Oikawa and Kyan, 1986; More O'Ferrall et al., 1989; Dijkstra et al., 1990; Van Veldhuizen et al., 1991; Van Vleck et al., 1992; Gregory et al., 1994b, 1995; Shackelford et al., 1994b; Barkhouse et al., 1996; Wheeler et al., 1996; Wulf et al., 1996; O'Connor et al., 1997; Kim et al., 1998; Splan et al., 1998.
[†]Number of studies.

objective indicators of tenderness, based on a limited number of studies, seems to be more heritable than subjective tenderness evaluated by sensory panellists. Other technological quality traits (water-binding capacity, pH and lean colour) appear to be slightly to moderately heritable. Subjective sensory traits appear to be considerably less heritable than carcass composition traits.

Genetic correlations with meat traits

Genetic correlations are important to consider in multiple-trait selection and in the design of breeding systems, because selection for one trait can cause a response in other traits. Genetic antagonisms tend to slow the rate of improvement or even cause undesirable change in some traits. When genetic antagonisms exist between dam traits and market calf traits, then terminal breeding systems offer a potential advantage. On an industry-wide basis, terminal systems could allow seedstock breeders to focus on fewer traits within a given breed, increasing the rate of change per trait.

Body composition
Genetic correlations among body composition traits are presented in Table 21.4. The genetic relationships of marbling score with fat thickness and lean yield are of particular interest, because in many markets these variables are important criteria in the determination of carcass price. Traditionally, it has been assumed that higher marbling scores were genetically associated with increased external fat and decreased lean yield, both within and between breeds, and the average genetic correlations found by Koots *et al.* (1994b) agree with that view. Recent estimates have generally ranged from moderately antagonistic to slightly favourable, suggesting that external fatness and

Table 21.4. Average values of genetic correlations (r_g) among carcass weight and composition traits.

Trait 1	Trait 2	Review estimates* (mean r_g)	Recent estimates[†] Mean r_g	(n)[‡]	Range
Carcass weight[§]	Lean yield	0.00	−0.06	(6)	−0.19 to 0.19
	Dressing percentage	0.04			
	Fat thickness	0.29	0.23	(7)	−0.01 to 0.39
	Longissimus muscle area	0.48	0.41	(7)	0.23 to 0.66
	Marbling score	0.25	0.09	(8)	−0.05 to 0.36
Dressing percentage	Longissimus muscle area	0.36	0.18	(1)	
	Marbling score	0.25	0.24	(2)	−0.20 to 0.68
Lean yield	Fat thickness	−0.56	−0.77	(6)	−0.86 to −0.62
	Longissimus muscle area	0.45	0.63	(5)	0.32 to 0.79
	Marbling score	−0.25	−0.19	(8)	−0.60 to 0.12
Fat thickness	Longissimus muscle area	0.01	−0.17	(7)	−0.43 to 0.01
	Marbling score	0.35	0.09	(9)	−0.12 to 0.44
Longissimus muscle area	Marbling score	−0.21	0.01	(10)	−0.40 to 0.49

*Source: Koots *et al.*, 1994b.
[†]Sources: Johnston *et al.*, 1992a; Van Vleck *et al.*, 1992; Gregory *et al.*, 1994b, 1995; Mukai *et al.*, 1995; Aass, 1996; Hirooka *et al.*, 1996; Moriya *et al.*, 1996; Wheeler *et al.*, 1996, 1997; Wulf *et al.*, 1996; AAA, 1997; Anderson, 1997; ASA, 1997; Kim *et al.*, 1998.
[‡]Number of estimates.
[§]Carcass weight adjusted to a common age or number of days in feedlot.

marbling are almost genetically independent in some populations. In a study involving a variety of breeds, Koch et al. (1982b) predicted that single-trait selection for reduced subcutaneous fat thickness would decrease marbling. Gwartney et al. (1996) demonstrated that selection based on expected progeny difference values of Angus bulls could decrease fat thickness while maintaining marbling.

Technological quality
The use of a subjective marbling score as an indicator of intramuscular fat percentage is confirmed by a high genetic correlation (Table 21.5). Fat thickness and lean yield appear to be more closely correlated with actual intramuscular fat percentage (Table 21.5) than with subjective marbling (Table 21.4), based on across-study averages. However, it should be noted that most of the estimates with intramuscular fat percentage were from studies reporting relatively high correlations with marbling, whereas most of the studies reporting low correlations with marbling did not measure intramuscular fat percentage.

Based on average genetic correlation estimates, improvement in shear force would be associated with increased intramuscular fat and decreased calpastatin activity and would have relatively little effect on muscling, ultimate pH or water-holding capacity (Table 21.5). Genetic correlation estimates of shear force with subcutaneous fatness or lean yield have varied considerably across studies, with the mean values indicating slight antagonisms.

Phenotypically, low pH in meat is often associated with pale colour, softness and low water-binding capacity. However, few estimates of genetic correlations for such traits in cattle are available. In contrast to the expected relationship between colour and firmness of lean, Shackelford et al. (1994a) reported a slight genetic tendency for darker meat to be softer and a moderate tendency for darker meat to be more coarsely textured. Renand (1985) reported genetic correlations of near zero among shear force, pH and water loss. Based on Hunter-L*a*b colour values, Aass (1996) reported that lower ultimate pH of meat was moderately genetically associated with increased degree of redness and yellowness, but was relatively unrelated to degree of lightness. Dinkel and Busch (1973) reported that increased desirability of lean colour score (pink = desirable versus dark red = undesirable) was slightly genetically associated with increased marbling and lower firmness scores and that increased marbling was moderately associated with increased firmness. Oikawa and Kyan (1986) found that higher (more desirable) scores for lean colour quality (Japanese system) were genetically associated with improved scores for marbling, quality of lean texture and firmness and quality and colour of fat. Several relatively large genetic correlations of carcass composition traits with meat colour have been reported (Dinkel and Busch, 1973; Oikawa and Kyan, 1986; Aass, 1996), although the numbers of estimates have been too small to make generalizations. The review of Renand (1988) reported relatively weak negative genetic correlations of carcass fatness with colour (reflectance), ultimate pH and water loss.

Table 21.5. Average values of genetic correlations (r_g) with technological quality traits.*

Trait 1	Trait 2	Mean r_g	No. estimates	Range
Intramuscular fat %	Carcass weight	0.32	2	0.26 to 0.38
	Lean yield	−0.47	5	−0.90 to −0.11
	Fat thickness	0.26	3	−0.06 to 0.71
	Longissimus muscle area	−0.10	3	−0.41 to 0.20
	Marbling score	0.81	2	0.65 to 0.96
	Shear force	−0.64	4	−0.93 to −0.05
	Calpastatin activity	−0.34	1	
	Meat colour – lightness[†]	0.05	1	
	Meat colour – redness[†]	−0.09	1	
	Meat colour – yellowness[†]	0.12	1	
	Ultimate meat pH	0.37	1	
Shear force	Carcass weight	−0.19	3	−0.47 to 0.00
	Lean yield	0.18	6	−0.19 to 0.70
	Fat thickness	−0.16	4	−0.40 to 0.33
	Longissimus muscle area	−0.26	5	−0.63 to 0.14
	Marbling score	−0.47	10	−1.00 to 0.28
	Calpastatin activity	0.63	4	0.35 to > 1
	Ultimate meat pH	−0.03	1	
	Water loss	−0.06	1	
Calpastatin activity	Lean yield	0.10	2	−0.25 to 0.44
	Longissimus muscle area	−0.30	1	
	Marbling score	−0.27	3	−0.75 to 0.61
Meat colour – reflectance	Carcass fatness	−0.15	3	
Meat colour – lightness[†]	Longissimus muscle area	0.92	1	
	Ultimate meat pH	−0.02	1	
	Meat colour – redness[†]	−0.36	1	
	Meat colour – yellowness[†]	0.39	1	
Meat colour – redness[†]	Longissimus muscle area	< −1	1	
	Ultimate meat pH	−0.37	1	
	Meat colour – yellowness[†]	0.77	1	
Meat colour – yellowness[†]	Longissimus muscle area	−0.44	1	
	Ultimate meat pH	−0.31	1	
Meat colour desirability[‡]	Marbling score	0.52	2	0.22 to 0.82
	Firmness of lean [‡]	−0.19	1	
	Texture and firmness of lean [‡]	0.86	1	
	Colour and quality of fat [‡]	0.51	1	
Meat colour darkness[§]	Softness of lean[§]	0.19	1	
	Coarseness of lean[§]	0.60	1	
Colour and quality of fat [‡]	Carcass weight	−0.20	1	
	Longissimus muscle area	−0.07	1	
	Marbling score	0.25	1	
	Texture and firmness of lean[‡]	0.48	1	
Ultimate pH	Dressing percentage	0.65	1	
	Longissimus muscle area	−0.62	1	
	Carcass fatness	−0.23	3	
	Water loss	−0.13	1	
Water loss	Carcass fat percentage	−0.35	1	
	Carcass fatness	−0.16	3	
Increased firmness of lean	Lean yield	0.60	1	
	Longissimus muscle area	0.61	1	
	Fat thickness	−0.34	1	
	Marbling score	0.47	1	
Decreased firmness of lean	Texture of lean	0.52	1	

*Sources: Dinkel and Busch, 1973; Koch *et al.*, 1982b; Renand, 1985, 1988; Oikawa and Kyan, 1986; Van Vleck *et al.*, 1992; Gregory *et al.*, 1994b, 1995; Shackelford *et al.*, 1994b; Aass, 1996; Barkhouse *et al.* 1996; Wheeler *et al.*, 1996, 1997; Wulf *et al.* 1996; O'Connor *et al.* 1997; Kim *et al.*, 1998.
[†]Colour evaluated by machine (Hunter-L*a*b).
[‡]Higher subjective scores assigned for increased desirability.
[§]Higher subjective scores for increased softness (less firm), darker colour or coarser (less fine) texture.

In summary, genetic correlations with technological quality attributes have not been widely studied. Limited evidence suggests that selection for leanness could be slightly antagonistic to water-binding capacity. Effects of selection for leanness on intramuscular fat content could range from negligible to moderately antagonistic. Effects of selection for increased intramuscular fat content or improved shear force on technological quality could range from slightly antagonistic to moderately favourable.

Sensory quality

Flavour intensity, juiciness and tenderness are the most commonly studied sensory traits of beef, and are highly genetically correlated to one another (Table 21.6). The bulk of evidence from sensory-panel evaluation indicates that selection for leanness could be slightly antagonistic to tenderness and juiciness. However, in a study of Charolais bulls, Renand *et al.* (1994) suggested

Table 21.6. Average values of genetic correlations (r_g) with sensory panel traits.*

Trait 1	Trait 2	Mean r_g	No. estimates	Range
Flavour intensity	Carcass weight	0.01	2	−0.12 to 0.13
	Lean yield	−0.06	4	−0.25 to 0.16
	Fat thickness	−0.07	3	−0.62 to 0.31
	Fat trim %	0.04	2	−0.11 to 0.19
	Longissimus muscle area	0.04	3	−0.25 to 0.22
	Marbling score	0.43	6	−0.19 to 1.00
	Intramuscular fat %	0.29	3	−0.14 to 0.48
	Intramuscular fat % (cooked)	0.35	1	
	Shear force	−0.71	5	−1.00 to 0.27
	Calpastatin activity	0.21	1	
Juiciness	Carcass weight	0.03	1	
	Lean yield	−0.26	2	−0.31 to −0.20
	Fat thickness	0.40	2	0.34 to 0.45
	Fat trim %	0.15	1	
	Longissimus muscle area	0.12	2	−0.01 to 0.24
	Marbling score	0.42	4	0.23 to 0.60
	Intramuscular fat %	0.33	3	0.29 to 0.41
	Shear force	−0.78	4	−0.96 to −0.23
	Flavour intensity	0.86	3	0.78 to 1.00
Tenderness	Carcass weight	0.24	2	0.15 to 0.32
	Lean yield	−0.19	4	−0.48 to 0.03
	Fat thickness	0.10	4	−0.14 to 0.30
	Fat trim %	0.20	2	−0.07 to 0.46
	Longissimus muscle area	0.21	4	−0.25 to 0.56
	Marbling score	0.38	9	0.00 to 0.90
	Intramuscular fat %	0.30	3	0.06 to 0.50
	Intramuscular fat % (cooked)	0.36	1	
	Shear force	−0.86	9	−1.00 to −0.64
	Calpastatin activity	−0.70	3	< −1.00 to 0.00
	Flavour intensity	0.86	5	0.63 to 1.00
	Juiciness	0.79	4	0.43 to 0.95

*Sources: Renand, 1988; Van Vleck *et al.*, 1992; Gregory *et al.*, 1994b, 1995; Barkhouse *et al.*, 1996; Wheeler *et al.*, 1996, 1997; Wulf *et al.*, 1996; O'Connor *et al.*, 1997; Kim *et al.*, 1998.

that selection for leanness would favourably affect tenderness, because of increased collagen solubility and proportion of white-type muscle fibres, even though intramuscular lipid content would be reduced. The genetic association between subcutaneous fat thickness and flavour intensity has been somewhat variable across studies, with the average estimate approximating zero. Longissimus muscle area is apparently not closely associated with flavour intensity or juiciness, although its genetic correlations with sensory tenderness have ranged from slightly negative to moderately positive. Genetic correlations of intramuscular fatness with flavour intensity and tenderness have ranged from slightly negative to moderately or highly positive. Recent estimates of the genetic correlation between intramuscular fatness and juiciness have been moderately positive, although Wilson *et al.* (1976) reported a value of –0.81 for marbling with juiciness (phenotypic correlation was 0.21). Myofibrillar fragmentation index and calpastatin activity might be genetically correlated to sensory tenderness, but have not been widely studied. Shear force appears to be highly genetically correlated to sensory quality, and seems to be the best indicator of genetic potential for sensory quality among all carcass composition or technological quality traits of beef that have been evaluated to date.

Reproduction and growth
The ease with which a cow maintains body condition may be important for her to begin cycling early after calving, at least in some environments. Therefore, there is concern among producers that selection for leanness could be antagonistic to cow rebreeding performance. Very little information exists regarding within-population genetic correlations between reproductive traits in the cow herd and carcass traits of market calves. The study of MacNeil *et al.* (1984) suggested that selection for reduced external fatness in steers could be associated with delayed puberty and reduced fertility in female relatives. In the study of Splan *et al.* (1998), heifer age at puberty was not genetically associated with subcutaneous fat depth or marbling score in steer relatives. Heifer calving rate was favourably associated with increased marbling, but unfavourably associated with increased leanness of steers. The impact on reproductive traits from genetic changes in body composition could be reduced through the use of terminal matings (Bennett and Williams, 1994).

Carcass external fat thickness is positively genetically correlated with live animal growth during pre- and postweaning (Koots *et al.*, 1994b; Marshall, 1994) and with age-adjusted carcass weight (Table 21.4). However, age-adjusted carcass weight appears to be unrelated to lean yield (Table 21.4). Thus, within-population selection for growth could be expected to increase carcass weight at a given age, with a corresponding increase in weight of both fat and lean. Renand *et al.* (1994) found that index selection for growth and improved feed efficiency in Charolais bulls resulted in progeny carcasses with increased lean percentage and decreased fat percentage.

Average genetic correlations with age-adjusted carcass weight were positive for marbling score, percentage intramuscular fat and longissimus muscle area (Tables 21.4 and 21.5). Recent studies indicate that age-adjusted carcass weight is poorly correlated with shear force (Table 21.5) and sensory tenderness (Table 21.6), although Renand (1988) reported moderately antagonistic genetic relationships of growth rate (daily gain or final live weight) with shear force ($r = 0.34$ from three estimates) and sensory tenderness ($r = -0.31$ from two estimates). Shackelford et al. (1994b) reported a favourable genetic relationship between postweaning gain and shear force ($r = -0.40$). Genetic correlations of carcass weight with other technological and sensory attributes appear to be quite low, although few estimates are available (Tables 21.5 and 21.6). The review of Renand (1988) reported low genetic correlations for growth rate with lean colour, ultimate pH and water loss. Selection for growth and improved feed efficiency in Charolais resulted in decreased intramuscular lipid content, but had little effect on several other technological measures of meat quality in the longissimus dorsi muscle (Renand et al., 1994). In general, there seems to be little basis for concern regarding genetic antagonisms between growth rate and meat quality within most populations.

Live-animal ultrasound evaluation

Historically, most cattle breeders considered carcass merit to be of lower economic importance than growth or reproductive traits, and assumed that the value of genetic improvement would be relatively low compared with the associated cost. More recently, meat quality has become more of a focal point for the industry, because of the realization that consumers were increasingly choosing alternative foods instead of beef. However, the development of genetic evaluation programmes for carcass traits has been slowed by the difficulty of measuring carcass traits in breeding animals and by the costly nature of progeny testing. Consequently, there has been much interest in the use of ultrasound technology to evaluate meat traits in the live animal, both to facilitate direct measurement on breeding animals and to expand data availability on non-breeding relatives.

In studies in which the average measurement age ranged from about 1 year to 470 days, heritability estimates of ultrasonic measurements have averaged approximately 0.30 (ranged from 0.04 to 0.56) for fat depth and 0.25 (ranged from 0.11 to 0.40) for longissimus muscle area (de Rose et al., 1988; Turner et al., 1990; Arnold et al., 1991; Johnson et al., 1993; Robinson et al., 1993; Nagamine et al., 1996; Shepard et al., 1996; Crump et al., 1998; Graser et al., 1998). Average heritability estimates of live-animal ultrasonic measures of subcutaneous fat depth and longissimus muscle area have been somewhat lower than those based on post-mortem carcass measurement, but sufficiently large to provide for effective genetic evaluation. Although ultrasound technology for evaluation of intramuscular fat is at a more preliminary stage of development, results have been reasonably promising (Izquierdo et al., 1997; Graser et al., 1998; Wilson et al., 1998).

Breeding animals typically are measured at a younger age, are leaner and have less intramuscular fat as compared with non-breeding cohorts. Thus, concern exists regarding whether actual differences in subcutaneous and intramuscular fatness of market progeny can be accurately predicted from estimated breeding values that are based on ultrasonic measurements of breeding animals. Although few estimates of genetic correlations between ultrasonic measures and actual carcass measures are available, a growing body of evidence suggests that ultrasound evaluation of seedstock should be encouraged (Bertrand et al., 1997; Robinson et al., 1998; Wilson et al., 1998). Future comprehensive genetic evaluation programmes for carcass traits will probably be based on a combination of ultrasonic measurements of seedstock and non-breeding animals and post-mortem carcass measurements of market animals.

Heterosis Effects in Breed Crosses

Body composition

Direct heterosis for body composition traits is related to an increased rate of maturing for crossbred animals (Marshall, 1994). At a given age, crossbred animals are generally heavier with increased marbling, fat cover and muscle, whereas overall lean yield is little affected by heterosis (Table 21.7). On a weight-constant basis, heterosis estimates for carcass composition traits tend to be relatively small (Gregory et al., 1978; Drewry et al., 1979; Johnston et al., 1992b).

Technological and sensory quality

Heterosis effects on technological quality or sensory traits of beef have not been widely studied with the exception of shear force. The majority of heterosis estimates for shear force have ranged from moderately favourable to slightly unfavourable (i.e. approximately −10 to 5%) (Winer et al., 1981; Peacock et al. 1982; Anderson et al., 1986; Marshall et al., 1987; Gregory et al., 1994a, b), although certain crosses between *Bos taurus* and *Bos indicus* may result in higher levels of favourable heterosis (DeRouen et al., 1992).

Table 21.7. Direct heterosis effects on carcass composition traits (age-constant or time-in-feedlot-constant basis).*

Trait	Heterosis (%)	No. studies
Fat cover	10.1	11
Longissimus muscle area	4.1	9
Lean yield or retail product %	−0.6	7
Marbling	3.8	7

*Values given are from the review of Marshall (1994). Heterosis estimates were averaged across specific crosses within study and across studies, and expressed as a percentage of the straightbred mean.

Non-significant heterosis estimates were reported by Winer et al. (1981) and Gregory et al. (1994b) for sensory evaluation of juiciness, tenderness and flavour and by Winer et al. (1981) for cooked colour and overall desirability.

Major Genes and Quantitative Trait Loci

Although quantitative traits have generally been assumed to be controlled by multiple genes, individual genes may account for a relatively large amount of variation for some traits. The muscle hypertrophy condition known as double muscling has for some time been thought to probably be controlled by a single major gene (Arthur, 1995). This condition has been observed in several breeds, and is predominant in some (e.g. Belgian Blue and Piedmontese). Recent evidence suggests that double muscling in these two breeds is caused by mutation of a gene located on bovine chromosome 2 that produces the protein myostatin (Grobet et al., 1997; Kambadur et al., 1997; Smith et al., 1997). Normally, myostatin serves to repress skeletal muscle growth, but the mutation apparently blocks this effect and permits extra muscle growth. Compared with normal cattle, double-muscled animals generally have an increased proportion of muscle relative to fat and bone, and have reduced organ weights at a given body weight (Hanset, 1981). Carcasses from double-muscled cattle tend to have a higher proportion of 'valuable' meat cuts, and, although results have varied across studies, the tenderness of meat from double-muscled cattle tends to be acceptable and often preferred compared with that of other breeds (Arthur, 1995). A recent study found that animals inheriting a single copy of the muscle hypertrophy (*mh*) allele from a crossbred Belgian Blue or crossbred Piedmontese sire had increased longissimus muscle area and retail yield and reduced external and intramuscular fatness compared to animals receiving no copies of the *mh* allele (Casas et al., 1998).

Currently, breeding animals can be evaluated for carcass composition by ultrasound or progeny testing, whereas sensory trait evaluation is limited to progeny testing only. The development of molecular marker-assisted methods of genetic evaluation could potentially allow direct evaluation of breeding animals and significantly reduce the time needed for evaluation. Preliminary data from genome-wide screening of DNA markers have revealed a number of putative quantitative trait loci (QTLs) associated with meat traits, although few results have been published to date (Hetzel and Davis, 1997; Stone, 1997; Sugimoto, 1997; Taylor and Davis, 1997). Beever et al. (1990) reported significant associations of genetic markers with carcass composition traits in a half-sib Angus family. Kim and Marshall (1998) reported a small association of lean yield with a polymorphism in the growth hormone gene. The calpastatin gene (Killefer and Koohmaraie, 1994) has been proposed as a candidate locus for marker-assisted selection, because of the role calpastatin plays in post-mortem tenderness as an inhibitory regulator of the calpain system (Koohmaraie, 1992). Significant associations between beef tenderness and calpastatin genotype, based on restriction fragment length polymorphisms, were detected by Green et al. (1996a, b), but not by Lonergan et al. (1995).

Breed Variation

Substantial between-breed variation exists for many carcass composition and meat quality characteristics. Thus, it would seem that beef producers could, with relative ease, identify an appropriate breed or blend of breeds to fit their particular market needs. However, the choice of breeds for beef production is often complicated by constraints in the cow herd. For example, much of the beef is produced in dairy herds in some countries. In many non-dairy situations, the breed type of the cow herd must be matched to available resources and environmental constraints of the individual production unit. In either case, the genetic type best suited for cow herd production efficiency might not be optimal for postweaning production or meat traits.

Body composition

Breed differences for body composition traits have been evaluated in numerous studies, and have been reviewed by Renand (1985) and Marshall (1994). Franke (1997) has reviewed carcass composition of subtropically adapted breeds in the USA. A general summary of approximate breed differences in body composition is presented in Table 21.8. It is important to recognize that breed differences and rankings can vary due to such factors as sampling effects, environmental effects and production system (including end-point criteria). Also, the performance of a given breed in crossbreeding can vary due to differences in heterosis, depending on which other breeds are used in the cross.

Several *B. taurus* breeds that rank high for marbling tend to rank low for lean-to-fat ratio and vice versa. In general, *B. indicus* breeds tend to have moderate lean-to-fat ratios and below average marbling, although Boran may be comparable to many *B. taurus* breeds for marbling.

Dairy breeds tend to be quite variable in carcass composition. Holsteins or Friesians are similar to many beef breeds in lean-to-fat ratio and marbling, although they tend to have more bone and thus less muscle per unit of carcass weight as compared with many beef and dual-purpose breeds (El-Hakim et al., 1986; Knapp et al., 1989). Jerseys tend to have lower lean-to-fat ratios and more marbling than most beef breeds. Terminal matings of dairy cows to bulls of breeds with high lean-to-fat ratios have been frequently used as a way to increase the value of meat from dairy herds.

Several of the beef breeds of British origin tend to have low lean-to-fat ratios and relatively high marbling scores. Among breeds originating in central Europe, the dual-purpose breeds tend to rank at or somewhat above average in lean-to-fat ratio, and about average for marbling, whereas several other

Table 21.8. Breed characterization for body composition.*†

Category/breed	Lean-to-fat ratio (age-constant)	Longissimus muscle area/carcass wt	Marbling (age-constant)
British beef			
Angus	– –	–	+ +
Devon	–	=	=
Galloway	=	+	+
Hereford	– –	–	=
Red Angus	– –	–	+ +
Red Poll	– –	=	+
Shorthorn	– –	–	+ +
South Devon	–	=	+
Continental beef			
Belgian Blue	+ + +	+ + +	– –
Blonde d'Aquitaine	+ + +	+ + +	– – –
Charolais	+ +	+ +	–
Chianina	+ + +	+ +	– – –
Limousin	+ +	+ +	– –
Piedmontese	+ + +	+ + +	–
Continental dual-purpose			
Salers	+	+	=/–
Maine Anjou	+	+	=/–
Gelbvieh	+	+	=/–
Pinzgauer	=	=	+
Simmental	+	+	=/–
Tarentaise	=	=	=
Braunvieh	+	+	=
Zebu			
Boran	=	=	–
Brahman	=	–	– –
Nellore	=	–	–
Sahiwal	=	=	– –
Brahman derivative			
Brangus	– –	– –	=
Santa Gertrudis	– –	– –	=
Non-zebu subtropical			
Tuli	=	=	=
Dairy			
Brown Swiss	+	+	=
Holstein	=	–	=
Jersey	– –	–	+ + +
Other			
Texas Longhorn	=	=	=

*Adapted from Renand, 1988; Cundiff, 1992; Marshall, 1994; Wheeler *et al.*, 1996; Franke, 1997; Cundiff *et al.*, 1997.
†Minus signs, equal signs and plus signs, respectively, indicate relatively lower levels, moderate levels and higher levels.

breeds (e.g. Belgian Blue, Piedmontese, Chianina) tend to have very high lean-to-fat ratios and relatively little marbling. Japanese Black, Japanese Brown and Wagyu have relatively high genetic potential for marbling, but a relatively low growth rate.

Technological quality

Homer et al. (1997) reported no significant sire breed differences among Limousin, Charolais, Belgian Blue, Piedmontese, Angus and Hereford in pH at 3 or 24 h. In that same study, meat (uncooked) from Angus crosses tended to be darker (colour reflectance) than meat from other breeds. Drip loss from uncooked sirloin steaks tended be slightly less for British crosses, but not significantly so.

Renand (1985) found no sire breed differences between carcasses of young bull progeny of Charolais, Blonde d'Aquitaine, Limousin, Coopelso 93 synthetic (Blonde d'Aquitaine × Charolais × Limousin) and Inra 95 double-muscled synthetic (Blonde d'Aquitaine × Charolais) in ultimate pH or water loss of longissimus dorsi muscle. Liboriussen et al. (1977) reported that colour (reflectance) of uncooked muscle was lighter for carcasses of crossbred bull calves from Limousin, Romagnola, Charolais and Blonde d'Aquitaine as compared with Simmental, Hereford, Danish Red and White and Chianina. More O'Ferrall et al. (1989) reported that drip loss from longissimus dorsi muscle was greater for carcasses of Charolais-sired progeny as compared with progeny of Friesian, Hereford or Simmental sires (dams were Friesian). Muscle pH was greater for Simmental than for Hereford at 48 h post-mortem, although there were no significant differences in pH at 3 h.

Shear force and sensory quality

Sire breed comparisons from the US Meat Animal Research Center germplasm evaluation (GPE) project for shear force and sensory quality traits are presented in Tables 21.9, 21.10 and 21.11. Different sire breeds have been compared over time in the GPE project, with Angus and Hereford sires used in each 'cycle' as a control. In cycles I, II and III, tenderness differences were quite small among *B. taurus* breeds (Table 21.9). Meat tenderness was slightly reduced for Brahman-sired cattle and significantly reduced for Sahiwal compared with most *B. taurus* breeds. In cycle IV, steaks from calves of Pinzgauer and Piedmontese sires were slightly more tender, whereas steaks from Nellore-sired calves were significantly less tender than average at age (Table 21.10), weight and fat end-points. In cycle V, steaks from the progeny of Belgian Blue, Piedmontese, Angus, Hereford and Tuli were more tender than steaks from progeny of Boran or Brahman (Table 21.11). In all cycles of the GPE project, breed differences for sensory juiciness and flavour were of little practical importance.

Table 21.9. Sire breed comparisons for shear force and sensory quality from cycles I, II and III of the MARC GPE study*.

Category/breed	Shear force (kg)	Tenderness[†]	Flavour[†]	Juiciness[†]
Hereford, Angus	3.3	7.3	7.3	7.3
Red Poll	3.4	7.3	7.4	7.1
South Devon	3.1	7.4	7.3	7.4
Charolais	3.3	7.3	7.4	7.3
Chianina	3.6	6.9	7.3	7.2
Limousin	3.5	6.9	7.4	7.3
Maine Anjou	3.4	7.1	7.3	7.2
Gelbvieh	3.5	6.9	7.4	7.2
Pinzgauer	3.4	7.1	7.4	7.2
Simmental	3.5	6.8	7.3	7.3
Tarentaise	3.7	6.7	7.3	7.0
Brahman	3.8	6.5	7.2	6.9
Sahiwal	4.1	5.8	7.1	7.0
Brown Swiss	3.5	7.2	7.4	7.2
Jersey	3.1	7.4	7.5	7.5

*Age-constant (457, 473 and 445 days for cycles I, II and III, respectively) values. Dam breeds were Angus and Hereford. Sources: Koch *et al.*, 1976, 1979, 1982a; Cundiff, 1992.
[†]Sensory panel scores: 1 = extremely undesirable to 9 = extremely desirable.
MARC, Meat Animal Research Center.

Table 21.10. Sire breed comparisons for shear force and sensory quality from cycle IV of the MARC GPE study*.

Category/breed	Shear force (kg)	Tenderness[†]	Flavour intensity[†]	Juiciness[†]
Hereford, Angus	5.7	4.7	4.8	5.1
Galloway	5.8	4.8	4.8	5.1
Shorthorn	5.9	4.7	4.8	5.1
Charolais	5.9	4.6	4.8	5.0
Piedmontese	5.4	5.0	4.7	5.1
Salers	6.3	4.5	4.8	5.0
Gelbvieh	5.6	4.7	4.7	5.0
Pinzgauer	5.1	5.1	4.8	5.1
Nellore	7.2	4.0	4.7	4.8
Texas Longhorn	6.1	4.8	4.8	5.1

*Age-constant (426 days) values. Dams were Angus and Hereford. Source: Wheeler *et al.*, 1996.
[†]Sensory panel scores: 1 = extremely tough, bland, or dry to 8 = extremely tender, intense or juicy.
MARC, Meat Animal Research Center.

Liboriussen *et al.* (1977) reported that sire breed means were similar among Simmental, Charolais, Danish Red and White, Romagnola, Chianina, Hereford, Blonde d'Aquitaine and Limousin for shear force, sensory juiciness and cooked colour. Sensory tenderness and flavour of the semitendinosus muscle were slightly more desirable for Blonde d'Aquitaine and Limousin than for Hereford progeny (other breeds were intermediate), although breed differences were non-significant for the longissimus dorsi muscle. Homer *et al.* (1997) reported no significant sire breed differences among Limousin,

Table 21.11. Breed characterization for shear force and sensory quality from cycle V of the MARC GPE study*.

Category/breed	Shear force (kg)	Tenderness†	Flavour†	Juiciness†
Hereford, Angus	5.4	5.3	4.9	5.3
Belgian Blue	5.9	4.9	4.9	5.0
Piedmontese	5.4	5.0	4.8	5.0
Brahman	7.3	4.0	4.8	4.8
Boran	6.6	4.5	4.8	5.0
Tuli	5.7	5.0	4.9	5.2

*Age-constant (447 days) values. Dams were Angus, Hereford and MARC III composite. Source: Cundiff et al., 1997.
†Sensory panel scores: 1 = extremely tough, bland, or dry to 8 = extremely tender, intense or juicy.
MARC, Meat Animal Research Center.

Charolais, Belgian Blue, Piedmontese, Angus and Hereford for tenderness, juiciness or flavour of sirloin steaks, although roasting joints from Belgian Blue crosses were more tender than those from other sire breed groups. More O'Ferrall et al. (1989) reported that shear force was lower for Simmental than for several other sire breeds, although sensory tenderness did not vary significantly across breeds. In that same study, sensory juiciness, flavour desirability and flavour intensity were slightly better for Hereford than for Charolais or Simmental, while overall acceptability was similar across sire breeds.

Skelley et al. (1980) reported no difference in subjective lean colour or firmness of steer carcasses from progeny of Holstein, Polled Hereford, Charolais or Simmental sires (cows were Angus), although carcass fat tended to be softer and more yellow for Holstein progeny. Although variation between breeds was not significant for shear force, Hereford crosses had slightly more desirable sensory ratings than Charolais crosses for tenderness, juiciness, flavour intensity and flavour desirability. Knapp et al. (1989) found that Holsteins were similar to English and Continental European beef breeds in most palatability traits of cooked beef and had fewer tough steaks than cattle of *B. indicus* breeding. May et al. (1993) reported no significant differences between steaks from Wagyu-cross and Angus steers in shear force or sensory palatability traits.

Shackelford et al. (1994a) found significant differences among 28 sire lines in subjective colour, texture and firmness of carcass lean, but concluded that genetic variation was small relative to environmental variation in the incidence of the unacceptable dark, firm and dry (DFD) condition. There was some evidence of genetic trend toward darker meat in some breeds. *B. indicus* breeds were similar to most *B. taurus* breeds in subjective colour score of uncooked lean.

Cattle of high percentage *B. indicus* breeding tend to have lower marbling scores at a given age and produce less tender steaks than *B. taurus* breeds (Table 21.8; Peacock et al., 1982; Crouse et al., 1989; DeRouen et al., 1992). Tenderness is often lower for *B. indicus* than for *B. taurus* at the same level of marbling (Koch *et al.*, 1988; Wheeler *et al.*, 1994), suggesting that factors other than marbling account for a significant portion of the tenderness difference. Increased activity of calpastatin in *B. indicus* may account for some of the reduced tenderness (Johnson *et al.*, 1990; Wheeler *et al.*, 1990; Whipple *et al.*, 1990; Shackelford *et al.*, 1991; O'Connor *et al.*, 1997).

Variation in tenderness is significant within as well as across breeds. In cycle IV of the GPE study, Wheeler *et al.* (1996) noted that the mean estimated purebred difference in shear force between the most and least tender breeds (i.e. Pinzgauer and Nellore) corresponded to 4.76 genetic standard deviations, whereas the total range within a breed is approximately 6 genetic standard deviations.

Genetic Effects on Attributes of Fat

A high level of cholesterol in the blood has been identified as a possible risk factor for coronary heart disease in humans. In recent years, health consultants in many countries have encouraged diets with decreased levels of cholesterol and fat and lower ratios of saturated-to-unsaturated fatty acids. Intake of cholesterol and most saturated fatty acids (stearic acid, for one, may be an exception) tends to raise blood cholesterol. Intake of unsaturated fatty acids in lieu of saturated fatty acids may decrease blood cholesterol. Compositional differences in fatty acids may also play a significant role in determining palatability and sensory characteristics of beef. For example, Melton *et al.* (1982) found that desirable flavour of ground beef was associated with increased oleic (C18 : 1) acid in the neutral lipids and with decreased concentrations of C18 : 0 and C18 : 3 in the neutral and polar lipids. Larick *et al.* (1989) suggested that the difference in flavour of longissimus muscle of American bison might be related to a greater proportion of polyunsaturated fatty acids as compared with Hereford or Brahman.

Cholesterol content

Subcutaneous fat clearly has a higher cholesterol content than does longissimus muscle on a wet-weight basis (Eichhorn *et al.*, 1986a, b; Wheeler *et al.*, 1987). Animals with increased subcutaneous fat would, in general, be expected to have more cholesterol in the total carcass. Interestingly, limited evidence suggests that intramuscular fat differences do not have a significant association with cholesterol content in longissimus muscle. Rhee *et al.* (1982) evaluated longissimus muscle steaks from carcasses of eight subjective marbling-score categories, ranging from 'practically devoid' to 'moderately abundant' and found no effect of marbling score on cholesterol content of cooked steaks. Among uncooked steaks, Rhee *et al.* (1982) reported that the only significant difference was that those with 'practically devoid' marbling had significantly less cholesterol than did those of all seven other marbling categories. Wheeler *et al.* (1987) reported that cholesterol content of longissimus muscle did not vary significantly as cattle increased in age, weight and intramuscular fat.

On a within-tissue basis, most studies have found non-significant or relatively small differences between breeds in cholesterol content. Eichhorn *et al.* (1986a) reported that breed type did not significantly affect cholesterol content of the longissimus or triceps brachii muscles in a study of meat from mature cows of 15 breeds and crosses. Wheeler *et al.* (1987) reported no difference between Chianina and British crosses in cholesterol content in the longissimus muscle or subcutaneous fat, even though Chianina had a lower content of ether-extractable fat in the uncooked longissimus muscle and lower serum cholesterol. Baker and Lunt (1990) reported that sire breed did not significantly affect cholesterol content in longissimus muscle tissue, even though total lipid content of muscle tissue was higher in Angus-sired calves than in calves sired by Charolais or Piedmontese. Koch *et al.* (1995) reported that *B. taurus* cattle had a higher content of cholesterol in subcutaneous fat but less in the longissimus muscle as compared with bison (American buffalo) or *B. taurus* × bison hybrids. Rule *et al.* (1997) reported higher cholesterol content of ground carcass (composite of muscle and adipose) tissue in moderate-growth as compared with high-growth steers, a difference which the authors suggested might be attributed to the increased overall fatness of the moderate-growth animals. There was no significant difference between genetic types in cholesterol content of defatted longissimus muscle of steers.

Fatty acid profile

Several studies have reported significant breed effects on fatty acid composition, although breed type has sometimes been confounded with other potential effects, such as age, weight and overall fatness. Evidence suggests that fat depots tend to become more unsaturated as an animal increases in age and/or fatness (Zembayashi and Nishimura, 1996; Malau-Aduli *et al.*, 1997), although these relationships do not hold up consistently across animal types (i.e. sexes and breeds) in various studies. There is evidence that fatty acid composition varies across tissue type or lipid source (Eichhorn *et al.*, 1985; Zembayashi and Nishimura, 1996), which is important because some tissues (e.g. subcutaneous fat) can be removed before eating whereas intramuscular fat is generally consumed.

Sumida *et al.* (1972) reported that breed type (Hereford, Angus and crosses among Hereford, Angus and Charolais steers) did not affect the ratio of total unsaturated to saturated fatty acids and that most breed-type comparisons for proportions of individual fatty acids were also non-significant. Gillis *et al.* (1973) reported that the proportions of several individual fatty acids varied significantly among breeds in both intramuscular and subcutaneous lipid, although the ratios of unsaturated to saturated fatty acids were quite similar across breed types. In a study of meat from mature cows of 15 breeds and crosses, Eichhorn *et al.* (1986a) reported that breed effects on total saturated, unsaturated and polyunsaturated fatty acid proportions were significant for the longissimus muscle and subcutaneous adipose tissue but not for the triceps brachii muscle or perinephric adipose tissue. In total lipid extracts of longissimus muscle, breed-type means ranged from 5.1 to 12.5% for total polyunsaturated fatty acids and from 40.9 to 48.3% for total saturated fatty acids. Mills *et al.* (1992) reported that muscle of Angus–Charolais–Simmental cross steers contained slightly more total saturated fat, including a higher content of stearic acid, than muscle from Holstein steers. Total saturated fat other than stearic acid was similar between muscle of Holsteins and crossbreds.

May *et al.* (1993) reported a higher ratio of monounsaturated to saturated fatty acids in subcutaneous and intramuscular adipose tissue for Wagyu-cross than for Angus steers, although it was unclear whether part of the difference might have been attributable to confounding effects of overall fatness. Zembayashi *et al.* (1995) found that Japanese Black (Wagyu) steers had a higher content of monounsaturated fatty acids in subcutaneous and intramuscular neutral lipids than Holstein, even when adjusting for difference in slaughter age and carcass fatness. Zembayashi and Nishimura (1996) found that dam-breed comparisons of Japanese Black versus F_1 Japanese Black × Holstein were usually non-significant for percentages of individual or group total fatty acids in subcutaneous and intramuscular lipids. The effect of individual sire (only one sire breed was used) was significant for total saturated, polyunsaturated and unsaturated fatty acids in subcutaneous neutral lipid (but not in intramuscular neutral lipid or phospholipid) and for some individual fatty acids in each lipid source, even after removing the effect of carcass fat percentage.

Average fat thickness and fatty acid content of subcutaneous adipose tissue were found to be less, whereas the degree of unsaturation of fatty acids was greater for Brahman than for Hereford both among mature cows (Huerta-Leidenz *et al.*, 1993) and in their male progeny (Huerta-Leidenz *et al.*, 1996). Covariate adjustment for body fatness did not remove the significant breed-type effects for most of the variables in the cow study. Perry *et al.* (1998) reported that subcutaneous fat from Brahman × Hereford steers had a higher proportion of unsaturated fatty acids and a lower melting-point than fat from Hereford, Simmental × Hereford, or Holstein × Hereford, independent of carcass fatness.

A series of recent studies was conducted in Australia to determine genetic effects on fatty acid profiles of beef. Heritability estimates of fatty acids were rather low in weaners, but quite promising in slaughter animals (0.23, 0.57 and 0.15 for total saturated, monounsaturated and polyunsaturated fatty acids, respectively) for the triacylglycerol fraction of adipose tissue (Malau-Aduli *et al.*, 1998). Among feedlot animals of several breed types, ranging from early- (Jersey × Hereford) to late- (Charolais × Simmental × Hereford) maturing, Jersey crosses had the highest concentration of monosaturates and the lowest level of polyunsaturates in intramuscular fat (Siebert *et al.*, 1996). Limousin cows had higher percentages of saturated fatty acids and lower percentages of mono- and polyunsaturated fatty acids than Jersey cows in fat extracted from shoulder muscle, although differences between these two breeds were non-significant among yearling heifers (Malau-Aduli et al., 1997). Comparisons among grass-fed animals were made between calves sired by Angus, Belgian Blue, Hereford, Jersey, Limousin, South Devon and Wagyu bulls mated to Hereford dams (Deland et al., 1998; Siebert et al., 1998). Concentration of monounsaturated fatty acids in longissimus muscle phospholipids and subcutaneous adipose tissue tended to be greatest for calves sired by Wagyu, Jersey and Belgian Blue, and least for calves sired by Angus, Hereford and Limousin. Proportion of total saturated fatty acids was lowest for Wagyu- and highest for Limousin-sired calves in subcutaneous adipose tissue, but did not differ significantly among sire breeds when measured in muscle phospholipids.

Fat colour

In some markets, white fat is preferred to yellow fat. The degree of yellowness is associated with concentration of β-carotene. Death et al. (1996) reported that the degree of yellowness was greater in the subcutaneous and intermuscular fats of Hereford × Jersey than Hereford × Friesian heifers. Kruk et al. (1998) found that β-carotene concentration and degree of yellowness in subcutaneous fat were higher as the proportion of Jersey relative to Limousin breeding increased. Their preliminary results indicated possible effects of a major gene, because individuals that had extreme values for β-carotene concentration and degree of yellowness of fat were observed among purebred Jersey, $\frac{3}{4}$ Jersey, and F_1 animals, but not among purebred or $\frac{3}{4}$ Limousin animals.

Prospects for Genetic Improvement in Meat Traits

Consumer demand for beef

The beef industry has been forced to give greater attention to consumer issues in response to negative publicity regarding health effects of meats and to increased competition from pork and poultry. General trends in consumer preferences are toward leaner meat products with predictable eating qualities and simple preparation requirements. Health consultants have encouraged diets with decreased levels of cholesterol and fat and lower ratios of saturated-to-unsaturated fatty acids. Consumers are also demanding improved consistency in sensory attributes of cooked beef, with tenderness probably being the most important of these, followed by flavour. The industry is responding by producing leaner animals and by trimming excess fat during processing, although consistency of tenderness remains less than desired.

General trends notwithstanding, consumer preferences remain variable for factors such as leanness, marbling and serving size. Consequently, alternative market categories with different target specifications continue to exist. An important challenge for the cattle industry is to maintain appropriate genetic variability to accommodate variable market demands, while providing improved product consistency within a given market category.

Consistency in eating quality

Excessive genetic variation has often been cited as an important contributor to a lack of consistency in sensory attributes of beef. However, considering that the market does have an array of demand categories to accommodate variable product specifications, it is more important to improve product consistency within a given market category than to reduce variation in general. The use of processing treatments (e.g. post-mortem storage, mechanical tenderization, electrical stimulation, addition of enzymes) has been suggested to improve consistency of palatability, but an accurate method of determining which carcasses need treatment is needed. Genetic improvement in palatability is limited by a lack of cost-effective methods for measurement. Improvement in the industry's ability to accurately classify animals and carcasses into eating quality categories is critically needed.

In some markets, marbling score (intramuscular fatness) has been used as an indicator of eating quality and as a determinant of carcass price, although such use has been controversial. Although intramuscular fat content is associated with consumer desirability of beef, the relationship is too weak to provide consistently accurate classification of carcasses into categories of eating quality (Smith *et al.*, 1987; Wheeler *et al.*, 1994). Higher levels of intramuscular fatness are desirably associated with shear force, calpastatin activity, meat colour, flavour intensity, juiciness and tenderness. Interestingly, estimates of the phenotypic correlations tend to be lower than the corresponding genetic correlations. Thus, it could be concluded that intramuscular fat content is more useful for genetic evaluation than for market classification, at least in populations in which its genetic correlation with lean yield is not significantly antagonistic. A more direct measure of palatability (e.g. shear force), or perhaps a combination of criteria, is needed to properly classify carcasses for market value and would be useful for genetic evaluation as well.

Shear-force evaluation of cooked meat is one of the most heritable traits associated with meat quality, it is closely genetically correlated to sensory attributes, and it has few, if any, genetic antagonisms with other traits. Thus far, routine measurement of shear force has been considered too costly to implement in commercial beef-packing facilities. A potential limitation of shear force, and probably other methods, is that measurement of one muscle might not accurately predict tenderness of other muscles (Shackelford *et al.*, 1995).

Multiple strategies, both genetic and non-genetic, might be needed to improve the consistency of eating quality. Methods to improve average palatability will also improve consistency, because fewer unacceptable products will be produced. Choice of breeds, seedstock selection, animal management and post-mortem treatment all represent potential methods of improving palatability of beef. An improved ability to accurately sort products into palatability categories is critical for enhancing consistency and will improve the effectiveness of genetic evaluation.

Genetic evaluation and improvement

Extensive genetic variation exists between and within cattle breeds for carcass composition and tenderness. Breed preferences tend to vary across regions and market categories because of differences in production systems and consumer demands. Heterosis effects tend to be relatively unimportant for most meat traits. However, crossbreeding has the potential to improve meat traits through complementary mating systems and blending of breeds. For example, Cundiff (1997) noted that certain crosses of Continental European × British inheritance are more likely to provide optimal levels of carcass weight, lean yield and marbling than most single breeds. Similarly, crosses between *B. indicus* and *B. taurus* can provide adaptability to tropical or subtropical climates, while maintaining a higher level of meat tenderness than purebred *B. indicus*. Terminal mating systems with specialized sire and dam lines are recommended when production constraints in the cow herd dictate the use of dam genetic types that are not optimal in terms of meat quality.

Cattle-industry programmes for genetic evaluation, many of which are sponsored by breed organizations, are increasingly expanding to include meat traits. The number of seedstock with estimated breeding values for body composition traits should increase rapidly with the adoption of ultrasound technology. Genetic improvement in these traits could occur relatively rapidly, considering that they are moderately to highly heritable, but this will depend on the intensity with which selection pressure is applied. Selection responses in sensory attributes of beef are expected to occur more slowly, because heritability is lower and evaluation methods are more difficult to implement.

Advances in molecular technology hold promise for genetic evaluation of meat traits, because marker-based selection has the potential to avoid some of the difficulties (e.g. trait measurement, long generation interval) associated with traditional evaluation methods. Results of genome studies in cattle are quite preliminary at present, but useful applications are anticipated in the future. This research should improve our basic knowledge of the underlying regulation of lean : fat deposition and muscle development. Marker-based genetic evaluation could be beneficial for improvement of sensory attributes, which are more costly to measure and less heritable than composition traits. Novel alleles with desirable effects could be introgressed into 'recipient' populations.

As the ability to genetically alter carcass merit in cattle populations increases, it is important to consider the effects of such changes on correlated responses in other traits. For most meat traits, relatively few significant genetic antagonisms are obvious when based on across-study averages, although estimates tend to vary considerably across studies. It is uncertain to what extent this across-study variation in genetic correlation estimates reflects sampling errors versus true population differences. In some populations, selection for leanness could have antagonistic effects on intramuscular fat and sensory perception of tenderness, flavour intensity and juiciness. Fortunately, the magnitude of these genetic relationships is not very strong within many breeds, indicating that simultaneous improvement in composition and sensory quality could be possible if selection pressure is exerted on each trait. Selection for carcass merit should be reasonably compatible with increased growth rate in most populations. Perhaps of greater concern, albeit based on limited evidence, is that selection for leanness could be antagonistic to reproductive efficiency. Increased use of terminal mating systems is suggested when it is desirable to maintain a different genetic potential for body composition in the cow herd from that in slaughter progeny.

Conclusions

Appropriate breeding system design, utilization of breed differences and seedstock selection are important if the beef industry is to improve product quality and consistency while maintaining production efficiency. The potential to genetically change lean : fat composition of beef is high, but the potential impact on other traits needs to be considered. Potential improvement of sensory attributes within a breed is currently limited by measurement cost and low heritability, and so the utilization of between breed differences is important. Development of a cost-effective strategy for accurate assessment of sensory attributes of beef would enhance genetic evaluation and improve the consistency with which product quality meets consumer expectations. The development of ultrasound and molecular technologies has the potential to significantly enhance genetic improvement of meat traits.

References

AAA (1997) *Spring 1997 American Angus Association Sire Summary.* American Angus Association, St Joseph, Missouri, USA.

Aass, L. (1996) Variation in carcass and meat quality traits and their relations to growth in dual purpose cattle. *Livestock Production Science* 46, 1–12.

Anderson, D.C., Kress, D.D., Burfening, P.J. and Blackwell, R.L. (1986) Heterosis among closed lines of Hereford cattle. III. Postweaning growth and carcass traits in steers. *Journal of Animal Science* 62, 950–957.

Anderson, K. (1997) *Interpreting and Using EPDs for Carcass Traits in Limousin Cattle.* Leaflet, North American Limousin Foundation, Englewood, Colorado, USA.

Arnold, J.W., Bertrand, J.K., Benyshek, L.L. and Ludwig, C. (1991) Estimates of genetic parameters for live animal ultrasound, actual carcass data, and growth traits in beef cattle. *Journal of Animal Science* 69, 985.

Arthur, P.F. (1995) Double muscling in cattle: a review. *Australian Journal of Agricultural Research* 46, 1493–1515.

ASA (1997) *National Simmental/Simbrah Spring Sire Summary.* American Simmental Association, Bozeman, Montana, USA.

Baker, J.F. and Lunt, D.K. (1990) Comparison of production characteristics from birth through slaughter of calves sired by Angus, Charolais or Piedmontese bulls. *Journal of Animal Science* 68, 1562–1568.

Barkhouse, K.L., Van Vleck, L.D., Cundiff, L.V., Koohmaraie, M., Lunstra, D.D. and Crouse, J.D. (1996) Prediction of breeding values for tenderness of market animals from measurements on bulls. *Journal of Animal Science* 74, 2612–2621.

Beever, J. E., George, P.D., Fernando, R.L., Stormont, C.J. and Lewin, H.A. (1990) Associations between genetic markers and growth and carcass traits in a paternal half-sib family of Angus cattle. *Journal of Animal Science* 68, 337–344.

Bennett, G.L. and Williams, C.B. (1994) Implications of genetic changes in body composition on beef production systems. *Journal of Animal Science* 72, 2756–2763.

Bertrand, J.K., Moser, D.W. and Herring, W.O. (1997) Selection for carcass traits. In: *Proceedings, 29th Beef Improvement Federation.* Dickinson, North Dakota, USA, pp. 93–98.

Casas, E., Keele, J.W., Shackelford, S.D., Koohmaraie, M., Sonstegard, T.S., Smith, T.P.L., Kappes, S.M. and Stone, R.T. (1998) Association of the muscle hypertrophy locus with carcass traits in beef cattle. *Journal of Animal Science* 76, 468–473.

Crouse, J.D., Cundiff, L.V., Koch, R.M., Koohmaraie, M. and Seideman, S.C. (1989) Comparisons of *Bos indicus* and *Bos taurus* inheritance for carcass beef characteristics and meat palatability. *Journal of Animal Science* 67, 2661.

Crump, R.E., Simm, G. and Thompson, R. (1998) Genetic parameters for live animal conformation measures of British pedigree beef cattle. In: *Proceedings, 6th World Congress on Genetics Applied to Livestock Production,* Vol. 23, University of New England, Armidale, Australia, pp. 181–184.

Cundiff, L.V. (1992) Genetic selection to improve the quality and composition of beef carcasses. In: *Proceedings 45th Annual Reciprocal Meat Conference.* National Livestock and Meat Board, Chicago, USA, p. 123.

Cundiff, L.V. (1997) How breeds common to Nebraska fit into various marketing alliances. In: *Proceedings, Integrated Resource Management: Managing for Profitability in the Beef Industry.* Nebraska Cattlemen and University of Nebraska, Lincoln, pp. 24–31.

Cundiff, L.V., Gregory, K.E., Wheeler, T.L., Shackelford, S.D., Koohmaraie, M., Freetly, H.C. and Lunstra, D.D. (1997) *Preliminary Results from Cycle V of the Cattle Germ Plasm Evaluation Program at the Roman L. Hruska U.S. Meat Animal Research Center.* Progress Report No. 16, Clay Center, Nebraska, USA.

Death, A.F., Knight, T.W., Purchas, R.W., Morris, S.T. and Burnham, D.L. (1996) Plasma carotenoid concentrations early in life can be used as a selection criterion for fat colour in heifers. *Proceedings, New Zealand Society of Animal Production* 56, 398–399.

Deland, M.P.B., Malau-Aduli, A.E.O., Siebert, B.D., Bottema, C.D.K. and Pitchford, W.S. (1998) Sex and breed differences in the fatty acid composition of muscle phospholipids in crossbred cattle. In: *Proceedings, 6th World Congress on Genetics Applied to Livestock Production,* Vol. 25. University of New England, Armidale, Australia, pp. 185–188.

de Rose, E.P., Wilton, J.E. and Schaeffer, L.R. (1988) Estimation of variance components for traits measured on station-tested beef bulls. *Journal of Animal Science* 66, 626–634.

DeRouen, S.M., Franke, D.E., Bidner, T.D. and Blouin, D.C. (1992) Direct and maternal genetic effects for carcass traits in beef cattle. *Journal of Animal Science* 70, 3677.

Dijkstra, J., Oldenbroek, J.K., Korver, S. and Van Der Werf, J.H.J. (1990) Breeding for veal and beef production in Dutch Red and White cattle. *Livestock Production Science* 25, 183–198.

Dinkel, C.A. and Busch, D.A. (1973) Genetic parameters among production, carcass composition, and carcass quality traits of beef cattle. *Journal of Animal Science* 36, 832–846.

Drewry, K.J., Becker, S.P., Martin, T.G. and Nelson, L.A. (1979) Crossing Angus and Milking Shorthorn cattle: steer carcass traits. *Journal of Animal Science* 48, 517.

Eichhorn, J.M., Bailey, C.M. and Blomquist, G.J. (1985) Fatty acid composition of muscle and adipose tissue from crossbred bulls and steers. *Journal of Animal Science* 61, 892–904.

Eichhorn, J.M., Coleman, L.J., Wakayama, E.J., Blomquist, G.J., Bailey, C.M. and Jenkins, T.G. (1986a) Effects of breed type and restricted versus *ad libitum* feeding on fatty acid composition and cholesterol content of muscle and adipose tissue from mature bovine females. *Journal of Animal Science* 63, 781–794.

Eichhorn, J.M., Wakayama, E.J., Blomquist, G.J. and Bailey, C.M. (1986b) Cholesterol content of muscle and adipose tissue from crossbred bulls and steers. *Meat Science* 16, 71–78.

El-Hakim, A., Eichinger, H. and Pirchner, F. (1986) Growth and carcass traits of bulls and veal calves of continental cattle breeds. 2. Carcass composition. *Animal Production* 43, 235–243.

Franke, D.E. (1997) Postweaning performance and carcass merit of F_1 steers sired by Brahman and alternative subtropically adapted breeds. *Journal of Animal Science* 75, 2604–2608.

Gillis, A.T., Eskin, N.A.M. and Cliplef, R.L. (1973) Fatty acid composition of bovine intramuscular and subcutaneous fat as related to breed and sex. *Journal of Food Science* 38(3), 408–411.

Graser, H.U., Reverter, A., Upton, W., Donoghue, K. and Wilson, D.E. (1998). Use of real-time ultrasonic measurements of fat thickness and percent intramuscular fat for the Angus breed in Australia. In: *Proceedings, 6th World Congress on Genetics Applied to Livestock Production*, Vol. 23, pp. 69–72.

Green, R.D., Cockett, N.E., Tatum, J.D., O'Connor, S.F., Hancock, D.L. and Smith, G.C. (1996a) Association of a Taq1 calpastatin polymorphism with postmortem measures of beef tenderness in *Bos taurus* and *Bos indicus–Bos taurus* steers and heifers. *Journal of Animal Science* 74(Suppl. 1), 111.

Green, R.D., Cockett, N.E., Tatum, J.D., Wulf, D.M., Hancock, D.L. and Smith, G.C. (1996b) Association of a Taq1 calpastatin polymorphism with postmortem measures of beef tenderness in Charolais- and Limousin-sired steers and heifers. *Journal of Animal Science* 74(Suppl. 1), 113.

Gregory, K.E., Crouse, J.D., Koch, R.M., Laster, D.B., Cundiff, L.V. and Smith, G.M. (1978) Heterosis and breed maternal and transmitted effects in beef cattle. IV. Carcass traits of steers. *Journal of Animal Science* 47, 1063.

Gregory, K.E., Cundiff, L.V., Koch, R.M., Dikeman, M.E. and Koohmaraie, M. (1994a) Breed effects and retained heterosis for growth, carcass, and meat traits in advanced generations of composite populations of beef cattle. *Journal of Animal Science* 72, 833–850.

Gregory, K.E., Cundiff, L.V., Koch, R.M., Dikeman, M.E. and Koohmaraie, M. (1994b) Breed effects, retained heterosis, and estimates of genetic and phenotypic

parameters for carcass and meat traits of beef cattle. *Journal of Animal Science* 72, 1174–1183.
Gregory, K.E., Cundiff, L.V. and Koch, R.M. (1995) Genetic and phenotypic (co)variances for growth and carcass traits of purebred and composite populations of beef cattle. *Journal of Animal Science* 73, 1920–1926.
Grobet, L., Martin, L.J.R., Poncelet, D., Pirottin, D., Brouwers, B., Riquet, J., Schoeberlein, A., Dunner, S., Menissier, F., Massabanda, J., Fries, R., Hanset, R. and Georges, M. (1997) A deletion in the bovine myostatin gene causes the double muscled phenotype in cattle. *Nature Genetics* 17, 71–74.
Gwartney, B.L., Calkins, C.R., Rasby, R.J., Stock, R.A., Vieselmeyer, B.A. and Gosey, J.A. (1996) Use of expected progeny differences for marbling in beef: II. Carcass and palatability traits. *Journal of Animal Science* 74, 1014–1022.
Hanset, R. (1981) Double muscling in cattle. In: Brackett, B.G., Seidel, G.E. and Seidel, S.M. (eds) *New Technologies in Animal Breeding*. Academic Press, New York, pp. 71–80.
Hetzel, J. and Davis, G. (1997) Mapping quantitative trait loci: a new paradigm. In: *Proceedings, Beef Cattle Genomics: Past, Present, and Future*. Texas A & M University, College Station, Texas, pp. 24–25.
Hirooka, H., Groen, A.F. and Matsumoto, M. (1996) Genetic parameters for growth and carcass traits in Japanese Brown cattle estimated from field records. *Journal of Animal Science* 74, 2112–2116.
Homer, D.B., Cuthbertson, A., Homer, D.L.M. and McMenamin, P. (1997). Eating quality of beef from different sire breeds. *Animal Science* 64, 403–408.
Huerta-Leidenz, N.O., Cross, H.R., Savell, J.W., Lunt, D.K., Baker, J.F., Pelton, L.S. and Smith, S.B. (1993) Comparison of the fatty acid composition of subcutaneous adipose tissue from mature Brahman and Hereford cows. *Journal of Animal Science* 71, 625–630.
Huerta-Leidenz, N.O., Cross, H.R., Savell, J.W., Lunt, D.K., Baker, J.F. and Smith, S.B. (1996) Fatty acid composition of subcutaneous adipose tissue from male calves at different stages of growth. *Journal of Animal Science* 74(6), 1256–1264.
Izquierdo, M.M., Wilson, D.E. and Rouse, G.H. (1997) Estimation of genetic parameters for fat composition traits measured in live beef animals. In: *Proceedings, 29th Beef Improvement Federation*. Beef Improvement Federation, Dickinson, North Dakota, USA, pp. 99–101.
Johnson, M.H., Calkins, C.R., Huffman, R.D., Johnson, D.D. and Hargrove, D.D. (1990) Differences in cathepsin B and L and calcium-dependent protease activities among breed type and their relationship to beef tenderness. *Journal of Animal Science* 68, 2371.
Johnson, M.Z., Schalles, R.R., Dikeman, M.E. and Golden, B.L. (1993) Genetic parameter estimates of ultrasound-measured longissimus muscle area and 12th rib fat thickness in Brangus cattle. *Journal of Animal Science* 71, 2623–2630.
Johnston, D.J., Benyshek, L.L. Bertrand, J.K., Johnson, M.H. and Weiss, G.M. (1992a) Estimates of genetic parameters for growth and carcass traits in Charolais cattle. *Canadian Journal of Animal Science* 72, 493–499.
Johnston, D.J., Thompson, J.M. and Hammond, K. (1992b) Additive and non additive differences in postweaning growth and carcass characteristics of Devon, Hereford, and reciprocal cross steers. *Journal of Animal Science* 70, 2688–2694.
Kambadur, R., Sharma, M., Smith, T.P.L. and Bass, J.J. (1997) Mutations in myostatin (gdf8) in double muscled Belgian Blue and Piedmontese cattle. *Genome Research* 7, 910–916.

Killefer, J. and Koohmaraie, M. (1994) Bovine skeletal muscle calpastatin: cloning, sequence analysis, and steady-state mRNA expression. *Journal of Animal Science* 72, 606–614.

Kim, J.H. and Marshall, D.M. (1998) Growth hormone genotypic effects on calf growth traits, carcass traits, and cow production traits. *Journal of Animal Science* 76 (Suppl. 1), 32.

Kim, J.J., Davis, S.K., Sanders, J.O., Turner, J.W., Miller, R.K., Savell, J.W., Smith, S.B. and Taylor, J.F. (1998) Estimation of genetic parameters for carcass and palatability traits in *Bos indicus/Bos taurus* cattle. In: *Proceedings, 6th World Congress on Genetics Applied to Livestock Production*, Vol. 25, University of New England, Armidale, Australia, pp. 173–176.

Knapp, R.H., Terry, C.A., Savell, J.W., Cross, H.R., Mies, W.L and Edwards, J.W. (1989) Characterization of cattle types to meet specific beef targets. *Journal of Animal Science* 67, 2294–2308.

Koch, R.M., Dikeman, M.E., Allen, D.M., May, M., Crouse, J.D. and Campion, D.R. (1976) Characterization of biological types of cattle. III. Carcass composition, quality and palatability. *Journal of Animal Science* 43, 48–62.

Koch, R.M., Dikeman, M.E., Lipsey, R.J., Allen, D.M. and Crouse, J.D. (1979) Characterization of biological types of cattle – cycle II: III. Carcass composition, quality and palatability. *Journal of Animal Science* 49, 448–460.

Koch, R.M., Dikeman, M.E. and Crouse, J.D. (1982a) Characterization of biological types of cattle (cycle III). III. Carcass composition, quality and palatability. *Journal of Animal Science* 54, 35–45.

Koch, R.M., Cundiff, L.V. and Gregory, K.E. (1982b) Heritabilities and genetic, environmental, and phenotypic correlations of carcass traits in a population of diverse biological types and their implications in selection programs. *Journal of Animal Science* 55, 1319–1329.

Koch, R.M., Crouse, J.D., Dikeman, M.E., Cundiff, L.V. and Gregory, K.E. (1988) Effects of marbling on sensory panel tenderness in *Bos taurus* and *Bos indicus* crosses. *Journal of Animal Science* 66(Suppl. 1), 305.

Koch, R.M., Jung, H.G., Crouse, J.D., Varel, V.H. and Cundiff, L.V. (1995) Growth, digestive capability, carcass, and meat characteristics of *Bison bison*, *Bos taurus*, and *Bos × Bison*. *Journal of Animal Science* 73, 1271–1281.

Koohmaraie, M. (1992) Role of the neutral proteinases in postmortem muscle protein degradation and meat tenderness. In: *Proceedings 45th Annual Reciprocal Meat Conference*, National Livestock and Meat Board, Chicago, USA, pp. 63–71.

Koots, K.R., Gibson, J.P., Smith, C. and Wilton, J.W. (1994a) Analyses of published genetic parameter estimates for beef production traits. 1. Heritability. *Animal Breeding Abstracts* 62, 309–338.

Koots, K.R., Gibson, J.P. and Wilton, J.W. (1994b) Analyses of published genetic parameter estimates for beef production traits. 2. Phenotypic and genetic correlations. *Animal Breeding Abstracts* 62, 825–853.

Kruk, Z.A., Malau-Aduli, A.E.O., Pitchford, W.S. and Bottema, C.D.K. (1998) Genetics of fat colour in cattle. In: *Proceedings, 6th World Congress on Genetics Applied to Livestock Production*, Vol. 23, pp. 121–124.

Larick, D.K., Turner, B.E., Koch, R.M. and Crouse, J.S. (1989) Influence of phospholipid content and fatty acid composition of individual phospholipids in muscle from Bison, Hereford and Brahman steers on flavour. *Journal of Food Science* 54, 521–526.

Lee, D.H., Seo, K.S., Park, Y.I., Park, C.J., Won, Y.S. and Cho, B.D. (1998) Estimates of heritabilities and expected progeny differences for carcass traits in Korean native cattle. In: *Proceedings, 6th World Congress on Genetics Applied to Livestock Production*, Vol. 23, University of New England, Armidale, Australia, pp. 209–212.

Liboriussen, T., Bech Anderson, B., Buchter, L., Kousgaard, K. and Moller, A.J. (1977) Crossbreeding experiment with beef and dual-purpose sire breeds on Danish dairy cows. IV. Physical, chemical and palatability characteristics of longissimus dorsi and semitendinosus muscles from crossbred young bulls. *Livestock Production Science* 4, 31–43.

Lonergan, S.M., Ernst, C.W., Bishop, M.D., Calkins, C.R. and Koohmaraie, M. (1995) Relationship of restriction fragment length polymorphisms (RFLP) at the bovine calpastatin locus to calpastatin activity and meat tenderness. *Journal of Animal Science* 73, 3608–3612.

MacNeil, M.D., Cundiff, L.V., Dinkel, C.A. and Koch, R.M. (1984) Genetic correlations among sex-limited traits in beef cattle. *Journal of Animal Science* 58, 1171.

Malau-Aduli, A.E.O., Siebert, B.D., Bottema, C.D.K. and Pitchford, W.S. (1997) A comparison of the fatty acid composition of triacylglycerols in adipose tissue from Limousin and Jersey cattle. *Australian Journal of Agricultural Research* 48(5), 715–722.

Malau-Aduli, A.E.O., Siebert, B.D., Bottema, C.D.K., Deland, M.P.B. and Pitchford, W.S. (1998) Heritabilities of triacylglycerol fatty acids from the adipose tissue of beef cattle at weaning and slaughter. In: *Proceedings, 6th World Congress on Genetics Applied to Livestock Production*, Vol. 25, University of New England, Armidale, Australia, pp. 181–184.

Marshall, D.M. (1994) Breed differences and genetic parameters for body composition traits in beef cattle. *Journal of Animal Science* 72, 2745–2755.

Marshall, T.T., Hargrove, D.D. and Olson, T.A. (1987) Heterosis and additive breed effects on feedlot and carcass traits from crossing Angus and Brown Swiss. *Journal of Animal Science* 64, 1332–1339.

May, S.G., Sturdivant, C.A., Lunt, D.K., Miller, R.K. and Smith, S.B. (1993) Comparison of sensory characteristics and fatty acid composition between Wagyu crossbred and Angus steers. *Meat Science* 35, 289–298.

Melton, S.L., Amiri, M., Davis, G.W. and Backus, W.R. (1982) Flavour and chemical characteristics of ground beef from grass-, forage-grain-, and grain-finished steers. *Journal of Animal Science* 55, 77–87.

Mills, E.W., Comerford, J.W., Hollender, R., Harpster, H.W., House, B. and Henning, W.R. (1992) Meat composition and palatability of Holstein and beef steers as influenced by forage type and protein source. *Journal of Animal Science* 70, 2446–2451.

More O'Ferrall, G.J., Joseph, R.L., Tarrant, P.V. and McGloughlin, P. (1989) Phenotypic and genetic parameters of carcass and meat-quality traits in cattle. *Livestock Production Science* 21, 35–47.

Moriya, K., Dohgo, T. and Sasaki, Y. (1996) Multiple-trait restricted maximum likelihood estimation of genetic and phenotypic correlations for carcass traits in the base and current populations of Japanese Black cattle. *Animal Science and Technology* 67, 53–57.

Mukai, F., Okanishi, T. and Yoshimura, T. (1993) Genetic relationships between body measurements of heifers and carcass traits of fattening cattle in Japanese Black. *Animal Science and Technology* 64, 865–872.

Mukai, F., Oyama, K. and Kohno, S. (1995) Genetic relationships between performance test traits and field carcass traits in Japanese Black cattle. *Livestock Production Science* 44, 199–205.

Nagamine, Y., Roehe, R., Awata, T. and Kalm, E. (1997) Genetic parameters for ultrasonic fat thickness of Japanese Shorthorn cattle. *Journal of Animal Breeding and Genetics* 114, 99–106.

O'Connor, S.F., Tatum, J.D., Wulf, D.M., Green, R.D. and Smith, G.C. (1997) Genetic effects on beef tenderness in *Bos indicus* composite and *Bos taurus* cattle. *Journal of Animal Science* 75, 1822–1830.

Oikawa, T. and Kyan, K. (1986) Examination of sire x environment interaction and estimate of genetic parameters for the carcass traits of Japanese Black steers. *Japanese Journal of Zootechnical Science* 57, 916–924.

Peacock, F.M., Koger, M., Palmer, A.Z., Carpenter, J.W. and Olson, T.A. (1982) Additive breed and heterosis effects for individual and maternal influences on feedlot gain and carcass traits of Angus, Brahman, Charolais, and Crossbred steers. *Journal of Animal Science* 55, 797.

Perry, D., Nicholls, P.J. and Thompson, J.M. (1998) The effect of sire breed on the melting point and fatty acid composition of subcutaneous fat in steers. *Journal of Animal Science* 76, 87–95.

Renand, G. (1985) Genetic parameters of French beef breeds used in crossbreeding for young bull production. II. Slaughter performance. *Génétique, Sélection, Evolution* 17, 265–281.

Renand, G. (1988) Genetic determinism of carcass and meat quality in cattle. In: *Proceedings, 3rd World Congress on Sheep and Beef Cattle Breeding, Paris.*, Vol. 1. Institute National de la Recherche Agronomique, Paris, France, pp. 381–394.

Renand, G., Berge, P., Picard, B., Robelin, J., Geay, Y., Krauss, D. and Menissier, F. (1994) Genetic parameters of beef production and meat quality traits of young Charolais bulls progeny of divergently selected sires. In: *Proceedings, 5th World Congress on Genetics Applied to Livestock Production*, Vol. 19, University of Guelph, Guelph, Canada, pp. 446–449.

Rhee, K.S., Dutson, T.R., Smith, G.C., Hostetler, R.L. and Reiser, R. (1982) Cholesterol content of raw and cooked beef longissimus muscles with different degrees of marbling. *Journal of Food Science* 47, 716–719.

Robinson, D.L., Hammond, K. and McDonald, C.A. (1993) Live animal measurement of carcass traits: estimation of genetic parameters for beef cattle. *Journal of Animal Science* 71, 1128–1135.

Robinson, J.A.B., Armstrong, S.L. and Kuehni, P.P. (1998) Across breed evaluation of beef cattle carcass traits from commercial carcass data and real-time ultrasound. In: *Proceedings, 6th World Congress on Genetics Applied to Livestock Production*, Vol. 23, University of New England, Armidale, Australia, pp. 89–92.

Rule, D.C., MacNeil, M.D. and Short, R.E. (1997) Influence of sire growth potential, time on feed, and growing-finishing strategy on cholesterol and fatty acids of the ground carcass and longissimus muscle of beef steers. *Journal of Animal Science* 75, 1525–1533.

Shackelford, S.D., Koohmaraie, M., Miller, M.F., Crouse, J.D. and Reagan, J.O. (1991) An evaluation of tenderness of the longissimus muscle of Angus by Hereford versus Brahman crossbred heifers. *Journal of Animal Science* 69, 171.

Shackelford, S.D., Koohmaraie, M., Wheeler, T.L., Cundiff, L.V. and Dikeman, M.E. (1994a) Effect of biological type of cattle on the incidence of the dark, firm, and dry condition in the longissimus muscle. *Journal of Animal Science* 72, 337–343.

Shackelford, S.D., Koohmaraie, M., Cundiff, L.V., Gregory, K.E., Rohrer, G.A. and Savell, J.W. (1994b) Heritabilities and phenotypic and genetic correlations for bovine postrigor calpastatin activity, intramuscular fat content, Warner-Bratzler shear force, retail product yield, and growth rate. *Journal of Animal Science* 72, 857–863.

Shackelford, S.D., Wheeler, T.L. and Koohmaraie, M. (1995) Relationship between shear force and trained sensory panel tenderness ratings of 10 major muscles from *Bos indicus* and *Bos taurus* cattle. *Journal of Animal Science* 73, 3333–3340.

Shepard, H.H., Green, R.D., Golden, B.L., Hamlin, K.E., Perkins, T.L. and Diles, J.B. (1996) Genetic parameter estimates of live animal ultrasonic measures of retail yield indicators in yearling breeding cattle. *Journal of Animal Science* 74, 761–768.

Siebert, B.D., Deland, M.P. and Pitchford, W.S. (1996) Breed differences in the fatty acid composition of subcutaneous and intramuscular lipid of early and late maturing, grain finished cattle. *Austrailian Journal of Agricultural Research* 47(6), 943–952.

Siebert, B.D., Malau-Aduli, A.E.O., Deland, M.P.B., Bottema, C.D.K. and Pitchford, W.S. (1998) Genetic variation between crossbred weaner calves in triacylglycerol fatty acid composition. In: *Proceedings, 6th World Congress on Genetics Applied to Livestock Production*, Vol. 25, University of New England, Armidale, Australia, pp. 177–180.

Skelley, G.C., Thompson, C.E., Cross, D.L. and Grimes, L.W. (1980) Carcass characteristics of Polled Hereford × Angus, Charolais × Angus, Simmental × Angus, and Holstein × Angus steers finished on high silage diets. *Journal of Animal Science* 51, 822–829.

Smith, G.C., Savell, J.W., Cross, H.R., Carpenter, Z.L., Murphey, C.E., Davis, G.W., Abraham, H.C., Parrish, F.C. and Berry, B.W. (1987) Relationship of USDA quality grades to palatability of cooked beef. *Journal of Food Quality* 10, 269–286.

Smith, T.P.L., Lopez-Corrales, N.L., Kappes, S.M. and Sonstegard, T.S. (1997) Myostatin maps to the interval containing the bovine *mh* locus. *Mammalian Genome* 8, 742–744.

Splan, R.K., Cundiff, L.V. and VanVleck, L.D. (1998) Genetic correlations between male carcass and female growth and reproductive traits in beef cattle. In: *Proceedings, 6th World Congress on Genetics Applied to Livestock Production*, Vol. 23, University of New England, Armidale, Australia, pp. 274–277.

Stone, R. (1997) Mapping quantitative trait loci: a new paradigm. In: *Proceedings, Beef Cattle Genomics: Past, Present, and Future*. Texas A & M University, College Station, Texas, pp. 22–23.

Sugimoto, Y. (1997) Mapping quantitative trait loci: a new paradigm. In: *Proceedings, Beef Cattle Genomics: Past, Present, and Future*. Texas A & M University, College Station, Texas, pp. 26–27.

Sumida, D.M., Vogt, D.W., Cobb, E.H., Iwanaga, I.I. and Reimer, D. (1972) Effect of breed type and feeding regime on fatty acid composition of certain bovine tissues. *Journal of Animal Science* 35, 1058–1063.

Taylor, J.F. and Davis, S.K. (1997) Mapping quantitative trait loci: a new paradigm. In: *Proceedings, Beef Cattle Genomics: Past, Present, and Future*. Texas A & M University, College Station, Texas, pp. 20–21.

Turner, J.W., Pelton, L.S. and Cross, H.R. (1990) Using live animal ultrasound measures of ribeye area and fat thickness in yearling Hereford bulls. *Journal of Animal Science* 68, 3502–3506.

Van Veldhuizen, A.E., Bekman, H., Oldenbroek, J.K., Van der Werf, J.H.J., Koorn, D.S. and Muller, J.S. (1991) Genetic parameters for beef and milk production in Dutch

Red and White dual-purpose cattle and their implications for a breeding program. *Livestock Production Science* 29, 17–30.

Van Vleck, L.D., Hakim, A.F., Cundiff, L.V., Koch, R.M., Crouse, J.D. and Boldman, K.G. (1992) Estimated breeding values for meat characteristics of crossbred cattle with an animal model. *Journal of Animal Science* 70, 363–371.

Wheeler, T.L., Davis, G.W., Stoecker, B.J. and Harmon, C.J. (1987) Cholesterol concentration of longissimus muscle, subcutaneous fat and serum of two beef cattle breed types. *Journal of Animal Science* 65, 1531–1537.

Wheeler, T.L., Savell, J.W., Cross, H.R., Lunt, D.K. and Smith, S.B. (1990) Mechanisms associated with the variation in tenderness of meat from Brahman and Hereford cattle. *Journal of Animal Science* 68, 4206.

Wheeler, T.L., Cundiff, L.V. and Koch, R.M. (1994) Effect of marbling degree on beef palatability in *Bos taurus* and *Bos indicus* cattle. *Journal of Animal Science* 72, 3145–3151.

Wheeler, T.L., Cundiff, L.V., Koch, R.M. and Crouse, J.D. (1996) Characterization of biological types of cattle (Cycle IV): carcass traits and longissimus palatability. *Journal of Animal Science* 74, 1023–1035.

Wheeler, T.L., Cundiff, L.V., Koch, R.M., Dikeman, M.E. and Crouse, J.D. (1997) Characterization of different biological types of steers (Cycle IV): wholesale, subprimal, and retail product yields. *Journal of Animal Science* 75, 2389–2403.

Whipple, G., Koohmaraie, M., Dikeman, M.E., Crouse, J.D., Hunt, M.C. and Klemm, R.D. (1990) Evaluation of attributes that affect longissimus muscle tenderness in *Bos taurus* and *Bos indicus* cattle. *Journal of Animal Science* 68, 2716.

Wilson, D.E., Graser, H.-U., Rouse, G.H. and Amin, V. (1998) Prediction of carcass traits using live animal ultrasound. In: *Proceedings, 6th World Congress on Genetics Applied to Livestock Production*, Vol. 23, pp. 61–68.

Wilson, L.L., McCurley, J.R., Ziegler, J.H. and Watkins, J.L. (1976) Genetic parameters of live and carcass characters from progeny of Polled Hereford sires and Angus–Holstein cows *Journal of Animal Science* 43, 569–576.

Winer, L.K., David, P.J., Bailey, C.M., Read, M., Ringkob, T.P. and Stevenson, M. (1981) Palatability characteristics of the longissimus muscle of young bulls representing divergent beef breeds and crosses. *Journal of Animal Science* 53, 387–394.

Wulf, D.M., Tatum, J.D., Green, R.D., Morgan, J.B., Golden, B.L. and Smith, G.C. (1996) Genetic influences on beef longissimus palatability in Charolais- and Limousin-sired steers and heifers. *Journal of Animal Science* 74, 2394–2405.

Zembayashi, M. and Nishimura, K. (1996) Genetic and nutritional effects on the fatty acid composition of subcutaneous and intramuscular lipids of steers. *Meat Science* 43(2), 83–92.

Zembayashi, M., Nishimura, K., Lunt, D.K. and Smith, S.B. (1995) Effect of breed type and sex on the fatty acid composition of subcutaneous and intramuscular lipids of finishing steers and heifers. *Journal of Animal Science* 73, 3325–3332.

Genetic Aspects of Cattle Adaptation in the Tropics

22

S. Newman and S.G. Coffey

Cooperative Research Centre for the Cattle and Beef Industry (Meat Quality), CSIRO Tropical Agriculture, Box 5545, Rockhampton Mail Centre, Queensland 4702, Australia

Introduction	637
What is Adaptation?	638
What Constitutes an Adaptive Genotype for the Tropics?	639
Adaptive Traits in Breeding Programmes	640
Breed utilization	641
Genetic parameters	643
Breeding objectives	647
Economic values in breeding programmes	649
The Role of Molecular Genetics	650
Practical Implications	651
References	653

Introduction

Beef cattle are produced for myriad markets, production situations and environments. For these reasons, no one of the estimated 750 or more breeds of cattle (Mason, 1989) will satisfy all production and marketing situations. Highly productive breeds have been introduced into environments where management inputs, such as improved pastures, supplementary feeding, dipping and drenching, housing and other veterinary services are used to achieve animal potential. However, breeds not suitably able to cope with heat, parasite, disease or nutrition stress often do not exhibit the expected production levels.

In tropical environments, high production is negatively associated with heat tolerance, survival and tolerance to parasites (Frisch and Vercoe, 1977, 1982). In many instances, coping mechanisms have been through application of chemicals to the livestock or attempted eradication of the pest from the environment. Increasingly, molecular genetics will allow the location and exploitation of genetic markers for innate resistance to be introduced into genetic backgrounds lacking resistance. Also, exotic adapted breeds and composite cattle have been introduced in an attempt to increase production but maintain adaptation. These technologies provide a 'clean' approach to increasing productivity in stressful environments.

Limitations to genetic improvement of cattle in the tropics include variable and most often small herd sizes (often related to land-tenure constraints), poor animal identification and record keeping, lack of infrastructure to enable controlled matings (for example, single-sire matings, use of artificial insemination or multiple ovulation and embryo transfer (MOET) schemes), inadequate reproduction levels, limited options to reduce or to effectively measure environmental effects, the inability of exotic-breed males to survive and successfully reproduce, and lack of understanding and thereby motivation among people involved (Davis and Arthur, 1994; Payne and Hodges 1997). This chapter investigates the process of adaptation as applied to breeding programme design in a tropical environment. The importance of adaptive traits in breeding programmes is reviewed and methods to accommodate adaptation in structured breeding programmes to optimize profitability rather than maximizing performance are demonstrated.

What is Adaptation?

As with many concepts in biology, there are numerous definitions of adaptation. Some definitions revolve around adaptation being a process or a trait. Roughgarden (1979), for example, defines an adaptation as a trait that permits an organism to function well in its environment, thus endowing an organism with capabilities especially appropriate in its particular environment. Conversely, Futuyma (1979) sees adaptation to be a process whereby members of a population become suited over generations to survive and reproduce. In general, we can define two types of adaptations (Black, 1983): evolutionary adaptations, which include changes in morphology, biochemistry and behaviour and are transmitted from parent to progeny, and physiological adaptations, which refer to the capacity of an individual animal to adjust, within its lifetime, to changes in the environment. In essence, adaptation defines the fitness (the contribution of offspring to the next generation) of an individual to its environment.

Adaptation is a primary process in the evolution and selection of domesticated species of livestock. Systems theory can be used to define a model for adaptation at a biological level. Systems scientists use the concept of adaptation to study changes in dynamic systems by the inclusion of new variables or removal of old, by changing initial states or by changing surrounding states of the system. Simply put, a form of behaviour is adaptive if it maintains biological processes within physiological limits. Some external disturbance tends to drive a physiological variable outside its normal limits, but the change itself activates a mechanism that opposes the external disturbance. By this mechanism, the physiological variable is maintained within limits much narrower than would occur if the external disturbance were unopposed. In systems terms, this would correspond to the concept of dynamic self-regulation (Ashby, 1960). This is akin to the concept of genetic homoeostasis, or the property of a population to equilibrate its genetic composition and to resist sudden changes (Lerner, 1954). Under natural selection, intermediate optima exist for many characteristics and high levels of heterozygosity often impart a buffering effect to a wide range of environments.

Homoeostatic mechanisms involved in adaptation can be observed through the reaction of the bovine to heat stress. Many breeds of cattle are adapted to hot environments. Their adaptation has been achieved through their ability to shed heat load and adjust themselves to prevailing high ambient temperatures. This is done by producing a lower metabolic heat as a result of their lower metabolic rate, lower feed intake per unit of metabolic body size, lower growth rate and reduced milk production (Oyenuga and Nestel, 1984). Thus, tropically adapted animals are better able to survive the nutritional and climatic stress associated with the harsh environment and have less metabolic heat to dissipate as ambient temperatures rise.

What Constitutes an Adaptive Genotype for the Tropics?

The best-adapted animals in a population are those which exhibit a congruent combination of all characters leading to maximum fitness. Natural selection tends to favour individuals exhibiting intermediate optima for most traits, and these optima may change with changes in gene frequency.

Previous studies of the evolution of adaptation in tropical environments have couched the adaptive genotype in general terms of disease resistance or tolerance, heat stress (homeothermy), nutrition stress and drought tolerance (Black, 1983; Baker and Rege, 1994). We have attempted to filter these categories based on the combined effects of genetics and the environment (Table 22.1).

Environmentally independent traits are those on which the environment has little, if any, influence on expression. Many of these traits are expressed as distinguishing features of individual breeds (e.g. coat colour). Other traits are those that are expressed exclusively due to being in a tropical environment. These include resistance or tolerance to endo- and ectoparasites and heat tolerance. Characteristics for cattle adapted to tropical heat include lower fasting metabolism, more and larger sweat glands, larger skin surface (e.g. dewlap), small amounts of subcutaneous fat and a smooth, short coat (Oyenuga and Nestel, 1984).

The traits appearing in the middle column of Table 22.1 ('Variable') reflect complexes that are most closely associated with the fitness of an animal to its environment. These traits are not considered discrete, in that they are comprised of several components. These might include reaction to heat stress,

Table 22.1. Desirable attributes of the tropically adapted genotype.

Environmentally independent	Variable	Environmentally dependent
Polled	Calving ease	Endoparasite tolerance
Coat colour	Calf viability	Trypanosomiasis
Hair coat	Calf survival	Gastrointestinal nematodes
Pigmentation	Growth rate to mature size	Ectoparasite tolerance
Sheath score	Product quality	Ticks (*Boophilus microplus*)
Predator avoidance	Longevity	Buffalo fly
Deposition of body fat	Maternal ability	Heat tolerance (respiratory rate, rectal temperature)
More and larger sweat glands		Homoeothermy
Skin surface		Water balance
		Caloric balance
		Resistance to bovine infectious keratitis (BIK)
		Ranging ability
		Fasting metabolism

disease resistance, endo- and ectoparasite resistance and the ability to survive and reproduce on low-quality forages.

Adaptive Traits in Breeding Programmes

Two broad options exist for the genetic improvement of tropically adapted livestock (Franklin, 1986). First, the breeding programme may simply concentrate on maximizing production and allow natural selection to maintain adaptation. The second approach is to simultaneously select for adaptation and production through some multiple-trait selection procedure. The former method will be successful if the environment in which the animals are selected is the same environment in which the animals will be used. This is often a difficult goal to achieve (Baker and Rege, 1994). Alternatively, the multiple-trait approach allows improvement of both adaptation and production but requires more information including indicators of the economics of production, genetic parameters and a genetic evaluation scheme to choose selection candidates.

We assume multiple-trait selection to be the optimum selection scheme, because it simultaneously accounts for both genetic and economic information (Harris and Newman, 1994). From this perspective, two choices are available for genetic improvement of beef cattle: between-breed or within-breed selection. The latter provides for sustained and permanent genetic improvement, while the former provides a way of exploiting breed differences to make very rapid genetic change. For many traits, breed differences are very large. By taking advantage of these differences, between-breed selection can produce genetic change at a greater rate than the gradual change from within-breed selection. However, genetic change under crossbreeding is not permanent and crosses must be repeated. An exception to this is the development of composite populations that exploit heterosis in the foundation breeds but utilize within-population selection once the composite has been established.

Breed utilization

Tremendous productivity gains have been achieved in less developed tropical countries through the introduction of more productive breeds from Europe and North America, rather than from selection within native adapted breeds. However, the use of unadapted, exotic breeds has either required increased management inputs or acceptance of reduced levels of production (Davis and Arthur, 1994). In spite of this, the greatest opportunities for improving production levels of commercial animals in the tropics will be those that exploit the large amount of heterosis in subspecies crosses in cattle (Davis and Arthur, 1994; Frisch and O'Neill, 1998a, b). Generation-to-generation differences in performance due to variation in adaptation makes some crossbreeding systems (e.g. rotational systems) prone to failure under fluctuating tropical environments (Plasse, 1988). Furthermore, genotype × environment interaction will play a role, especially when either new genotypes are introduced into tropical regions from completely different environments, or genotypes tested in experiment stations are used in the field (Payne and Hodges, 1997). As noted by Barlow (1981), genotype × environment interaction is greatest among fitness traits and in stressful environments.

Frisch and O'Neill (1998b) evaluated cattle breeds of African, Indian and European origins for their suitability for beef production in northern Australia. The zebu breeds (*Bos indicus*) used were the Brahman and the Boran, the Sanga breed (African *B. taurus*) was the Tuli, and the European breeds (*B.

Table 22.2. Mean differences in tick counts (per animal day^{-1}) and in worm-egg counts (egg g^{-1}) from crosses of African, European and Indian origin as a deviation from Brahman (from Frisch and O'Neill, 1998b).

Dam genotype	Sire genotype						
	HS	AX	Ch	Tu	BX	Bo	B
Tick counts							
HS	28			14		0	3
AX		28		24		8	3
BX				12	14	3	1
B	9	4	20	12	5	3	0
Worm-egg counts							
HS	234			301		160	24
AX		207		258		191	81
BX				200	166	96	58
B	34	121	45	81	19	39	0

HS, Belmont Adaptaur; AX, Belmont Red; Ch, Charolais; Tu, Tuli; BX, Brahman × HS; Bo, Boran; B, Brahman.

taurus) were the Charolais and the Belmont Adaptaur (HS), a composite breed nominally ½ Hereford and ½ Shorthorn. The Belmont Red (AX), a *B. taurus* composite nominally ½ Africander, ¼ Hereford and ¼ Shorthorn, and the Belmont BX, a composite nominally ½ Brahman, ¼ Hereford and ¼ Shorthorn, were also used. Table 22.2 presents differences in tick and worm-egg counts for each genotype from the study as a deviation from Brahman. No straightbred or crossbred progeny exhibited lower tick counts or worm-egg counts than the Brahman. Indeed, most Brahman-sired crosses exhibited reduced tick and egg counts compared with other crosses. Crosses with Tuli and Boran expressed lower tick counts than crosses involving unadapted European breeds but not superior to Brahman crosses. Tuli and Belmont Red were similar. Thus, crosses involving other adapted genotypes (e.g. Boran and Tuli) do widen the choices for producers who desire to improve productivity while at least maintaining adaptation.

Bortolussi *et al.* (1999) compiled a database of environmental stress effects on beef productivity. These data were used to develop prediction equations of stress effects on production in tropical beef cattle as part of an effort to develop decision-support aids for crossbreeding in the tropics (Newman *et al.*, 1997). This was done to gauge variability in response to environmental stress between multibreed genotypes. Fifty-three citations were found in the literature covering production losses due to ticks (22), ambient temperature (5), rectal temperature (7) and worms (19). For illustrative purposes, genotype composition of calves was classified as purebred or cross among *B. taurus* breeds (T; European origin), zebu breeds (Z; Indian or African origin) or adapted *B. taurus* or Sanga (A; Tuli, Belmont Red, Belmont Adaptaur). Table 22.3 presents least-squares means for breed type for levels of environmental stressors. Purebred *B. taurus* cattle had significantly ($P < 0.05$) greater numbers of ticks per side than all other breed types. There was a consistent pattern to the means of zebu and adapted purebreds and crosses between them having lower numbers of ticks per side. Crosses between *B. taurus* and zebu or adapted *B. taurus* had intermediate numbers of ticks and purebred *B. taurus* cattle had the greatest number. The results suggest an additive mode of gene action rather than a dominance model. Heterosis was not significant ($P > 0.05$).

Table 22.3. Means for various genotype groupings for three measures of environmental stress (from Bortolussi *et al.*, 1998).

Breed type	Ticks per side (n)	Rectal temp. (°C)	Intestinal worms (eggs g^{-1})
AA	20^a ± 5.1	38.84^a ± 0.27	279^a ± 76
AT	17^a ± 5.6		258^{ab} ± 101
TT	35^b ± 2.7	39.53^b ± 0.08	339^a ± 39
ZA	12^a ± 6.2		148^{ab} ± 70
ZT	18^a ± 3.2	39.24^a ± 0.12	170^a ± 41
ZZ	14^a ± 3.9	38.86^a ± 0.18	34^b ± 60

A, Adapted *B. taurus*; T, *B. taurus*; Z, zebu.
Means in the same column that do not have a common superscript differ ($P < 0.05$).

In the analysis of rectal temperature, there was indication of a similar adaptive pattern among breed types to that indicated by the tick data. Purebred *B. taurus* cattle had the highest temperatures ($P < 0.05$), with ZT crosses intermediate and marginally different ($P > 0.11$) from purebred zebu or adapted *B. taurus* purebreds, which did not differ ($P > 0.3$).

Purebred zebu (ZZ) genotypes expressed significantly fewer intestinal worms than adapted (AA) or unadapted (TT) genotypes ($P < 0.02$). All other comparisons among breed types were not significant. However, there was a trend for zebu crosses to have lower infestations than purebred *B. taurus*, whether supposedly adapted or not. Results indicate that worm resistance has not been acquired among the adapted breeds reported in these literature sources.

Davis and Arthur (1994) summarized the use of crossbreeding in the tropics. Relative superiority of F_1 *B. taurus* × *B. indicus* crosses compared with local adapted breeds was reported in Latin America, tropical Australia, South-East Asia and tropical Africa. Crosses among zebu breeds gave varying results. Furthermore, composite or *inter se*-mated *B. taurus* × *B. indicus* expressed lower relative performance than the F_1 (Plasse, 1988; Arthur *et al.*, 1994; Frisch and O'Neill, 1998a). However, use of composites in some environments provided superior performance compared with the local breed. Composite populations can produce large amounts of heterosis, utilize breed complementarity in the formation of the synthetic, possess consistency of performance, produce their own replacements and are simple to manage once the composite population is stabilized and established. Composite breeds can be 'customized' for specific environments through judicious selection of breeds. Many composite populations lack access to accurate estimated breeding values (EBV), because full genetic models (accounting for non-additive genetic effects) have not been developed for large-scale genetic evaluation. In spite of this, accurate EBV can be calculated once composite populations have stabilized.

These results support contentions that crossbreeding can be used to exploit breed differences in adaptation while improving overall productivity. However, due to genetic relationships between traits of economic importance, selection for adaptation cannot be done in isolation of growth, reproduction and product quantity and quality. For this reason, it is useful to study relationships between production traits and adaptive traits with respect to the development of structured breeding programmes for the tropics.

Genetic parameters

A primary consideration in the development of structured breeding programmes is the breeding objective and its predictor, the selection index. Genetic parameters like (co)variances, correlations and heritabilities are essential to their construction. Other uses of genetic parameters arise as a function of our ability to model larger and more complex systems and our attempts to integrate information across herds, breeds or enterprises. Examples include multibreed EBV and modelling of crossbreeding systems.

Extensive reviews of genetic parameter estimates covering both tropical and temperate production systems have been reported by Koots *et al.* (1994a, b). However, there remains a paucity of published information on genetic parameters estimated in tropical areas, the most recent being that of Davis (1993). Tables 22.4, 22.5 and 22.6 present estimates of parameters from that research, and recent work by Burrow (1998).

Table 22.4 presents heritability estimates for growth from northern Australia-produced progeny from *B. taurus* (Hereford × Shorthorn, HS), zebu-

Table 22.4. Estimates of heritability for growth traits in tropical and temperate environments (from Davis, 1993; Meyer, 1995; Burrow, 1999).

	Northern Australia								Temperate	
	Bos taurus		Zebu-derived		Sanga-derived		Brahman			
Trait	Mean	SE	Mean	SE	Mean	SE	Mean	SE	Mean	SE
Birth weight	23	5	53	9	61				37	1
Weaning (direct)	20	23	30	9	41		39	6	23	1
Weaning (maternal)	12	17	24	6	34		5	3	8	1
Yearling			24	6	47		39	6	31	1
550-day	32	41	25	6	46		39	6	32	2
700-day			32	8			45	8		
900-day			38	10			40	11		

Table 22.5. Estimates of heritability (%) of other traits of economic importance.

Trait	Heritability*	SE	Burrow (1999)
Preweaning gain	16	4	36
Postweaning gain	31	8	22
Dry-season gain	30	8	17
Wet-season gain	18	7	19
Calving success	14	12	5
Number of calves per lifetime of cow	36	14	
Calving rate	17	14	
Days to calving	9	4	7
Bull fertility (cows conceiving per cow exposed)	5	7	
Scrotal circumference (550 days)	31	12	41
Testosterone concentration (550 days)	32	16	
Tick resistance (count)	34	6	42
Buffalo fly resistance (count)	26	11	30
Internal parasite resistance (eggs g^{-1})	28	3	36
Heat tolerance (rectal temperature)	20	2	17
Flight speed			40

*Davis, 1993.

Table 22.6. Estimates of genetic correlations (%) between growth, reproduction and adaptation traits (from Burrow, 1999).

	Wwt	Ywt	Fwt	Pre	Pst	Tck	Wms	Fls	Tmp	Fsd	Scs	Dtc	Csc
Bwt	53	55	55	29	26	1	0	30	-11	-3	20	22	8
Wwt		90	86	96	28	0	-6	50	-23	0	32	-18	-14
Ywt			95	86	55	-8	-2	68	-26	1	36	-34	-30
Fwt				81	72	-13	-15	74	-40	0	37	-43	-35
Pre					20	-1	-8	47	-20	0	28	-27	-24
Pst						-22	-14	70	-37	1	30	-48	-41
Tck							30	5	22	-1	9	10	1
Wms								-2	-3	0	9	6	1
Fls									-17	0	39	-16	1
Tmp										-6	1	16	26
Fsd											11	15	-1
Scs												32	2
Dtc													93

Bwt, birth weight; Wwt, weaning weight; Ywt, yearling weight; Fwt, final weight (~18 months); Pre, preweaning gain; Pst, postweaning gain; Tck, ticks; Wms, worms; Fls, buffalo flies; Tmp, temperature; Fsd, flight speed; Scs, scrotal size; Dtc, days to calving; Csc, calving success.

derived (based on Brahman or Sahiwal), Sanga-derived (Africander × Hereford-Shorthorn, or AX; Burrow 1999) and Brahman sires. It also includes comparisons from the Angus breed produced in temperate Australia (Meyer, 1995). The zebu-derived values also include early data from Sanga-derived breeds. However, the values in the Sanga-derived column are more current and represent AX and AXBX (AX × Brahman–HS) animals. In general, heritabilities are moderate in magnitude and are higher compared with *B. taurus* breeds. Heritabilities for the direct component of weaning, yearling and 550-day weights are higher in *B. indicus* cattle than in temperate breeds. The higher estimates of Burrow (1999) are probably due to the use of more contemporary data-analytic techniques, using animal models with maternal genetic effects and repeated records, a larger number of sires and only two breed types (AX and AXBX). Davis (1993) noted that the maternal component of heritability for weaning weight differed across classes of cattle. This might indicate that variation in growth up to weaning of calves of Brahman and Brahman-derived breeds is determined to a larger extent by an individual's own breeding value than in Africander-derived breeds, where the maternal genetic component was larger.

Heritability of other traits of economic importance in beef cattle in the tropics is presented in Table 22.5. These are average estimates from published reports (where possible) and are compared with results by Burrow (1999). Heritabilites of traits associated with reproduction remain low in all studies, although scrotal circumference has an estimate as high as 41%. The way in which reproductive success is measured differs between temperate and tropical areas. Temperate areas tend to utilize restricted mating seasons, where most cows calve annually. Thus traits like days to calving will be more indicative of fertility than calving success. On the other hand, in northern Australia and other tropical areas where large herds are managed, a lower number of cows calve annually and mating and calving seasons are often long and can be year-round. Thus, calving success may be a more useful indicator of fertility (Davis, 1993).

Traits associated with adaptation are also presented in Table 22.5. In general the estimates for these traits are moderate, showing genetic variation and opportunities for selection. The estimates of Burrow (1999) are higher than the average values, because different analytical procedures were used for analysis of resistance traits.

Presented in Table 22.6 are genetic correlations derived from the work of Burrow (1999) relevant to this discussion. Genetic correlations between weights at different ages are moderate to high and in agreement with the results of Davis (1993) and Mackinnon *et al.* (1991a). However, the relationship between pre- and postweaning gain is small (20%) and may show that weight gain under a relatively benign environment with respect to stress (preweaning) is under separate genetic control from that postweaning, where environmental stress may be higher (Mackinnon *et al.*,1991a).

Genetic correlations between growth and tick counts were in general negative (favourable) and small. As the number of ticks decrease, weights and weight gains will increase. Correlations with worm counts were similar to those for ticks, in that increasing numbers of worms will be reflected in lower weight gains. Alternatively, relationships between buffalo fly counts and growth were positive and therefore unfavourable. Mackinnon *et al.* (1991a) and Davis (1993) found variable responses as well, and attributed this to the level of resistance of the breeds under study. Selection for growth in breeds with relatively high resistance might lead to either no change or increased parasite counts (reduced resistance), while selection for growth in a less resistant genotype would result in reduced parasite counts (Davis, 1993).

As temperature decreases, weight and weight gains increase (Table 22.4). Mackinnon *et al.* (1991a) found a small correlated response in growth from animals selected on repeated measures of rectal temperature. In this study, use of resistance to heat stress as an indirect selection criterion was not useful for improving growth. This would be true for ticks as well. Thus, selection for growth would be more successful by direct selection for it as opposed to indirect selection for stress resistance. Indirect selection might be of more value in breeds with low levels of resistance because genetic correlations between growth and resistance traits are higher in these breeds (Davis, 1993).

Scrotal circumference in general was positively related to all other traits, displaying moderate genetic correlations. Both days to calving and calving success were negatively correlated with weight and weight gain. Higher weights and gains were associated with fewer days to calving but fewer calves. Mackinnon *et al.* (1991b) reported largely unfavourable correlations between scrotal circumference and endo- and ectoparasite resistance, and a negative relationship between worm resistance and female conception rate (Mackinnon *et al.*, 1990). This may be a general result indicating negative correlations between reproductive performance in the presence of parasites and parasite resistance (Davis, 1993). At this stage, some evidence exists for a favourable genetic relationship between fertility traits attained in the presence of heat stress and the resistance to such stress.

Breeding objectives

Independent of environment, the development of structured beef cattle breeding programmes should proceed through a logical sequence of steps (Ponzoni and Newman, 1989).

1. Define the breeding objective.
2. Choose selection criteria.
3. Develop a pedigree and performance recording scheme.
4. Conduct the genetic evaluation.
5. Use selected individuals.

The breeding objective is simply a statement (model) describing the relationship between various biological traits and income and expenditure in the livestock industries. The objective is developed to maximize socioeconomic benefit in the most populous unit, be it smallholder, village or large-scale property (Newman and Davis, 1995). The breeding objective is often described as a profit equation $P = I - E$, where P is profit, I is income and E is expenditure. The profit equation is normally expressed as a function of biological traits influencing income and expenditure. These traits might be calves weaned, carcass weight and food intake. At this stage, interest is usually confined to the economics of the production systems and not how expensive or difficult it might be to measure economically important traits.

Once the breeding objective has been developed, the next step is to decide on the selection criteria that will be used to estimate the value of animals as parents of future generations and to select replacements. Selection criteria should be heritable, be measured relatively early in life, be inexpensive to measure and, most importantly, be genetically correlated with traits in the breeding objective. The characters used as selection criteria need not be the same as the traits in the breeding objective. Information required at this point in the development of the breeding programme includes phenotypic and genetic parameters for the traits in the breeding objective and for the characters chosen as selection criteria.

Formal breeding objectives for livestock in tropical areas of the world are scarce (Annor, 1996; Amer et al., 1998). Through the process of natural selection, animals tolerant of various diseases and other stresses have been developed. However, some of these adaptive characters (e.g. small body size and fat carcasses) tend to devalue native breeds for commercial production purposes (Moyo, 1990; Amer et al., 1998). Use of previously discussed technologies such as crossbreeding will improve production but sacrifice adaptation and increase maintenance requirements. Thus, it is of primary importance to expand the range of traits used as selection criteria in the development of breeding programmes. This has been achieved on a very small scale for many adaptability traits, due to lack of estimates of genetic correlations with other traits of economic importance. However, results from two studies have been published to provide clues to the importance of adaptation traits in breeding programmes.

Davis (1994) reported on the refinement of genetic improvement programmes for cattle in northern Australia. The environment can be described as dry, with seasonable but unreliable rainfall. Challenge from endo- and ectoparasites was low to moderate. Among other objectives of the study, the one germane to the present chapter was to evaluate new traits (tick resistance, heat tolerance and worm resistance) for incorporation into BREEDPLAN, the national beef cattle genetic evaluation scheme for Australia (Graser *et al.*, 1995). The proposed systems were either the production of premium export steers from central Queensland or production of manufacturing beef for the American market. The breeding objective contained sale live weight, dressing percentage, saleable meat percentage, fat depth (rump), cow weaning rate, bull fertility, cow survival rate and cow weight. The selection criteria (EBV) included 200-, 550-, 700- and 900-day weight, calving success, days to calving and scrotal circumference. The evaluation of these traits plus the resistance traits was carried out utilizing BREEDOBJECT, a breeding objectives and selection index package (Barwick *et al.*, 1997), and utilizing genetic parameters estimated from studies previously discussed (summarized by Davis, 1993). Evaluations were made for the case of a relatively resistant and a relatively susceptible zebu-derived breed. In this study, susceptibility was defined by the genetic correlation between sale weight and the resistance trait in the presence of the stress. For the resistant breed, this correlation was negligible (−0.2), but for the susceptible breed the correlation was moderate (−0.4). For *B. indicus* breeds these correlations are zero and there is negligible genetic variance. Thus, EBV for these traits are unlikely to be of use in breeding programmes using pure *B. indicus* cattle.

Their conclusions were as follows.

1. There was negligible merit in having or using EBV for resistance traits when a breed is relatively resistant because there was only a small gain in accuracy (correlation between objective and index) and negligible gain in sale weight.

2. It is unlikely that the marginal improvement in genetic gain from including EBV for resistance traits would be sufficient to offset the expense associated with measurement of the traits in zebu-derived breeds. Adaptation traits require repeated measurements for sufficient accuracy and require expertise in measurement.

3. For susceptible breeds there is a gain from including EBV for resistance traits in the index. While this may be indicative of having EBV for adaptation traits available in genetic evaluation, the most effective method for improving resistance in susceptible breeds would be to increase Brahman content or utilize other adapted breeds. Of course, in some countries (e.g. Australia, the USA), Brahman content is perceived to be associated with meat-quality issues (Kemp,

1994). Thus, increases in Brahman content as a means to increasing adaptation might provide a financial disincentive in more developed economies.

Annor (1996) developed a breeding objective for beef cattle production in Ghana. Life-cycle modelling was used to estimate economic values for calving rate, age at first calving, calving interval, survival rate of calves from birth to weaning, survival rate of calves from weaning to maturity, preweaning gain, postweaning gain and food intake. Profit equations were used to derive economic values. Selection indices were not developed in the study, so there was no test of the importance of adaptation traits in either genetic or economic response to selection. However, the breeding objective was developed using data from the N'Dama and zebu breeds of Ghana, assumed to be adapted to the environment in which they are produced. Thus, in a similar conclusion to the work of Davis (1994), there might not be a need to use EBV for adaptation.

Annor (1996) found that, under most circumstances, reducing mortality from birth to maturity (increasing survival) made the greatest positive contribution to profit, followed by improving reproductive performance. Growth rate made the lowest positive contribution. This makes sense if we consider that, in many subsistence tropical-farming systems, survival under environmental stress is the primary economic trait to improve (Baker and Rege, 1994). The conclusions of Davis (1994) are similar from the perspective that cow weaning rate (the additional profit accruing from an extra 1% of calves), of which survival is a part, was the largest contributor to profit in the production systems they defined.

Economic values in breeding programmes

In many tropical areas of the world, it is lack of resistance to diseases and parasites that is the most important contributor to mortality and reduction of genetic progress in cattle production systems. As pointed out by Annor (1996), disease traits have an impact on profitability through their effects on expenditure for control, or their effects on production if left untreated, and the benefits of control may be difficult to quantify. Furthermore, the future cost of control may be very different from current values.

The presence of genetic variation in resistance to disease, coupled with increased consumer pressure against the use of drugs and drenches, makes genetic solutions to animal health problems increasingly attractive. The non-permanent effectiveness of chemical agents (due to development of resistance by the pathogen) further contributes to this interest. When animals are individually treated against an illness, calculation of the corresponding economic value is relatively straightforward (e.g. Eriksson and Solbu, 1993). There are instances, however, in which treatments are applied to groups of animals as a whole, thus leaving no between-animal variation in the cost of treatment. In such cases there would be no immediate economic gain resulting from response to selection against the disease.

Breeding objectives are normally defined with 'explicit' economic values (Newman and Ponzoni 1994). That is, as an aggregate of breeding values for all traits influencing income and/or expenditure, with each breeding value weighted by an appropriately derived economic value. Another method of defining a breeding objective is by stating target production levels for the population in question, sometimes within a given time frame (Newman and Melton, 1996). A target level is the desired level of a trait. As such, it is the optimal level of the trait for some objective defined by the individual breeder. Hence target levels (like economic values, when properly estimated) are individualized values, which may differ considerably from one breeder to another. However, the biggest advantage of target levels over selection indices lies in their intuitive appeal. A target level reflects the breeder's objective and may thus include any number of considerations important to the individual breeder. Furthermore, the breeder is free to adopt any means to rework the target, once it has been specified. Thus, where a selection index has historically been intended for within-breed or within-herd selection, target levels can also be achieved by selection across breeds and herds. This approach is akin to the 'desired gains' approach advocated by Pesek and Baker (1969).

The target-level approach is useful in the derivation of economic values for disease resistance and perhaps can be extended to other adaptation traits. Using theory developed by Brascamp (1984), Woolaston (1994) developed a breeding objective in which economic values were used for the production traits and a desired-gains approach used for resistance to internal parasites. Thus, the breeding objective is partly defined using the economic (explicit) approach, and partly defined using the target production-level approach. Some have argued that desired-gains approaches may compromise the economic effectiveness of the index approach (Gibson and Kennedy, 1990). However, such approaches may be the best alternative when economic values are difficult to specify or are at risk of being erroneous (Newman and Ponzoni, 1994).

The Role of Molecular Genetics

Techniques in molecular genetics offer the opportunity to isolate markers and genes associated with adaptation, especially disease resistance. Applications of molecular genetics in adaptation include quantification of genetic distance among, and genetic diversity within, populations, use of gene markers to speed up the selection process, and transgenesis.

The theory of breed utilization is based on the notion that, the more distantly related lines or breeds are, the greater the degree of heterosis that will be displayed in crossbred progeny. The genetic distance between two breeds defines the extent to which the breeds differ in their gene frequencies. A phylogenetic tree can be drawn from the allele data collected and analysed. The length of the branches separating any two populations is proportional to the genetic distance between them (Nicholas, 1996). Microsatellites are useful in calculating genetic distances, because they are polymorphic and are interspersed throughout the genome (see Chapter 11). Genetic-distance studies provide a guide to genetic diversity within and between species and serve as a foundation for decisions about conservation of germplasm. Genetic-distance studies can also be an aid to understanding domestication, breed origins and their history and evolution, in order to identify genetically unique breeds and to aid in the formulation of breeding programmes. Indeed, due to the process of adaptation, traits that confer survival advantages have evolved in certain cattle breeds. Thus, there are a variety of traits distributed among cattle breeds around the world with differing potential to benefit livestock production.

Molecular genetics will make a great contribution to the study of adaptation through the exploitation of genetic markers linked to quantitative trait loci (QTL) to enhance the artificial selection process. By targeting cattle populations that have been maintained in harsh tropical environments, it is likely that genes of major effect influencing heat and drought tolerance, disease resistance and parasite resistance will be found. This was the case in the detection of a putative major gene for tick resistance in the Belmont Adaptaur, an adapted *B. taurus* research population, nominally $\frac{1}{2}$ Hereford and $\frac{1}{2}$ Shorthorn (Kerr et al., 1994). A search for a marker associated with the putative gene is under way (see Chapter 11).

Trypanosomiasis-tolerant animals have been identified in populations of N'Dama cattle in Africa, and preliminary screening suggests evidence of gene(s) contributing to trypanotolerance on bovine chromosome 23 in the region of the major histocompatibility complex (MHC; see Chapter 8).

The use of comparative mapping techniques (the comparison of genetic maps of different species) has greatly accelerated the accurate identification of QTL (Georges, 1997). The basis for comparative mapping lies in the remarkable conservation of DNA (in terms of groups of loci) across widely divergent species (Nicholas, 1996). This technique has been used to search for QTL associated with trypanosomiasis resistance in mice (see Chapter 8). Three loci of large effect have been localized, indicating that an apparently quantitative infectious-disease resistance trait is in large part controlled by only two or three regions of the mouse genome. Making use of cattle populations segregating for trypanosomiasis resistance, comparative mapping can be used to search for similar QTL in cattle populations.

Practical Implications

Adaptation is a complex process, involving a suite of traits acting in consort to regulate and maintain an animal's physiological functions in the environment in which it is being raised. Often, adaptation is negatively associated with increased production in cattle and is affected by the ability of the animal to maintain itself under nutrition, heat, parasite and disease stress. This coping under environmental stress might be reflected in reduced feed intake, delayed calving, shorter lactation and post-partum anoestrus. Thus, the process involves various mechanisms to maintain a physiological equilibrium. Because of the negative relationships between stress and production, single trait selection will not be successful in balancing productivity with environmental challenges. Optimization in the form of a multiple-trait selection approach is the only logical method available to achieve this goal.

Defining breeding objectives in economic terms is difficult in the tropics, because of greater environmental and managerial complexity (Franklin, 1986). Breeding objectives should be based on body size, feed intake and nutrient partitioning, longevity, fertility, fecundity, heat tolerance, disease resistance and structural soundness, all of which are physiologically interrelated. Detailed economic assessments of costs and returns from tropical cattle-production systems, so basic to the development of structured breeding programmes, are rare, but becoming more available (e.g. Annor, 1996).

The keys to the development of breeding programmes to maintain or enhance adaptation of cattle need to incorporate sustainability issues. Production levels need to be increased to ensure food security at the household, regional, national and continental level, while ensuring the efficient and stable utilization and effective management and conservation of natural resources (Amer *et al.*, 1998). To achieve these goals, Payne and Hodges (1997) outline the necessary characteristics of an effective tropical breeding strategy, which include implementation with local resources and packaging the genetic improvement programme with options for improved management, animal health, marketing and extension support with trained local people. Amer *et al.* (1998) describe the following protocol for incorporating sustainability issues into breeding programme design.

1. Construct a key list of animal traits that influence the value of an animal to all potential stakeholders.
2. Establish a breeding objective with quantified economic values for existing livestock managers in the production system of interest. This should take account of all livestock uses outlined in the descriptions of production systems.
3. Establish an additional breeding objective with quantified economic values which reflect the interests of the external stakeholders concerned with broader sustainability issues than those directly concerning livestock managers.
4. Use the two (or more) sets of economic values to compare the effects of alternative breeding strategies (e.g. breed substitution or specific trait recording) on the benefits of animal breeding to the livestock manager and to the additional stakeholder(s).
5. Identify technological and sociological limitations to breeding strategies which when overcome result in significant benefits to both the livestock manager and the external stakeholder(s).
6. Encourage the external stakeholder(s) to support sociological and technological developments which overcome limitations to mutually beneficial breeding strategies.

7. Ensure that technological and sociological developments are followed up with appropriate genetic changes in the target production system.

To maximize genetic improvement, either the range of economically important traits needs to be increased (or reduced), so that the results of selection can contribute directly to the sustainability of the production system, or the recording costs reduced, using improved methods of trait recording, direct subsidies or genetic markers (Amer et al., 1998). There is still a dearth of information on the effect of adaptation on overall productivity from a multiple-trait selection perspective. It is likely that, because of antagonistic relationships between adaptation and production, breeding objectives combining adaptation and production will be feasible for the development of composite populations and for within-breed selection programmes to improve indigenous and already adapted breeds. Indeed, Payne and Hodges (1997) point out that basic cattle-breeding strategies for the tropics in the future will capitalize on the generation of F_1 crossbred bulls (indigenous and adapted cow breeds, with temperate sire breeds not to exceed 50% temperate genes) for use at the local level. Enhanced strategies for genetic improvement will also include genetic evaluation of indigenous breeds, progressive breed substitution with another indigenous breed, crossbreeding with temperate breeds when artificial insemination is available, and composite breed formation.

References

Amer, P.R., Mpofu, M. and Bondoc, O. (1998) Definition of breeding objectives for sustainable production systems. In: Barker, S. (ed.) *Proceedings of the Sixth World Congress on Genetics Applied to Livestock Production*, Vol. 28. University of New England, Armidale, pp. 97–106.

Annor, S.Y. (1996) Development of a breeding objective for beef cattle in Ghana. MSc thesis, Massey University, Palmerston North, New Zealand, 191 pp.

Arthur, P.F., Hearnshaw, H., Kohun, P.J. and Barlow, R. (1994) Evaluation of *Bos indicus* and *Bos taurus* straightbreds and crosses. *Australian Journal of Agriculture Research* 45, 783–794.

Ashby, W.R. (1960) *Design for a Brain.* Chapman & Hall, London, 286.pp.

Baker, R.L. and Rege, J.E.O. (1994) Genetic resistance to diseases and other stresses in improvement of ruminant livestock in the tropics. In: Smith, C. (ed.) *Proceedings of the Fifth World Congress on Genetics Applied to Livestock Production*, Vol. 20. University of Guelph, Guelph, Ontario, pp. 405–412.

Barlow, R. (1981) Experimental evidence for interaction between heterosis and environment in animals. *Animal Breeding Abstracts* 49, 715–737.

Barwick, S.A., Henzell, A.L., Graser, H.U., Upton W.H., Johnston, D.J., Allen, J., Sundstrom, B. and Goddard, M.E. (1997) *BREEDOBJECT: Training and Accreditation School Handbook.* Animal Genetics and Breeding Unit, University of New England, Armidale, 85 pp.

Black, J.L. (1983) Evolutionary adaptations and their significance in animal production. In: Peel, L. and Tribe, D.E. (eds) *Domestication, Conservation, and Use of Animal Resources.* Elsevier, Amsterdam, pp. 107–132.

Bortolussi, G., Stewart, T.S. and Newman, S. (1999) The effects of environmental stress on beef cattle production in the tropics: a synthesis of published work. *Australian Journal of Agricultural Research* (in press).

Brascamp, E.W. (1984) Selection indices with constraints. *Animal Breeding Abstracts* 52, 645–654.

Burrow H.M. (1999) Genetic analysis of temperament and its relationships with productive and adaptive traits in tropical beef cattle. PhD thesis, University of New England, Armidale, Australia.

Davis, G.P. (1993) Genetic parameters for tropical beef cattle in northern Australia: a review. *Australian Journal of Agricultural Research* 44, 179–198.

Davis, G.P. (1994) *Refining Genetic Selection Programs for Cattle in Northern Australia*. Final report on Project CS123, Meat Research Corporation, Sydney, 64 pp.

Davis, G.P. and Arthur, P.F. (1994) Crossbreeding large ruminants in the tropics: current knowledge and future directions. In: Smith, C. (ed.) *Proceedings of the Fifth World Congress on Genetics Applied to Livestock Production*, Vol. 20. University of Guelph, Guelph, Ontario, pp. 332–339.

Eriksson, J.-Å. and Solbu, H. (1993) Practical experience of breeding for health traits in Scandinavia. *Proceedings European Association of Animal Production, Denmark* 44, 1–11.

Franklin, I.R. (1986) Breeding ruminants for the tropics. In: Dickerson, G.E. and Johnson, R.K. (eds) *Proceedings of the Third World Congress on Genetics Applied to Livestock Production*, Vol. 11. University of Nebraska, Lincoln, pp. 451–461.

Frisch, J.E. and O'Neill, C.J. (1998a) Comparative evaluation of beef cattle breeds of African, European, and Indian origins. (i) Liveweights and heterosis at birth, weaning, and 18 months. *Animal Science* 67, 27–38.

Frisch, J.E. and O'Neill, C.J. (1998b) Comparative evaluation of beef cattle breeds of African, European, and Indian origins. (ii) Resistance to cattle ticks and gastrointestinal nematodes. *Animal Science* 67, 39–48.

Frisch, J.E. and Vercoe, J.E. (1977) Food intake, eating rate, weight gains, metabolic rate and efficiency of feed utilisation in *Bos taurus* and *Bos indicus* crossbred cattle. *Animal Production* 25, 343–358.

Frisch, J.E. and Vercoe, J.E. (1982) Consideration of adaptive and productive components of productivity in breeding beef for tropical Australia. In: Garsi (ed.) *Proceedings of the Second World Congress on Genetics Applied to Livestock Production*, Vol. 4. Madrid, Spain, pp. 307–321.

Futuyma, D.J. (1979) *Evolutionary Biology*. Sinauer Associates, Sunderland, Massachusetts, 565 pp.

Georges, M. (1997) QTL mapping to QTL cloning: mice to the rescue. *Genome Research* 7, 663–665.

Gibson, J.P. and Kennedy, B.W. (1990) The use of constrained selection indices in breeding for economic merit. *Theoretical and Applied Genetics* 80, 801–805.

Graser, H.-U., Goddard, M.E. and Allen, J. (1995) Better genetic technology for the beef industry. *Proceedings of the Australian Association for Animal Breeding and Genetics* 11, 56–64.

Harris, D.L. and Newman, S. (1994) Breeding for profit: synergism between genetic improvement and livestock production. *Journal of Animal Science* 72, 2178–2201.

Kemp, R.A. (1994) Genetics of meat quality in cattle. In: Smith, C. (ed.) *Proceedings of the Fifth World Congress on Genetics Applied to Livestock Production*, Vol. 19. University of Guelph, Guelph, Ontario, pp. 439–445.

Kerr, R.J., Frisch, J.E. and Kinghorn, B.P. (1994) Evidence of a major gene for tick resistance in cattle. In: Smith, C. (ed.) *Proceedings of the Fifth World Congress on Genetics Applied to Livestock Production*, Vol. 20. University of Guelph, Guelph, Ontario, pp. 265–268.

Koots, K.R., Gibson, J.P., Smith, C. and Wilton, J.W. (1994a) Analyses of published genetic parameter estimates for beef production traits. 1. Heritability. *Animal Breeding Abstracts* 62, 309–337.

Koots, K.R., Gibson, J.P., Smith, C. and Wilton, J.W. (1994b) Analyses of published genetic parameter estimates for beef production traits. 2. Phenotypic and genetic correlations. *Animal Breeding Abstracts* 62, 826–853.

Lerner, I.M. (1954) *Genetic Homeostasis*. Oliver & Boyd, London, 134 pp.

Mackinnon, M.J., Taylor, J.F. and Hetzel, D.J.S. (1990) Genetic variation and covariation in beef cow and bull fertility. *Journal of Animal Science* 68, 1208–1214.

Mackinnon, M.J., Meyer, K. and Hetzel, D.J.S. (1991a) Genetic variation and covariation for growth, parasite resistance and heat tolerance in tropical cattle. *Livestock Production Science* 27, 105–122.

Mackinnon, M.J., Corbet, N.J., Meyer, K., Burrow, H.M., Bryan, R.P. and Hetzel, D.J.S. (1991b) Genetic parameters for testosterone response to GnRH stimulation and scrotal circumference in tropical beef bulls. *Livestock Production Science* 29, 297–309.

Mason, I.L. (1989) *World Dictionary of Livestock Breeds*, 3rd edn. CAB International, Wallingford, UK, 348 pp.

Meyer, K. (1995) Estimates of genetic parameters and breeding values for New Zealand and Australian Angus cattle. *Australian Journal of Agricultural Research* 46, 1219–1229.

Moyo, S. (1990) *Evaluation of the Productivity of Indigenous Cattle and Some Exotic Beef Breeds and Their Crosses in Zimbabwe*. International Livestock Centre for Africa, Addis Ababa, 139 pp.

Newman, S. and Davis, G.P. (1995) Design of structured livestock breeding programs. In: Pryor, W.J. (ed.) *Exploring Approaches to Research in the Animal Sciences in Vietnam*. Publication 68, Australian Centre for International Agricultural Research, Canberra, pp. 127–131.

Newman, S. and Melton, B. (1996) Multi-trait selection for the US beef industry: a question of balance. In: Wilson, D.E. (ed.) *Fifth Genetic Prediction Workshop*, Vol. 5, Beef Improvement Federation, Kansas City, pp. 92–113.

Newman, S. and Ponzoni, R.W. (1994) Experience with economic weights. In: Smith, C. (ed.) *Proceedings of the Fifth World Congress on Genetics Applied to Livestock Production*, Vol. 18. University of Guelph, Guelph, Ontario, pp. 217–223.

Newman, S., Stewart, T.S., Goddard, M.E. and Gregory, M. (1997) HotCross – a decision support aid for crossbreeding of beef cattle in tropical and subtropical environments. *Proceedings of the Association for the Advancement of Animal Breeding and Genetics* 12, 400–403.

Nicholas, F.W. (1996) *Introduction to Veterinary Genetics*. Oxford University Press, Oxford, UK, 317 pp.

Oyenuga, V.A. and Nestel, B. (1984) Development of animal production systems in humid Africa. In: Nestel, B. (ed.) *World Animal Science, Development of Animal Production Systems*. Elsevier, London, pp. 189–199.

Payne, W.J.A. and Hodges, J. (1997) *Tropical Cattle: Origins, Breeds and Breeding Policies*. Blackwell Scientific, London, 328 pp.

Pesek, J. and Baker, R.J. (1969) Desired improvement in relation to selection indices. *Canadian Journal of Plant Science* 49, 803–804.

Plasse, D. (1988) Results from crossbreeding *Bos taurus* and *Bos indicus* in tropical Latin America. In: *Proceedings of the 3rd World Congress on Sheep and Beef Cattle Breeding*, Vol. 2, INRA, Paris, France, pp. 73–92.

Ponzoni, R.W. and Newman, S (1989) Developing breeding objectives for Australian beef production. *Animal Production* 49, 35–47.

Roughgarden, J. (1979) *Theory of Population Genetics and Evolutionary Ecology: An Introduction.* Macmillan, New York, 634 pp.

Woolaston, R.R. (1994) Preliminary evaluation of strategies to breed Merinos for resistance to roundworms. In: Smith, C. (ed.) *Proceedings of the Fifth World Congress on Genetics Applied to Livestock Production*, Vol. 20. University of Guelph, Guelph, Ontario, pp. 281–284.

Standardized Genetic Nomenclature for Cattle

23

C.H.S. Dolling
COGNOSAG, Box 74, McLaren Vale, South Australia 5171, Australia

Introduction	657
Development of Nomenclature	658
The first guidelines for sheep and goats	658
The 1991 guidelines for ruminants	658
The 1993 guidelines for ruminants	658
Categories of loci in sheep	659
Categories of loci in cattle	659
The Guidelines for Genetic Nomenclature for Cattle – Locus Categories 1–4	660
Locus	660
Alleles	661
Genotype terminology	662
Phenotype terminology	663
Additional Comments for Categories 3 and 4	663
Listings of Loci of Cattle, Sheep and Goats by the Committee on Genetic Nomenclature of Sheep and Goats	664
Acknowledgements	665
References	665

Introduction

The Committee on Genetic Nomenclature of Sheep and Goats (COGNOSAG) was founded with the express purpose of drawing up guidelines for the nomenclature of loci and alleles in sheep and goats. It was launched at Palmerston North, New Zealand, in 1984, and is an association registered in France under the Loi de 1901 as the Comité de Nomenclature Génétique des Ovins et Caprins (COGOVICA).

The need for guidelines became manifest during the National Congress on Breeding Coloured Sheep and Using Coloured Wool in Adelaide, South

Australia, in 1979, at which there were almost as many systems of nomenclature for coat colour and pattern as there were speakers from half a dozen countries.

Development of Nomenclature

The first guidelines for sheep and goats

A project for standardizing genetic nomenclature in sheep was developed by Lauvergne (1984) and presented to the COGNOSAG Workshop of 1986. At the same workshop, Searle (1988) reviewed the mouse nomenclature rules and gene nomenclature in the human and in other mammals. The rules proposed for sheep by the workshop were a compromise between nomenclature regulations for mice and those for humans.

The rules permitted the use of mainly human rules for loci defined by biochemical variants and tissue or blood groups, and retained useful rules from the mouse nomenclature for the loci where dominance and recessivity could be distinguished in the various genotypes (Lauvergne, 1988).

The 1986 Workshop guidelines for sheep and goats were reviewed at the COGNOSAG Workshop of 1987 (COGNOSAG, 1989).

The 1991 guidelines for ruminants

The guidelines from the 1987 Workshop were revised during the 1988 and 1989 Workshops. A further development was the revision at the 1991 Workshop to accommodate all ruminants and to facilitate the development of a comparative genome nomenclature. This development had been stimulated by the interest expressed by workers with cattle, at the 1990 Workshop, for the guidelines to be suitable for use with cattle as well as with sheep and goats. The ruminant guidelines were published by COGNOSAG (1991).

The 1993 guidelines for ruminants

Revised guidelines for gene nomenclature in ruminants were prepared by the COGNOSAG *ad hoc* committee at the 1993 Workshop (COGNOSAG *ad hoc* committee, 1995). The core of these revised guidelines consists of the 1991 proposals of COGNOSAG (1991). Their rewording was undertaken to reduce their length and to increase their clarity. The earlier recommendations to limit the length of symbols of loci and alleles to a maximum of five and four characters, respectively, were relaxed. The designation of top dominant and codominant alleles by a capital initial letter was undertaken to assist in the recognition of alleles with visible effects. Thus, the changes made are intended to render the guidelines more permissive and user-friendly, while retaining consistency with the human and mouse systems of nomenclature. The new recommendations include the use of species prefixes – for example, *BTA* or *BBO* for cattle, *OAR* or *OOV* for sheep – and the adoption of the nomenclature for keratins and keratin-associated proteins proposed by Powell and Rogers (1993). An additional proposal for provisionally assigning symbols and for listing newly reported deoxyribonucleic acid (DNA) segments and proteins that have no known homologues, official names or symbols was also outlined.

Some minor changes in wording, made to the 1993 guidelines at the 1997 and 1998 Workshops, have been incorporated here.

As far as possible, COGNOSAG respects the names of loci and alleles proposed by authors, and COGNOSAG will propose new names only in the light of new knowledge or to maintain consistency with the existing nomenclature.

The use of names and symbols in italics for loci and their alleles is preferred. However, if it is not possible to comply with this, then those names and symbols should be underlined. Authors are asked to be consistent in the underlining or use of italics for the names and symbols within a document and/or file. Locus and allele symbols need not be in italics or underlined in databases, but should be in italics in hard copy from these databases.

Categories of loci in sheep

In *Mendelian Inheritance in Sheep 1996 (MIS96)* (Lauvergne *et al.*, 1996), COGNOSAG presents reports on loci, which have been classified into one of four categories.

Category 1. Coat colour loci.
Category 2. Loci for visible traits other than coat colour.
Category 3. Loci controlling blood and milk polymorphisms.
Category 4. Mapped loci and other genetic systems.

Categories of loci in cattle

In *Mendelian Inheritance in Cattle (MIC)* (Lauvergne *et al.*, 1998), COGNOSAG presents reports on loci that have been classified into one of the following three categories.

Category 1. Coat colour loci.
Category 2. Loci for visible traits other than coat colour.
Category 3. Loci controlling blood and milk polymorphisms.

Sources of listings of category 4 loci are given.

In the next section the guidelines for genetic nomenclature for cattle loci in categories 1–4 are presented. Additional comments for categories 3 and 4 are given in the subsequent section.

The Guidelines for Genetic Nomenclature for Cattle – Locus Categories 1–4

Locus

Locus name
Choice of name
- The name in English should be as brief as possible but not consist of a single letter, and should convey as accurately as possible the character affected or the function by which the locus is recognized. The name may indicate, for example, a morphological character (*Achondroplastic Dwarfism 1*), a disease character *(Epidemolysis)* or a biochemical property *(Albumin)*.
- As far as possible the locus name should reflect interspecies homology.
- All Greek symbols should be spelt out in roman letters and placed after the name, e.g. β *Haemoglobin* becomes *Haemoglobin Beta*.
- If a newly described locus has an effect similar to that of a locus which has already been named, it may be named according to the breed, geographical location or population of origin.
- Should a new locus be identified later as being the same as a locus already named, the name invoking breed, geographical location or population of origin should be abandoned.

Printing the name
- The locus name should be in roman letters or a combination of roman letters and arabic numerals. Wherever possible, the locus name should be printed in italics; otherwise it should be underlined.
- The initial letter of the locus name should be a capital roman letter.
- Locus names should begin with capital roman letters, e.g. *Atresia Coli*, *Haemoglobin Beta*; Atresia Coli, Haemoglobin Beta.

Locus symbol
Choice of symbol
- For newly reported loci, unmapped DNA segments and proteins which have no known homologues or official names or symbols, special care should be exercised in selecting an appropriate symbol to avoid duplication and confusion with existing nomenclature. Every effort should be made to ensure that the symbols selected conform with those in current use for homologous loci.
- The locus symbol should consist of as few roman letters as possible, or a combination of roman letters and arabic numerals.
- The initial character should always be a capital roman letter which, if possible, should be the initial letter of the name of the locus.
- For loci other than those for coat colour and visible traits, upper-case roman letters only, or upper-case letters combined with arabic numerals, should be used.

- If the locus name is of two or more words and the initial letters are used in the locus symbol, then these letters should be in capitals.
- All characters in a locus symbol should be written on the same line; no superscripts or subscripts should be used, nor should roman numerals or Greek letters.
- Where appropriate, the symbol should indicate the biochemical property or designate a particular nucleotide segment.
- The rules of mammalian interspecific homology already used in the choice of the name of the locus should be applied to the choice of the symbol.
- The designation of prefixes denoting mammalian species of origin, when being used to distinguish between the species homologues of a locus (e.g. *BBO* or *BTA* for cattle and *HSA* for humans), should follow the recommendations of the Human Genome Nomenclature Committee.

Currently, two systems of nomenclature are in use for DNA segments. The two systems have not been considered in detail by COGNOSAG, but a proposal for their integration is set out here.

- For unmapped DNA segments and proteins that have no known homologues or official names or symbols, the symbols used will, wherever possible, be the same as those first reported, except that the letters used will all be in capital roman letters, e.g. RM095.
- For mapped DNA segments that have no known homologues or official names or symbols, the symbols used will, wherever possible, consist of two components: the first component will be the same as that assigned to the unmapped DNA segment (see above), whose letters will be in capital roman letters, e.g. RM095, and the second component, written in brackets after the first component, will be the D number, designated by the BovGBase, BovMap or Cattle Genome Database. Thus, for RM095 which has been mapped on to cattle chromosome 1, the D number designated is D1S13. The full symbol for this marker is therefore suggested to be *RM095(D1S13)*, e.g. see BovMap accession number BTA000851.

Printing the symbol Wherever possible, the locus symbol should be in italics; otherwise it should be underlined, e.g. the symbol of the *Agouti* locus: *A* or A.

Alleles

Allele name
Choice of name
- The name should be as brief as possible, but should convey the variation associated with the allele. If not given names, alleles should be given symbols as described in the 'Allele symbol' section below.
- If a newly described allele is similar to one that is already named, it should be named according to the breed, geographical location or population of

origin. The names of new alleles at a recognized locus should conform to the nomenclature established for that locus.
- Should a new allele be identified later as being the same as an allele already named, the name invoking breed, geographical location or population of origin should be abandoned.

Printing the name Wherever possible, the allele name should be in italics; otherwise it should be underlined. A lower-case initial letter of the allele name is preferred. This does not apply when a symbol is used instead of an allele name; for example, an allele at the *Haemoglobin Beta* locus: B or B.

Allele symbol
Choice of symbol
- The allele symbol should be as brief as possible, consisting of roman letters and/or arabic numerals.
- As far as possible, the allele symbol should be an abbreviation of the allele name and should start with the same letter. In the loci detected by biochemical, serological or nucleotide methods, the allele name and symbol may be identical.
- Greek letters and roman numerals should not be used.
- The symbol + can be used alone for identification of the standard allele ('wild type') for alleles having visible effects. Neither + nor − symbols should be used in alleles detected by biochemical, serological or nucleotide methods. Null alleles should be designated by the number zero.
- The initial letter of the symbol of the top dominant allele should be a capital letter. When there are codominant alleles only, they should each have a capital initial letter. The initial letter of all other alleles should be lower-case.

Printing the symbol
- The allele symbol should always be written with the locus symbol. It may be written as a superscript following the locus symbol or it may be written following an asterisk on the same line as the locus symbol. The allele symbol should be printed immediately adjacent to the locus symbol, i.e. with no gaps.
- Wherever possible, the allele symbol should be in italics; otherwise it should be underlined. For example, the recessive allele *alopecia 1* at the *Alopecia 1* locus in cattle will be printed in italics: $ALOP1^{alop1}$ or *ALOP1*alop1*, or underlined: ALOP1^{alop1} or ALOP1*alop1.

Genotype terminology

- The genotype of an individual should be shown by printing the relevant locus and allele symbols for the two homologous chromosomes

concerned, separated by a solidus, e.g. *ALOP1*^{*alop1*}/*ALOP1*^{*alop1*}, or <u>ALOP1</u>^{alop1}/<u>ALOP1</u>^{alop1} – which would be the genotype of a hairless calf.
- Unlinked loci should be separated by semicolons.
- Linked or syntenic loci should be separated by a space and listed in alphabetical order when gene order and/or phase are not known.
- For X-linked loci, the hemizygous case should be designated by /Y following the locus and allele symbols.
- Y-linked loci should be designated by /X following the locus and allele symbols.

Phenotype terminology

The phenotype symbol should be in the same characters as are the locus and allele symbols. The difference is that the characters should not be in italics, should not be underlined and should be written with a space between locus characters and allele characters instead of with an asterisk. Square brackets [ ] may also be used. For example, the dominant genotype *ALOP1*^{*Alop1*}/*ALOP1*^{*alop1*} is equivalent to the phenotype ALOP1 Alop1 or [ALOP1^{Alop1}] – the phenotype of a calf with a normal phenotype.

Additional Comments for Categories 3 and 4

The group of COGNOSAG members working on category 3 – blood and milk polymorphisms – has given detailed consideration to the nomenclature for blood and milk polymorphisms in cattle, sheep and goats. To comply with the decision made at the 21st International Conference on Animal Genetics, Michigan, 1990, to change existing gene nomenclature closer to that used for humans, additional proposals were put forward for cattle, sheep and goats (Larsen *et al.*, 1992). These were that the locus symbols would be written in capital letters but that lower-case letters would be retained to designate recessive alleles. Lower-case letters were also retained for sheep and goat blood-group factors.

The guidelines for gene nomenclature for category 4 – mapped loci and other genetic systems – are as described in the previous section, but, in addition, COGNOSAG recommends that, if an unnamed, newly mapped cattle locus is identified, contact be made with Dr Sue Povey, University College, London, Medical Research Council (MRC) Biochemical Genetics Unit, Wolfson House, 4 Stephenson Way, London NW1 2HE; phone +44-171-380 7410; fax +44-171-387 3496; email mpovey@hgmp.mrc.ac.uk, before the locus is named to ensure that the symbol proposed for cattle has not already been adopted for the human. Alternatively, researchers may contact their nearest available COGNOSAG member, or the cattle genome databases – BovMAP of Institut National de la Recherche Agronomique (INRA), France, or BOVGBASE of Texas A & M, USA – for assistance.

Listings of the loci of category 4 in cattle have been prepared in the following:

BOVGBASE, of Texas A & M, USA
http://bos.cvm.tamu.edu/bovgbase.html
BovMAP, of INRA, France
http://locus.jouy.inra.fr/cgi-bin/bovmap/Bovmap/main.pl
Cattle Genome Database, Commonwealth Scientific and Industrial Research Organisation (CSIRO), Australia
http://spinal.tag.csiro.au/
Animal Genome Database, Japan
http://ws4.niai.affrc.go.jp/jgbase.html
Online Mendelian Inheritance in Animals (OMIA), Australia
http://www.angis.org.au/Databases/BIRX/omia/
Meat Animal Research Center (MARC), Nebraska, USA
http://sol.marc.usda.gov/marc/html/gene1.html

Listings of Loci of Cattle, Sheep and Goats by the Committee on Genetic Nomenclature of Sheep and Goats

The following listings by COGNOSAG members document and record the loci and alleles of cattle, sheep and goats reported in the scientific literature and illustrate the implementation of COGNOSAG's guidelines for gene nomenclature.

- Loci for category 2 – visible traits other than coat colour – have been published for sheep and goats (COGNOSAG, 1989).
- Loci for category 3 – blood and milk polymorphisms – have been published for sheep and goats (COGNOSAG, 1989) and for cattle, sheep and goats (Larsen et al., 1992).
- Loci for category 1 – coat colour – have been published for sheep and goats (COGNOSAG, 1990).
- In *Mendelian Inheritance in Sheep 1996 (MIS96)* (Lauvergne et al., 1996), published by the University of Camerino, Camerino, Italy, and COGNOSAG, a full listing of loci and alleles in categories 1, 2 and 3 is given and of loci in category 4. This serves as a link between the gene mapper and the sheep breeder.
- Listings of loci and alleles in sheep in categories 1–3 and of loci of category 4 – mapped loci and other genetic systems – are available on the Internet through the agency of the University of Sydney (contact person is Dr F.W. Nicholas, Department of Animal Science, University of Sydney, New South Wales 2006, Australia, fax: +61-2-9351-8097, email: frankn@vetsci.usyd.edu.au).
- In *Mendelian Inheritance in Cattle (MIC)* (Lauvergne et al., 1999), published by the Mendel University of Agriculture and Forestry Brno, in the Czech Republic, the Laboratoire de Génétique Factorielle du Département

de Génétique Animale de l'INRA in France and by COGNOSAG, a listing of loci and alleles in categories 1–3 is given. This will serve as a link between the gene mapper and the cattle breeder.

Acknowledgements

This chapter has reported work on the genetic nomenclature of cattle, sheep and goats which has been done by members of COGNOSAG from countries around the world, both during and between COGNOSAG Workshops. The chronology of COGNOSAG activities emphasizes the time and effort which members have put into the development of COGNOSAG guidelines for nomenclature and procedures for listing loci (Table 23.1)

Grateful acknowledgement is made to my co-editors of *Mendelian Inheritance in Cattle (MIC)* – J.J. Lauvergne and P. Millar – and to those attending the COGNOSAG 1997 Workshop, to which the manuscript was presented. For the preparation of the chapter, it is a pleasure to acknowledge the assistance of three further members of COGNOSAG, Tom Broad, of New Zealand, and Frank Nicholas, of New South Wales, Australia, for helpful comments on the manuscript, and Verle Wood, of South Australia, for the preparation of the typescript.

Table 23.1. Chronology of COGNOSAG activities.

Year	Activities
1979	Original concept in Adelaide (South Australia)
1984	Foundation at Massey University (New Zealand)
1986	1st Workshop, Manosque (France)
1987	Registration in Antony Sub-Prefecture (France)
	2nd Workshop, Manosque (France)
1988	3rd Workshop, Manosque (France)
1989	4th Workshop, Eugene (Oregon, USA)
1990	5th Workshop, Edinburgh (Scotland)
1991	6th Workshop, Manosque (France)
1992	7th Workshop, Manosque (France)
1993	8th Workshop, Turretfield (South Australia)
	Seminar, Australian Society of Animal Production/COGNOSAG (Adelaide)
1994	9th Workshop, Rome (Italy)
1995	10th Workshop, Matelica (Italy)
1996	11th Workshop, Nouzilly (France)
1997	12th Workshop, Lednice (Czech Republic)
1998	13th Workshop, Lednice (Czech Republic)

References

COGNOSAG (Andresen, E., Broad, T., di Stasio, L., Dolling, C.H.S., Hill, D., Huston, K., Larsen B., Lauvergne, J.J., Levéziel, H., Malher, X., Millar, P., Rae, A.L., Renieri, C. and Tucker, E.M.) (1991) Guidelines for gene nomenclature in ruminants 1991. *Genetics, Selection, Evolution* 23, 461–466.

COGNOSAG (1989) Gene nomenclature in sheep and goats 1987. In: Lauvergne, J.J. (ed.) *Standardized Genetic Nomenclature for Sheep and Goats 1987. Loci for visible traits other than colour and blood and milk polymorphisms. Proceedings of the COGNOSAG Workshop, 1987.* TEC and DOC – LAVOISIER, Paris, pp. 17–21.

COGNOSAG (1990) Loci for coat colour of sheep and goats 1989. In: Lauvergne J.J. (ed.) *Proceedings of the COGNOSAG Workshops of 1988 and 1989.* COGOVICA/COGNOSAG, Clamart, France.

COGNOSAG ad hoc committee (1995) Revised guidelines for gene nomenclature in ruminants 1993. *Genetics, Selection, Evolution* 27, 89–93.

Larsen, B., di Stasio L. and Tucker, E.M. (1992) List of alleles for blood and milk polymorphisms in cattle, sheep and goats. *Animal Genetics* 23, 188–192.

Lauvergne, J.J. (1984) *A Project for Standardizing Genetic Nomenclature in Sheep.* Bulletin technique du Département de Génétique animale No. 38, INRA Département de Génétique animale, CNRZ, Jouy-en-Josas, France, 59 pp.

Lauvergne, J.J. (1988) The project of genic nomenclature for the COGNOSAG. In: Lauvergne, J.J. (ed.) *Proceedings of the COGNOSAG Workshop, 1986.* TEC and DOC – LAVOISIER, Paris, pp. 21–25.

Lauvergne, J.J., Dolling, C.H.S. and Renieri, C. (eds) (1996) *Mendelian Inheritance in Sheep 1996 (MIS 96).* COGOVICA/COGNOSAG, Clamart, France, and University of Camerino, Camerino, Italy.

Lauvergne, J.J., Dolling, C.H.S. and Millar, P (eds). (1999) *Mendelian Inheritance in Cattle (MIC).* Mendel University of Agriculture and Forestry Brno, Czech Republic; Laboratoire de Génétique Factorielle de l'Institut National de la Recherche Agronomique. France; and COGOVICA/COGNOSAG, Clamart, France.

Powell, B.C. and Rogers, G.E. (1993) Differentiation in hard keratin tissues, hair and related structures. In: Leigh, I., Watt, F. and Lane, E.B., (eds) *Keratinocyte Handbook.* Cambridge University Press, Cambridge, UK.

Searle, A.G. (1988) The genetic nomenclatures of the mammals. In: Lauvergne, J.J. (ed.) *Proceedings of the COGNOSAG Workshop 1986.* TEC and DOC – LAVOISIER, Paris, pp. 15–20.

Breeds of Cattle

D.S. Buchanan and S.L. Dolezal
Animal Science Department, Oklahoma State University, Stillwater, OK 74078, USA

Introduction	667
Bos taurus* and *Bos indicus	668
Classification of Breeds	668
References	686

Introduction

Since domestication of cattle began more than 9000 years ago, humans have attempted, through various means, to identify superior animals and retain them. The widely varying geographical areas in which cattle existed and the multiplicity of uses (meat, dairy, draught, hides, ceremonial, etc.) meant that cattle would develop in many diverse ways. It was inevitable that cattle would begin to fall into groupings which we have come to refer to as 'breeds'. The term 'breed' is a difficult one to define precisely, because it means different things to different people. Breed might be defined as a group of animals with similar physical characteristics (such as colour, horns, body type, etc.). However, there are breeds that contain wide variation in such characteristics, while members of different breeds may be quite similar. There is general agreement that the concept of a breed denotes common ancestry and yet some organizations that protect the purity of a breed choose, periodically, to allow entry of animals from exotic ancestry. Lush (1994), quoting from Lloyd-Jones (1915), makes the following observation:

> A breed is a group of domestic animals, termed such by common consent of the breeders, a term which arose among breeders of livestock, created one might say for their own use, and no one is warranted in assigning to this word a scientific definition and in calling the breeders wrong when they deviated from the formulated definition. It is their word and the breeders' common usage is what we must accept as the correct definition.

Wright (1977), in his description of breed formation, describes a breed as something which arises more rapidly than normal evolutionary processes would dictate but more slowly than would be true in the laboratory. Breed development probably covers almost the entire range of rates in that spectrum. Some breeds arise almost entirely through natural forces, while others are developed by human managers in a highly directed fashion.

Bos taurus and *Bos indicus*

Worldwide, cattle fall into two reasonably distinct groups (Felius, 1985). All cattle are contained within the genus *Bos*. However, most cattle can be assigned to either the species *Bos taurus* or the species *Bos indicus*. There is no uniform opinion about whether these should be considered as separate species. They freely interbreed, so there is no reproductive barrier. However, there are obvious physical differences and, while apparently derived from a common progenitor species, they evolved quite separately for several thousand years. The most obvious physical difference between *B. indicus* and *B. taurus* cattle is that *B. indicus* (zebu) cattle generally have a very pronounced hump on their shoulders, while *B. taurus* cattle are humpless. *Bos indicus* cattle remained on the Indian subcontinent for many generations and then began to migrate along the east coast of Africa and toward South East Asia.

Bos taurus cattle evolved in more northern areas of Asia and in Europe. There were migrations of *B. taurus* cattle along western Africa and the Americas, with the explorations of the Spaniards. Although it is generally true that *B. indicus* cattle are tropically adapted and *B. taurus* cattle are adapted to temperate regions, the migration of *B. taurus* cattle along western Africa resulted in some tropically adapted *B. taurus* breeds. *Bos taurus* cattle brought to the Americas by the Spaniards were left to adapt to their environment for several hundred years. These are referred to as the Criollo cattle and several breeds arose in South and Central America, as well as the Texas Longhorn and Florida Cracker breeds in areas now contained in the USA. Some early crossing of *B. taurus* and *B. indicus* cattle in Africa resulted in a subgroup referred to as Sanga cattle. In the last century, numerous breeds have been developed which take advantage of the complementary characteristics of *B. taurus* and *B. indicus* cattle.

Classification of Breeds

Another way to subdivide cattle breeds is by utility. Artificial selection within many breeds has caused them to excel for either meat or milk production. This is especially true in Great Britain and North America, where there is a fairly clear delineation between such categories. Breeds such as the Holstein produce quantities of milk far in excess of that which could ever be consumed by a calf and have become well adapted to a highly intensive schedule of being milked twice, or thrice, daily. Other breeds give only enough milk to sustain a calf but have highly developed muscularity, possibly from use as a draught animal, which is important for meat production.

Breeds are easy to recognize in many of the developed countries, because organizations have arisen to protect the purity of the breed and to pursue its improvement. These 'breed societies' originated in Great Britain during the early part of the 19th century (Willham, 1987) and spread to other countries, most notably the USA. Some breed societies are large businesses with millions of cattle registered, while others number their registrations with three digits and are organized by a single individual.

A difficulty associated with describing breeds of cattle is identifying the number of breeds to include. One source includes more than 1000 different breeds (Mason, 1996), although many of these are national derivatives of a breed that is imported from its native country. It would be desirable to identify all of the 'important' breeds around the world. This task is rendered nearly impossible because of the difficulty in defining 'important'. Breeds with high census numbers are likely to be considered important, but there may be breeds with low numbers which are important either historically or as a source of unique genetic material for some future use. This raises the issue of conservation of genetic material. Breeds may be conserved due to economic, scientific or cultural reasons (Committee on Managing Global Genetic Resources, 1993). It also seems likely that some breeds have acquired new names as more is learned about breeds in developing countries, and this may confuse the identification of important breeds.

Renewed interest in developing new breeds has arisen in recent years. 'Composites', 'synthetics' or 'hybrids' are labels used to signify new breeds or newly formed lines developed from crossing. The Composite Cattle Breeders International Alliance is an organization that meets annually to address the needs and issues of these cattle.

Ideally, a description of breeds in a publication would include those breeds with a well-understood origin and well-researched characteristics. Many breeds are important, at least in some parts of the world, even though they fit neither of these characteristics. The origin of many breeds has been lost due to inadequate historical records or is irrelevant due to large-scale introduction of individuals from outside the breed. There is considerable research information for some breeds, but resources for such research are limited and may not be applicable to all environments around the world.

We are left with only imperfect methods for identifying breeds to include in a publication such as this one. We have chosen the following approach. Breeds in common use in North America were included if there was research information available or if they were included in one or more of the following sources of information: Briggs and Briggs (1980), Walker (1989) or one of several popular wall posters of breed information distributed by organizations like Better Beef Business. Breeds from other parts of the world were included based upon information from the Food and Agriculture Organization of the United Nations (http://dad.fao.org/dad-is/data/index.htm), Rouse

(1970a, b, 1973), Felius (1985) or the Committee on Managing Global Genetic Resources (1993). Most of the breeds are included in the Breeds of Livestock Website maintained by Oklahoma State University (http://www.ansi.okstate.edu/breeds/cattle/).

The breeds are described in a series of tables. Table 24.1 includes several breeds that have been selected primarily for milk production. All of the breeds, including those shown in Table 24.1, are listed in Tables 24.2–24.8, by region of origin, with beef production characteristics. Where information is available, breeds are described for size, age at puberty, lean-to-fat ratio and milk production. These four traits have been used by the scientists at the US Meat Animal Research Center to characterize breeds included in the germplasm evaluation (GPE) experiment (Cundiff et al., 1986, 1993, 1997). This experiment has been conducted since the late 1960s to evaluate breeds in use in North America. The descriptors, for breeds other than those in the GPE, are highly subjective. They probably reflect performance that is dependent upon the environment in which the breeds are used and may not indicate the performance levels that would be achieved if all of the breeds were managed in a uniform environment.

Table 24.9 is included to provide references for research information concerning the breeds. Numerous research papers, published in refereed journals, are included, with a list of the breeds evaluated in the project described in each paper.

The concept of a breed is likely to remain rather fluid. The number of breeds developed in North America during the last half of the 20th century is indicative of a general effort to identify combinations of germplasm for use in the varied environments in which cattle are raised. These developments are, apparently, continuing unabated. It is tempting to assume that the important breeds of today will continue to be important in the future. One has only to examine the history of breeds during the 20th century, in cattle and in other species of livestock, to see the fallacy of this assumption. Improved techniques for identification of superior genetic material, including techniques from molecular biology, will probably speed the evolutionary pace in cattle. This will mean even more rapid assembly and recombination of genetic stocks.

Table 24.1. Breeds of cattle used primarily for milk production.

Breed	Place of origin	Species of *Bos*	Distribution	Colour	Other	Size and growth	Milk production
Ayrshire	Ayr, Scotland	*taurus*	North America, Europe	Red and white		Moderate	Moderate
Braunvieh	Switzerland	*taurus*	Europe	Brown		Moderate to large	Moderate
Brown Swiss	Switzerland	*taurus*	North America, Europe	Light brown		Moderate to large	Moderate to high
Danish Red	Denmark	*taurus*	Europe	Red		Moderate to large	Moderate
Dexter	Ireland	*taurus*	North America, Europe	Black		Small	Moderate
Dutch Belted	Netherlands	*taurus*	Europe	Black and white		Moderate	Small to moderate
Flamande	France	*taurus*	Europe	Dark brown to black		Moderate to large	Moderate
Guernsey	Guernsey (Channel Islands)	*taurus*	North America, Europe	Fawn and white	Moderately high in butterfat	Moderate	Moderate
Holstein	Netherlands	*taurus*	North America, Europe	Black and white	Beef strain referred to as Beef Friesian	Moderate to large	High
Illawarra	New South Wales, Australia	*taurus*	Australia, Asia	Red, some roans or whites	Ayrshire, Devon, Milking Shorthorn hybrid	Moderate	Moderate
Javanese	Indonesia	*indicus*	Oceania	Tan		Small to moderate	Moderate

Breed	Origin	Type	Distribution	Color	Special characteristics	Size	Milk production
Jersey	Jersey (Channel Islands)	*taurus*	North America, Europe	Fawn	High in butterfat	Small	Low to moderate
Local Indian Dairy	Malaysia	*indicus*	Asia, Oceania	White		Small	Low to moderate
Montbéliarde	France	*taurus*	Europe	Red and white		Moderate to large	Moderate
Norwegian Red	Norway	*taurus*	Europe	Red and white		Moderate	Moderate
Red Sindhi	Pakistan	*indicus*	Asia, Africa, Australia	Red		Small	Moderate
Russian Black Pied	Russia	*taurus*	Europe, Asia	Black and white		Moderate	Moderate to high
Sahiwal	Punjab region, India	*indicus*	Asia, Africa, North America, Australia	Red		Small to moderate	Moderate to high
Shorthorn	Northumberland and Durham, England	*taurus*	North America, Europe	Dark red, white or roan		Moderate	Moderate
Xinjiang Brown	Xinjiang Uygur Region, China	*taurus*	Asia	Variable		Small to moderate	Small to moderate

Table 24.2. Breeds of cattle with origin in Asia.

Breed	Place of origin	Species of *Bos*	Distribution	Colour	Other	Size and growth	Lean-to-fat ratio	Age at puberty	Milk production (for calf)
Bengali	Bangladesh and Bengal, India	*indicus*	Asia	Light brown		Small			
Chinese Yellow	China	Transitional	Asia	Yellow		Small			
Dhanni	Pakistan	*indicus*	Asia	White with black spots		Moderate			
Gir	Gujerat, India	*indicus*	Asia, South America, North America	Variable, red to white		Moderate			
Guzerat	India	*indicus*	Asia, South America, North America	White	Kankrej (in India)	Moderate to large			Moderate
Hissar	India	*indicus*	Asia	White		Moderate to large			
Krishna Valley	India	*indicus*	Asia	White		Moderate to large			
Mongolian	Mongolia	*taurus*	Asia	Red and white		Small			
Nellore	India	*indicus*	Asia, Africa, South America, Australia	White		Moderate to large*	Moderate*	Late*	Moderate*
Ongole	India	*indicus*	Asia	White		Moderate to large			Moderate
Red Sindhi	Pakistan	*indicus*	Asia, Africa, Australia	Red		Small			Moderate to high
Sahiwal	Punjab region, India	*indicus*	Asia, Africa, North America, Australia	Red		Small to moderate*	Moderate*	Late*	Moderate
Tharparkar	India	*indicus*	Asia	White		Small to moderate			
Wagyu	Japan	*taurus*	Asia, North America	Both black and red strains		Small			Moderate
Xinjiang Brown	Xinjiang Uygur Region, China	*taurus*	Asia	Variable		Small to moderate			

*Cundiff *et al.*, 1993.

Table 24.3. Breeds of cattle with origin in Africa.

Breed	Place of origin	Species of *Bos*	Distribution	Colour	Other	Size and growth	Lean-to-fat ratio	Age at puberty	Milk production (for calf)
Abyssinian Shorthorned Zebu	Ethiopia	*indicus*	Africa	Variable		Small			
Adamawa	Nigeria	*indicus*	Africa	Variable		Small	Moderate	Moderate	
Africander	South Africa	Sanga	Africa, Australia	Red		Moderate		to late	
Ankole-Watusi	Egypt	*indicus*	Africa, North America	Red or black with white spots	Produce high-fat milk	Moderate	Moderate to high		
Arsi	Ethiopia	*indicus*	Africa	Variable		Small			
Bonsmara	South Africa	Hybrid	Africa, Australia	Red	Afrikander–Hereford–Shorthorn hybrid	Moderate	Moderate		Moderate
Boran	Ethiopia	*indicus*	Africa	Red		Small to moderate	Moderate to high	Moderate to late	Moderate
Brown Atlas	Algeria	*taurus*	Africa	Brown		Small			
Butana	Sudan	*indicus*	Africa	Dark red		Small			Moderate to high
Danakil	Ethiopia	Sanga	Africa	Variable		Moderate to large			Moderate
Dinka (Nilotic)	Sudan	Sanga	Africa	White		Small to moderate			
Fogera	Ethiopia	Sanga–*indicus*	Asia	Variable		Moderate			
Kenana	Sudan	*indicus*	Africa	White		Small to moderate			
Keteku	Nigeria	Hybrid	Africa	Black and white	Zebu–Shorthorn hybrid	Small			
Menufi (Baladi)	Egypt	*taurus*	Africa	Red		Small			
Muturu	Nigeria	*taurus*	Africa	Black and white		Small			
N'Dama	Guinea, West Africa	*taurus*	Africa	Fawn	Trypanotolerant	Moderate			
Nguni	South Africa	Sanga	Africa	Variable		Small to moderate	Moderate to high		
Tuli	Zimbabwe	Sanga	Africa	Yellow		Moderate			
White Fulani	Nigeria	*indicus*	Africa	White		Small to moderate			

Breeds of Cattle

Table 24.4. Breeds of cattle with origin in Europe.

Breed	Place of origin	Species of *Bos*	Distribution	Colour	Other	Size and growth	Lean-to-fat ratio	Age at puberty	Milk production (for calf)
Alentejana	Portugal	*taurus*	Europe	Red		Moderate			
Aubrac	France	*taurus*	Europe	Brown		Small to moderate			
Belgian Blue	Belgium	*taurus*	North America, Europe	White or blue roan		Moderate to large	High	Moderate	Moderate to high
Blonde d'Aquitaine	Garonne, France	*taurus*	North America, Europe	Yellow		Moderate to large	Moderate to high		Low to moderate
Braunvieh	Switzerland	*taurus*	Europe	Brown		Moderate to large*	Moderate to high*	Early to moderate*	Moderate to high*
Brown Swiss	Switzerland	*taurus*	North America, Europe	Light brown		Moderate to large			
Charolais	Charolais and Nievre, France	*taurus*	North America, Europe	White		Large*	High*	Moderate to late*	Low*
Chianina	Chiana Valley, Italy	*taurus*	North America, Europe	White		Large*	High*	Moderate to late*	Low*
Danish Red	Denmark	*taurus*	Europe	Red		Moderate to large			Moderate to high
Danish Red and White	Denmark	*taurus*	Europe	Red and white		Moderate			Moderate
Dutch Belted	Netherlands	*taurus*	Europe	Black and white		Moderate			
Fighting Bull (Toro de Lidia)	Spain	*taurus*	Europe, Latin American countries	Variable	Bred for bullfighting	Small to moderate			
Flamande	France	*taurus*	Europe	Dark brown to black		Moderate to large			Moderate to large
Fleckvieh	Germany	*taurus*	Europe	Red and white		Moderate to large			
Gelbvieh	Bavaria, Germany	*taurus*	North America, Europe	Red		Moderate to large*	Moderate to high*	Early to moderate*	Moderate to high*

Table 24.4. Continued

Breed	Place of origin	Species of Bos	Distribution	Colour	Other	Size and growth	Lean-to-fat ratio	Age at puberty	Milk production (for calf)
Holstein	Netherlands	taurus	North America, Europe	Black and white	Beef strain referred to as Beef Friesian	Moderate to large*	Moderate to high*	Early to moderate*	High*
Hungarian Spotted	Hungary	taurus	Europe	Red, white markings		Moderate			
Icelandic	Iceland	taurus	Iceland	Variable		Small to moderate			
Limousin	Aquitaine region of France	taurus	North America, Europe	Red		Moderate*	High*	Moderate to late*	Low*
Maine Anjou	Mancelle region of France	taurus	North America, Europe	Red, white markings		Large*	Moderate to high*	Moderate*	Moderate*
Marchigiana	Marche, Italy	taurus	North America, Europe	White		Moderate to large	Moderate to high	Early to moderate	
Meuse-Rhine-Yssel	Netherlands	taurus	Europe	Red with white markings		Moderate to large			
Montbéliarde	France	taurus	Europe	Red and white		Moderate to large			Moderate to high
Normande	Manche and Calvados, France	taurus	Europe, North America	Yellow to dark brown, white markings		Moderate to large	Moderate to high		Moderate to high
Norwegian Red	Norway	taurus	Europe	Red and white		Moderate			
Piedmontese	North-western Italy	taurus	North America, Europe	White	High-frequency 'double-muscled'	Moderate*	Very high*	Early to moderate*	Low to moderate*
Pinzgauer	Pinz Valley, Austria	taurus	North America, Europe	Red, white markings		Moderate*	Moderate*	Early to moderate*	Moderate*
Polish Red	Poland	taurus	Europe	Red		Small to moderate			
Polish Red and White	Poland	taurus	Europe	Red and white		Moderate			

Breeds of Cattle

Breed	Origin	Species	Distribution	Color	Markings	Size			
Romagnola	North-eastern Italy	taurus	North America, Europe	Grey		Large	High	Moderate	
Rotvieh (German Red)	Germany	taurus	Europe	Red		Moderate			
Russian Black Pied	Russia	taurus	Europe, Asia	Black and white		Moderate			High
Salers	Salers District, France	taurus	North America, Europe	Red, some white markings		Large*	Moderate to high*	Moderate*	Moderate*
Simmental	Simme Valley, Switzerland	taurus	North America, Europe, Asia	Red with white markings	White face	Large*	Moderate to high*	Moderate*	Moderate to high*
Swedish Red and White	Sweden	taurus	Europe	Red and white		Moderate			
Swedish Red Polled	Sweden	taurus	Europe	Red		Small			
Tarentaise	Moutiers, France	taurus	North America, Europe	Red		Moderate	Moderate*	Early to moderate*	Moderate*

*Cundiff et al., 1993.

Table 24.5. Breeds of cattle with origin in the British Isles.

Breed	Place of origin	Species of Bos	Distribution	Colour	Other	Size and growth	Lean-to-fat ratio	Age at puberty	Milk production (for calf)
Angus	Aberdeen and Angus Counties, Scotland	taurus	North America, Europe	Black, also red strain	Polled	Moderate*	Low to moderate*	Moderate*	Moderate*
Ayrshire	Ayr, Scotland	taurus	North America, Europe	Red and white		Moderate			
Belted Galloway	Scotland	taurus	North America, Europe	Black with white belt	Polled	Small to moderate			
British White	England	taurus	North America, Europe	White	Polled	Small to moderate			
Devon	Devon, England	taurus	North America, Europe	Red		Small to moderate*	Low to moderate*	Moderate*	Low to moderate*
Dexter	Ireland	taurus	North America, Europe	Black		Small			
Galloway	Galloway, Scotland	taurus	North America, Europe	Black	Long, curly hair; polled	Small to moderate*		Moderate*	Low to moderate*
Guernsey	Guernsey (Channel Islands)	taurus	North America, Europe	Fawn and white	Moderately high in butterfat	Moderate			
Hereford	Herefordshire, England	taurus	North America, Europe	Red, white markings	White face	Moderate*	Low to moderate*	Moderate*	Low to moderate*
Jersey	Jersey (Channel Islands)	taurus	North America, Europe	Fawn	High in butterfat	Small		Early	High
Lincoln Red	England	taurus	Europe	Red		Moderate			
Longhorn (English)	England	taurus	Europe	Red, grey or brindle		Moderate			

Red Angus	Scotland	taurus	North America, Europe	Red	Polled	Moderate	Moderate	Early to moderate	Moderate
Red Poll	Suffolk and Norfolk counties, England	taurus	North America, Europe	Red	Polled	Small to moderate*	Moderate*	Early to moderate*	Moderate*
Scotch Highland	Western Scotland	taurus	North America, Europe	Brown, black or red	Long hair	Small			
Shorthorn	Northumberland and Durham, England	taurus	North America, Europe	Dark red, white or roan		Moderate*	Low to moderate*	Moderate*	Moderate*
South Devon	Devon and Cornwall, England	taurus	North America, Europe	Red		Moderate*	Moderate*	Early to Moderate*	Moderate*
Sussex	South-east England	taurus	Europe, Africa	Red		Small			
Welsh Black	Wales	taurus	Europe	Black		Moderate			
White Park	England	taurus	Europe, North America	White		Moderate			

*Cundiff et al., 1993.

Table 24.6. Breeds of cattle with origin in North America.

Breed	Place of origin	Species of *Bos*	Distribution	Colour	Other	Size and growth	Lean-to-fat ratio	Age at puberty	Milk production (for calf)
American White Park	United States	*taurus*	North America	White		Moderate			
Amerifax	United States	*taurus*	North America	Red or black	Polled, Angus–Beef Friesian hybrid	Moderate to large			Moderate to high
Ankina	United States	*taurus*	North America	Black	Polled	Moderate to large	Moderate to high		
Barzona	Arizona, United States	Hybrid	North America	Dark red	Africander–Hereford–Angus–Santa Gertrudis hybrid	Moderate	Moderate		
Beefmaster	Texas, United States	Hybrid	North America, Africa	Red and other colours		Moderate to large	Moderate		Moderate
Braford	Florida, United States	Hybrid	North America, Australia	Red, white markings	Brahman–Hereford hybrid	Moderate	Moderate	Early to moderate	Moderate
Brah-Maine	United States	Hybrid	North America	Red with white markings	Brahman–Maine Anjou hybrid	Moderate to large	Moderate to high	Moderate	Moderate
Brahman	United States	*indicus*	North America, Africa	Grey strains, red strains	Blending of Gir, Guzerat and Nellore	Moderate to large*	Moderate*	Late*	Moderate*
Brahmousin	United States	Hybrid	North America	Red	Limousin–Brahman hybrid	Moderate	Moderate to high		
Bralers	United States	Hybrid	North America	Red	Brahman–Salers hybrid	Moderate	Moderate		Moderate
Brangus	Louisiana, United States	Hybrid	North America, Africa	Black	Angus–Brahman hybrid	Moderate*	Low to moderate*	Moderate to late*	Low to moderate*
Canadienne	Canada	*taurus*	North America	Red		Small			

Breeds of Cattle

Breed	Country	Type	Region	Color	Composition	Size				
Charbray	United States	Hybrid	North America, Australia	White to tan	Charolais–Brahman hybrid	Moderate to large	Moderate to high	Moderate to late	Moderate	Moderate
Corriente	Northern Mexico	taurus	North America	Variable	Criollo cattle of northern Mexico	Small				
Florida Cracker	Florida, United States	taurus	North America	Variable	Criollo cattle of Florida	Small		Early		
Gelbray	United States	Hybrid	North America	Red	Gelbvieh–Brahman hybrid	Moderate to large	Moderate to high		Moderate to high	
Hays Converter	Canada	taurus	North America	Black or red with white markings	Holstein–Hereford hybrid	Moderate to large			Moderate to high	
Red Brangus	United States	Hybrid	North America	Red	Red Angus–Brahman hybrid	Moderate				
RX3	United States	taurus	North America	Red	Hereford–Holstein–Red Angus cross	Moderate	Moderate		Moderate to high	
Salorn	United States	taurus	North America	Red	Salers–Texas Longhorn synthetic	Moderate	Moderate to high	Early to moderate		
Santa Cruz	Texas, United States	Hybrid	North America	Red	Santa Gertrudis–Red Angus–Gelbvieh hybrid	Moderate	Moderate to high	Early to moderate		
Santa Gertrudis	Texas, United States	Hybrid	North America, Africa	Red	Shorthorn–Brahman hybrid	Moderate*	Low to moderate*	Moderate to late*	Moderate to late*	Low to moderate*
Senepol	Virgin Islands	taurus	North America	Red		Small to moderate				
Simbrah	United States	Hybrid	North America	Red with white markings	Simmental–Brahman hybrid	Moderate to large				
Texas Longhorn	Mexico and southern United States	taurus	North America	Variable	Criollo cattle of South-western United States	Small*	Moderate*	Moderate*	Low to moderate*	

*Cundiff et al., 1993.

Table 24.7. Breeds of cattle with origin in Australia and Oceania.

Breed	Place of origin	Species of *Bos*	Distribution	Colour	Other	Size and growth	Lean-to-fat ratio	Age at puberty	Milk production (for calf)
Bali Cattle	Bali	*taurus*	Australia, Oceania	Grey or tan		Small			
Banteng	Java	*taurus*	Australia, Oceania	Grey or tan		Small			
Grati	Indonesia	*taurus*	Oceania	Red or black and white		Moderate			
Illawarra	New South Wales, Australia	*taurus*	Australia, Asia	Red, some roans or whites	Ayrshire–Devon–Milking Shorthorn hybrid	Moderate			
Javanese	Indonesia	*indicus*	Oceania	Tan		Small to moderate			Moderate to high
Kelantan	Malaysia	*indicus*	Oceania	Tan to brown		Small			
Local Indian Dairy	Malaysia	*indicus*	Asia, Oceania	White		Small			
Madura	Indonesia	*indicus*	Oceania	Tan		Small			
Mandalong	New South Wales, Australia	Hybrid		Yellow to brown	Charolais–Chianina–Shorthorn–British White–Brahman hybrid	Moderate to large			
Murray Grey	New South Wales, Australia	*taurus*	Australia	Grey	Polled	Moderate	Moderate to high		Moderate

Table 24.8. Breeds of cattle with origin in South America.

Breed	Place of origin	Species	Distribution	Colour	Other	Size and Growth
Caracu	Brazil	*taurus*	South America	Variable	Criollo cattle of Brazil	Moderate
Indo-Brazil	Brazil	*indicus*	South America, North America	White or grey		Moderate
Blanco Orejinegro	Colombia	*taurus*	South America	White	Criollo cattle of Colombia	Moderate

Table 24.9. References for research information concerning the breeds

Authors	Breeds
Adams *et al.* (1973)	Hereford, Simmental, Limousin, Maine Anjou, Lincoln Red, Brown Swiss, Charolais and Angus
Adams *et al.* (1977)	Hereford, Angus, Lincoln Red, Brown Swiss, Simmental, Limousin, Maine Anjou and Charolais
Alenda and Martin (1981)	Angus, Charolais and Hereford
Alenda *et al.* (1980a, b)	Angus, Charolais and Hereford
Anderson *et al.* (1978)	Angus, Chianina, Holstein, Charolais and Simmental
Bailey and Moore (1980)	Hereford, Red Poll, Angus, Charolais, Brahman
Berndtson *et al.* (1987)	Angus, Hereford
Bond *et al.* (1972)	Holstein, Jersey, Milking Shorthorn, Angus and Hereford
Brown and Dinkel (1982)	Angus, Charolais, Salers, Limousin and Polled Hereford
Brown *et al.* (1972)	Angus, Hereford
Browning *et al.* (1995)	Angus, Brahman and Tuli
Butts *et al.* (1980a, b)	Angus, Hereford and Charolais
Chapman *et al.* (1970)	Angus, Polled Hereford, Santa Gertrudis, Brahman and Shorthorn
Chapman *et al.* (1971)	Angus, Polled Hereford and Santa Gertrudis
Chapman *et al.* (1978)	Angus, Hereford, Limousin and Simmental
Charles and Johnson (1976)	Hereford, Angus, Friesian and Charolais
Cianzio *et al.* (1982)	Limousin, Maine Anjou, Simmental, Angus, Hereford, Holstein and Brown Swiss
Comerford *et al.* (1987)	Brahman, Limousin, Polled Hereford and Simmental
Coulter *et al.* (1987)	Angus, Hereford
Crockett *et al.* (1978a, b)	Angus, Brahman and Hereford
Crockett *et al.* (1979)	Brahman, Brangus, Beefmaster, Limousin, Simmental, Maine Anjou, Angus and Hereford
Crouse *et al.* (1975)	Hereford, Angus, Limousin, Charolais, Simmental, South Devon and Jersey
Cundiff (1970)	Hereford, Angus, Shorthorn and Charolais
Cundiff *et al.* (1974a, b)	Angus, Hereford and Shorthorn
Dean *et al.* (1976)	Hereford, Holstein, Angus and Charolais
Deutscher and Slyter (1978)	Angus, Hereford and Charolais
Deutscher and Whiteman (1971)	Angus, Holstein
Dhuyvetter *et al.* (1985)	Charolais, Limousin, Angus, Hereford, Simmental, Brown Swiss and Jersey

Table 24.9. *Continued*

Authors	Breeds
Dikeman and Crouse (1975)	Hereford, Limousin, Angus and Simmental
Dillard *et al.* (1980)	Angus, Charolais and Hereford
Drewry *et al.* (1979a, b)	Angus, Milking Shorthorn
Dunn *et al.* (1969)	Angus, Hereford
Fortin *et al.* (1980a, b)	Holstein, Angus
Fortin *et al.* (1981a, b)	Holstein, Angus
Gaines *et al.* (1967)	Angus, Hereford and Shorthorn
Gaines *et al.* (1970)	Angus, Hereford and Shorthorn
Garcia-de-Siles *et al.* (1977)	Hereford, Holstein
Garrett (1971)	Holstein, Hereford
Glimp *et al.* (1971)	Angus, Hereford
Gray *et al.* (1978)	Angus, Hereford
Gregory and Cundiff (1980)	Angus, Brahman, Charolais and Hereford
Gregory *et al.* (1978a, b, c, d, e)	Angus, Hereford, Red Poll, Brown Swiss, Gelbvieh, Maine Anjou and Chianina
Gregory *et al.* (1979a, b)	Angus, Hereford, Brahman, Sahiwal, Pinzgauer and Tarentaise
Hedrick *et al.* (1970)	Angus, Charolais, Holstein, Brahman, Shorthorn and Hereford
Hedrick *et al.* (1975)	Angus, Charolais, Hereford
Holloway *et al.* (1975a, b)	Hereford, Holstein
Hooven *et al.* (1972)	Angus, Hereford, Holstein Friesian, Jersey and Milking Shorthorn
Jain *et al.* (1971)	Angus, Charolais and Hereford
Jenkins *et al.* (1981)	Angus, Brahman, Holstein, Hereford and Jersey
Koch and Dikeman (1977)	Hereford, Angus, Jersey, South Devon, Limousin, Charolais and Simmental
Koch *et al.* (1976)	Hereford, Angus, Jersey, South Devon, Limousin, Charolais and Simmental
Koch *et al.* (1979)	Hereford, Angus, Red Poll, Brown Swiss, Gelbvieh, Maine Anjou and Chianina
Koch *et al.* (1981)	Angus, Hereford, Red Poll, Brown Swiss, Gelbvieh, Maine Anjou and Chianina
Koger (1980)	Zebu, Brahman, Santa Gertrudis, Beefmaster, Brangus, Braford, Barzona, Charbray, Simbrah and Bramousin
Kress *et al.* (1995)	Tarentaise, Hereford
Kroger *et al.* (1975)	Brahman, Shorthorn
Kropp *et al.* (1973a, b)	Hereford, Holstein
Lasley *et al.* (1971)	Angus, Charolais and Hereford
Lasley *et al.* (1973)	Angus, Charolais and Hereford
Laster *et al.* (1972)	Hereford, Angus, Charolais, Simmental, South Devon, Jersey and Limousin
Laster *et al.* (1973a, b)	Red Poll, Brown Swiss, Hereford, Angus, Jersey, South Devon, Limousin, Simmental and Charolais
Laster *et al.* (1976)	Hereford, Angus, Jersey, South Devon, Limousin, Charolais and Simmental
Laster *et al.* (1979)	Hereford, Angus, Red Poll, Brown Swiss, Gelbvieh, Maine Anjou and Chianina
Lemka *et al.* (1973)	Hariana, Deshi, Blanco Orejinegro and Costeno Con Cuernos
LeVan *et al.* (1979)	Angus, Charolais

Table 24.9. *Continued*

Authors	Breeds
Long (1980)	Angus, Hereford, Charolais, Jersey, Limousin, Simmental, South Devon, Brown Swiss, Chianina, Gelbvieh, Maine Anjou, Red Poll, Brahman, Pinzgauer, Sahiwal, Tarentaise, Argentine Holstein, Blonde D'Aquitaine, German Black and White, German Red and White, Marchigiana, Normandie, Piedmontese, Romagnola, Santa Gertrudis and Shorthorn
Long and Gregory (1974)	Angus, Hereford
Long and Gregory (1975a, b)	Angus, Hereford
Long *et al.* (1979a, b)	Angus, Brahman, Hereford, Holstein and Jersey
Luckett *et al.* (1975)	Angus, Brahman, Hereford and Charolais
McAllister *et al.* (1976)	Polled Hereford, Charolais, Limousin, Simmental and Angus–Holstein
McDonald and Turner (1972)	Angus, Brahman, Brangus and Hereford
Marshall *et al.* (1976)	Angus, Charolais and Polled Hereford
Melton *et al.* (1967)	Angus, Charolais and Hereford
Nadarahaj *et al.* (1985)	Hereford, Angus, Shorthorn, Charolais, Simmental, Brown Swiss and Holstein
Nelsen *et al.* (1982a, b)	Angus, Brahman, Hereford, Holstein and Jersey
Northcutt *et al.* (1990)	Angus, Hereford and Brahman
Notter *et al.* (1978a, b)	Hereford, Angus, Charolais, Simmental, Limousin, Jersey, Brahman, Holstein, Maine Anjou, Chianina, Gelbvieh and South Devon
Nour *et al.* (1981)	Angus, Holstein
Ohlson *et al.* (1981)	Hereford, Simmental
O'Mary *et al.* (1979)	Angus, Charolais
Pahnish *et al.* (1969)	Charolais, Brown Swiss, Angus, Hereford and Brahman
Pahnish *et al.* (1971)	Charolais, Brown Swiss, Hereford and Angus
Peacock and Koger (1980)	Angus, Brahman and Charolais
Peacock *et al.* (1971)	Brahman, Shorthorn
Peacock *et al.* (1977)	Angus, Brahman and Charolais
Peacock *et al.* (1978)	Angus, Brahman and Charolais
Plasse *et al.* (1968)	Brahman, Shorthorn
Reynolds *et al.* (1978)	Angus, Brahman, Brangus and Africander–Angus
Reynolds *et al.* (1979)	Angus, Zebu, Brangus, Brahman and Africander–Angus
Reynolds *et al.* (1980)	Angus, Brahman, Brangus and Africander–Angus
Reynolds *et al.* (1982)	Angus, Brahman, Brangus and Africander–Angus
Rodriguez-Almeida *et al.* (1995a, b)	Angus, Charolais, Gelbvieh, Hereford, Limousin, Pinzgauer, Polled Hereford and Simmental
Rollins *et al.* (1969)	Angus, Hereford and Shorthorn
Rutledge (1975)	Hereford, Angus, Brown Swiss and Holstein
Sagebiel *et al.* (1969)	Angus, Charolais and Hereford
Sagebiel *et al.* (1973)	Angus, Charolais and Hereford
Sagebiel *et al.* (1974)	Angus, Charolais and Hereford
Sanders (1980)	Zebu, Guzerat, Nellore, Gir, Indu-Brazil, Brahman, Red Brahman and Grey Brahman
Scarth *et al.* (1973)	Angus, Hereford and Shorthorn
Smith (1976)	Hereford, Angus, Jersey, South Devon, Limousin, Charolais and Simmental
Smith and Cundiff (1976)	Angus, Hereford and Shorthorn
Smith *et al.* (1976a, b)	Angus, Hereford and Shorthorn

Table 24.9. *Continued*

Authors	Breeds
Smith et al. (1976c, d)	Hereford, Angus, Jersey, South Devon, Limousin, Charolais and Simmental
Stewart et al. (1980)	Angus, Brahman, Hereford, Holstein and Jersey
Thonney et al. (1981)	Holstein, Angus
Thrift et al. (1978)	Angus, Hereford, Charolais, Maine Anjou, Simmental and Holstein
Turner and McDonald (1969)	Angus, Brahman, Brangus, Charolais and Hereford
Turner et al. (1968)	Angus, Brahman, Brangus, Shorthorn, Charolais and Hereford
Urick et al. (1971)	Angus, Charolais, and Hereford
Urick et al. (1974)	Angus, Hereford, Charolais and Brown Swiss
Vogt et al. (1967)	Aberdeen-Angus, Hereford and Beef Shorthorn
Wettemann et al. (1982)	Brahman, Hereford and Holstein
Wiltbank et al. (1967)	Angus, Hereford and Shorthorn
Wiltbank et al. (1969)	Angus, Hereford
Wilton and Batra (1972)	Angus, Charolais, Hereford
Winer et al. (1981)	Hereford, Red Poll, Angus, Charolais and Brahman
Wyatt et al. (1977)	Hereford, Holstein and Charolais
Young et al. (1978a, b)	Hereford, Angus, Jersey, South Devon, Simmental, Limousin, Brahman, Charolais and Holstein
Ziegler et al. (1971)	Angus, Charolais, Holstein and Hereford

References

Adams, N.J., Garrett, W.N. and Elings, J.T. (1973) Performance and carcass characteristics of crosses from imported breeds. *Journal of Animal Science* 37, 623–628.

Adams, N.J., Smith, G.C. and Carpenter, Z.L. (1977) Carcass and palatability characteristics of Hereford and crossbred steers. *Journal of Animal Science* 46, 438–448.

Alenda, R. and Martin, T.G. (1981) Estimation of genetic and maternal effects in crossbred cattle of Angus, Charolais, and Hereford parentage. III. Optimal breed composition of crossbreds. *Journal of Animal Science* 53, 347–353.

Alenda, R., Martin, T.G., Lasley, J.F. and Ellersieck, M.R. (1980a) Estimation of genetic and maternal effects in crossbred cattle of Angus, Charolais and Hereford parentage. I. Birth and weaning weights. *Journal of Animal Science* 50, 226–234.

Alenda, R., Martin, T.G., Lasley, J.F. and Ellersieck, M.R. (1980b) Estimation of genetic and maternal effects in crossbred cattle of Angus, Charolais, and Hereford parentage. II. Postweaning growth, ribeye area and fat cover. *Journal of Animal Science* 50, 235–241.

Anderson, D.C., O'Mary, C.C. and Martin, E.L. (1978) Birth, preweaning and postweaning traits of Angus, Holstein, Simmental, and Chianina sired calves. *Journal of Animal Science* 46, 362–369.

Bailey, C.M. and Moore, J.D. (1980) Reproductive performance and birth characters of divergent breeds and crosses of beef cattle. *Journal of Animal Science* 50, 645–652.

Berndtson, W.E., Igboeli, G. and Pickett, B.W. (1987) Relationship of absolute numbers of Sertoli cells to testicular size and spermatogenesis in young beef bulls. *Journal of Animal Science* 64, 241–246.

Bond, J., Hooven, N.W., Jr, Warick, E.J., Hiner, R.L. and Richardson, G.V. (1972) Influence of breed and plane of nutrition on performance of dairy, dual-purpose and beef steers. II. From 180 days of age to slaughter. *Journal of Animal Science* 34, 1046–1053.

Briggs, H.M. and Briggs, D.M. (1980) *Modern Breeds of Livestock*, 4th edn. Macmillan, New York, 802 pp.

Brown, J.E., Brown, C.J. and Butts, W.T. (1972) A discussion of the genetic aspects of weight, mature weight and rate of maturing in Hereford and Angus cattle. *Journal of Animal Science* 34, 525–537.

Brown, M.A. and Dinkel, C.A. (1982) Efficiency to slaughter of calves from Angus, Charolais and reciprocal cross cows. *Journal of Animal Science* 55, 254–262.

Browning, R., Leite-Browning, M.L., Neuendorff, D.A. and Randel, R.D. (1995) Preweaning growth of Angus- (*Bos taurus*), Brahman- (*Bos indicus*), and Tuli- (Sanga) sired calves and reproductive performance of their Brahman dams. *Journal of Animal Science* 73, 2558–2563.

Butts, W.T., Jr, Backus, W.R., Lidvall, E.R., Corrick, J.A. and Montgomery, R.F. (1980a) Relationships among definable characteristics of feeder calves, subsequent performance and carcass traits. I. Objective measurements. *Journal of Animal Science* 51, 1297–1305.

Butts, W.T., Jr, Lidvall, E.R., Backus, W.R. and Corrick J.A. (1980b) Relationships among definable characteristics of feeder calves, subsequent performance and carcass traits. II. Subjective scores. *Journal of Animal Science* 51, 1306–1313.

Chapman, H.D., Clyburn, T.M. and McCormick, W.C. (1970) Grading, two- and three-breed rotational crossing as systems for production of calves to weaning. *Journal of Animal Science* 31, 642–651.

Chapman, H.D., Clyburn, T.M. and McCormick, W.C. (1971) Grading and two- and three-breed rotational crossing as systems for production of slaughter steers. *Journal of Animal Science* 32, 1062–1068.

Chapman, H.D., Morrison, E.G. and Edwards, N.C., Jr (1978) Limousin and Simmental sires mated with Angus and Hereford cows. *Journal of Animal Science* 46, 341–344.

Charles, D.D. and Johnson, E.R. (1976) Breed differences in amount and distribution of bovine carcass dissectible fat. *Journal of Animal Science* 42, 332–341.

Cianzio, D.S., Topel, D.G., Whitehurst, G.B., Beitz, D.C. and Self, H.L. (1982) Adipose tissue growth in cattle representing two frame sizes: distribution among depots. *Journal of Animal Science* 55, 305–312.

Comerford, J.W., Bertrand, J.K., Benyshek, L.L. and Johnson, M.H. (1987) Reproductive rates, birth weight, calving ease and 24-h calf survival in a four-breed diallel among Simmental, Limousin, Polled Hereford and Brahman beef cattle. *Journal of Animal Science* 64, 65–76.

Committee on Managing Global Genetic Resources (1993) *Managing Global Genetic Resources – Livestock*. National Academy Press, Washington, DC, 276 pp.

Coulter, G.H., Carruthers, T.D., Amann, R.P. and Kozub, G.C. (1987) Testicular development, daily sperm production and epididymal sperm reserves in 15-mo-old Angus and Hereford bulls: effects of bull strain plus dietary energy. *Journal of Animal Science* 64, 254–260.

Crockett, J.R., Koger, M. and Franke, D.E. (1978a) Rotational crossbreeding of beef cattle: reproduction by generation. *Journal of Animal Science* 46, 1163–1169.

Crockett, J.R., Koger, M. and Franke, D.E. (1978b) Rotational crossbreeding of beef cattle: preweaning traits by generation. *Journal of Animal Science* 46, 1170–1177.

Crockett, J.R., Baker, F.S. Jr, Carpenter, J.W. and Koger, M. (1979) Preweaning, feedlot and carcass characteristics of calves sired by Continental, Brahman and

Brahman-derivative sires in subtropical Florida. *Journal of Animal Science* 49, 900–907.

Crouse, J.D., Dikeman, M.E., Koch, R,M. and Murphey, C.E. (1975) Evaluation of traits in the U.S.D.A. yield grade equation for predicting beef carcass cutability in breed groups differing in growth and fattening characteristics. *Journal of Animal Science* 41, 548–553.

Cundiff, L.V. (1970) Experimental results on crossbreeding cattle for beef production. *Journal of Animal Science* 30, 694–705.

Cundiff, L.V., Gregory, K.E. and Koch, R.M. (1974a) Effects of heterosis on reproduction in Hereford, Angus and Shorthorn cattle. *Journal of Animal Science* 38, 711–727.

Cundiff, L.V., Gregory, K.E., Schwulst, F.J. and Koch, R.M. (1974b) Effects of heterosis on maternal performance and milk production in Hereford, Angus and Shorthorn cattle. *Journal of Animal Science* 38, 728–745.

Cundiff, L.V., Gregory, K.E., Koch, R.M. and Dickerson, G.E. (1986) Genetic diversity among cattle breeds and its use to increase beef production efficiency in a temperate environment. In: Dickerson, G.E. and Johnson, R.K. (eds) *Proceedings 3rd World Congress on Genetics Applied to Livestock Production: IX. Breeding Programs for Dairy and Beef Cattle, Water Buffalo, Sheep, and Goats*. University of Nebraska, Lincoln, pp. 271–282.

Cundiff, L.V., Szabo, F., Gregory, K.E., Koch, R.M., Dikeman, M.E. and Crouse, J.D. (1993) Breed comparisons in the germplasm evaluation program at MARC. In: Bolze, R. (ed.) *Proceedings of Beef Improvement Federation Research Symposium and Annual Meeting*. Beef Improvement Federation, Colby, Kansas, pp. 124–138.

Cundiff, L.V., Gregory, K.E., Wheeler, T.L., Shackelford, S.D., Koohmaraie, M., Freetly, H.C. and Lunstra, D.D. (1997) Preliminary results from Cycle V of the cattle germplasm evaluation program at the Roman L. Hruska U.S. Meat Animal Research Center. In: *Germplasm Evaluation Program Progress Report No. 16*. US Department of Agriculture, Clay Center, Nebraska, pp. 2–11.

Dean, R.A., Walters, L.E., Whiteman, J.V., Stephens, D.F. and Totusek, R. (1976) Carcass traits of progeny of Hereford, Hereford × Holstein and Holstein cows. *Journal of Animal Science* 42, 1427–1433.

Deutscher, G.H. and Slyter, A.L. (1978) Crossbreeding and management systems for beef production. *Journal of Animal Science* 47, 19–28.

Deutscher, G.H. and Whiteman, J.V. (1971) Productivity as two-year-olds of Angus–Holstein crossbreds compared to Angus heifers under range conditions. *Journal of Animal Science* 33, 337–342.

Dhuyvetter, J.M., Frahm, R.R. and Marshall, D.M. (1985) Comparison of Charolais and Limousin as terminal cross sire breeds. *Journal of Animal Science* 60, 935–941.

Dikeman, M.E. and Crouse, J.D. (1975) Chemical composition of carcasses from Hereford, Limousin and Simmental crossbred cattle as related to growth and meat palatability. *Journal of Animal Science* 40, 463–467.

Dillard, E.U., Rodriquez, O. and Robison, O.W. (1980) Estimation of additive and nonadditive direct and maternal genetic effects from crossbreeding beef cattle. *Journal of Animal Science* 50, 653–663.

Drewry, K.J., Becker, S.P., Martin, T.G. and Nelson, L.A. (1979a) Crossing Angus and Milking Shorthorn cattle: feedlot performance of steers. *Journal of Animal Science* 48, 313–318.

Drewry, K.J., Becker, S.P., Martin, T.G. and Nelson, L.A. (1979b) Crossing Angus and Milking Shorthorn cattle: steer carcass traits. *Journal of Animal Science* 48, 517–524.

Dunn, T.G., Ingalls, J.E., Zimmerman, D.R, and Wiltbank, J.N. (1969) Reproductive performance of 2-year-old Hereford and Angus heifers as influenced by pre- and post-calving energy intake. *Journal of Animal Science* 29, 719–726.

Felius, M. (1985) *Cattle Breeds of the World*. Merck, Rahway, New Jersey, 234 pp.

Fortin, A., Simpfendorfer, S., Reid, J.T., Ayala, H.J., Anrique, R. and Kertz, A.F. (1980a) Effect of level of energy intake and influence of breed and sex on the chemical composition of cattle. *Journal of Animal Science* 51, 604–614.

Fortin, A., Reid, J.T., Maiga, A.M., Sim, D.W. and Wellington, G.H. (1980b) Effect of level of energy intake and influence of breed and sex on muscle growth and distribution in the bovine carcass. *Journal of Animal Science* 51, 1288–1296.

Fortin, A., Reid, J.T., Maiga, A.M., Sim, D.W. and Wellington, G.H. (1981a) Effect of energy intake level and influence of breed and sex on the physical composition of the carcass of cattle. *Journal of Animal Science* 51, 331–339.

Fortin, A., Reid, T.J., Maiga, A.M., Sim, D.W. and Wellington, G.H. (1981b) Effect of level of energy intake and influence of breed and sex on growth of fat tissue and distribution in the bovine carcass. *Journal of Animal Science* 53, 982–991.

Gaines, J.A., Richardson, G.V., McClure, W.H., Vogt, D.W. and Carter, R.C. (1967) Heterosis from crosses among British breeds of beef cattle: carcass characteristics. *Journal of Animal Science* 26, 1217–1225.

Gaines, J.A., Richardson, G.V., Carter, R.C. and McClure, W.H. (1970) General combining ability and maternal effects in crossing three British breeds of beef cattle. *Journal of Animal Science* 31, 19–26.

Garcia-de-Siles, J.L., Ziegler, J.H., Wilson, L.L. and Sink, J.D. (1977) Growth, carcass and muscle characters of Hereford and Holstein steers. *Journal of Animal Science* 44, 973–984.

Garrett, W.N. (1971) Energetic efficiency of beef and dairy steers. *Journal of Animal Science* 32, 451–456.

Glimp, H.A., Dikeman, M.E., Tuma, H.J., Gregory, K.E. and Cundiff, L.V. (1971) Effect of sex condition on growth and carcass traits of male Hereford and Angus cattle. *Journal of Animal Science* 33, 1242–1247.

Gray, E.F., Thrift, F.A. and Absher, C.W. (1978) Heterosis expression for preweaning traits under commercial beef cattle conditions. *Journal of Animal Science* 47, 370–374.

Gregory, K.E. and Cundiff, L.V. (1980) Crossbreeding in beef cattle: evaluation of systems. *Journal of Animal Science* 51, 1224–1242.

Gregory, K.E., Cundiff, L.V., Smith, G.M., Laster, D.B. and Fitzhugh, H.A., Jr (1978a) Characterization of biological types of cattle – cycle II: I. Birth and weaning traits. *Journal of Animal Science* 47, 1022–1030.

Gregory, K.E., Cundiff, L.V., Koch, R.M., Laster, D.B. and Smith, G.M. (1978b) Heterosis and breed maternal and transmitted effects in beef cattle. I. Preweaning traits. *Journal of Animal Science* 47, 1031–1041.

Gregory, K.E., Laster, D.B., Cundiff L.V., Koch, R.M. and Smith, G.M. (1978c) Heterosis and breed maternal and transmitted effects in beef cattle. II. Growth rate and puberty in females. *Journal of Animal Science* 47, 1042–1053.

Gregory, K.E., Koch, R.M., Laster, D.B., Cundiff, L.V. and Smith G.M. (1978d) Heterosis and breed maternal and transmitted effects in beef cattle. III. Growth traits of steers. *Journal of Animal Science* 47, 1054–1062.

Gregory, K.E., Crouse, J.D., Koch, R.M., Laster, D.B., Cundiff, L.V. and Smith, G.M. (1978e) Heterosis and breed maternal and transmitted effects in beef cattle. IV. Carcass traits of steers. *Journal of Animal Science* 47, 1063–1079.

Gregory, K.E., Smith, G.M., Cundiff, L.V., Koch, R.M. and Laster, D.B. (1979a) Characterization of biological types of cattle – cycle III: I. Birth and weaning traits. *Journal of Animal Science* 48, 271–279.

Gregory, K.E., Laster, D.B., Cundiff, L.V., Smith, G.M. and Koch, R.M (1979b) Characterization of biological types of cattle – cycle III: II. Growth rate and puberty in females. *Journal of Animal Science* 49, 461–471.

Hedrick, H.B., Lasley, J.F, Jain, J.P., Krause, G.F., Sibbit, B., Langford, L., Comfort, J.E. and Dyer, A.J. (1970) Quantitative carcass characteristics of reciprocally crossed Angus, Charolais and Hereford heifers. *Journal of Animal Science* 31, 633–641.

Hedrick, H.B., Krause, G.F., Lasley, J.F., Sibbit, B., Langford, L. and Dyer, A.J. (1975) Quantitative and qualitative carcass characteristics of straightbred and reciprocally crossed Angus, Charolais and Hereford steers. *Journal of Animal Science* 41, 1581–1591.

Holloway, J.W., Stephens, D.F., Whiteman, J.V. and Totusek, R. (1975a) Performance of 3-year-old Hereford, Hereford × Holstein and Holstein cows on range and in drylot. *Journal of Animal Science* 40, 114–125.

Holloway, J.W., Stephens, D.F., Whiteman, J.V. and Totusek, R. (1975b) Efficiency of production of 2- and 3-year-old Hereford, Hereford × Holstein and Holstein cows. *Journal of Animal Science* 41, 855–867.

Hooven, N.W., Jr., Bond, J., Warwick, E.J., Hiner, R.L. and Richardson, G.V. (1972) Influence of breed and plane of nutrition on the performance of dairy, dual-purpose and beef steers. I. Birth to 180 days of age. *Journal of Animal Science* 34, 1037–1045.

Jain, J.P., Lasley, J.F., Sibbit, B., Langford, L., Comfort, J.E., Dyer, A.J., Krause, G.F. and Hedrick, H.B. (1971) Growth traits of reciprocally crossed Angus, Hereford and Charolais heifers. *Journal of Animal Science* 32, 399–405.

Jenkins, T.G., Long, C.R., Cartwright, T.C. and Smith, G.C. (1981) Characterization of cattle of a five-breed diallel. IV. Slaughter and carcass characters of serially slaughtered bulls. *Journal of Animal Science* 53, 62–79.

Koch, R.M. and Dikeman, M.E. (1977) Characterization of biological types of cattle. V. Carcass wholesale cut composition. *Journal of Animal Science* 45, 30–42.

Koch, R.M., Dikeman, M.E., Allen, D.M., May, M., Crouse, J.D. and Campion, D.R. (1976) Characterization of biological types of cattle. III. Carcass composition, quality, and palatability. *Journal of Animal Science* 43, 48–62.

Koch, R.M., Dikeman, M.E., Lipse, J., Allen, D.M. and Crouse, J.D. (1979) Characterization of biological types of cattle – cycle II: III. Carcass composition, quality and palatability. *Journal of Animal Science* 49, 448–460.

Koch, R.M., Dikeman, M.E. and Cundiff, L.V. (1981) Characterization of biological types of cattle (cycle II). V. Carcass wholesale cut composition. *Journal of Animal Science* 53, 992–999.

Koger, M. (1980) Effective crossbreeding systems utilizing Zebu cattle. *Journal of Animal Science* 50, 1215–1220.

Kress, D.D., Doornbos, D.E., Anderson, D.C. and Davis, K.C. (1995) Tarentase and Hereford breed effects on cow and calf traits and estimates of individual heterosis. *Journal of Animal Science* 73, 2574–2578.

Kroger, M., Peacock, F.M., Kirk, W.G. and Crockett, J.R. (1975) Heterosis effects on weaning performance of Brahman–Shorthorn calves. *Journal of Animal Science* 40, 826–833.

Kropp, J.R., Holloway, J.W., Stephens, D.F., Knori, L., Morrison, R.D. and Totusek, R. (1973a) Range behavior of Hereford, Hereford × Holstein and Holstein non-lactating heifers. *Journal of Animal Science* 36, 797–802.

Kropp, J.R., Stephens, D.F., Holloway, J.W., Whiteman, J.V., Knori, L. and Totusek, R. (1973b) Performance on range and in drylot of two-year-old Hereford, Hereford × Holstein and Holstein females as influenced by level of winter supplementation. *Journal of Animal Science* 37, 1222–1232.

Lasley, J.F., Krause, G.F., Jain, J.P., Hedrick, H.B., Sibbit, B., Langford, L., Comfort, J.E. and Dyer, A.J. (1971) Carcass quality characteristics in heifers of reciprocal crosses of the Angus, Charolais and Hereford breeds. *Journal of Animal Science* 32, 406–412.

Lasley, J.F., Sibbit, B., Langford, L., Comfort, J.E., Dyer, A.J., Krause, G.F. and Hedrick, H.B. (1973) Growth traits in straightbred and reciprocally crossed Angus, Hereford and Charolais steers. *Journal of Animal Science* 36, 1044–1056.

Laster, D.B., Glimp, H.A. and Gregory, K.E. (1972) Age and weight at puberty and conception in different breeds and breed-crosses of beef heifers. *Journal of Animal Science* 34, 1031–1036.

Laster, D.B., Glimp, H.A., Cundiff, L.V. and Gregory, K.E. (1973a) Factors affecting dystocia and the effects of dystocia on subsequent reproduction in beef cattle. *Journal of Animal Science* 36, 695–705.

Laster, D.B., Glimp, H.A. and Gregory, K.E. (1973b) Effects of early weaning on postpartum reproduction of cows. *Journal of Animal Science* 36, 734–740.

Laster, D.B., Smith G.M. and Gregory, K.E. (1976) Characterization of biological types of cattle. IV. Postweaning growth and puberty of heifers. *Journal of Animal Science* 43, 63–70.

Laster, D.B., Smith, G.M., Cundiff, L.V. and Gregory, K.E. (1979) Characterization of biological types of cattle (cycle II). II. Postweaning growth and puberty of heifers. *Journal of Animal Science* 48, 500–508.

Lemka, L., McDowell, R.E., Van Vleck, L.D., Guha, H. and Salazar, J.J. (1973) Reproductive efficiency and viability in two *Bos indicus* and two *Bos taurus* breeds in the tropics of India and Colombia. *Journal of Animal Science* 36, 644–652.

LeVan, P.J., Wilson, L.L., Watkins, J.L., Grieco, C.K., Ziegler, J.H. and Barber, K.A. (1979) Retail lean, bone and fat distribution of Angus and Charolais steers slaughtered at similar stages of physiological maturity. *Journal of Animal Science* 49, 683–692.

Lloyd-Jones, O. (1915) What is a breed? *Journal of Heredity* 6, 531.

Long, C.R. (1980) Crossbreeding for beef production: experimental results. *Journal of Animal Science* 51, 1197–1223.

Long, C.R. and Gregory, K.E. (1974) Heterosis and breed effects in preweaning traits of Angus, Hereford and reciprocal cross calves. *Journal of Animal Science* 39, 11–17.

Long, C.R. and Gregory, K.E. (1975a) Heterosis and management effects in postweaning growth of Angus, Hereford and reciprocal cross cattle. *Journal of Animal Science* 41, 1563–1571.

Long, C.R. and Gregory, K.E. (1975b) Heterosis and management effects in carcass characters of Angus, Hereford and reciprocal cross cattle. *Journal of Animal Science* 41, 1572–1580.

Long, C.R., Stewart, T.S., Cartwright, T.C. and Jenkins, T.G. (1979a) Characterization of cattle of a five breed diallel: I. Measures of size, condition and growth in bulls. *Journal of Animal Science* 49, 418–431.

Long, C.R., Stewart, T.S., Cartwright, T.C. and Baker, J.F. (1979b) Characterization of cattle of a five breed diallel: II. Measures of size, condition, and growth in heifers. *Journal of Animal Science* 49, 432–447.

Luckett, R.L., Bidner, T.D., Icaza, E.A. and Turner, J.W. (1975) Tenderness studies in straightbred and crossbred steers. *Journal of Animal Science* 40, 468–475.

Lush, J.L. (1994) Making new breeds. In: *The Genetics of Populations*. Iowa State University, Ames, pp. 816–860.

McAllister, T.J., Wilson, L.L., Ziegler, J.H. and Sink, J.D. (1976) Growth rate, carcass quality and fat, lean and bone distribution of British- and Continental-sired crossbred steers. *Journal of Animal Science* 42, 324–331.

McDonald, R.P. and Turner, J.W. (1972) Estimation of maternal heterosis in preweaning traits of beef cattle. *Journal of Animal Science* 35, 1146–1153.

Marshall, D.A., Parker, W.R. and Dinkel, C.A. (1976) Factors affecting efficiency to weaning in Angus, Charolais and reciprocal cross cows. *Journal of Animal Science* 43, 1176–1187.

Mason, I.L. (1996) *A World Dictionary of Livestock Breeds, Types and Varieties*, 4th edn. CAB International, Wallingford, UK, 273 pp.

Melton, A.A., Riggs, J.K., Nelson, L.A. and Cartwright, T.C. (1967) Milk production, composition and calf gains of Angus, Charolais and Hereford cows. *Journal of Animal Science* 26, 804–809.

Nadarahaj, K., Marlowe, T.J. and Notter, D.R. (1985) Growth patterns of cows sired by British and Continental beef and American dairy bulls and out of Hereford dams. *Journal of Animal Science* 60, 890–901.

Nelsen, T.C., Long, C.R. and Cartwright, T.C. (1982a) Postinflection growth in straightbred and crossbred cattle. I. Heterosis for weight, height and maturing rate. *Journal of Animal Science* 55, 280–292.

Nelsen, T.C., Long, C.R. and Cartwright, T.C. (1982b) Postinflection growth in straightbred and crossbred cattle. II. Relationships among weight, height and pubertal characters. *Journal of Animal Science* 55, 293–304.

Northcutt, S.L., Aaron, D.K. and Thrift, F.A. (1990) Influence of specific genotype × environment interactions on preweaning beef cattle traits in the southern region. *Applied Agricultural Research* 5, 63–69.

Notter, D.R., Cundiff, L.V., Smith, G.M., Laster, D.B. and Gregory, K.E. (1978a) Characterization of biological types of cattle. VI. Transmitted and maternal effects on birth and survival traits in progeny of young cows. *Journal of Animal Science* 46, 892–907.

Notter, D.R., Cundiff, L.V., Smith, G.M., Laster, D.B. and Gregory, K.E. (1978b) Characterization of biological types of cattle. VII. Milk production in young cows and transmitted and maternal effects on preweaning growth of progeny. *Journal of Animal Science* 46, 908–921.

Nour, A.Y.M., Thonney, M.L., Stouffer, J.R. and White, W.R.C., Jr (1981) Muscle, fat and bone in serially slaughtered large dairy or small beef cattle fed corn or corn silage diets in one of two locations. *Journal of Animal Science* 52, 512–521.

Ohlson, D.L., Davis, S.L., Ferrell, C.L. and Jenkins, T.G. (1981) Plasma growth hormone prolactin and thyrotropin secretory patterns in Hereford and Simmental calves. *Journal of Animal Science* 53, 371–375.

O'Mary, C.C., Martin, E.L. and Anderson, D.C. (1979) Production and carcass characteristics of Angus and Charolais × Angus steers. *Journal of Animal Science* 48, 239–245.

Pahnish, O.F., Brinks, J.S., Urick, J.J., Knapp, B.W. and Riley, T.M. (1969) Results from crossing beef × beef and beef × dairy breeds: calf performance to weaning. *Journal of Animal Science* 28, 291–299.

Pahnish, O.F., Knapp, B.W., Urick, J.J., Brinks, J.S. and Willson, F.S. (1971) Results from crossing beef × beef and beef × dairy breeds: postweaning performance of heifers. *Journal of Animal Science* 33, 736–743.

Peacock, F.M. and Koger, M. (1980) Reproductive performance of Angus, Brahman, Charolais and crossbred dams. *Journal of Animal Science* 50, 689–693.

Peacock, F.M., Koger, M., Kirk, W.G., Hodges, E.M. and Warnick, A.C. (1971) Reproduction in Brahman, Shorthorn and crossbred cows on different pasture programs. *Journal of Animal Science* 33, 458–465.

Peacock, F.M., Koger, M., Crockett, J.R. and Warnick, A.C. (1977) Reproductive performance and crossbreeding Angus, Brahman and Charolais cattle. *Journal of Animal Science* 44, 729–733.

Peacock, F.M., Koger, M. and Hodges, E.M. (1978) Weaning traits of Angus, Brahman, Charolais and F_1 crosses of these breeds. *Journal of Animal Science* 47, 366–369.

Plasse, D., Warnick, A.C. and Koger, M. (1968) Reproductive behavior of *Bos indicus* females in a subtropical environment. I. Puberty and ovulation frequency in Brahman and Brahman × British heifers. *Journal of Animal Science* 27, 94–100.

Reynolds, W.L., DeRouen, T.M. and Bellows, R.A. (1978) Relationships of milk yield of dam to early growth rate of straightbred and crossbred calves. *Journal of Animal Science* 47, 584–594.

Reynolds, W.L., DeRouen, T.M., Moin, S. and Koonce, K.L. (1979) Factors affecting pregnancy rate of Angus, Zebu and Zebu-cross cattle. *Journal of Animal Science* 48, 1312–1321.

Reynolds, W.L., DeRouen, T.M., Moin, S. and Koonce, K.L. (1980) Factors influencing gestation length, birth weight and calf survival of Angus, Zebu and Zebu cross beef cattle. *Journal of Animal Science* 51, 860–867.

Reynolds, W.L., DeRouen, T.M. and Koonce, K.L. (1982) Preweaning growth rate and weaning traits of Angus, Zebu and Zebu-cross cattle. *Journal of Animal Science* 54, 241–247.

Rodriguez-Almeida, F.A., Van Vleck, L.D., Cundiff, L.V. and Kachman, S.D. (1995a) Heterogeneity of variance by sire, breed, sex, and dam breed in 200- and 365-day weights of beef cattle from a top cross experiment. *Journal of Animal Science* 73, 2579–2588.

Rodriguez-Almeida, F.A., Van Vleck, L.D. and Cundiff, L.V. (1995b) Effect of accounting for different phenotypic variances by sire breed and sex on selection of sires based on expected progeny differences for 200- and 365-day weights. *Journal of Animal Science* 73, 2589–2599.

Rollins, W.C., Loy, R.G., Carroll, F.D. and Wagnon, K.A. (1969) Heterotic effects in reproduction and growth to weaning in crosses of the Angus, Hereford and Shorthorn breeds. *Journal of Animal Science* 28, 431–436.

Rouse, J.E. (1970a) *World Cattle*, Vol. I. *Cattle of Europe, South America, Australia, and New Zealand*. University of Oklahoma Press, Norman, 485 pp.

Rouse, J.E. (1970b) *World Cattle*, Vol. II. *Cattle of Africa and Asia*. University of Oklahoma Press, Norman, 557 pp.

Rouse, J.E. (1973) *World Cattle*, Vol. III. *Cattle of North America*. University of Oklahoma Press, Norman, 650 pp.

Rutledge, J.J. (1975) Twinning in cattle. *Journal of Animal Science* 40, 803–815.

Sagebiel, J.A., Krause, G.F., Sibbit, B., Langford, L., Comfort, J.E., Dyer, A.J. and Lasley, J.F. (1969) Dystocia in reciprocally crossed Angus, Hereford and Charolais cattle. *Journal of Animal Science* 29, 245–250.

Sagebiel, J.A., Krause, G.F., Sibbit, B., Langford, L., Dyer, A.J. and Lasley, J.F. (1973) Effect of heterosis and maternal influence on gestation length and birth weight in reciprocal crosses among Angus, Charolais and Hereford cattle. *Journal of Animal Science* 37, 1273–1278.

Sagebiel, J.A., Krause, G.F., Sibbit, B., Langford, L., Dyer, A.J. and Lasley, J.F. (1974) Effect of heterosis and maternal influence on weaning traits in reciprocal crosses among Angus, Charolais and Hereford cattle. *Journal of Animal Science* 39, 471–479.

Sanders, J.O. (1980) History and development of Zebu cattle in the United States. *Journal of Animal Science* 50, 1188–1200.

Scarth, R.D., Kauffman, R.G. and Bray, R.W. (1973) Effects of breed and age classification on live weight and carcass traits of steers shown at International Quality Beef Show. *Journal of Animal Science* 36, 653–657.

Smith, G.M. (1976) Sire breed effects on economic efficiency of a terminal-cross beef production system. *Journal of Animal Science* 43, 1163–1170.

Smith, G.M. and Cundiff, L.V. (1976) Genetic analysis of relative growth rate in crossbreed and straightbred Hereford, Angus and Shorthorn steers. *Journal of Animal Science* 43, 1171–1175.

Smith, G.M., Laster, D.B. and Gregory, K.E. (1976a) Characterization of biological types of cattle I. Dystocia and preweaning growth. *Journal of Animal Science* 43, 27–36.

Smith, G.M., Laster, D.B., Cundiff, L.V. and Gregory, K.E. (1976b) Characterization of biological types of cattle II. Postweaning growth and feed efficiency of steers. *Journal of Animal Science* 43, 37–47.

Smith, G.M., Fitzhugh, H.A., Jr, Cundiff, L.V., Cartwright, T.C. and Gregory, K.E. (1976c) Heterosis for maturing patterns in Hereford, Angus and Shorthorn cattle. *Journal of Animal Science* 43, 380–388.

Smith, G.M., Fitzhugh, H.A., Jr, Cundiff, L.V., Cartwright, T.C. and Gregory, K.E. (1976d) A genetic analysis of maturing patterns in straightbred and crossbred Hereford, Angus and Shorthorn cattle. *Journal of Animal Science* 43, 389–395.

Stewart, T.S., Long, C.R. and Cartwright, T.C. (1980) Characterization of cattle of a five-breed diallel. III. Puberty in bulls and heifers. *Journal of Animal Science* 50, 808–820.

Thonney, M.L., Heide, E.K., Duhaime, D.J, Nour, A.Y.M, and Oltenacu, P.A. (1981) Growth and feed efficiency of cattle of different mature sizes. *Journal of Animal Science* 53, 354–362.

Thrift, F.A., Gallion, S.R. and Absher, C.W. (1978) Breed of sire and dam comparisons for preweaning traits under commercial beef cattle conditions. *Journal of Animal Science* 46, 977–982.

Turner, J.W. and McDonald, R.P. (1969) Mating-type comparisons among crossbred beef cattle for preweaning traits. *Journal of Animal Science* 29, 389–397.

Turner, J.W., Farthing, B.R. and Robertson, G.L. (1968) Heterosis in reproductive performance of beef cows. *Journal of Animal Science* 27, 336–338.

Urick, J.J., Knapp, B.W., Brinks, J.S., Pahnish, O.F. and Riley, T.M. (1971) Relationships between cow weights and calf weaning weights in Angus, Charolais and Hereford breeds. *Journal of Animal Science* 33, 343–348.

Urick, J.J., Knapp, B.W., Hiner, R.L., Pahnish, O.F., Brinks, J.S. and Blackwell, R.L. (1974) Results from crossing beef × beef and beef × Brown Swiss: carcass quantity and quality traits. *Journal of Animal Science* 39, 292–302.

Vogt, D.W., Gaines, J.A., Carter, R.C., McClure, W.H. and Kincaid, C.M. (1967) Heterosis from crosses among British breeds of beef cattle: post-weaning performance to slaughter. *Journal of Animal Science* 26, 443–452.

Walker, H., III (1989) *Blue Book of Beef Breeds*. PAW Publishing, Allen, Kansas, 136 pp.

Wettemann, R.P., Tucker, H.A., Beck, T.W. and Meyerhoeffer, D.C. (1982) Influence of ambient temperature on prolactin concentrations in serum of Holstein and Brahman × Hereford heifers. *Journal of Animal Science* 55, 391–394.

Willham, R.L. (1987) *Taking Stock*. Iowa State University, Ámes.

Wiltbank, J.N., Gregory, K.E., Rothlisberger, J.A., Ingalls, J.E. and Kasson, C.W. (1967) Fertility in beef cows bred to produce straightbred and crossbred calves. *Journal of Animal Science* 26, 1005–1010.

Wiltbank, J.N., Kasson, C.W. and Ingalls, J.E. (1969) Puberty in crossbred and straightbred beef heifers on two levels of feed. *Journal of Animal Science* 29, 602–605.

Wilton, J.W. and Batra, T.R. (1972) Variance of gain of beef bulls in central stations. *Journal of Animal Science* 538–540.

Winer, L.K., David, P.J., Curtiss, M.B., Read, M., Ringkob, T.P. and Stevenson, M. (1981) Palatability characteristics of the *Longissimus* muscle of young bulls representing divergent beef breeds and crosses. *Journal of Animal Science* 53, 387–394.

Wright, S. (1977) *Evolution and the Genetics of Populations*, Vol. 3. University of Chicago Press, Chicago.

Wyatt, R.D., Lusby, K.S., Gould, M.B., Walters, L.E., Whiteman, J.V. and Totusek R. (1977) Feedlot performance and carcass traits of progeny of Hereford, Hereford × Holstein and Holstein cows. *Journal of Animal Science* 45, 1131–1137.

Young, L.D., Cundiff, L.V., Crouse, J.D., Smith, G.M. and Gregory, K.E. (1978a) Characterization of biological types of cattle. VIII. Postweaning growth and carcass traits of three-way cross steers. *Journal of Animal Science* 46, 1178–1191.

Young, L.D., Laster, D.B., Cundiff, L.V., Smith, G.M. and Gregory, K.E. (1978b) Characterization of biological types of cattle. IX. Postweaning growth and puberty of three-breed cross heifers. *Journal of Animal Science* 47, 843–852.

Ziegler, J.H., Wilson, L.L. and Coble, D.S. (1971) Comparisons of certain carcass traits of several breeds and crosses of cattle. *Journal of Animal Science* 32, 446–450.

Index

Note: all breeds are entered under the heading 'breeds'

abortion and twinning 394, 395
acid phosphatase polymorphism 95
acrodermatitis enteropathica 65
acrosome reaction 413
adaptation
    definition 638–639
    evolutionary 638
    physiological 638
    in the Tropics 637–656
additive genetic value 499–500
adenosine deaminase polymorphism 98
Aepycerotinae 4
African buffalo 5
age
    and dystocia 402
    and gestation length 400
    and twinning rate 393–394
Agg locus 88
aggression 368, 370–371
albumin polymorphism 90
alkaline phosphatase polymorphism 90–92
alkaline ribonuclease polymorphism 98
allele 336, 340, 661–662
    discrimination 342
    frequency 334
    maternal/paternal 445-446
allele-specific oligonucleotide (ASO) 341, 342, 343
allo-grooming 370
allotypes 93
α-lactalbumin 544–545

downregulation 562
    polymorphism 100
    transgenes 561
α-mannosidosis 60
Amblyomma 210, 216
amelogenin 457–458
American bison 6–7, 10, 11, 35
amplification-created restriction sites (ACRS) 341
amplified fragment length polymorphism (AFLP) 341, 342, 343
amylase polymorphism 92
Ancelaphine 3
androgen insensitivity syndrome 67
androgen receptor deficiency 67
androstenedione 456
aneuploid heteroploidy 256
Angleton research herd 337
aniridia 453
anoa 5–6, 10
anthelminthic resistance 201
antibodies 172
antigen-presenting cells (APCs) 166
anti-J 85
Antilopinae 3
anti-Müllerian hormone 456, 459
aplasia cutis 64
approach/avoidance behaviour 372
argininosuccinate synthetase 58
ark-farm projects 479–480
arni 5, 6, 10

697

artificial insemination (AI) 375
   beef cattle 594–595
   and breeding programmes 528–530
   selection of bulls for 502
artiodactyls 2, 3
associative behaviour 369–370
auroch 9, 15, 16
   coat colour 35
   population bottleneck during domestication 25–26
   sexual dimorphism 23
   social behaviour 367
autosomes 124, 248
average effect of gene/allele substitution 500
Axin 449

B-cell receptor (BCR) 172
B cells 164, 172, 174
bacterial artificial chromosomes (BACs) 132, 263, 264
bail test 372
Banteng 8, 11, 35, 681
baulking rating 372
BCS 127, 128
beef, consumer demand for 625–626
beef cattle
   adaptation in the Tropics 637–656
   breeding goals 578–584
   breeding replacement females in specialized herds 583–584
   breeds 587–588
   cloning 599–600
   cooperative breeding schemes 593
   crossbreeding 582, 587–588, 627
      theory 584–587
   Europe 19
   evaluation of breed resources 584
   evaluation across herds, breeds and countries 594–595
   evaluation of individuals 588–592
   evidence and value of genetic improvement 596–597
   formulation of breeding objectives 579–581
   genetic improvement 577–603
   indices of overall economic merit 595–596
   multiple ovulation and embryo transfer 598–599
   performance testing 592
   progeny testing 593
   quantitative trait loci 597–598
   selection within breeds 588–597
   sexing 599
   terminal sire breeds 582–583
   testing systems 592–593
   traits recorded 593–594
beef production without brood cows 413
behaviour 365–366, 384–385
   feeding 381–384
   innate and learned traits 366
   maternal 380–381
   sexual 375, 378–380
   social 367–371
behavioural tests 372
Bennett–Goodspeed phenomenon 86
best linear unbiased prediction (BLUP) 590–591
β-carotene 625
β-casein element-1 (BCE-1) 558–559
β-lactoglobulin 545–546
   endocrine control of expression 548
   polymorphism 100, 523–524
β-mannosidosis 65
Bibos frontalis 8, 15, 20
Bibos gaurus 8, 11, 20
Bibos sauveli 8
biochemical polymorphisms 87, 90, 91, 334
   and breed characteristics 102–103
   and economically important traits 103–105
   milk proteins 99–100, 523–524
   and parentage verification 101
   plasma proteins 90–95
   red-cell proteins 95–97
   and twin determinations 101–102
   utilization 100–105
   white-cell proteins 98–99
birth weight and dystocia 401
Bison bison 6–7, 10, 11, 35
Bison bonasus 7, 10, 11
blastocoele 439
blastocyst 438–439, 440–446, 449
blood groups
   antigens 88–90
   early studies 77–79
   factors 79–80
   loci 80
   monoclonal antibodies with specificity for factors 87–88

nomenclature and definitions 79–80
standardization of typing results 87
systems 80, 81–87
utilization 100–105
Wisconsin contribution 78–79
blood typing 101–105
blood-typing reagent 80
BMF 127, 128
body composition 606–608
    breed variation 617–619
    genetic correlations 609–610
    heterosis effects 615
Boophilus decoloratus 216
Boophilus microplus 216
Bos frontalis 8, 15, 20
Bos gaurus 8, 11, 20
Bos grunniens see yak
Bos indicus see zebu
Bos primigenius see auroch
Bos sauveli 8
Bos taurus 9, 15, 16–22, 668
Boselaphini 5
Bov-A 127, 128
Bov-A2 127, 128
Bov-B 127, 128
Bov-tA 127, 128
Bovidae 1, 3–4
Bovinae 1, 3, 4, 5
bovine leucocyte antigens (BoLA) 86, 166
    Class I genes and proteins 168–169
    Class I peptide binding motifs 171–172
    Class IIa genes and proteins 170–171
    Class IIb genes and proteins 171
    genomic organization 167–168
bovine leucosis resistance 207–208
bovine spongiform encephalopathy (BSE) 229–230, 241
    clinical signs and pathology 230
    experimental 233
    genetics 235–241
    natural 232
    pathogenesis 232–233
    preclininal diagnosis 233
    prion protein in 231–232
    transmission characteristics 233–234
Bovini 1, 5–9
    chromosome numbers 5
    phylogeny 9–11
BovMap consortium 263
branchygnatia 256

breed societies 669
breeding programmes 528–532, 598–599, 637–651
breeding value 499, 588
    see also estimated breeding values
BREEDOBJECT 648
BREEDPLAN 648
breeds 9
    Abyssinian 673
    Adamawa 673
    Africander 673
    Alentejana 674
    Alpine Brown 480
    American White Park 49, 679
    Amerifax 679
    Angler 18
    Angus 19, 21, 22, 587, 677
    Ankina 679
    Ankole 21
    Ankole-Watusi 673
    as appropriate units of conservation 484
    Arsi 673
    Atlas Brown 20–21
    Aubrac 674
    Ayrshire 45, 50, 317, 516, 670, 677
    Baggara 21
    Baladi 19, 20, 673
    Bali cattle 8, 15, 20, 681
    Baoulé 21
    Bapedi 51
    Barotse 21
    Barzona 679
    Beefmaster 45, 679
    Belgian Blue 45, 48, 49, 61, 454, 587, 616, 641, 642, 651, 674
    Belmont Adaptaur 641, 642, 651
    Belmont BX 641, 642
    Belmont Red 641, 642
    Belted Galloway 45, 50, 677
    Bengali 672
    Black Pied 208, 371, 480
    Blanco Orejinegro 49, 682
    Blonde d'Aquitaine 19, 62, 618, 619, 620, 674
    Bonsmara 673
    Boran 21, 213, 673
    Braford 45, 679
    Brah-Maine 679
    Brahman 679
    Brahmousin 679

breeds *continued*
   Bralers 679
   Brangus 382, 396, 397, 618, 679
   Braunvieh 397, 618, 670, 674
   British White 45, 677
   Brown Atlas 673
   Brown Swiss 18, 371, 670, 674
   Bruna Alpina 369
   Butana 21, 29, 673
   Canadienne 679
   Caracu 682
   characteristics 102–103
   Charbray 679
   Charolais 19, 21, 587, 674
   Chianina 674
   Chinese Yellow 672
   classification 668–686
   Corriente 680
   criollo 22, 210, 478, 668
   Danakil 673
   Danish Red 18, 38, 619, 620, 670, 674
   Danish Red and White 674
   Danish White 619, 620
   definition 667
   Devon 397, 618, 677
   Dexter 63, 670, 677
   Dexter bulldog dwarf 63
   Dhanni 51, 672
   Dinka 673
   Dutch Belted 45, 50, 670, 674
   endangered 476, 482–489
   Eringer 371
   Fighting Bull 674
   Finnish Ayrshire 18
   Flamande 670, 674
   Fleckvieh 18, 674
   Florida Cracker 45, 46–47, 48, 49, 50, 680
   Fogera 673
   Friesian 215, 371, 617
   Fulani 21
   Galloway 373, 383, 618, 620, 677
   Gelbray 680
   Gelbvieh 674
   and genetic resources 475–476
   genetic uniqueness 488
   Gir 20, 22, 672
   Grati 681
   Grey Mountain 480
   Grey Steppe 37, 41
   Grigia delle Alpi 369
   Gronigen 45
   Guernsey 670, 677
   Guzerat 22, 672
   Hariana 20
   Hereford 19, 21, 22, 587, 677
   Hereford–Angus 397
   Hissar 672
   Holstein 515, 670, 675
   Holstein Friesian 18, 21, 22
   Horro 217
   Hungarian Spotted 675
   Iberian Brown 480
   Icelandic 40, 675
   Illawarra 670, 681
   Indian bison 8, 11, 20
   Indian wild buffalo 5, 6, 10
   Indo-Brazil 682
   Japanese Black 619
   Japanese Brown 619
   Javanese 670, 681
   Jersey 21, 515–516, 671, 677
   Kankrej 20
   Kapsiki 21
   Karan Fries 375
   Karan Swiss 375
   Kelantan 681
   Kenana 21, 29, 673
   Keteku 673
   Krishna Valley 672
   Landim 51
   Limousin 19, 587, 675
   Lincoln Red 677
   Local Indian Dairy 671, 681
   Lohani 51
   Longhorn 45, 48, 677
   Madura 8, 681
   Maine Anjou 675
   Mandalong 681
   Marchigiana 675
   Mashona 21
   Maure 21, 29
   Menufi 673
   Meuse-Rhine-Yssel 675
   Mongolian 672
   Montbéliarde 18, 671, 675
   MRY 18
   Murrah 375
   Murray Grey 37, 681
   Muturu 673

Index

Namchi 21
N'Dama 21, 29, 673
Nellore 22, 618, 619, 620, 672
New Zealand 21
Nguni 21, 51, 217, 673
Normande 18, 38, 45, 50, 675
Norwegian Red 50, 671, 675
Ongole 20, 672
Parda 371
Pecora 2
Piedmontese 675
Pinzgauer 675
Pirenaica 371
Polish Red 675
Polish Red and White 675
Rathi 369
Red Angus 37, 38, 618, 678
Red Brahman 51
Red Brangus 680
Red Holstein 18
Red Poll 678
Red Sindhi 20, 671, 672
references for research
information 682–685
Rendena 369
Rojhan 51
Romagnola 619, 620, 676
Romosinuano 378
Rotbunte 18
Rotvieh 676
Russian Black Pied 671, 676
Sahiwal 20, 671, 672
Salers 618, 620, 676
Salorn 680
Sanga 21, 668
Santa Cruz 680
Santa Gertrudis 396, 397, 400, 618, 680
Scottish Highland 37, 678
Senepol 378, 680
Shorthorn 671, 678
shorthorned zebu 673
Siboney 382
Simbrah 680
Simmental 18, 21, 676
social hierarchies 368–369
South Devon 397, 618, 620, 678
Sussex 678
Swedish Red and White 676
Swedish Red Polled 676
Tarentaise 397, 618, 620, 676

Texas Longhorn 680
Tharparkar 20, 672
Toro de Lidia 674
in tropical breeding programmes 641–643
Tswana 21
Tuli 21, 673
Wagyu 619, 624, 672
Welsh Black 678
West African Shorthorn 21, 51, 212
White Fulani 673
White Galloway 49
White Lineback 480
White Park 45, 49, 678
Xinjiang Brown 671, 672
brucellosis resistance 208–209
Bubalus arnee 5, 6, 10
Bubalus bubalis 5, 6
Bubalus depressicornis 5–6, 10
Bubalus mindorensis 5, 6, 10
buffalo species 5–6, 10

C-bands 248
cachectin 185
E-cadherin 442, 443
calf, gender 400, 402, 415–416
calpastatin 607, 610, 622
calving ease 514, 518, 582
Cambodian wild ox 8
cancer eye 34
Caprinae 3
carbonic anhydrase polymorphism 95
carcass
    composition traits 606–608, 609–610, 615, 617–619
    merit 583
casein locus 90, 99, 138–144, 541–543
caseins 541
    characteristics 138–139
    endocrine control of gene expression 548
    evolutionary comparisons 142–144
    extracellular matrix control of gene expression 549
    gene 543–544
    gene transcription activation 551–555
    microsatellites 141–142
    origins 139–140
    polymorphisms 99, 100, 140–141, 524
    protein and DNA sequences 139–140
    transgenes 561, 563
catalase polymorphism 95

cathepsin B 454
Cephalophinae 3
ceroid lipofuscinosis 63
ceruloplasmin polymorphism 92
cetaceans 2
Chediak–Higashi syndrome 63
chevrotains 2
chiasmata 330
chimeras 102, 459
cholesterol 622–623
chorion 439
chromatids 330
chromosome mapping 260, 266, 298–301, 302
    comparative 265, 651
    cytogenetic 267–297
    in situ hybridization techniques 262–264
    linkage 265, 329–330, 331–335, 343–360
    radiation-hybrid panel 264, 265
    rationale 260, 262
    somatic-cell hybrid 262
chromosome painting technique 133–134
chromosomes 124, 126, 248–256
    abnormalities 256–260, 261
    gene loci listing 267–297
    and genetic linkage 330–332
    karyotype standardization 248–252, 253
    mechanically stretched 263
    meiotic 254–256
    nucleolar organizer regions 248, 252–254, 259
    RBG-banded 253
    sorting and microdissection 265–266
chute test 372
cingulin 443
citrullinaemia 58, 353
CJD 233, 235
class II-associated invariant chain peptide (CLIP) 166
cleavage 438, 440–446
cloning 421–424
    applications in agriculture 427–428
    beef cattle 599–600
    and breeding programmes 531
    from cultured embryonic stem cells 424–426
    from differentiated cells 426–427, 460–461
    and totipotency 425–427, 460–461
coat colour
    Agouti (*A*) locus 36, 37, 40, 41, 63, 455
    Albino (*C*) locus 42, 63
    basis of pigmentation 35
    Belting (*Bt*) locus 45, 50
    Blaze (*Bl*) locus 45, 47
    blue roan 48
    Brindle (*Br*) locus 37, 38–39, 42
    brockling (*Bc*) locus 45, 50
    Charolais dilution (*Dc*) locus 43
    Chinchilla mutant 41
    colour-sided (*Cs*) locus 45, 48–50
    developmental effects of mutations 455
    dun (*Dn*) locus 37, 41–42
    Extension (*E*) locus 36, 37, 38–39, 41, 42, 58, 455
    Gloucester pattern 46
    mutations affecting 36–44, 58, 63
    patterned blackish gene 40
    Roan (*R*) locus 45, 48, 63, 353, 455, 458
    S locus 45, 46–47
    Simmental dilution (*Ds*) locus 43–44
    trait localization 353
    variation 33–53
    white park pattern 49–50
    white-spotting mutants 44–51
    wild type 35
coefficient of inbreeding 486
colony-stimulating factors (CSFs) 186
Committee on Genetic Nomenclature of Sheep and Goats (COGNOSAG) 57, 657–658, 664–665
complement 78
Composite Cattle Breeders International Alliance 669
conformation traits in dairy cattle 518, 519
congenital erythropoietic porphyria 66
Conservation and Use of Animal Genetic Resources in Asia and the Pacific project 478
conservation of genetic resources 477–480, 489–490, 492–493
    costs 481–482
    developing countries 477–478
    Europe 479
    objectives 482–484
    strategies 484–490, 491
continuous traits 351–352, 353
cortisol and lactogenesis 547–548

cre–lox system 419–420
Creutzfeld-Jakob disease (CJD) 233, 235
crossbreeding
    beef production systems 582, 584–587,
        587–588, 627
    dairy cattle 516
    and estimated breeding values 591–592
crush test 372
cryopreservation of embryos and
    semen 480, 481–482, 489–490
cytogenetics 247–260, 261
cytokine receptors 186, 187
cytokines 180–186
    embryonic expression 444
    in maternal recognition of
        pregnancy 447–448

dairy cattle 512, 515–525, 670–671
    breeding programme design 528–532
    economic indices 528
    genetic evaluation 525–528
    genetic improvement 511–537
    heterosis 520–522
    profit functions 512–515
    reproduction traits 518
    workability traits 518–519
    yield traits 516–518, 522–525
dairy temperament score 372, 377
daughter–dam comparison 526
dendritic cells 164, 166
dermatophilosis resistance 209–211
developmental genetics 437–473
dihydrotestosterone receptor deficiency 67
discrete traits 353
disease control options 200–201, 202
disease resistance 200–227
diversity function 489
DNA
    centromeric satellite *see* satellites
    introduction into cultured cells 419
    introduction into sperm 420
    large-insert libraries 131–133, 264
    mitochondrial 26–27
    polymorphisms 334, 337–338, 488
        detection 339–343
        identification and
            mapping 343–353
        types 338–339
    segment nomenclature 661
DNA oligonucleotide hybridization 339, 340

DNA visualization 339, 340
domestication
    and behaviour 365–366
    genetic effects 22–26
dominance 377, 585
double-muscle trait 61–62, 147, 353, 454, 616
double-stranded DNA-binding complex
    (DSI) 556
dual-purpose breeds 477–478, 578, 588
    beef breeding goals 581–582
dulong 20
duplication 125
dwarfism 63–64
dynamic self-regulation 639
dyserythropoiesis 64
dystocia 401–402, 514
    and twinning 394, 395

economic indices, dairy cattle 528
economic trait loci (ETLs) 124
ectoderm 439
effective population size 485–486
Ehlers–Danlos syndrome, type VII 64
embryo 439-440
    cryopreservation 480, 481–482, 489–490
    developmental stages 438–440
    gene expression 442–444
    genetic control of morphogenesis 449–455
    genome activation 441–442
    in vitro production 412–415, 530
    molecular diagnoses 144–147
    organogenesis 452–453
    ovum/preimplantation stage 438–439
    sexing 144, 145–146, 416, 599
embryoblast 439
embryonic lethals 59
embryonic stem cells 424–426
epidermal growth factor 444
epistasis and heterosis 585
epitheliogenesis imperfecta 64
epitopes 172
erythrocytic protein antigen
    polymorphism 95–96
erythrocytic protein polymorphisms 96
esterase D polymorphism 96
estimated breeding values (EBVs) 528, 530,
    588–591
    across breeds 591–592
euchromatin 254, 263, 266, 423
eumelanin 35, 36, 58, 455

European Association of Animal Production, Animal Genetic Data Bank 479
evolution
    adaptation 638–639
    of the casein locus 138, 141, 142–144
    of the genome 24–26, 128, 130, 133
    of mitochondria 26–27
    of the phenotype 22, 24, 33–34
    of zebu 28–29
expected progeny differences (EPDs) 399
expressed sequence tags (ESTs) 130–131
extracellular matrix
    in regulation of milk protein expression 549–550
    transcriptional regulation by 558–559
eyelid pigmentation 34
ezrin 442–443

factor XI deficiency 65
familial goitre 59
FAO
    Global Domestic Animal Diversity Information System 479
    World Watch List 477, 478
fat 622–625
fatty acids 623–625
feed intake 514
feeding behaviour 381–384
feeding methods, factors influencing aggression 370–371
fertility 379, 514
fertilization 413, 414, 445
fetal development
    muscle development 453–454
    sexual differentiation 145, 256–257, 456
    timing and duration 439, 440
fibroblast growth factor 444
fibronectin 443
flight distance test 372
flow karyotype 265
fluorescence *in situ* hybridization (FISH) 263–264
    fibre 263–264
    multiple-colour 263
freemartinism 256–257
    diagnosis 460
    phenomenology 458–459
    role of placenta 459
    and sex-chromosome chimerism 459

G-bands 248
gametic imprinting 445–446
gangliosidosis 65
gastrulation 449–450
gaur 8, 11, 20
gayal 8, 15, 20
gene loci lists 267–297, 343–360
genetic distance studies 650–651
genetic drift 366
genetic evaluation of dairy cattle 525–526, 527
genetic homoeostasis 639
genetic markers
    from microdissected chromosomes 134
    from somatic cell hybrids 134–135
    from sorted chromosomes 133–134
genetic resources 475–476, 477–493
genome, cattle 124–130
genotype, effects of domestication on 24–26
germ cells 255, 398, 425, 426, 459
    chimera 257
Gerstmann–Straussler syndrome 235
gestation length 397, 399–400
Ggf-4 449
Gli-1 450
glial growth factor 451
globin genes 125
Glossina 212
glucocorticoid receptor 554–555
glucocorticoid response elements (GREs) 554
glucocorticoids 547–548
glutamic oxalacetic transaminase polymorphism 98
glycogen storage diseases 59, 65, 353
goitre, familial 59
gonadal development 145, 456–458
    in freemartins 459
    SRY gene 457–460
granddaughter design 136
grooming behaviour 370
group breeding schemes, beef cattle 593
growth
    and gestation length 400
    and meat quality 613–614
growth factors 444, 451, 454
growth hormone-receptor deficiency 64
gsc 449

H-Y antigen 145
haemoglobin 82, 96–97
    chromosome location 349

Haemphysalis longicornis 216
hairlessness 65
Hays Converter 680
head process 450
heat stress adaptation 639
Hed 445
helminthiasis resistance 214–216
hemizygosity 331
herd-book 17
hereditary parakeratosis 65
hereditary thymus hypoplasia 65
hereditary zinc deficiency 65
heritability 515
    meat traits 606–608
    temperament 376–377
heterochromatin 265
heteroduplex (HD) 341, 342
heterosis
    beef cattle 584–586, 615–616
    dairy cattle 520–522
    genetic basis 585–586
heterozygosity 331
Hippotraginae 3
HNF-3beta 449
homoeobox 451
homozygosity 331
Hox gene family 451, 452
*Hpa*II tiny fragments (HTF) islands 128
humoral immunity 164, 172–177
humped cattle *see* zebu
humpless cattle *see Bos taurus*
Hxt 445
hybrid viral vector 418–419
hypervariable loci 338
hypotrichosis 65

Illinois Reference Resource families 337
immune system 163–197
immunoglobulins 93, 172–177
implantation 439, 443, 446–447, 450
imprinted genes 445–446
in vitro fertilization (IVF) 412–415, 530, 599
inbreeding 366
    dairy cattle 520, 521
    minimization in breeding programmes 531
inbreeding coefficient 520
inbreeding depression 531
inherited disorders 55–76

inner cell mass (ICM) 439
insulin and lactogenesis 547–548
insulin-like growth factors 444, 454
integrins 549
Interbull Centre 527
interferons 184–185, 447–448
interleukins 181–184
International Bovine Reference Panel (IBRP) family 337
International System for Cytogenetic Nomenclature of Domestic Animals (ISCNDA) 248–249
    revision 250–251
introgression and marker-assisted selection 505
isoimmunization 78
Ixodes rubicundus 216

karyotype 248–252, 253, 265
khainag 8
kouprey 8

lactation
    genetic markers of traits 103–105
    and twinning rate 392
lactoferrin 547, 549–550, 562–563
lactogenesis 547–550
lactose 544, 562
lactose synthase complex 544
laminin 549
lethal trait A46 65
leucocyte adhesion deficiency 60, 353, 498
leucocytic protein 2 polymorphism 98
leukaemia inhibitory factor 444
Leydig cells 456
linkage 329–330
    cattle as experimental species 139, 335–337
    and chromosomes 330–332
    mapping 265, 329–330, 331–335, 343–360
    with milk production traits 103–105
    reference families 337
    Rh genes 83
linkage disequilibrium 352, 353, 507
lod score 333
long interspersed elements (LINEs) 125, 126, 127
long terminal repeat-type retrotransposons 125–126
luteinizing hormone (LH) 398
lysozyme 125, 562–563

macroglobulin antigen polymorphism 93
macrophages 164, 166
major histocompatibility complex
    (MHC) 25, 86, 164, 165–166
    class I molecules and genes 166–167,
        168–169
    class I peptide binding motifs 171–172
    class II molecules and genes 166–167,
        170–171, 200
    genomic organization 167–168
    structural gene location 178
    structure and function of
        molecules 166–167
malate dehydrogenase polymorphism 98
mammary cell-activating factor
    (MAF) 557–558
mammospheres 549
mannose phosphate isomerase
    polymorphism 98
mapping 131, 134–135, 260–266, 331–332,
    334
    blood group B 82, 83
    of casein locus 138–144, 551–554
    comparative 265
    by lod score 136, 333–334
    microsatellites 127
maps 136, 137, 266–302, 334
    anonymous loci 344, 345
    blood groups 82–87, 91
    casein locus 143, 551
    coding sequence 347–351
    β-lactoglobulin 558
maple syrup urine disease 61
marbling 583
marbling score 607, 609–610, 626
marker-assisted selection (MAS) 490–491,
    501–507, 532, 598
Mash-1 451
Mash-2 445, 451
mastitis 211–212, 514, 519
maternal behaviour 380–381
mean herd life 515
meat
    breed variation 606–608, 617–622
    eating quality 583, 605–636, 626–627
    heterosis effects in breed
        crosses 615–616
    live-animal ultrasound
        evaluation 614–615
    quantitative trait loci 616–617

sensory quality 608, 612–613, 615–618
shear force 613, 619–622, 626
technological quality 607–608
    genetic correlations 610–612
    heterosis effects 615–616
traits
    genetic correlations 609–614
    prospects for genetic
        improvement 625–628
    see also body composition
Meat Animal Research Center 337, 393, 396
meiosis 254–256, 330–331
melanins 35
melanoblasts 455
*melanocyte-stimulating hormone (MSH)
    receptor* locus 36, 37, 38–39, 58
Mendelian Inheritance in Animals 57, 68
Mendelian Inheritance in Cattle 659, 664–665
Mendelian Inheritance in Man 57
Mendelian Inheritance in Sheep 659, 664
Mendelian sampling 490–491
mesoderm 439
microsatellites 127–128, 338, 488, 650–651
    casein 141–142
    phylogeny based on 28–29
    production 134
milk 540–541, 541–544, 544–547
    alteration of composition via
        transgenesis 561–563
    protein polymorphisms 99–100, 523–524
    regulation of protein gene
        expression 547–550, 550–559
milk yield 516
    control of 548–549
    genes affecting 522–525
    within-breed variations 516–518
milking speed 515, 518–519
minisatellites 338
mithan 8, 15, 20
mitochondrial DNA variation 26–27
MMp9 445
molecular genetics 123–161
    application to tropical breeding
        programmes 650–651
    immune system 163–197
molecular markers 135, 136–138
molecular tools 130–135
monoclonal antibody specificity for
    blood-group factors 87–88
monosomies 256

morphogenesis 449–455
morphological traits 55–76
morula 438
mtDNA variation 26–27
multiple ovulation and embryo transfer (MOET)
    beef cattle 598–599
    dairy cattle 530
multitrait across-country evaluation (MACE) 527
muscle
    genetic control of development 453–454
    hypertrophy 61–62, 147, 353, 454, 616
Mycobacterium bovis 209
myoclonus 66
myogenesis 453–454
myosin 453–454
myostatin 454, 616

nanismus 256
necrosin 185
neonatal isoerythrolysis 105–106
neural crest 451
neural tube 450
NF-1 family 557
nomenclature 248–252, 660–664
    1991 ruminant guidelines 658
    1993 ruminant guidelines 658–659
    blood and milk polymorphisms 663
    published listings of loci 664–665
non-restrained tests 372, 376
notochord 450
nuclear transfer 421–424
    use of cultured embryonic stem cells 424–426
    use of differentiated cells 426–427
nucleolar organizer regions (NORs) 248, 252–254, 259
numerator relationship matrix 590
nutrition and age at puberty 398

Oc substance 85
oestrus behaviour 379
1/29 Robertsonian translocation 259–260
Online Mendelian Inheritance in Animals (OMIA) 57, 68
oocytes, transvaginal ultrasound-guided recovery 414–415, 599
open field test 372

orotic acid 59
ovulation rate
    quantitative trait loci 396, 404
    and twinning 392–393
ovum pick-up 414–415, 599

P1-derived artificial chromosomes (PACs) 132
parentage verification 101
Pax genes 453
Peleinae 4
peptidase polymorphisms 97
performance testing of beef cattle 592
perinatal mortality 394, 401
phaeomelanin 35, 36, 58, 455
pharmaceutical production in milk 417–418
phenogroups 80
phenotype, effects of domestication on 22–24
phosphoglucomutase polymorphism 97, 98
phosphogluconate dehydrogenase polymorphism 99
phospholigands 177
phylogenetic analysis 102–103
phylogeny 1–14
pigmentation 35
    see also coat colour
pink tooth 66
placenta
    development 440
    in freemartinism 459
    gene expression 448–449
    retained 394
plasma protein polymorphisms 90–95
plasma thromboplastin antecedent (PTA) deficiency 65
platelet-derived growth factor 444
point mutations 338
polled condition 65, 353
polyandry 256
polygyny 256
polymerase chain reaction (PCR) 339, 340–341, 342
polymorphism information coefficient (PIC) 337–338
polymorphisms *see* biochemical polymorphisms; DNA, polymorphisms
polyploidy 256
Pompe's disease 65, 353
population bottlenecks 25–26
porphyria, congenital erythropoietic 66
porphyrins 66

positional cloning 131, 136–138, 147
post-transferrin polymorphisms 93–94
pound test 372
pregnancy, maternal recognition 447–448
pregnancy-specific mammary nuclear factor (PMF) 555–556
primitive streak 449
prion protein (PrP) 230, 231–232
processed retropseudogenes 125
procollagen III 448, 449
productive life of dairy cattle 519
profit functions, dairy cattle 512–515
progeny testing 528–529, 593
progesterone 547
progressive degenerative myeloencephalopathy 66, 353
prolactin 547–548, 551–554
prolactin-related proteins 448
protamine-2 deficiency 66
protease inhibitor alpha polymorphism 94
proto-oncogenes 443
protoporphyria 62
PrP gene 230, 235–241
    human 235
    sheep 234–235
pseudogenes 125
    alpha-lactalbumin 545
    beta-lactoglobulin 546
pseudotyped viral vector 418–419
puberty 396–399
purine nucleoside phosphorylase polymorphism 97

Q-bands 248
quantitative trait loci (QTL)
    beef cattle 597–598
    detection 506
    mapping 497
    and marker-assisted selection 532
    meat quality 616–617
    milk yield 524–525
    ovulation rate 396, 404

R-bands 248
radiation-hybrid panel analysis 264, 265
radioactive *in situ* hybridization 262–263
RAG-1 protein 174
RAG-2 protein 174
random amplified polymorphic DNA (RAPD) 341, 342, 343

Rare Breeds Survival Trust 487
Reading Conference 248
Reading nomenclature system 248
recombination fraction 331, 332, 333, 336
recombination loss 521, 522
red-cell protein polymorphisms 95–97
red-eyed condition 50
Reduncinae 3
reference families, genetic linkage mapping 337
relatives, use of information from 589–591
relaxin 402
renal coloboma syndrome 453
repeatability 515
replication-defective retroviral vectors 418
reproduction 391–410
reproductive technology 411–436
restrained tests 372, 376–377
restriction endonuclease 338
restriction fragment length polymorphism (RFLP) 341, 342, 343
retinol 545–546
retroposons 125–126
retroviruses 126
Rhipicephalus appendiculatus 216
RNA
    genes 125
    in prenatal development 440, 443–444
    transcription of polymorphisms 338
Robertsonian translocation 257–260
Rosenthal syndrome 65
rumination 384
RX3 680

Salmonella dublin 209
satellites 128–130, 338
scrapie 230, 234–235
scrapie-associated fibrils (SAF) 231
scrotal circumference 398, 399
selection indices 590, 591, 595–596, 643, 650
selembu 8
self-recognition 79
semen cryopreservation 480, 481–482, 489–490
Sertoli cells 456
sex
    defects of determination 458
    determination and differentiation 456–458
    predetermination in calves 415–416
sex chromosomes 124, 248, 330–331
    chimerism and freemartinism 459

numerical abnormalities 256
sex reversal 62
sexual behaviour
    female 379–380
    male 375, 378–379
sexual dimorphism 23
short interspersed elements (SINEs) 125, 126, 127, 128
sialic acid 213–214
simple sequence repeated polymorphisms *see* microsatellites
simple tandem repeat polymorphism (STRP) 341, 342, 343, 345–346
SINE-PCR primers 128
single nucleotide polymorphism (SNP) 341, 342, 343
single-strand conformational polymorphism (SSCP) 240, 341, 342
single-stranded DNA activator region-binding protein (SARP) 556
single-stranded DNA-binding complex (SS) 556
single-stranded DNA-binding transcriptional regulator (STR) 556
sire referencing schemes, beef cattle 593
size, effects of domestication on 22–23
small-eye syndrome 453
snorter dwarfism 64
social behaviour
    aggression 368, 370–371
    association 369–370
    dominance 367–369
    group structure 367–369
somatic cell score/count (SCS) 211, 514, 519
somitogenesis 450
sonic hedgehog (Shh) 450
sperm
    capacitation 413
    introduction of DNA into 420
    sexing 415, 599
spherocytosis 62
spinal muscular atrophy 67
SRY gene 456, 457–458
SRY protein 145, 457
starch-gel electrophoresis 90
Stat5 (MGF) 551, 553–554, 555, 557, 559
StockMarks panel 346
Strepsocerotini 5
suckler herds 578
suckling 381

Syncerus caffer 5
syndactyly 67
synteny 92, 123–124, 126, 128–130, 136, 137, 138–144, 171, 175, 182, 183, 249–251, 262–263, 265, 267–300, 546, 551
systematics 1–14

T-box gene family 450, 452–453
T-cell receptors 165, 177
    diversification 179
    gene organization 179
    gene segments 179–180
    structure 179
    and T-cell subsets 177–178
T cells 164
    cytotoxic 165, 177
    $\gamma\delta$ 177–178, 200
    helper 164–165, 177
    subsets 164–165, 177–178
T gene 449–450
T protein 450
tamarao 5, 6, 10
tameness 365–366
target levels of traits 650
temperament 515, 519
    and breed 372–373
    and growth 374
    heritability 376–377
    and milk yield 374–375
    and sex 374
    tests of 372
territorial behaviour 370
test-day model 526
testicular development 456
testicular feminization 67
testosterone 456
Texas standard chromosome nomenclature 249–252
Theileria parva 200
tibial hemimelia 67
tick resistance 216–217, 651
totipotency and cloning 425–427, 460–461
Tragelaphini 5
transferrin polymorphism 94–95
transforming growth factors (TGFs) 185–186, 444, 451
transgenics 144, 417–421, 540
    applications 416
    and disease resistance 205
    history 416

transgenics *continued*
    limitations 416–417
    marker-assisted selection 505
    molecular diagnosis 146–147
    utilization of milk-protein gene
        promoters 559–563
translocations *see* Robertsonian
    translocations
transmissible spongiform encephalopathies
    (TSEs) 230, 234–235
trisomies 256
trophoblast 439, 444–445
trophoblast protein 1 (TP-1) 447
trophoectoderm 439
Tropics
    adaptive genotype 639–640
    breeding programmes 637–651
trypanocide resistance 201
trypanotolerance 21, 204, 212–214, 651
TSPY 457
tumour necrosis factors (TNFs) 185
twins 391–396
    determinations 101–102
    monozygotic/dizygotic 102, 392
    and ovulation rate 392–393
tyrosinase 36, 58, 455

ulnar–mammary syndrome 453
uridine monophosphate synthase
    deficiency 59, 353, 498

variable number of tandem repeats
    (VNTR) 341, 342
vertical-fibre hide defect 67
vitamin D-binding protein 90, 92–93

water-buffalo 5, 6
weaning weight 591

weaver disease 66, 353
whey acidic protein 541, 549
whey proteins 544–547
white-cell protein polymorphisms 98–99
white heifer disease 48, 63, 455, 458
wisent 7, 10, 11
workability traits in dairy cattle 518–519

X chromosome 248
    cycle 458
    and meiosis 255–256
XY female 62

Y chromosome 248, 456
    polymorphism 29
yak 7–8, 10, 11, 16, 47
yakow 8
yeast artificial chromosomes (YACs) 131–132,
    263, 264

zebu 9, 15, 16, 668
    Africa 28–29
    Australia 21–22
    in beef breeding systems 588
    body composition 618
    breeds 19–20
    coat colour 34, 36, 38, 40, 51
    composite breeds based on 477–478
    meat quality 621–622
    North America 22
    sexual behaviour 378
    tick resistance 217
    in tropical breeding programmes 641, 642,
        643
    white-spotted 51
ZFX 146
ZFY 145, 146
Zoo-FISH painting 265